Thank you for purchasing a new copy of *Mechanics of Materials, Eighth Edition in SI* [...] C. Hibbeler. The one-time password below provides access to the animations and video solutions found on the C[...] Website.

For students only:

To access the Companion Website:

1. Go to http://www.pearsoned-asia.com/hibbeler
2. Click on the title *Mechanics of Materials, Eighth Edition in SI Units*
3. Click on "Companion Website"
4. Register your access code given below and create an account

ISWHMM-SCOFF-ARTIC-FLAMB-STASH-TUNES

5. Follow the instructions on screen

For instructors only:

To access the suite of Instructor's Resources available for your use, please follow steps 1 and 2 above and click on "Instructor's Resources" instead of "Companion Website". Follow the instructions on screen, and you will be prompted to enter a user ID and password. Please proceed if you are already registered and have existing instructor's access.

If you do not have instructor's access, please contact your Pearson Education representative. He/She will then register for you and provide you with a user ID and password.

The animations and video solutions can be viewed on the Companion Website.

MECHANICS OF MATERIALS

EIGHTH EDITION IN SI UNITS

R. C. Hibbeler

SI Conversion by

S. C. Fan
Nanyang Technological University

Prentice Hall

Singapore   London   New York   Toronto   Sydney   Tokyo   Madrid
Mexico City   Munich   Paris   Capetown   Hong Kong   Montreal

Published in 2011 by
Prentice Hall
Pearson Education South Asia Pte Ltd
23/25 First Lok Yang Road, Jurong
Singapore 629733

Publishing Director: *Joy Tan*
Publishing Manager: *Gan Meng Meng*
Acquisitions Editor: *Gloria Seow*
Project Editor: *Jeremy Wong*
Prepress Executive: *Kimberly Yap*

Pearson Education offices in Asia: *Bangkok, Beijing, Hong Kong, Jakarta, Kuala Lumpur, Manila, New Delhi, Seoul, Singapore, Taipei, Tokyo*

Authorized adaptation from the United States edition, entitled *Mechanics of Materials, Eighth Edition*, ISBN: 0136022308 by HIBBELER, RUSSELL C., published by Pearson Education, Inc., publishing as Prentice Hall, Copyright © 2011 R. C. Hibbeler.

WORLD WIDE adaptation edition published by PEARSON EDUCATION SOUTH ASIA PTE LTD, Copyright © 2011.

Printed in Singapore

4   3   2   1
14   13   12   11

ISBN-13   978-981-06-8509-6
ISBN-10   981-06-8509-2

Prentice Hall
is an imprint of

www.pearson-asia.com

TO THE STUDENT

With the hope that this work will stimulate an interest in Mechanics of Materials and provide an acceptable guide to its understanding

Mechanics of Materials *in SI Units, Eighth Edition* was developed in order to provide students with a clear and thorough presentation of the theory and applications of this subject. Renowned for its clarity in explanation and robust problem sets, Hibbeler is currently one of the best-selling course texts in this field.

In this edition, visualization aids such as animations, real-life photos with vectors and video solutions are provided to enable students to better understand concepts taught and problems posed. The ample problems provided in this text are arranged in increasing level of difficulty to build students' analytical and problem-solving skills, as well as give them the practice they require.

MasteringEngineering, the most technologically advanced and effective online system that offers customized coaching, has also been introduced with this edition to aid in student tutoring. With MasteringEngineering and Hibbeler's sound pedagogical approach, this combination will form the definitive authority in the teaching and learning of Mechanics of Materials that grace lecture halls today.

This book contains the following new elements:

1. **Animations that help the visualization of concepts**
2. **24/7 video solution walkthroughs that enable independent revision**
3. **A wide variety of problems for practice and problem solving**
4. **MasteringEngineering which offers students customized coaching**
5. **Realistic diagrams with vectors to demonstrate real-world applications**

1   Animations That Help the Visualization of Concepts

Animations

Located on the Companion Website are 8 animations identified as fundamental engineering mechanics concepts. Flagged by an "interactive animation" icon, the animations help students visualize the relation between mathematical explanation and real structure, breaking down complicated sequences and showing how free-body diagrams can be derived. These animations lend a graphic component to tutorials and lectures, assisting instructors in demonstrating the teaching of concepts with greater ease and clarity.

Each animation is flagged by an "interactive animation" icon.

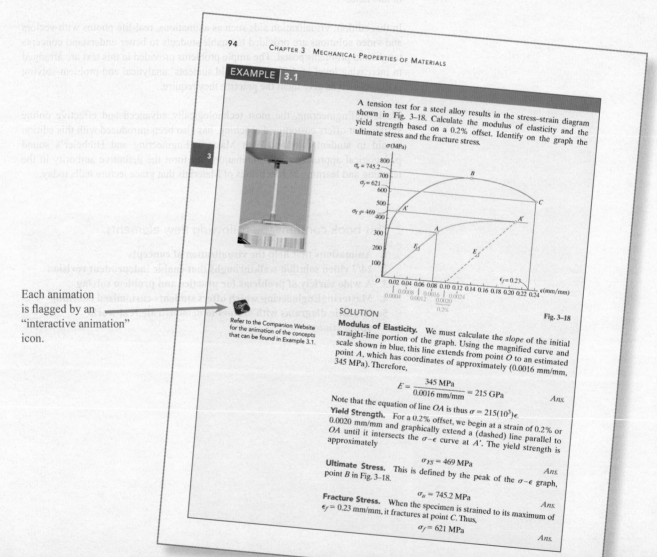

94   CHAPTER 3   MECHANICAL PROPERTIES OF MATERIALS

EXAMPLE 3.1

A tension test for a steel alloy results in the stress–strain diagram shown in Fig. 3–18. Calculate the modulus of elasticity and the yield strength based on a 0.2% offset. Identify on the graph the ultimate stress and the fracture stress.

Refer to the Companion Website for the animation of the concepts that can be found in Example 3.1.

Fig. 3–18

SOLUTION

Modulus of Elasticity. We must calculate the *slope* of the initial straight-line portion of the graph. Using the magnified curve and scale shown in blue, this line extends from point O to an estimated point A, which has coordinates of approximately (0.0016 mm/mm, 345 MPa). Therefore,

$$E = \frac{345 \text{ MPa}}{0.0016 \text{ mm/mm}} = 215 \text{ GPa} \qquad Ans.$$

Note that the equation of line OA is thus $\sigma = 215(10^3)\epsilon$.

Yield Strength. For a 0.2% offset, we begin at a strain of 0.2% or 0.0020 mm/mm and graphically extend a (dashed) line parallel to OA until it intersects the $\sigma-\epsilon$ curve at A'. The yield strength is approximately

$$\sigma_{YS} = 469 \text{ MPa} \qquad Ans.$$

Ultimate Stress. This is defined by the peak of the $\sigma-\epsilon$ graph, point B in Fig. 3–18.

$$\sigma_u = 745.2 \text{ MPa} \qquad Ans.$$

Fracture Stress. When the specimen is strained to its maximum of $\epsilon_f = 0.23$ mm/mm, it fractures at point C. Thus,

$$\sigma_f = 621 \text{ MPa} \qquad Ans.$$

Instructors can demonstrate the different methods of analysis step-by-step.

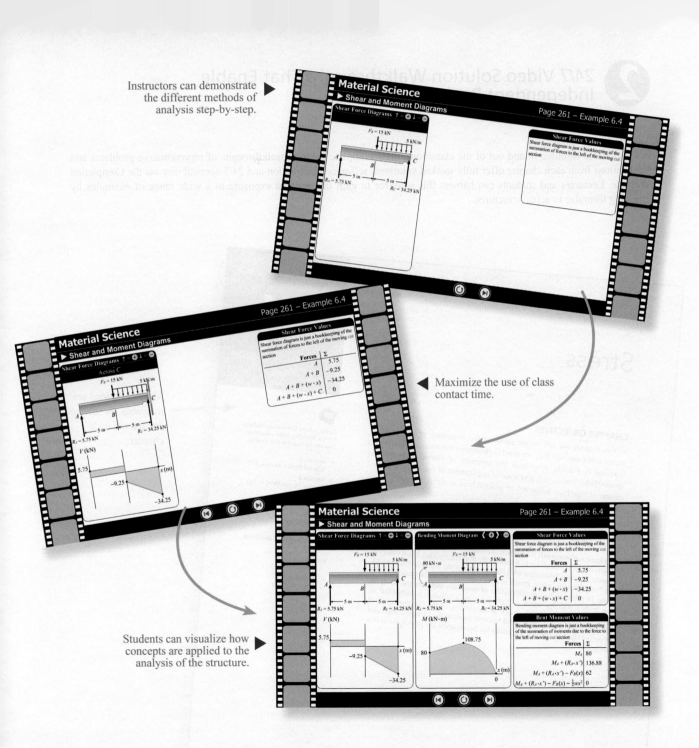

Maximize the use of class contact time.

Students can visualize how concepts are applied to the analysis of the structure.

2 **24/7 Video Solution Walkthroughs That Enable Independent Revision**

Video Solutions

An invaluable resource in and out of the classroom, these complete solution walkthroughs of representative problems and applications from each chapter offer fully worked solutions, self-paced instruction and 24/7 accessibility via the Companion Website. Lecturers and students can harness this resource to gain independent exposure to a wide range of examples by applying formulae to actual structures.

Stress

1

CHAPTER OBJECTIVES

In this chapter we will review some of the important principles of statics and show how they are used to determine the internal resultant loadings in a body. Afterwards the concepts of normal and shear stress will be introduced, and specific applications of the analysis and design of members subjected to an axial load or direct shear will be discussed.

1.1 Introduction

Mechanics of materials is a branch of mechanics that studies the internal effects of stress and strain in a solid body that is subjected to an external loading. Stress is associated with the strength of the material from which the body is made, while strain is a measure of the deformation of the body. In addition to this, mechanics of materials includes the study of the body's stability when a body such as a column is subjected to compressive loading. A thorough understanding of the fundamentals of this subject is of vital importance because many of the formulas and rules of design cited in engineering codes are based upon the principles of this subject.

Check out the Companion Website for the following video solution(s) that can be found in this chapter.

- Section 2
  Internal Forces Review, 2-D
- Section 2
  Internal Forces Review, 3-D
- Section 3
  Force Resultant Review
- Section 4
  Average Normal Stress
- Section 4, 5
  Average Normal and Shear Stress
- Section 6
  Allowable Stress
- Section 7
  Factor of Safety

Video solutions are available for selected sections in each chapter.

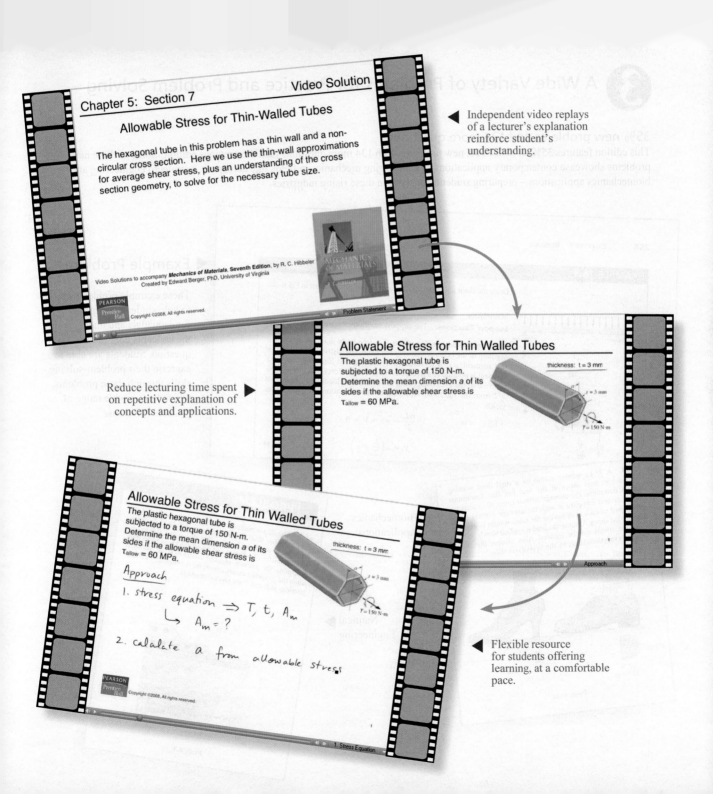

◀ Independent video replays of a lecturer's explanation reinforce student's understanding.

Reduce lecturing time spent on repetitive explanation of concepts and applications.

◀ Flexible resource for students offering learning, at a comfortable pace.

3 A Wide Variety of Problems for Practice and Problem Solving

35% new problems and 134 more questions

This edition features 35%, or about 550, new problems with 134 more questions than the previous edition! Some of these new problems showcase contemporary applications of engineering mechanics in fields like aerospace, petroleum engineering and biomechanics applications—preparing students for work in these rising industries.

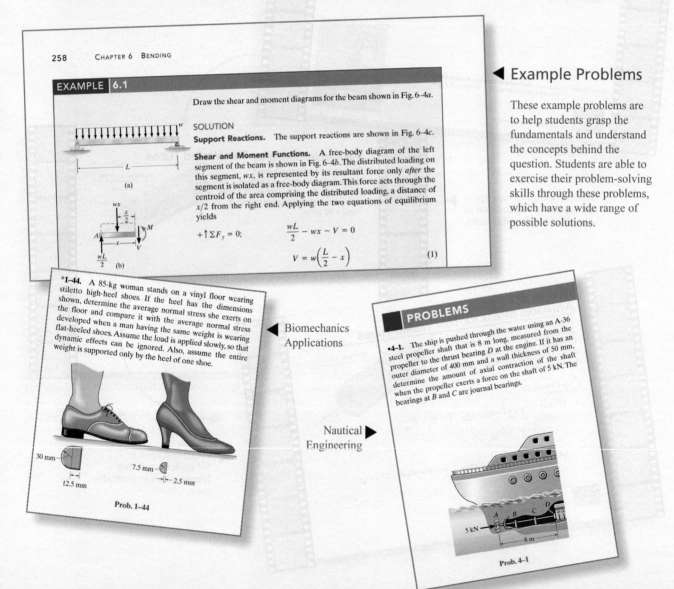

258   CHAPTER 6   BENDING

EXAMPLE 6.1

Draw the shear and moment diagrams for the beam shown in Fig. 6–4a.

SOLUTION

Support Reactions. The support reactions are shown in Fig. 6–4c.

Shear and Moment Functions. A free-body diagram of the left segment of the beam is shown in Fig. 6–4b. The distributed loading on this segment, wx, is represented by its resultant force only *after* the segment is isolated as a free-body diagram. This force acts through the centroid of the area comprising the distributed loading, a distance of $x/2$ from the right end. Applying the two equations of equilibrium yields

$$+\uparrow \Sigma F_y = 0; \qquad \frac{wL}{2} - wx - V = 0$$

$$V = w\left(\frac{L}{2} - x\right) \qquad (1)$$

(a)

(b)

◀ **Example Problems**

These example problems are to help students grasp the fundamentals and understand the concepts behind the question. Students are able to exercise their problem-solving skills through these problems, which have a wide range of possible solutions.

*1–44. A 85-kg woman stands on a vinyl floor wearing stiletto high-heel shoes. If the heel has the dimensions shown, determine the average normal stress she exerts on the floor and compare it with the average normal stress developed when a man having the same weight is wearing flat-heeled shoes. Assume the load is applied slowly, so that dynamic effects can be ignored. Also, assume the entire weight is supported only by the heel of one shoe.

◀ Biomechanics
Applications

30 mm

7.5 mm

12.5 mm   2.5 mm

Prob. 1–44

PROBLEMS

•4–1. The ship is pushed through the water using an A-36 steel propeller shaft that is 8 m long, measured from the propeller to the thrust bearing D at the engine. If it has an outer diameter of 400 mm and a wall thickness of 50 mm, determine the amount of axial contraction of the shaft when the propeller exerts a force on the shaft of 5 kN. The bearings at B and C are journal bearings.

Nautical ▶
Engineering

5 kN

8 m

Prob. 4–1

Fundamental ▶ Problems

A new feature in this edition, these problems are located after the example problems. They offer students simple applications of concepts to ensure they have understood the chapter before attempting to tackle any other questions.

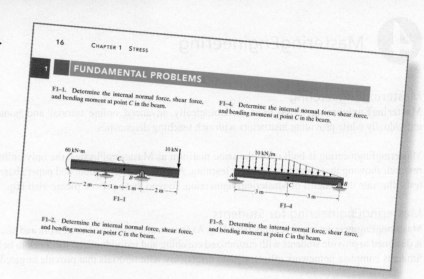

FUNDAMENTAL PROBLEMS

F1–1. Determine the internal normal force, shear force, and bending moment at point C in the beam.

F1–1

F1–4. Determine the internal normal force, shear force, and bending moment at point C in the beam.

F1–4

F1–2. Determine the internal normal force, shear force, and bending moment at point C in the beam.

F1–5. Determine the internal normal force, shear force, and bending moment at point C in the beam.

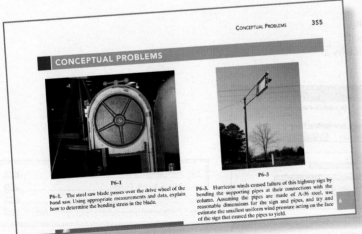

CONCEPTUAL PROBLEMS

P6-1

P6-1. The steel saw blade passes over the drive wheel of the band saw. Using appropriate measurements and data, explain how to determine the bending stress in the blade.

P6-3

P6-3. Hurricane winds caused failure of this highway sign by bending the supporting pipes at their connections with the column. Assuming the pipes are made of A-36 steel, use reasonable dimensions for the sign and pipes, and try and estimate the smallest uniform wind pressure acting on the face of the sign that caused the pipes to yield.

◀ Conceptual Problems

These problems depict realistic situations encountered in practice. They are developed to test a student's ability to apply concepts learnt. Students are also kept up to date on the latest applications of engineering mechanics. Lecturers too, have a wide range of questions from which to select, modify and add, as new questions to their resources.

In every set of conceptual problems, an attempt has been made to arrange the problems in order of increasing difficulty. The answers to all but every fourth problem are listed in the back of the book. To alert the user to a problem without a reported answer, an asterisk (*) is placed before the problem number. Answers are reported to three significant figures, even though the data for material properties may be known with less accuracy. Although this might appear to be a poor practice, it is done simply to be consistent and to allow the student a better chance to validate his or her solution. A solid square (■) is used to identify problems that require a numerical analysis or a computer application.

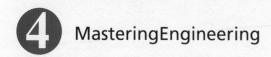 ④ MasteringEngineering

MasteringEngineering

MasteringEngineering is the most technologically advanced online tutorial and homework system. It tutors students individually while providing instructors with rich teaching diagnostics.

MasteringEngineering is built upon the same platform as MasteringPhysics, the only online physics homework system with research showing that it improves student learning. A wide variety of published papers based on NSF-sponsored research and tests illustrate the benefits of MasteringEngineering. To read these papers, please visit http://www.masteringengineering.com.

MasteringEngineering for Students

MasteringEngineering improves understanding. As an instructor-assigned homework and tutorial system, MasteringEngineering is designed to provide students with customized coaching and individualized feedback to help improve problem-solving skills. Students complete homework efficiently and effectively with tutorials that provide targeted help.

Immediate and specific feedback on wrong answers coach students individually. Specific feedback on common errors helps explain why a ▼ particular answer is incorrect.

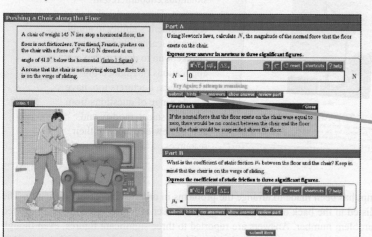

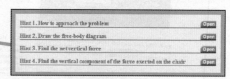

▲ Hints provide individualized coaching. Skip the hints you don't need and access only the ones that you need, for the most efficient path to the correct solution.

MasteringEngineering for Instructors

Incorporate dynamic homework into your course with automatic grading and adaptive tutoring. Choose from a wide variety of stimulating problems, including free-body diagram drawing capabilities, algorithmically generated problem sets, and more.

MasteringEngineering emulates the instructor's office-hour environment, coaching students on problem-solving techniques by asking students simpler sub-questions.

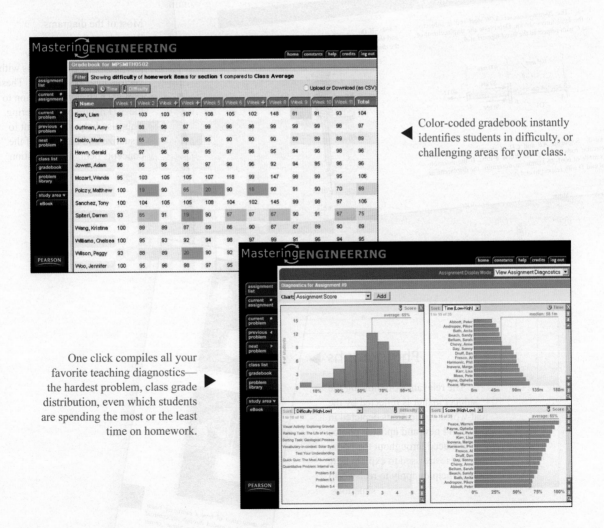

Color-coded gradebook instantly identifies students in difficulty, or challenging areas for your class.

One click compiles all your favorite teaching diagnostics—the hardest problem, class grade distribution, even which students are spending the most or the least time on homework.

 5 Realistic Diagrams with Vectors to Demonstrate Real-World Applications

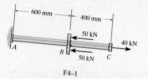

130 CHAPTER 4 AXIAL LOAD

FUNDAMENTAL PROBLEMS

F4–1. The 20-mm-diameter A-36 steel rod is subjected to the axial forces shown. Determine the displacement of end C with respect to the fixed support at A.

600 mm 400 mm

50 kN

40 kN

A B 50 kN C

F4–1

F4–2. Segments AB and CD of the assembly are solid circular rods, and segment BC is a tube. If the assembly is made of 6061-T6 aluminum, determine the displacement of end D with respect to end A.

20 mm 20 mm

F4–4. If the 20-mm-diameter rod is made of A-36 steel and the stiffness of the spring is k = 50 MN/m, determine the displacement of end A when the 60-kN force is applied.

B

400 mm k = 50 MN/m

400 mm

A

60 kN

F4–4

4

◀ Illustrations with Vectors

Most of the diagrams throughout the book are in full color art and many photorealistic illustrations with vectors have been added. These provide a strong connection to the 3-D nature of engineering. This also helps the student to visualize and be aware of the concepts behind the question.

Photographs ▶

The relevance of knowing the subject matter is reflected by the real-world applications depicted in over 44 new and updated photos placed throughout the book. These are used to explain how the principles apply to real-world situations.

7

Demonstration of how a cantilever beam deflects when loaded through the centroid (above) and through the shear center (below).

Resources for Instructors

Instructor's Solutions Manual

This supplement provides complete solutions supported by problem statements and problem figures. The eighth edition Solutions Manual includes homework assignment lists and was also checked as part of the accuracy checking program.

Teaching Slides

Fully editable powerpoint slides that instructors can use for frontal teaching are available under "Instructor's Resources" at http://www.pearsoned-asia.com/hibbeler. Professor Fan Sau Cheong of Nanyang Technological University Singapore developed the example-led focus of the slides, which contain information and diagrams from the respective chapters. Chapters 9 and 10 feature an equation simulator, a Flash-based tool that instantly calculates the results, letting lecturers demonstrate how the formulae work without going through the unnecessary math.

Image Bank

The image bank comprises illustrations, diagrams and photos found in the textbook. Instructors will find this particularly useful in customizing their lesson materials such as creating of worksheets or modification of questions.

Video Solutions

Developed by Professor Edward Berger, University of Virginia, video solutions are located on the Companion Website and they offer step-by-step solution walkthroughs of representative problems and applications from the text. Make efficient use of class time and office hours by showing students the complete and concise problem-solving approaches that they can access anytime and view at their own pace. The videos are designed to be a flexible resource to be used however each instructor and student prefers. A valuable tutorial resource, the videos are also helpful for student self-evaluation as students can pause the videos to check their understanding and work alongside the video. Access the videos at http://www.pearsoned-asia.com/hibbeler and follow the links under *Mechanics of Materials, Eighth Edition in SI Units*.

Resources for Students

Companion Website

The Companion Website, located at http://www.pearsoned-asia.com/hibbeler, includes opportunities for practice and review including:

- **Animations** — Eight specially selected animations help students visualize and simplify the relation between mathematical explanation and real structure.
- **Video solutions** — Complete, step-by-step solution walkthroughs from selected sections. Videos offer:
  - **Fully worked solutions**—Showing every step of representative problems and applications, to help students make vital connections between concepts.
  - **Self-paced instruction**—Students can navigate each problem and select, play, rewind, fast-forward, stop, and jump to sections within each problem's solution.
  - **24/7 access**—Help whenever students need it with over 20 hours of helpful review.
  - **Differentiated system of units**—Video solutions are categorized according to the system of unit, allowing students to improve problem solving by practicing in either imperial or SI units.

MECHANICS OF MATERIALS

Over the years, this text has been shaped by the suggestions and comments of many of my colleagues in the teaching profession. Their encouragement and willingness to provide constructive criticism are very much appreciated and it is hoped that they will accept this anonymous recognition. A note of thanks is given to the reviewers.

Akthem Al-Manaseer, *San Jose State University*
Yabin Liao, *Arizona State University*
Cliff Lissenden, *Penn State*
Gregory M. Odegard, *Michigan Technological University*
John Oyler, *University of Pittsburgh*
Roy Xu, *Vanderbilt University*
Paul Ziehl, *University of South Carolina*

There are a few people that I feel deserve particular recognition. A longtime friend and associate, Kai Beng Yap, was of great help to me in checking the entire manuscript and helping to prepare the problem solutions. A special note of thanks also goes to Kurt Norlin of Laurel Tech Integrated Publishing Services in this regard. During the production process I am thankful for the assistance of Rose Kernan, my production editor for many years, and to my wife, Conny, and daughter, Mary Ann, for their help in proofreading and typing, that was needed to prepare the manuscript for publication.

I would also like to thank all my students who have used the previous edition and have made comments to improve its contents.

I would greatly appreciate hearing from you if at any time you have any comments or suggestions regarding the contents of this edition.

RUSSELL CHARLES HIBBELER
hibbeler@bellsouth.net

About the Adaptor

Fan Sau Cheong, from Nanyang Technolgical University (NTU), Singapore, received his PhD from the University of Hong Kong. Professor Fan's research interest lies in the studies of responses of structures due to environmental effect or man-made events. His industrial experience includes work and research in bridges, tall buildings, shell structures, jetties, pavements, cable structures, and glass diaphragm walls. Professor Fan was also the adaptor for the fifth and sixth SI editions of Hibbeler's *Mechanics of Materials*, and the eleventh SI edition of Hibbeler's *Engineering Mechanics: Statics and Dynamics*.

CONTENTS

12

Deflection of Beams and Shafts 569

13

Buckling of Columns 657

14

Energy Methods 715

Appendices

PHOTOCREDITS

Chapter 1   Close up of iron girders. Jack Sullivan\Alamy Images.

Chapter 2   Photoelastic phenomena: tension in a screw mount. Alfred Pasieka\Alamy Images.

Chapter 3   A woman stands near a collapsed bridge in one of the worst earthquake-hit areas of Yingxiu town in Wenchuan county, in China's southwestern province of Sichuan on June 2, 2008. UN Secretary of State Condoleezza Rice on June 29 met children made homeless by the devastating earthquake that hit southwest China last month and praised the country's response to the disaster. LIU JIN/Stringer\Getty Images, Inc.AFP.

Chapter 3 text   Cup and cone steel. Alamy Images.

Chapter 4   Rotary bit on portable oil drilling rig. © Lowell Georgia/CORBIS. All Rights Reserved.

Chapter 5   Steam rising from soils and blurred spinning hollow stem auger. Alamy Images.

Chapter 6   Steel framework at construction site. Corbis RF.

Chapter 7   Train wheels on track. Jill Stephenson\Alamy Images.

Chapter 7 text   Highway flyover. Gari Wyn Williams\Alamy Images.

Chapter 8   Ski lift with snow covered mountain in background. Shutterstock.

Chapter 9   Turbine blades. Chris Pearsall\Alamy Images.

Chapter 10   Complex stresses developed within an airplane wing. Courtesy of Measurements Group, Inc. Raleigh, North Carolina, 27611, USA.

Chapter 11   Metal frame and yellow crane. Stephen Finn\Alamy Images.

Chapter 12   Man pole vaulting in desert. © Patrick Giardino/CORBIS. All Rights Reserved.

Chapter 13   Water storage tower. John Dorado\Shutterstock.

Chapter 14   Shot of jack-up-pile-driver and floating crane. John MacCooey\Alamy Images.

Other images provided by the author.

MECHANICS OF MATERIALS

EIGHTH EDITION IN SI UNITS

The bolts used for the connections of this steel framework are subjected to stress. In this chapter we will discuss how engineers design these connections and their fasteners.

Stress

1

CHAPTER OBJECTIVES

In this chapter we will review some of the important principles of statics and show how they are used to determine the internal resultant loadings in a body. Afterwards the concepts of normal and shear stress will be introduced, and specific applications of the analysis and design of members subjected to an axial load or direct shear will be discussed.

1.1 Introduction

Mechanics of materials is a branch of mechanics that studies the internal effects of stress and strain in a solid body that is subjected to an external loading. Stress is associated with the strength of the material from which the body is made, while strain is a measure of the deformation of the body. In addition to this, mechanics of materials includes the study of the body's stability when a body such as a column is subjected to compressive loading. A thorough understanding of the fundamentals of this subject is of vital importance because many of the formulas and rules of design cited in engineering codes are based upon the principles of this subject.

Check out the Companion Website for the following video solution(s) that can be found in this chapter.

- Section 2
  Internal Forces Review, 2-D
- Section 2
  Internal Forces Review, 3-D
- Section 3
  Force Resultant Review
- Section 4
  Average Normal Stress
- Section 4, 5
  Average Normal and Shear Stress
- Section 6
  Allowable Stress
- Section 7
  Factor of Safety

Historical Development.   The origin of mechanics of materials dates back to the beginning of the seventeenth century, when Galileo performed experiments to study the effects of loads on rods and beams made of various materials. However, at the beginning of the eighteenth century, experimental methods for testing materials were vastly improved, and at that time many experimental and theoretical studies in this subject were undertaken primarily in France, by such notables as Saint-Venant, Poisson, Lamé, and Navier.

Over the years, after many of the fundamental problems of mechanics of materials had been solved, it became necessary to use advanced mathematical and computer techniques to solve more complex problems. As a result, this subject expanded into other areas of mechanics, such as the *theory of elasticity* and the *theory of plasticity*. Research in these fields is ongoing, in order to meet the demands for solving more advanced problems in engineering.

1.2   Equilibrium of a Deformable Body

Since statics has an important role in both the development and application of mechanics of materials, it is very important to have a good grasp of its fundamentals. For this reason we will review some of the main principles of statics that will be used throughout the text.

External Loads.   A body is subjected to only two types of external loads; namely, surface forces or body forces, Fig. 1–1.

Surface Forces.   *Surface forces* are caused by the direct contact of one body with the surface of another. In all cases these forces are distributed over the *area* of contact between the bodies. If this area is small in comparison with the total surface area of the body, then the surface force can be *idealized* as a single *concentrated force*, which is applied to a *point* on the body. For example, the force of the ground on the wheels of a bicycle can be considered as a concentrated force. If the surface loading is applied along a narrow strip of area, the loading can be *idealized* as a *linear distributed load*, $w(s)$. Here the loading is measured as having an intensity of force/length along the strip and is represented graphically by a series of arrows along the line s. *The resultant force $\mathbf{F}_R$ of $w(s)$ is equivalent to the area under the distributed loading curve, and this resultant acts through the centroid C or geometric center of this area.* The loading along the length of a beam is a typical example of where this idealization is often applied.

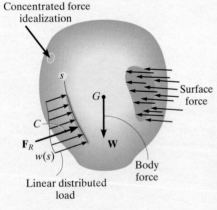

Concentrated force idealization

Surface force

Body force

Linear distributed load

Fig. 1–1

Body Forces.   A *body force* is developed when one body exerts a force on another body without direct physical contact between the bodies. Examples include the effects caused by the earth's gravitation or its electromagnetic field. Although body forces affect each of the particles composing the body, these forces are normally represented by a single concentrated force acting on the body. In the case of gravitation, this force is called the **weight** of the body and acts through the body's center of gravity.

Support Reactions.   The surface forces that develop at the supports or points of contact between bodies are called **reactions**. For two-dimensional problems, i.e., bodies subjected to coplanar force systems, the supports most commonly encountered are shown in Table 1–1. Note carefully the symbol used to represent each support and the type of reactions it exerts on its contacting member. As a general rule, *if the support prevents translation in a given direction, then a force must be developed on the member in that direction. Likewise, if rotation is prevented, a couple moment must be exerted on the member*. For example, the roller support only prevents translation perpendicular or normal to the surface. Hence, the roller exerts a normal force **F** on the member at its point of contact. Since the member can freely rotate about the roller, a couple moment cannot be developed on the member.

Many machine elements are pin connected in order to enable free rotation at their connections. These supports exert a force on a member, but no moment.

TABLE 1–1			
Type of connection	Reaction	Type of connection	Reaction
Cable	One unknown: F	External pin	Two unknowns: F_x, F_y
Roller	One unknown: F	Internal pin	Two unknowns: F_x, F_y
Smooth support	One unknown: F	Fixed support	Three unknowns: F_x, F_y, M

1

Equations of Equilibrium.

Equilibrium of a body requires both a **balance of forces**, to prevent the body from translating or having accelerated motion along a straight or curved path, and a **balance of moments**, to prevent the body from rotating. These conditions can be expressed mathematically by two vector equations

$$\Sigma F = 0$$
$$\Sigma M_O = 0 \qquad (1-1)$$

Here, $\Sigma \mathbf{F}$ represents the sum of all the forces acting on the body, and $\Sigma \mathbf{M}_O$ is the sum of the moments of all the forces about any point O either on or off the body. If an x, y, z coordinate system is established with the origin at point O, the force and moment vectors can be resolved into components along each coordinate axis and the above two equations can be written in scalar form as six equations, namely,

$$\Sigma F_x = 0 \qquad \Sigma F_y = 0 \qquad \Sigma F_z = 0$$
$$\Sigma M_x = 0 \qquad \Sigma M_y = 0 \qquad \Sigma M_z = 0 \qquad (1-2)$$

Often in engineering practice the loading on a body can be represented as a system of *coplanar forces*. If this is the case, and the forces lie in the x–y plane, then the conditions for equilibrium of the body can be specified with only three scalar equilibrium equations; that is,

$$\Sigma F_x = 0$$
$$\Sigma F_y = 0$$
$$\Sigma M_O = 0 \qquad (1-3)$$

Here all the moments are summed about point O and so they will be directed along the z axis.

Successful application of the equations of equilibrium requires complete specification of all the known and unknown forces that act *on* the body, and so **the best way to account for all these forces is to draw the body's free-body diagram.**

In order to design the horizontal members of this building frame, it is first necessary to find the internal loadings at various points along their length.

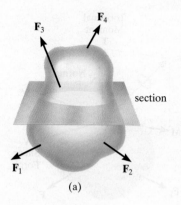

section

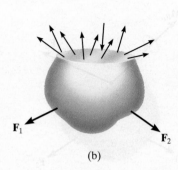

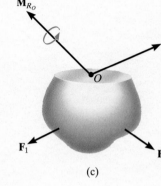

(a)

(b)

(c)

Fig. 1–2

Internal Resultant Loadings.

In mechanics of materials, statics is primarily used to determine the resultant loadings that act within a body. For example, consider the body shown in Fig. 1–2a, which is held in equilibrium by the four external forces.* In order to obtain the internal loadings acting on a specific region within the body, it is necessary to pass an imaginary section or "cut" through the region where the internal loadings are to be determined. The two parts of the body are then separated, and a free-body diagram of one of the parts is drawn, Fig. 1–2b. Notice that there is actually a distribution of internal force acting on the "exposed" area of the section. These forces represent the effects of the material of the top part of the body acting on the adjacent material of the bottom part.

Although the exact distribution of this internal loading may be *unknown*, we can use the equations of equilibrium to relate the external forces on the bottom part of the body to the distribution's *resultant force and moment*, $\mathbf{F}_R$ and $\mathbf{M}_{R_O}$, *at any specific point O* on the sectioned area, Fig. 1–2c. It will be shown in later portions of the text that point O is most often chosen at the *centroid* of the sectioned area, and so we will always choose this location for O, unless otherwise stated. Also, if a member is long and slender, as in the case of a rod or beam, the section to be considered is generally taken *perpendicular* to the longitudinal axis of the member. This section is referred to as the ***cross section***.

*The body's weight is not shown, since it is assumed to be quite small, and therefore negligible compared with the other loads.

1

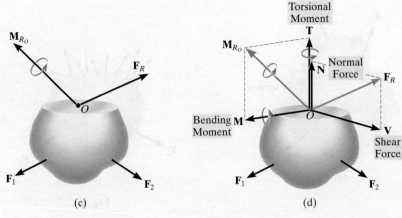

Fig. 1–2 (cont.)

Three Dimensions.   Later in this text we will show how to relate the resultant loadings, $\mathbf{F}_R$ and $\mathbf{M}_{R_O}$, to the *distribution of force* on the sectioned area, and thereby develop equations that can be used for analysis and design. To do this, however, the components of $\mathbf{F}_R$ and $\mathbf{M}_{R_O}$ acting both normal and perpendicular to the sectioned area must be considered, Fig. 1–2d. Four different types of resultant loadings can then be defined as follows:

Normal force, N.   This force acts perpendicular to the area. It is developed whenever the external loads tend to push or pull on the two segments of the body.

Shear force, V.   The shear force lies in the plane of the area and it is developed when the external loads tend to cause the two segments of the body to slide over one another.

Torsional moment or torque, T.   This effect is developed when the external loads tend to twist one segment of the body with respect to the other about an axis perpendicular to the area.

Bending moment, M.   The bending moment is caused by the external loads that tend to bend the body about an axis lying within the plane of the area.

In this text, note that graphical representation of a moment or torque is shown in three dimensions as a vector with an associated curl. By the *right-hand rule*, the thumb gives the arrowhead sense of this vector and the fingers or curl indicate the tendency for rotation (twisting or bending).

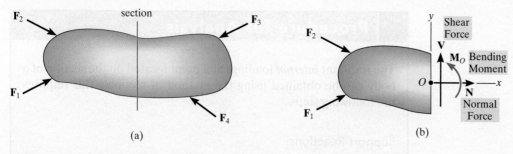

Fig. 1–3

Coplanar Loadings.   If the body is subjected to a *coplanar system of forces*, Fig. 1–3*a*, then only normal-force, shear-force, and bending- moment components will exist at the section, Fig. 1–3*b*. If we use the x, y, z coordinate axes, as shown on the left segment, then $\mathbf{N}$ can be obtained by applying $\Sigma F_x = 0$, and $\mathbf{V}$ can be obtained from $\Sigma F_y = 0$. Finally, the bending moment $\mathbf{M}_O$ can be determined by summing moments about point O (the z axis), $\Sigma M_O = 0$, in order to eliminate the moments caused by the unknowns $\mathbf{N}$ and $\mathbf{V}$.

Important Points

- *Mechanics of materials* is a study of the relationship between the external loads applied to a body and the stress and strain caused by the internal loads within the body.

- External forces can be applied to a body as *distributed* or *concentrated surface loadings*, or as *body forces* that act throughout the volume of the body.

- Linear distributed loadings produce a *resultant force* having a *magnitude* equal to the *area* under the load diagram, and having a *location* that passes through the *centroid* of this area.

- A support produces a *force* in a particular direction on its attached member if it *prevents translation* of the member in that direction, and it produces a *couple moment* on the member if it *prevents rotation*.

- The equations of equilibrium $\Sigma \mathbf{F} = \mathbf{0}$ and $\Sigma \mathbf{M} = \mathbf{0}$ must be satisfied in order to prevent a body from translating with accelerated motion and from rotating.

- When applying the equations of equilibrium, it is important to first draw the free-body diagram for the body in order to account for all the terms in the equations.

- The method of sections is used to determine the internal resultant loadings acting on the surface of the sectioned body. In general, these resultants consist of a normal force, shear force, torsional moment, and bending moment.

1

Procedure for Analysis

The resultant *internal* loadings at a point located on the section of a body can be obtained using the method of sections. This requires the following steps.

Support Reactions.

- First decide which segment of the body is to be considered. If the segment has a support or connection to another body, then *before* the body is sectioned, it will be necessary to determine the reactions acting on the chosen segment. To do this draw the free-body diagram of the *entire body* and then apply the necessary equations of equilibrium to obtain these reactions.

Free-Body Diagram.

- Keep all external distributed loadings, couple moments, torques, and forces in their *exact locations*, before passing an imaginary section through the body at the point where the resultant internal loadings are to be determined.
- Draw a free-body diagram of one of the "cut" segments and indicate the unknown resultants **N**, **V**, **M**, and **T** at the section. These resultants are normally placed at the point representing the geometric center or *centroid* of the sectioned area.
- If the member is subjected to a *coplanar* system of forces, only **N**, **V**, and **M** act at the centroid.
- Establish the x, y, z coordinate axes with origin at the centroid and show the resultant internal loadings acting along the axes.

Equations of Equilibrium.

- Moments should be summed at the section, about each of the coordinate axes where the resultants act. Doing this eliminates the unknown forces **N** and **V** and allows a direct solution for **M** (and **T**).
- If the solution of the equilibrium equations yields a negative value for a resultant, the assumed *directional sense* of the resultant is *opposite* to that shown on the free-body diagram.

The following examples illustrate this procedure numerically and also provide a review of some of the important principles of statics.

EXAMPLE 1.1

Determine the resultant internal loadings acting on the cross section at C of the cantilevered beam shown in Fig. 1–4a.

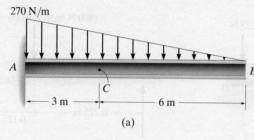

(a)

Fig. 1–4

SOLUTION

Support Reactions.   The support reactions at A do not have to be determined if segment CB is considered.

Free-Body Diagram.   The free-body diagram of segment CB is shown in Fig. 1–4b. It is important to keep the distributed loading on the segment until *after* the section is made. Only then should this loading be replaced by a single resultant force. Notice that the intensity of the distributed loading at C is found by proportion, i.e., from Fig. 1–4a, $w/6 \text{ m} = (270 \text{ N/m})/9 \text{ m}$, $w = 180 \text{ N/m}$. The magnitude of the resultant of the distributed load is equal to the area under the loading curve (triangle) and acts through the centroid of this area. Thus, $F = \frac{1}{2}(180 \text{ N/m})(6 \text{ m}) = 540 \text{ N}$, which acts $\frac{1}{3}(6 \text{ m}) = 2 \text{ m}$ from C as shown in Fig. 1–4b.

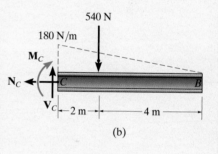

(b)

Equations of Equilibrium.   Applying the equations of equilibrium we have

$$\xrightarrow{+} \Sigma F_x = 0; \qquad -N_C = 0$$
$$N_C = 0 \qquad \qquad Ans.$$
$$+\uparrow \Sigma F_y = 0; \qquad V_C - 540 \text{ N} = 0$$
$$V_C = 540 \text{ N} \qquad \qquad Ans.$$
$$\zeta + \Sigma M_C = 0; \qquad -M_C - 540 \text{ N}(2 \text{ m}) = 0$$
$$M_C = -1080 \text{ N} \cdot \text{m} \qquad \qquad Ans.$$

NOTE: The negative sign indicates that $\mathbf{M}_C$ acts in the opposite direction to that shown on the free-body diagram. Try solving this problem using segment AC, by first obtaining the support reactions at A, which are given in Fig. 1–4c.

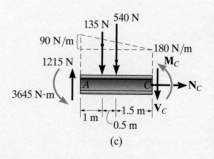

(c)

EXAMPLE 1.2

Determine the resultant internal loadings acting on the cross section at C of the machine shaft shown in Fig. 1–5a. The shaft is supported by journal bearings at A and B, which only exert vertical forces on the shaft.

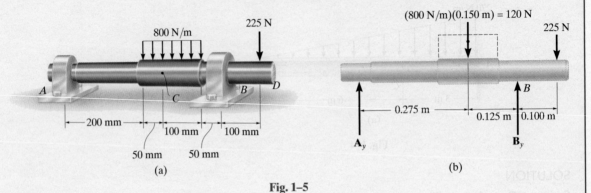

Fig. 1–5

SOLUTION

We will solve this problem using segment AC of the shaft.

Support Reactions. The free-body diagram of the entire shaft is shown in Fig. 1–5b. Since segment AC is to be considered, only the reaction at A has to be determined. Why?

$$\zeta + \Sigma M_B = 0; \quad -A_y(0.400 \text{ m}) + 120 \text{ N}(0.125 \text{ m}) - 225 \text{ N}(0.100 \text{ m}) = 0$$

$$A_y = -18.75 \text{ N}$$

The negative sign indicates that $\mathbf{A}_y$ acts in the *opposite sense* to that shown on the free-body diagram.

Free-Body Diagram. The free-body diagram of segment AC is shown in Fig. 1–5c.

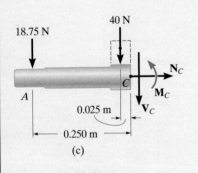

Equations of Equilibrium.

$$\xrightarrow{+} \Sigma F_x = 0; \qquad\qquad N_C = 0 \qquad\qquad\qquad Ans.$$

$$+\uparrow \Sigma F_y = 0; \qquad -18.75 \text{ N} - 40 \text{ N} - V_C = 0$$

$$V_C = -58.8 \text{ N} \qquad\qquad Ans.$$

$$\zeta + \Sigma M_C = 0; \quad M_C + 40 \text{ N}(0.025 \text{ m}) + 18.75 \text{ N}(0.250 \text{ m}) = 0$$

$$M_C = -5.69 \text{ N} \cdot \text{m} \qquad\qquad Ans.$$

NOTE: The negative signs for V_C and M_C indicate they act in the opposite directions on the free-body diagram. As an exercise, calculate the reaction at B and try to obtain the same results using segment CBD of the shaft.

EXAMPLE 1.3

The 500-kg engine is suspended from the crane boom in Fig. 1–6a. Determine the resultant internal loadings acting on the cross section of the boom at point E.

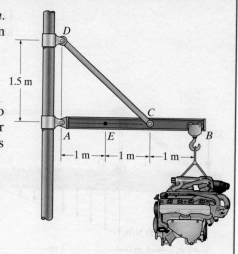

(a)

SOLUTION

Support Reactions.   We will consider segment AE of the boom so we must first determine the pin reactions at A. Notice that member CD is a two-force member. The free-body diagram of the boom is shown in Fig. 1–6b. Applying the equations of equilibrium,

$$\zeta + \Sigma M_A = 0; \qquad F_{CD}\left(\tfrac{3}{5}\right)(2\text{ m}) - [500(9.81)\text{ N}](3\text{ m}) = 0$$

$$F_{CD} = 12\ 262.5 \text{ N}$$

$$\xrightarrow{+}\Sigma F_x = 0; \qquad A_x - (12\ 262.5 \text{ N})\left(\tfrac{4}{5}\right) = 0$$

$$A_x = 9810 \text{ N}$$

$$+\uparrow\Sigma F_y = 0; \qquad -A_y + (12\ 262.5 \text{ N})\left(\tfrac{3}{5}\right) - 500(9.81)\text{ N} = 0$$

$$A_y = 2452.5 \text{ N}$$

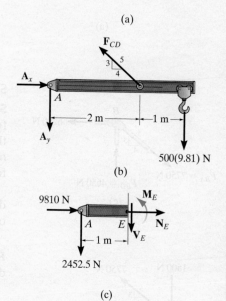

(b)

Free-Body Diagram.   The free-body diagram of segment AE is shown in Fig. 1–6c.

Equations of Equilibrium.

$$\xrightarrow{+}\Sigma F_x = 0; \qquad N_E + 9810 \text{ N} = 0$$

$$N_E = -9810 \text{ N} = -9.81 \text{ kN} \qquad Ans.$$

$$+\uparrow\Sigma F_y = 0; \qquad -V_E - 2452.5 \text{ N} = 0$$

$$V_E = -2452.5 \text{ N} = -2.45 \text{ kN} \qquad Ans.$$

$$\zeta + \Sigma M_E = 0; \qquad M_E + (2452.5 \text{ N})(1\text{ m}) = 0$$

$$M_E = -2452.5 \text{ N}\cdot\text{m} = -2.45 \text{ kN}\cdot\text{m} \qquad Ans.$$

(c)

Fig. 1–6

1

EXAMPLE   **1.4**

Determine the resultant internal loadings acting on the cross section at G of the beam shown in Fig. 1–7a. Each joint is pin connected.

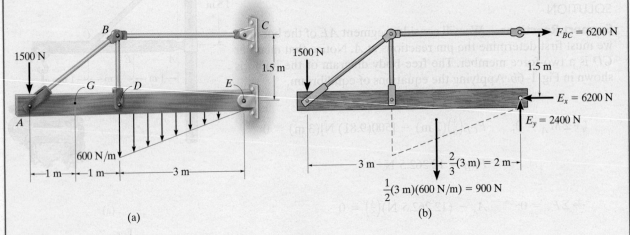

Fig. 1–7

SOLUTION

Support Reactions. Here we will consider segment AG. The free-body diagram of the *entire* structure is shown in Fig. 1–7b. Verify the calculated reactions at E and C. In particular, note that BC is a *two-force member* since only two forces act on it. For this reason the force at C must act along BC, which is horizontal as shown.

Since BA and BD are also two-force members, the free-body diagram of joint B is shown in Fig. 1–7c. Again, verify the magnitudes of forces $\mathbf{F}_{BA}$ and $\mathbf{F}_{BD}$.

Free-Body Diagram. Using the result for $\mathbf{F}_{BA}$, the free-body diagram of segment AG is shown in Fig. 1–7d.

Equations of Equilibrium.

$$\xrightarrow{+}\Sigma F_x = 0; \qquad 7750\text{ N}\left(\tfrac{4}{5}\right) + N_G = 0 \qquad N_G = -6200\text{ N} \qquad \textit{Ans.}$$

$$+\uparrow\Sigma F_y = 0; \qquad -1500\text{ N} + 7750\text{ N}\left(\tfrac{3}{5}\right) - V_G = 0$$

$$V_G = 3150\text{ N} \qquad \textit{Ans.}$$

$$\downarrow{+}\Sigma M_G = 0; \qquad M_G - (7750\text{ N})\left(\tfrac{3}{5}\right)(1\text{ m}) + 1500\text{ N}(1\text{ m}) = 0$$

$$M_G = 3150\text{ N}\cdot\text{m} \qquad \textit{Ans.}$$

EXAMPLE | 1.5

Determine the resultant internal loadings acting on the cross section at B of the pipe shown in Fig. 1–8a. The pipe has a mass of 2 kg/m and is subjected to both a vertical force of 50 N and a couple moment of 70 N $\cdot$ m at its end A. It is fixed to the wall at C.

SOLUTION

The problem can be solved by considering segment AB, so we do not need to calculate the support reactions at C.

Free-Body Diagram.   The x, y, z axes are established at B and the free-body diagram of segment AB is shown in Fig. 1–8b. The resultant force and moment components at the section are assumed to act in the positive coordinate directions and to pass through the *centroid* of the cross-sectional area at B. The weight of each segment of pipe is calculated as follows:

$$W_{BD} = (2 \text{ kg/m})(0.5 \text{ m})(9.81 \text{ N/kg}) = 9.81 \text{ N}$$

$$W_{AD} = (2 \text{ kg/m})(1.25 \text{ m})(9.81 \text{ N/kg}) = 24.525 \text{ N}$$

These forces act through the center of gravity of each segment.

Equations of Equilibrium.   Applying the six scalar equations of equilibrium, we have*

$\Sigma F_x = 0;$ $\qquad\qquad (F_B)_x = 0$ $\qquad\qquad$ *Ans.*

$\Sigma F_y = 0;$ $\qquad\qquad (F_B)_y = 0$ $\qquad\qquad$ *Ans.*

$\Sigma F_z = 0;$ $\quad (F_B)_z - 9.81 \text{ N} - 24.525 \text{ N} - 50 \text{ N} = 0$

$\qquad\qquad\qquad (F_B)_z = 84.3 \text{ N}$ $\qquad\qquad$ *Ans.*

$\Sigma (M_B)_x = 0;$ $\quad (M_B)_x + 70 \text{ N} \cdot \text{m} - 50 \text{ N} (0.5 \text{ m})$

$\qquad\qquad - 24.525 \text{ N} (0.5 \text{ m}) - 9.81 \text{ N} (0.25 \text{ m}) = 0$

$\qquad\qquad\qquad (M_B)_x = -30.3 \text{ N} \cdot \text{m}$ $\qquad\qquad$ *Ans.*

$\Sigma (M_B)_y = 0;$ $\;(M_B)_y + 24.525 \text{ N} (0.625 \text{ m}) + 50 \text{ N} (1.25 \text{ m}) = 0$

$\qquad\qquad\qquad (M_B)_y = -77.8 \text{ N} \cdot \text{m}$ $\qquad\qquad$ *Ans.*

$\Sigma (M_B)_z = 0;$ $\qquad\qquad (M_B)_z = 0$ $\qquad\qquad$ *Ans.*

NOTE: What do the negative signs for $(M_B)_x$ and $(M_B)_y$ indicate? Note that the normal force $N_B = (F_B)_y = 0$, whereas the shear force is $V_B = \sqrt{(0)^2 + (84.3)^2} = 84.3 \text{ N}$. Also, the torsional moment is $T_B = (M_B)_y = 77.8 \text{ N} \cdot \text{m}$ and the bending moment is $M_B = \sqrt{(30.3)^2 + (0)^2} = 30.3 \text{ N} \cdot \text{m}$.

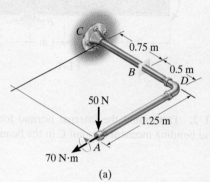

(a)

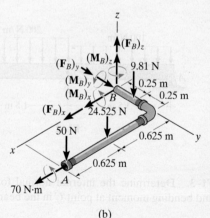

(b)

Fig. 1–8

---

*The *magnitude* of each moment about an axis is equal to the magnitude of each force times the perpendicular distance from the axis to the line of action of the force. The *direction* of each moment is determined using the right-hand rule, with positive moments (thumb) directed along the positive coordinate axes.

FUNDAMENTAL PROBLEMS

F1–1. Determine the internal normal force, shear force, and bending moment at point C in the beam.

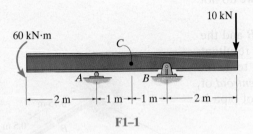

F1–1

F1–2. Determine the internal normal force, shear force, and bending moment at point C in the beam.

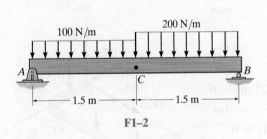

F1–2

F1–3. Determine the internal normal force, shear force, and bending moment at point C in the beam.

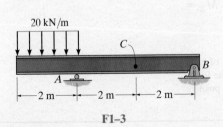

F1–3

F1–4. Determine the internal normal force, shear force, and bending moment at point C in the beam.

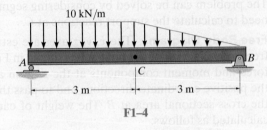

F1–4

F1–5. Determine the internal normal force, shear force, and bending moment at point C in the beam.

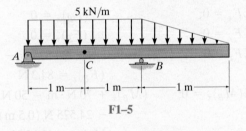

F1–5

F1–6. Determine the internal normal force, shear force, and bending moment at point C in the beam.

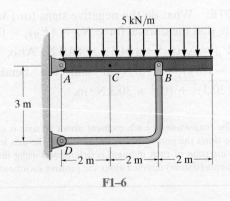

F1–6

PROBLEMS

1–1. Determine the resultant internal normal force acting on the cross section through point A in each column. In (a), segment BC weighs 300 kg/m and segment CD weighs 400 kg/m. In (b), the column has a mass of 200 kg/m.

1–3. Determine the resultant internal torque acting on the cross sections through points B and C.

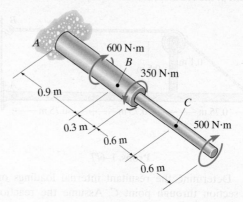

Prob. 1–3

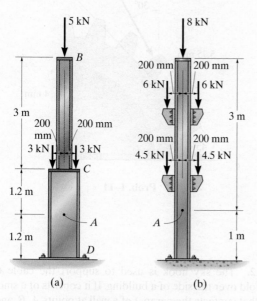

(a)   (b)

Prob. 1–1

*1–4. A force of 80 N is supported by the bracket as shown. Determine the resultant internal loadings acting on the section through point A.

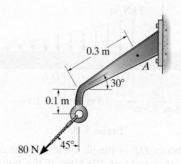

Prob. 1–4

1–2. Determine the resultant internal torque acting on the cross sections through points C and D. The support bearings at A and B allow free turning of the shaft.

•1–5. Determine the resultant internal loadings in the beam at cross sections through points D and E. Point E is just to the right of the 15-kN load.

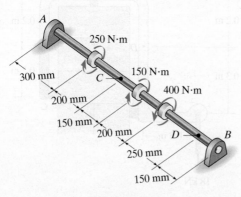

Prob. 1–2

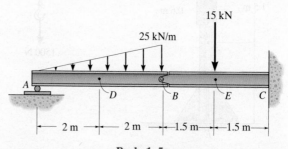

Prob. 1–5

1–6. Determine the normal force, shear force, and moment at a section through point *C*. Take *P* = 8 kN.

1–7. The cable will fail when subjected to a tension of 2 kN. Determine the largest vertical load *P* the frame will support and calculate the internal normal force, shear force, and moment at the cross section through point *C* for this loading.

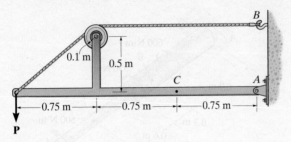

Probs. 1–6/7

*1–8. Determine the resultant internal loadings on the cross section through point *C*. Assume the reactions at the supports *A* and *B* are vertical.

•1–9. Determine the resultant internal loadings on the cross section through point *D*. Assume the reactions at the supports *A* and *B* are vertical.

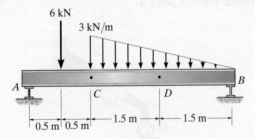

Probs. 1–8/9

1–10. The boom *DF* of the jib crane and the column *DE* have a uniform weight of 750 N/m. If the hoist and load weigh 1500 N, determine the resultant internal loadings in the crane on cross sections through points *A*, *B*, and *C*.

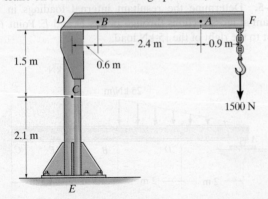

Prob. 1–10

1–11. The force 400 N acts on the gear tooth. Determine the resultant internal loadings on the root of the tooth, i.e., at the centroid point *A* of section *a–a*.

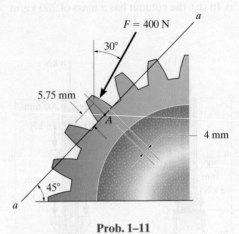

Prob. 1–11

*1–12. The sky hook is used to support the cable of a scaffold over the side of a building. If it consists of a smooth rod that contacts the parapet of a wall at points *A*, *B*, and *C*, determine the normal force, shear force, and moment on the cross section at points *D* and *E*.

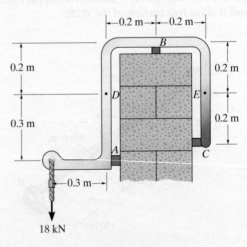

Prob. 1–12

•1–13. The 4-kN load is being hoisted at a constant speed using the motor *M*, which has a weight of 0.45 kN. Determine the resultant internal loadings acting on the cross section through point *B* in the beam. The beam has a weight of 0.6 kN/m and is fixed to the wall at *A*.

1–14. Determine the resultant internal loadings acting on the cross section through points *C* and *D* of the beam in Prob. 1–13.

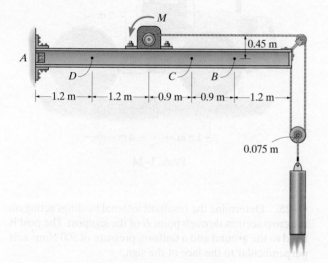

Probs. 1–13/14

1–15. Determine the resultant internal loading on the cross section through point *C* of the pliers. There is a pin at *A*, and the jaws at *B* are smooth.

*1–16. Determine the resultant internal loading on the cross section through point *D* of the pliers.

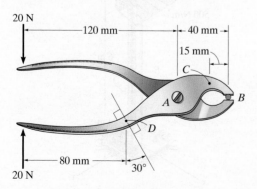

Probs. 1–15/16

•1–17. Determine resultant internal loadings acting on section *a–a* and section *b–b*. Each section passes through the centerline at point *C*.

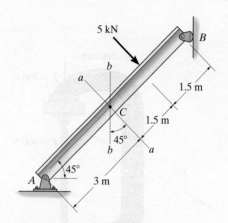

Prob. 1–17

1–18. The bolt shank is subjected to a tension of 400 N. Determine the resultant internal loadings acting on the cross section at point *C*.

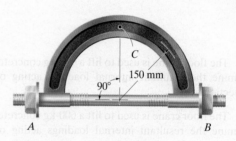

Prob. 1–18

1–19. Determine the resultant internal loadings acting on the cross section through point *C*. Assume the reactions at the supports *A* and *B* are vertical.

*1–20. Determine the resultant internal loadings acting on the cross section through point *D*. Assume the reactions at the supports *A* and *B* are vertical.

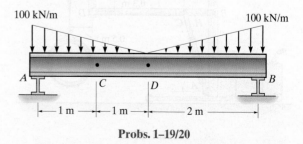

Probs. 1–19/20

•1–21.   The forged steel clamp exerts a force of $F = 900$ N on the wooden block. Determine the resultant internal loadings acting on section a–a passing through point A.

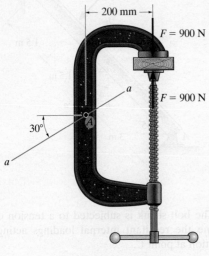

200 mm

$F = 900$ N

a

$F = 900$ N

30°

A

a

Prob. 1–21

1–22.   The floor crane is used to lift a 600-kg concrete pipe. Determine the resultant internal loadings acting on the cross section at G.

1–23.   The floor crane is used to lift a 600-kg concrete pipe. Determine the resultant internal loadings acting on the cross section at H.

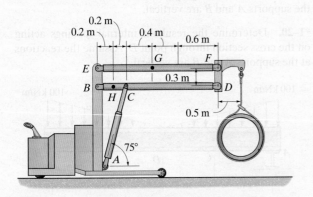

0.2 m
0.2 m   0.4 m
0.6 m
G F
E 0.3 m
B
H C D
0.5 m

75°
A

Probs. 1–22/23

*1–24.   The machine is moving with a constant velocity. It has a total mass of 20 Mg, and its center of mass is located at G, excluding the front roller. If the front roller has a mass of 5 Mg, determine the resultant internal loadings acting on point C of each of the two side members that support the roller. Neglect the mass of the side members. The front roller is free to roll.

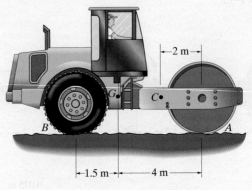

2 m

G C

B A

1.5 m   4 m

Prob. 1–24

•1–25.   Determine the resultant internal loadings acting on the cross section through point B of the signpost. The post is fixed to the ground and a uniform pressure of 500 N/m² acts perpendicular to the face of the sign.

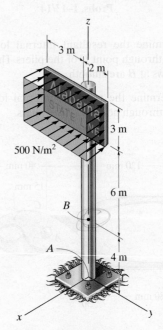

z

3 m

2 m

3 m

500 N/m²

B 6 m

A 4 m

x y

Prob. 1–25

1–26. The shaft is supported at its ends by two bearings A and B and is subjected to the forces applied to the pulleys fixed to the shaft. Determine the resultant internal loadings acting on the cross section located at point C. The 300-N forces act in the $-z$ direction and the 500-N forces act in the $+x$ direction. The journal bearings at A and B exert only x and z components of force on the shaft.

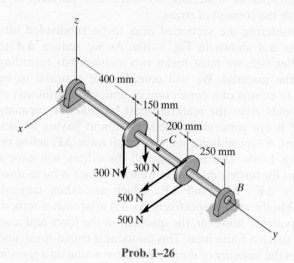

Prob. 1–26

1–27. The pipe has a mass of 12 kg/m. If it is fixed to the wall at A, determine the resultant internal loadings acting on the cross section at B. Neglect the weight of the wrench CD.

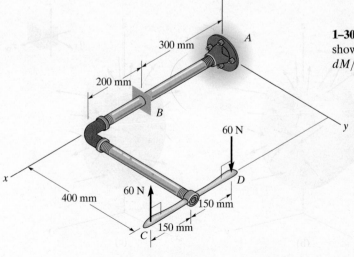

Prob. 1–27

*1–28. The brace and drill bit is used to drill a hole at O. If the drill bit jams when the brace is subjected to the forces shown, determine the resultant internal loadings acting on the cross section of the drill bit at A.

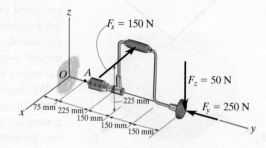

Prob. 1–28

•1–29. The curved rod has a radius r and is fixed to the wall at B. Determine the resultant internal loadings acting on the cross section through A which is located at an angle θ from the horizontal.

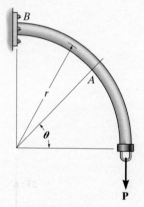

Prob. 1–29

1–30. A differential element taken from a curved bar is shown in the figure. Show that $dN/d\theta = V$, $dV/d\theta = -N$, $dM/d\theta = -T$, and $dT/d\theta = M$.

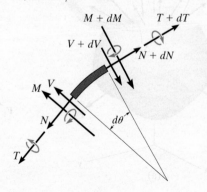

Prob. 1–30

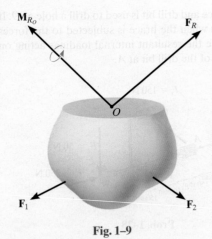

Fig. 1–9

1.3 Stress

It was stated in Section 1.2 that the force and moment acting at a specified point O on the sectioned area of the body, Fig. 1–9, represents the resultant effects of the actual *distribution of loading* acting over the sectioned area, Fig. 1–10a. Obtaining this *distribution* is of primary importance in mechanics of materials. To solve this problem it is necessary to establish the concept of stress.

We begin by considering the sectioned area to be subdivided into small areas, such as ΔA shown in Fig. 1–10a. As we reduce ΔA to a smaller and smaller size, we must make two assumptions regarding the properties of the material. We will consider the material to be ***continuous***, that is, to consist of a *continuum* or uniform distribution of matter having no voids. Also, the material must be ***cohesive***, meaning that all portions of it are connected together, without having breaks, cracks, or separations. A typical finite yet very small force $\Delta \mathbf{F}$, acting on ΔA, is shown in Fig. 1–10a. This force, like all the others, will have a unique direction, but for further discussion we will replace it by its *three components*, namely, $\Delta \mathbf{F}_x$, $\Delta \mathbf{F}_y$, and $\Delta \mathbf{F}_z$, which are taken tangent, tangent, and normal to the area, respectively. As ΔA approaches zero, so do $\Delta \mathbf{F}$ and its components; however, the quotient of the force and area will, in general, approach a finite limit. This quotient is called *stress*, and as noted, it describes the *intensity of the internal force* acting on a *specific plane* (area) passing through a point.

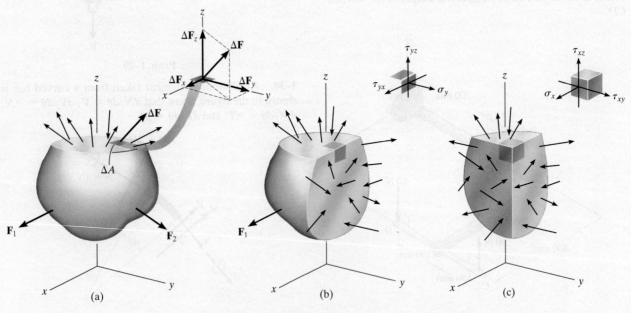

Fig. 1–10

Normal Stress.

The *intensity* of the force acting normal to ΔA is defined as the **normal stress**, σ (sigma). Since $\Delta \mathbf{F}_z$ is normal to the area then

$$\sigma_z = \lim_{\Delta A \to 0} \frac{\Delta F_z}{\Delta A} \qquad (1-4)$$

If the normal force or stress "pulls" on ΔA as shown in Fig. 1–10a, it is referred to as *tensile stress*, whereas if it "pushes" on ΔA it is called *compressive stress*.

Shear Stress.

The intensity of force acting tangent to ΔA is called the **shear stress**, τ (tau). Here we have shear stress components,

$$\tau_{zx} = \lim_{\Delta A \to 0} \frac{\Delta F_x}{\Delta A}$$

$$\tau_{zy} = \lim_{\Delta A \to 0} \frac{\Delta F_y}{\Delta A} \qquad (1-5)$$

Note that in this subscript notation z specifies the orientation of the area ΔA, Fig. 1–11, and x and y indicate the axes along which each shear stress acts.

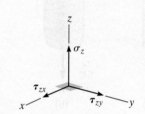

Fig. 1–11

General State of Stress.

If the body is further sectioned by planes parallel to the $x-z$ plane, Fig. 1–10b, and the $y-z$ plane, Fig. 1–10c, we can then "cut out" a cubic volume element of material that represents the **state of stress** acting around the chosen point in the body. This state of stress is then characterized by three components acting on each face of the element, Fig. 1–12.

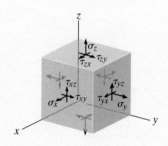

Fig. 1–12

Units.

Since stress represents a force per unit area, in the International Standard or SI system, the magnitudes of both normal and shear stress are specified in the basic units of newtons per square meter (N/m^2). This unit, called a pascal $(1 \, \text{Pa} = 1 \, \text{N/m}^2)$ is rather small, and in engineering work prefixes such as kilo- (10^3), symbolized by k, mega- (10^6), symbolized by M, or giga- (10^9), symbolized by G, are used to represent larger, more realistic values of stress.*

---

*Sometimes stress is expressed in units of N/mm^2, where $1 \, \text{mm} = 10^{-3}$ m. However, in the SI system, prefixes are not allowed in the denominator of a fraction and therefore it is better to use the equivalent $1 \, \text{N/mm}^2 = 1 \, \text{MN/m}^2 = 1 \, \text{MPa}$.

P

P

P

P

(a)

P

Region of
uniform
deformation
of bar

P

(b)

1.4   Average Normal Stress in an Axially Loaded Bar

In this section we will determine the average stress distribution acting on the cross-sectional area of an axially loaded bar such as the one shown in Fig. 1–13*a*. This bar is **prismatic** since all cross sections are the same throughout its length. When the load P is applied to the bar through the centroid of its cross-sectional area, then the bar will deform uniformly throughout the central region of its length, as shown in Fig. 1–13*b*, provided the material of the bar is both homogeneous and isotropic.

Homogeneous material has the same physical and mechanical properties throughout its volume, and ***isotropic material*** has these same properties in all directions. Many engineering materials may be approximated as being both homogeneous and isotropic as assumed here. Steel, for example, contains thousands of randomly oriented crystals in each cubic millimeter of its volume, and since most problems involving this material have a physical size that is very much larger than a single crystal, the above assumption regarding its material composition is quite realistic.

Note that anisotropic materials such as wood have different properties in different directions, and although this is the case, like wood if the anisotropy is oriented along the bar's axis, then the bar will also deform uniformly when subjected to the axial load P.

Average Normal Stress Distribution.   If we pass a section through the bar, and separate it into two parts, then equilibrium requires the resultant normal force at the section to be P, Fig. 1–13*c*. Due to the *uniform* deformation of the material, it is necessary that the cross section be subjected to a *constant normal stress distribution*, Fig. 1–13*d*.

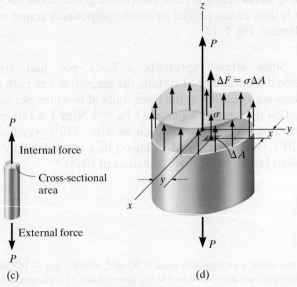

z

P

$\Delta F = \sigma \Delta A$

σ

x

y

ΔA

P

y

x

P

Internal force

Cross-sectional
area

External force

P

(c)

(d)

Fig. 1–13

As a result, each small area ΔA on the cross section is subjected to a force $\Delta F = \sigma \, \Delta A$, and the *sum* of these forces acting over the entire cross-sectional area must be equivalent to the internal resultant force **P** at the section. If we let $\Delta A \rightarrow dA$ and therefore $\Delta F \rightarrow dF$, then, recognizing σ is *constant*, we have

$$+\uparrow F_{Rz} = \Sigma F_z; \qquad \int dF = \int_A \sigma \, dA$$

$$P = \sigma A$$

$$\boxed{\sigma = \frac{P}{A}} \qquad\qquad (1\text{–}6)$$

Here

σ = average normal stress at any point on the cross-sectional area

P = *internal resultant normal force*, which acts through the *centroid* of the cross-sectional area. P is determined using the method of sections and the equations of equilibrium

A = cross-sectional area of the bar where σ is determined

Since the internal load P passes through the centroid of the cross-section the uniform stress distribution will produce zero moments about the x and y axes passing through this point, Fig. 1–13d. To show this, we require the moment of P about each axis to be equal to the moment of the stress distribution about the axes, namely,

$$(M_R)_x = \Sigma M_x; \qquad 0 = \int_A y \, dF = \int_A y\sigma \, dA = \sigma \int_A y \, dA$$

$$(M_R)_y = \Sigma M_y; \qquad 0 = -\int_A x \, dF = -\int_A x\sigma \, dA = -\sigma \int_A x \, dA$$

These equations are indeed satisfied, since by definition of the centroid, $\int y \, dA = 0$ and $\int x \, dA = 0$. (See Appendix A.)

Equilibrium.   It should be apparent that only a normal stress exists on any small volume element of material located at each point on the cross section of an axially loaded bar. If we consider vertical equilibrium of the element, Fig. 1–14, then apply the equation of force equilibrium,

$$\Sigma F_z = 0; \qquad\qquad \sigma(\Delta A) - \sigma'(\Delta A) = 0$$

$$\sigma = \sigma'$$

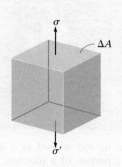

Fig. 1–14

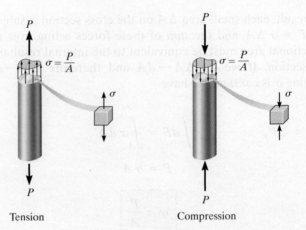

Tension   Compression

Fig. 1–15

In other words, the two normal stress components on the element must be equal in magnitude but opposite in direction. This is referred to as *uniaxial stress*.

The previous analysis applies to members subjected to either tension or compression, as shown in Fig. 1–15. As a graphical interpretation, the ***magnitude*** of the internal resultant force **P** is ***equivalent*** to the ***volume*** under the stress diagram; that is, $P = \sigma A$ (volume = height × base). Furthermore, as a consequence of the balance of moments, ***this resultant passes through the centroid of this volume***.

Although we have developed this analysis for *prismatic* bars, this assumption can be relaxed somewhat to include bars that have a *slight taper*. For example, it can be shown, using the more exact analysis of the theory of elasticity, that for a tapered bar of rectangular cross section, for which the angle between two adjacent sides is 15°, the average normal stress, as calculated by $\sigma = P/A$, is only 2.2% *less* than its value found from the theory of elasticity.

Maximum Average Normal Stress.

In our analysis both the internal force P and the cross-sectional area A were *constant* along the longitudinal axis of the bar, and as a result the normal stress $\sigma = P/A$ is also *constant* throughout the bar's length. Occasionally, however, the bar may be subjected to *several* external loads along its axis, or a change in its cross-sectional area may occur. As a result, the normal stress within the bar could be different from one section to the next, and, if the *maximum* average normal stress is to be determined, then it becomes important to find the location where the ratio P/A is a *maximum*. To do this it is necessary to determine the internal force P at various sections along the bar. Here it may be helpful to show this variation by drawing an ***axial or normal force diagram***. Specifically, this diagram is a plot of the normal force P versus its position x along the bar's length. As a sign convention, P will be positive if it causes tension in the member, and negative if it causes compression. Once the internal loading throughout the bar is known, the maximum ratio of P/A can then be identified.

This steel tie rod is used as a hanger to suspend a portion of a staircase, and as a result it is subjected to tensile stress.

Important Points

- When a body subjected to external loads is sectioned, there is a distribution of force acting over the sectioned area which holds each segment of the body in equilibrium. The intensity of this internal force at a point in the body is referred to as *stress*.

- Stress is the limiting value of force per unit area, as the area approaches zero. For this definition, the material is considered to be continuous and cohesive.

- The magnitude of the stress components at a point depends upon the type of loading acting on the body, and the orientation of the element at the point.

- When a prismatic bar is made from homogeneous and isotropic material, and is subjected to an axial force acting through the centroid of the cross-sectional area, then the center region of the bar will deform uniformly. As a result, the material will be subjected *only to normal stress*. This stress is uniform or *averaged* over the cross-sectional area.

Procedure for Analysis

The equation $\sigma = P/A$ gives the *average* normal stress on the cross-sectional area of a member when the section is subjected to an internal resultant normal force **P**. For axially loaded members, application of this equation requires the following steps.

Internal Loading.

- Section the member *perpendicular* to its longitudinal axis at the point where the normal stress is to be determined and use the necessary free-body diagram and force equation of equilibrium to obtain the *internal axial force* **P** at the section.

Average Normal Stress.

- Determine the member's cross-sectional area at the section and calculate the average normal stress $\sigma = P/A$.

- It is suggested that σ be shown acting on a small volume element of the material located at a point on the section where stress is calculated. To do this, first draw σ on the face of the element coincident with the sectioned area A. Here σ acts in the *same direction* as the internal force **P** since all the normal stresses on the cross section develop this resultant. The normal stress σ on the other face of the element acts in the opposite direction.

EXAMPLE | 1.6

The bar in Fig. 1–16a has a constant width of 35 mm and a thickness of 10 mm. Determine the maximum average normal stress in the bar when it is subjected to the loading shown.

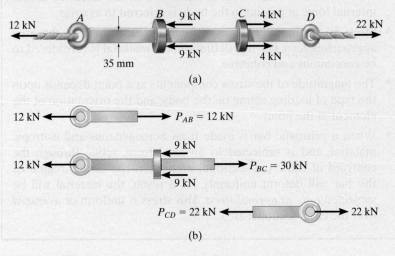

(a)

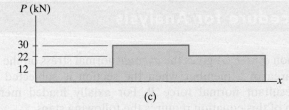

(b)

(c)

SOLUTION

Internal Loading. By inspection, the internal axial forces in regions *AB*, *BC*, and *CD* are all constant yet have different magnitudes. Using the method of sections, these loadings are determined in Fig. 1–16b; and the normal force diagram which represents these results graphically is shown in Fig. 1–16c. The largest loading is in region *BC*, where $P_{BC} = 30$ kN. Since the cross-sectional area of the bar is *constant*, the largest average normal stress also occurs within this region of the bar.

Average Normal Stress. Applying Eq. 1–6, we have

$$\sigma_{BC} = \frac{P_{BC}}{A} = \frac{30(10^3)\,\text{N}}{(0.035\,\text{m})(0.010\,\text{m})} = 85.7\,\text{MPa} \qquad Ans.$$

NOTE: The stress distribution acting on an arbitrary cross section of the bar within region *BC* is shown in Fig. 1–16d. Graphically the *volume* (or "block") represented by this distribution of stress is equivalent to the load of 30 kN; that is, 30 kN = (85.7 MPa)(35 mm)(10 mm).

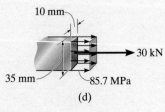

(d)

Fig. 1–16

EXAMPLE 1.7

The 80-kg lamp is supported by two rods AB and BC as shown in Fig. 1–17a. If AB has a diameter of 10 mm and BC has a diameter of 8 mm, determine the average normal stress in each rod.

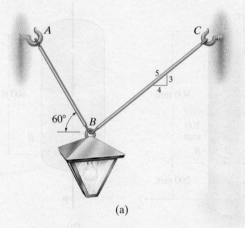

(a)

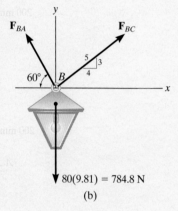

(b)

Fig. 1–17

SOLUTION

Internal Loading.   We must first determine the axial force in each rod. A free-body diagram of the lamp is shown in Fig. 1–17b. Applying the equations of force equilibrium,

$$\xrightarrow{+} \Sigma F_x = 0; \qquad F_{BC}\left(\tfrac{4}{5}\right) - F_{BA}\cos 60° = 0$$

$$+\uparrow \Sigma F_y = 0; \quad F_{BC}\left(\tfrac{3}{5}\right) + F_{BA}\sin 60° - 784.8 \text{ N} = 0$$

$$F_{BC} = 395.2 \text{ N}, \qquad F_{BA} = 632.4 \text{ N}$$

By Newton's third law of action, equal but opposite reaction, these forces subject the rods to tension throughout their length.

Average Normal Stress.   Applying Eq. 1–6,

$$\sigma_{BC} = \frac{F_{BC}}{A_{BC}} = \frac{395.2 \text{ N}}{\pi(0.004 \text{ m})^2} = 7.86 \text{ MPa} \qquad Ans.$$

$$\sigma_{BA} = \frac{F_{BA}}{A_{BA}} = \frac{632.4 \text{ N}}{\pi(0.005 \text{ m})^2} = 8.05 \text{ MPa} \qquad Ans.$$

NOTE: The average normal stress distribution acting over a cross section of rod AB is shown in Fig. 1–17c, and at a point on this cross section, an element of material is stressed as shown in Fig. 1–17d.

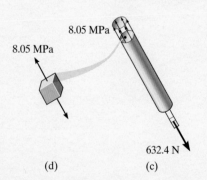

(d)

(c)

EXAMPLE | 1.8

The casting shown in Fig. 1–18a is made of steel having a specific weight of 80 kN/m³. Determine the average compressive stress acting at points A and B.

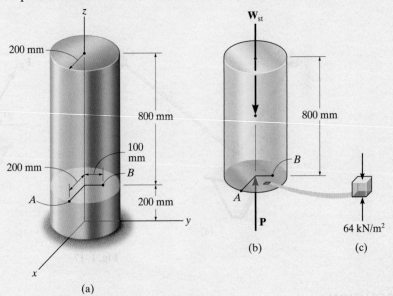

Fig. 1–18

SOLUTION

Internal Loading. A free-body diagram of the top segment of the casting where the section passes through points A and B is shown in Fig. 1–18b. The weight of this segment is determined from $W_{st} = \gamma_{st} V_{st}$. Thus the internal axial force P at the section is

$$+\uparrow \Sigma F_z = 0; \qquad P - W_{st} = 0$$

$$P - (80 \text{ kN/m}^3)(0.8 \text{ m})[\pi(0.2 \text{ m})^2] = 0$$

$$P = 8.042 \text{ kN}$$

Average Compressive Stress. The cross-sectional area at the section is $A = \pi(0.2 \text{ m})^2$, and so the average compressive stress becomes

$$\sigma = \frac{P}{A} = \frac{8.042 \text{ kN}}{\pi(0.2 \text{ m})^2} = 64.0 \text{ kN/m}^2 \qquad \qquad Ans.$$

NOTE: The stress shown on the volume element of material in Fig. 1–18c is representative of the conditions at either point A or B. Notice that this stress acts *upward* on the bottom or shaded face of the element since this face forms part of the bottom surface area of the section, and on this surface, the resultant internal force **P** is pushing upward.

EXAMPLE 1.9

Member AC shown in Fig. 1–19a is subjected to a vertical force of 3 kN. Determine the position x of this force so that the average compressive stress at the smooth support C is equal to the average tensile stress in the tie rod AB. The rod has a cross-sectional area of 400 mm² and the contact area at C is 650 mm².

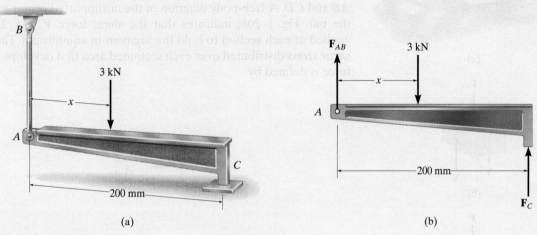

(a)   (b)

Fig. 1–19

SOLUTION

Internal Loading.   The forces at A and C can be related by considering the free-body diagram for member AC, Fig. 1–19b. There are three unknowns, namely, F_{AB}, F_C, and x. To solve this problem we will work in units of newtons and millimeters.

$+\uparrow \Sigma F_y = 0;$ $\qquad\qquad F_{AB} + F_C - 3000\text{ N} = 0 \qquad (1)$

$\zeta+\Sigma M_A = 0;$ $\qquad -3000\text{ N}(x) + F_C(200\text{ mm}) = 0 \qquad (2)$

Average Normal Stress.   A necessary third equation can be written that requires the tensile stress in the bar AB and the compressive stress at C to be equivalent, i.e.,

$$\sigma = \frac{F_{AB}}{400\text{ mm}^2} = \frac{F_C}{650\text{ mm}^2}$$

$$F_C = 1.625F_{AB}$$

Substituting this into Eq. 1, solving for F_{AB}, then solving for F_C, we obtain

$$F_{AB} = 1143\text{ N}$$
$$F_C = 1857\text{ N}$$

The position of the applied load is determined from Eq. 2,

$$x = 124\text{ mm} \qquad\qquad\qquad Ans.$$

NOTE: $0 < x < 200$ mm, as required.

1

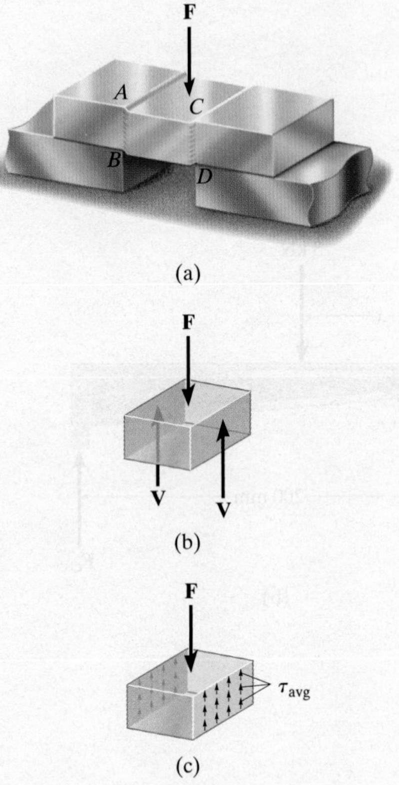

F

A   C

B   D

(a)

F

V   **V**

(b)

F

τ_{avg}

(c)

Fig. 1–20

1.5   Average Shear Stress

Shear stress has been defined in Section 1.3 as the stress component that acts *in the plane* of the sectioned area. To show how this stress can develop, consider the effect of applying a force **F** to the bar in Fig. 1–20*a*. If the supports are considered rigid, and **F** is large enough, it will cause the material of the bar to deform and fail along the planes identified by *AB* and *CD*. A free-body diagram of the unsupported center segment of the bar, Fig. 1–20*b*, indicates that the shear force $V = F/2$ must be applied at each section to hold the segment in equilibrium. The *average shear stress* distributed over each sectioned area that develops this shear force is defined by

$$\tau_{avg} = \frac{V}{A} \qquad (1\text{–}7)$$

Here

τ_{avg} = average shear stress at the section, which is assumed to be the *same* at each point located on the section

V = internal resultant shear force on the section determined from the equations of equilibrium

A = area at the section

The distribution of average shear stress acting over the sections is shown in Fig. 1–20*c*. Notice that τ_{avg} is in the *same direction* as **V**, since the shear stress must create associated forces all of which contribute to the internal resultant force **V** at the section.

The loading case discussed here is an example of *simple or direct shear*, since the shear is caused by the *direct action* of the applied load **F**. This type of shear often occurs in various types of simple connections that use bolts, pins, welding material, etc. In all these cases, however, application of Eq. 1–7 is *only approximate*. A more precise investigation of the shear-stress distribution over the section often reveals that much larger shear stresses occur in the material than those predicted by this equation. Although this may be the case, application of Eq. 1–7 is generally acceptable for many problems in engineering design and analysis. For example, engineering codes allow its use when considering design sizes for fasteners such as bolts and for obtaining the bonding strength of glued joints subjected to shear loadings.

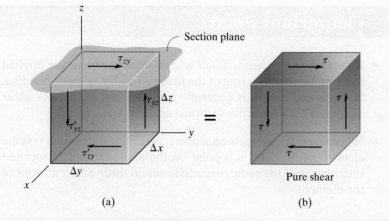

Fig. 1–21

Shear Stress Equilibrium.

Figure 1–21a shows a volume element of material taken at a point located on the surface of a sectioned area which is subjected to a shear stress τ_{zy}. Force and moment equilibrium requires the shear stress acting on this face of the element to be accompanied by shear stress acting on three other faces. To show this we will first consider force equilibrium in the y direction. Then

$$\overset{\overbrace{\text{force}}}{\underset{\underbrace{\text{stress area}}}{}}$$

$$\Sigma F_y = 0; \qquad \tau_{zy}(\Delta x\,\Delta y) - \tau'_{zy}\,\Delta x\,\Delta y = 0$$

$$\tau_{zy} = \tau'_{zy}$$

In a similar manner, force equilibrium in the z direction yields $\tau_{yz} = \tau'_{yz}$. Finally, taking moments about the x axis,

$$\overset{\text{moment}}{\overset{\overbrace{\text{force} \qquad \text{arm}}}{\underset{\underbrace{\text{stress area}}}{}}}$$

$$\Sigma M_x = 0; \qquad -\tau_{zy}(\Delta x\,\Delta y)\,\Delta z + \tau_{yz}(\Delta x\,\Delta z)\,\Delta y = 0$$

$$\tau_{zy} = \tau_{yz}$$

so that

$$\tau_{zy} = \tau'_{zy} = \tau_{yz} = \tau'_{yz} = \tau$$

In other words, ***all four shear stresses must have equal magnitude and be directed either toward or away from each other at opposite edges of the element***, Fig. 1–21b. This is referred to as the *complementary property of shear*, and under the conditions shown in Fig. 1–21, the material is subjected to *pure shear*.

1

Important Points

- If two parts are *thin or small* when joined together, the applied loads may cause shearing of the material with negligible bending. If this is the case, it is generally assumed that an *average shear stress* acts over the cross-sectional area.

- When shear stress τ acts on a plane, then equilibrium of a volume element of material at a point on the plane requires associated shear stress of the same magnitude act on three adjacent sides of the element.

Procedure for Analysis

The equation $\tau_{\mathrm{avg}} = V/A$ is used to determine the *average shear stress* in the material. Application requires the following steps.

Internal Shear.

- Section the member at the point where the average shear stress is to be determined.

- Draw the necessary free-body diagram, and calculate the internal shear force $\mathbf{V}$ acting at the section that is necessary to hold the part in equilibrium.

Average Shear Stress.

- Determine the sectioned area A, and determine the average shear stress $\tau_{\mathrm{avg}} = V/A$.

- It is suggested that τ_{avg} be shown on a small volume element of material located at a point on the section where it is determined. To do this, first draw τ_{avg} on the face of the element, coincident with the sectioned area A. This stress acts in the *same direction* as $\mathbf{V}$. The shear stresses acting on the three adjacent planes can then be drawn in their appropriate directions following the scheme shown in Fig. 1–21.

EXAMPLE | 1.10

Determine the average shear stress in the 20-mm-diameter pin at A and the 30-mm-diameter pin at B that support the beam in Fig. 1–22a.

SOLUTION

Internal Loadings. The forces on the pins can be obtained by considering the equilibrium of the beam, Fig. 1–22b.

$$\zeta + \Sigma M_A = 0; \quad F_B\left(\frac{4}{5}\right)(6\text{ m}) - 30\text{ kN}(2\text{ m}) = 0 \quad F_B = 12.5\text{ kN}$$

$$\xrightarrow{+} \Sigma F_x = 0; \quad (12.5\text{ kN})\left(\frac{3}{5}\right) - A_x = 0 \quad A_x = 7.50\text{ kN}$$

$$+\uparrow \Sigma F_y = 0; \quad A_y + (12.5\text{ kN})\left(\frac{4}{5}\right) - 30\text{ kN} = 0$$

$$A_y = 20\text{ kN}$$

Thus, the resultant force acting on pin A is

$$F_A = \sqrt{A_x^2 + A_y^2} = \sqrt{(7.50\text{ kN})^2 + (20\text{ kN})^2} = 21.36\text{ kN}$$

The pin at A is supported by two fixed "leaves" and so the free-body diagram of the center segment of the pin shown in Fig. 1–22c has *two* shearing surfaces between the beam and each leaf. The force of the beam (21.36 kN) acting on the pin is therefore supported by shear force on each of these surfaces. This case is called *double shear*. Thus,

$$V_A = \frac{F_A}{2} = \frac{21.36\text{ kN}}{2} = 10.68\text{ kN}$$

In Fig. 1–22a, note that pin B is subjected to *single shear*, which occurs on the section between the cable and beam, Fig. 1–22d. For this pin segment,

$$V_B = F_B = 12.5\text{ kN}$$

Average Shear Stress.

$$(\tau_A)_{\text{avg}} = \frac{V_A}{A_A} = \frac{10.68(10^3)\text{ N}}{\frac{\pi}{4}(0.02\text{ m})^2} = 34.0\text{ MPa} \qquad Ans.$$

$$(\tau_B)_{\text{avg}} = \frac{V_B}{A_B} = \frac{12.5(10^3)\text{ N}}{\frac{\pi}{4}(0.03\text{ m})^2} = 17.7\text{ MPa} \qquad Ans.$$

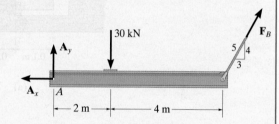

(a)

(b)

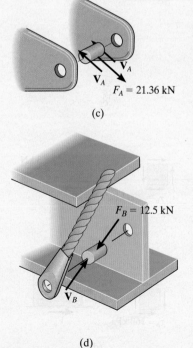

(c)

(d)

Fig. 1–22

EXAMPLE 1.11

If the wood joint in Fig. 1–23a has a width of 150 mm, determine the average shear stress developed along shear planes a–a and b–b. For each plane, represent the state of stress on an element of the material.

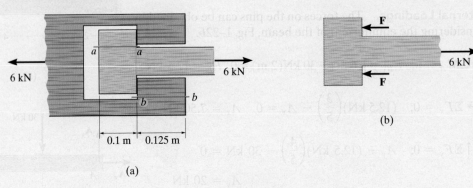

(a)

(b)

Fig. 1–23

SOLUTION

Internal Loadings. Referring to the free-body diagram of the member, Fig. 1–23b,

$$\xrightarrow{+} \Sigma F_x = 0; \qquad 6 \text{ kN} - F - F = 0 \qquad F = 3 \text{ kN}$$

Now consider the equilibrium of segments cut across shear planes a–a and b–b, shown in Figs. 1–23c and 1–23d.

$$\xrightarrow{+} \Sigma F_x = 0; \qquad V_a - 3 \text{ kN} = 0 \qquad V_a = 3 \text{ kN}$$

$$\xrightarrow{+} \Sigma F_x = 0; \qquad 3 \text{ kN} - V_b = 0 \qquad V_b = 3 \text{ kN}$$

Average Shear Stress.

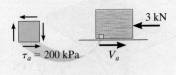

(c)

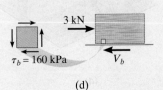

(d)

$$(\tau_a)_{\text{avg}} = \frac{V_a}{A_a} = \frac{3(10^3) \text{ N}}{(0.1 \text{ m})(0.15 \text{ m})} = 200 \text{ kPa} \qquad \textit{Ans.}$$

$$(\tau_b)_{\text{avg}} = \frac{V_b}{A_b} = \frac{3(10^3) \text{ N}}{(0.125 \text{ m})(0.15 \text{ m})} = 160 \text{ kPa} \qquad \textit{Ans.}$$

The state of stress on elements located on sections a–a and b–b is shown in Figs. 1–23c and 1–23d, respectively.

EXAMPLE | 1.12

The inclined member in Fig. 1–24a is subjected to a compressive force of 3000 N. Determine the average compressive stress along the smooth areas of contact defined by *AB* and *BC*, and the average shear stress along the horizontal plane defined by *DB*.

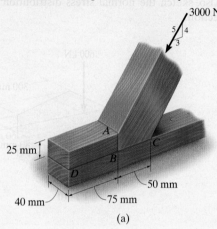

(a)

Fig. 1–24

SOLUTION

Internal Loadings. The free-body diagram of the inclined member is shown in Fig. 1–24b. The compressive forces acting on the areas of contact are

$$\xrightarrow{+} \Sigma F_x = 0; \qquad F_{AB} - 3000\ \text{N}\left(\tfrac{3}{5}\right) = 0 \qquad F_{AB} = 1800\ \text{N}$$

$$+\uparrow \Sigma F_y = 0; \qquad F_{BC} - 3000\ \text{N}\left(\tfrac{4}{5}\right) = 0 \qquad F_{BC} = 2400\ \text{N}$$

Also, from the free-body diagram of the top segment *ABD* of the bottom member, Fig. 1–24c, the shear force acting on the sectioned horizontal plane *DB* is

$$\xrightarrow{+} \Sigma F_x = 0; \qquad\qquad V = 1800\ \text{N}$$

Average Stress. The average compressive stresses along the horizontal and vertical planes of the inclined member are

$$\sigma_{AB} = \frac{P_{AB}}{A_{AB}} = \frac{1800\ \text{N}}{(25\ \text{mm})(40\ \text{mm})} = 1.80\ \text{N/mm}^2 \qquad \textit{Ans.}$$

$$\sigma_{BC} = \frac{P_{BC}}{A_{BC}} = \frac{2400\ \text{N}}{(25\ \text{mm})(40\ \text{mm})} = 1.20\ \text{N/mm}^2 \qquad \textit{Ans.}$$

These stress distributions are shown in Fig. 1–24d.

The average shear stress acting on the horizontal plane defined by *DB* is

$$\tau_{\text{avg}} = \frac{1800\ \text{N}}{(75\ \text{mm})(40\ \text{mm})} = 0.60\ \text{N/mm}^2 \qquad \textit{Ans.}$$

This stress is shown uniformly distributed over the sectioned area in Fig. 1–24e.

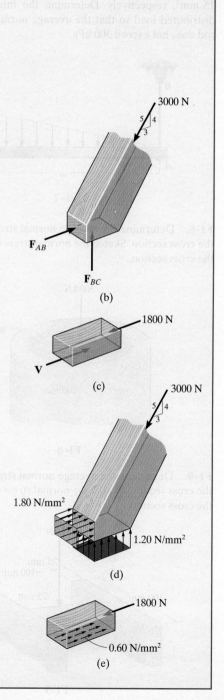

(b)

(c)

(d)

(e)

FUNDAMENTAL PROBLEMS

F1–7. The uniform beam is supported by two rods AB and CD that have cross-sectional areas of 10 mm² and 15 mm², respectively. Determine the intensity w of the distributed load so that the average normal stress in each rod does not exceed 300 kPa.

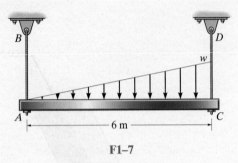

F1–7

F1–8. Determine the average normal stress developed on the cross section. Sketch the normal stress distribution over the cross section.

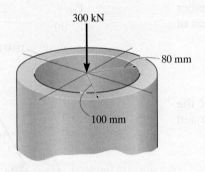

F1–8

F1–9. Determine the average normal stress developed on the cross section. Sketch the normal stress distribution over the cross section.

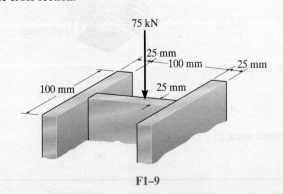

F1–9

F1–10. If the 600-kN force acts through the centroid of the cross section, determine the location $\bar{y}$ of the centroid and the average normal stress developed on the cross section. Also, sketch the normal stress distribution over the cross section.

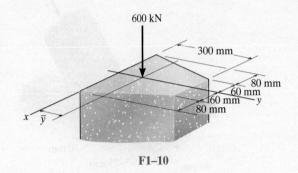

F1–10

F1–11. Determine the average normal stress developed at points A, B, and C. The diameter of each segment is indicated in the figure.

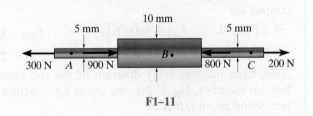

F1–11

F1–12. Determine the average normal stress developed in rod AB if the load has a mass of 50 kg. The diameter of rod AB is 8 mm.

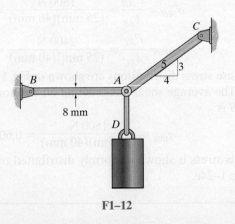

F1–12

PROBLEMS

1–31. The column is subjected to an axial force of 8 kN, which is applied through the centroid of the cross-sectional area. Determine the average normal stress acting at section *a–a*. Show this distribution of stress acting over the area's cross section.

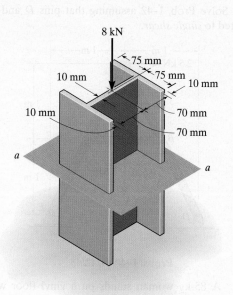

Prob. 1–31

*1–32. The lever is held to the fixed shaft using a tapered pin *AB*, which has a mean diameter of 6 mm. If a couple is applied to the lever, determine the average shear stress in the pin between the pin and lever.

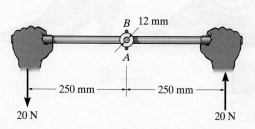

Prob. 1–32

•1–33. The bar has a cross-sectional area *A* and is subjected to the axial load *P*. Determine the average normal and average shear stresses acting over the shaded section, which is oriented at θ from the horizontal. Plot the variation of these stresses as a function of θ ($0 \le \theta \le 90°$).

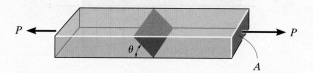

Prob. 1–33

1–34. The built-up shaft consists of a pipe *AB* and solid rod *BC*. The pipe has an inner diameter of 20 mm and outer diameter of 28 mm. The rod has a diameter of 12 mm. Determine the average normal stress at points *D* and *E* and represent the stress on a volume element located at each of these points.

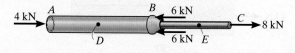

Prob. 1–34

1–35. The bars of the truss each have a cross-sectional area of 780 mm². Determine the average normal stress in each member due to the loading *P* = 40 kN. State whether the stress is tensile or compressive.

*1–36. The bars of the truss each have a cross-sectional area of 780 mm². If the maximum average normal stress in any bar is not to exceed 140 MPa, determine the maximum magnitude *P* of the loads that can be applied to the truss.

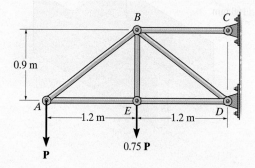

Probs. 1–35/36

•1–37. The plate has a width of 0.5 m. If the stress distribution at the support varies as shown, determine the force **P** applied to the plate and the distance d to where it is applied.

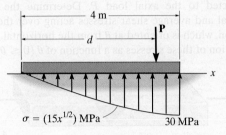

$\sigma = (15x^{1/2})$ MPa   30 MPa

Prob. 1–37

1–38. The two members used in the construction of an aircraft fuselage are joined together using a 30° fish-mouth weld. Determine the average normal and average shear stress on the plane of each weld. Assume each inclined plane supports a horizontal force of 2 kN.

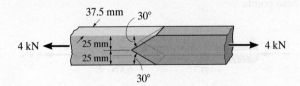

37.5 mm   30°

4 kN ◄—   25 mm / 25 mm   —► 4 kN

30°

Prob. 1–38

1–39. If the block is subjected to the centrally applied force of 600 kN, determine the average normal stress in the material. Show the stress acting on a differential volume element of the material.

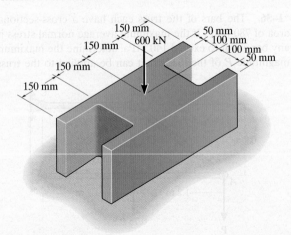

150 mm
150 mm   600 kN
150 mm
150 mm
150 mm
50 mm
100 mm
100 mm
50 mm

Prob. 1–39

***1–40.** The pins on the frame at B and C each have a diameter of 6 mm. If these pins are subjected to *double shear*, determine the average shear stress in each pin.

•1–41. Solve Prob. 1–40 assuming that pins B and C are subjected to *single shear*.

1–42. The pins on the frame at D and E each have a diameter of 6 mm. If these pins are subjected to *double shear,* determine the average shear stress in each pin.

1–43. Solve Prob. 1–42 assuming that pins D and E are subjected to *single shear*.

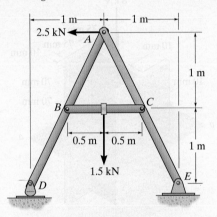

1 m   1 m
2.5 kN ◄—
A
1 m
B C
0.5 m   0.5 m
1 m
1.5 kN
D E

Probs. 1–40/41/42/43

***1–44.** A 85-kg woman stands on a vinyl floor wearing stiletto high-heel shoes. If the heel has the dimensions shown, determine the average normal stress she exerts on the floor and compare it with the average normal stress developed when a man having the same weight is wearing flat-heeled shoes. Assume the load is applied slowly, so that dynamic effects can be ignored. Also, assume the entire weight is supported only by the heel of one shoe.

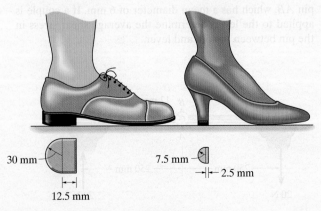

30 mm   7.5 mm
2.5 mm
12.5 mm

Prob. 1–44

•1–45.   The truss is made from three pin-connected members having the cross-sectional areas shown in the figure. Determine the average normal stress developed in each member when the truss is subjected to the load shown. State whether the stress is tensile or compressive.

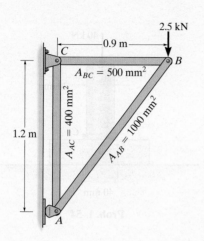

Prob. 1–45

1–46.   Determine the average normal stress developed in links AB and CD of the smooth two-tine grapple that supports the log having a mass of 3 Mg. The cross-sectional area of each link is 400 mm².

1–47.   Determine the average shear stress developed in pins A and B of the smooth two-tine grapple that supports the log having a mass of 3 Mg. Each pin has a diameter of 25 mm and is subjected to double shear.

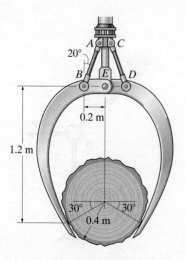

Probs. 1–46/47

*1–48.   The beam is supported by a pin at A and a short link BC. If $P = 15$ kN, determine the average shear stress developed in the pins at A, B, and C. All pins are in double shear as shown, and each has a diameter of 18 mm.

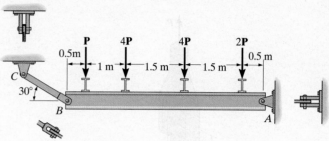

Prob. 1–48

•1–49.   The beam is supported by a pin at A and a short link BC. Determine the maximum magnitude P of the loads the beam will support if the average shear stress in each pin is not to exceed 80 MPa. All pins are in double shear as shown, and each has a diameter of 18 mm.

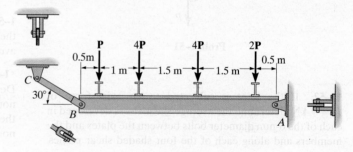

Prob. 1–49

1–50.   The block is subjected to a compressive force of 2 kN. Determine the average normal and average shear stress developed in the wood fibers that are oriented along section a–a at 30° with the axis of the block.

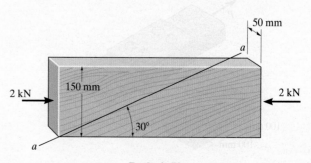

Prob. 1–50

1–51. During the tension test, the wooden specimen is subjected to an average normal stress of 15 MPa. Determine the axial force **P** applied to the specimen. Also, find the average shear stress developed along section *a–a* of the specimen.

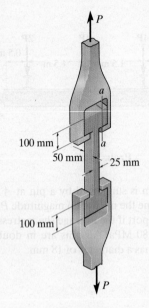

100 mm

50 mm

25 mm

100 mm

Prob. 1–51

*1–52.** If the joint is subjected to an axial force of $P = 9$ kN, determine the average shear stress developed in each of the 6-mm diameter bolts between the plates and the members and along each of the four shaded shear planes.

•**1–53.** The average shear stress in each of the 6-mm diameter bolts and along each of the four shaded shear planes is not allowed to exceed 80 MPa and 500 kPa, respectively. Determine the maximum axial force **P** that can be applied to the joint.

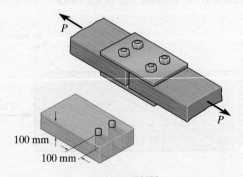

100 mm

100 mm

Probs. 1–52/53

1–54. The shaft is subjected to the axial force of 40 kN. Determine the average bearing stress acting on the collar C and the normal stress in the shaft.

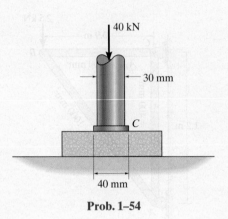

40 kN

30 mm

C

40 mm

Prob. 1–54

1–55. Rods AB and BC each have a diameter of 5 mm. If the load of $P = 2$ kN is applied to the ring, determine the average normal stress in each rod if $\theta = 60°$.

*1–56.** Rods AB and BC each have a diameter of 5 mm. Determine the angle θ of rod BC so that the average normal stress in rod AB is 1.5 times that in rod BC. What is the load **P** that will cause this to happen if the average normal stress in each rod is not allowed to exceed 100 MPa?

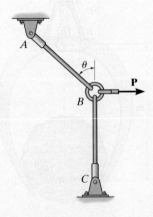

A

θ

P

B

C

Probs. 1–55/56

•1–57. The specimen failed in a tension test at an angle of 52° when the axial load was 100 kN. If the diameter of the specimen is 12 mm, determine the average normal and average shear stress acting on the area of the inclined failure plane. Also, what is the average normal stress acting on the *cross section* when failure occurs?

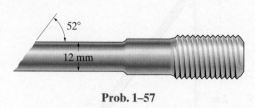

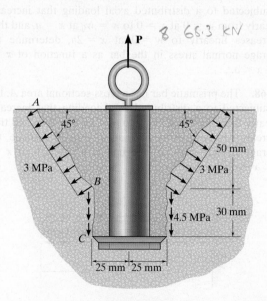

Prob. 1–57

1–58. The anchor bolt was pulled out of the concrete wall and the failure surface formed part of a frustum and cylinder. This indicates a shear failure occurred along the cylinder *BC* and tension failure along the frustum *AB*. If the shear and normal stresses along these surfaces have the magnitudes shown, determine the force **P** that must have been applied to the bolt.

8 68.3 KN

Prob. 1–58

1–59. The open square butt joint is used to transmit a force of 250 kN from one plate to the other. Determine the average normal and average shear stress components that this loading creates on the face of the weld, section *AB*.

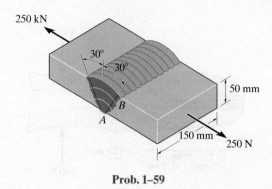

Prob. 1–59

*1–60. If *P* = 20 kN, determine the average shear stress developed in the pins at *A* and *C*. The pins are subjected to double shear as shown, and each has a diameter of 18 mm.

•1–61. Determine the maximum magnitude *P* of the load the beam will support if the average shear stress in each pin is not to allowed to exceed 60 MPa. All pins are subjected to double shear as shown, and each has a diameter of 18 mm.

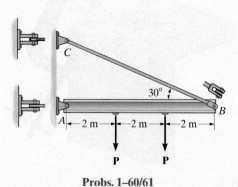

Probs. 1–60/61

1–62. The crimping tool is used to crimp the end of the wire E. If a force of 100 N is applied to the handles, determine the average shear stress in the pin at A. The pin is subjected to double shear and has a diameter of 5 mm. Only a vertical force is exerted on the wire.

1–63. Solve Prob. 1–62 for pin B. The pin is subjected to double shear and has a diameter of 5 mm.

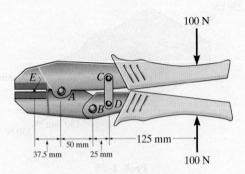

Probs. 1–62/63

***1–64.** The triangular blocks are glued along each side of the joint. A C-clamp placed between two of the blocks is used to draw the joint tight. If the glue can withstand a maximum average shear stress of 800 kPa, determine the maximum allowable clamping force **F**.

•1–65. The triangular blocks are glued along each side of the joint. A C-clamp placed between two of the blocks is used to draw the joint tight. If the clamping force is $F = 900$ N, determine the average shear stress developed in the glued shear plane.

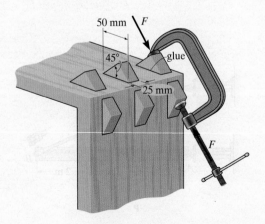

Probs. 1–64/65

1–66. Determine the largest load P that can be a applied to the frame without causing either the average normal stress or the average shear stress at section a–a to exceed $\sigma = 150$ MPa and $\tau = 60$ MPa, respectively. Member CB has a square cross section of 25 mm on each side.

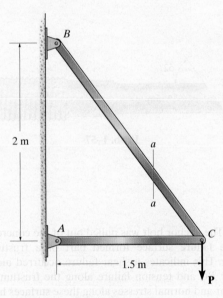

Prob. 1–66

1–67. The prismatic bar has a cross-sectional area A. If it is subjected to a distributed axial loading that increases linearly from $w = 0$ at $x = 0$ to $w = w_0$ at $x = a$, and then decreases linearly to $w = 0$ at $x = 2a$, determine the average normal stress in the bar as a function of x for $0 \le x < a$.

***1–68.** The prismatic bar has a cross-sectional area A. If it is subjected to a distributed axial loading that increases linearly from $w = 0$ at $x = 0$ to $w = w_0$ at $x = a$, and then decreases linearly to $w = 0$ at $x = 2a$, determine the average normal stress in the bar as a function of x for $a < x \le 2a$.

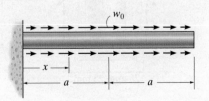

Probs. 1–67/68

•**1–69.** The tapered rod has a radius of $r = (50 - \frac{x}{6})$ mm and is subjected to the distributed loading of $w = (12 + 0.32x)$ N/mm. Determine the average normal stress at the center of the rod, B.

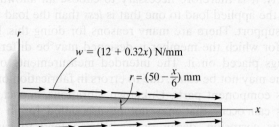

$w = (12 + 0.32x)$ N/mm

$r = (50 - \frac{x}{6})$ mm

75 mm   75 mm

Prob. 1–69

1–70. The pedestal supports a load **P** at its center. If the material has a mass density ρ, determine the radial dimension r as a function of z so that the average normal stress in the pedestal remains constant. The cross section is circular.

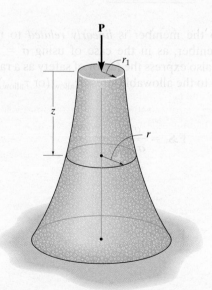

Prob. 1–70

1–71. Determine the average normal stress at section a–a and the average shear stress at section b–b in member AB. The cross section is square, 12 mm on each side.

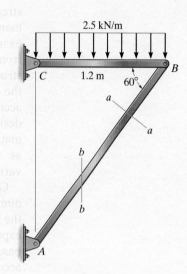

2.5 kN/m

1.2 m

60°

B

C

a

a

b

b

A

Prob. 1–71

*1–72. Consider the general problem of a bar made from m segments, each having a constant cross-sectional area A_m and length L_m. If there are n loads on the bar as shown, write a computer program that can be used to determine the average normal stress at any specified location x. Show an application of the program using the values $L_1 = 2$ m, $d_1 = 1$ m, $P_1 = 1200$ N, $A_1 = 1800$ mm^2, $L_2 = 1$ m, $d_2 = 3$ m, $P_2 = -1000$ N, $A_2 = 600$ mm^2.

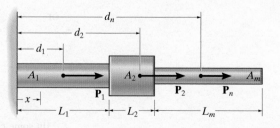

d_n

d_2

d_1

A_1 A_2 A_m

P_1 P_2 P_n

x

L_1 L_2 L_m

Prob. 1–72

1.6 Allowable Stress

To properly *design* a structural member or mechanical element it is necessary to restrict the stress in the material to a level that will be safe. To ensure this safety, it is therefore necessary to choose an allowable stress that restricts the applied load to one that is *less* than the load the member can fully support. There are many reasons for doing this. For example, the load for which the member is designed may be different from actual loadings placed on it. The intended measurements of a structure or machine may not be exact, due to errors in fabrication or in the assembly of its component parts. Unknown vibrations, impact, or accidental loadings can occur that may not be accounted for in the design. Atmospheric corrosion, decay, or weathering tend to cause materials to deteriorate during service. And lastly, some materials, such as wood, concrete, or fiber-reinforced composites, can show high variability in mechanical properties.

One method of specifying the allowable load for a member is to use a number called the factor of safety. The *factor of safety* (F.S.) is a ratio of the failure load F_{fail} to the allowable load F_{allow}. Here F_{fail} is found from experimental testing of the material, and the factor of safety is selected based on experience so that the above mentioned uncertainties are accounted for when the member is used under similar conditions of loading and geometry. Stated mathematically,

$$\text{F.S.} = \frac{F_{\text{fail}}}{F_{\text{allow}}} \tag{1-8}$$

If the load applied to the member is *linearly related* to the stress developed within the member, as in the case of using $\sigma = P/A$ and $\tau_{\text{avg}} = V/A$, then we can also express the factor of safety as a ratio of the failure stress σ_{fail} (or τ_{fail}) to the allowable stress σ_{allow} (or τ_{allow});* that is,

$$\text{F.S.} = \frac{\sigma_{\text{fail}}}{\sigma_{\text{allow}}} \tag{1-9}$$

or

$$\text{F.S.} = \frac{\tau_{\text{fail}}}{\tau_{\text{allow}}} \tag{1-10}$$

*In some cases, such as columns, the applied load is not linearly related to stress and therefore only Eq. 1–8 can be used to determine the factor of safety. See Chapter 13.

In any of these equations, the factor of safety must be *greater* than 1 in order to avoid the potential for failure. Specific values depend on the types of materials to be used and the intended purpose of the structure or machine. For example, the F.S. used in the design of aircraft or space-vehicle components may be close to 1 in order to reduce the weight of the vehicle. Or, in the case of a nuclear power plant, the factor of safety for some of its components may be as high as 3 due to uncertainties in loading or material behavior. In many cases, the factor of safety for a specific case can be found in design codes and engineering handbooks. These values are intended to form a balance of ensuring public and environmental safety and providing a reasonable economic solution to design.

1.7 Design of Simple Connections

By making simplifying assumptions regarding the behavior of the material, the equations $\sigma = P/A$ and $\tau_{avg} = V/A$ can often be used to analyze or design a simple connection or mechanical element. In particular, if a member is subjected to *normal force* at a section, its required area at the section is determined from

$$A = \frac{P}{\sigma_{\text{allow}}} \tag{1–11}$$

On the other hand, if the section is subjected to an average *shear force*, then the required area at the section is

$$A = \frac{V}{\tau_{\text{allow}}} \tag{1–12}$$

As discussed in Sec. 1.6, the allowable stress used in each of these equations is determined either by applying a factor of safety to the material's normal or shear failure stress or by finding these stresses directly from an appropriate design code.

Three examples of where the above equations apply are shown in Fig. 1–25.

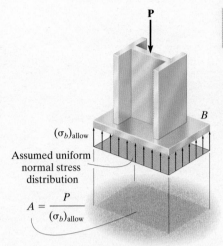

$$A = \frac{P}{(\sigma_b)_{\text{allow}}}$$

The area of the column base plate B is determined from the allowable bearing stress for the concrete.

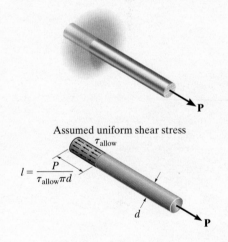

$$l = \frac{P}{\tau_{\text{allow}} \pi d}$$

The embedded length l of this rod in concrete can be determined using the allowable shear stress of the bonding glue.

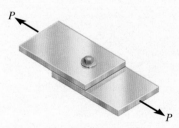

The area of the bolt for this lap joint is determined from the shear stress, which is largest between the plates.

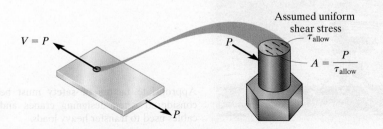

$$A = \frac{P}{\tau_{\text{allow}}}$$

Fig. 1–25

Important Point

- Design of a member for strength is based on selecting an allowable stress that will enable it to safely support its intended load. Since there are many unknown factors that can influence the actual stress in a member, then depending upon the intended use of the member, a *factor of safety* is applied to obtain the allowable load the member can support.

Procedure for Analysis

When solving problems using the average normal and shear stress equations, a careful consideration should first be made as to choose the section over which the critical stress is acting. Once this section is determined, the member must then be designed to have a sufficient area at the section to resist the stress that acts on it. This area is determined using the following steps.

Internal Loading.

- Section the member through the area and draw a free-body diagram of a segment of the member. The internal resultant force at the section is then determined using the equations of equilibrium.

Required Area.

- Provided the allowable stress is known or can be determined, the required area needed to sustain the load at the section is then determined from $A = P/\sigma_{\text{allow}}$ or $A = V/\tau_{\text{allow}}$.

Appropriate factors of safety must be considered when designing cranes and cables used to transfer heavy loads.

EXAMPLE | 1.13

The control arm is subjected to the loading shown in Fig. 1–26a. Determine to the nearest 5 mm the required diameter of the steel pin at C if the allowable shear stress for the steel is τ_{allow} = 55 MPa.

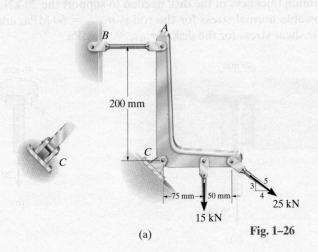

(a)

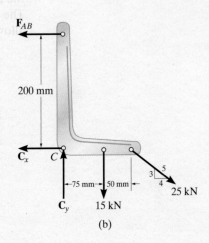

(b)

Fig. 1–26

SOLUTION

Internal Shear Force.   A free-body diagram of the arm is shown in Fig. 1–26b. For equilibrium we have

$$\zeta+\Sigma M_C = 0; \quad F_{AB}(0.2 \text{ m}) - 15 \text{ kN}(0.075 \text{ m}) - 25 \text{ kN}\left(\tfrac{3}{5}\right)(0.125 \text{ m}) = 0$$

$$F_{AB} = 15 \text{ kN}$$

$$\xrightarrow{+}\Sigma F_x = 0; \quad -15 \text{ kN} - C_x + 25 \text{ kN}\left(\tfrac{4}{5}\right) = 0 \quad C_x = 5 \text{ kN}$$

$$+\uparrow\Sigma F_y = 0; \quad C_y - 15 \text{ kN} - 25 \text{ kN}\left(\tfrac{3}{5}\right) = 0 \quad C_y = 30 \text{ kN}$$

The pin at C resists the resultant force at C, which is

$$F_C = \sqrt{(5 \text{ kN})^2 + (30 \text{ kN})^2} = 30.41 \text{ kN}$$

Since the pin is subjected to double shear, a shear force of 15.205 kN acts over its cross-sectional area *between* the arm and each supporting leaf for the pin, Fig. 1–26c.

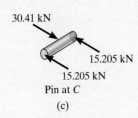

Pin at C

(c)

Required Area.   We have

$$A = \frac{V}{\tau_{allow}} = \frac{15.205 \text{ kN}}{55(10^3) \text{ kN/m}^2} = 276.45(10^{-6}) \text{ m}^2 = 276.45 \text{ mm}^2$$

$$\pi\left(\frac{d}{2}\right)^2 = 276.45 \text{ mm}^2$$

$$d = 18.8 \text{ mm}$$

Use a pin having a diameter of

$$d = 20 \text{ mm} \qquad\qquad Ans.$$

EXAMPLE | 1.14

The suspender rod is supported at its end by a fixed-connected circular disk as shown in Fig. 1–27a. If the rod passes through a 40-mm-diameter hole, determine the minimum required diameter of the rod and the minimum thickness of the disk needed to support the 20-kN load. The allowable normal stress for the rod is $\sigma_{allow} = 60$ MPa, and the allowable shear stress for the disk is $\tau_{allow} = 35$ MPa.

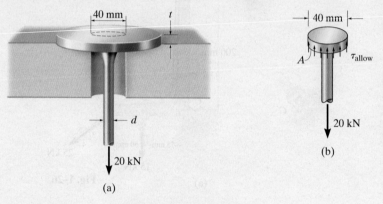

(a)

(b)

Fig. 1–27

SOLUTION

Diameter of Rod.   By inspection, the axial force in the rod is 20 kN. Thus the required cross-sectional area of the rod is

$$A = \frac{P}{\sigma_{allow}}; \qquad \frac{\pi}{4}d^2 = \frac{20(10^3) \text{ N}}{60(10^6) \text{ N/m}^2}$$

so that

$$d = 0.0206 \text{ m} = 20.6 \text{ mm} \qquad\qquad Ans.$$

Thickness of Disk.   As shown on the free-body diagram in Fig. 1–27b, the material at the sectioned area of the disk must resist *shear stress* to prevent movement of the disk through the hole. If this shear stress is *assumed* to be uniformly distributed over the sectioned area, then, since $V = 20$ kN, we have

$$A = \frac{V}{\tau_{allow}}; \qquad 2\pi(0.02 \text{ m})(t) = \frac{20(10^3) \text{ N}}{35(10^6) \text{ N/m}^2}$$

$$t = 4.55(10^{-3}) \text{ m} = 4.55 \text{ mm} \qquad\qquad Ans.$$

EXAMPLE | 1.15

The shaft shown in Fig. 1–28a is supported by the collar at C, which is attached to the shaft and located on the right side of the bearing at B. Determine the largest value of P for the axial forces at E and F so that the bearing stress on the collar does not exceed an allowable stress of $(\sigma_b)_{\text{allow}} = 75$ MPa and the average normal stress in the shaft does not exceed an allowable stress of $(\sigma_t)_{\text{allow}} = 55$ MPa.

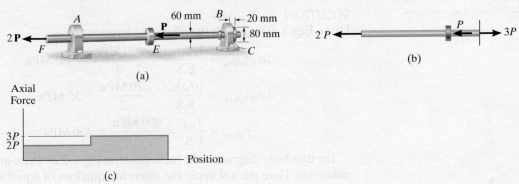

(a)

(b)

(c)

Fig. 1–28

SOLUTION

To solve the problem we will determine P for each possible failure condition. Then we will choose the *smallest* value. Why?

Normal Stress. Using the method of sections, the axial load within region FE of the shaft is 2P, whereas the *largest* axial force, 3P, occurs within region EC, Fig. 1–28b. The variation of the internal loading is clearly shown on the normal-force diagram, Fig. 1–28c. Since the cross-sectional area of the entire shaft is constant, region EC is subjected to the maximum average normal stress. Applying Eq. 1–11, we have

$$A = \frac{P}{\sigma_{\text{allow}}}; \qquad \pi(0.03 \text{ m})^2 = \frac{3P}{55(10^6) \text{ N/m}^2}$$

$$P = 51.8 \text{ kN} \qquad \qquad Ans.$$

Bearing Stress. As shown on the free-body diagram in Fig. 1–28d, the collar at C must resist the load of 3P, which acts over a bearing area of $A_b = [\pi(0.04 \text{ m})^2 - \pi(0.03 \text{ m})^2] = 2.199(10^{-3}) \text{ m}^2$. Thus,

$$A = \frac{P}{\sigma_{\text{allow}}}; \qquad 2.199(10^{-3}) \text{ m}^2 = \frac{3P}{75(10^6) \text{ N/m}^2}$$

$$P = 55.0 \text{ kN}$$

(d)

By comparison, the largest load that can be applied to the shaft is $P = 51.8$ kN, since any load larger than this will cause the allowable normal stress in the shaft to be exceeded.

NOTE: Here we have not considered a possible shear failure of the collar as in Example 1.14.

1

EXAMPLE | 1.16

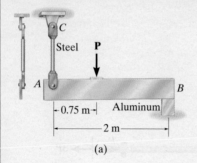

(a)

The rigid bar AB shown in Fig. 1–29a is supported by a steel rod AC having a diameter of 20 mm and an aluminum block having a cross-sectional area of 1800 mm^2. The 18-mm-diameter pins at A and C are subjected to *single shear*. If the failure stress for the steel and aluminum is $(\sigma_{st})_{fail} = 680$ MPa and $(\sigma_{al})_{fail} = 70$ MPa, respectively, and the failure shear stress for each pin is $\tau_{fail} = 900$ MPa, determine the largest load P that can be applied to the bar. Apply a factor of safety of F.S. = 2.

SOLUTION

Using Eqs. 1–9 and 1–10, the allowable stresses are

$$(\sigma_{st})_{allow} = \frac{(\sigma_{st})_{fail}}{F.S.} = \frac{680 \text{ MPa}}{2} = 340 \text{ MPa}$$

$$(\sigma_{al})_{allow} = \frac{(\sigma_{al})_{fail}}{F.S.} = \frac{70 \text{ MPa}}{2} = 35 \text{ MPa}$$

$$\tau_{allow} = \frac{\tau_{fail}}{F.S.} = \frac{900 \text{ MPa}}{2} = 450 \text{ MPa}$$

The free-body diagram of the bar is shown in Fig. 1–29b. There are three unknowns. Here we will apply the moment equations of equilibrium in order to express F_{AC} and F_B in terms of the applied load P. We have

$$\zeta + \Sigma M_B = 0; \qquad P(1.25 \text{ m}) - F_{AC}(2 \text{ m}) = 0 \qquad (1)$$

$$\zeta + \Sigma M_A = 0; \qquad F_B(2 \text{ m}) - P(0.75 \text{ m}) = 0 \qquad (2)$$

We will now determine each value of P that creates the allowable stress in the rod, block, and pins, respectively.

Rod AC. This requires

$$F_{AC} = (\sigma_{st})_{allow}(A_{AC}) = 340(10^6) \text{ N/m}^2 [\pi(0.01 \text{ m})^2] = 106.8 \text{ kN}$$

Using Eq. 1,

$$P = \frac{(106.8 \text{ kN})(2 \text{ m})}{1.25 \text{ m}} = 171 \text{ kN}$$

Block B. In this case,

$$F_B = (\sigma_{al})_{allow} A_B = 35(10^6) \text{ N/m}^2 [1800 \text{ mm}^2 (10^{-6}) \text{ m}^2/\text{mm}^2] = 63.0 \text{ kN}$$

Using Eq. 2,

$$P = \frac{(63.0 \text{ kN})(2 \text{ m})}{0.75 \text{ m}} = 168 \text{ kN}$$

Pin A or C. Due to single shear,

$$F_{AC} = V = \tau_{allow} A = 450(10^6) \text{ N/m}^2 [\pi(0.009 \text{ m})^2] = 114.5 \text{ kN}$$

From Eq. 1,

$$P = \frac{114.5 \text{ kN } (2 \text{ m})}{1.25 \text{ m}} = 183 \text{ kN}$$

By comparison, as P reaches its *smallest value* (168 kN), the allowable normal stress will first be developed in the aluminum block. Hence,

$$P = 168 \text{ kN} \qquad \qquad Ans.$$

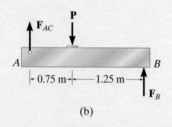

(b)

Fig. 1–29

FUNDAMENTAL PROBLEMS

F1–13. Rods *AC* and *BC* are used to suspend the 200-kg mass. If each rod is made of a material for which the average normal stress can not exceed 150 MPa, determine the minimum required diameter of each rod to the nearest mm.

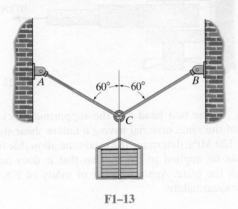

F1–13

F1–14. The frame supports the loading shown. The pin at *A* has a diameter of 50 mm. If it is subjected to double shear, determine the average shear stress in the pin.

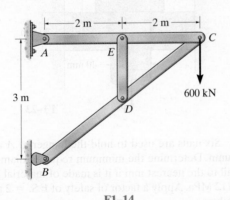

F1–14

F1–15. Determine the maximum average shear stress developed in each 12-mm-diameter bolt.

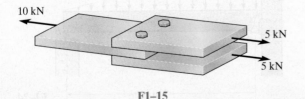

F1–15

F1–16. If each of the three nails has a diameter of 4 mm and can withstand an average shear stress of 60 MPa, determine the maximum allowable force **P** that can be applied to the board.

F1–16

F1–17. The strut is glued to the horizontal member at surface *AB*. If the strut has a thickness of 25 mm and the glue can withstand an average shear stress of 600 kPa, determine the maximum force **P** that can be applied to the strut.

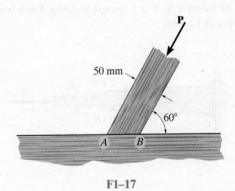

F1–17

F1–18. Determine the maximum average shear stress developed in the 30-mm-diameter pin.

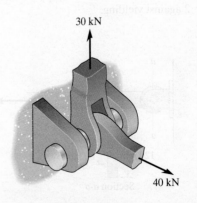

F1–18

F1–19. If the eyebolt is made of a material having a yield stress of $\sigma_Y = 250$ MPa, determine the minimum required diameter d of its shank. Apply a factor of safety F.S. = 1.5 against yielding.

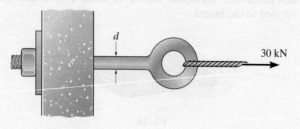

30 kN

F1–19

F1–20. If the bar assembly is made of a material having a yield stress of $\sigma_Y = 350$ MPa, determine the minimum required dimensions h_1 and h_2 to the nearest mm. Apply a factor of safety F.S. = 1.5 against yielding. Each bar has a thickness of 12 mm.

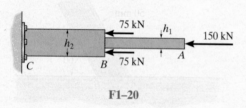

75 kN

h_1

150 kN

h_2

A

C   B   75 kN

F1–20

F1–21. Determine the maximum force **P** that can be applied to the rod if it is made of material having a yield stress of $\sigma_Y = 250$ MPa. Consider the possibility that failure occurs in the rod and at section a–a. Apply a factor of safety of F.S. = 2 against yielding.

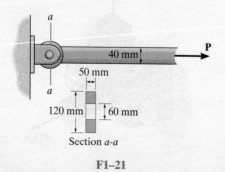

a

40 mm

P

50 mm

a

120 mm   60 mm

Section a–a

F1–21

F1–22. The pin is made of a material having a failure shear stress of $\tau_{fail} = 100$ MPa. Determine the minimum required diameter of the pin to the nearest mm. Apply a factor of safety of F.S. = 2.5 against shear failure.

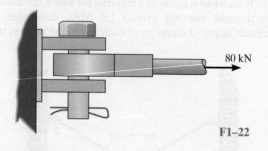

80 kN

F1–22

F1–23. If the bolt head and the supporting bracket are made of the same material having a failure shear stress of $\tau_{fail} = 120$ MPa, determine the maximum allowable force **P** that can be applied to the bolt so that it does not pull through the plate. Apply a factor of safety of F.S. = 2.5 against shear failure.

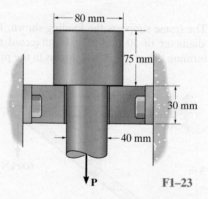

80 mm

75 mm

30 mm

40 mm

P   F1–23

F1–24. Six nails are used to hold the hanger at A against the column. Determine the minimum required diameter of each nail to the nearest mm if it is made of material having $\tau_{fail} = 112$ MPa. Apply a factor of safety of F.S. = 2 against shear failure.

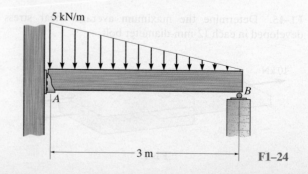

5 kN/m

A

B

3 m   F1–24

PROBLEMS

•1–73. Member B is subjected to a compressive force of 4 kN. If A and B are both made of wood and are 10 mm thick, determine to the nearest multiples of 5 mm the smallest dimension h of the horizontal segment so that it does not fail in shear. The average shear stress for the segment is $\tau_{\text{allow}} = 2.1$ MPa.

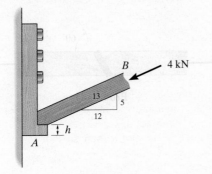

Prob. 1–73

1–74. The lever is attached to the shaft A using a key that has a width d and length of 25 mm. If the shaft is fixed and a vertical force of 200 N is applied perpendicular to the handle, determine the dimension d if the allowable shear stress for the key is $\tau_{\text{allow}} = 35$ MPa.

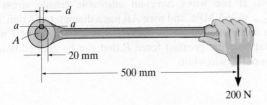

Prob. 1–74

1–75. The joint is fastened together using two bolts. Determine the required diameter of the bolts if the failure shear stress for the bolts is $\tau_{\text{fail}} = 350$ MPa. Use a factor of safety for shear of F.S. $= 2.5$.

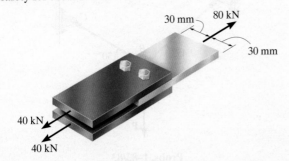

Prob. 1–75

*1–76. The lapbelt assembly is to be subjected to a force of 800 N. Determine (a) the required thickness t of the belt if the allowable tensile stress for the material is $(\sigma_t)_{\text{allow}} = 10$ MPa, (b) the required lap length d_l if the glue can sustain an allowable shear stress of $(\tau_{\text{allow}})_g = 0.75$ MPa, and (c) the required diameter d_r of the pin if the allowable shear stress for the pin is $(\tau_{\text{allow}})_p = 30$ MPa.

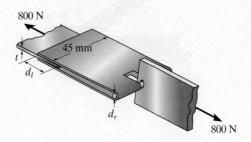

Prob. 1–76

•1–77. The wood specimen is subjected to the pull of 10 kN in a tension testing machine. If the allowable normal stress for the wood is $(\sigma_t)_{\text{allow}} = 12$ MPa and the allowable shear stress is $\tau_{\text{allow}} = 1.2$ MPa, determine the required dimensions b and t so that the specimen reaches these stresses simultaneously. The specimen has a width of 25 mm.

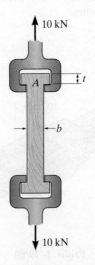

Prob. 1–77

1–78. Member B is subjected to a compressive force of 3 kN. If A and B are both made of wood and are 40 mm thick, determine to the nearest multiples of 5 mm the smallest dimension a of the support so that the average shear stress along the blue line does not exceed $\tau_{allow} = 0.35$ MPa. Neglect friction.

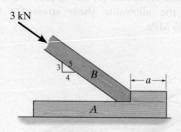

Prob. 1–78

1–79. The joint is used to transmit a torque of $T = 3$ kN·m. Determine the required minimum diameter of the shear pin A if it is made from a material having a shear failure stress of $\tau_{fail} = 150$ MPa. Apply a factor of safety of 3 against failure.

*1–80. Determine the maximum allowable torque $\mathbf{T}$ that can be transmitted by the joint. The shear pin A has a diameter of 25 mm, and it is made from a material having a failure shear stress of $\tau_{fail} = 150$ MPa. Apply a factor of safety of 3 against failure.

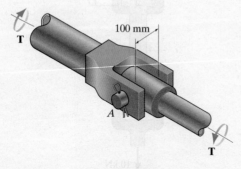

Probs. 1–79/80

•1–81. The tension member is fastened together using *two* bolts, one on each side of the member as shown. Each bolt has a diameter of 7.5 mm. Determine the maximum load P that can be applied to the member if the allowable shear stress for the bolts is $\tau_{allow} = 84$ MPa. and the allowable average normal stress is $\sigma_{allow} = 140$ MPa.

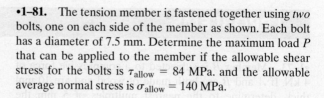

Prob. 1–81

1–82. The three steel wires are used to support the load. If the wires have an allowable tensile stress of $\sigma_{allow} = 165$ MPa, determine the required diameter of each wire if the applied load is $P = 6$ kN.

1–83. The three steel wires are used to support the load. If the wires have an allowable tensile stress of $\sigma_{allow} = 165$ MPa, and wire AB has a diameter of 6 mm, BC has a diameter of 5 mm, and BD has a diameter of 7 mm, determine the greatest force P that can be applied before one of the wires fails.

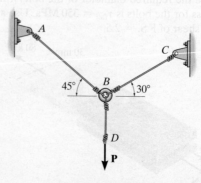

Probs. 1–82/83

*1–84. The assembly consists of three disks A, B, and C that are used to support the load of 140 kN. Determine the smallest diameter d_1 of the top disk, the diameter d_2 within the support space, and the diameter d_3 of the hole in the bottom disk. The allowable bearing stress for the material is $(\sigma_{\text{allow}})_b = 350$ MPa and allowable shear stress is $\tau_{\text{allow}} = 125$ MPa.

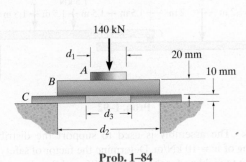

Prob. 1–84

•1–85. The boom is supported by the winch cable that has a diameter of 6 mm and an allowable normal stress of $\sigma_{\text{allow}} = 168$ MPa. Determine the greatest load that can be supported without causing the cable to fail when $\theta = 30°$ and $\phi = 45°$. Neglect the size of the winch.

1–86. The boom is supported by the winch cable that has an allowable normal stress of $\sigma_{\text{allow}} = 168$ MPa. If it is required that it be able to slowly lift 25 kN, from $\theta = 20°$ to $\theta = 50°$, determine the smallest diameter of the cable to the nearest multiples of 5 mm. The boom AB has a length of 6 m. Neglect the size of the winch. Set $d = 3.6$ m.

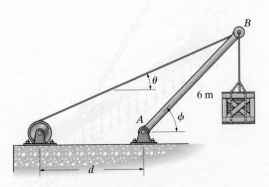

Probs. 1–85/86

1–87. The 60 mm × 60 mm oak post is supported on the pine block. If the allowable bearing stresses for these materials are $\sigma_{\text{oak}} = 43$ MPa and $\sigma_{\text{pine}} = 25$ MPa, determine the greatest load P that can be supported. If a rigid bearing plate is used between these materials, determine its required area so that the maximum load P can be supported. What is this load?

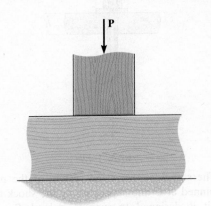

Prob. 1–87

*1–88. The frame is subjected to the load of 4 kN which acts on member ABD at D. Determine the required diameter of the pins at D and C if the allowable shear stress for the material is $\tau_{\text{allow}} = 40$ MPa. Pin C is subjected to double shear, whereas pin D is subjected to single shear.

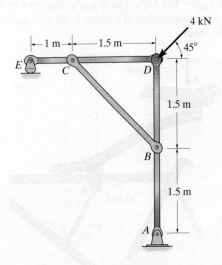

Prob. 1–88

•1–89.   The eye bolt is used to support the load of 25 kN. Determine its diameter d to the nearest multiples of 5 mm and the required thickness h to the nearest multiples of 5 mm of the support so that the washer will not penetrate or shear through it. The allowable normal stress for the bolt is $\sigma_{allow} = 150$ MPa and the allowable shear stress for the supporting material is $\tau_{allow} = 35$ MPa.

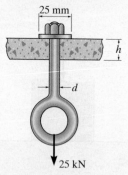

Prob. 1–89

1–90.   The soft-ride suspension system of the mountain bike is pinned at C and supported by the shock absorber BD. If it is designed to support a load $P = 1500$ N, determine the required minimum diameter of pins B and C. Use a factor of safety of 2 against failure. The pins are made of material having a failure shear stress of $\tau_{fail} = 150$ MPa, and each pin is subjected to double shear.

1–91.   The soft-ride suspension system of the mountain bike is pinned at C and supported by the shock absorber BD. If it is designed to support a load of $P = 1500$ N, determine the factor of safety of pins B and C against failure if they are made of a material having a shear failure stress of $\tau_{fail} = 150$ MPa. Pin B has a diameter of 7.5 mm, and pin C has a diameter of 6.5 mm. Both pins are subjected to double shear.

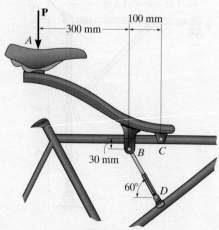

Probs. 1–90/91

*1–92.   The compound wooden beam is connected together by a bolt at B. Assuming that the connections at A, B, C, and D exert only vertical forces on the beam, determine the required diameter of the bolt at B and the required outer diameter of its washers if the allowable tensile stress for the bolt is $(\sigma_t)_{allow} = 150$ MPa and the allowable bearing stress for the wood is $(\sigma_b)_{allow} = 28$ MPa. Assume that the hole in the washers has the same diameter as the bolt.

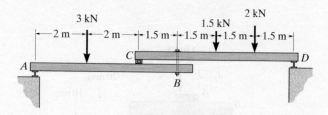

Prob. 1–92

•1–93.   The assembly is used to support the distributed loading of $w = 10$ kN/m. Determine the factor of safety with respect to yielding for the steel rod BC and the pins at B and C if the yield stress for the steel in tension is $\sigma_y = 250$ MPa and in shear $\tau_y = 125$ MPa. The rod has a diameter of 13 mm, and the pins each have a diameter of 10 mm.

1–94.   If the allowable shear stress for each of the 10-mm-diameter steel pins at A, B, and C is $\tau_{allow} = 90$ MPa, and the allowable normal stress for the 13-mm-diameter rod is $\sigma_{allow} = 150$ MPa, determine the largest intensity w of the uniform distributed load that can be suspended from the beam.

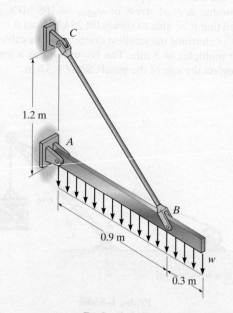

Probs. 1–93/94

1–95. If the allowable bearing stress for the material under the supports at A and B is $(\sigma_b)_{\text{allow}} = 1.5$ MPa, determine the size of *square* bearing plates A' and B' required to support the load. Dimension the plates to the nearest mm. The reactions at the supports are vertical. Take $P = 100$ kN.

*1–96. If the allowable bearing stress for the material under the supports at A and B is $(\sigma_b)_{\text{allow}} = 1.5$ MPa, determine the maximum load P that can be applied to the beam. The bearing plates A' and B' have square cross sections of $150 \text{ mm} \times 150 \text{ mm}$ and $250 \text{ mm} \times 250 \text{ mm}$, respectively.

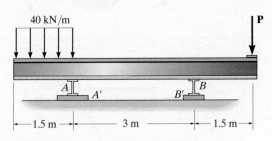

Probs. 1–95/96

•1–97. The rods AB and CD are made of steel having a failure tensile stress of $\sigma_{\text{fail}} = 510$ MPa. Using a factor of safety of F.S. $= 1.75$ for tension, determine their smallest diameter so that they can support the load shown. The beam is assumed to be pin connected at A and C.

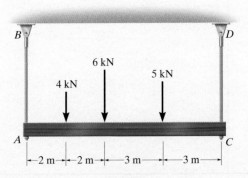

Prob. 1–97

1–98. The aluminum bracket A is used to support the centrally applied load of 40 kN. If it has a constant thickness of 12 mm, determine the smallest height h in order to prevent a shear failure. The failure shear stress is $\tau_{\text{fail}} = 160$ MPa. Use a factor of safety for shear of F.S. $= 2.5$.

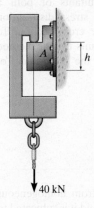

Prob. 1–98

1–99. The hanger is supported using the rectangular pin. Determine the magnitude of the allowable suspended load P if the allowable bearing stress is $(\sigma_b)_{\text{allow}} = 220$ MPa, the allowable tensile stress is $(\sigma_t)_{\text{allow}} = 150$ MPa, and the allowable shear stress is $\tau_{\text{allow}} = 130$ MPa. Take $t = 6$ mm, $a = 5$ mm, and $b = 25$ mm.

*1–100. The hanger is supported using the rectangular pin. Determine the required thickness t of the hanger, and dimensions a and b if the suspended load is $P = 60$ kN. The allowable tensile stress is $(\sigma_t)_{\text{allow}} = 150$ MPa, the allowable bearing stress is $(\sigma_b)_{\text{allow}} = 290$ MPa, and the allowable shear stress is $\tau_{\text{allow}} = 125$ MPa.

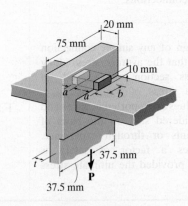

Probs. 1–99/100

CHAPTER REVIEW

The internal loadings in a body consist of a normal force, shear force, bending moment, and torsional moment. They represent the resultants of both a normal and shear stress distribution that acts over the cross section. To obtain these resultants, use the method of sections and the equations of equilibrium.	$\Sigma F_x = 0$ $\Sigma F_y = 0$ $\Sigma F_z = 0$ $\Sigma M_x = 0$ $\Sigma M_y = 0$ $\Sigma M_z = 0$ Ref.: Section 1.2	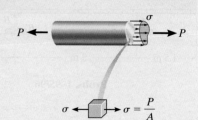
If a bar is made from homogeneous isotropic material and it is subjected to a series of external axial loads that pass through the centroid of the cross section, then a uniform normal stress distribution will act over the cross section. This average normal stress can be determined from $\sigma = P/A$, where P is the internal axial load at the section.	$\sigma = \dfrac{P}{A}$ Ref.: Section 1.4	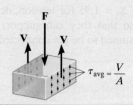
The average shear stress can be determined using $\tau_{avg} = V/A$, where V is the shear force acting on the cross-sectional area A. This formula is often used to find the average shear stress in fasteners or in parts used for connections.	$\tau_{avg} = \dfrac{V}{A}$ Ref.: Section 1.5	
The design of any simple connection requires that the average stress along any cross section not exceed an allowable stress of σ_{allow} or τ_{allow}. These values are reported in codes and are considered safe on the basis of experiments or through experience. Sometimes a factor of safety is reported provided the ultimate stress is known.	$\text{F.S.} = \dfrac{\sigma_{fail}}{\sigma_{allow}} = \dfrac{\tau_{fail}}{\tau_{allow}}$ Ref.: Section 1.6	

CONCEPTUAL PROBLEMS

P1–1

P1–3

P1–1. Here hurricane winds caused the failure of this highway sign. Assuming the wind creates a uniform pressure on the sign of 2 kPa, use reasonable dimensions for the sign and determine the resultant shear and moment at the two connections where the failure occurred.

P1–3. The hydraulic cylinder H applies a horizontal force F on the pin at A. Draw the free-body diagram of the pin and show the forces acting on it. Using the method of sections, explain why the average shear stress in the pin is largest at sections through the gaps D and E and not at some intermediate section.

P1–2

P1–4

P1–2. The two structural tubes are connected by the pin which passes through them. If the vertical load being supported is 100 kN, draw a free-body diagram of the pin and then use the method of sections to find the maximum average shear force in the pin. If the pin has a diameter of 50 mm, what is the maximum average shear stress in the pin?

P1–4. The vertical load on the hook is 5 kN. Draw the appropriate free-body diagrams and determine the maximum average shear force on the pins at A, B, and C. Note that due to symmetry four wheels are used to support the loading on the railing.

1

REVIEW PROBLEMS

•1–101. The 200-mm-diameter aluminum cylinder supports a compressive load of 300 kN. Determine the average normal and shear stress acting on section *a–a*. Show the results on a differential element located on the section.

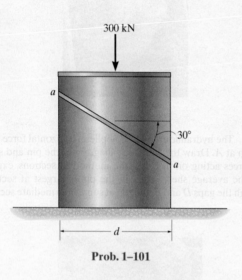

Prob. 1–101

1–102. The long bolt passes through the 30-mm-thick plate. If the force in the bolt shank is 8 kN, determine the average normal stress in the shank, the average shear stress along the cylindrical area of the plate defined by the section lines *a–a*, and the average shear stress in the bolt head along the cylindrical area defined by the section lines *b–b*.

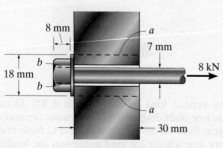

Prob. 1–102

1–103. Determine the required thickness of member *BC* and the diameter of the pins at *A* and *B* if the allowable normal stress for member *BC* is $\sigma_{allow} = 200$ MPa and the allowable shear stress for the pins is $\tau_{allow} = 70$ MPa.

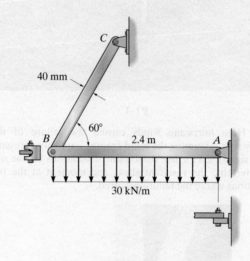

Prob. 1–103

*1–104. Determine the resultant internal loadings acting on the cross sections located through points *D* and *E* of the frame.

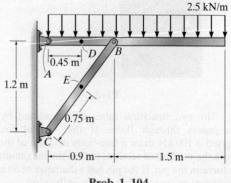

Prob. 1–104

•1–105. The pulley is held fixed to the 20-mm-diameter shaft using a key that fits within a groove cut into the pulley and shaft. If the suspended load has a mass of 50 kg, determine the average shear stress in the key along section a–a. The key is 5 mm by 5 mm square and 12 mm long.

1–107. The yoke-and-rod connection is subjected to a tensile force of 5 kN. Determine the average normal stress in each rod and the average shear stress in the pin A between the members.

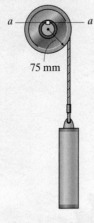

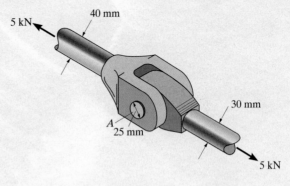

Prob. 1–105

Prob. 1–107

1–106. The bearing pad consists of a 150 mm by 150 mm block of aluminum that supports a compressive load of 6 kN. Determine the average normal and shear stress acting on the plane through section a–a. Show the results on a differential volume element located on the plane.

*1–108. The cable has a specific weight γ (weight/volume) and cross-sectional area A. If the sag s is small, so that its length is approximately L and its weight can be distributed uniformly along the horizontal axis, determine the average normal stress in the cable at its lowest point C.

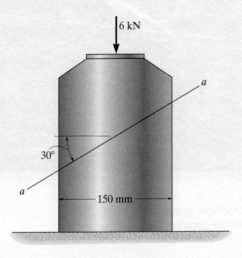

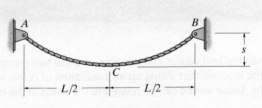

Prob. 1–106

Prob. 1–108

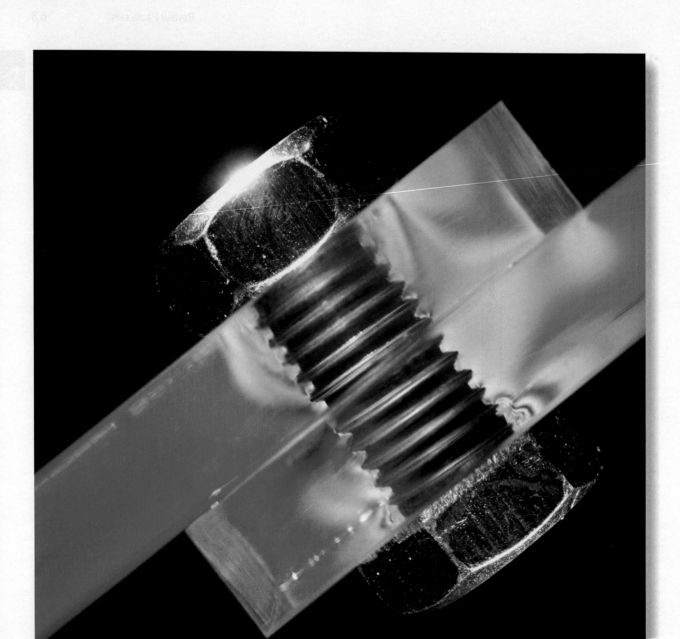

When the bolt causes compression of these two transparent plates it produces strains in the material that shows up as a spectrum of colors when displayed under polarized light. These strains can be related to the stress in the material.

Strain

CHAPTER OBJECTIVES

In engineering the deformation of a body is specified using the concepts of normal and shear strain. In this chapter we will define these quantities and show how they can be determined for various types of problems.

Check out the Companion Website for the following video solution(s) that can be found in this chapter.

- Section 1
  Normal Strain

- Section 2
  Shear Strain

2.1 Deformation

Whenever a force is applied to a body, it will tend to change the body's shape and size. These changes are referred to as *deformation*, and they may be either highly visible or practically unnoticeable. For example, a rubber band will undergo a very large deformation when stretched, whereas only slight deformations of structural members occur when a building is occupied by people walking about. Deformation of a body can also occur when the temperature of the body is changed. A typical example is the thermal expansion or contraction of a roof caused by the weather.

In a general sense, the deformation of a body will not be uniform throughout its volume, and so the change in geometry of any line segment within the body may vary substantially along its length. Hence, to study deformational changes in a more uniform manner, we will consider line segments that are very short and located in the neighborhood of a point. Realize, however, that these changes will also depend on the orientation of the line segment at the point. For example, a line segment may elongate if it is oriented in one direction, whereas it may contract if it is oriented in another direction.

Note the before and after positions of three different line segments on this rubber membrane which is subjected to tension. The vertical line is lengthened, the horizontal line is shortened, and the inclined line changes its length and rotates.

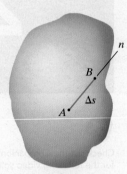

Undeformed body
(a)

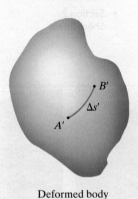

Deformed body
(b)

Fig. 2–1

2.2 Strain

In order to describe the deformation of a body by changes in length of line segments and the changes in the angles between them, we will develop the concept of strain. Strain is actually measured by experiments, and once the strain is obtained, it will be shown in the next chapter how it can be related to the stress acting within the body.

Normal Strain. If we define the normal strain as the change in length of a line per unit length, then we will not have to specify the *actual length* of any particular line segment. Consider, for example, the line AB, which is contained within the undeformed body shown in Fig. 2–1a. This line lies along the n axis and has an original length of Δs. After deformation, points A and B are displaced to A' and B', and the line becomes a curve having a length of $\Delta s'$, Fig. 2–1b. The change in length of the line is therefore $\Delta s' - \Delta s$. If we define the *average normal strain* using the symbol ϵ_{avg} (epsilon), then

$$\epsilon_{avg} = \frac{\Delta s' - \Delta s}{\Delta s} \qquad (2\text{–}1)$$

As point B is chosen closer and closer to point A, the length of the line will become shorter and shorter, such that $\Delta s \rightarrow 0$. Also, this causes B' to approach A', such that $\Delta s' \rightarrow 0$. Consequently, in the limit the normal strain at point A and in the direction of n is

$$\epsilon = \lim_{B \rightarrow A \text{ along } n} \frac{\Delta s' - \Delta s}{\Delta s} \qquad (2\text{–}2)$$

Hence, when ϵ (or ϵ_{avg}) is positive the initial line will elongate, whereas if ϵ is negative the line contracts.

Note that normal strain is a *dimensionless quantity*, since it is a ratio of two lengths. Although this is the case, it is sometimes stated in terms of a ratio of length units. If the SI system is used, then the basic unit for length is the meter (m). Ordinarily, for most engineering applications ϵ will be very small, so measurements of strain are in micrometers per meter (μm/m), where $1 \, \mu$m $= 10^{-6}$ m. Sometimes for experimental work, strain is expressed as a percent, e.g., 0.001 m/m = 0.1%. As an example, a normal strain of $480(10^{-6})$ can be reported as $480(10^{-6})$ in./in., $480 \mu m/m$, or 0.0480%. Also, one can state this answer as simply 480μ (480 "micros").

Shear Strain. Deformations not only cause line segments to elongate or contract, but they also cause them to change direction. If we select two line segments that are originally perpendicular to one another, then the change in angle that occurs between these two line segments is referred to as **shear strain**. This angle is denoted by γ (gamma) and is always measured in radians (rad), which are dimensionless. For example, consider the line segments AB and AC originating from the same point A in a body, and directed along the perpendicular n and t axes, Fig. 2–2a. After deformation, the ends of both lines are displaced, and the lines themselves become curves, such that the angle between them at A is θ', Fig. 2–2b. Hence the shear strain at point A associated with the n and t axes becomes

$$\gamma_{nt} = \frac{\pi}{2} - \lim_{\substack{B \to A \text{ along } n \\ C \to A \text{ along } t}} \theta' \qquad (2–3)$$

Notice that if θ' is smaller than $\pi/2$ the shear strain is positive, whereas if θ' is larger than $\pi/2$ the shear strain is negative.

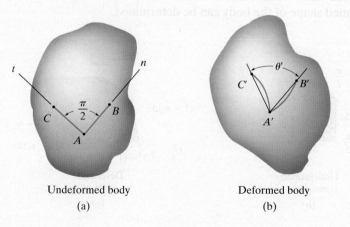

Undeformed body   Deformed body
(a)   (b)

Fig. 2–2

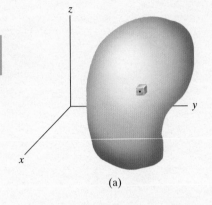

(a)

Cartesian Strain Components.

Using the definitions of normal and shear strain, we will now show how they can be used to describe the deformation of the body in Fig. 2–3a. To do so, imagine the body is subdivided into small elements such as the one shown in Fig. 2–3b. This element is rectangular, has undeformed dimensions Δx, Δy, and Δz, and is located in the neighborhood of a point in the body, Fig. 2–3a. If the element's dimensions are very small, then its deformed shape will be a parallelepiped, Fig. 2–3c, since very small line segments will remain approximately straight after the body is deformed. In order to achieve this deformed shape, we will first consider how the normal strain changes the lengths of the sides of the rectangular element, and then how the shear strain changes the angles of each side. For example, Δx elongates $\epsilon_x \Delta x$, so its new length is $\Delta x + \epsilon_x \Delta x$. Therefore, the approximate lengths of the three sides of the parallelepiped are

$$(1 + \epsilon_x)\, \Delta x \qquad (1 + \epsilon_y)\, \Delta y \qquad (1 + \epsilon_z)\, \Delta z$$

And the approximate angles between these sides are

$$\frac{\pi}{2} - \gamma_{xy} \qquad \frac{\pi}{2} - \gamma_{yz} \qquad \frac{\pi}{2} - \gamma_{xz}$$

Notice that the ***normal strains cause a change in volume*** of the element, whereas the ***shear strains cause a change in its shape***. Of course, both of these effects occur simultaneously during the deformation.

In summary, then, the *state of strain* at a point in a body requires specifying three normal strains, ϵ_x, ϵ_y, ϵ_z, and three shear strains, γ_{xy}, γ_{yz}, γ_{xz}. These strains completely describe the deformation of a rectangular volume element of material located at the point and oriented so that its sides are originally parallel to the x, y, z axes. Provided these strains are defined at all points in the body, then the deformed shape of the body can be determined.

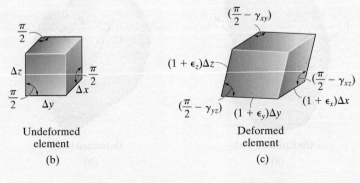

Undeformed element

(b)

Deformed element

(c)

Fig. 2–3

Small Strain Analysis. Most engineering design involves applications for which only *small deformations* are allowed. In this text, therefore, we will assume that the deformations that take place within a body are almost infinitesimal. In particular, the *normal strains* occurring within the material are *very small* compared to 1, so that $\epsilon \ll 1$. This assumption has wide practical application in engineering, and it is often referred to as a *small strain analysis*. It can be used, for example, to approximate $\sin \theta = \theta$, $\cos \theta = 1$, and $\tan \theta = \theta$, provided θ is very small.

The rubber bearing support under this concrete bridge girder is subjected to both normal and shear strain. The normal strain is caused by the weight and bridge loads on the girder, and the shear strain is caused by the horizontal movement of the girder due to temperature changes.

Important Points

- Loads will cause all material bodies to deform and, as a result, points in a body will undergo *displacements or changes in position*.

- *Normal strain* is a measure per unit length of the elongation or contraction of a small line segment in the body, whereas *shear strain* is a measure of the change in angle that occurs between two small line segments that are originally perpendicular to one another.

- The state of strain at a point is characterized by six strain components: three normal strains ϵ_x, ϵ_y, ϵ_z and three shear strains γ_{xy}, γ_{yz}, γ_{xz}. These components depend upon the original orientation of the line segments and their location in the body.

- Strain is the geometrical quantity that is measured using experimental techniques. Once obtained, the stress in the body can then be determined from material property relations, as discussed in the next chapter.

- Most engineering materials undergo very small deformations, and so the normal strain $\epsilon \ll 1$. This assumption of "small strain analysis" allows the calculations for normal strain to be simplified, since first-order approximations can be made about their size.

EXAMPLE | 2.1

The slender rod shown in Fig. 2–4 is subjected to an increase of temperature along its axis, which creates a normal strain in the rod of $\epsilon_z = 40(10^{-3})z^{1/2}$, where z is measured in meters. Determine (a) the displacement of the end B of the rod due to the temperature increase, and (b) the average normal strain in the rod.

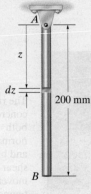

Fig. 2–4

SOLUTION

Part (a). Since the normal strain is reported at each point along the rod, a differential segment dz, located at position z, Fig. 2–4, has a deformed length that can be determined from Eq. 2–1; that is,

$$dz' = dz + \epsilon_z \, dz$$

$$dz' = \left[1 + 40(10^{-3})z^{1/2}\right] dz$$

The sum of these segments along the axis yields the *deformed length* of the rod, i.e.,

$$z' = \int_0^{0.2 \text{ m}} \left[1 + 40(10^{-3})z^{1/2}\right] dz$$

$$= \left[z + 40(10^{-3})\tfrac{2}{3}z^{3/2}\right]\Big|_0^{0.2 \text{ m}}$$

$$= 0.20239 \text{ m}$$

The displacement of the end of the rod is therefore

$$\Delta_B = 0.20239 \text{ m} - 0.2 \text{ m} = 0.00239 \text{ m} = 2.39 \text{ mm} \downarrow \quad \textit{Ans.}$$

Part (b). The average normal strain in the rod is determined from Eq. 2–1, which assumes that the rod or "line segment" has an original length of 200 mm and a change in length of 2.39 mm. Hence,

$$\epsilon_{\text{avg}} = \frac{\Delta s' - \Delta s}{\Delta s} = \frac{2.39 \text{ mm}}{200 \text{ mm}} = 0.0119 \text{ mm/mm} \quad \textit{Ans.}$$

EXAMPLE 2.2

When force **P** is applied to the rigid lever arm ABC in Fig. 2–5a, the arm rotates counterclockwise about pin A through an angle of 0.05°. Determine the normal strain developed in wire BD.

SOLUTION I

Geometry. The orientation of the lever arm after it rotates about point A is shown in Fig. 2–5b. From the geometry of this figure,

$$\alpha = \tan^{-1}\left(\frac{400 \text{ mm}}{300 \text{ mm}}\right) = 53.1301°$$

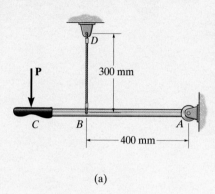

(a)

Then

$$\phi = 90° - \alpha + 0.05° = 90° - 53.1301° + 0.05° = 36.92°$$

For triangle ABD the Pythagorean theorem gives

$$L_{AD} = \sqrt{(300 \text{ mm})^2 + (400 \text{ mm})^2} = 500 \text{ mm}$$

Using this result and applying the law of cosines to triangle $AB'D$,

$$
\begin{aligned}
L_{B'D} &= \sqrt{L_{AD}^2 + L_{AB'}^2 - 2(L_{AD})(L_{AB'})\cos\phi} \\
&= \sqrt{(500 \text{ mm})^2 + (400 \text{ mm})^2 - 2(500 \text{ mm})(400 \text{ mm})\cos 36.92°} \\
&= 300.3491 \text{ mm}
\end{aligned}
$$

Normal Strain.

$$\epsilon_{BD} = \frac{L_{B'D} - L_{BD}}{L_{BD}} = \frac{300.3491 \text{ mm} - 300 \text{ mm}}{300 \text{ mm}} = 0.00116 \text{ mm/mm} \quad Ans.$$

(b)

Fig. 2–5

SOLUTION II

Since the strain is small, this same result can be obtained by approximating the elongation of wire BD as ΔL_{BD}, shown in Fig. 2–5b. Here,

$$\Delta L_{BD} = \theta L_{AB} = \left[\left(\frac{0.05°}{180°}\right)(\pi \text{ rad})\right](400 \text{ mm}) = 0.3491 \text{ mm}$$

Therefore,

$$\epsilon_{BD} = \frac{\Delta L_{BD}}{L_{BD}} = \frac{0.3491 \text{ mm}}{300 \text{ mm}} = 0.00116 \text{ mm/mm} \quad Ans.$$

EXAMPLE 2.3

Due to a loading, the plate is deformed into the dashed shape shown in Fig. 2–6a. Determine (a) the average normal strain along the side AB, and (b) the average shear strain in the plate at A relative to the x and y axes.

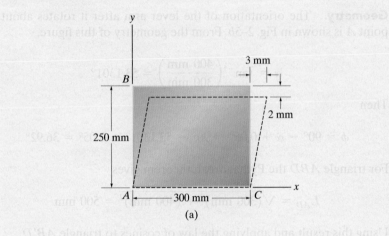

(a)

Fig. 2–6

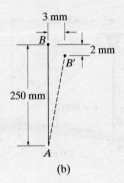

(b)

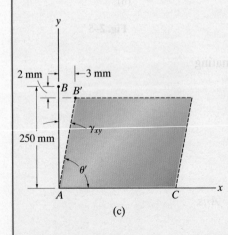

(c)

SOLUTION

Part (a).   Line AB, coincident with the y axis, becomes line AB' after deformation, as shown in Fig. 2–6b. The length of AB' is

$$AB' = \sqrt{(250 \text{ mm} - 2 \text{ mm})^2 + (3 \text{ mm})^2} = 248.018 \text{ mm}$$

The average normal strain for AB is therefore

$$(\epsilon_{AB})_{\text{avg}} = \frac{AB' - AB}{AB} = \frac{248.018 \text{ mm} - 250 \text{ mm}}{250 \text{ mm}}$$

$$= -7.93(10^{-3}) \text{ mm/mm} \qquad Ans.$$

The negative sign indicates the strain causes a contraction of AB.

Part (b).   As noted in Fig. 2–6c, the once 90° angle BAC between the sides of the plate at A changes to θ' due to the displacement of B to B'. Since $\gamma_{xy} = \pi/2 - \theta'$, then γ_{xy} is the angle shown in the figure. Thus,

$$\gamma_{xy} = \tan^{-1}\left(\frac{3 \text{ mm}}{250 \text{ mm} - 2 \text{ mm}}\right) = 0.0121 \text{ rad} \qquad Ans.$$

EXAMPLE | 2.4

The plate shown in Fig. 2–7a is fixed connected along AB and held in the horizontal guides at its top and bottom, AD and BC. If its right side CD is given a uniform horizontal displacement of 2 mm, determine (a) the average normal strain along the diagonal AC, and (b) the shear strain at E relative to the x, y axes.

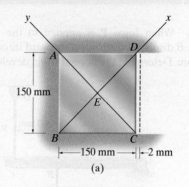

(a)

SOLUTION

Part (a). When the plate is deformed, the diagonal AC becomes AC', Fig. 2–7b. The length of diagonals AC and AC' can be found from the Pythagorean theorem. We have

$$AC = \sqrt{(0.150 \text{ m})^2 + (0.150 \text{ m})^2} = 0.21213 \text{ m}$$

$$AC' = \sqrt{(0.150 \text{ m})^2 + (0.152 \text{ m})^2} = 0.21355 \text{ m}$$

Therefore the average normal strain along the diagonal is

$$(\epsilon_{AC})_{\text{avg}} = \frac{AC' - AC}{AC} = \frac{0.21355 \text{ m} - 0.21213 \text{ m}}{0.21213 \text{ m}}$$

$$= 0.00669 \text{ mm/mm} \qquad \qquad Ans.$$

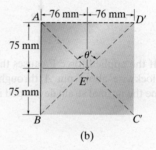

(b)

Fig. 2–7

Part (b). To find the shear strain at E relative to the x and y axes, it is first necessary to find the angle θ' after deformation, Fig. 2–7b. We have

$$\tan\left(\frac{\theta'}{2}\right) = \frac{76 \text{ mm}}{75 \text{ mm}}$$

$$\theta' = 90.759° = \left(\frac{\pi}{180°}\right)(90.759°) = 1.58404 \text{ rad}$$

Applying Eq. 2–3, the shear strain at E is therefore

$$\gamma_{xy} = \frac{\pi}{2} - 1.58404 \text{ rad} = -0.0132 \text{ rad} \qquad Ans.$$

The *negative sign* indicates that the angle θ' is *greater than* 90°.

NOTE: If the x and y axes were horizontal and vertical at point E, then the 90° angle between these axes would not change due to the deformation, and so $\gamma_{xy} = 0$ at point E.

FUNDAMENTAL PROBLEMS

F2–1. When force **P** is applied to the rigid arm *ABC*, point *B* displaces vertically downward through a distance of 0.2 mm. Determine the normal strain developed in wire *CD*.

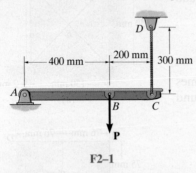

F2–1

F2–2. If the applied force **P** causes the rigid arm *ABC* to rotate clockwise about pin *A* through an angle of 0.02°, determine the normal strain developed in wires *BD* and *CE*.

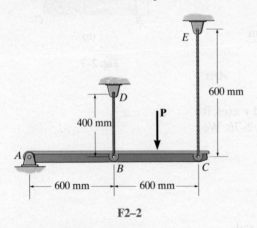

F2–2

F2–3. The rectangular plate is deformed into the shape of a rhombus shown by the dashed line. Determine the average shear strain at corner *A* with respect to the *x* and *y* axes.

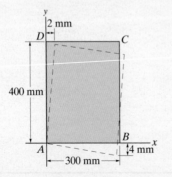

F2–3

F2–4. The triangular plate is deformed into the shape shown by the dashed line. Determine the normal strain developed along edge *BC* and the average shear strain at corner *A* with respect to the *x* and *y* axes.

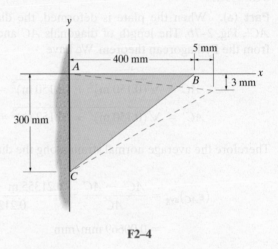

F2–4

F2–5. The square plate is deformed into the shape shown by the dashed line. Determine the average normal strain along diagonal *AC* and the shear strain of point *E* with respect to the *x* and *y* axes.

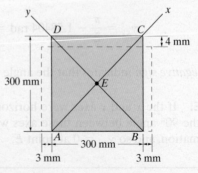

F2–5

PROBLEMS

2–1. An air-filled rubber ball has a diameter of 150 mm. If the air pressure within it is increased until the ball's diameter becomes 175 mm, determine the average normal strain in the rubber.

2–2. A thin strip of rubber has an unstretched length of 375 mm. If it is stretched around a pipe having an outer diameter of 125 mm, determine the average normal strain in the strip.

2–3. The rigid beam is supported by a pin at A and wires BD and CE. If the load **P** on the beam causes the end C to be displaced 10 mm downward, determine the normal strain developed in wires CE and BD.

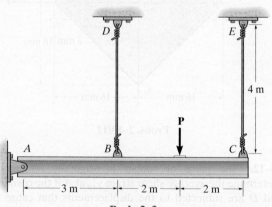

Prob. 2–3

*2–4. The two wires are connected together at A. If the force **P** causes point A to be displaced horizontally 2 mm, determine the normal strain developed in each wire.

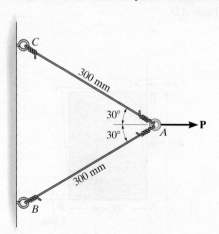

Prob. 2–4

•2–5. The rigid beam is supported by a pin at A and wires BD and CE. If the distributed load causes the end C to be displaced 10 mm downward, determine the normal strain developed in wires CE and BD.

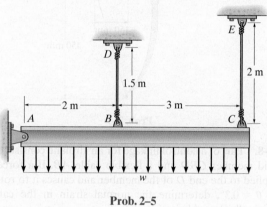

Prob. 2–5

2–6. Nylon strips are fused to glass plates. When moderately heated the nylon will become soft while the glass stays approximately rigid. Determine the average shear strain in the nylon due to the load **P** when the assembly deforms as indicated.

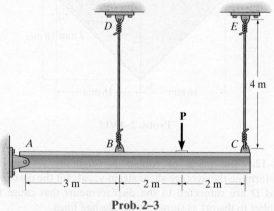

Prob. 2–6

2–7. If the unstretched length of the bowstring is 887.5 mm, determine the average normal strain in the string when it is stretched to the position shown.

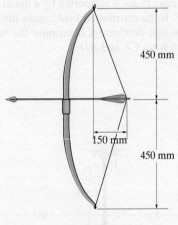

450 mm

150 mm

450 mm

Prob. 2–7

*2–8. Part of a control linkage for an airplane consists of a rigid member *CBD* and a flexible cable *AB*. If a force is applied to the end *D* of the member and causes it to rotate by $\theta = 0.3°$, determine the normal strain in the cable. Originally the cable is unstretched.

•2–9. Part of a control linkage for an airplane consists of a rigid member *CBD* and a flexible cable *AB*. If a force is applied to the end *D* of the member and causes a normal strain in the cable of 0.0035 mm/mm, determine the displacement of point *D*. Originally the cable is unstretched.

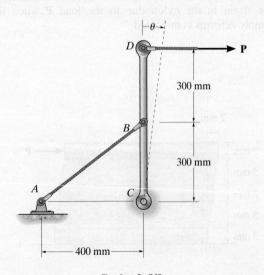

D

θ

P

300 mm

B

300 mm

A

C

400 mm

Probs. 2–8/9

2–10. The corners *B* and *D* of the square plate are given the displacements indicated. Determine the shear strains at *A* and *B*.

2–11. The corners *B* and *D* of the square plate are given the displacements indicated. Determine the average normal strains along side *AB* and diagonal *DB*.

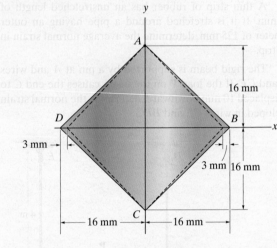

y

A

16 mm

D

B

x

3 mm

3 mm 16 mm

16 mm   C   16 mm

Probs. 2–10/11

*2–12. The piece of rubber is originally rectangular. Determine the average shear strain γ_{xy} at *A* if the corners *B* and *D* are subjected to the displacements that cause the rubber to distort as shown by the dashed lines.

•2–13. The piece of rubber is originally rectangular and subjected to the deformation shown by the dashed lines. Determine the average normal strain along the diagonal *DB* and side *AD*.

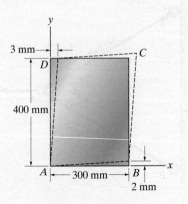

y

3 mm

D   C

400 mm

A   300 mm   B   x

2 mm

Probs. 2–12/13

2–14. Two bars are used to support a load. When unloaded, AB is 125 mm long, AC is 200 mm long, and the ring at A has coordinates $(0, 0)$. If a load **P** acts on the ring at A, the normal strain in AB becomes $\epsilon_{AB} = 0.02$ mm/mm, and the normal strain in AC becomes $\epsilon_{AC} = 0.035$ mm/mm. Determine the coordinate position of the ring due to the load.

2–15. Two bars are used to support a load **P**. When unloaded, AB is 125 mm long, AC is 200 mm long, and the ring at A has coordinates $(0,0)$. If a load is applied to the ring at A, so that it moves it to the coordinate position $(6.25$ mm, -18.25 mm$)$, determine the normal strain in each bar.

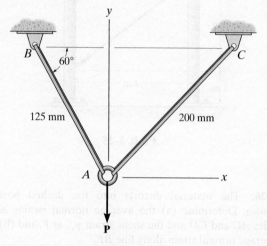

Probs. 2–14/15

*2–16. The square deforms into the position shown by the dashed lines. Determine the average normal strain along each diagonal, AB and CD. Side $D'B'$ remains horizontal.

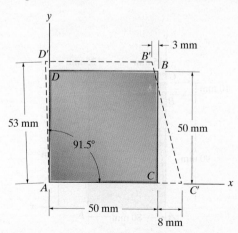

Prob. 2–16

•2–17. The three cords are attached to the ring at B. When a force is applied to the ring it moves it to point B', such that the normal strain in AB is ϵ_{AB} and the normal strain in CB is ϵ_{CB}. Provided these strains are small, determine the normal strain in DB. Note that AB and CB remain horizontal and vertical, respectively, due to the roller guides at A and C.

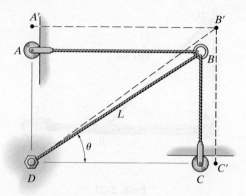

Prob. 2–17

2–18. The piece of plastic is originally rectangular. Determine the shear strain γ_{xy} at corners A and B if the plastic distorts as shown by the dashed lines.

2–19. The piece of plastic is originally rectangular. Determine the shear strain γ_{xy} at corners D and C if the plastic distorts as shown by the dashed lines.

*2–20. The piece of plastic is originally rectangular. Determine the average normal strain that occurs along the diagonals AC and DB.

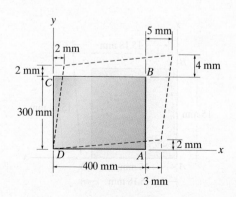

Probs. 2–18/19/20

•2–21. The force applied to the handle of the rigid lever arm causes the arm to rotate clockwise through an angle of 3° about pin A. Determine the average normal strain developed in the wire. Originally, the wire is unstretched.

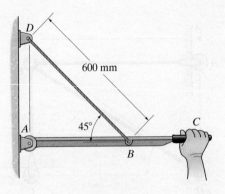

Prob. 2–21

2–22. A square piece of material is deformed into the dashed position. Determine the shear strain γ_{xy} at A.

2–23. A square piece of material is deformed into the dashed parallelogram. Determine the average normal strain that occurs along the diagonals AC and BD.

*2–24. A square piece of material is deformed into the dashed position. Determine the shear strain γ_{xy} at C.

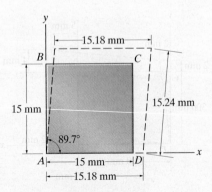

Probs. 2–22/23/24

•2–25. The guy wire AB of a building frame is originally unstretched. Due to an earthquake, the two columns of the frame tilt $\theta = 2°$. Determine the approximate normal strain in the wire when the frame is in this position. Assume the columns are rigid and rotate about their lower supports.

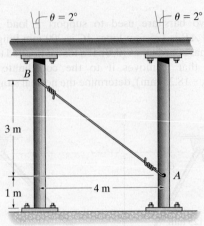

Prob. 2–25

2–26. The material distorts into the dashed position shown. Determine (a) the average normal strains along sides AC and CD and the shear strain γ_{xy} at F, and (b) the average normal strain along line BE.

2–27. The material distorts into the dashed position shown. Determine the average normal strain that occurs along the diagonals AD and CF.

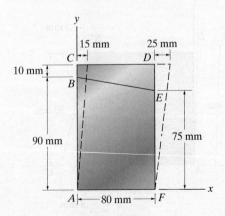

Probs. 2–26/27

*2–28. The wire is subjected to a normal strain that is defined by $\epsilon = xe^{-x^2}$, where x is in millimeters. If the wire has an initial length L, determine the increase in its length.

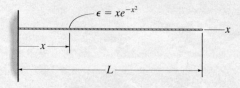

Prob. 2–28

•2–29. The curved pipe has an original radius of 0.6 m. If it is heated nonuniformly, so that the normal strain along its length is $\epsilon = 0.05 \cos \theta$, determine the increase in length of the pipe.

2–30. Solve Prob. 2–29 if $\epsilon = 0.08 \sin \theta$.

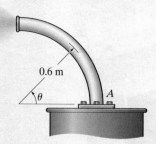

Probs. 2–29/30

2–31. The rubber band AB has an unstretched length of 1 m. If it is fixed at B and attached to the surface at point A', determine the average normal strain in the band. The surface is defined by the function $y = (x^2)$ m, where x is in meters.

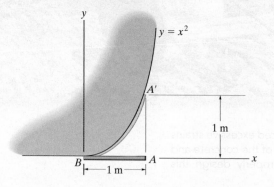

Prob. 2–31

*2–32. The bar is originally 300 mm long when it is flat. If it is subjected to a shear strain defined by $\gamma_{xy} = 0.02x$, where x is in meters, determine the displacement Δy at the end of its bottom edge. It is distorted into the shape shown, where no elongation of the bar occurs in the x direction.

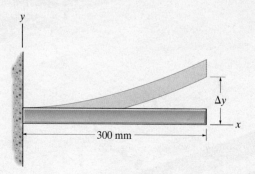

Prob. 2–32

•2–33. The fiber AB has a length L and orientation θ. If its ends A and B undergo very small displacements u_A and v_B, respectively, determine the normal strain in the fiber when it is in position $A'B'$.

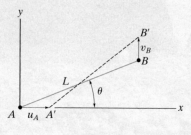

Prob. 2–33

2–34. If the normal strain is defined in reference to the final length, that is,

$$\epsilon'_n = \lim_{p \to p'} \left(\frac{\Delta s' - \Delta s}{\Delta s'} \right)$$

instead of in reference to the original length, Eq. 2–2, show that the difference in these strains is represented as a second-order term, namely, $\epsilon_n - \epsilon'_n = \epsilon_n \epsilon'_n$.

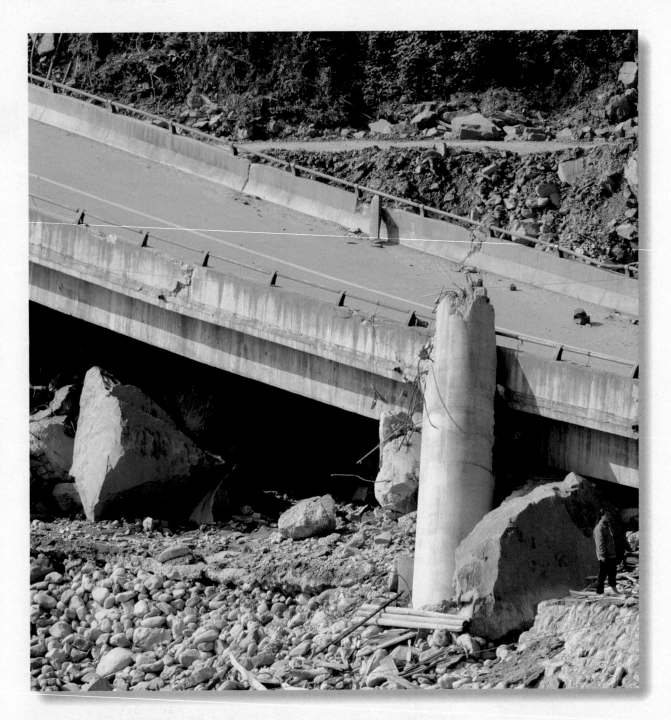

Horizontal ground displacements caused by an earthquake produced excessive strains in these bridge piers until they fractured. The material properties of the concrete and steel reinforcement must be known so that engineers can properly design this structure and thereby avoid such failures.

Mechanical Properties of Materials

3

CHAPTER OBJECTIVES

Having discussed the basic concepts of stress and strain, we will in this chapter show how stress can be related to strain by using experimental methods to determine the stress–strain diagram for a specific material. The behavior described by this diagram will then be discussed for materials that are commonly used in engineering. Also, mechanical properties and other tests that are related to the development of mechanics of materials will be discussed.

Check out the Companion Website for the following video solution(s) that can be found in this chapter.

- Section 1
  Mechanical Properties
- Section 3,4
  Strain and Elongation
- Section 5
  Elastic Recovery
- Section 6
  Poisson's Ratio - Length and Diameter Change

3.1 The Tension and Compression Test

The strength of a material depends on its ability to sustain a load without undue deformation or failure. This property is inherent in the material itself and must be determined by *experiment*. One of the most important tests to perform in this regard is the **tension or compression test**. Although several important mechanical properties of a material can be determined from this test, it is used primarily to determine the relationship between the average normal stress and average normal strain in many engineering materials such as metals, ceramics, polymers, and composites.

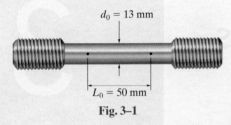

$d_0 = 13$ mm

$L_0 = 50$ mm

Fig. 3–1

Typical steel specimen with attached strain gauge.

To perform a tension or compression test a specimen of the material is made into a "standard" shape and size. It has a constant circular cross section with enlarged ends, so that failure will not occur at the grips. Before testing, two small punch marks are placed along the specimen's uniform length. Measurements are taken of both the specimen's initial cross-sectional area, A_0, and the **gauge-length** distance L_0 between the punch marks. For example, when a metal specimen is used in a tension test it generally has an initial diameter of $d_0 = 13$ mm and a gauge length of $L_0 = 50$ mm, Fig. 3–1. In order to apply an axial load with no bending of the specimen, the ends are usually seated into ball-and-socket joints. A testing machine like the one shown in Fig. 3–2 is then used to stretch the specimen at a very slow, constant rate until it fails. The machine is designed to read the load required to maintain this uniform stretching.

At frequent intervals during the test, data is recorded of the applied load P, as read on the dial of the machine or taken from a digital readout. Also, the elongation $\delta = L - L_0$ between the punch marks on the specimen may be measured using either a caliper or a mechanical or optical device called an **extensometer**. This value of δ (delta) is then used to calculate the average normal strain in the specimen. Sometimes, however, this measurement is not taken, since it is also possible to read the strain *directly* by using an **electrical-resistance strain gauge**, which looks like the one shown in Fig. 3–3. The operation of this gauge is based on the change in electrical resistance of a very thin wire or piece of metal foil under strain. Essentially the gauge is cemented to the specimen along its length. If the cement is very strong in comparison to the gauge, then the gauge is in effect an integral part of the specimen, so that when the specimen is strained in the direction of the gauge, the wire and specimen will experience the same strain. By measuring the electrical resistance of the wire, the gauge may be calibrated to read values of normal strain directly.

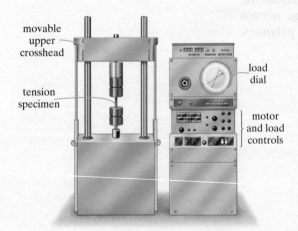

movable upper crosshead

tension specimen

load dial

motor and load controls

Fig. 3–2

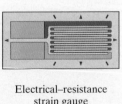

Electrical–resistance strain gauge

Fig. 3–3

3.2   The Stress–Strain Diagram

It is not feasible to prepare a test specimen to match the size, A_0 and L_0, of each structural member. Rather, the test results must be reported so they apply to a member of *any size*. To achieve this, the load and corresponding deformation data are used to calculate various values of the stress and corresponding strain in the specimen. A plot of the results produces a curve called the ***stress–strain diagram***. There are two ways in which it is normally described.

Conventional Stress–Strain Diagram.
We can determine the ***nominal*** or ***engineering stress*** by dividing the applied load P by the specimen's *original* cross-sectional area A_0. This calculation assumes that the stress is *constant* over the cross section and throughout the gauge length. We have

$$\sigma = \frac{P}{A_0} \tag{3–1}$$

Likewise, the ***nominal*** or ***engineering strain*** is found directly from the strain gauge reading, or by dividing the change in the specimen's gauge length, δ, by the specimen's original gauge length L_0. Here the strain is assumed to be constant throughout the region between the gauge points. Thus,

$$\epsilon = \frac{\delta}{L_0} \tag{3–2}$$

If the corresponding values of σ and ϵ are plotted so that the vertical axis is the stress and the horizontal axis is the strain, the resulting curve is called a ***conventional stress–strain diagram***. Realize, however, that two stress–strain diagrams for a particular material will be quite similar, but will never be exactly the same. This is because the results actually depend on variables such as the material's composition, microscopic imperfections, the way it is manufactured, the rate of loading, and the temperature during the time of the test.

We will now discuss the characteristics of the conventional stress–strain curve as it pertains to *steel*, a commonly used material for fabricating both structural members and mechanical elements. Using the method described above, the characteristic stress–strain diagram for a steel specimen is shown in Fig. 3–4. From this curve we can identify four different ways in which the material behaves, depending on the amount of strain induced in the material.

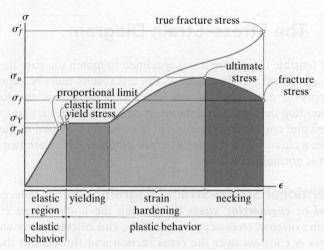

Conventional and true stress-strain diagrams
for ductile material (steel) (not to scale)

Fig. 3–4

Elastic Behavior.   Elastic behavior of the material occurs when the strains in the specimen are within the light orange region shown in Fig. 3–4. Here the curve is actually a *straight line* throughout most of this region, so that the stress is *proportional* to the strain. The material in this region is said to be *linear elastic*. The upper stress limit to this linear relationship is called the ***proportional limit***, σ_{pl}. If the stress slightly exceeds the proportional limit, the curve tends to bend and flatten out as shown. This continues until the stress reaches the ***elastic limit***. Upon reaching this point, if the load is removed the specimen will still return back to its original shape. Normally for steel, however, the elastic limit is seldom determined, since it is very close to the proportional limit and therefore rather difficult to detect.

Yielding.   A slight increase in stress above the elastic limit will result in a breakdown of the material and cause it to *deform permanently*. This behavior is called ***yielding***, and it is indicated by the rectangular dark orange region of the curve. The stress that causes yielding is called the ***yield stress*** or ***yield point***, σ_Y, and the deformation that occurs is called ***plastic deformation***. Although not shown in Fig. 3–4, for low-carbon steels or those that are hot rolled, the yield point is often distinguished by two values. The ***upper yield point*** occurs first, followed by a sudden decrease in load-carrying capacity to a ***lower yield point***. Notice that once the yield point is reached, then as shown in Fig. 3–4, the specimen will continue to elongate (strain) *without any increase in load*. When the material is in this state, it is often referred to as being ***perfectly plastic***.

Strain Hardening.   When yielding has ended, an increase in load can be supported by the specimen, resulting in a curve that rises continuously but becomes flatter until it reaches a maximum stress referred to as the *ultimate stress*, σ_u. The rise in the curve in this manner is called *strain hardening*, and it is identified in Fig. 3–4 as the region in light green.

Necking.   Up to the ultimate stress, as the specimen elongates, its cross-sectional area will decrease. This decrease is fairly *uniform* over the specimen's entire gauge length; however, just after, at the ultimate stress, the cross-sectional area will begin to decrease in a *localized* region of the specimen. As a result, a constriction or "neck" tends to form in this region as the specimen elongates further, Fig. 3–5a. This region of the curve due to necking is indicated in dark green in Fig. 3–4. Here the stress–strain diagram tends to curve downward until the specimen breaks at the *fracture stress*, σ_f, Fig. 3–5b.

True Stress–Strain Diagram.   Instead of always using the *original* cross-sectional area and specimen length to calculate the (engineering) stress and strain, we could have used the *actual* cross-sectional area and specimen length at the *instant* the load is measured. The values of stress and strain found from these measurements are called *true stress* and *true strain*, and a plot of their values is called the ***true stress–strain diagram***. When this diagram is plotted it has a form shown by the light-blue curve in Fig. 3–4. Note that the conventional and true σ–ϵ diagrams are practically coincident when the strain is small. The differences between the diagrams begin to appear in the strain-hardening range, where the magnitude of strain becomes more significant. In particular, there is a large divergence within the necking region. Here it can be seen from the conventional σ–ϵ diagram that the specimen *actually* supports a *decreasing load*, since A_0 is constant when calculating engineering stress, $\sigma = P/A_0$. However, from the true σ–ϵ diagram, the actual area A within the necking region is always decreasing until fracture, σ'_f, and so the material actually sustains *increasing stress*, since $\sigma = P/A$.

Typical necking pattern which has occurred on this steel specimen just before fracture.

This steel specimen clearly shows the necking that occurred just before the specimen failed. This resulted in the formation of a "cup-cone" shape at the fracture location, which is characteristic of ductile materials.

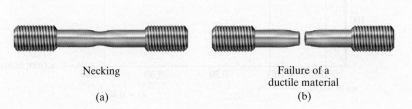

Necking	Failure of a ductile material
(a)	(b)

Fig. 3–5

Although the true and conventional stress–strain diagrams are different, most engineering design is done so that the material supports a stress within the elastic range. This is because the deformation of the material is generally not severe and the material will restore itself when the load is removed. The true strain up to the elastic limit will remain small enough so that the error in using the engineering values of σ and ϵ is very small (about 0.1%) compared with their true values. This is one of the primary reasons for using conventional stress–strain diagrams.

The above concepts can be summarized with reference to Fig. 3–6, which shows an actual conventional stress–strain diagram for a mild steel specimen. In order to enhance the details, the elastic region of the curve has been shown in light blue color using an exaggerated strain scale, also shown in light blue. Tracing the behavior, the proportional limit is reached at $\sigma_{pl} = 241$ MPa, where $\epsilon_{pl} = 0.0012$ mm/mm. This is followed by an upper yield point of $(\sigma_Y)_u = 262$ MPa, then suddenly a lower yield point of $(\sigma_Y)_l = 248$ MPa. The end of yielding occurs at a strain of $\epsilon_Y = 0.030$ mm/mm, which is 25 times greater than the strain at the proportional limit! Continuing, the specimen undergoes strain hardening until it reaches the ultimate stress of $\sigma_u = 434$ MPa, then it begins to neck down until a fracture occurs, $\sigma_f = 324$ MPa. By comparison, the strain at failure, $\epsilon_f = 0.380$ mm/mm is 317 times greater than ϵ_{pl}!

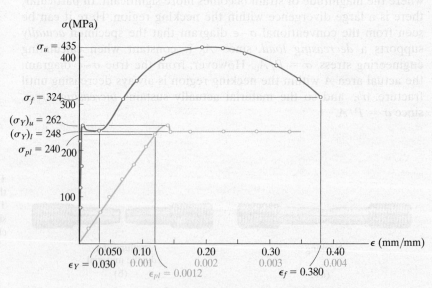

Stress-strain diagram for mild steel

Fig. 3–6

3.3   Stress–Strain Behavior of Ductile and Brittle Materials

Materials can be classified as either being ductile or brittle, depending on their stress–strain characteristics.

Ductile Materials.   Any material that can be subjected to large strains before it fractures is called a *ductile material*. Mild steel, as discussed previously, is a typical example. Engineers often choose ductile materials for design because these materials are capable of absorbing shock or energy, and if they become overloaded, they will usually exhibit large deformation before failing.

One way to specify the ductility of a material is to report its percent elongation or percent reduction in area at the time of fracture. The *percent elongation* is the specimen's fracture strain expressed as a percent. Thus, if the specimen's original gauge length is L_0 and its length at fracture is L_f, then

$$\text{Percent elongation} = \frac{L_f - L_0}{L_0}(100\%) \qquad (3\text{–}3)$$

As seen in Fig. 3–6, since $\epsilon_f = 0.380$, this value would be 38% for a mild steel specimen.

The *percent reduction in area* is another way to specify ductility. It is defined within the region of necking as follows:

$$\text{Percent reduction of area} = \frac{A_0 - A_f}{A_0}(100\%) \qquad (3\text{–}4)$$

Here A_0 is the specimen's original cross-sectional area and A_f is the area of the neck at fracture. Mild steel has a typical value of 60%.

Besides steel, other metals such as brass, molybdenum, and zinc may also exhibit ductile stress–strain characteristics similar to steel, whereby they undergo elastic stress–strain behavior, yielding at constant stress, strain hardening, and finally necking until fracture. In most metals, however, constant yielding will *not occur* beyond the elastic range. One metal for which this is the case is aluminum. Actually, this metal often does not have a well-defined *yield point*, and consequently it is standard practice to define a *yield strength* using a graphical procedure called the *offset method*. Normally a 0.2% strain (0.002 mm/mm) is chosen, and from this point on the ϵ axis, a line parallel to the initial straight-line portion of the stress–strain diagram is drawn. The point where this line intersects the curve defines the yield strength. An example of the construction for determining the yield strength for an aluminum alloy is shown in Fig. 3–7. From the graph, the yield strength is $\sigma_{YS} = 352$ MPa.

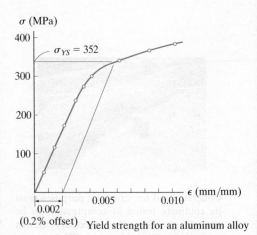

Yield strength for an aluminum alloy

Fig. 3–7

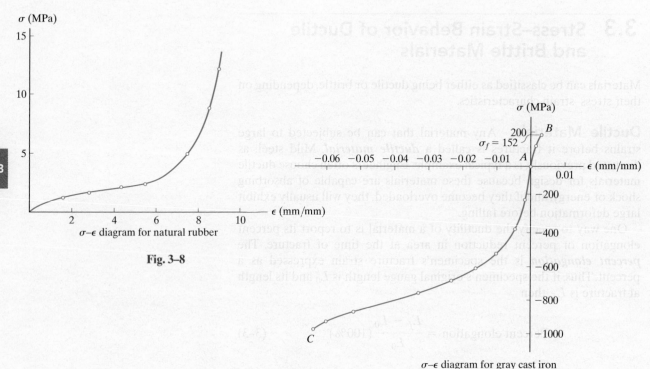

σ–ε diagram for natural rubber

Fig. 3–8

σ–ε diagram for gray cast iron

Fig. 3–9

Concrete used for structural purposes must be routinely tested in compression to be sure it provides the necessary design strength for this bridge deck. The concrete cylinders shown are compression tested for ultimate stress after curing for 30 days.

Realize that the yield strength is not a physical property of the material, since it is a stress that causes a *specified* permanent strain in the material. In this text, however, we will assume that the yield strength, yield point, elastic limit, and proportional limit all *coincide* unless otherwise stated. An exception would be natural rubber, which in fact does not even have a proportional limit, since stress and strain are *not* linearly related. Instead, as shown in Fig. 3–8, this material, which is known as a polymer, exhibits *nonlinear elastic behavior*.

Wood is a material that is often moderately ductile, and as a result it is usually designed to respond only to elastic loadings. The strength characteristics of wood vary greatly from one species to another, and for each species they depend on the moisture content, age, and the size and arrangement of knots in the wood. Since wood is a fibrous material, its tensile or compressive characteristics will differ greatly when it is loaded either parallel or perpendicular to its grain. Specifically, wood splits easily when it is loaded in tension perpendicular to its grain, and consequently tensile loads are almost always intended to be applied parallel to the grain of wood members.

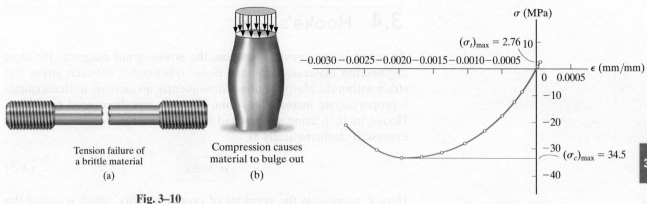

Tension failure of
a brittle material
(a)

Compression causes
material to bulge out
(b)

Fig. 3–10

$\sigma - \epsilon$ diagram for typical concrete mix

Fig. 3–11

Brittle Materials.   Materials that exhibit little or no yielding before failure are referred to as ***brittle materials***. Gray cast iron is an example, having a stress–strain diagram in tension as shown by portion AB of the curve in Fig. 3–9. Here fracture at $\sigma_f = 152$ MPa took place initially at an imperfection or microscopic crack and then spread rapidly across the specimen, causing complete fracture. Since the appearance of initial cracks in a specimen is quite random, brittle materials do not have a well-defined tensile fracture stress. Instead the *average* fracture stress from a set of observed tests is generally reported. A typical failed specimen is shown in Fig. 3–10a.

Compared with their behavior in tension, brittle materials, such as gray cast iron, exhibit a much higher resistance to axial compression, as evidenced by portion AC of the curve in Fig. 3–9. For this case any cracks or imperfections in the specimen tend to close up, and as the load increases the material will generally bulge or become barrel shaped as the strains become larger, Fig. 3–10b.

Like gray cast iron, concrete is classified as a brittle material, and it also has a low strength capacity in tension. The characteristics of its stress–strain diagram depend primarily on the mix of concrete (water, sand, gravel, and cement) and the time and temperature of curing. A typical example of a "complete" stress–strain diagram for concrete is given in Fig. 3–11. By inspection, its maximum compressive strength is almost 12.5 times greater than its tensile strength, $(\sigma_c)_{max} = 34.5$ MPa versus $(\sigma_t)_{max} = 2.76$ MPa. For this reason, concrete is almost always reinforced with steel bars or rods whenever it is designed to support tensile loads.

It can generally be stated that most materials exhibit both ductile and brittle behavior. For example, steel has brittle behavior when it contains a high carbon content, and it is ductile when the carbon content is reduced. Also, at low temperatures materials become harder and more brittle, whereas when the temperature rises they become softer and more ductile. This effect is shown in Fig. 3–12 for a methacrylate plastic.

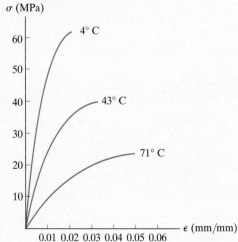

Steel rapidly loses its strength when heated. For this reason engineers often require main structural members to be insulated in case of fire.

$\sigma - \epsilon$ diagrams for a methacrylate plastic

Fig. 3–12

3.4 Hooke's Law

As noted in the previous section, the stress–strain diagrams for most engineering materials exhibit a *linear relationship* between stress and strain within the elastic region. Consequently, an increase in stress causes a proportionate increase in strain. This fact was discovered by Robert Hooke in 1676 using springs and is known as *Hooke's law*. It may be expressed mathematically as

$$\sigma = E\epsilon \qquad (3\text{–}5)$$

Here E represents the constant of proportionality, which is called the **modulus of elasticity** or **Young's modulus**, named after Thomas Young, who published an account of it in 1807.

Equation 3–5 actually represents the equation of the *initial straight-lined portion* of the stress–strain diagram up to the proportional limit. Furthermore, the modulus of elasticity represents the *slope* of this line. Since strain is dimensionless, from Eq. 3–5, E will have the same units as stress, such as psi, ksi, or pascals. As an example of its calculation, consider the stress–strain diagram for steel shown in Fig. 3–6. Here $\sigma_{pl} = 240$ MPa and $\epsilon_{pl} = 0.0012$ mm/mm, so that

$$E = \frac{\sigma_{pl}}{\epsilon_{pl}} = \frac{240 \text{ MPa}}{0.0012 \text{ mm}} = 200 \text{ GPa}$$

As shown in Fig. 3–13, the proportional limit for a particular type of steel alloy depends on its carbon content; however, most grades of steel, from the softest rolled steel to the hardest tool steel, have about the

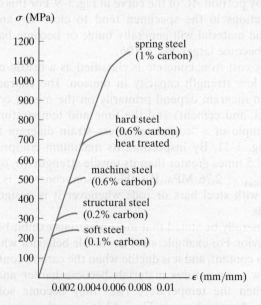

Fig. 3–13

same modulus of elasticity, generally accepted to be $E_{st} = 200$ GPa. Values of E for other commonly used engineering materials are often tabulated in engineering codes and reference books. Representative values are also listed on the inside back cover of this book. It should be noted that the modulus of elasticity is a mechanical property that indicates the *stiffness* of a material. Materials that are very stiff, such as steel, have large values of $E[E_{st} = 200$ GPa], whereas spongy materials such as vulcanized rubber may have low values [$E_r = 0.70$ MPa].

The modulus of elasticity is one of the most important mechanical properties used in the development of equations presented in this text. It must always be remembered, though, that E can be used only if a material has *linear elastic behavior*. Also, if the stress in the material is *greater* than the proportional limit, the stress–strain diagram ceases to be a straight line and so Eq. 3–5 is no longer valid.

Strain Hardening.

If a specimen of ductile material, such as steel, is loaded into the *plastic region* and then unloaded, *elastic strain is recovered* as the material returns to its equilibrium state. The *plastic strain remains*, however, and as a result the material is subjected to a **permanent set**. For example, a wire when bent (plastically) will spring back a little (elastically) when the load is removed; however, it will not fully return to its original position. This behavior can be illustrated on the stress–strain diagram shown in Fig. 3–14a. Here the specimen is first loaded beyond its yield point A to point A'. Since interatomic forces have to be overcome to elongate the specimen *elastically*, then these same forces pull the atoms back together when the load is removed, Fig. 3–14a. Consequently, the modulus of elasticity, E, is the same, and therefore the slope of line $O'A'$ is the same as line OA.

If the load is reapplied, the atoms in the material will again be displaced until yielding occurs at or near the stress A', and the stress–strain diagram continues along the same path as before, Fig. 3–14b. It should be noted, however, that this new stress–strain diagram, defined by $O'A'B$, now has a *higher* yield point (A'), a consequence of strain-hardening. In other words, the material now has a *greater elastic region*; however, it has *less ductility*, a smaller plastic region, than when it was in its original state.

This pin was made from a hardened steel alloy, that is, one having a high carbon content. It failed due to brittle fracture.

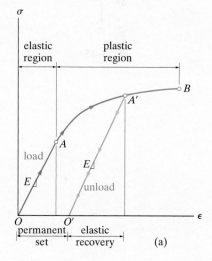

(a)

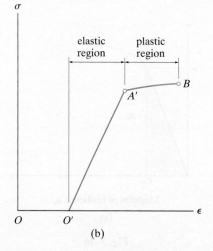

(b)

Fig. 3–14

3.5 Strain Energy

As a material is deformed by an external loading, it tends to store energy *internally* throughout its volume. Since this energy is related to the strains in the material, it is referred to as **strain energy**. To obtain this strain energy consider a volume element of material from a tension test specimen. It is subjected to uniaxial stress as shown in Fig. 3–15. This stress develops a force $\Delta F = \sigma \, \Delta A = \sigma(\Delta x \, \Delta y)$ on the top and bottom faces of the element *after* the element of length Δz undergoes a vertical displacement $\epsilon \, \Delta z$. By definition, *work* is determined by the product of the force and displacement in the direction of the force. Since the force is increased uniformly from zero to its final magnitude ΔF when the displacement $\epsilon \, \Delta z$ is attained, the work done on the element by the force is equal to the *average* force magnitude $(\Delta F/2)$ times the displacement $\epsilon \, \Delta z$. This "external work" on the element is equivalent to the "internal work" or strain energy stored in the element—assuming that no energy is lost in the form of heat. Consequently, the strain energy ΔU is $\Delta U = (\frac{1}{2} \Delta F) \, \epsilon \, \Delta z = (\frac{1}{2} \sigma \, \Delta x \, \Delta y) \, \epsilon \, \Delta z$. Since the volume of the element is $\Delta V = \Delta x \, \Delta y \, \Delta z$, then $\Delta U = \frac{1}{2} \sigma \epsilon \, \Delta V$.

For applications, it is sometimes convenient to specify the strain energy per unit volume of material. This is called the **strain-energy density**, and it can be expressed as

$$u = \frac{\Delta U}{\Delta V} = \frac{1}{2} \sigma \epsilon \qquad (3\text{–}6)$$

If the material behavior is *linear elastic*, then Hooke's law applies, $\sigma = E\epsilon$, and therefore we can express the elastic strain-energy density in terms of the uniaxial stress as

$$u = \frac{1}{2} \frac{\sigma^2}{E} \qquad (3\text{–}7)$$

Modulus of Resilience.   In particular, when the stress σ reaches the proportional limit, the strain-energy density, as calculated by Eq. 3–6 or 3–7, is referred to as the **modulus of resilience**, i.e.,

$$u_r = \frac{1}{2} \sigma_{pl} \epsilon_{pl} = \frac{1}{2} \frac{\sigma_{pl}^2}{E} \qquad (3\text{–}8)$$

From the elastic region of the stress–strain diagram, Fig. 3–16a, notice that u_r is equivalent to the shaded *triangular area* under the diagram. Physically a material's resilience represents the ability of the material to absorb energy without any permanent damage to the material.

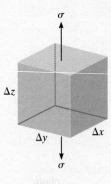

Fig. 3–15

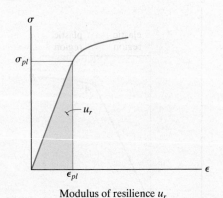

Modulus of resilience u_r

(a)

Fig. 3–16

Modulus of Toughness.

Another important property of a material is the **modulus of toughness**, u_t. This quantity represents the *entire area* under the stress–strain diagram, Fig. 3–16b, and therefore it indicates the strain-energy density of the material just before it fractures. This property becomes important when designing members that may be accidentally overloaded. Alloying metals can also change their resilience and toughness. For example, by changing the percentage of carbon in steel, the resulting stress–strain diagrams in Fig. 3–17 show how the degrees of resilience and toughness can be changed.

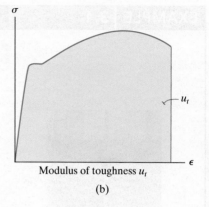

Modulus of toughness u_t

(b)

Fig. 3–16 (cont.)

Important Points

- A *conventional stress–strain diagram* is important in engineering since it provides a means for obtaining data about a material's tensile or compressive strength without regard for the material's physical size or shape.

- *Engineering stress and strain* are calculated using the *original* cross-sectional area and gauge length of the specimen.

- A *ductile material*, such as mild steel, has four distinct behaviors as it is loaded. They are *elastic behavior, yielding, strain hardening,* and *necking*.

- A material is *linear elastic* if the stress is proportional to the strain within the elastic region. This behavior is described by *Hooke's law*, $\sigma = E\epsilon$, where the *modulus of elasticity E* is the slope of the line.

- Important points on the stress–strain diagram are the *proportional limit, elastic limit, yield stress, ultimate stress,* and *fracture stress.*

- The *ductility* of a material can be specified by the specimen's *percent elongation* or the *percent reduction in area.*

- If a material does not have a distinct yield point, a *yield strength* can be specified using a graphical procedure such as the *offset method.*

- *Brittle materials*, such as gray cast iron, have very little or no yielding and so they can fracture suddenly.

- *Strain hardening* is used to establish a higher yield point for a material. This is done by straining the material beyond the elastic limit, then releasing the load. The modulus of elasticity remains the same; however, the material's ductility *decreases.*

- *Strain energy* is energy stored in a material due to its deformation. This energy per unit volume is called *strain-energy density*. If it is measured up to the proportional limit, it is referred to as the *modulus of resilience*, and if it is measured up to the point of fracture, it is called the *modulus of toughness*. It can be determined from the area under the σ–ϵ diagram.

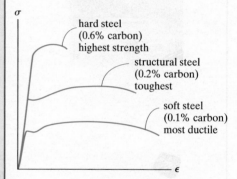

hard steel
(0.6% carbon)
highest strength

structural steel
(0.2% carbon)
toughest

soft steel
(0.1% carbon)
most ductile

Fig. 3–17

This nylon specimen exhibits a high degree of toughness as noted by the large amount of necking that has occurred just before fracture.

EXAMPLE 3.1

A tension test for a steel alloy results in the stress–strain diagram shown in Fig. 3–18. Calculate the modulus of elasticity and the yield strength based on a 0.2% offset. Identify on the graph the ultimate stress and the fracture stress.

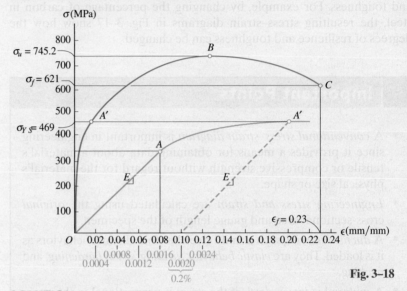

Fig. 3–18

Refer to the Companion Website for the animation of the concepts that can be found in Example 3.1.

SOLUTION

Modulus of Elasticity. We must calculate the *slope* of the initial straight-line portion of the graph. Using the magnified curve and scale shown in blue, this line extends from point O to an estimated point A, which has coordinates of approximately (0.0016 mm/mm, 345 MPa). Therefore,

$$E = \frac{345 \text{ MPa}}{0.0016 \text{ mm/mm}} = 215 \text{ GPa} \qquad Ans.$$

Note that the equation of line OA is thus $\sigma = 215(10^3)\epsilon$.

Yield Strength. For a 0.2% offset, we begin at a strain of 0.2% or 0.0020 mm/mm and graphically extend a (dashed) line parallel to OA until it intersects the $\sigma-\epsilon$ curve at A'. The yield strength is approximately

$$\sigma_{YS} = 469 \text{ MPa} \qquad Ans.$$

Ultimate Stress. This is defined by the peak of the $\sigma-\epsilon$ graph, point B in Fig. 3–18.

$$\sigma_u = 745.2 \text{ MPa} \qquad Ans.$$

Fracture Stress. When the specimen is strained to its maximum of $\epsilon_f = 0.23$ mm/mm, it fractures at point C. Thus,

$$\sigma_f = 621 \text{ MPa} \qquad Ans.$$

EXAMPLE 3.2

The stress–strain diagram for an aluminum alloy that is used for making aircraft parts is shown in Fig. 3–19. If a specimen of this material is stressed to 600 MPa, determine the permanent strain that remains in the specimen when the load is released. Also, find the modulus of resilience both before and after the load application.

SOLUTION

Permanent Strain.   When the specimen is subjected to the load, it strain-hardens until point B is reached on the $\sigma-\epsilon$ diagram. The strain at this point is approximately 0.023 mm/mm. When the load is released, the material behaves by following the straight line BC, which is parallel to line OA. Since both lines have the same slope, the strain at point C can be determined analytically. The slope of line OA is the modulus of elasticity, i.e.,

$$E = \frac{450 \text{ MPa}}{0.006 \text{ mm/mm}} = 75.0 \text{ GPa}$$

From triangle CBD, we require

$$E = \frac{BD}{CD}; \qquad 75.0(10^9) \text{ Pa} = \frac{600(10^6) \text{ Pa}}{CD}$$

$$CD = 0.008 \text{ mm/mm}$$

This strain represents the amount of *recovered elastic strain*. The permanent strain, ϵ_{OC}, is thus

$$\epsilon_{OC} = 0.023 \text{ mm/mm} - 0.008 \text{ mm/mm}$$

$$= 0.0150 \text{ mm/mm} \qquad \qquad Ans.$$

Note: If gauge marks on the specimen were originally 50 mm apart, then after the load is *released* these marks will be 50 mm + $(0.0150)(50 \text{ mm}) = 50.75 \text{ mm}$ apart.

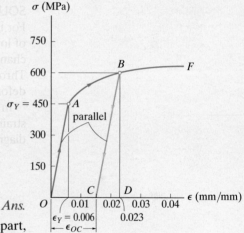

Fig. 3–19

Modulus of Resilience.   Applying Eq. 3–8, we have[*]

$$(u_r)_{\text{initial}} = \frac{1}{2}\sigma_{pl}\epsilon_{pl} = \frac{1}{2}(450 \text{ MPa})(0.006 \text{ mm/mm})$$

$$= 1.35 \text{ MJ/m}^3 \qquad \qquad Ans.$$

$$(u_r)_{\text{final}} = \frac{1}{2}\sigma_{pl}\epsilon_{pl} = \frac{1}{2}(600 \text{ MPa})(0.008 \text{ mm/mm})$$

$$= 2.40 \text{ MJ/m}^3 \qquad \qquad Ans.$$

NOTE: By comparison, the effect of strain-hardening the material has caused an increase in the modulus of resilience; however, note that the modulus of toughness for the material has decreased since the area under the original curve, $OABF$, is larger than the area under curve CBF.

[*] Work in the SI system of units is measured in joules, where $1 \text{ J} = 1 \text{ N} \cdot \text{m}$.

EXAMPLE | **3.3**

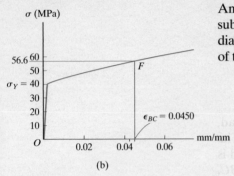

(b)

An aluminum rod shown in Fig. 3–20a has a circular cross section and is subjected to an axial load of 10 kN. If a portion of the stress–strain diagram is shown in Fig. 3–20b, determine the approximate elongation of the rod when the load is applied. Take $E_{al} = 70$ GPa.

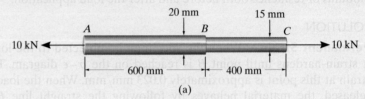

(a)

Fig. 3–20

SOLUTION

For the analysis we will neglect the *localized deformations* at the point of load application and where the rod's cross-sectional area suddenly changes. (These effects will be discussed in Sections 4.1 and 4.7.) Throughout the midsection of each segment the normal stress and deformation are uniform.

In order to find the elongation of the rod, we must first obtain the strain. This is done by calculating the stress, then using the stress–strain diagram. The normal stress within each segment is

$$\sigma_{AB} = \frac{P}{A} = \frac{10(10^3)\text{ N}}{\pi(0.01\text{ m})^2} = 31.83\text{ MPa}$$

$$\sigma_{BC} = \frac{P}{A} = \frac{10(10^3)\text{ N}}{\pi(0.0075\text{ m})^2} = 56.59\text{ MPa}$$

From the stress–strain diagram, the material in segment AB is strained *elastically* since $\sigma_{AB} < \sigma_Y = 40$ MPa. Using Hooke's law,

$$\epsilon_{AB} = \frac{\sigma_{AB}}{E_{al}} = \frac{31.83(10^6)\text{ Pa}}{70(10^9)\text{ Pa}} = 0.0004547\text{ mm/mm}$$

The material within segment BC is strained plastically, since $\sigma_{BC} > \sigma_Y = 40$ MPa. From the graph, for $\sigma_{BC} = 56.59$ MPa, $\epsilon_{BC} \approx 0.045$ mm/mm. The approximate elongation of the rod is therefore

$$\delta = \Sigma\epsilon L = 0.0004547(600\text{ mm}) + 0.0450(400\text{ mm})$$

$$= 18.3\text{ mm} \qquad\qquad Ans.$$

FUNDAMENTAL PROBLEMS

F3–1. Define homogeneous material.

F3–2. Indicate the points on the stress-strain diagram which represent the proportional limit and the ultimate stress.

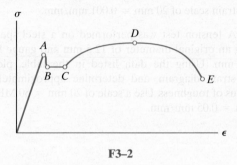

F3–2

F3–3. Define the modulus of elasticity E.

F3–4. At room temperature, mild steel is a ductile material. True or false?

F3–5. Engineering stress and strain are calculated using the *actual* cross-sectional area and length of the specimen. True or false?

F3–6. As the temperature increases the modulus of elasticity will increase. True or false?

F3–7. A 100-mm long rod has a diameter of 15 mm. If an axial tensile load of 100 kN is applied, determine its change is length. $E = 200$ GPa.

F3–8. A bar has a length of 200 mm and cross-sectional area of 7500 mm^2. Determine the modulus of elasticity of the material if it is subjected to an axial tensile load of 50 kN and stretches 0.075 mm. The material has linear-elastic behavior.

F3–9. A 10-mm-diameter brass rod has a modulus of elasticity of $E = 100$ GPa. If it is 4 m long and subjected to an axial tensile load of 6 kN, determine its elongation.

F3–10. The material for the 50-mm-long specimen has the stress–strain diagram shown. If $P = 100$ kN, determine the elongation of the specimen.

F3–11. The material for the 50-mm-long specimen has the stress–strain diagram shown. If $P = 150$ kN is applied and then released, determine the permanent elongation of the specimen.

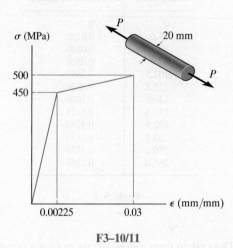

F3–10/11

F3–12. If the elongation of wire BC is 0.2 mm after the force **P** is applied, determine the magnitude of **P**. The wire is A-36 steel and has a diameter of 3 mm.

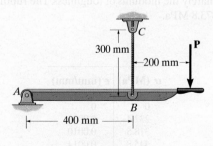

F3–12

PROBLEMS

•3–1. A concrete cylinder having a diameter of 150 mm and gauge length of 300 mm is tested in compression. The results of the test are reported in the table as load versus contraction. Draw the stress–strain diagram using scales of 10 mm = 2 MPa and 10 mm = 0.1(10^{-3}) mm/mm. From the diagram, determine approximately the modulus of elasticity.

Load (kN)	Contraction (mm)
0	0
25.0	0.0150
47.5	0.0300
82.5	0.0500
102.5	0.0650
127.5	0.0850
150.0	0.1000
172.5	0.1125
192.5	0.1250
232.5	0.1550
250.0	0.1750
265.0	0.1850

Prob. 3–1

3–2. Data taken from a stress–strain test for a ceramic are given in the table. The curve is linear between the origin and the first point. Plot the diagram, and determine the modulus of elasticity and the modulus of resilience.

3–3. Data taken from a stress–strain test for a ceramic are given in the table. The curve is linear between the origin and the first point. Plot the diagram, and determine approximately the modulus of toughness. The rupture stress is $\sigma_r = 373.8$ MPa.

σ (MPa)	ϵ (mm/mm)
0	0
232.4	0.0006
318.5	0.0010
345.8	0.0014
360.5	0.0018
373.8	0.0022

Probs. 3–2/3

*3–4. A tension test was performed on a specimen having an original diameter of 12.5 mm and a gauge length of 50 mm. The data are listed in the table. Plot the stress–strain diagram, and determine approximately the modulus of elasticity, the ultimate stress, and the fracture stress. Use a scale of 20 mm = 50 MPa and 20 mm = 0.05 mm/mm. Redraw the linear-elastic region, using the same stress scale but a strain scale of 20 mm = 0.001 mm/mm.

3–5. A tension test was performed on a steel specimen having an original diameter of 12.5 mm and gauge length of 50 mm. Using the data listed in the table, plot the stress–strain diagram, and determine approximately the modulus of toughness. Use a scale of 20 mm = 50 MPa and 20 mm = 0.05 mm/mm.

Load (kN)	Elongation (mm)
0	0
11.1	0.0175
31.9	0.0600
37.8	0.1020
40.9	0.1650
43.6	0.2490
53.4	1.0160
62.3	3.0480
64.5	6.3500
62.3	8.8900
58.8	11.9380

Probs. 3–4/5

3–6. A specimen is originally 300 mm long, has a diameter of 12 mm, and is subjected to a force of 2.5 kN. When the force is increased from 2.5 kN to 9 kN, the specimen elongates 0.225 mm. Determine the modulus of elasticity for the material if it remains linear elastic.

3–7. A structural member in a nuclear reactor is made of a zirconium alloy. If an axial load of 20 kN is to be supported by the member, determine its required cross-sectional area. Use a factor of safety of 3 relative to yielding. What is the load on the member if it is 1 m long and its elongation is 0.5 mm? $E_{zr} = 100$ GPa, $\sigma_Y = 400$ MPa. The material has elastic behavior.

*3–8. The strut is supported by a pin at C and an A-36 steel guy wire AB. If the wire has a diameter of 5 mm, determine how much it stretches when the distributed load acts on the strut.

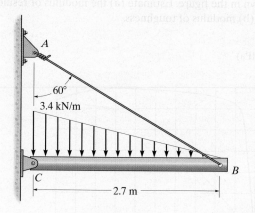

Prob. 3–8

•3–9. The $\sigma-\epsilon$ diagram for a collagen fiber bundle from which a human tendon is composed is shown. If a segment of the Achilles tendon at A has a length of 165 mm and an approximate cross-sectional area of 145 mm², determine its elongation if the foot supports a load of 625 N, which causes a tension in the tendon of 1718.75 N.

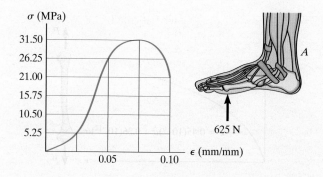

Prob. 3–9

3–10. The stress–strain diagram for a metal alloy having an original diameter of 12 mm and a gauge length of 50 mm is given in the figure. Determine approximately the modulus of elasticity for the material, the load on the specimen that causes yielding, and the ultimate load the specimen will support.

3–11. The stress–strain diagram for a steel alloy having an original diameter of 12 mm and a gauge length of 50 mm is given in the figure. If the specimen is loaded until it is stressed to 500 MPa, determine the approximate amount of elastic recovery and the increase in the gauge length after it is unloaded.

*3–12. The stress–strain diagram for a steel alloy having an original diameter of 12 mm and a gauge length of 50 mm is given in the figure. Determine approximately the modulus of resilience and the modulus of toughness for the material.

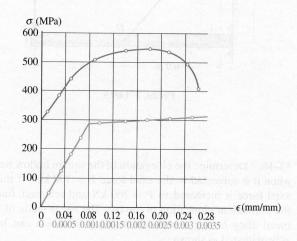

Probs. 3–10/11/12

•3–13. A bar having a length of 125 mm and cross-sectional area of 437.5 mm² is subjected to an axial force of 40 kN. If the bar stretches 0.05 mm, determine the modulus of elasticity of the material. The material has linear-elastic behavior.

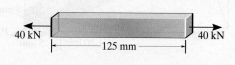

Prob. 3–13

3–14. The rigid pipe is supported by a pin at A and an A-36 steel guy wire BD. If the wire has a diameter of 6 mm, determine how much it stretches when a load of $P = 3$ kN acts on the pipe.

3–15. The rigid pipe is supported by a pin at A and an A-36 guy wire BD. If the wire has a diameter of 6 mm, determine the load P if the end C is displaced 1.875 mm downward.

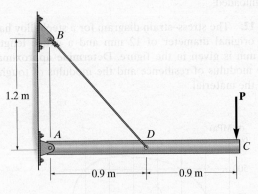

Probs. 3–14/15

***3–16.** Determine the elongation of the square hollow bar when it is subjected to the axial force $P = 100$ kN. If this axial force is increased to $P = 360$ kN and released, find the permanent elongation of the bar. The bar is made of a metal alloy having a stress–strain diagram which can be approximated as shown.

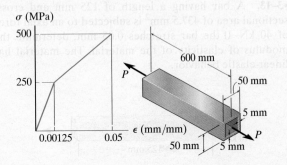

Prob. 3–16

3–17. A tension test was performed on an aluminum 2014-T6 alloy specimen. The resulting stress–strain diagram is shown in the figure. Estimate (a) the proportional limit, (b) the modulus of elasticity, and (c) the yield strength based on a 0.2% strain offset method.

3–18. A tension test was performed on an aluminum 2014-T6 alloy specimen. The resulting stress–strain diagram is shown in the figure. Estimate (a) the modulus of resilience; and (b) modulus of toughness.

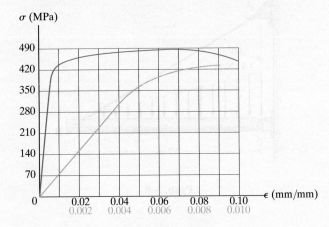

Probs. 3–17/18

3–19. The stress–strain diagram for a bone is shown, and can be described by the equation $\epsilon = 0.45(10^{-6}) \sigma + 0.36(10^{-12}) \sigma^3$, where σ is in kPa. Determine the yield strength assuming a 0.3% offset.

***3–20.** The stress–strain diagram for a bone is shown and can be described by the equation $\epsilon = 0.45(10^{-6}) \sigma + 0.36(10^{-12}) \sigma^3$, where σ is in kPa. Determine the modulus of toughness and the amount of elongation of a 200-mm-long region just before it fractures if failure occurs at $\epsilon = 0.12$ mm/mm.

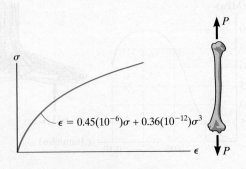

Probs. 3–19/20

•3–21. The stress–strain diagram for a polyester resin is given in the figure. If the rigid beam is supported by a strut AB and post CD, both made from this material, and subjected to a load of $P = 80$ kN, determine the angle of tilt of the beam when the load is applied. The diameter of the strut is 40 mm and the diameter of the post is 80 mm.

3–22. The stress–strain diagram for a polyester resin is given in the figure. If the rigid beam is supported by a strut AB and post CD made from this material, determine the largest load P that can be applied to the beam before it ruptures. The diameter of the strut is 12 mm and the diameter of the post is 40 mm.

3–23. By adding plasticizers to polyvinyl chloride, it is possible to reduce its stiffness. The stress–strain diagrams for three types of this material showing this effect are given below. Specify the type that should be used in the manufacture of a rod having a length of 125 mm and a diameter of 50 mm, that is required to support at least an axial load of 100 kN and also be able to stretch at most 6 mm.

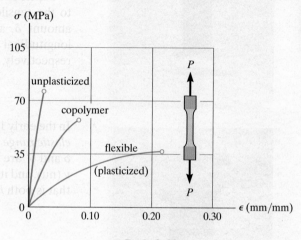

Prob. 3–23

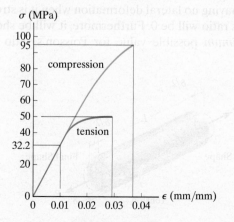

Probs. 3–21/22

*3–24. The stress–strain diagram for many metal alloys can be described analytically using the Ramberg-Osgood three parameter equation $\epsilon = \sigma/E + k\sigma^n$, where E, k, and n are determined from measurements taken from the diagram. Using the stress–strain diagram shown in the figure, take $E = 210$ GPa and determine the other two parameters k and n and thereby obtain an analytical expression for the curve.

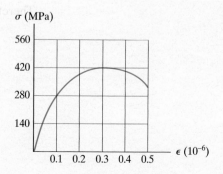

Prob. 3–24

When the rubber block is compressed (negative strain) its sides will expand (positive strain). The ratio of these strains remains constant.

3.6 Poisson's Ratio

When a deformable body is subjected to an axial tensile force, not only does it elongate but it also contracts laterally. For example, if a rubber band is stretched, it can be noted that both the thickness and width of the band are decreased. Likewise, a compressive force acting on a body causes it to contract in the direction of the force and yet its sides expand laterally.

Consider a bar having an original radius r and length L and subjected to the tensile force P in Fig. 3–21. This force elongates the bar by an amount δ, and its radius contracts by an amount δ'. Strains in the longitudinal or axial direction and in the lateral or radial direction are, respectively,

$$\epsilon_{\text{long}} = \frac{\delta}{L} \quad \text{and} \quad \epsilon_{\text{lat}} = \frac{\delta'}{r}$$

In the early 1800s, the French scientist S. D. Poisson realized that within the *elastic range* the *ratio* of these strains is a *constant*, since the deformations δ and δ' are proportional. This constant is referred to as **Poisson's ratio**, ν (nu), and it has a numerical value that is unique for a particular material that is both *homogeneous and isotropic*. Stated mathematically it is

$$\nu = -\frac{\epsilon_{\text{lat}}}{\epsilon_{\text{long}}} \tag{3–9}$$

The negative sign is included here since *longitudinal elongation* (positive strain) causes *lateral contraction* (negative strain), and vice versa. Notice that these strains are caused only by the axial or longitudinal force P; i.e., no force or stress acts in a lateral direction in order to strain the material in this direction.

Poisson's ratio is a *dimensionless* quantity, and for most nonporous solids it has a value that is generally between $\frac{1}{4}$ and $\frac{1}{3}$. Typical values of ν for common engineering materials are listed on the inside back cover. For an "ideal material" having no lateral deformation when it is stretched or compressed Poisson's ratio will be 0. Furthermore, it will be shown in Sec. 10.6 that the *maximum* possible value for Poisson's ratio is 0.5. Therefore $0 \leq \nu \leq 0.5$.

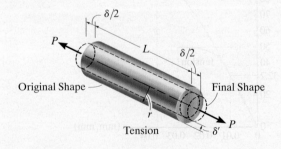

Fig. 3–21

EXAMPLE 3.4

A bar made of A-36 steel has the dimensions shown in Fig. 3–22. If an axial force of $P = 80$ kN is applied to the bar, determine the change in its length and the change in the dimensions of its cross section after applying the load. The material behaves elastically.

Refer to the Companion Website for the animation of the concepts that can be found in Example 3.4.

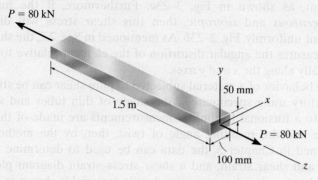

Fig. 3–22

SOLUTION

The normal stress in the bar is

$$\sigma_z = \frac{P}{A} = \frac{80(10^3)\ \text{N}}{(0.1\ \text{m})(0.05\ \text{m})} = 16.0(10^6)\ \text{Pa}$$

From the table on the inside back cover for A-36 steel $E_{st} = 200$ GPa, and so the strain in the z direction is

$$\epsilon_z = \frac{\sigma_z}{E_{st}} = \frac{16.0(10^6)\ \text{Pa}}{200(10^9)\ \text{Pa}} = 80(10^{-6})\ \text{mm/mm}$$

The axial elongation of the bar is therefore

$$\delta_z = \epsilon_z L_z = [80(10^{-6})](1.5\ \text{m}) = 120\ \mu\text{m} \qquad Ans.$$

Using Eq. 3–9, where $\nu_{st} = 0.32$ as found from the inside back cover, the lateral contraction strains in *both* the x and y directions are

$$\epsilon_x = \epsilon_y = -\nu_{st}\epsilon_z = -0.32[80(10^{-6})] = -25.6\ \mu\text{m/m}$$

Thus the changes in the dimensions of the cross section are

$$\delta_x = \epsilon_x L_x = -[25.6(10^{-6})](0.1\ \text{m}) = -2.56\ \mu\text{m} \qquad Ans.$$

$$\delta_y = \epsilon_y L_y = -[25.6(10^{-6})](0.05\ \text{m}) = -1.28\ \mu\text{m} \qquad Ans.$$

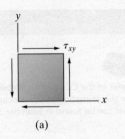

(a)

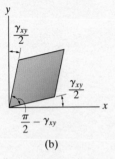

(b)

Fig. 3–23

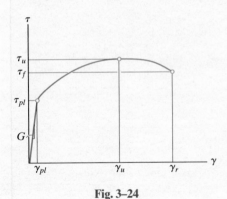

Fig. 3–24

3.7 The Shear Stress–Strain Diagram

In Sec. 1.5 it was shown that when a small element of material is subjected to *pure shear*, equilibrium requires that equal shear stresses must be developed on four faces of the element. These stresses τ_{xy} must be directed toward or away from diagonally opposite corners of the element, as shown in Fig. 3–23a. Furthermore, if the material is *homogeneous* and *isotropic*, then this shear stress will distort the element uniformly, Fig. 3–23b. As mentioned in Sec. 2.2, the shear strain γ_{xy} measures the angular distortion of the element relative to the sides originally along the *x* and *y* axes.

The behavior of a material subjected to pure shear can be studied in a laboratory using specimens in the shape of thin tubes and subjecting them to a torsional loading. If measurements are made of the applied torque and the resulting angle of twist, then by the methods to be explained in Chapter 5, the data can be used to determine the shear stress and shear strain, and a shear stress–strain diagram plotted. An example of such a diagram for a ductile material is shown in Fig. 3–24. Like the tension test, this material when subjected to shear will exhibit linear-elastic behavior and it will have a defined *proportional limit* τ_{pl}. Also, strain hardening will occur until an *ultimate shear stress* τ_u is reached. And finally, the material will begin to lose its shear strength until it reaches a point where it fractures, τ_f.

For most engineering materials, like the one just described, the elastic behavior is *linear*, and so Hooke's law for shear can be written as

$$\tau = G\gamma \qquad (3\text{–}10)$$

Here G is called the **shear modulus of elasticity** or the **modulus of rigidity**. Its value represents the slope of the line on the $\tau{-}\gamma$ diagram, that is, $G = \tau_{pl}/\gamma_{pl}$. Typical values for common engineering materials are listed on the inside back cover. Notice that the units of measurement for G will be the *same* as those for τ (Pa or psi), since γ is measured in radians, a dimensionless quantity.

It will be shown in Sec. 10.6 that the three material constants, E, ν, and G are actually *related* by the equation

$$G = \frac{E}{2(1 + \nu)} \qquad (3\text{–}11)$$

Provided E and G are known, the value of ν can then be determined from this equation rather than through experimental measurement. For example, in the case of A-36 steel, $E_{st} = 200$ GPa and $G_{st} = 76$ GPa, so that, from Eq. 3–11, $\nu_{st} = 0.32$.

EXAMPLE 3.5

A specimen of titanium alloy is tested in torsion and the shear stress–strain diagram is shown in Fig. 3–25a. Determine the shear modulus G, the proportional limit, and the ultimate shear stress. Also, determine the maximum distance d that the top of a block of this material, shown in Fig. 3–25b, could be displaced horizontally if the material behaves elastically when acted upon by a shear force $\mathbf{V}$. What is the magnitude of $\mathbf{V}$ necessary to cause this displacement?

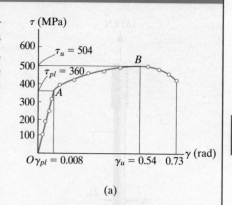

(a)

SOLUTION

Shear Modulus.   This value represents the slope of the straight-line portion OA of the τ–γ diagram. The coordinates of point A are (0.008 rad, 360 MPa). Thus,

$$G = \frac{360 \text{ MPa}}{0.008 \text{ rad}} = 45(10^3) \text{ MPa} = 45 \text{ GPa} \qquad Ans.$$

The equation of line OA is therefore $\tau = G\gamma = 45(10^3)\gamma$, which is Hooke's law for shear.

Proportional Limit.   By inspection, the graph ceases to be linear at point A. Thus,

$$\tau_{pl} = 360 \text{ MPa} \qquad Ans.$$

Ultimate Stress.   This value represents the maximum shear stress, point B. From the graph,

$$\tau_u = 504 \text{ MPa} \qquad Ans.$$

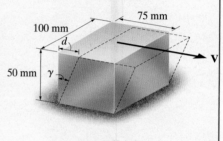

(b)

Fig. 3–25

Maximum Elastic Displacement and Shear Force.   Since the maximum elastic shear strain is 0.008 rad, a very small angle, the top of the block in Fig. 3–25b will be displaced horizontally:

$$\tan(0.008 \text{ rad}) \approx 0.008 \text{ rad} = \frac{d}{50 \text{ mm}}$$

$$d = 0.4 \text{ mm} \qquad Ans.$$

The corresponding *average* shear stress in the block is $\tau_{pl} = 360$ MPa. Thus, the shear force V needed to cause the displacement is

$$\tau_{avg} = \frac{V}{A}; \qquad 360 \text{ MPa} = \frac{V}{(75 \text{ mm})(100 \text{ mm})}$$

$$V = 2700 \text{ kN} \qquad Ans.$$

EXAMPLE 3.6

165 kN

d_0 ← → L_0

165 kN

Fig. 3–26

An aluminum specimen shown in Fig. 3–26 has a diameter of $d_0 = 25$ mm and a gauge length of $L_0 = 250$ mm. If a force of 165 kN elongates the gauge length 1.20 mm, determine the modulus of elasticity. Also, determine by how much the force causes the diameter of the specimen to contract. Take $G_{al} = 26$ GPa and $\sigma_Y = 440$ MPa.

SOLUTION

Modulus of Elasticity.   The average normal stress in the specimen is

$$\sigma = \frac{P}{A} = \frac{165(10^3)\ \text{N}}{(\pi/4)(0.025\ \text{m})^2} = 336.1\ \text{MPa}$$

and the average normal strain is

$$\epsilon = \frac{\delta}{L} = \frac{1.20\ \text{mm}}{250\ \text{mm}} = 0.00480\ \text{mm/mm}$$

Since $\sigma < \sigma_Y = 440$ MPa, the material behaves elastically. The modulus of elasticity is therefore

$$E_{al} = \frac{\sigma}{\epsilon} = \frac{336.1(10^6)\ \text{Pa}}{0.00480} = 70.0\ \text{GPa} \qquad Ans.$$

Contraction of Diameter.   First we will determine Poisson's ratio for the material using Eq. 3–11.

$$G = \frac{E}{2(1 + \nu)}$$

$$26\ \text{GPa} = \frac{70.0\ \text{GPa}}{2(1 + \nu)}$$

$$\nu = 0.347$$

Since $\epsilon_{long} = 0.00480$ mm/mm, then by Eq. 3–9,

$$\nu = -\frac{\epsilon_{lat}}{\epsilon_{long}}$$

$$0.347 = -\frac{\epsilon_{lat}}{0.00480\ \text{mm/mm}}$$

$$\epsilon_{lat} = -0.00166\ \text{mm/mm}$$

The contraction of the diameter is therefore

$$\delta' = (0.00166)(25\ \text{mm})$$

$$= 0.0416\ \text{mm} \qquad Ans.$$

*3.8   Failure of Materials Due to Creep and Fatigue

The mechanical properties of a material have up to this point been discussed only for a static or slowly applied load at constant temperature. In some cases, however, a member may have to be used in an environment for which loadings must be sustained over long periods of time at elevated temperatures, or in other cases, the loading may be repeated or cycled. We will not consider these effects in this book, although we will briefly mention how one determines a material's strength for these conditions, since they are given special treatment in design.

Creep.   When a material has to support a load for a very long period of time, it may continue to deform until a sudden fracture occurs or its usefulness is impaired. This time-dependent permanent deformation is known as *creep*. Normally creep is considered when metals and ceramics are used for structural members or mechanical parts that are subjected to high temperatures. For some materials, however, such as polymers and composite materials—including wood or concrete—temperature is *not* an important factor, and yet creep can occur strictly from long-term load application. As a typical example, consider the fact that a rubber band will not return to its original shape after being released from a stretched position in which it was held for a very long period of time. In the general sense, therefore, both *stress and/or temperature* play a significant role in the *rate* of creep.

For practical purposes, when creep becomes important, a member is usually designed to resist a specified creep strain for a given period of time. An important mechanical property that is used in this regard is called the *creep strength*. This value represents the highest stress the material can withstand during a specified time without exceeding an allowable creep strain. The creep strength will vary with temperature, and for design, a given temperature, duration of loading, and allowable creep strain must all be specified. For example, a creep strain of 0.1% per year has been suggested for steel in bolts and piping.

Several methods exist for determining an allowable creep strength for a particular material. One of the simplest involves testing several specimens simultaneously at a constant temperature, but with each subjected to a *different axial stress*. By measuring the length of time needed to produce either an allowable strain or the fracture strain for each specimen, a curve of stress versus time can be established. Normally these tests are run to a maximum of 1000 hours. An example of the results for stainless steel at a temperature of 650°C and prescribed creep strain of 1% is shown in Fig. 3–27. As noted, this material has a yield strength of 276 MPa at room temperature (0.2% offset) and the creep strength at 1000 h is found to be approximately $\sigma_c = 138$ MPa.

The long-term application of the cable loading on this pole has caused the pole to deform due to creep.

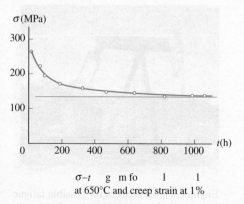

σ–t g m fo   l   l
at 650°C and creep strain at 1%

Fig. 3–27

The design of members used for amusement park rides requires careful consideration of cyclic loadings that can cause fatigue.

Engineers must account for possible fatigue failure of the moving parts of this oil-pumping rig.

In general, the creep strength will *decrease* for *higher temperatures* or for *higher applied stresses*. For longer periods of time, extrapolations from the curves must be made. To do this usually requires a certain amount of experience with creep behavior, and some supplementary knowledge about the creep properties of the material. Once the material's creep strength has been determined, however, a factor of safety is applied to obtain an appropriate allowable stress for design.

Fatigue.   When a metal is subjected to repeated *cycles* of stress or strain, it causes its structure to break down, ultimately leading to fracture. This behavior is called *fatigue*, and it is usually responsible for a large percentage of failures in connecting rods and crankshafts of engines; steam or gas turbine blades; connections or supports for bridges, railroad wheels, and axles; and other parts subjected to cyclic loading. In all these cases, fracture will occur at a stress that is *less* than the material's yield stress.

The nature of this failure apparently results from the fact that there are microscopic imperfections, usually on the surface of the member, where the localized stress becomes *much greater* than the average stress acting over the cross section. As this higher stress is cycled, it leads to the formation of minute cracks. Occurrence of these cracks causes a further increase of stress at their tips or boundaries, which in turn causes a further extension of the cracks into the material as the stress continues to be cycled. Eventually the cross-sectional area of the member is reduced to the point where the load can no longer be sustained, and as a result sudden fracture occurs. The material, even though known to be ductile, behaves as if it were *brittle*.

In order to specify a safe strength for a metallic material under repeated loading, it is necessary to determine a limit below which no evidence of failure can be detected after applying a load for a specified number of cycles. This limiting stress is called the *endurance* or *fatigue limit*. Using a testing machine for this purpose, a series of specimens are each subjected to a specified stress and cycled to failure. The results are plotted as a graph representing the stress S (or σ) on the vertical axis and the number of cycles-to-failure N on the horizontal axis. This graph is called an *S–N diagram* or *stress–cycle diagram*, and most often the values of N are plotted on a logarithmic scale since they are generally quite large.

Examples of *S–N* diagrams for two common engineering metals are shown in Fig. 3–28. The endurance limit is usually identified as the stress for which the *S–N* graph becomes horizontal or asymptotic. As noted, it has a well-defined value of $(S_{el})_{st} = 186$ MPa for steel. For aluminum, however, the endurance limit is not well defined, and so it is normally specified as the stress having a limit of 500 million cycles, $(S_{el})_{al} = 131$ MPa. Once a particular value is obtained, it is often assumed that for any stress below this value the fatigue life is infinite, and therefore the number of cycles to failure is no longer given consideration.

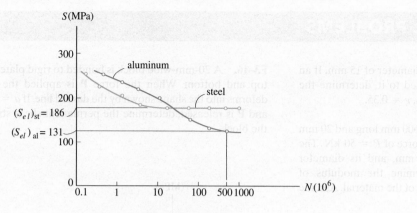

S–N diagram for steel and aluminum alloys
(N axis has a logarithmic scale)

Fig. 3–28

Important Points

- *Poisson's ratio*, ν, is a ratio of the lateral strain of a homogeneous and isotropic material to its longitudinal strain. Generally these strains are of opposite signs, that is, if one is an elongation, the other will be a contraction.

- The *shear stress–strain diagram* is a plot of the shear stress versus the shear strain. If the material is homogeneous and isotropic, and is also linear elastic, the slope of the straight line within the elastic region is called the modulus of rigidity or the shear modulus, G.

- There is a mathematical relationship between G, E, and ν.

- *Creep* is the time-dependent deformation of a material for which stress and/or temperature play an important role. Members are designed to resist the effects of creep based on their material creep strength, which is the largest initial stress a material can withstand during a specified time without exceeding a specified creep strain.

- *Fatigue* occurs in metals when the stress or strain is cycled. This phenomenon causes brittle fracture of the material. Members are designed to resist fatigue by ensuring that the stress in the member does not exceed its *endurance* or *fatigue limit*. This value is determined from an *S–N* diagram as the maximum stress the material can resist when subjected to a specified number of cycles of loading.

FUNDAMENTAL PROBLEMS

F3–13. A 100-mm long rod has a diameter of 15 mm. If an axial tensile load of 10 kN is applied to it, determine the change in its diameter. $E = 70$ GPa, $\nu = 0.35$.

F3–14. A solid circular rod that is 600 mm long and 20 mm in diameter is subjected to an axial force of $P = 50$ kN. The elongation of the rod is $\delta = 1.40$ mm, and its diameter becomes $d' = 19.9837$ mm. Determine the modulus of elasticity and the modulus of rigidity of the material. Assume that the material does not yield.

F3–16. A 20-mm-wide block is bonded to rigid plates at its top and bottom. When the force **P** is applied the block deforms into the shape shown by the dashed line. If $a = 3$ mm and **P** is released, determine the permanent shear strain in the block.

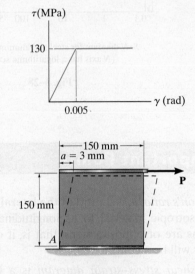

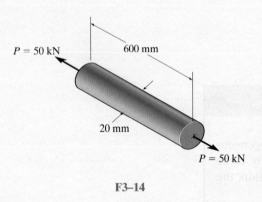

F3–14

F3–16

F3–15. A 20-mm-wide block is firmly bonded to rigid plates at its top and bottom. When the force **P** is applied the block deforms into the shape shown by the dashed line. Determine the magnitude of **P**. The block's material has a modulus of rigidity of $G = 26$ GPa. Assume that the material does not yield and use small angle analysis.

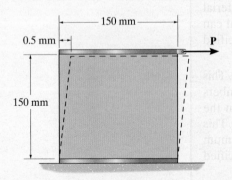

F3–15

PROBLEMS

•3–25. The acrylic plastic rod is 200 mm long and 15 mm in diameter. If an axial load of 300 N is applied to it, determine the change in its length and the change in its diameter. $E_p = 2.70$ GPa, $\nu_p = 0.4$.

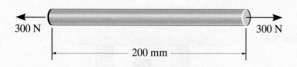

Prob. 3–25

3–26. The short cylindrical block of 2014-T6 aluminum, having an original diameter of 12 mm and a length of 37.5 mm, is placed in the smooth jaws of a vise and squeezed until the axial load applied is 4 kN. Determine (a) the decrease in its length and (b) its new diameter.

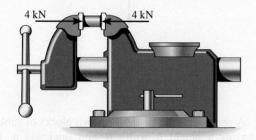

Prob. 3–26

3–27. The elastic portion of the stress–strain diagram for a steel alloy is shown in the figure. The specimen from which it was obtained had an original diameter of 13 mm and a gauge length of 50 mm. When the applied load on the specimen is 50 kN, the diameter is 12.99265 mm. Determine Poisson's ratio for the material.

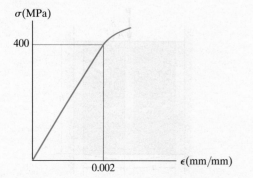

Prob. 3–27

*3–28. The elastic portion of the stress–strain diagram for a steel alloy is shown in the figure. The specimen from which it was obtained had an original diameter of 13 mm and a gauge length of 50 mm. If a load of $P = 20$ kN is applied to the specimen, determine its diameter and gauge length. Take $\nu = 0.4$.

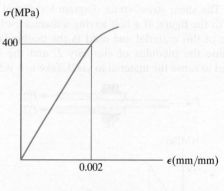

Prob. 3–28

•3–29. The aluminum block has a rectangular cross section and is subjected to an axial compressive force of 40 kN. If the 37.5-mm side changed its length to 37.5033 mm, determine Poisson's ratio and the new length of the 50-mm. side. $E_{al} = 70$ GPa.

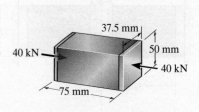

Prob. 3–29

3–30. The block is made of titanium Ti-6A1-4V and is subjected to a compression of 1.5 mm along the y axis, and its shape is given a tilt of $\theta = 89.7°$. Determine ϵ_x, ϵ_y, and γ_{xy}.

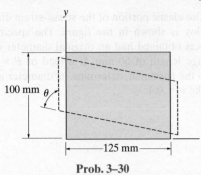

Prob. 3–30

3–31. The shear stress–strain diagram for a steel alloy is shown in the figure. If a bolt having a diameter of 20 mm is made of this material and used in the double lap joint, determine the modulus of elasticity E and the force P required to cause the material to yield. Take $\nu = 0.3$.

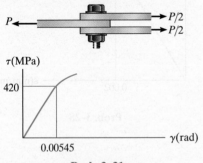

Prob. 3–31

*3–32. A shear spring is made by bonding the rubber annulus to a rigid fixed ring and a plug. When an axial load **P** is placed on the plug, show that the slope at point y in the rubber is $dy/dr = -\tan\gamma = -\tan(P/(2\pi h Gr))$. For small angles we can write $dy/dr = -P/(2\pi h Gr)$. Integrate this expression and evaluate the constant of integration using the condition that $y = 0$ at $r = r_o$. From the result compute the deflection $y = \delta$ of the plug.

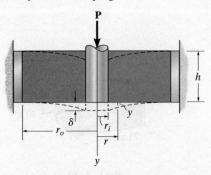

Prob. 3–32

•3–33. The support consists of three rigid plates, which are connected together using two symmetrically placed rubber pads. If a vertical force of 5 N is applied to plate A, determine the approximate vertical displacement of this plate due to shear strains in the rubber. Each pad has cross-sectional dimensions of 30 mm and 20 mm. $G_r = 0.20$ MPa.

Ref.: Section 3.3

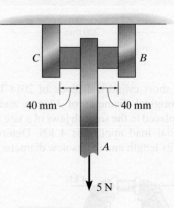

Prob. 3–33

3–34. A shear spring is made from two blocks of rubber, each having a height h, width b, and thickness a. The blocks are bonded to three plates as shown. If the plates are rigid and the shear modulus of the rubber is G, determine the displacement of plate A if a vertical load **P** is applied to this plate. Assume that the displacement is small so that $\delta = a \tan\gamma \approx a\gamma$.

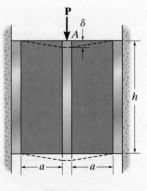

Prob. 3–34

CHAPTER REVIEW

One of the most important tests for material strength is the tension test. The results, found from stretching a specimen of known size, are plotted as normal stress on the vertical axis and normal strain on the horizontal axis.

Many engineering materials exhibit initial linear elastic behavior, whereby stress is proportional to strain, defined by Hooke's law, $\sigma = E\epsilon$. Here E, called the modulus of elasticity, is the slope of this straight line on the stress–strain diagram.

$$\sigma = E\epsilon$$

Ref.: Section 3.4

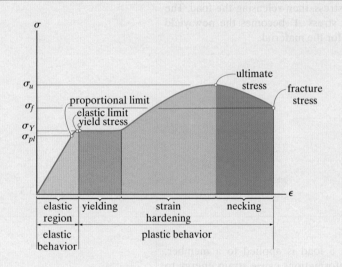

ductile material

When the material is stressed beyond the yield point, permanent deformation will occur. In particular, steel has a region of yielding, whereby the material will exhibit an increase in strain with no increase in stress. The region of strain hardening causes further yielding of the material with a corresponding increase in stress. Finally, at the ultimate stress, a localized region on the specimen will begin to constrict, forming a neck. It is after this that the fracture occurs.

Ref.: Section 3.2

Ductile materials, such as most metals, exhibit both elastic and plastic behavior. Wood is moderately ductile. Ductility is usually specified by the permanent elongation to failure or by the percent reduction in the cross-sectional area.

$$\text{Percent elongation} = \frac{L_f - L_0}{L_0}(100\%)$$

$$\text{Percent reduction of area} = \frac{A_0 - A_f}{A_0}(100\%)$$

Ref.: Section 3.3

3

Brittle materials exhibit little or no yielding before failure. Cast iron, concrete, and glass are typical examples.

Ref.: Section 3.3

The yield point of a material at A can be increased by strain hardening. This is accomplished by applying a load that causes the stress to be greater than the yield stress, then releasing the load. The larger stress A' becomes the new yield point for the material.

Ref.: Section 3.3

When a load is applied to a member, the deformations cause strain energy to be stored in the material. The strain energy per unit volume or strain energy density is equivalent to the area under the stress–strain curve. This area up to the yield point is called the modulus of resilience. The entire area under the stress–strain diagram is called the modulus of toughness.

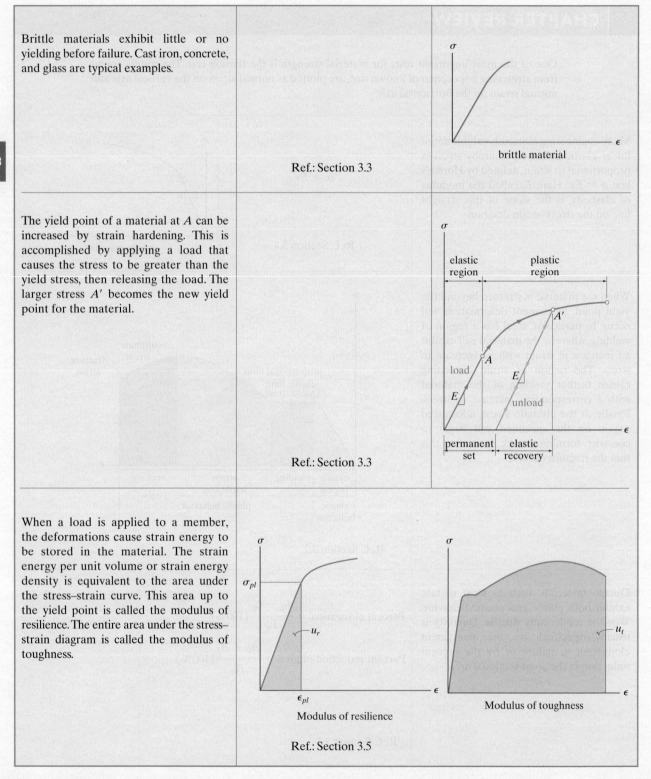

brittle material

elastic region   plastic region

load

unload

permanent set   elastic recovery

Modulus of resilience

Modulus of toughness

Ref.: Section 3.5

3

Poisson's ratio ν is a dimensionless material property that relates the lateral strain to the longitudinal strain. Its range of values is $0 \leq \nu \leq 0.5$.

$$\nu = -\frac{\epsilon_{lat}}{\epsilon_{long}}$$

Ref.: Section 3.6

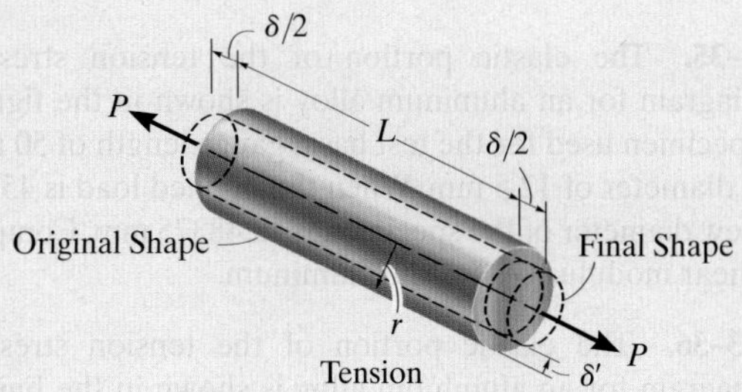

Original Shape

Final Shape

Tension

Shear stress versus shear strain diagrams can also be established for a material. Within the elastic region, $\tau = G\gamma$, where G is the shear modulus, found from the slope of the line. The value of ν can be obtained from the relationship that exists between G, E and ν.

$$G = \frac{E}{2(1 + \nu)}$$

Ref.: Section 3.7

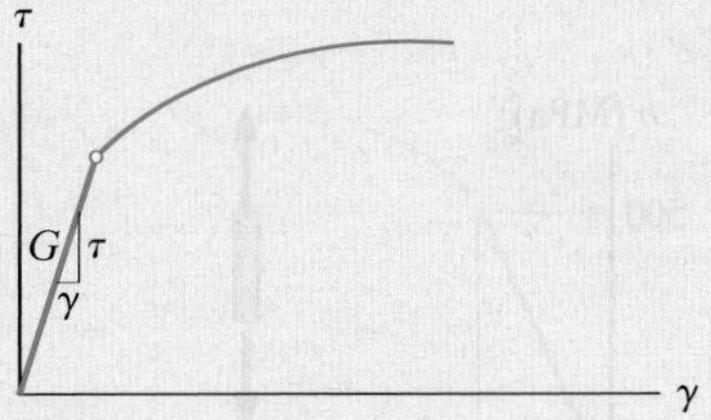

When materials are in service for long periods of time, considerations of creep become important. Creep is the time rate of deformation, which occurs at high stress and/or high temperature. Design requires that the stress in the material not exceed an allowable stress which is based on the material's creep strength.

Fatigue can occur when the material undergoes a large number of cycles of loading. This effect will cause microscopic cracks to form, leading to a brittle failure. To prevent fatigue, the stress in the material must not exceed a specified endurance or fatigue limit.

Ref.: Section 3.8

REVIEW PROBLEMS

3–35. The elastic portion of the tension stress–strain diagram for an aluminum alloy is shown in the figure. The specimen used for the test has a gauge length of 50 mm and a diameter of 12.5 mm. When the applied load is 45 kN, the new diameter of the specimen is 12.48375 mm. Compute the shear modulus G_{al} for the aluminum.

*3–36. The elastic portion of the tension stress–strain diagram for an aluminum alloy is shown in the figure. The specimen used for the test has a gauge length of 50 mm and a diameter of 12.5 mm. If the applied load is 50 kN, determine the new diameter of the specimen. The shear modulus is $G_{al} = 28$ GPa.

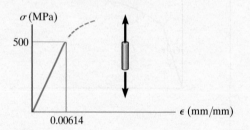

Probs. 3–35/36

3–37. The σ–ϵ diagram for elastic fibers that make up human skin and muscle is shown. Determine the modulus of elasticity of the fibers and estimate their modulus of toughness and modulus of resilience.

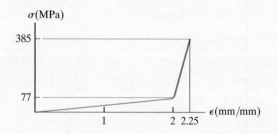

Prob. 3–37

3–38. A short cylindrical block of 6061-T6 aluminum, having an original diameter of 20 mm and a length of 75 mm, is placed in a compression machine and squeezed until the axial load applied is 5 kN. Determine (a) the decrease in its length and (b) its new diameter.

3–39. The rigid beam rests in the horizontal position on two 2014-T6 aluminum cylinders having the *unloaded* lengths shown. If each cylinder has a diameter of 30 mm, determine the placement x of the applied 80-kN load so that the beam remains horizontal. What is the new diameter of cylinder A after the load is applied? $\nu_{al} = 0.35$.

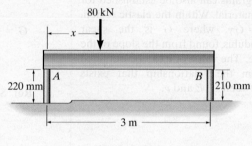

Prob. 3–39

*3–40. The head H is connected to the cylinder of a compressor using six steel bolts. If the clamping force in each bolt is 4 kN, determine the normal strain in the bolts. Each bolt has a diameter of 5 mm. If $\sigma_Y = 280$ MPa and $E_{st} = 200$ GPa, what is the strain in each bolt when the nut is unscrewed so that the clamping force is released?

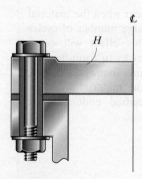

Prob. 3–40

•**3–41.** The stone has a mass of 800 kg and center of gravity at G. It rests on a pad at A and a roller at B. The pad is fixed to the ground and has a compressed height of 30 mm, a width of 140 mm, and a length of 150 mm. If the coefficient of static friction between the pad and the stone is $\mu_s = 0.8$, determine the approximate horizontal displacement of the stone, caused by the shear strains in the pad, before the stone begins to slip. Assume the normal force at A acts 1.5 m from G as shown. The pad is made from a material having $E = 4$ MPa and $\nu = 0.35$.

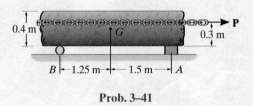

Prob. 3–41

3–42. The bar DA is rigid and is originally held in the horizontal position when the weight W is supported from C. If the weight causes B to be displaced downward 0.625 mm, determine the strain in wires DE and BC. Also, if the wires are made of A-36 steel and have a cross-sectional area of 1.25 mm^2, determine the weight W.

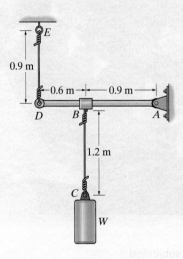

Prob. 3–42

3–43. The 8-mm-diameter bolt is made of an aluminum alloy. It fits through a magnesium sleeve that has an inner diameter of 12 mm and an outer diameter of 20 mm. If the original lengths of the bolt and sleeve are 80 mm and 50 mm, respectively, determine the strains in the sleeve and the bolt if the nut on the bolt is tightened so that the tension in the bolt is 8 kN. Assume the material at A is rigid. $E_{al} = 70$ GPa, $E_{mg} = 45$ GPa.

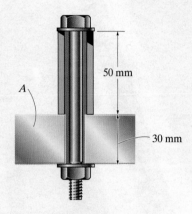

Prob. 3–43

3–44. The A-36 steel wire AB has a cross-sectional area of 10 mm^2 and is unstretched when $\theta = 45.0°$. Determine the applied load P needed to cause $\theta = 44.9°$.

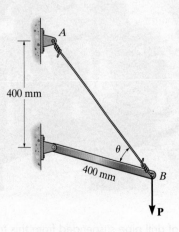

Prob. 3–44

The string of drill pipe suspended from this traveling block on an oil rig is subjected to extremely large loadings and axial deformations.

Axial Load

4

CHAPTER OBJECTIVES

In Chapter 1, we developed the method for finding the normal stress in axially loaded members. In this chapter we will discuss how to determine the deformation of these members, and we will also develop a method for finding the support reactions when these reactions cannot be determined strictly from the equations of equilibrium. An analysis of the effects of thermal stress, stress concentrations, inelastic deformations, and residual stress will also be discussed.

4.1 Saint-Venant's Principle

In the previous chapters, we have developed the concept of *stress* as a means of measuring the force distribution within a body and *strain* as a means of measuring a body's deformation. We have also shown that the mathematical relationship between stress and strain depends on the type of material from which the body is made. In particular, if the material behaves in a linear elastic manner, then Hooke's law applies, and there is a proportional relationship between stress and strain.

Check out the Companion Website for the following video solution(s) that can be found in this chapter.

- Section 2
  Axial Deformation – Rod
- Section 2
  Axial Deformation – Truss
- Section 4
  Static Indeterminacy – Similar Triangles
- Section 4
  Static Indeterminacy – Prestressed Wires
- Section 4
  Static Indeterminacy – Load Location
- Section 6
  Static Indeterminacy – Thermal Stress

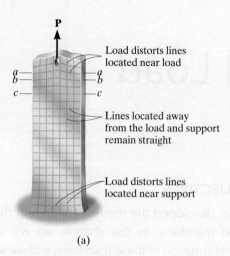

P

Load distorts lines
located near load

a
b

a
b

c — — c

Lines located away
from the load and support
remain straight

Load distorts lines
located near support

(a)

Fig. 4–1

Using this idea, consider the manner in which a rectangular bar will deform elastically when the bar is subjected to a force **P** applied along its centroidal axis, Fig. 4–1a. Here the bar is fixed connected at one end, with the force applied through a hole at its other end. Due to the loading, the bar deforms as indicated by the once horizontal and vertical grid lines drawn on the bar. Notice how the *localized deformation* that occurs at each end tends to even out and become uniform throughout the midsection of the bar.

If the material remains elastic then the strains caused by this deformation are directly related to the stress in the bar. As a result, the stress will be distributed more uniformly throughout the cross-sectional area when the section is taken farther and farther from the point where any external load is applied. For example, consider a profile of the variation of the stress distribution acting at sections *a–a*, *b–b*, and *c–c*, each of which is shown in Fig. 4–1b. By comparison, the stress tends to reach a uniform value at section *c–c*, which is sufficiently removed from the end since the localized deformation caused by **P** *vanishes*. The minimum distance from the bar's end where this occurs can be determined using a mathematical analysis based on the theory of elasticity.

It has been found that this distance should at least be equal to the *largest dimension* of the loaded cross section. Hence, section *c–c* should be located at a distance at least equal to the width (not the thickness) of the bar.*

---

*When section *c–c* is so located, the theory of elasticity predicts the maximum stress to be $\sigma_{max} = 1.02\sigma_{avg}$.

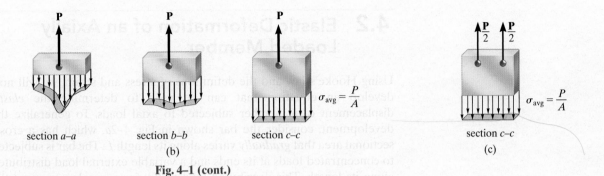

section *a–a*   section *b–b*   section *c–c* $\sigma_{avg} = \dfrac{P}{A}$ section *c–c* $\sigma_{avg} = \dfrac{P}{A}$

(b)   (c)

Fig. 4–1 (cont.)

In the same way, the stress distribution at the support will also even out and become uniform over the cross section located the same distance away from the support.

The fact that stress and deformation behave in this manner is referred to as *Saint-Venant's principle*, since it was first noticed by the French scientist Barré de Saint-Venant in 1855. Essentially it states that the *stress and strain produced at points in a body sufficiently removed from the region of load application will be the same as the stress and strain produced by any applied loadings that have the same statically equivalent resultant, and are applied to the body within the same region.* For example, if two symmetrically applied forces $P/2$ act on the bar, Fig. 4–1c, the stress distribution at section *c–c* will be uniform and therefore equivalent to $\sigma_{avg} = P/A$ as in Fig. 4–1b.

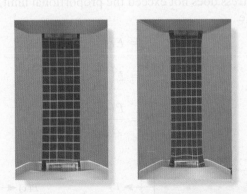

Notice how the lines on this rubber membrane distort after it is stretched. The localized distortions at the grips smooth out as stated by Saint-Venant's principle.

4.2   Elastic Deformation of an Axially Loaded Member

Using Hooke's law and the definitions of stress and strain, we will now develop an equation that can be used to determine the *elastic* displacement of a member subjected to axial loads. To generalize the development, consider the bar shown in Fig. 4–2a, which has a cross-sectional area that *gradually* varies along its length L. The bar is subjected to concentrated loads at its ends and a variable external load distributed along its length. This distributed load could, for example, represent the weight of the bar if it does not remain horizontal, or friction forces acting on the bar's surface. Here we wish to find the **relative displacement** δ (delta) of one end of the bar with respect to the other end as caused by this loading. We will *neglect* the localized deformations that occur at points of concentrated loading and where the cross section suddenly changes. From Saint-Venant's principle, these effects occur within small regions of the bar's length and will therefore have only a slight effect on the final result. For the most part, the bar will deform uniformly, so the normal stress will be uniformly distributed over the cross section.

Using the method of sections, a differential element (or wafer) of length dx and cross-sectional area $A(x)$ is isolated from the bar at the arbitrary position x. The free-body diagram of this element is shown in Fig. 4–2b. The resultant internal axial force will be a function of x since the external distributed loading will cause it to vary along the length of the bar. This load, $P(x)$, will deform the element into the shape indicated by the dashed outline, and therefore the displacement of one end of the element with respect to the other end is $d\delta$. The stress and strain in the element are

$$\sigma = \frac{P(x)}{A(x)} \quad \text{and} \quad \epsilon = \frac{d\delta}{dx}$$

Provided the stress does not exceed the proportional limit, we can apply Hooke's law; i.e.,

$$\sigma = E\epsilon$$

$$\frac{P(x)}{A(x)} = E\left(\frac{d\delta}{dx}\right)$$

$$d\delta = \frac{P(x)\,dx}{A(x)E}$$

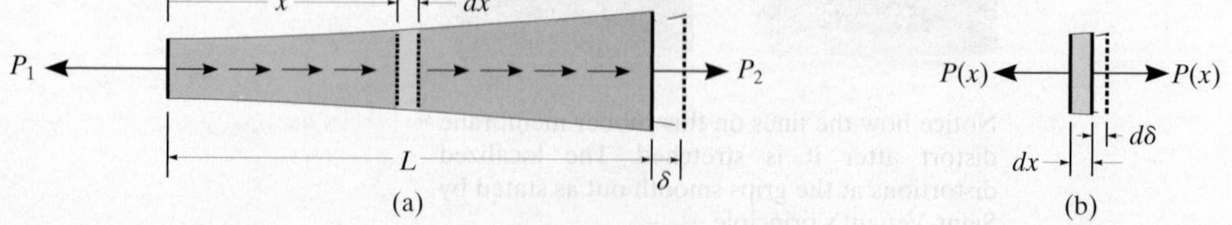

(a)

(b)

Fig. 4–2

For the entire length L of the bar, we must integrate this expression to find δ. This yields

$$\delta = \int_0^L \frac{P(x)\,dx}{A(x)E} \qquad (4\text{--}1)$$

where

δ = displacement of one point on the bar relative to the other point

L = original length of bar

$P(x)$ = internal axial force at the section, located a distance x from one end

$A(x)$ = cross-sectional area of the bar, expressed as a function of x

E = modulus of elasticity for the material

Constant Load and Cross-Sectional Area.

In many cases the bar will have a constant cross-sectional area A; and the material will be homogeneous, so E is constant. Furthermore, if a constant external force is applied at each end, Fig. 4–3, then the internal force P throughout the length of the bar is also constant. As a result, Eq. 4–1 can be integrated to yield

$$\delta = \frac{PL}{AE} \qquad (4\text{--}2)$$

If the bar is subjected to several different axial forces along its length, or the cross-sectional area or modulus of elasticity changes abruptly from one region of the bar to the next, the above equation can be applied to each *segment* of the bar where these quantities remain *constant*. The displacement of one end of the bar with respect to the other is then found from the *algebraic addition* of the relative displacements of the ends of each segment. For this general case,

$$\delta = \sum \frac{PL}{AE} \qquad (4\text{--}3)$$

The vertical displacement at the top of these building columns depends upon the loading applied on the roof and to the floor attached to their midpoint.

$$
\begin{array}{c}
\end{array}
$$

Fig. 4–3

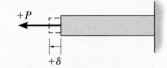

Positive sign convention for P and δ

Fig. 4–4

Sign Convention.

In order to apply Eq. 4–3, we must develop a sign convention for the internal axial force and the displacement of one end of the bar with respect to the other end. To do so, we will consider both the force and displacement to be *positive* if they cause *tension and elongation*, respectively, Fig. 4–4; whereas a *negative* force and displacement will cause *compression* and *contraction*, respectively.

For example, consider the bar shown in Fig. 4–5a. The *internal axial forces* "P," are determined by the method of sections for each segment, Fig. 4–5b. They are $P_{AB} = +5$ kN, $P_{BC} = -3$ kN, $P_{CD} = -7$ kN. This variation in axial load is shown on the axial or *normal force diagram* for the bar, Fig. 4–5c. Since we now know how the *internal* force varies throughout the bar's length, the displacement of end A relative to end D is determined from

$$\delta_{A/D} = \sum \frac{PL}{AE} = \frac{(5 \text{ kN})L_{AB}}{AE} + \frac{(-3 \text{ kN})L_{BC}}{AE} + \frac{(-7 \text{ kN})L_{CD}}{AE}$$

If the other data are substituted and a positive answer is calculated, it means that end A will move away from end D (the bar elongates), whereas a negative result would indicate that end A moves toward end D (the bar shortens). The double subscript notation is used to indicate this relative displacement ($\delta_{A/D}$); however, if the displacement is to be determined relative to a *fixed point*, then only a single subscript will be used. For example, if D is located at a *fixed* support, then the displacement will be denoted as simply δ_A.

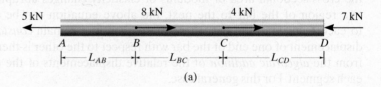

(a)

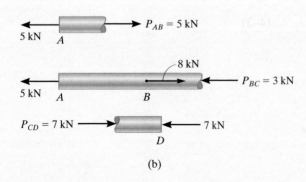

(b)

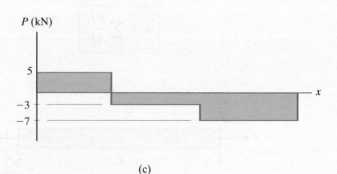

(c)

Fig. 4–5

Important Points

- *Saint-Venant's principle* states that both the localized deformation and stress which occur within the regions of load application or at the supports tend to "even out" at a distance sufficiently removed from these regions.

- The displacement of one end of an axially loaded member relative to the other end is determined by relating the applied *internal* load to the stress using $\sigma = P/A$ and relating the displacement to the strain using $\epsilon = d\delta/dx$. Finally these two equations are combined using Hooke's law, $\sigma = E\epsilon$, which yields Eq. 4–1.

- Since Hooke's law has been used in the development of the displacement equation, it is important that no internal load causes yielding of the material, and that the material is homogeneous and behaves in a linear elastic manner.

Procedure for Analysis

The relative displacement between any two points A and B on an axially loaded member can be determined by applying Eq. 4–1 (or Eq. 4–2). Application requires the following steps.

Internal Force.

- Use the method of sections to determine the internal axial force P within the member.

- If this force varies along the member's length due to an *external distributed loading*, a section should be made at the arbitrary location x from one end of the member and the force represented as a function of x, i.e., $P(x)$.

- If several *constant external forces* act on the member, the internal force in each *segment* of the member, between any two external forces, must be determined.

- For any segment, an internal *tensile force* is *positive* and an internal *compressive force* is *negative*. For convenience, the results of the internal loading can be shown graphically by constructing the normal-force diagram.

Displacement.

- When the member's cross-sectional area *varies* along its length, the area must be expressed as a function of its position x, i.e., $A(x)$.

- If the cross-sectional area, the modulus of elasticity, or the internal loading *suddenly changes*, then Eq. 4–2 should be applied to each segment for which these quantities are constant.

- When substituting the data into Eqs. 4–1 through 4–3, be sure to account for the proper sign for the internal force **P**. Tensile loadings are positive and compressive loadings are negative. Also, use a consistent set of units. For any segment, if the result is a *positive* numerical quantity, it indicates *elongation*; if it is *negative*, it indicates a *contraction*.

EXAMPLE | 4.1

The A-36 steel bar shown in Fig. 4–6a is made from two segments having cross-sectional areas of $A_{AB} = 600$ mm^2 and $A_{BD} = 1200$ mm^2. Determine the vertical displacement of end A and the displacement of B relative to C.

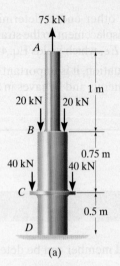

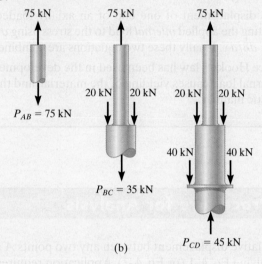

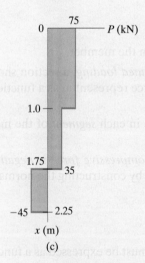

Fig. 4–6

(a)

(b)

(c)

SOLUTION

Internal Force. Due to the application of the external loadings, the internal axial forces in regions AB, BC, and CD will all be different. These forces are obtained by applying the method of sections and the equation of vertical force equilibrium as shown in Fig. 4–6b. This variation is plotted in Fig. 4–6c.

Displacement. From the inside back cover, $E_{st} = 200(10^6)$ kPa. Using the sign convention, i.e., internal tensile forces are positive and compressive forces are negative, the vertical displacement of A relative to the *fixed* support D is

$$\delta_A = \sum \frac{PL}{AE} = \frac{[+75\text{ kN}](1\text{ m})}{[600(10^{-6})\text{ m}^2][200(10^6)\text{ kN/m}^2]} + \frac{[+35\text{ kN}](0.75\text{ m})}{[1200(10^{-6})\text{ m}^2][200(10^6)\text{ kN/m}^2]}$$

$$+ \frac{[-45\text{ kN}](0.5\text{ m})}{[1200(10^{-6})\text{ m}^2][200(10^6)\text{ kN/m}^2]}$$

$$= +0.641\text{ mm} \qquad\qquad\qquad\qquad Ans.$$

Since the result is *positive*, the bar *elongates* and so the displacement at A is upward.

Applying Eq. 4–2 between points B and C, we obtain,

$$\delta_{B/C} = \frac{P_{BC}L_{BC}}{A_{BC}E} = \frac{[+35\text{ kN}](0.75\text{ m})}{[1200(10^{-6})\text{ m}^2][200(10^6)\text{ kN/m}^2]} = +0.109\text{ mm} \quad Ans.$$

Here B moves away from C, since the segment elongates.

EXAMPLE | 4.2

The assembly shown in Fig. 4–7a consists of an aluminum tube AB having a cross-sectional area of 400 mm². A steel rod having a diameter of 10 mm is attached to a rigid collar and passes through the tube. If a tensile load of 80 kN is applied to the rod, determine the displacement of the end C of the rod. Take $E_{st} = 200$ GPa, $E_{al} = 70$ GPa.

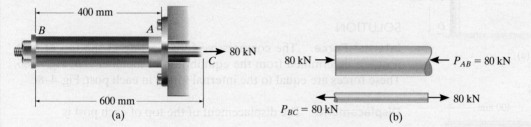

Fig. 4–7

SOLUTION

Internal Force.   The free-body diagram of the tube and rod segments in Fig. 4–7b, shows that the rod is subjected to a tension of 80 kN and the tube is subjected to a compression of 80 kN.

Displacement.   We will first determine the displacement of end C with respect to end B. Working in units of newtons and meters, we have

$$\delta_{C/B} = \frac{PL}{AE} = \frac{[+80(10^3)\ N](0.6\ m)}{\pi(0.005\ m)^2[200(10^9)\ N/m^2]} = +0.003056\ m \rightarrow$$

The positive sign indicates that end C moves *to the right* relative to end B, since the bar elongates.

The displacement of end B with respect to the *fixed* end A is

$$\delta_B = \frac{PL}{AE} = \frac{[-80(10^3)\ N](0.4\ m)}{[400\ mm^2(10^{-6})\ m^2/mm^2][70(10^9)\ N/m^2]}$$

$$= -0.001143\ m = 0.001143\ m \rightarrow$$

Here the negative sign indicates that the tube shortens, and so B moves to the *right* relative to A.

Since both displacements are to the right, the displacement of C relative to the fixed end A is therefore

$$(\xrightarrow{+}) \qquad \delta_C = \delta_B + \delta_{C/B} = 0.001143\ m + 0.003056\ m$$

$$= 0.00420\ m = 4.20\ mm \rightarrow \qquad\qquad Ans.$$

EXAMPLE | **4.3**

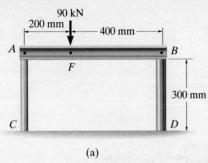

(a)

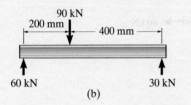

(b)

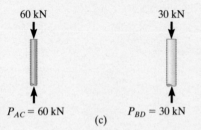

$P_{AC} = 60$ kN $P_{BD} = 30$ kN

(c)

Rigid beam AB rests on the two short posts shown in Fig. 4–8a. AC is made of steel and has a diameter of 20 mm, and BD is made of aluminum and has a diameter of 40 mm. Determine the displacement of point F on AB if a vertical load of 90 kN is applied over this point. Take $E_{st} = 200$ GPa, $E_{al} = 70$ GPa.

SOLUTION

Internal Force. The compressive forces acting at the top of each post are determined from the equilibrium of member AB, Fig. 4–8b. These forces are equal to the internal forces in each post, Fig. 4–8c.

Displacement. The displacement of the top of each post is

Post AC:

$$\delta_A = \frac{P_{AC}L_{AC}}{A_{AC}E_{st}} = \frac{[-60(10^3)\,\text{N}](0.300\,\text{m})}{\pi(0.010\,\text{m})^2[200(10^9)\,\text{N/m}^2]} = -286(10^{-6})\,\text{m}$$

$$= 0.286\,\text{mm} \downarrow$$

Post BD:

$$\delta_B = \frac{P_{BD}L_{BD}}{A_{BD}E_{al}} = \frac{[-30(10^3)\,\text{N}](0.300\,\text{m})}{\pi(0.020\,\text{m})^2[70(10^9)\,\text{N/m}^2]} = -102(10^{-6})\,\text{m}$$

$$= 0.102\,\text{mm} \downarrow$$

A diagram showing the centerline displacements at A, B, and F on the beam is shown in Fig. 4–8d. By proportion of the blue shaded triangle, the displacement of point F is therefore

$$\delta_F = 0.102\,\text{mm} + (0.184\,\text{mm})\left(\frac{400\,\text{mm}}{600\,\text{mm}}\right) = 0.225\,\text{mm} \downarrow \quad Ans.$$

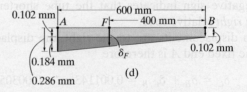

(d)

Fig. 4–8

EXAMPLE | 4.4

A member is made from a material that has a specific weight γ and modulus of elasticity E. If it is in the form of a *cone* having the dimensions shown in Fig. 4–9a, determine how far its end is displaced due to gravity when it is suspended in the vertical position.

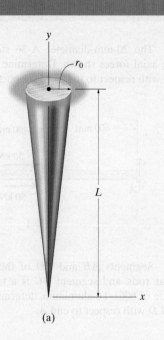

SOLUTION

Internal Force.   The internal axial force varies along the member since it is dependent on the weight $W(y)$ of a segment of the member below any section, Fig. 4–9b. Hence, to calculate the displacement, we must use Eq. 4–1. At the section located a distance y from its free end, the radius x of the cone as a function of y is determined by proportion; i.e.,

$$\frac{x}{y} = \frac{r_0}{L}; \qquad x = \frac{r_0}{L}y$$

The volume of a cone having a base of radius x and height y is

$$V = \frac{1}{3}\pi y x^2 = \frac{\pi r_0^2}{3L^2}y^3$$

(a)

Since $W = \gamma V$, the internal force at the section becomes

$+\uparrow \Sigma F_y = 0;$ $\qquad P(y) = \dfrac{\gamma \pi r_0^2}{3L^2}y^3$

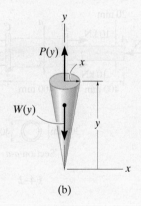

Displacement.   The area of the cross section is also a function of position y, Fig. 4–9b. We have

$$A(y) = \pi x^2 = \frac{\pi r_0^2}{L^2}y^2$$

Applying Eq. 4–1 between the limits of $y = 0$ and $y = L$ yields

$$\delta = \int_0^L \frac{P(y)\,dy}{A(y)E} = \int_0^L \frac{\left[(\gamma \pi r_0^2 / 3L^2)\,y^3\right]dy}{\left[(\pi r_0^2 / L^2)\,y^2\right]E}$$

$$= \frac{\gamma}{3E}\int_0^L y\,dy$$

$$= \frac{\gamma L^2}{6E} \qquad\qquad\qquad Ans.$$

(b)

Fig. 4–9

NOTE: As a partial check of this result, notice how the units of the terms, when canceled, give the displacement in units of length as expected.

FUNDAMENTAL PROBLEMS

F4–1. The 20-mm-diameter A-36 steel rod is subjected to the axial forces shown. Determine the displacement of end C with respect to the fixed support at A.

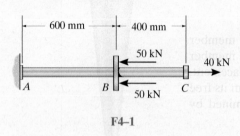

F4–1

F4–2. Segments AB and CD of the assembly are solid circular rods, and segment BC is a tube. If the assembly is made of 6061-T6 aluminum, determine the displacement of end D with respect to end A.

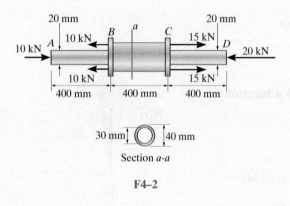

Section *a-a*

F4–2

F4–3. The 30-mm-diameter A-36 steel rod is subjected to the loading shown. Determine the displacement of end A with respect to end C.

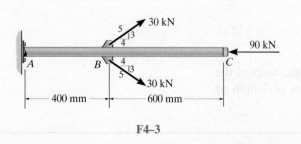

F4–3

F4–4. If the 20-mm-diameter rod is made of A-36 steel and the stiffness of the spring is $k = 50$ MN/m, determine the displacement of end A when the 60-kN force is applied.

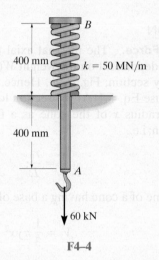

F4–4

F4–5. The 20-mm-diameter 2014-T6 aluminum rod is subjected to the uniform distributed axial load. Determine the displacement of end A.

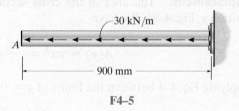

F4–5

F4–6. The 20-mm-diameter 2014-T6 aluminum rod is subjected to the triangular distributed axial load. Determine the displacement of end A.

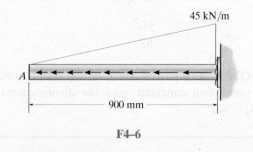

F4–6

PROBLEMS

•4–1. The ship is pushed through the water using an A-36 steel propeller shaft that is 8 m long, measured from the propeller to the thrust bearing D at the engine. If it has an outer diameter of 400 mm and a wall thickness of 50 mm, determine the amount of axial contraction of the shaft when the propeller exerts a force on the shaft of 5 kN. The bearings at B and C are journal bearings.

4–3. The A-36 steel rod is subjected to the loading shown. If the cross-sectional area of the rod is 50 mm^2, determine the displacement of its end D. Neglect the size of the couplings at B, C, and D.

*4–4. The A-36 steel rod is subjected to the loading shown. If the cross-sectional area of the rod is 50 mm^2, determine the displacement of C. Neglect the size of the couplings at B, C, and D.

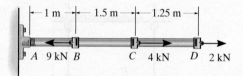

Probs. 4–3/4

4–5. The assembly consists of a steel rod CB and an aluminum rod BA, each having a diameter of 12 mm. If the rod is subjected to the axial loadings at A and at the coupling B, determine the displacement of the coupling B and the end A. The unstretched length of each segment is shown in the figure. Neglect the size of the connections at B and C, and assume that they are rigid. $E_{st} = 200$ GPa, $E_{al} = 70$ GPa.

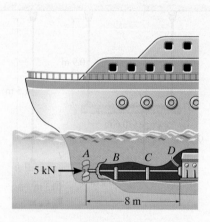

Prob. 4–1

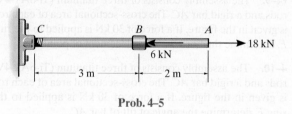

Prob. 4–5

4–2. The copper shaft is subjected to the axial loads shown. Determine the displacement of end A with respect to end D. The diameters of each segment are $d_{AB} = 75$ mm, $d_{BC} = 50$ mm, and $d_{CD} = 25$ mm. Take $E_{cu} = 126$ GPa.

4–6. The bar has a cross-sectional area of 1800 mm^2, and $E = 250$ GPa. Determine the displacement of its end A when it is subjected to the distributed loading.

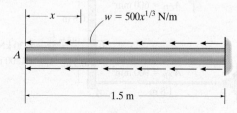

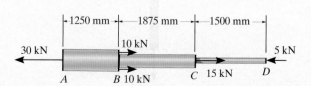

Prob. 4–2

Prob. 4–6

4–7. The load of 4 kN is supported by the four 304 stainless steel wires that are connected to the rigid members AB and DC. Determine the vertical displacement of the load if the members were horizontal before the load was applied. Each wire has a cross-sectional area of 30 mm^2.

*4–8. The load of 4 kN is supported by the four 304 stainless steel wires that are connected to the rigid members AB and DC. Determine the angle of tilt of each member after the load is applied. The members were originally horizontal, and each wire has a cross-sectional area of 30 mm^2.

4–11. The load is supported by the four 304 stainless steel wires that are connected to the rigid members AB and DC. Determine the vertical displacement of the 2.5-kN load if the members were originally horizontal when the load was applied. Each wire has a cross-sectional area of 16 mm^2.

*4–12. The load is supported by the four 304 stainless steel wires that are connected to the rigid members AB and DC. Determine the angle of tilt of each member after the 2.5 kN load is applied. The members were originally horizontal, and each wire has a cross-sectional area of 16 mm^2.

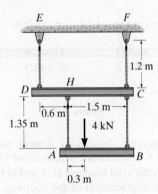

Probs. 4–7/8

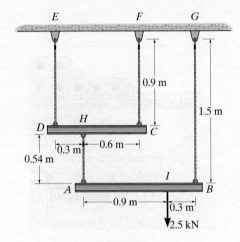

Probs. 4–11/12

•4–9. The assembly consists of three titanium (Ti-6A1-4V) rods and a rigid bar AC. The cross-sectional area of each rod is given in the figure. If a force of 30 kN is applied to the ring F, determine the horizontal displacement of point F.

4–10. The assembly consists of three titanium (Ti-6A1-4V) rods and a rigid bar AC. The cross-sectional area of each rod is given in the figure. If a force of 30 kN is applied to the ring F, determine the angle of tilt of bar AC.

•4–13. The bar has a length L and cross-sectional area A. Determine its elongation due to the force $\mathbf{P}$ and its own weight. The material has a specific weight γ (weight/volume) and a modulus of elasticity E.

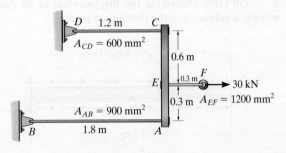

Probs. 4–9/10

Prob. 4–13

4–14. The post is made of Douglas fir and has a diameter of 60 mm. If it is subjected to the load of 20 kN and the soil provides a frictional resistance that is uniformly distributed along its sides of $w = 4$ kN/m, determine the force **F** at its bottom needed for equilibrium. Also, what is the displacement of the top of the post A with respect to its bottom B? Neglect the weight of the post.

4–15. The post is made of Douglas fir and has a diameter of 60 mm. If it is subjected to the load of 20 kN and the soil provides a frictional resistance that is distributed along its length and varies linearly from $w = 0$ at $y = 0$ to $w = 3$ kN/m at $y = 2$ m, determine the force **F** at its bottom needed for equilibrium. Also, what is the displacement of the top of the post A with respect to its bottom B? Neglect the weight of the post.

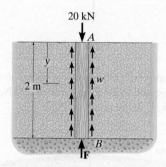

Probs. 4–14/15

*4–16. The linkage is made of two pin-connected A-36 steel members, each having a cross-sectional area of 1000 mm². If a vertical force of $P = 250$ kN is applied to point A, determine its vertical displacement at A.

•4–17. The linkage is made of two pin-connected A-36 steel members, each having a cross-sectional area of 1000 mm². Determine the magnitude of the force **P** needed to displace point A 0.625 mm downward.

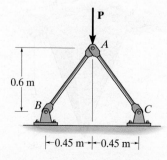

Probs. 4–16/17

4–18. The assembly consists of two A-36 steel rods and a rigid bar BD. Each rod has a diameter of 20 mm. If a force of 50 kN is applied to the bar as shown, determine the vertical displacement of the load.

4–19. The assembly consists of two A-36 steel rods and a rigid bar BD. Each rod has a diameter of 20 mm. If a force of 50 kN is applied to the bar, determine the angle of tilt of the bar.

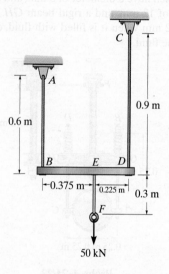

Probs. 4–18/19

*4–20. The rigid bar is supported by the pin-connected rod CB that has a cross-sectional area of 500 mm² and is made of A-36 steel. Determine the vertical displacement of the bar at B when the load is applied.

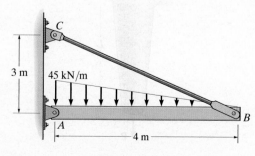

Prob. 4–20

•4–21. A spring-supported pipe hanger consists of two springs which are originally unstretched and have a stiffness of $k = 60$ kN/m, three 304 stainless steel rods, AB and CD, which have a diameter of 5 mm, and EF, which has a diameter of 12 mm, and a rigid beam GH. If the pipe and the fluid it carries have a total weight of 4 kN, determine the displacement of the pipe when it is attached to the support.

4–22. A spring-supported pipe hanger consists of two springs, which are originally unstretched and have a stiffness of $k = 60$ kN/m, three 304 stainless steel rods, AB and CD, which have a diameter of 5 mm, and EF, which has a diameter of 12 mm, and a rigid beam GH. If the pipe is displaced 82 mm when it is filled with fluid, determine the weight of the fluid.

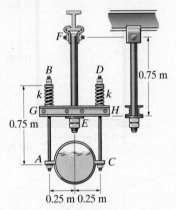

Probs. 4–21/22

4–23. The rod has a slight taper and length L. It is suspended from the ceiling and supports a load $\mathbf{P}$ at its end. Show that the displacement of its end due to this load is $\delta = PL/(\pi E r_2 r_1)$. Neglect the weight of the material. The modulus of elasticity is E.

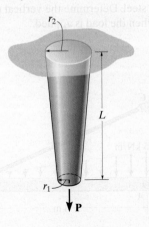

Prob. 4–23

*4–24. Determine the relative displacement of one end of the tapered plate with respect to the other end when it is subjected to an axial load P.

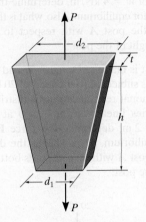

Prob. 4–24

4–25. Determine the elongation of the A-36 steel member when it is subjected to an axial force of 30 kN. The member is 10 mm thick. Use the result of Prob. 4–24.

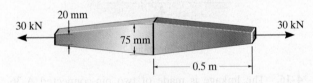

Prob. 4–25

4–26. The casting is made of a material that has a specific weight γ and modulus of elasticity E. If it is formed into a pyramid having the dimensions shown, determine how far its end is displaced due to gravity when it is suspended in the vertical position.

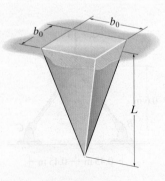

Prob. 4–26

4–27. The circular bar has a variable radius of $r = r_0 e^{ax}$ and is made of a material having a modulus of elasticity of E. Determine the displacement of end A when it is subjected to the axial force **P**.

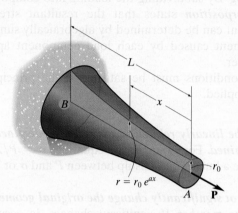

Prob. 4–27

*4–28. The pedestal is made in a shape that has a radius defined by the function $r = 2/(2 + y^{1/2})$ m, where y is in meters. If the modulus of elasticity for the material is $E = 100$ GPa, determine the displacement of its top when it supports the 5-kN load.

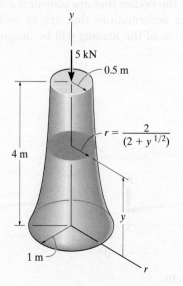

Prob. 4–28

•4–29. The support is made by cutting off the two opposite sides of a sphere that has a radius r_0. If the original height of the support is $r_0/2$, determine how far it shortens when it supports a load **P**. The modulus of elasticity is E.

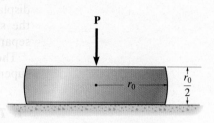

Prob. 4–29

4–30. The weight of the kentledge exerts an axial force of $P = 1500$ kN on the 300-mm diameter high strength concrete bore pile. If the distribution of the resisting skin friction developed from the interaction between the soil and the surface of the pile is approximated as shown, and the resisting bearing force **F** is required to be zero, determine the maximum intensity p_0 kN/m for equilibrium. Also, find the corresponding elastic shortening of the pile. Neglect the weight of the pile.

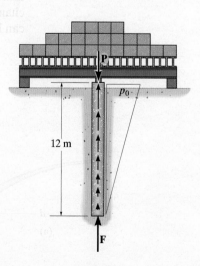

Prob. 4–30

4.3 Principle of Superposition

The principle of superposition is often used to determine the stress or displacement at a point in a member when the member is subjected to a complicated loading. By subdividing the loading into components, the ***principle of superposition*** states that the resultant stress or displacement at the point can be determined by algebraically summing the stress or displacement caused by each load component applied separately to the member.

The following two conditions must be satisfied if the principle of superposition is to be applied.

1. ***The loading must be linearly related to the stress or displacement that is to be determined.*** For example, the equations $\sigma = P/A$ and $\delta = PL/AE$ involve a linear relationship between P and σ or δ.

2. ***The loading must not significantly change the original geometry or configuration of the member.*** If significant changes do occur, the direction and location of the applied forces and their moment arms will change. For example, consider the slender rod shown in Fig. 4–10a, which is subjected to the load **P**. In Fig. 4–10b, **P** is replaced by two of its components, $\mathbf{P} = \mathbf{P}_1 + \mathbf{P}_2$. If **P** causes the rod to deflect a large amount, as shown, the moment of the load about its support, Pd, will not equal the sum of the moments of its component loads, $Pd \neq P_1d_1 + P_2d_2$, because $d_1 \neq d_2 \neq d$.

This principle will be used throughout this text whenever we assume Hooke's law applies and also, the bodies that are considered will be such that the loading will produce deformations that are so small that the change in position and direction of the loading will be insignificant and can be neglected.

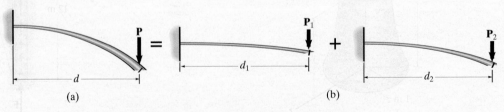

(a)   (b)

Fig. 4–10

4.4   Statically Indeterminate Axially Loaded Member

Consider the bar shown in Fig. 4–11a which is fixed supported at both of its ends. From the free-body diagram, Fig. 4–11b, equilibrium requires

$$+\uparrow \Sigma F = 0; \qquad F_B + F_A - P = 0$$

This type of problem is called ***statically indeterminate***, since the equilibrium equation(s) are not sufficient to determine the two reactions on the bar.

In order to establish an additional equation needed for solution, it is necessary to consider how points on the bar displace. Specifically, an equation that specifies the conditions for displacement is referred to as a ***compatibility*** or ***kinematic condition***. In this case, a suitable compatibility condition would require the displacement of one end of the bar with respect to the other end to be equal to zero, since the end supports are fixed. Hence, the compatibility condition becomes

$$\delta_{A/B} = 0$$

This equation can be expressed in terms of the applied loads by using a *load–displacement relationship*, which depends on the material behavior. For example, if linear-elastic behavior occurs, $\delta = PL/AE$ can be used. Realizing that the internal force in segment AC is $+F_A$, and in segment CB the internal force is $-F_B$, Fig. 4–11c, the above equation can be written as

$$\frac{F_A L_{AC}}{AE} - \frac{F_B L_{CB}}{AE} = 0$$

Assuming that AE is constant, then $F_A = F_B(L_{CB}/L_{AC})$, so that using the equilibrium equation, the equations for the reactions become

$$F_A = P\left(\frac{L_{CB}}{L}\right) \quad \text{and} \quad F_B = P\left(\frac{L_{AC}}{L}\right)$$

Since both of these results are positive, the direction of the reactions is shown correctly on the free-body diagram.

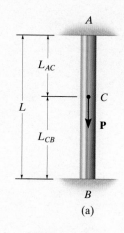

(a)

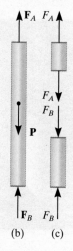

(b)   (c)

Fig. 4–11

Important Points

- The *principle of superposition* is sometimes used to simplify stress and displacement problems having complicated loadings. This is done by subdividing the loading into components, then algebracially adding the results.
- Superposition requires that the loading be linearly related to the stress or displacement, and the loading does not significantly change the original geometry of the member.
- A problem is *statically indeterminate* if the equations of equilibrium are not sufficient to determine all the reactions on a member.
- *Compatibility conditions* specify the displacement constraints that occur at the supports or other points on a member.

Procedure for Analysis

The support reactions for statically indeterminate problems are determined by satisfying equilibrium, compatibility, and force-displacement requirements for the member.

Equilibrium.

- Draw a free-body diagram of the member in order to identify all the forces that act on it.
- The problem can be classified as statically indeterminate if the number of unknown reactions on the free-body diagram is greater than the number of available equations of equilibrium.
- Write the equations of equilibrium for the member.

Compatibility.

- Consider drawing a displacement diagram in order to investigate the way the member will elongate or contract when subjected to the external loads.
- Express the compatibility conditions in terms of the displacements caused by the loading.
- Use a load–displacement relation, such as $\delta = PL/AE$, to relate the unknown displacements to the reactions.
- Solve the equilibrium and compatibility equations for the reactions. If any of the results has a negative numerical value, it indicates that this force acts in the opposite sense of direction to that indicated on the free-body diagram.

Most concrete columns are reinforced with steel rods; and since these two materials work together in supporting the applied load, the forces in each material become statically indeterminate.

EXAMPLE | 4.5

The steel rod shown in Fig. 4–12a has a diameter of 10 mm. It is fixed to the wall at A, and before it is loaded, there is a gap of 0.2 mm between the wall at B' and the rod. Determine the reactions at A and B' if the rod is subjected to an axial force of $P = 20$ kN as shown. Neglect the size of the collar at C. Take $E_{st} = 200$ GPa.

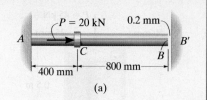

(a)

SOLUTION

Equilibrium. As shown on the free-body diagram, Fig. 4–12b, we will *assume* that force P is large enough to cause the rod's end B to contact the wall at B'. The problem is statically indeterminate since there are two unknowns and only one equation of equilibrium.

$$\stackrel{+}{\rightarrow} \Sigma F_x = 0; \qquad -F_A - F_B + 20(10^3) \text{ N} = 0 \qquad (1)$$

(b)

Compatibility. The force P causes point B to move to B', with no further displacement. Therefore the compatibility condition for the rod is

$$\delta_{B/A} = 0.0002 \text{ m}$$

(c)

This displacement can be expressed in terms of the unknown reactions using the load–displacement relationship, Eq. 4–2, applied to segments AC and CB, Fig. 4–12c. Working in units of newtons and meters, we have

Fig. 4–12

$$\delta_{B/A} = 0.0002 \text{ m} = \frac{F_A L_{AC}}{AE} - \frac{F_B L_{CB}}{AE}$$

$$0.0002 \text{ m} = \frac{F_A(0.4 \text{ m})}{\pi(0.005 \text{ m})^2[200(10^9) \text{ N/m}^2]}$$

$$- \frac{F_B(0.8 \text{ m})}{\pi(0.005 \text{ m})^2[200(10^9) \text{ N/m}^2]}$$

or

$$F_A(0.4 \text{ m}) - F_B(0.8 \text{ m}) = 3141.59 \text{ N} \cdot \text{m} \qquad (2)$$

Solving Eqs. 1 and 2 yields

$$F_A = 16.0 \text{ kN} \qquad F_B = 4.05 \text{ kN} \qquad \qquad Ans.$$

Since the answer for F_B is *positive*, indeed end B contacts the wall at B' as originally assumed.

NOTE: If F_B were a negative quantity, the problem would be statically determinate, so that $F_B = 0$ and $F_A = 20$ kN.

EXAMPLE 4.6

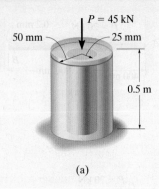

(a)

(b)

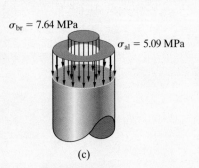

(c)

Fig. 4–13

The aluminum post shown in Fig. 4–13a is reinforced with a brass core. If this assembly supports an axial compressive load of $P = 45$ kN, applied to the rigid cap, determine the average normal stress in the aluminum and the brass. Take $E_{al} = 70(10^3)$ MPa and $E_{br} = 105(10^3)$ MPa.

SOLUTION

Equilibrium. The free-body diagram of the post is shown in Fig. 4–13b. Here the resultant axial force at the base is represented by the unknown components carried by the aluminum, $\mathbf{F}_{al}$, and brass, $\mathbf{F}_{br}$. The problem is statically indeterminate. Why?

Vertical force equilibrium requires

$$+\uparrow \Sigma F_y = 0; \qquad -45 \text{ kN} + F_{al} + F_{br} = 0 \qquad (1)$$

Compatibility. The rigid cap at the top of the post causes both the aluminum and brass to displace the same amount. Therefore,

$$\delta_{al} = \delta_{br}$$

Using the load–displacement relationships,

$$\frac{F_{al}L}{A_{al}E_{al}} = \frac{F_{br}L}{A_{br}E_{br}}$$

$$F_{al} = F_{br}\left(\frac{A_{al}}{A_{br}}\right)\left(\frac{E_{al}}{E_{br}}\right)$$

$$F_{al} = F_{br}\left[\frac{\pi[(0.05 \text{ m})^2 - (0.025 \text{ m})^2]}{\pi(0.025 \text{ m})^2}\right]\left[\frac{70(10^3) \text{ MPa}}{105(10^3) \text{ MPa}}\right]$$

$$F_{al} = 2F_{br} \qquad (2)$$

Solving Eqs. 1 and 2 simultaneously yields

$$F_{al} = 30 \text{ kN} \qquad F_{br} = 15 \text{ kN}$$

Since the results are positive, indeed the stress will be compressive. The average normal stress in the aluminum and brass is therefore

$$\sigma_{al} = \frac{30 \text{ kN}}{\pi[(0.05 \text{ m})^2 - (0.025 \text{ m})^2]} = 5.09 \text{ MPa} \qquad \textit{Ans.}$$

$$\sigma_{br} = \frac{15 \text{ kN}}{\pi(0.025 \text{ m})^2} = 7.64 \text{ MPa} \qquad \textit{Ans.}$$

NOTE: Using these results, the stress distributions are shown in Fig. 4–13c.

EXAMPLE 4.7

The three A-36 steel bars shown in Fig. 4–14a are pin connected to a *rigid* member. If the applied load on the member is 15 kN, determine the force developed in each bar. Bars *AB* and *EF* each have a cross-sectional area of 50 mm², and bar *CD* has a cross-sectional area of 30 mm².

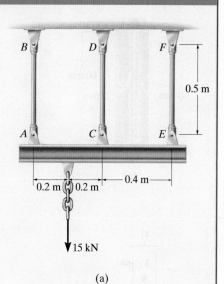

SOLUTION

Equilibrium. The free-body diagram of the rigid member is shown in Fig. 4–14b. This problem is statically indeterminate since there are three unknowns and only two available equilibrium equations.

$$+\uparrow \Sigma F_y = 0; \qquad F_A + F_C + F_E - 15 \text{ kN} = 0 \qquad (1)$$

$$\downarrow + \Sigma M_C = 0; \quad -F_A(0.4 \text{ m}) + 15 \text{ kN}(0.2 \text{ m}) + F_E(0.4 \text{ m}) = 0 \quad (2)$$

Compatibility. The applied load will cause the horizontal line *ACE* shown in Fig. 4–14c to move to the inclined line *A'C'E'*. The displacements of points *A*, *C*, and *E* can be related by similar triangles. Thus the compatibility equation that relates these displacements is

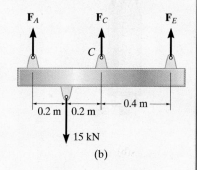

$$\frac{\delta_A - \delta_E}{0.8 \text{ m}} = \frac{\delta_C - \delta_E}{0.4 \text{ m}} \qquad \frac{\delta_C}{0.4} = \frac{\delta_A}{0.8} + \frac{\delta_E}{0.8}$$

$$\delta_C = \frac{1}{2}\delta_A + \frac{1}{2}\delta_E$$

Using the load–displacement relationship, Eq. 4–2, we have

$$\frac{F_C L}{(30 \text{ mm}^2)E_{st}} = \frac{1}{2}\left[\frac{F_A L}{(50 \text{ mm}^2)E_{st}}\right] + \frac{1}{2}\left[\frac{F_E L}{(50 \text{ mm}^2)E_{st}}\right]$$

$$F_C = 0.3F_A + 0.3F_E \qquad (3)$$

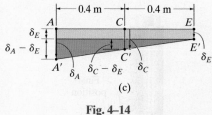

Fig. 4–14

Solving Eqs. 1–3 simultaneously yields

$$F_A = 9.52 \text{ kN} \qquad \qquad Ans.$$

$$F_C = 3.46 \text{ kN} \qquad \qquad Ans.$$

$$F_E = 2.02 \text{ kN} \qquad \qquad Ans.$$

EXAMPLE | 4.8

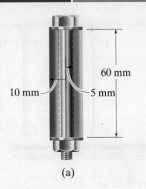

(a)

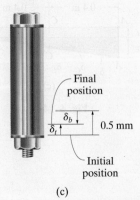

(b)

Final position

δ_b
δ_t

0.5 mm

Initial position

(c)

Fig. 4–15

The bolt shown in Fig. 4–15a is made of 2014-T6 aluminum alloy and is tightened so it compresses a cylindrical tube made of Am 1004-T61 magnesium alloy. The tube has an outer radius of 10 mm and it is assumed that both the inner radius of the tube and the radius of the bolt are 5 mm. The washers at the top and bottom of the tube are considered to be rigid and have a negligible thickness. Initially the nut is hand tightened snugly; then, using a wrench, the nut is further tightened one-half turn. If the bolt has 25 threads per 25 mm, determine the stress in the bolt.

SOLUTION

Equilibrium. The free-body diagram of a section of the bolt and the tube, Fig. 4–15b, is considered in order to relate the force in the bolt F_b to that in the tube, F_t. Equilibrium requires

$$+\uparrow \Sigma F_y = 0; \qquad F_b - F_t = 0 \qquad (1)$$

Compatibility. When the nut is tightened on the bolt, the tube will shorten δ_t, and the bolt will *elongate* δ_b, Fig. 4–15c. Since the nut undergoes one-half turn, it advances a distance of $(\frac{1}{2})(\frac{25\text{ mm}}{25}) = 0.5$ mm along the bolt. Thus, the compatibility of these displacements requires

$$(+\uparrow) \qquad\qquad \delta_t = 0.5\text{ mm} - \delta_b$$

Taking the moduli of elasticity from the table on the inside back cover, and applying Eq. 4–2, yields

$$\frac{F_t(60\text{ mm})}{\pi[(10\text{ mm})^2 - (5\text{ mm})^2][44.7(10^3)\text{ MPa}]} =$$
$$0.5\text{ mm} - \frac{F_b(60\text{ mm})}{\pi(5\text{ mm})^2[73.1(10^3)\text{ MPa}]}$$

$$0.017897F_t = 0.5\pi(10^3) - 0.032832F_b \qquad (2)$$

Solving Eqs. 1 and 2 simultaneously, we get

$$F_b = F_t = 30.96(10^3)\text{ N} = 30.96\text{ kN}$$

The stresses in the bolt and tube are therefore

$$\sigma_b = \frac{F_b}{A_b} = \frac{30.96\text{ kN}}{\pi(5\text{ mm})^2} = 394.2\text{ N/mm}^2 = 394.2\text{ MPa} \qquad Ans.$$

$$\sigma_b = \frac{F_t}{A_t} = \frac{30.96\text{ kN}}{\pi[(10\text{ mm})^2 - (5\text{ mm})^2]} = 131.4\text{ N/mm}^2 = 131.4\text{ MPa}$$

These stresses are less than the reported yield stress for each material, $(\sigma_Y)_{al} = 414$ MPa and $(\sigma_Y)_{mg} = 152$ MPa (see the inside back cover), and therefore this "elastic" analysis is valid.

4.5   The Force Method of Analysis for Axially Loaded Members

It is also possible to solve statically indeterminate problems by writing the compatibility equation using the principle of superposition. This method of solution is often referred to as the *flexibility or force method of analysis*. To show how it is applied, consider again the bar in Fig. 4–16a. If we choose the support at B as "redundant" and *temporarily* remove its effect on the bar, then the bar will become statically determinate as in Fig. 4–16b. By using the principle of superposition, we must add back the unknown redundant load F_B, as shown in Fig. 4–16c.

If load P causes B to be displaced *downward* by an amount δ_P, the reaction F_B must displace end B of the bar *upward* by an amount δ_B, such that no displacement occurs at B when the two loadings are superimposed. Thus,

$$(+\downarrow) \qquad\qquad 0 = \delta_P - \delta_B$$

This equation represents the compatibility equation for displacements at point B, for which we have assumed that displacements are positive downward.

Applying the load–displacement relationship to each case, we have $\delta_P = PL_{AC}/AE$ and $\delta_B = F_B L/AE$. Consequently,

$$0 = \frac{PL_{AC}}{AE} - \frac{F_B L}{AE}$$

$$F_B = P\left(\frac{L_{AC}}{L}\right)$$

From the free-body diagram of the bar, Fig. 4–11b, the reaction at A can now be determined from the equation of equilibrium,

$$+\uparrow \Sigma F_y = 0; \qquad P\left(\frac{L_{AC}}{L}\right) + F_A - P = 0$$

Since $L_{CB} = L - L_{AC}$, then

$$F_A = P\left(\frac{L_{CB}}{L}\right)$$

These results are the same as those obtained in Sec. 4.4, except that here we have applied the condition of compatibility to obtain one reaction *and then* the equilibrium condition to obtain the other.

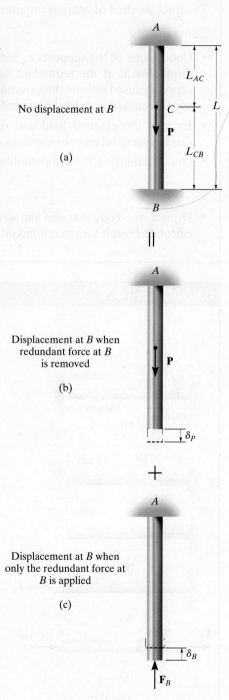

No displacement at B

(a)

Displacement at B when redundant force at B is removed

(b)

Displacement at B when only the redundant force at B is applied

(c)

Fig. 4–16

Procedure for Analysis

The force method of analysis requires the following steps.

Compatibility.

- Choose one of the supports as redundant and write the equation of compatibility. To do this, the known displacement at the redundant support, which is usually zero, is equated to the displacement at the support caused *only* by the external loads acting on the member *plus* (vectorially) the displacement at this support caused *only* by the redundant reaction acting on the member.
- Express the external load and redundant displacements in terms of the loadings by using a load–displacement relationship, such as $\delta = PL/AE$.
- Once established, the compatibility equation can then be solved for the magnitude of the redundant force.

Equilibrium.

- Draw a free-body diagram and write the appropriate equations of equilibrium for the member using the calculated result for the redundant. Solve these equations for any other reactions.

EXAMPLE 4.9

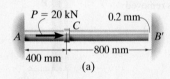

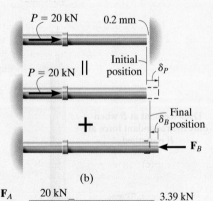

(a)

(b)

(c)

Fig. 4–17

The A-36 steel rod shown in Fig. 4–17a has a diameter of 10 mm. It is fixed to the wall at A, and before it is loaded there is a gap between the wall at B' and the rod of 0.2 mm. Determine the reactions at A and B'. Neglect the size of the collar at C. Take $E_{st} = 200$ GPa.

SOLUTION

Compatibility. Here we will consider the support at B' as redundant. Using the principle of superposition, Fig. 4–17b, we have

$$(\overset{+}{\rightarrow}) \qquad\qquad 0.0002 \text{ m} = \delta_P - \delta_B \qquad\qquad (1)$$

The deflections δ_P and δ_B are determined from Eq. 4–2.

$$\delta_P = \frac{PL_{AC}}{AE} = \frac{[20(10^3) \text{ N}](0.4 \text{ m})}{\pi(0.005 \text{ m})^2[200(10^9) \text{ N/m}^2]} = 0.5093(10^{-3}) \text{ m}$$

$$\delta_B = \frac{F_B L_{AB}}{AE} = \frac{F_B(1.20 \text{ m})}{\pi(0.005 \text{ m})^2[200(10^9) \text{ N/m}^2]} = 76.3944(10^{-9})F_B$$

Substituting into Eq. 1, we get

$$0.0002 \text{ m} = 0.5093(10^{-3}) \text{ m} - 76.3944(10^{-9})F_B$$

$$F_B = 4.05(10^3) \text{ N} = 4.05 \text{ kN} \qquad\qquad Ans.$$

Equilibrium. From the free-body diagram, Fig. 4–17c,

$$\overset{+}{\rightarrow} \Sigma F_x = 0; \quad -F_A + 20 \text{ kN} - 4.05 \text{ kN} = 0 \quad F_A = 16.0 \text{ kN} \qquad Ans.$$

PROBLEMS

4–31. The column is constructed from high-strength concrete and six A-36 steel reinforcing rods. If it is subjected to an axial force of 150 kN, determine the average normal stress in the concrete and in each rod. Each rod has a diameter of 20 mm.

***4–32.** The column is constructed from high-strength concrete and six A-36 steel reinforcing rods. If it is subjected to an axial force of 150 kN, determine the required diameter of each rod so that one-fourth of the load is carried by the concrete and three-fourths by the steel.

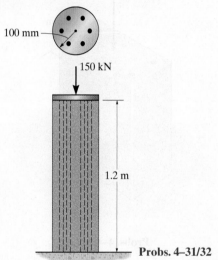

Probs. 4–31/32

•4–33. The steel pipe is filled with concrete and subjected to a compressive force of 80 kN. Determine the average normal stress in the concrete and the steel due to this loading. The pipe has an outer diameter of 80 mm and an inner diameter of 70 mm. E_{st} = 200 GPa, E_c = 24 GPa.

Prob. 4–33

4–34. The 304 stainless steel post A has a diameter of d = 50 mm and is surrounded by a red brass C83400 tube B. Both rest on the rigid surface. If a force of 25 kN is applied to the rigid cap, determine the average normal stress developed in the post and the tube.

4–35. The 304 stainless steel post A is surrounded by a red brass C83400 tube B. Both rest on the rigid surface. If a force of 25 kN is applied to the rigid cap, determine the required diameter d of the steel post so that the load is shared equally between the post and tube.

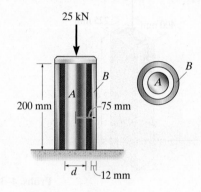

Probs. 4–34/35

***4–36.** The composite bar consists of a 20-mm-diameter A-36 steel segment AB and 50-mm-diameter red brass C83400 end segments DA and CB. Determine the average normal stress in each segment due to the applied load.

•4–37. The composite bar consists of a 20-mm-diameter A-36 steel segment AB and 50-mm-diameter red brass C83400 end segments DA and CB. Determine the displacement of A with respect to B due to the applied load.

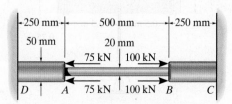

Probs. 4–36/37

4–38. The A-36 steel column, having a cross-sectional area of 11 250 mm², is encased in high-strength concrete as shown. If an axial force of 300 kN is applied to the column, determine the average compressive stress in the concrete and in the steel. How far does the column shorten? It has an original length of 2.4 m.

4–39. The A-36 steel column is encased in high-strength concrete as shown. If an axial force of 300 kN is applied to the column, determine the required area of the steel so that the force is shared equally between the steel and concrete. How far does the column shorten? It has an original length of 2.4 m.

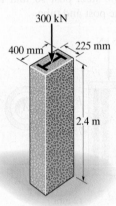

Probs. 4–38/39

*4–40. The rigid member is held in the position shown by three A-36 steel tie rods. Each rod has an unstretched length of 0.75 m and a cross-sectional area of 125 mm². Determine the forces in the rods if a turnbuckle on rod *EF* undergoes one full turn. The lead of the screw is 1.5 mm. Neglect the size of the turnbuckle and assume that it is rigid. *Note:* The lead would cause the rod, when *unloaded*, to shorten 1.5 mm when the turnbuckle is rotated one revolution.

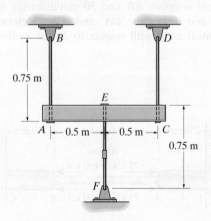

Prob. 4–40

•4–41. The concrete post is reinforced using six steel reinforcing rods, each having a diameter of 20 mm. Determine the stress in the concrete and the steel if the post is subjected to an axial load of 900 kN. $E_{st} = 200$ GPa, $E_c = 25$ GPa.

4–42. The post is constructed from concrete and six A-36 steel reinforcing rods. If it is subjected to an axial force of 900 kN, determine the required diameter of each rod so that one-fifth of the load is carried by the steel and four-fifths by the concrete. $E_{st} = 200$ GPa, $E_c = 25$ GPa.

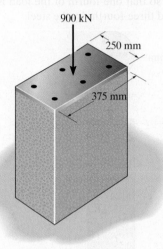

Probs. 4–41/42

4–43. The assembly consists of two red brass C83400 copper alloy rods *AB* and *CD* of diameter 30 mm, a stainless 304 steel alloy rod *EF* of diameter 40 mm, and a rigid cap *G*. If the supports at *A*, *C* and *F* are rigid, determine the average normal stress developed in rods *AB*, *CD* and *EF*.

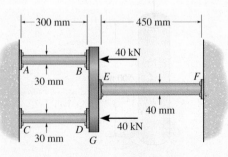

Prob. 4–43

*4–44. The two pipes are made of the same material and are connected as shown. If the cross-sectional area of BC is A and that of CD is $2A$, determine the reactions at B and D when a force $\mathbf{P}$ is applied at the junction C.

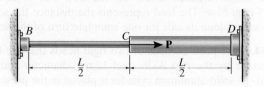

Prob. 4–44

•4–45. The bolt has a diameter of 20 mm and passes through a tube that has an inner diameter of 50 mm and an outer diameter of 60 mm. If the bolt and tube are made of A-36 steel, determine the normal stress in the tube and bolt when a force of 40 kN is applied to the bolt. Assume the end caps are rigid.

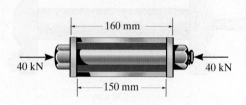

Prob. 4–45

4–46. If the gap between C and the rigid wall at D is initially 0.15 mm, determine the support reactions at A and D when the force $\mathbf{P} = 200$ kN is applied. The assembly is made of A36 steel.

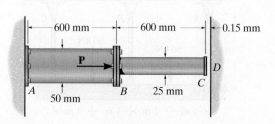

Prob. 4–46

4–47. Two A-36 steel wires are used to support the 3.25-kN ($\approx$ 325-kg) engine. Originally, AB is 800 mm long and $A'B'$ is 800.2 mm long. Determine the force supported by each wire when the engine is suspended from them. Each wire has a cross-sectional area of 6.25 mm^2.

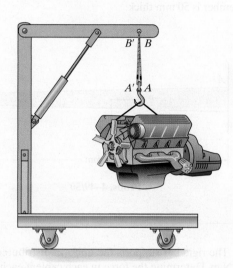

Prob. 4–47

*4–48. Rod AB has a diameter d and fits snugly between the rigid supports at A and B when it is unloaded. The modulus of elasticity is E. Determine the support reactions at A and B if the rod is subjected to the linearly distributed axial load.

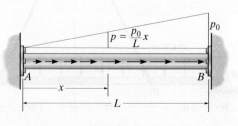

Prob. 4–48

•**4–49.** The tapered member is fixed connected at its ends A and B and is subjected to a load $P = 35$ kN at $x = 750$ mm. Determine the reactions at the supports. The material is 50 mm thick and is made from 2014-T6 aluminum.

4–50. The tapered member is fixed connected at its ends A and B and is subjected to a load **P**. Determine the location x of the load and its greatest magnitude so that the average normal stress in the bar does not exceed $\sigma_{allow} = 28$ MPa. The member is 50 mm thick.

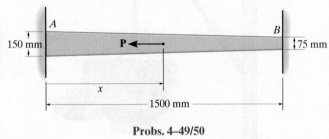

Probs. 4–49/50

4–51. The rigid bar supports the uniform distributed load of 90 kN/m. Determine the force in each cable if each cable has a cross-sectional area of 36 mm^2 and $E = 200$ GPa.

*__4–52.__ The rigid bar is originally horizontal and is supported by two cables each having a cross-sectional area of 36 mm^2, and $E = 200$ GPa. Determine the slight rotation of the bar when the uniform load is applied.

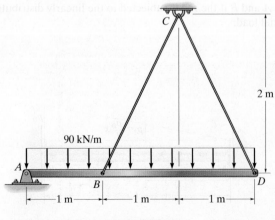

Probs. 4–51/52

•**4–53.** The press consists of two rigid heads that are held together by the two A-36 steel 12-mm-diameter rods. A 6061-T6-solid-aluminum cylinder is placed in the press and the screw is adjusted so that it just presses up against the cylinder. If it is then tightened one-half turn, determine the average normal stress in the rods and in the cylinder. The single-threaded screw on the bolt has a lead of 0.25 mm. *Note:* The lead represents the distance the screw advances along its axis for one complete turn of the screw.

4–54. The press consists of two rigid heads that are held together by the two A-36 steel 12-mm-diameter rods. A 6061-T6-solid-aluminum cylinder is placed in the press and the screw is adjusted so that it just presses up against the cylinder. Determine the angle through which the screw can be turned before the rods or the specimen begin to yield. The single-threaded screw on the bolt has a lead of 0.25 mm. *Note:* The lead represents the distance the screw advances along its axis for one complete turn of the screw.

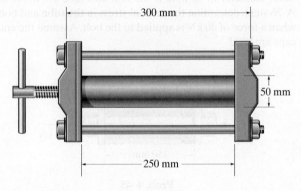

Probs. 4–53/54

4–55. The three suspender bars are made of A-36 steel and have equal cross-sectional areas of 450 mm^2. Determine the average normal stress in each bar if the rigid beam is subjected to the loading shown.

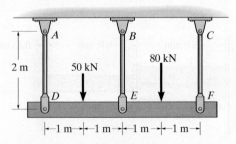

Prob. 4–55

*4–56. The rigid bar supports the 4-kN load. Determine the normal stress in each A-36 steel cable if each cable has a cross-sectional area of 25 mm².

•4–57. The rigid bar is originally horizontal and is supported by two A-36 steel cables each having a cross-sectional area of 25 mm². Determine the rotation of the bar when the 4-kN load is applied.

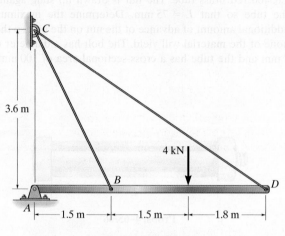

Probs. 4–56/57

*4–60. The assembly consists of two posts AD and CF made of A-36 steel and having a cross-sectional area of 1000 mm², and a 2014-T6 aluminum post BE having a cross-sectional area of 1500 mm². If a central load of 400 kN is applied to the rigid cap, determine the normal stress in each post. There is a small gap of 0.1 mm between the post BE and the rigid member ABC.

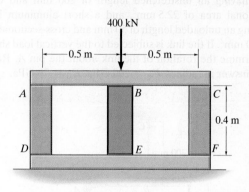

Prob. 4–60

4–58. The horizontal beam is assumed to be rigid and supports the distributed load shown. Determine the vertical reactions at the supports. Each support consists of a wooden post having a diameter of 120 mm and an unloaded (original) length of 1.40 m. Take $E_w = 12$ GPa.

4–59. The horizontal beam is assumed to be rigid and supports the distributed load shown. Determine the angle of tilt of the beam after the load is applied. Each support consists of a wooden post having a diameter of 120 mm and an unloaded (original) length of 1.40 m. Take $E_w = 12$ GPa.

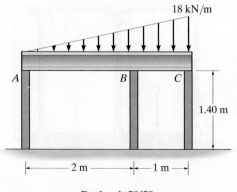

Probs. 4–58/59

•4–61. The distributed loading is supported by the three suspender bars. AB and EF are made of aluminum and CD is made of steel. If each bar has a cross-sectional area of 450 mm², determine the maximum intensity w of the distributed loading so that an allowable stress of $(\sigma_{allow})_{st} = 180$ MPa in the steel and $(\sigma_{allow})_{al} = 94$ MPa in the aluminum is not exceeded. $E_{st} = 200$ GPa, $E_{al} = 70$ GPa. Assume ACE is rigid.

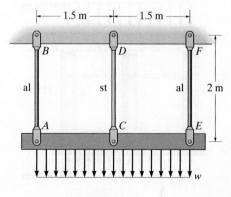

Prob. 4–61

4-62. The rigid link is supported by a pin at A, a steel wire BC having an unstretched length of 200 mm and cross-sectional area of 22.5 mm², and a short aluminum block having an unloaded length of 50 mm and cross-sectional area of 40 mm². If the link is subjected to the vertical load shown, determine the average normal stress in the wire and the block. E_{st} = 200 GPa, E_{al} = 70 GPa.

4-63. The rigid link is supported by a pin at A, a steel wire BC having an unstretched length of 200 mm and cross-sectional area of 22.5 mm², and a short aluminum block having an unloaded length of 50 mm and cross-sectional area of 40 mm². If the link is subjected to the vertical load shown, determine the rotation of the link about the pin A. Report the answer in radians. E_{st} = 200 GPa, E_{al} = 70 GPa.

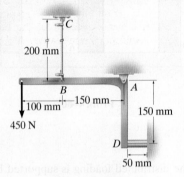

Probs. 4–62/63

***4-64.** The center post B of the assembly has an original length of 124.7 mm, whereas posts A and C have a length of 125 mm. If the caps on the top and bottom can be considered rigid, determine the average normal stress in each post. The posts are made of aluminum and have a cross-sectional area of 400 mm². E_{al} = 70 GPa.

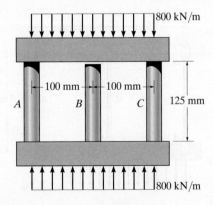

Prob. 4–64

•4-65. The assembly consists of an A-36 steel bolt and a C83400 red brass tube. If the nut is drawn up snug against the tube so that L = 75 mm, then turned an additional amount so that it advances 0.02 mm on the bolt, determine the force in the bolt and the tube. The bolt has a diameter of 7 mm and the tube has a cross-sectional area of 100 mm².

4-66. The assembly consists of an A-36 steel bolt and a C83400 red brass tube. The nut is drawn up snug against the tube so that L = 75 mm. Determine the maximum additional amount of advance of the nut on the bolt so that none of the material will yield. The bolt has a diameter of 7 mm and the tube has a cross-sectional area of 100 mm².

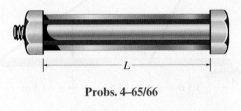

Probs. 4–65/66

4-67. The three suspender bars are made of the same material and have equal cross-sectional areas A. Determine the average normal stress in each bar if the rigid beam ACE is subjected to the force **P**.

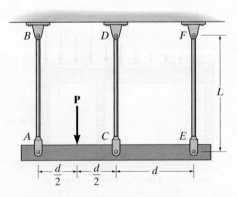

Prob. 4–67

4.6   Thermal Stress

A change in temperature can cause a body to change its dimensions. Generally, if the temperature increases, the body will expand, whereas if the temperature decreases, it will contract. Ordinarily this expansion or contraction is *linearly* related to the temperature increase or decrease that occurs. If this is the case, and the material is homogeneous and isotropic, it has been found from experiment that the displacement of a member having a length L can be calculated using the formula

$$\delta_T = \alpha\, \Delta T L \qquad (4\text{--}4)$$

where

Most traffic bridges are designed with expansion joints to accommodate the thermal movement of the deck and thus avoid any thermal stress.

    α = a property of the material, referred to as the **linear coefficient of thermal expansion**. The units measure strain per degree of temperature. They are 1/°C (Celsius) or 1/K (Kelvin) in the SI system. Typical values are given on the inside back cover

ΔT = the algebraic change in temperature of the member

  L = the original length of the member

  δ_T = the algebraic change in the length of the member

The change in length of a *statically determinate* member can easily be calculated using Eq. 4–4, since the member is free to expand or contract when it undergoes a temperature change. However, in a *statically indeterminate* member, these thermal displacements will be constrained by the supports, thereby producing **thermal stresses** that must be considered in design. Determining these thermal stresses is possible using the methods outlined in the previous sections. The following examples illustrate some applications.

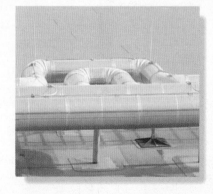

Long extensions of ducts and pipes that carry fluids are subjected to variations in climate that will cause them to expand and contract. Expansion joints, such as the one shown, are used to mitigate thermal stress in the material.

EXAMPLE | 4.10

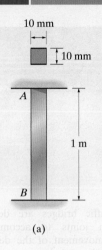

10 mm

10 mm

A

1 m

B

(a)

F

F

(b)

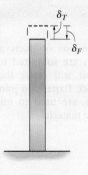

δ_T

δ_F

(c)

Fig. 4–18

The A-36 steel bar shown in Fig. 4–18a is constrained to just fit between two fixed supports when $T_1 = 30°C$. If the temperature is raised to $T_2 = 60°C$, determine the average normal thermal stress developed in the bar.

SOLUTION

Equilibrium. The free-body diagram of the bar is shown in Fig. 4–18b. Since there is no external load, the force at A is equal but opposite to the force at B; that is,

$$+\uparrow \Sigma F_y = 0; \qquad\qquad F_A = F_B = F$$

The problem is statically indeterminate since this force cannot be determined from equilibrium.

Compatibility. Since $\delta_{A/B} = 0$, the thermal displacement δ_T at A that occurs, Fig. 4–18c, is counteracted by the force $\mathbf{F}$ that is required to push the bar δ_F back to its original position. The compatibility condition at A becomes

$$(+\uparrow) \qquad\qquad \delta_{A/B} = 0 = \delta_T - \delta_F$$

Applying the thermal and load–displacement relationships, we have

$$0 = \alpha \Delta T L - \frac{FL}{AE}$$

Thus, from the data on the inside back cover,

$$F = \alpha \Delta T A E$$
$$= [12(10^{-6})/°C](60°C - 30°C)(0.010\ \text{m})^2[200(10^6)\ \text{kN/m}^2]$$
$$= 7.2\ \text{kN}$$

Since $\mathbf{F}$ also represents the internal axial force within the bar, the average normal compressive stress is thus

$$\sigma = \frac{F}{A} = \frac{7.2\ \text{kN}}{(0.010\ \text{m})^2} = 72\ 000\ \text{kPa} = 72\ \text{MPa} \qquad Ans.$$

NOTE: From the magnitude of $\mathbf{F}$, it should be apparent that changes in temperature can cause large reaction forces in statically indeterminate members.

EXAMPLE | 4.11

The rigid beam shown in Fig. 4–19a is fixed to the top of the three posts made of A-36 steel and 2014-T6 aluminum. The posts each have a length of 250 mm when no load is applied to the beam, and the temperature is $T_1 = 20°C$. Determine the force supported by each post if the bar is subjected to a uniform distributed load of 150 kN/m and the temperature is raised to $T_2 = 80°C$.

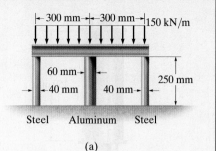

(a)

SOLUTION

Equilibrium. The free-body diagram of the beam is shown in Fig. 4–19b. Moment equilibrium about the beam's center requires the forces in the steel posts to be equal. Summing forces on the free-body diagram, we have

$$+\uparrow \Sigma F_y = 0; \qquad 2F_{st} + F_{al} - 90(10^3)\ N = 0 \qquad (1)$$

Compatibility. Due to load, geometry, and material symmetry, the top of each post is displaced by an equal amount. Hence,

$$(+\downarrow) \qquad \qquad \delta_{st} = \delta_{al} \qquad \qquad (2)$$

The final position of the top of each post is equal to its displacement caused by the temperature increase, plus its displacement caused by the internal axial compressive force, Fig. 4–19c. Thus, for the steel and aluminum post, we have

$$(+\downarrow) \qquad \qquad \delta_{st} = -(\delta_{st})_T + (\delta_{st})_F$$
$$(+\downarrow) \qquad \qquad \delta_{al} = -(\delta_{al})_T + (\delta_{al})_F$$

Applying Eq. 2 gives

$$-(\delta_{st})_T + (\delta_{st})_F = -(\delta_{al})_T + (\delta_{al})_F$$

Using Eqs. 4–2 and 4–4 and the material properties on the inside back cover, we get

$$-[12(10^{-6})/°C](80°C - 20°C)(0.250\ m) + \frac{F_{st}(0.250\ m)}{\pi(0.020\ m)^2[200(10^9)\ N/m^2]}$$

$$= -[23(10^{-6})/°C](80°C - 20°C)(0.250\ m) + \frac{F_{al}(0.250\ m)}{\pi(0.030\ m)^2[73.1(10^9)\ N/m^2]}$$

$$F_{st} = 1.216F_{al} - 165.9(10^3) \qquad (3)$$

To be *consistent*, all numerical data has been expressed in terms of newtons, meters, and degrees Celsius. Solving Eqs. 1 and 3 simultaneously yields

$$F_{st} = -16.4\ kN \quad F_{al} = 123\ kN \qquad \qquad Ans.$$

The negative value for F_{st} indicates that this force acts opposite to that shown in Fig. 4–19b. In other words, the steel posts are in tension and the aluminum post is in compression.

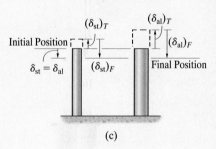

(c)

Fig. 4–19

EXAMPLE | 4.12

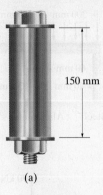

(a)

A 2014-T6 aluminum tube having a cross-sectional area of 600 mm² is used as a sleeve for an A-36 steel bolt having a cross-sectional area of 400 mm², Fig. 4–20a. When the temperature is $T_1 = 15°C$, the nut holds the assembly in a snug position such that the axial force in the bolt is negligible. If the temperature increases to $T_2 = 80°C$, determine the force in the bolt and sleeve.

SOLUTION

Equilibrium. The free-body diagram of a top segment of the assembly is shown in Fig. 4–20b. The forces F_b and F_s are produced since the sleeve has a higher coefficient of thermal expansion than the bolt, and therefore the sleeve will expand more when the temperature is increased. It is required that

$$+\uparrow \Sigma F_y = 0; \qquad\qquad F_s = F_b \qquad\qquad (1)$$

(b)

Compatibility. The temperature increase causes the sleeve and bolt to expand $(\delta_s)_T$ and $(\delta_b)_T$, Fig. 4–20c. However, the redundant forces F_b and F_s elongate the bolt and shorten the sleeve. Consequently, the end of the assembly reaches a final position, which is not the same as its initial position. Hence, the compatibility condition becomes

$$(+\downarrow) \qquad \delta = (\delta_b)_T + (\delta_b)_F = (\delta_s)_T - (\delta_s)_F$$

Applying Eqs. 4–2 and 4–4, and using the mechanical properties from the table on the inside back cover, we have

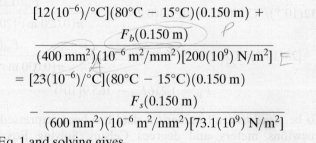

Initial position

$$[12(10^{-6})/°C](80°C - 15°C)(0.150\ m) +$$
$$\frac{F_b(0.150\ m)}{(400\ mm^2)(10^{-6}\ m^2/mm^2)[200(10^9)\ N/m^2]}$$
$$= [23(10^{-6})/°C](80°C - 15°C)(0.150\ m)$$
$$- \frac{F_s(0.150\ m)}{(600\ mm^2)(10^{-6}\ m^2/mm^2)[73.1(10^9)\ N/m^2]}$$

Final position

Using Eq. 1 and solving gives

$$F_s = F_b = 20.3\ kN \qquad\qquad Ans.$$

(c)

Fig. 4–20

NOTE: Since linear elastic material behavior was assumed in this analysis, the average normal stresses should be checked to make sure that they do not exceed the proportional limits for the material.

PROBLEMS

*4–68.   A steel surveyor's tape is to be used to measure the length of a line. The tape has a rectangular cross section of 1.25 mm by 5 mm and a length of 30 m when $T_1 = 20°C$ and the tension or pull on the tape is 100 N. Determine the true length of the line if the tape shows the reading to be 139 m when used with a pull of 175 N at $T_2 = 40°C$. The ground on which it is placed is flat. $\alpha_{st} = 17(10^{-6})/°C$, $E_{st} = 200$ GPa.

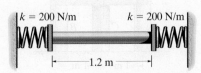

P ◄——————————————————————► P

5 mm

1.25 mm

Prob. 4–68

•4–69.   Three bars each made of different materials are connected together and placed between two walls when the temperature is $T_1 = 12°C$. Determine the force exerted on the (rigid) supports when the temperature becomes $T_2 = 18°C$. The material properties and cross-sectional area of each bar are given in the figure.

Steel	Brass	Copper
$E_{st} = 200$ GPa	$E_{br} = 100$ GPa	$E_{cu} = 120$ GPa
$\alpha_{st} = 12(10^{-6})/°C$	$\alpha_{br} = 21(10^{-6})/°C$	$\alpha_{cu} = 17(10^{-6})/°C$

$A_{cu} = 515$ mm²

$A_{st} = 200$ mm² $A_{br} = 450$ mm²

|←———300 mm———→|←—200 mm—→|

100 mm

Prob. 4–69

4–70.   The rod is made of A-36 steel and has a diameter of 6 mm. If the rod is 1.2 m long when the springs are compressed 12 mm and the temperature of the rod is $T = 10°C$, determine the force in the rod when its temperature is $T = 75°C$.

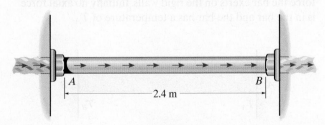

$k = 200$ N/m $k = 200$ N/m

|←———1.2 m———→|

Prob. 4–70

4–71.   A 1.8-m-long steam pipe is made of A-36 steel with $\sigma_Y = 280$ MPa. It is connected directly to two turbines A and B as shown. The pipe has an outer diameter of 100 mm and a wall thickness of 6 mm. The connection was made at $T_1 = 20°C$. If the turbines' points of attachment are assumed rigid, determine the force the pipe exerts on the turbines when the steam and thus the pipe reach a temperature of $T_2 = 135°C$.

*4–72.   A 1.8-m-long steam pipe is made of A-36 steel with $\sigma_Y = 280$ MPa. It is connected directly to two turbines A and B as shown. The pipe has an outer diameter of 100 mm and a wall thickness of 6 mm. The connection was made at $T_1 = 20°C$. If the turbines' points of attachment are assumed to have a stiffness of $k = 16$ MN/mm, determine the force the pipe exerts on the turbines when the steam and thus the pipe reach a temperature of $T_2 = 135°C$.

|←—1.8 m—→|

A   B

Probs. 4–71/72

•4–73.   The pipe is made of A-36 steel and is connected to the collars at A and B. When the temperature is 15°C, there is no axial load in the pipe. If hot gas traveling through the pipe causes its temperature to rise by $\Delta T = (20 + 30x)°C$, where x is in feet, determine the average normal stress in the pipe. The inner diameter is 50 mm, the wall thickness is 4 mm.

4–74.   The bronze C86100 pipe has an inner radius of 12.5 mm and a wall thickness of 5 mm. If the gas flowing through it changes the temperature of the pipe uniformly from $T_A = 60°C$ at A to $T_B = 15°C$ at B, determine the axial force it exerts on the walls. The pipe was fitted between the walls when $T = 15°C$.

A   B

|←————————2.4 m————————→|

Probs. 4–73/74

4–75. The 12-m-long A-36 steel rails on a train track are laid with a small gap between them to allow for thermal expansion. Determine the required gap δ so that the rails just touch one another when the temperature is increased from $T_1 = -30°C$ to $T_2 = 30°C$. Using this gap, what would be the axial force in the rails if the temperature were to rise to $T_3 = 40°C$? The cross-sectional area of each rail is 3200 mm^2.

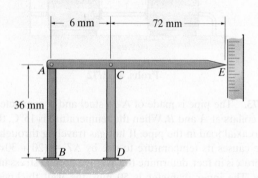

Prob. 4–75

***4–76.** The device is used to measure a change in temperature. Bars AB and CD are made of A-36 steel and 2014-T6 aluminum alloy respectively. When the temperature is at 40°C, ACE is in the horizontal position. Determine the vertical displacement of the pointer at E when the temperature rises to 80°C.

Prob. 4–76

•4–77. The bar has a cross-sectional area A, length L, modulus of elasticity E, and coefficient of thermal expansion α. The temperature of the bar changes uniformly along its length from T_A at A to T_B at B so that at any point x along the bar $T = T_A + x(T_B - T_A)/L$. Determine the force the bar exerts on the rigid walls. Initially no axial force is in the bar and the bar has a temperature of T_A.

Prob. 4–77

4–78. The A-36 steel rod has a diameter of 50 mm and is lightly attached to the rigid supports at A and B when $T_1 = 80°C$. If the temperature becomes $T_2 = 20°C$ and an axial force of $P = 200$ kN is applied to its center, determine the reactions at A and B.

4–79. The A-36 steel rod has a diameter of 50 mm and is lightly attached to the rigid supports at A and B when $T_1 = 50°C$. Determine the force P that must be applied to the collar at its midpoint so that, when $T_2 = 30°C$, the reaction at B is zero.

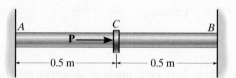

Probs. 4–78/79

***4–80.** The rigid block has a weight of 400 kN and is to be supported by posts A and B, which are made of A-36 steel, and the post C, which is made of C83400 red brass. If all the posts have the same original length before they are loaded, determine the average normal stress developed in each post when post C is heated so that its temperature is increased by 10°C. Each post has a cross-sectional area of 5000 mm^2.

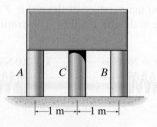

Prob. 4–80

•4–81. The three bars are made of A-36 steel and form a pin-connected truss. If the truss is constructed when $T_1 = 10°C$, determine the force in each bar when $T_2 = 43°C$. Each bar has a cross-sectional area of 1250 mm².

4–82. The three bars are made of A-36 steel and form a pin-connected truss. If the truss is constructed when $T_1 = 10°C$, determine the vertical displacement of joint A when $T_2 = 65.5°C$. Each bar has a cross-sectional area of 1250 mm².

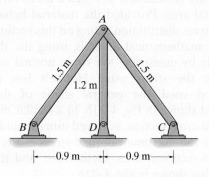

Probs. 4–81/82

4–83. The wires AB and AC are made of steel, and wire AD is made of copper. Before the 750-N force is applied, AB and AC are each 1500 mm long and AD is 1000 mm long. If the temperature is increased by 25°C, determine the force in each wire needed to support the load. Take $E_{st} = 200$ GPa, $E_{cu} = 120$ GPa, $\alpha_{st} = 4.5(10^{-6})°C$, $\alpha_{cu} = 5.4(10^{-6})°C$. Each wire has a cross-sectional area of 7.68 mm².

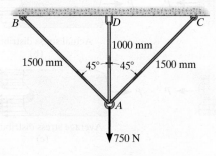

Prob. 4–83

*4–84. The AM1004-T61 magnesium alloy tube AB is capped with a rigid plate E. The gap between E and end C of the 6061-T6 aluminum alloy solid circular rod CD is 0.2 mm when the temperature is at 30°C. Determine the normal stress developed in the tube and the rod if the temperature rises to 80°C. Neglect the thickness of the rigid cap.

•4–85. The AM1004-T61 magnesium alloy tube AB is capped with a rigid plate. The gap between E and end C of the 6061-T6 aluminum alloy solid circular rod CD is 0.2 mm when the temperature is at 30°C. Determine the highest temperature to which it can be raised without causing yielding either in the tube or the rod. Neglect the thickness of the rigid cap.

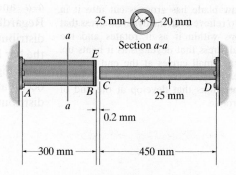

Probs. 4–84/85

4–86. The steel bolt has a diameter of 7 mm and fits through an aluminum sleeve as shown. The sleeve has an inner diameter of 8 mm and an outer diameter of 10 mm. The nut at A is adjusted so that it just presses up against the sleeve. If the assembly is originally at a temperature of $T_1 = 20°C$ and then is heated to a temperature of $T_2 = 100°C$, determine the average normal stress in the bolt and the sleeve. $E_{st} = 200$ GPa, $E_{al} = 70$ GPa, $\alpha_{st} = 14(10^{-6})/°C$, $\alpha_{al} = 23(10^{-6})/°C$.

Prob. 4–86

This saw blade has grooves cut into it in order to relieve both the dynamic stress that develops within it as it rotates and the thermal stress that develops as it heats up. Note the small circles at the end of each groove. These serve to reduce the stress concentrations that develop at the end of each groove.

4.7 Stress Concentrations

In Sec. 4.1, it was pointed out that when an axial force is applied to a member, it creates a complex stress distribution within the localized region of the point of load application. Not only do complex stress distributions arise just under a concentrated loading, they can also arise at sections where the member's cross-sectional area changes. For example, consider the bar in Fig. 4–21a, which is subjected to an axial force P. Here the once horizontal and vertical grid lines deflect into an irregular pattern around the hole centered in the bar. The maximum normal stress in the bar occurs on section a–a, which is taken through the bar's *smallest* cross-sectional area. Provided the material behaves in a linear-elastic manner, the stress distribution acting on this section can be determined either from a mathematical analysis, using the theory of elasticity, or experimentally by measuring the strain normal to section a–a and then calculating the stress using Hooke's law, $\sigma = E\epsilon$. Regardless of the method used, the general shape of the stress distribution will be like that shown in Fig. 4–21b. In a similar manner, if the bar has a reduction in its cross section, achieved using shoulder fillets as in Fig. 4–22a, then again the maximum normal stress in the bar will occur at the *smallest* cross-sectional area, section a–a, and the stress distribution will look like that shown in Fig. 4–22b.

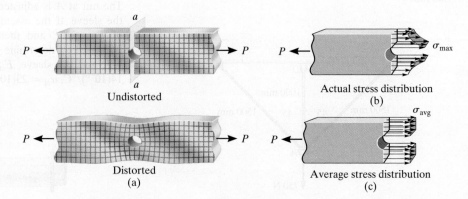

Undistorted

Distorted
(a)

Actual stress distribution
(b)

Average stress distribution
(c)

Fig. 4–21

In both of these cases, *force equilibrium* requires the magnitude of the *resultant force* developed by the stress distribution to be equal to P. In other words,

$$P = \int_A \sigma \, dA \qquad (4\text{--}5)$$

This integral *graphically* represents the total *volume* under each of the stress-distribution diagrams shown in Fig. 4–21b or Fig. 4–22b. The resultant **P** must act through the *centroid* of each *volume*.

In engineering practice, the actual stress distributions in Fig. 4–21b and Fig. 4–22b do *not* have to be determined. Instead, only the *maximum stress* at these sections must be known, and the member is then designed to resist this stress when the axial load **P** is applied. Specific values of this maximum normal stress can be determined by experimental methods or by advanced mathematical techniques using the theory of elasticity. The results of these investigations are usually reported in graphical form using a **stress-concentration factor** K. We define K as a ratio of the maximum stress to the average normal stress acting at the cross section; i.e.,

$$\boxed{K = \frac{\sigma_{\max}}{\sigma_{\text{avg}}}} \qquad (4\text{--}6)$$

Provided K is known, and the average normal stress has been calculated from $\sigma_{\text{avg}} = P/A$, where A is the *smallest* cross-sectional area, Figs. 4–21c and 4–22c, then the maximum normal stress at the cross section is $\sigma_{\max} = K(P/A)$.

Stress concentrations often arise at sharp corners on heavy machinery. Engineers can mitigate this effect by using stiffeners welded to the corners.

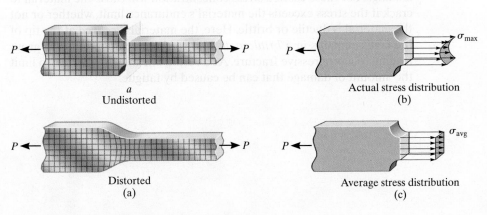

Undistorted

Actual stress distribution
(b)

Distorted
(a)

Average stress distribution
(c)

Fig. 4–22

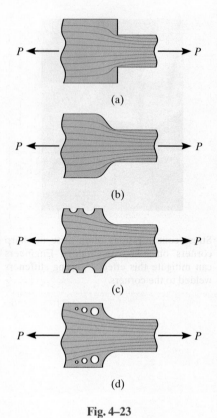

(a)

(b)

(c)

(d)

Fig. 4–23

Specific values of K are generally reported in handbooks related to stress analysis.* Examples are given in Figs. 4–24 and 4–25. Note that K is independent of the bar's material properties; rather, it *depends only on the bar's geometry* and the type of discontinuity. As the size r of the discontinuity is *decreased*, the stress concentration is increased. For example, if a bar requires a change in cross section, it has been determined that a sharp corner, Fig. 4–23a, produces a stress-concentration factor greater than 3. In other words, the maximum normal stress will be three times greater than the average normal stress on the smallest cross section. However, this can be reduced to, say, 1.5 by introducing a fillet, Fig. 4–23b. A further reduction can be made by means of small grooves or holes placed at the transition, Fig. 4–23c and 4–23d. In all of these cases these designs help to reduce the rigidity of the material surrounding the corners, so that both the strain and the stress are more evenly spread throughout the bar.

The stress-concentration factors given in Figs. 4–24 and 4–25 were determined on the basis of a static loading, with the assumption that the stress in the material does not exceed the proportional limit. If the material is *very brittle*, the proportional limit may be at the fracture stress, and so for this material, failure will begin *at* the point of stress concentration. Essentially a crack begins to form at this point, and a higher stress concentration will develop at the *tip* of this crack. This, in turn, causes the crack to propagate over the cross section, resulting in sudden fracture. For this reason, it is very important to use stress-concentration factors in design when using brittle materials. On the other hand, if the material is ductile and subjected to a static load, it is often not necessary to use stress-concentration factors since any stress that exceeds the proportional limit will not result in a crack. Instead, the material will have reserve strength due to yielding and strain-hardening. In the next section we will discuss the effects caused by this phenomenon.

Stress concentrations are also responsible for many failures of structural members or mechanical elements subjected to *fatigue loadings*. For these cases, a stress concentration will cause the material to crack if the stress exceeds the material's endurance limit, whether or not the material is ductile or brittle. Here, the material *localized* at the tip of the crack remains in a *brittle state*, and so the crack continues to grow, leading to a progressive fracture. As a result, one must seek ways to limit the amount of damage that can be caused by fatigue.

*See Lipson, C. and R. C. Juvinall, *Handbook of Stress and Strength*, Macmillan.

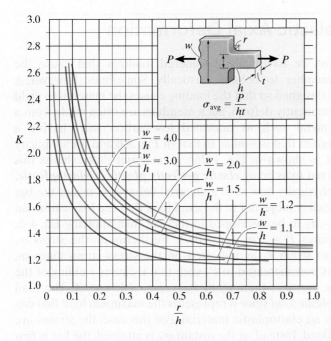

Fig. 4–24

Fig. 4–25

Important Points

- *Stress concentrations* occur at sections where the cross-sectional area suddenly changes. The more severe the change, the larger the stress concentration.

- For design or analysis, it is only necessary to determine the maximum stress acting on the smallest cross-sectional area. This is done using a *stress concentration factor*, K, that has been determined through experiment and is only a function of the geometry of the specimen.

- Normally the stress concentration in a ductile specimen that is subjected to a static loading will *not* have to be considered in design; however, if the material is *brittle*, or subjected to *fatigue* loadings, then stress concentrations become important.

Failure of this steel pipe in tension occurred at its smallest cross-sectional area, which is through the hole. Notice how the material yielded around the fractured surface.

*4.8   Inelastic Axial Deformation

Up to this point we have only considered loadings that cause the material of a member to behave elastically. Sometimes, however, a member may be designed so that the loading causes the material to yield and thereby permanently deform. Such members are often made from a highly ductile metal such as annealed low-carbon steel having a stress–strain diagram that is similar to that of Fig. 3–6 and for simplicity can be *modeled* as shown in Fig. 4–26b. A material that exhibits this behavior is referred to as being ***elastic perfectly plastic*** or ***elastoplastic.***

To illustrate physically how such a material behaves, consider the bar in Fig. 4–26a, which is subjected to the axial load **P**. If the load causes an *elastic stress* $\sigma = \sigma_1$ to be developed in the bar, then applying Eq. 4–5, *equilibrium* requires $P = \int \sigma_1 \, dA = \sigma_1 A$. Furthermore, the stress σ_1 causes the bar to strain ϵ_1 as indicated on the stress–strain diagram, Fig. 4–26b. If P is now increased to P_p such that it causes yielding of the material, that is, $\sigma = \sigma_Y$, then again $P_p = \int \sigma_Y \, dA = \sigma_Y A$. The load P_p is called the *plastic load* since it represents the maximum load that can be supported by an elastoplastic material. For this case, the strains are *not* uniquely defined. Instead, at the instant σ_Y is attained, the bar is *first* subjected to the yield strain ϵ_Y, Fig. 4–26b, after which the bar will *continue to yield* (or elongate) such that the strains ϵ_2, then ϵ_3, etc., are generated. Since our "model" of the material exhibits perfectly plastic material behavior, this elongation will continue indefinitely with no increase in load. In reality, however, the material will, after some yielding, actually begin to strain-harden so that the extra strength it attains will *stop* any further straining. As a result, any design based on this behavior will be safe, since strain-hardening provides the potential for the material to support an *additional* load if necessary.

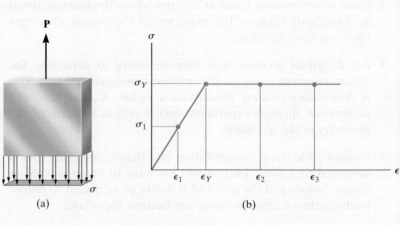

Fig. 4–26

Consider now the case of a bar having a hole through it as shown in Fig. 4–27a. As the magnitude of **P** is increased, a stress concentration occurs in the material at the edge of the hole, on section a–a. The stress here will reach a maximum value of, say, $\sigma_{max} = \sigma_1$ and have a corresponding *elastic strain* of ϵ_1, Fig. 4–27b. The stresses and corresponding strains at other points on the cross section will be smaller, as indicated by the stress distribution shown in Fig. 4–27c. Equilibrium requires $P = \int \sigma\, dA$. In other words, P is geometrically equivalent to the "volume" contained within the stress distribution. If the load is now increased to P', so that $\sigma_{max} = \sigma_Y$, then the material will begin to yield outward from the hole, until the equilibrium condition $P' = \int \sigma\, dA$ is satisfied, Fig. 4–27d. As shown, this produces a stress distribution that has a geometrically *greater* "volume" than that shown in Fig. 4–27c. A further increase in load will cause the material over the *entire cross section* to yield eventually. When this happens, *no greater load* can be sustained by the bar. This *plastic load* P_p is shown in Fig. 4–27e. It can be calculated from the equilibrium condition

$$P_p = \int_A \sigma_Y\, dA = \sigma_Y A$$

where A is the bar's cross-sectional area at section a–a.

The following examples illustrate numerically how these concepts apply to other types of problems for which the material has elastoplastic behavior.

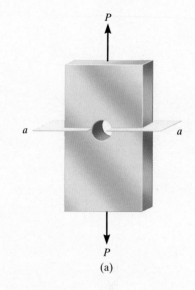

(a)

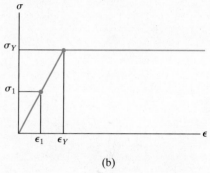

(b)

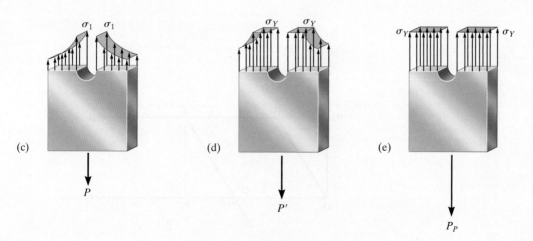

(c)

(d)

(e)

Fig. 4–27

*4.9   Residual Stress

If an axially loaded member or group of such members forms a statically indeterminate system that can support both tensile and compressive loads, then excessive external loadings, which cause yielding of the material, will create *residual stresses* in the members when the loads are removed. The reason for this has to do with the elastic recovery of the material that occurs during unloading. To show this, consider a prismatic member made from an elastoplastic material having the stress–strain diagram shown in Fig. 4–28. If an axial load produces a stress σ_Y in the material and a corresponding plastic strain ϵ_C, then when the load is *removed*, the material will respond elastically and follow the line CD in order to recover some of the plastic strain. A recovery to zero stress at point O' will be possible only if the member is statically determinate, since the support reactions for the member must be zero when the load is removed. Under these circumstances the member will be permanently deformed so that the permanent set or strain in the member is $\epsilon_{O'}$.

If the member is *statically indeterminate*, however, removal of the external load will cause the support forces to respond to the elastic recovery CD. Since these forces will constrain the member from full recovery, they will induce **residual stresses** in the member. To solve a problem of this kind, the complete cycle of loading and then unloading of the member can be considered as the *superposition* of a positive load (loading) on a negative load (unloading). The loading, O to C, results in a plastic stress distribution, whereas the unloading, along CD, results only in an elastic stress distribution. Superposition requires the loads to cancel; however, the stress distributions will not cancel, and so residual stresses will remain.

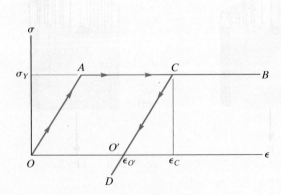

Fig. 4–28

EXAMPLE | 4.13

The bar in Fig. 4–29a is made of steel that is assumed to be elastic perfectly plastic, with $\sigma_Y = 250$ MPa. Determine (a) the maximum value of the applied load P that can be applied without causing the steel to yield and (b) the maximum value of P that the bar can support. Sketch the stress distribution at the critical section for each case.

SOLUTION

Part (a). When the material behaves elastically, we must use a stress-concentration factor determined from Fig. 4–24 that is unique for the bar's geometry. Here

$$\frac{r}{h} = \frac{4 \text{ mm}}{(40 \text{ mm} - 8 \text{ mm})} = 0.125$$

$$\frac{w}{h} = \frac{40 \text{ mm}}{(40 \text{ mm} - 8 \text{ mm})} = 1.25$$

From the figure $K \approx 1.75$. The maximum load, without causing yielding, occurs when $\sigma_{max} = \sigma_Y$. The average normal stress is $\sigma_{avg} = P/A$. Using Eq. 4–6, we have

$$\sigma_{max} = K\sigma_{avg}; \qquad \sigma_Y = K\left(\frac{P_Y}{A}\right)$$

$$250(10^6) \text{ Pa} = 1.75\left[\frac{P_Y}{(0.002 \text{ m})(0.032 \text{ m})}\right]$$

$$P_Y = 9.14 \text{ kN} \qquad\qquad Ans.$$

This load has been calculated using the *smallest* cross section. The resulting stress distribution is shown in Fig. 4–29b. For equilibrium, the "volume" contained within this distribution must equal 9.14 kN.

Part (b). The maximum load sustained by the bar will cause *all the material* at the smallest cross section to yield. Therefore, as P is increased to the *plastic load* P_p, it gradually changes the stress distribution from the elastic state shown in Fig. 4–29b to the plastic state shown in Fig. 4–29c. We require

$$\sigma_Y = \frac{P_p}{A}$$

$$250(10^6) \text{ Pa} = \frac{P_p}{(0.002 \text{ m})(0.032 \text{ m})}$$

$$P_p = 16.0 \text{ kN} \qquad\qquad Ans.$$

Here P_p equals the "volume" contained within the stress distribution, which in this case is $P_p = \sigma_Y A$.

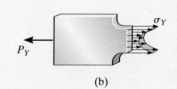

(b)

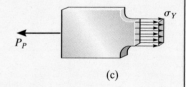

(c)

Fig. 4–29

EXAMPLE 4.14

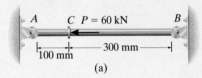

(a)

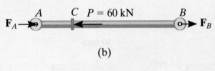

(b)

Fig. 4–30

The rod shown in Fig. 4–30a has a radius of 5 mm and is made of an elastic perfectly plastic material for which $\sigma_Y = 420$ MPa, $E = 70$ GPa, Fig. 4–30c. If a force of $P = 60$ kN is applied to the rod and then removed, determine the residual stress in the rod.

SOLUTION

The free-body diagram of the rod is shown in Fig. 4–30b. Application of the load **P** will cause one of three possibilities, namely, both segments AC and CB remain elastic, AC is plastic while CB is elastic, or both AC and CB are plastic.*

An *elastic analysis*, similar to that discussed in Sec. 4.4, will produce $F_A = 45$ kN and $F_B = 15$ kN at the supports. However, this results in a stress of

$$\sigma_{AC} = \frac{45 \text{ kN}}{\pi (0.005 \text{ m})^2} = 573 \text{ MPa (compression)} > \sigma_Y = 420 \text{ MPa}$$

$$\sigma_{CB} = \frac{15 \text{ kN}}{\pi (0.005 \text{ m})^2} = 191 \text{ MPa (tension)}$$

Since the material in segment AC will yield, we will assume that AC becomes plastic, while CB remains elastic.

For this case, the maximum possible force developed in AC is

$$(F_A)_Y = \sigma_Y A = 420(10^3) \text{ kN/m}^2 [\pi (0.005 \text{ m})^2] = 33.0 \text{ kN}$$

and from the equilibrium of the rod, Fig. 4–31b,

$$F_B = 60 \text{ kN} - 33.0 \text{ kN} = 27.0 \text{ kN}$$

The stress in each segment of the rod is therefore

$$\sigma_{AC} = \sigma_Y = 420 \text{ MPa (compression)}$$

$$\sigma_{CB} = \frac{27.0 \text{ kN}}{\pi (0.005 \text{ m})^2} = 344 \text{ MPa (tension)} < 420 \text{ MPa (OK)}$$

*The possibility of CB becoming plastic before AC will not occur because when point C moves, the *strain* in AC (since it is shorter) will always be larger than the strain in CB.

Residual Stress.   In order to obtain the residual stress, it is also necessary to know the strain in each segment due to the loading. Since CB responds elastically,

$$\delta_C = \frac{F_B L_{CB}}{AE} = \frac{(27.0 \text{ kN})(0.300 \text{ m})}{\pi(0.005 \text{ m})^2[70(10^6) \text{ kN/m}^2]} = 0.001474 \text{ m}$$

$$\epsilon_{CB} = \frac{\delta_C}{L_{CB}} = \frac{0.001474 \text{ m}}{0.300 \text{ m}} = +0.004913$$

$$\epsilon_{AC} = \frac{\delta_C}{L_{AC}} = -\frac{0.001474 \text{ m}}{0.100 \text{ m}} = -0.01474$$

Here the yield strain is

$$\epsilon_Y = \frac{\sigma_Y}{E} = \frac{420(10^6) \text{ N/m}^2}{70(10^9) \text{ N/m}^2} = 0.006$$

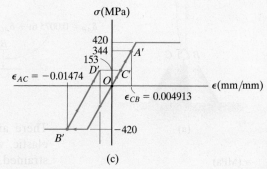

Fig. 4–30 (cont.)

Therefore, when **P** is *applied*, the stress–strain behavior for the material in segment CB moves from O to A', Fig. 4–30c, and the stress–strain behavior for the material in segment AC moves from O to B'. If the load **P** is applied in the *reverse direction*, in other words, the load is removed, then an elastic response occurs and a reverse force of $F_A = 45$ kN and $F_B = 15$ kN must be applied to each segment. As calculated previously, these forces now produce stresses $\sigma_{AC} = 573$ MPa (tension) and $\sigma_{CB} = 191$ MPa (compression), and as a result the residual stress in each member is

$$(\sigma_{AC})_r = -420 \text{ MPa} + 573 \text{ MPa} = 153 \text{ MPa} \qquad Ans.$$

$$(\sigma_{CB})_r = 344 \text{ MPa} - 191 \text{ MPa} = 153 \text{ MPa} \qquad Ans.$$

This residual stress is the *same* for both segments, which is to be expected. Also note that the stress–strain behavior for segment AC moves from B' to D' in Fig. 4–30c, while the stress–strain behavior for the material in segment CB moves from A' to C' when the load is removed.

EXAMPLE | **4.15**

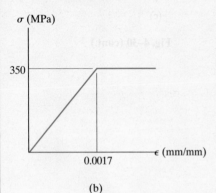

(a)

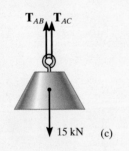

(b)

T$_{AB}$ **T**$_{AC}$

15 kN   (c)

Fig. 4–31

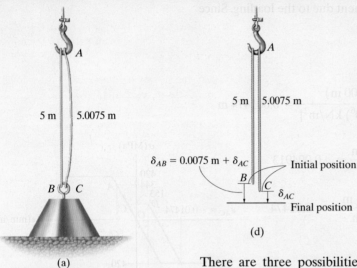

$\delta_{AB} = 0.0075\ \text{m} + \delta_{AC}$ — Initial position

δ_{AC}

Final position

(d)

Two steel wires are used to lift the weight of 15 kN ($\approx$ 1500 kg), Fig. 4–31a. Wire AB has an unstretched length of 5 m and wire AC has an unstretched length of 5.0075 m. If each wire has a cross-sectional area of 30 mm^2, and the steel can be considered elastic perfectly plastic as shown by the σ–ϵ graph in Fig. 4–31b, determine the force in each wire and its elongation.

SOLUTION

Once the weight is supported by both wires, then the stress in the wires depends on the corresponding strain. There are three possibilities, namely, the strains in both wires are elastic, wire AB is plastically strained while wire AC is elastically strained, or both wires are plastically strained. We will assume that AC remains *elastic* and AB is plastically strained.

Investigation of the free-body diagram of the suspended weight, Fig. 4–31c, indicates that the problem is statically indeterminate. The equation of equilibrium is

$$+\uparrow\Sigma F_y = 0; \qquad T_{AB} + T_{AC} - 15\ \text{kN} = 0 \qquad (1)$$

Since AB becomes plastically strained then it must support its maximum load.

$$T_{AB} + \sigma_Y A_{AB} = 350\ \text{N/mm}^2(30\ \text{mm}^2) = 10500\ \text{N} = 10.5\ \text{kN} \quad Ans.$$

Therefore, from Eq. 1,

$$T_{AC} = 4.5\ \text{kN} \qquad\qquad Ans.$$

Note that wire AC remains elastic as assumed since the stress in the wire is $\sigma_{AC} = 4.5\ \text{kN}/30\ \text{mm}^2 = 150\ \text{MPa} < 350\ \text{MPa}$. The corresponding elastic strain is determined by proportion, Fig. 4–31b; i.e.,

$$\frac{\epsilon_{AC}}{150\ \text{MPa}} = \frac{0.0017}{350\ \text{MPa}}$$

$$\epsilon_{AC} = 0.000729$$

The elongation of AC is thus

$$\delta_{AC} = (0.000729)(5.0075\ \text{m}) = 0.00365\ \text{m} \qquad Ans.$$

And from Fig. 4–31d, the elongation of AB is then

$$\delta_{AB} = 0.0075\ \text{m} + 0.00365\ \text{m} = 0.01115\ \text{m} \qquad Ans.$$

PROBLEMS

4–87. Determine the maximum normal stress developed in the bar when it is subjected to a tension of $P = 8$ kN.

*4–88. If the allowable normal stress for the bar is $\sigma_{allow} = 120$ MPa, determine the maximum axial force P that can be applied to the bar.

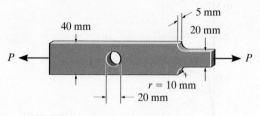

Probs. 4–87/88

•4–89. The member is to be made from a steel plate that is 6 mm thick. If a 25-mm hole is drilled through its center, determine the approximate width w of the plate so that it can support an axial force of 16.75 kN. The allowable stress is $\sigma_{allow} = 150$ MPa.

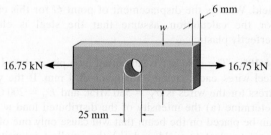

Prob. 4–89

4–90. The A-36 steel plate has a thickness of 12 mm. If there are shoulder fillets at B and C, and $\sigma_{allow} = 150$ MPa, determine the maximum axial load P that it can support. Calculate its elongation, neglecting the effect of the fillets.

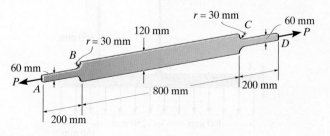

Prob. 4–90

4–91. Determine the maximum axial force P that can be applied to the bar. The bar is made from steel and has an allowable stress of $\sigma_{allow} = 147$ MPa.

*4–92. Determine the maximum normal stress developed in the bar when it is subjected to a tension of $P = 8$ kN.

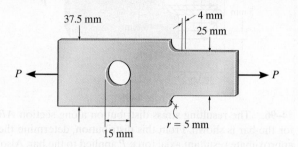

Probs. 4–91/92

•4–93. Determine the maximum normal stress developed in the bar when it is subjected to a tension of $P = 8$ kN.

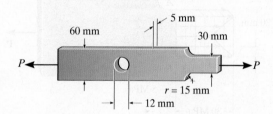

Prob. 4–93

4–94. The resulting stress distribution along section AB for the bar is shown. From this distribution, determine the approximate resultant axial force P applied to the bar. Also, what is the stress-concentration factor for this geometry?

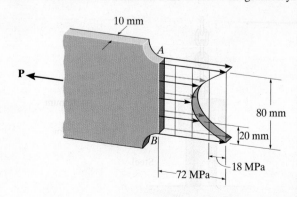

Prob. 4–94

4–95. The resulting stress distribution along section AB for the bar is shown. From this distribution, determine the approximate resultant axial force P applied to the bar. Also, what is the stress-concentration factor for this geometry?

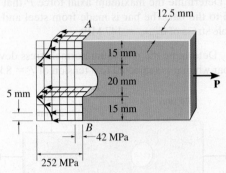

Prob. 4–95

*4–96. The resulting stress distribution along section AB for the bar is shown. From this distribution, determine the approximate resultant axial force P applied to the bar. Also, what is the stress-concentration factor for this geometry?

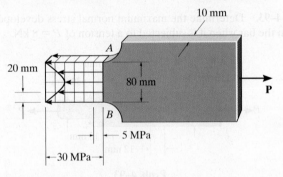

Prob. 4–96

•4–97. The 1500-kN weight is slowly set on the top of a post made of 2014-T6 aluminum with an A-36 steel core. If both materials can be considered elastic perfectly plastic, determine the stress in each material.

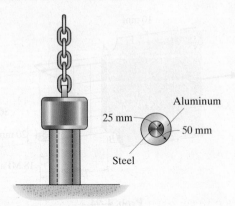

Prob. 4–97

4–98. The bar has a cross-sectional area of 300 mm² and is made of a material that has a stress–strain diagram that can be approximated by the two line segments shown. Determine the elongation of the bar due to the applied loading.

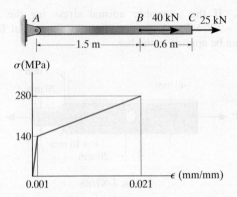

Prob. 4–98

4–99. The rigid bar is supported by a pin at A and two steel wires, each having a diameter of 4 mm. If the yield stress for the wires is $\sigma_Y = 530$ MPa, and $E_{st} = 200$ GPa, determine the intensity of the distributed load w that can be placed on the beam and will just cause wire EB to yield. What is the displacement of point G for this case? For the calculation, assume that the steel is elastic perfectly plastic.

*4–100. The rigid bar is supported by a pin at A and two steel wires, each having a diameter of 4 mm. If the yield stress for the wires is $\sigma_Y = 530$ MPa, and $E_{st} = 200$ GPa, determine (a) the intensity of the distributed load w that can be placed on the beam that will cause only one of the wires to start to yield and (b) the smallest intensity of the distributed load that will cause both wires to yield. For the calculation, assume that the steel is elastic perfectly plastic.

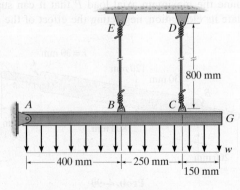

Probs. 4–99/100

•4–101. The rigid lever arm is supported by two A-36 steel wires having the same diameter of 4 mm. If a force of P = 3 kN is applied to the handle, determine the force developed in both wires and their corresponding elongations. Consider A-36 steel as an elastic-perfectly plastic material.

4–102. The rigid lever arm is supported by two A-36 steel wires having the same diameter of 4 mm. Determine the smallest force **P** that will cause (a) only one of the wires to yield; (b) both wires to yield. Consider A-36 steel as an elastic-perfectly plastic material.

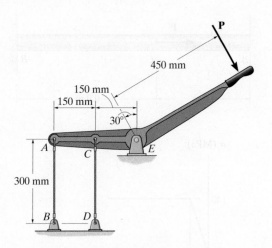

Probs. 4–101/102

4–103. The three bars are pinned together and subjected to the load **P**. If each bar has a cross-sectional area A, length L, and is made from an elastic perfectly plastic material, for which the yield stress is σ_Y, determine the largest load (ultimate load) that can be supported by the bars, i.e., the load P that causes all the bars to yield. Also, what is the horizontal displacement of point A when the load reaches its ultimate value? The modulus of elasticity is E.

*4–104. The rigid beam is supported by three 25-mm diameter A-36 steel rods. If the beam supports the force of P = 230 kN, determine the force developed in each rod. Consider the steel to be an elastic perfectly-plastic material.

•4–105. The rigid beam is supported by three 25-mm diameter A-36 steel rods. If the force of P = 230 kN is applied on the beam and removed, determine the residual stresses in each rod. Consider the steel to be an elastic perfectly-plastic material.

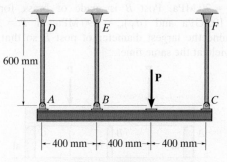

Probs. 4–104/105

4–106. The distributed loading is applied to the rigid beam, which is supported by the three bars. Each bar has a cross-sectional area of 780 mm² and is made from a material having a stress–strain diagram that can be approximated by the two line segments shown. If a load of w = 400 kN/m is applied to the beam, determine the stress in each bar and the vertical displacement of the beam.

4–107. The distributed loading is applied to the rigid beam, which is supported by the three bars. Each bar has a cross-sectional area of 468 mm² and is made from a material having a stress–strain diagram that can be approximated by the two line segments shown. Determine the intensity of the distributed loading w needed to cause the beam to be displaced downward 37.5 mm.

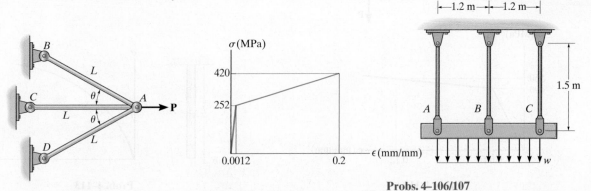

Probs. 4–106/107

Prob. 4–103

*4–108. The rigid beam is supported by the three posts A, B, and C of equal length. Posts A and C have a diameter of 75 mm and are made of aluminum, for which $E_{al} = 70$ GPa and $(\sigma_Y)_{al} = 20$ MPa. Post B has a diameter of 20 mm and is made of brass, for which $E_{br} = 100$ GPa and $(\sigma_Y)_{br} = 590$ MPa. Determine the smallest magnitude of **P** so that (a) only rods A and C yield and (b) all the posts yield.

•4–109. The rigid beam is supported by the three posts A, B, and C. Posts A and C have a diameter of 60 mm and are made of aluminum, for which $E_{al} = 70$ GPa and $(\sigma_Y)_{al} = 20$ MPa. Post B is made of brass, for which $E_{br} = 100$ GPa and $(\sigma_Y)_{br} = 590$ MPa. If $P = 130$ kN, determine the largest diameter of post B so that all the posts yield at the same time.

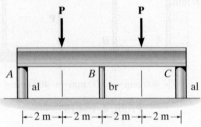

Probs. 4–108/109

4–110. The wire BC has a diameter of 3.4 mm and the material has the stress–strain characteristics shown in the figure. Determine the vertical displacement of the handle at D if the pull at the grip is slowly increased and reaches a magnitude of (a) $P = 2250$ N, (b) $P = 3000$ N.

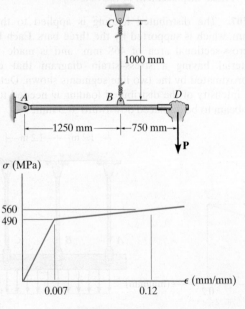

Prob. 4–110

4–111. The bar having a diameter of 50 mm is fixed connected at its ends and supports the axial load **P**. If the material is elastic perfectly plastic as shown by the stress–strain diagram, determine the smallest load P needed to cause segment CB to yield. If this load is released, determine the permanent displacement of point C.

*4–112. Determine the elongation of the bar in Prob. 4–111 when both the load **P** and the supports are removed.

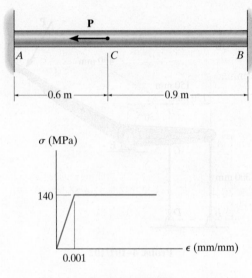

Probs. 4–111/112

•4–113. A material has a stress–strain diagram that can be described by the curve $\sigma = c\epsilon^{1/2}$. Determine the deflection δ of the end of a rod made from this material if it has a length L, cross-sectional area A, and a specific weight γ.

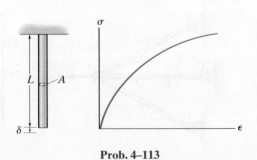

Prob. 4–113

CHAPTER REVIEW

When a loading is applied at a point on a body, it tends to create a stress distribution within the body that becomes more uniformly distributed at regions removed from the point of application of the load. This is called Saint-Venant's principle.

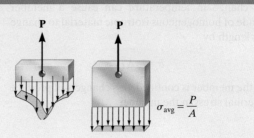

$$\sigma_{avg} = \frac{P}{A}$$

Ref.: Section 4.1

The relative displacement at the end of an axially loaded member relative to the other end is determined from

$$\delta = \int_0^L \frac{P(x)\,dx}{AE}$$

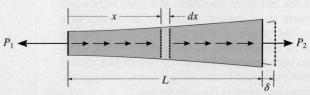

Ref.: Section 4.2

If a series of concentrated external axial forces are applied to a member and AE is also constant for the member, then

$$\delta = \Sigma \frac{PL}{AE}$$

For application, it is necessary to use a sign convention for the internal load P and displacement δ. We considered tension and elongation as positive values. Also, the material must not yield, but rather it must remain linear elastic.

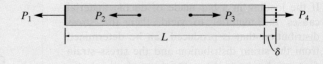

Ref.: Section 4.2

Superposition of load and displacement is possible provided the material remains linear elastic and no significant changes in the geometry of the member occur after loading.

Ref.: Section 4.3

The reactions on a statically indeterminate bar can be determined using the equilibrium equations and compatibility conditions that specify the displacement at the supports. These displacements are related to the loads using a load–displacement relationship such as $\delta = PL/AE$.

Ref.: Sections 4.4 & 4.5

A change in temperature can cause a member made of homogeneous isotropic material to change its length by

$$\delta = \alpha \Delta T L$$

If the member is confined, this change will produce thermal stress in the member.

Ref.: Section 4.6

Holes and sharp transitions at a cross section will create stress concentrations. For the design of a member made of brittle material one obtains the stress concentration factor K from a graph, which has been determined from experiment. This value is then multiplied by the average stress to obtain the maximum stress at the cross section.

$$\sigma_{\max} = K\sigma_{\text{avg}}$$

Ref.: Section 4.7

If the loading in a bar made of ductile material causes the material to yield, then the stress distribution that is produced can be determined from the strain distribution and the stress–strain diagram. Assuming the material is perfectly plastic, yielding will cause the stress distribution at the cross section of a hole or transition to even out and become uniform.

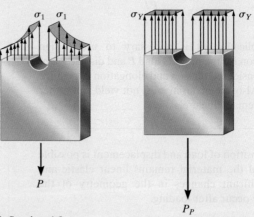

Ref.: Section 4.8

If a member is constrained and an external loading causes yielding, then when the load is released, it will cause residual stress in the member.

Ref.: Section 4.9

CONCEPTUAL PROBLEMS

P4–1

P4–2

P4–1. The concrete footing A was poured when this column was put in place. Later the rest of the foundation slab was poured. Can you explain why the 45° cracks occurred at each corner? Can you think of a better design that would avoid such cracks?

P4–2. The row of bricks, along with mortar and an internal steel reinforcing rod, was intended to serve as a lintel beam to support the bricks above this ventilation opening on an exterior wall of a building. Explain what may have caused the bricks to fail in the manner shown.

REVIEW PROBLEMS

4–114. The 2014-T6 aluminum rod has a diameter of 12 mm and is lightly attached to the rigid supports at A and B when $T_1 = 25°C$. If the temperature becomes $T_2 = -20°C$, and an axial force of $P = 80$ N is applied to the rigid collar as shown, determine the reactions at A and B.

4–115. The 2014-T6 aluminum rod has a diameter of 12 mm and is lightly attached to the rigid supports at A and B when $T_1 = 40°C$. Determine the force P that must be applied to the collar so that, when $T = 0°C$, the reaction at B is zero.

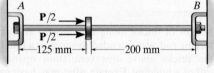
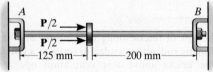

Probs. 114/115

*4–116. The rods each have the same 25-mm diameter and 600-mm length. If they are made of A-36 steel, determine the forces developed in each rod when the temperature increases to 50°C.

•4–117. Two A-36 steel pipes, each having a cross-sectional area of 200 mm², are screwed together using a union at B as shown. Originally the assembly is adjusted so that no load is on the pipe. If the union is then tightened so that its screw, having a lead of 3.75 mm, undergoes two full turns, determine the average normal stress developed in the pipe. Assume that the union at B and couplings at A and C are rigid. Neglect the size of the union. *Note:* The lead would cause the pipe, when *unloaded*, to shorten 3.75 mm when the union is rotated one revolution.

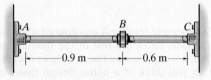

Prob. 4–117

4–118. The brass plug is force-fitted into the rigid casting. The uniform normal bearing pressure on the plug is estimated to be 15 MPa. If the coefficient of static friction between the plug and casting is $\mu_s = 0.3$, determine the axial force P needed to pull the plug out. Also, calculate the displacement of end B relative to end A just before the plug starts to slip out. $E_{br} = 98$ GPa.

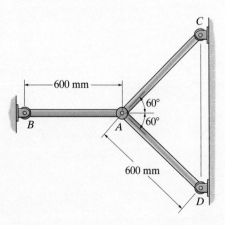

Prob. 4–116

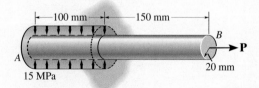

Prob. 4–118

4–119. The assembly consists of two bars AB and CD of the same material having a modulus of elasticity E_1 and coefficient of thermal expansion α_1, and a bar EF having a modulus of elasticity E_2 and coefficient of thermal expansion α_2. All the bars have the same length L and cross-sectional area A. If the rigid beam is originally horizontal at temperature T_1, determine the angle it makes with the horizontal when the temperature is increased to T_2.

*4–120. The rigid link is supported by a pin at A and two A-36 steel wires, each having an unstretched length of 300 mm and cross-sectional area of 7.8 mm^2. Determine the force developed in the wires when the link supports the vertical load of 1.75 kN.

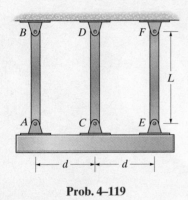

Prob. 4–119

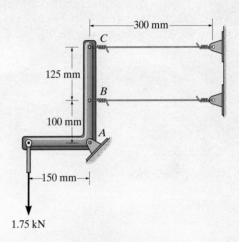

Prob. 4–120

The torsional stress and angle of twist of this soil auger depend upon the output of the machine turning the bit as well as the resistance of the soil in contact with the shaft.

Torsion

5

CHAPTER OBJECTIVES

In this chapter we will discuss the effects of applying a torsional loading to a long straight member such as a shaft or tube. Initially we will consider the member to have a circular cross section. We will show how to determine both the stress distribution within the member and the angle of twist when the material behaves in a linear elastic manner and also when it is inelastic. Statically indeterminate analysis of shafts and tubes will also be discussed, along with special topics that include those members having noncircular cross sections. Lastly, stress concentrations and residual stress caused by torsional loadings will be given special consideration.

Check out the Companion Website for the following video solution(s) that can be found in this chapter.

- Section 2
  Shear Stress in Torsion
- Section 2
  Allowable Torque
- Section 5
  Static Indeterminacy in Torsion
- Section 7
  Allowable Stress for Thin-Walled Tubes
- Section 9
  Inelastic Torsion

5.1 Torsional Deformation of a Circular Shaft

Torque is a moment that tends to twist a member about its longitudinal axis. Its effect is of primary concern in the design of axles or drive shafts used in vehicles and machinery. We can illustrate physically what happens when a torque is applied to a circular shaft by considering the shaft to be made of a highly deformable material such as rubber, Fig. 5–1*a*. When the torque is applied, the circles and longitudinal grid lines originally marked on the shaft tend to distort into the pattern shown in Fig. 5–1*b*. Note that twisting causes the circles to *remain circles*, and each longitudinal grid line deforms into a helix that intersects the circles at equal angles. Also, the cross sections from the ends along the shaft will remain *flat*—that is, they do not warp or bulge in or out—and radial lines *remain straight* during the deformation, Fig. 5–1*b*. From these observations we can assume that if the angle of twist is *small*, the *length of the shaft* and its *radius* will *remain unchanged*.

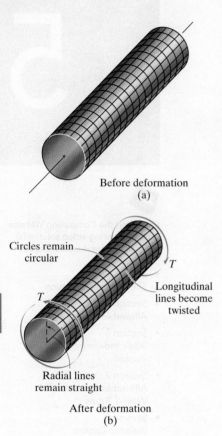

Before deformation
(a)

Circles remain
circular

T

T

Longitudinal
lines become
twisted

Radial lines
remain straight

After deformation
(b)

Fig. 5–1

If the shaft is fixed at one end and a torque is applied to its other end, the dark green shaded plane in Fig. 5–2 will distort into a skewed form as shown. Here a radial line located on the cross section at a distance x from the fixed end of the shaft will rotate through an angle $\phi(x)$. The angle $\phi(x)$, so defined, is called the *angle of twist*. It depends on the position x and will vary along the shaft as shown.

In order to understand how this distortion strains the material, we will now isolate a small element located at a radial distance ρ (rho) from the axis of the shaft, Fig. 5–3. Due to the deformation as noted in Fig. 5–2, the front and rear faces of the element will undergo a rotation—the back face by $\phi(x)$, and the front face by $\phi(x) + \Delta\phi$. As a result, the *difference* in these rotations, $\Delta\phi$, causes the element to be subjected to a *shear strain*. To calculate this strain, note that before deformation the angle between the edges AB and AC is 90°; after deformation, however, the edges of the element are AD and AC and the angle between them is θ'. From the definition of shear strain, Eq. 2–4, we have

$$\gamma = \frac{\pi}{2} - \theta'$$

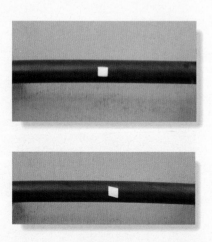

Notice the deformation of the rectangular element when this rubber bar is subjected to a torque.

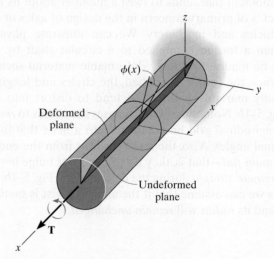

$\phi(x)$

y

x

Deformed
plane

Undeformed
plane

T

x

The angle of twist $\phi(x)$ increases as x increases.

Fig. 5–2

This angle, γ, is indicated on the element. It can be related to the length Δx of the element and the angle $\Delta\phi$ between the shaded planes by considering the length of arc BD, that is

$$BD = \rho\Delta\phi = \Delta x\,\gamma$$

Therefore, if we let $\Delta x \rightarrow dx$ and $\Delta\phi \rightarrow d\phi$,

$$\gamma = \rho\frac{d\phi}{dx} \qquad (5\text{–}1)$$

Since dx and $d\phi$ are the *same* for *all elements* located at points on the cross section at x, then $d\phi/dx$ is constant over the cross section, and Eq. 5–1 states that the magnitude of the shear strain for any of these elements varies only with its radial distance ρ from the axis of the shaft. In other words, the shear strain within the shaft varies linearly along any radial line, from zero at the axis of the shaft to a maximum γ_{max} at its outer boundary, Fig. 5–4. Since $d\phi/dx = \gamma/\rho = \gamma_{max}/c$, then

$$\gamma = \left(\frac{\rho}{c}\right)\gamma_{max} \qquad (5\text{–}2)$$

The results obtained here are also valid for circular tubes. They depend only on the assumptions regarding the deformations mentioned above.

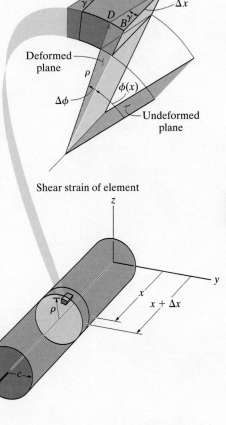

Fig. 5–3

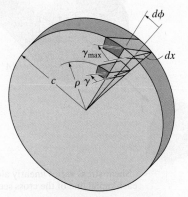

The shear strain at points on the cross section increases linearly with ρ, i.e., $\gamma = (\rho/c)\gamma_{max}$.

Fig. 5–4

5

5.2   The Torsion Formula

When an external torque is applied to a shaft it creates a corresponding internal torque within the shaft. In this section, we will develop an equation that relates this internal torque to the shear stress distribution on the cross section of a circular shaft or tube.

If the material is linear-elastic, then Hooke's law applies, $\tau = G\gamma$, and consequently a *linear variation in shear strain*, as noted in the previous section, leads to a corresponding *linear variation in shear stress* along any radial line on the cross section. Hence, τ will vary from zero at the shaft's longitudinal axis to a maximum value, τ_{max}, at its outer surface. This variation is shown in Fig. 5–5 on the front faces of a selected number of elements, located at an intermediate radial position ρ and at the outer radius c. Due to the proportionality of triangles, we can write

$$\tau = \left(\frac{\rho}{c}\right)\tau_{max} \tag{5–3}$$

This equation expresses the shear-stress distribution over the cross section in terms of the radial position ρ of the element. Using it, we can now apply the condition that requires the torque produced by the stress distribution over the entire cross section to be equivalent to the resultant internal torque T at the section, which holds the shaft in equilibrium, Fig. 5–5.

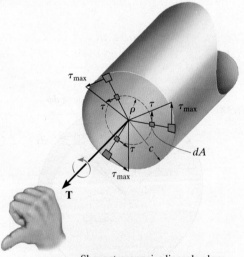

Shear stress varies linearly along each radial line of the cross section.

Fig. 5–5

Specifically, each element of area dA, located at ρ, is subjected to a force of $dF = \tau\, dA$. The torque produced by this force is $dT = \rho(\tau\, dA)$. We therefore have for the entire cross section

$$T = \int_A \rho(\tau\, dA) = \int_A \rho\left(\frac{\rho}{c}\right)\tau_{max}\, dA \qquad (5\text{–}4)$$

Since τ_{max}/c is constant,

$$T = \frac{\tau_{max}}{c} \int_A \rho^2\, dA \qquad (5\text{–}5)$$

The integral depends only on the geometry of the shaft. It represents the *polar moment of inertia* of the shaft's cross-sectional area about the shaft's longitudinal axis. We will symbolize its value as J, and therefore the above equation can be rearranged and written in a more compact form, namely,

$$\boxed{\tau_{max} = \frac{Tc}{J}} \qquad (5\text{–}6)$$

Here

τ_{max} = the maximum shear stress in the shaft, which occurs at the outer surface

T = the resultant *internal torque* acting at the cross section. Its value is determined from the method of sections and the equation of moment equilibrium applied about the shaft's longitudinal axis

J = the polar moment of inertia of the cross-sectional area

c = the outer radius of the shaft

Combining Eqs. 5–3 and 5–6, the shear stress at the intermediate distance ρ can be determined from

$$\boxed{\tau = \frac{T\rho}{J}} \qquad (5\text{–}7)$$

Either of the above two equations is often referred to as the *torsion formula*. Recall that it is used only if the shaft is circular and the material is homogeneous and behaves in a linear elastic manner, since the derivation is based on Hooke's law.

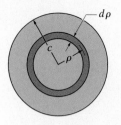

Fig. 5–6

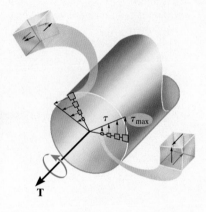

T

(a)

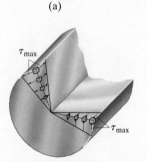

Shear stress varies linearly along
each radial line of the cross section.

(b)

Fig. 5–7

Solid Shaft. If the shaft has a solid circular cross section, the polar moment of inertia J can be determined using an area element in the form of a *differential ring* or annulus having a thickness $d\rho$ and circumference $2\pi\rho$, Fig. 5–6. For this ring, $dA = 2\pi\rho \, d\rho$, and so

$$J = \int_A \rho^2 \, dA = \int_0^c \rho^2(2\pi\rho \, d\rho) = 2\pi \int_0^c \rho^3 \, d\rho = 2\pi\left(\frac{1}{4}\right)\rho^4 \Big|_0^c$$

$$\boxed{J = \frac{\pi}{2}c^4} \tag{5–8}$$

Note that J is a *geometric property* of the circular area and is always positive. Common units used for its measurement are mm^4.

The shear stress has been shown to vary linearly along each radial line of the cross section of the shaft. However, if an element of material on the cross section is isolated, then due to the complementary property of shear, equal shear stresses must also act on four of its adjacent faces as shown in Fig. 5–7a. Hence, ***not only does the internal torque T develop a linear distribution of shear stress along each radial line in the plane of the cross-sectional area, but also an associated shear-stress distribution is developed along an axial plane***, Fig. 5–7b. It is interesting to note that because of this axial distribution of shear stress, shafts made from wood tend to *split* along the axial plane when subjected to excessive torque, Fig. 5–8. This is because wood is an anisotropic material. Its shear resistance parallel to its grains or fibers, directed along the axis of the shaft, is much less than its resistance perpendicular to the fibers, directed in the plane of the cross section.

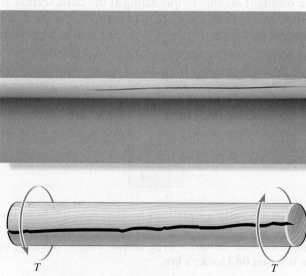

Failure of a wooden shaft due to torsion.

Fig. 5–8

Tubular Shaft.

If a shaft has a tubular cross section, with inner radius c_i and outer radius c_o, then from Eq. 5–8 we can determine its polar moment of inertia by subtracting J for a shaft of radius c_i from that determined for a shaft of radius c_o. The result is

$$J = \frac{\pi}{2}(c_o^4 - c_i^4) \qquad (5\text{–}9)$$

Like the solid shaft, the shear stress distributed over the tube's cross-sectional area varies linearly along any radial line, Fig. 5–9a. Furthermore, the shear stress varies along an axial plane in this same manner, Fig. 5–9b.

This tubular drive shaft for a truck was subjected to an excessive torque, resulting in failure caused by yielding of the material.

Absolute Maximum Torsional Stress.

If the absolute maximum torsional stress is to be determined, then it becomes important to find the location where the ratio Tc/J is a maximum. In this regard, it may be helpful to show the variation of the internal torque T at each section along the axis of the shaft by drawing a ***torque diagram***, which is a plot of the internal torque T versus its position x along the shaft's length. As a sign convention, T will be positive if by the right-hand rule the thumb is directed outward from the shaft when the fingers curl in the direction of twist as caused by the torque, Fig. 5–5. Once the internal torque throughout the shaft is determined, the maximum ratio of Tc/J can then be identified.

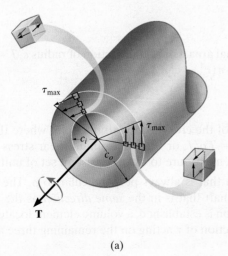

(a)

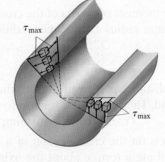

Shear stress varies linearly along each radial line of the cross section.

(b)

Fig. 5–9

Important Points

- When a shaft having a *circular cross section* is subjected to a torque, the cross section *remains plane* while radial lines rotate. This causes a *shear strain* within the material that *varies linearly* along any radial line, from zero at the axis of the shaft to a maximum at its outer boundary.

- For linear elastic homogeneous material the *shear stress* along any radial line of the shaft also *varies linearly*, from zero at its axis to a maximum at its outer boundary. This maximum shear stress *must not* exceed the proportional limit.

- Due to the complementary property of shear, the linear shear stress distribution within the plane of the cross section is also distributed along an adjacent axial plane of the shaft.

- The torsion formula is based on the requirement that the resultant torque on the cross section is equal to the torque produced by the shear stress distribution about the longitudinal axis of the shaft. It is required that the shaft or tube have a *circular* cross section and that it is made of *homogeneous* material which has *linear-elastic* behavior.

Procedure for Analysis

The torsion formula can be applied using the following procedure.

Internal Loading.

- Section the shaft perpendicular to its axis at the point where the shear stress is to be determined, and use the necessary free-body diagram and equations of equilibrium to obtain the internal torque at the section.

Section Property.

- Calculate the polar moment of inertia of the cross-sectional area. For a solid section of radius c, $J = \pi c^4/2$, and for a tube of outer radius c_o and inner radius c_i, $J = \pi(c_o^4 - c_i^4)/2$.

Shear Stress.

- Specify the radial distance ρ, measured from the center of the cross section to the point where the shear stress is to be found. Then apply the torsion formula $\tau = T\rho/J$, or if the maximum shear stress is to be determined use $\tau_{max} = Tc/J$. When substituting the data, make sure to use a consistent set of units.

- The shear stress acts on the cross section in a direction that is always perpendicular to ρ. The force it creates must contribute a torque about the axis of the shaft that is in the *same direction* as the internal resultant torque **T** acting on the section. Once this direction is established, a volume element located at the point where τ is determined can be isolated, and the direction of τ acting on the remaining three adjacent faces of the element can be shown.

EXAMPLE | 5.1

The *solid* shaft of radius c is subjected to a torque **T**, Fig. 5–10a. Determine the fraction of T that is resisted by the material contained within the outer region of the shaft, which has an inner radius of $c/2$ and outer radius c.

(a)

SOLUTION

The stress in the shaft varies linearly, such that $\tau = (\rho/c)\tau_{max}$, Eq. 5–3. Therefore, the torque dT' on the ring (area) located within the lighter-shaded region, Fig. 5–10b, is

$$dT' = \rho(\tau\, dA) = \rho(\rho/c)\tau_{max}(2\pi\rho\, d\rho)$$

For the entire lighter-shaded area the torque is

$$T' = \frac{2\pi\tau_{max}}{c} \int_{c/2}^{c} \rho^3\, d\rho$$

$$= \frac{2\pi\tau_{max}}{c} \frac{1}{4}\rho^4 \Big|_{c/2}^{c}$$

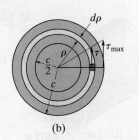

(b)

Fig. 5–10

So that

$$T' = \frac{15\pi}{32}\tau_{max}c^3 \qquad\qquad (1)$$

   This torque T' can be expressed in terms of the applied torque T by first using the torsion formula to determine the maximum stress in the shaft. We have

$$\tau_{max} = \frac{Tc}{J} = \frac{Tc}{(\pi/2)c^4}$$

or

$$\tau_{max} = \frac{2T}{\pi c^3}$$

Substituting this into Eq. 1 yields

$$T' = \frac{15}{16}T \qquad\qquad Ans.$$

NOTE: Here, approximately 94% of the torque is resisted by the lighter-shaded region, and the remaining 6% (or $\frac{1}{16}$) of T is resisted by the inner "core" of the shaft, $\rho = 0$ to $\rho = c/2$. As a result, the material located at the *outer region* of the shaft is highly effective in resisting torque, which justifies the use of tubular shafts as an efficient means for transmitting torque, and thereby saving material.

EXAMPLE 5.2

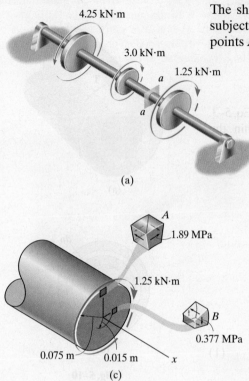

The shaft shown in Fig. 5–11a is supported by two bearings and is subjected to three torques. Determine the shear stress developed at points A and B, located at section a–a of the shaft, Fig. 5–11c.

(a)

(b)

1.89 MPa

1.25 kN·m

B

0.377 MPa

0.075 m   0.015 m   x

(c)

Fig. 5–11

SOLUTION

Internal Torque. The bearing reactions on the shaft are zero, provided the shaft's weight is neglected. Furthermore, the applied torques satisfy moment equilibrium about the shaft's axis.

The internal torque at section a–a will be determined from the free-body diagram of the left segment, Fig. 5–11b. We have

$$\Sigma M_x = 0; \quad 4.25 \text{ kN} \cdot \text{m} - 3.0 \text{ kN} \cdot \text{m} - T = 0 \quad T = 1.25 \text{ kN} \cdot \text{m}$$

Section Property. The polar moment of inertia for the shaft is

$$J = \frac{\pi}{2}(0.075 \text{ m})^4 = 4.97(10^{-5})\text{m}^4$$

Shear Stress. Since point A is at $\rho = c = 0.075$ m,

$$\tau_A = \frac{Tc}{J} = \frac{(1.25 \text{ kN} \cdot \text{m})(0.075 \text{ m})}{4.97(10^{-5})\text{m}^4} = 1886 \text{ kPa} = 1.89 \text{ MPa} \quad Ans.$$

Likewise for point B, at $\rho = 0.015$ m, we have

$$\tau_B = \frac{T\rho}{J} = \frac{(1.25 \text{ kN} \cdot \text{m})(0.015 \text{ m})}{4.97(10^{-5})\text{m}^4} = 377.3 \text{ kPa} = 0.377 \text{ MPa} \quad Ans.$$

NOTE: The directions of these stresses on each element at A and B, Fig. 5–11c, are established from the direction of the resultant internal torque T, shown in Fig. 5–11b. Note carefully how the shear stress acts on the planes of each of these elements.

EXAMPLE 5.3

The pipe shown in Fig. 5–12a has an inner diameter of 80 mm and an outer diameter of 100 mm. If its end is tightened against the support at A using a torque wrench at B, determine the shear stress developed in the material at the inner and outer walls along the central portion of the pipe when the 80-N forces are applied to the wrench.

SOLUTION

Internal Torque. A section is taken at an intermediate location C along the pipe's axis, Fig. 5–12b. The only unknown at the section is the internal torque $\mathbf{T}$. We require

$$\Sigma M_y = 0; \quad 80\ \text{N}\ (0.3\ \text{m}) + 80\ \text{N}\ (0.2\ \text{m}) - T = 0$$

$$T = 40\ \text{N} \cdot \text{m}$$

Section Property. The polar moment of inertia for the pipe's cross-sectional area is

$$J = \frac{\pi}{2}[(0.05\ \text{m})^4 - (0.04\ \text{m})^4] = 5.796(10^{-6})\ \text{m}^4$$

Shear Stress. For any point lying on the outside surface of the pipe, $\rho = c_o = 0.05\ \text{m}$, we have

$$\tau_o = \frac{Tc_o}{J} = \frac{40\ \text{N} \cdot \text{m}\ (0.05\ \text{m})}{5.796(10^{-6})\ \text{m}^4} = 0.345\ \text{MPa} \qquad \textit{Ans.}$$

And for any point located on the inside surface, $\rho = c_i = 0.04\ \text{m}$, so that

$$\tau_i = \frac{Tc_i}{J} = \frac{40\ \text{N} \cdot \text{m}\ (0.04\ \text{m})}{5.796(10^{-6})\ \text{m}^4} = 0.276\ \text{MPa} \qquad \textit{Ans.}$$

NOTE: To show how these stresses act at representative points D and E on the cross-section, we will first view the cross section from the front of segment CA of the pipe, Fig. 5–12a. On this section, Fig. 5–12c, the resultant internal torque is equal but opposite to that shown in Fig. 5–12b. The shear stresses at D and E contribute to this torque and therefore act on the shaded faces of the elements in the directions shown. As a consequence, notice how the shear-stress components act on the other three faces. Furthermore, since the top face of D and the inner face of E are in stress-free regions taken from the pipe's outer and inner walls, no shear stress can exist on these faces or on the other corresponding faces of the elements.

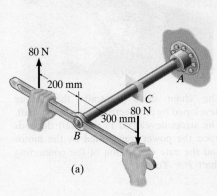

(a)

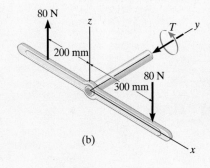

(b)

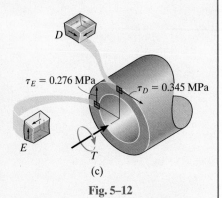

(c)

Fig. 5–12

The chain drive transmits the torque developed by the electric motor to the shaft. The stress developed in the shaft depends upon the power transmitted by the motor and the rate of rotation of the connecting shaft. $P = T\omega$.

5.3 Power Transmission

Shafts and tubes having circular cross sections are often used to transmit power developed by a machine. When used for this purpose, they are subjected to a torque that depends on the power generated by the machine and the angular speed of the shaft. *Power* is defined as the work performed per unit of time. Also, the work transmitted by a rotating shaft equals the torque applied times the angle of rotation. Therefore, if during an instant of time dt an applied torque $\mathbf{T}$ causes the shaft to rotate $d\theta$, then the instantaneous power is

$$P = \frac{T\,d\theta}{dt}$$

Since the shaft's angular velocity is $\omega = d\theta/dt$, we can express the power as

$$\boxed{P = T\omega} \tag{5–10}$$

In the SI system, power is expressed in *watts* when torque is measured in newton-meters ($\text{N}\cdot\text{m}$) and ω is in radians per second (rad/s) ($1\ \text{W} = 1\ \text{N}\cdot\text{m/s}$).

For machinery, the *frequency* of a shaft's rotation, f, is often reported. This is a measure of the number of revolutions or cycles the shaft makes per second and is expressed in hertz (1 Hz = 1 cycle/s). Since 1 cycle = 2π rad, then $\omega = 2\pi f$, and so the above equation for power becomes

$$\boxed{P = 2\pi f T} \tag{5–11}$$

Shaft Design. When the power transmitted by a shaft and its frequency of rotation are known, the torque developed in the shaft can be determined from Eq. 5–11, that is, $T = P/2\pi f$. Knowing T and the allowable shear stress for the material, τ_{allow}, we can determine the size of the shaft's cross section using the torsion formula, provided the material behavior is linear elastic. Specifically, the design or geometric parameter J/c becomes

$$\frac{J}{c} = \frac{T}{\tau_{\text{allow}}} \tag{5–12}$$

For a *solid shaft*, $J = (\pi/2)c^4$, and thus, upon substitution, a *unique value* for the shaft's radius c can be determined. If the shaft is *tubular*, so that $J = (\pi/2)(c_o^4 - c_i^4)$, design permits a wide range of possibilities for the solution. This is because an *arbitrary choice* can be made for either c_o or c_i and the other radius can then be determined from Eq. 5–12.

EXAMPLE 5.4

A solid steel shaft AB shown in Fig. 5–13 is to be used to transmit 3750 W from the motor M to which it is attached. If the shaft rotates at $\omega = 175$ rpm and the steel has an allowable shear stress of $\tau_{allow} = 100$ MPa determine the required diameter of the shaft to the nearest mm.

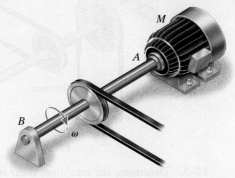

Fig. 5–13

SOLUTION

The torque on the shaft is determined from Eq. 5–10, that is, $P = T\omega$. Expressing P in foot-pounds per second and ω in radians/second, we have

$$P = 3750 \text{ W} = 3750 \text{ N} \cdot \text{m/s}$$

$$\omega = \frac{175 \text{ rev}}{\text{min}}\left(\frac{2\pi \text{ rad}}{1 \text{ rev}}\right)\left(\frac{1 \text{ min}}{60 \text{ s}}\right) = 18.33 \text{ rad/s}$$

Thus,

$$P = T\omega; \qquad 3750 \text{ N} \cdot \text{m/s} = T(18.33 \text{ rad/s})$$

$$T = 204.6 \text{ N} \cdot \text{m}$$

Applying Eq. 5–12 yields

$$\frac{J}{c} = \frac{\pi}{2}\frac{c^4}{c} = \frac{T}{\tau_{allow}}$$

$$c = \left(\frac{2T}{\pi\tau_{allow}}\right)^{1/3} = \left(\frac{2(204.6 \text{ N} \cdot \text{m})(1000 \text{ mm/m})}{\pi(100 \text{ N/mm}^2)}\right)^{1/3}$$

$$c = 10.92 \text{ mm}$$

Since $2c = 21.84$ mm, select a shaft having a diameter of

$$d = 22 \text{ mm} \qquad\qquad Ans.$$

FUNDAMENTAL PROBLEMS

F5–1. The solid circular shaft is subjected to an internal torque of $T = 5$ kN·m. Determine the shear stress developed at points A and B. Represent each state of stress on a volume element.

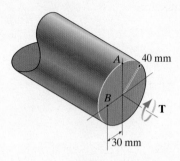

40 mm

30 mm

F5–1

F5–2. The hollow circular shaft is subjected to an internal torque of $T = 10$ kN·m. Determine the shear stress developed at points A and B. Represent each state of stress on a volume element.

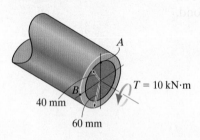

A

B

40 mm

60 mm

$T = 10$ kN·m

F5–2

F5–3. The shaft is hollow from A to B and solid from B to C. Determine the maximum shear stress developed in the shaft. The shaft has an outer diameter of 80 mm, and the thickness of the wall of the hollow segment is 10 mm.

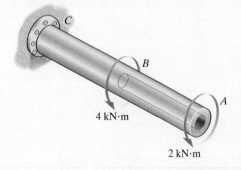

C

B

A

4 kN·m

2 kN·m

F5–3

F5–4. Determine the maximum shear stress developed in the 40-mm diameter shaft.

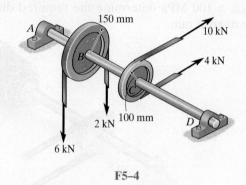

150 mm

A

B

10 kN

4 kN

C

2 kN

100 mm

6 kN

D

F5–4

F5–5. Determine the maximum shear stress developed in the shaft at section a–a.

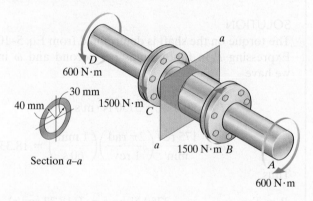

a

D

600 N·m

30 mm

40 mm

1500 N·m

C

1500 N·m B

a

A

Section a–a

600 N·m

F5–5

F5–6. Determine the shear stress developed at point A on the surface of the shaft. Represent the state of stress on a volume element at this point. The shaft has a radius of 40 mm.

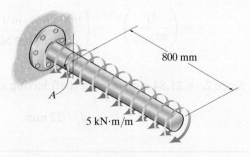

800 mm

A

5 kN·m/m

F5–6

PROBLEMS

•5–1. A shaft is made of a steel alloy having an allowable shear stress of $\tau_{allow} = 84$ MPa. If the diameter of the shaft is 37.5 mm, determine the maximum torque **T** that can be transmitted. What would be the maximum torque **T′** if a 25-mm-diameter hole is bored through the shaft? Sketch the shear-stress distribution along a radial line in each case.

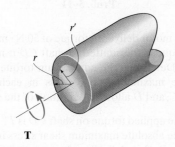

Prob. 5–1

5–2. The solid shaft of radius r is subjected to a torque **T**. Determine the radius r' of the inner core of the shaft that resists one-half of the applied torque ($T/2$). Solve the problem two ways: (a) by using the torsion formula, (b) by finding the resultant of the shear-stress distribution.

Prob. 5–2

5–3. The solid shaft is fixed to the support at C and subjected to the torsional loadings shown. Determine the shear stress at points A and B and sketch the shear stress on volume elements located at these points.

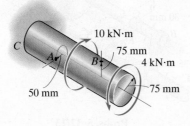

10 kN·m
75 mm
4 kN·m
75 mm
50 mm

Prob. 5–3

*5–4. The tube is subjected to a torque of 750 N·m. Determine the amount of this torque that is resisted by the gray shaded section. Solve the problem two ways: (a) by using the torsion formula, (b) by finding the resultant of the shear-stress distribution.

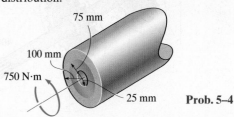

75 mm
100 mm
750 N·m
25 mm

Prob. 5–4

5–5. The copper pipe has an outer diameter of 40 mm and an inner diameter of 37 mm. If it is tightly secured to the wall at A and three torques are applied to it as shown, determine the absolute maximum shear stress developed in the pipe.

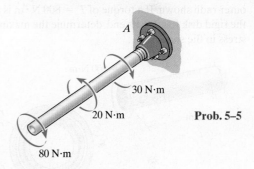

A
30 N·m
20 N·m
80 N·m

Prob. 5–5

5–6. The solid shaft has a diameter of 16 mm. If it is subjected to the torques shown, determine the maximum shear stress developed in regions BC and DE of the shaft. The bearings at A and F allow free rotation of the shaft.

5–7. The solid shaft has a diameter of 16 mm. If it is subjected to the torques shown, determine the maximum shear stress developed in regions CD and EF of the shaft. The bearings at A and F allow free rotation of the shaft.

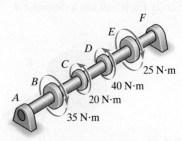

F
E
D
C
25 N·m
B
40 N·m
A
20 N·m
35 N·m

Probs. 5–6/7

*5–8. The solid 30-mm-diameter shaft is used to transmit the torques applied to the gears. Determine the absolute maximum shear stress on the shaft.

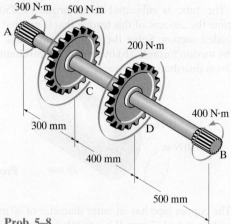

Prob. 5–8

•5–9. The shaft consists of three concentric tubes, each made from the same material and having the inner and outer radii shown. If a torque of $T = 800 \text{ N} \cdot \text{m}$ is applied to the rigid disk fixed to its end, determine the maximum shear stress in the shaft.

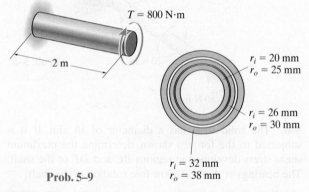

$T = 800 \text{ N·m}$

2 m

$r_i = 20 \text{ mm}$
$r_o = 25 \text{ mm}$

$r_i = 26 \text{ mm}$
$r_o = 30 \text{ mm}$

$r_i = 32 \text{ mm}$
$r_o = 38 \text{ mm}$

Prob. 5–9

5–10. The coupling is used to connect the two shafts together. Assuming that the shear stress in the bolts is *uniform*, determine the number of bolts necessary to make the maximum shear stress in the shaft equal to the shear stress in the bolts. Each bolt has a diameter d.

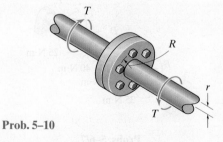

Prob. 5–10

5–11. The assembly consists of two sections of galvanized steel pipe connected together using a reducing coupling at B. The smaller pipe has an outer diameter of 18.75 mm and an inner diameter of 17 mm, whereas the larger pipe has an outer diameter of 25 mm and an inner diameter of 21.5 mm. If the pipe is tightly secured to the wall at C, determine the maximum shear stress developed in each section of the pipe when the couple shown is applied to the handles of the wrench.

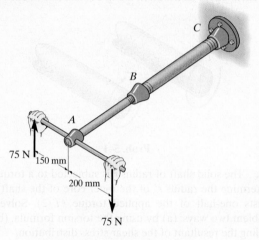

Prob. 5–11

*5–12. The motor delivers a torque of 50 N · m to the shaft AB. This torque is transmitted to shaft CD using the gears at E and F. Determine the equilibrium torque $\mathbf{T}'$ on shaft CD and the maximum shear stress in each shaft. The bearings B, C, and D allow free rotation of the shafts.

•5–13. If the applied torque on shaft CD is $T' = 75 \text{ N} \cdot \text{m}$, determine the absolute maximum shear stress in each shaft. The bearings B, C, and D allow free rotation of the shafts, and the motor holds the shafts fixed from rotating.

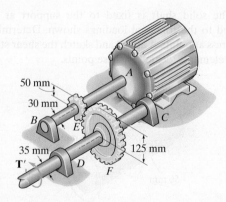

Probs. 5–12/13

5–14. The solid 50-mm-diameter shaft is used to transmit the torques applied to the gears. Determine the absolute maximum shear stress in the shaft.

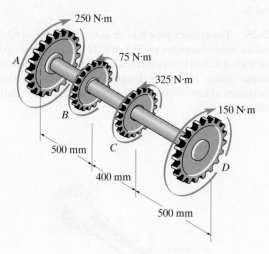

Prob. 5–14

5–15. The solid shaft is made of material that has an allowable shear stress of $\tau_{allow} = 10$ MPa. Determine the required diameter of the shaft to the nearest mm.

*5–16. The solid shaft has a diameter of 40 mm. Determine the absolute maximum shear stress in the shaft and sketch the shear-stress distribution along a radial line of the shaft where the shear stress is maximum.

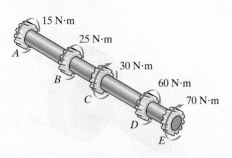

Probs. 5–15/16

•5–17. The rod has a diameter of 25 mm and a weight of 150 N/m. Determine the maximum torsional stress in the rod at a section located at A due to the rod's weight.

5–18. The rod has a diameter of 25 mm and a weight of 225 N/m. Determine the maximum torsional stress in the rod at a section located at B due to the rod's weight.

Probs. 5–17/18

5–19. Two wrenches are used to tighten the pipe. If $P = 300$ N is applied to each wrench, determine the maximum torsional shear stress developed within regions AB and BC. The pipe has an outer diameter of 25 mm and inner diameter of 20 mm. Sketch the shear stress distribution for both cases.

*5–20. Two wrenches are used to tighten the pipe. If the pipe is made from a material having an allowable shear stress of $\tau_{allow} = 85$ MPa, determine the allowable maximum force **P** that can be applied to each wrench. The pipe has an outer diameter of 25 mm and inner diameter of 20 mm.

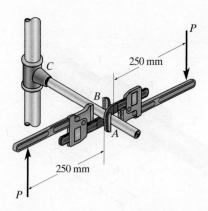

Probs. 5–19/20

•5–21. The 60-mm-diameter solid shaft is subjected to the distributed and concentrated torsional loadings shown. Determine the absolute maximum and minimum shear stresses on the outer surface of the shaft and specify their locations, measured from the fixed end A.

5–22. The solid shaft is subjected to the distributed and concentrated torsional loadings shown. Determine the required diameter d of the shaft to the nearest mm if the allowable shear stress for the material is $\tau_{allow} = 50$ MPa.

2 kN·m/m

1200 N·m

1.5 m

B

0.8 m

C

A

Probs. 5–21/22

5–23. Consider the general problem of a circular shaft made from m segments each having a radius of c_m. If there are n torques on the shaft as shown, write a computer program that can be used to determine the maximum shear stress at any specified location x along the shaft. Show an application of the program using the values $L_1 = 0.6$ m, $c_1 = 50$ mm, $L_2 = 1.2$ m, $c_2 = 25$ mm, $T_1 = 1200$ N · m, $d_1 = 0, T_2 = -900$ N · m, $d_2 = 1.5$ m.

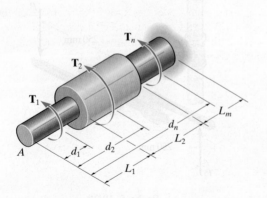

T_n

T_2

T_1

L_m

d_n

L_2

A

d_1

d_2

L_1

Prob. 5–23

*5–24. The copper pipe has an outer diameter of 62.5 mm and an inner diameter of 57.5 mm. If it is tightly secured to the wall at C and a uniformly distributed torque is applied to it as shown, determine the shear stress developed at points A and B. These points lie on the pipe's outer surface. Sketch the shear stress on volume elements located at A and B.

•5–25. The copper pipe has an outer diameter of 62.5 mm and an inner diameter of 57.5 mm. If it is tightly secured to the wall at C and it is subjected to the uniformly distributed torque along its entire length, determine the absolute maximum shear stress in the pipe. Discuss the validity of this result.

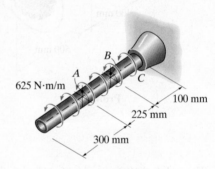

B

A

C

625 N·m/m

100 mm

225 mm

300 mm

Probs. 5–24/25

5–26. A cylindrical spring consists of a rubber annulus bonded to a rigid ring and shaft. If the ring is held fixed and a torque **T** is applied to the shaft, determine the maximum shear stress in the rubber.

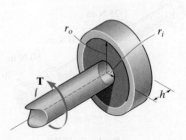

r_o

r_i

T

h

Prob. 5–26

5–27. The A-36 steel shaft is supported on smooth bearings that allow it to rotate freely. If the gears are subjected to the torques shown, determine the maximum shear stress developed in the segments AB and BC. The shaft has a diameter of 40 mm.

*5–28. The A-36 steel shaft is supported on smooth bearings that allow it to rotate freely. If the gears are subjected to the torques shown, determine the required diameter of the shaft to the nearest mm if $\tau_{allover} = 60$ MPa.

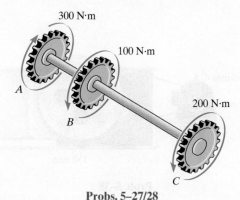

Probs. 5–27/28

•5–29. When drilling a well at constant angular velocity, the bottom end of the drill pipe encounters a torsional resistance T_A. Also, soil along the sides of the pipe creates a distributed frictional torque along its length, varying uniformly from zero at the surface B to t_A at A. Determine the minimum torque T_B that must be supplied by the drive unit to overcome the resisting torques, and compute the maximum shear stress in the pipe. The pipe has an outer radius r_o and an inner radius r_i.

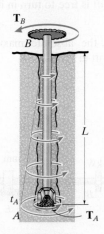

Prob. 5–29

5–30. The shaft is subjected to a distributed torque along its length of $t = (10x^2)$ N·m/m, where x is in meters. If the maximum stress in the shaft is to remain constant at 80 MPa, determine the required variation of the radius c of the shaft for $0 \leq x \leq 3$ m.

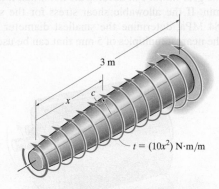

Prob. 5–30

5–31. The solid steel shaft AC has a diameter of 25 mm and is supported by smooth bearings at D and E. It is coupled to a motor at C, which delivers 3 kW of power to the shaft while it is turning at 50 rev/s. If gears A and B remove 1 kW and 2 kW, respectively, determine the maximum shear stress developed in the shaft within regions AB and BC. The shaft is free to turn in its support bearings D and E.

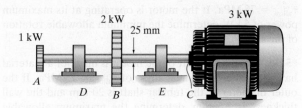

Prob. 5–31

*5–32. The pump operates using the motor that has a power of 85 W. If the impeller at B is turning at 150 rev/min, determine the maximum shear stress developed in the 20-mm-diameter transmission shaft at A.

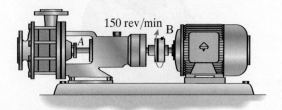

Prob. 5–32

•5–33.   The gear motor can develop 1.6 kW when it turns at 450 rev/min. If the shaft has a diameter of 25 mm, determine the maximum shear stress developed in the shaft.

5–34.   The gear motor can develop 2.4 kW when it turns at 150 rev/min. If the allowable shear stress for the shaft is $\tau_{\text{allow}} = 84$ MPa, determine the smallest diameter of the shaft to the nearest multiples of 5 mm that can be used.

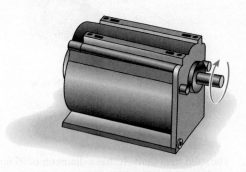

Probs. 5–33/34

5–35.   The 25-mm-diameter shaft on the motor is made of a material having an allowable shear stress of $\tau_{\text{allow}} = 75$ MPa. If the motor is operating at its maximum power of 5 kW, determine the minimum allowable rotation of the shaft.

*5–36.   The drive shaft of the motor is made of a material having an allowable shear stress of $\tau_{\text{allow}} = 75$ MPa. If the outer diameter of the tubular shaft is 20 mm and the wall thickness is 2.5 mm, determine the maximum allowable power that can be supplied to the motor when the shaft is operating at an angular velocity of 1500 rev/min.

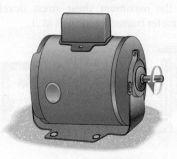

Probs. 5–35/36

•5–37.   A ship has a propeller drive shaft that is turning at 1500 rev/min while developing 1500 kW. If it is 2.4 m long and has a diameter of 100 mm, determine the maximum shear stress in the shaft caused by torsion.

5–38.   The motor A develops a power of 300 W and turns its connected pulley at 90 rev/min. Determine the required diameters of the steel shafts on the pulleys at A and B if the allowable shear stress is $\tau_{\text{allow}} = 85$ MPa.

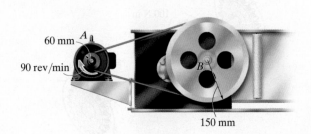

60 mm
90 rev/min
150 mm

Prob. 5–38

5–39.   The solid steel shaft DF has a diameter of 25 mm and is supported by smooth bearings at D and E. It is coupled to a motor at F, which delivers 12 kW of power to the shaft while it is turning at 50 rev/s. If gears A, B, and C remove 3 kW, 4 kW, and 5 kW respectively, determine the maximum shear stress developed in the shaft within regions CF and BC. The shaft is free to turn in its support bearings D and E.

*5–40.   Determine the absolute maximum shear stress developed in the shaft in Prob. 5–39.

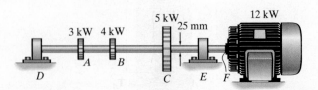

3 kW   4 kW   5 kW   25 mm   12 kW
D   A   B   C   E   F

Probs. 5–39/40

•5–41. The A-36 steel tubular shaft is 2 m long and has an outer diameter of 50 mm. When it is rotating at 40 rad/s, it transmits 25 kW of power from the motor M to the pump P. Determine the smallest thickness of the tube if the allowable shear stress is $\tau_{allow} = 80$ MPa.

5–42. The A-36 solid tubular steel shaft is 2 m long and has an outer diameter of 60 mm. It is required to transmit 60 kW of power from the motor M to the pump P. Determine the smallest angular velocity the shaft can have if the allowable shear stress is $\tau_{allow} = 80$ MPa.

*5–44. The drive shaft AB of an automobile is made of a steel having an allowable shear stress of $\tau_{allow} = 56$ MPa. If the outer diameter of the shaft is 62.5 mm and the engine delivers 165 kW to the shaft when it is turning at 1140 rev/min, determine the minimum required thickness of the shaft's wall.

•5–45. The drive shaft AB of an automobile is to be designed as a thin-walled tube. The engine delivers 125 kW when the shaft is turning at 1500 rev/min. Determine the minimum thickness of the shaft's wall if the shaft's outer diameter is 62.5 mm. The material has an allowable shear stress of $\tau_{allow} = 50$ MPa.

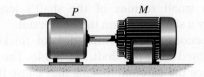

Probs. 5–41/42

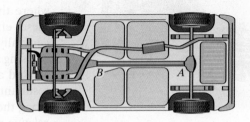

Probs. 5–44/45

5–43. A steel tube having an outer diameter of 65 mm is used to transmit 30 kW when turning at 2700 rev/min. Determine the inner diameter d of the tube to the nearest multiples of 5 mm if the allowable shear stress is $\tau_{allow} = 70$ MPa.

5–46. The motor delivers 12 kW to the pulley at A while turning at a constant rate of 1800 rpm. Determine to the nearest multiples of 5 mm the smallest diameter of shaft BC if the allowable shear stress for steel is $\tau_{allow} = 84$ MPa. The belt does not slip on the pulley.

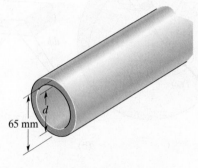

Prob. 5–43

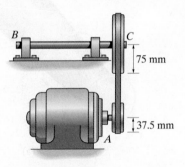

Prob. 5–46

5.4   Angle of Twist

Occasionally the design of a shaft depends on restricting the amount of rotation or twist that may occur when the shaft is subjected to a torque. Furthermore, being able to compute the angle of twist for a shaft is important when analyzing the reactions on statically indeterminate shafts.

In this section we will develop a formula for determining the **angle of twist** ϕ (phi) of one end of a shaft with respect to its other end. The shaft is assumed to have a circular cross section that can gradually vary along its length, Fig. 5–14a. Also, the material is assumed to be homogeneous and to behave in a linear-elastic manner when the torque is applied. Like the case of an axially loaded bar, we will neglect the localized deformations that occur at points of application of the torques and where the cross section changes abruptly. By Saint-Venant's principle, these effects occur within small regions of the shaft's length and generally they will have only a slight effect on the final result.

Using the method of sections, a differential disk of thickness dx, located at position x, is isolated from the shaft, Fig. 5–14b. The internal resultant torque is $T(x)$, since the external loading may cause it to vary along the axis of the shaft. Due to $T(x)$, the disk will twist, such that the *relative rotation* of one of its faces with respect to the other face is $d\phi$, Fig. 5–14b. As a result an element of material located at an arbitrary radius ρ within the disk will undergo a shear strain γ. The values of γ and $d\phi$ are related by Eq. 5–1, namely,

$$d\phi = \gamma \frac{dx}{\rho} \qquad (5\text{–}13)$$

Oil wells are commonly drilled to depths exceeding a thousand meters. As a result, the total angle of twist of a string of drill pipe can be substantial and must be determined.

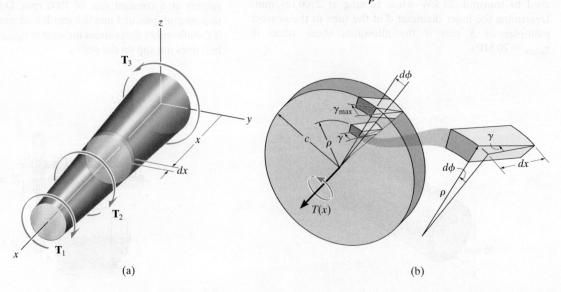

(a)   (b)

Fig. 5–14

Since Hooke's law, $\gamma = \tau/G$, applies and the shear stress can be expressed in terms of the applied torque using the torsion formula $\tau = T(x)\rho/J(x)$, then $\gamma = T(x)\rho/J(x)G$. Substituting this into Eq. 5–13, the angle of twist for the disk is

$$d\phi = \frac{T(x)}{J(x)G}\,dx$$

Integrating over the entire length L of the shaft, we obtain the angle of twist for the entire shaft, namely,

$$\phi = \int_0^L \frac{T(x)\,dx}{J(x)G} \qquad (5\text{–}14)$$

When computing both the stress and the angle of twist of this soil auger, it is necessary to consider the variable torsional loading which acts along its length.

Here

ϕ = the angle of twist of one end of the shaft with respect to the other end, measured in radians

$T(x)$ = the *internal torque* at the arbitrary position x, found from the method of sections and the equation of moment equilibrium applied about the shaft's axis

$J(x)$ = the shaft's polar moment of inertia expressed as a function of position x

G = the shear modulus of elasticity for the material

Constant Torque and Cross-Sectional Area.

Usually in engineering practice the material is homogeneous so that G is constant. Also, the shaft's cross-sectional area and the external torque are constant along the length of the shaft, Fig. 5–15. If this is the case, the internal torque $T(x) = T$, the polar moment of inertia $J(x) = J$, and Eq. 5–14 can be integrated, which gives

$$\phi = \frac{TL}{JG} \qquad (5\text{–}15)$$

The similarities between the above two equations and those for an axially loaded bar ($\delta = \int P(x)\,dx/A(x)E$ and $\delta = PL/AE$) should be noted.

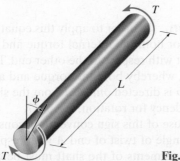

Fig. 5–15

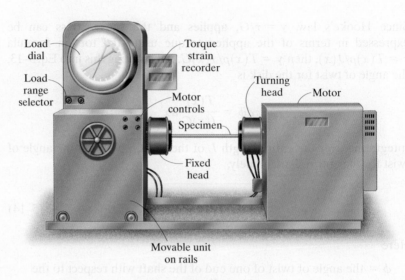

Fig. 5–16

Equation 5–15 is often used to determine the shear modulus of elasticity G of a material. To do so, a specimen of known length and diameter is placed in a torsion testing machine like the one shown in Fig. 5–16. The applied torque T and angle of twist ϕ are then measured along the length L. Using Eq. 5–15, $G = TL/J\phi$. Usually, to obtain a more reliable value of G, several of these tests are performed and the average value is used.

Multiple Torques. If the shaft is subjected to several different torques, or the cross-sectional area or shear modulus changes abruptly from one region of the shaft to the next, Eq. 5–15 can be applied to each segment of the shaft where these quantities are all constant. The angle of twist of one end of the shaft with respect to the other is then found from the vector addition of the angles of twist of each segment. For this case,

$$\phi = \sum \frac{TL}{JG} \qquad (5–16)$$

Sign Convention. In order to apply this equation, we must develop a sign convention for both the internal torque and the angle of twist of one end of the shaft with respect to the other end. To do this, we will use the right-hand rule, whereby both the torque and angle will be *positive*, provided the *thumb* is directed *outward* from the shaft when the fingers curl to give the tendency for rotation, Fig. 5–17.

To illustrate the use of this sign convention, consider the shaft shown in Fig. 5–18a. The angle of twist of end A with respect to end D is to be determined. Three segments of the shaft must be considered, since the

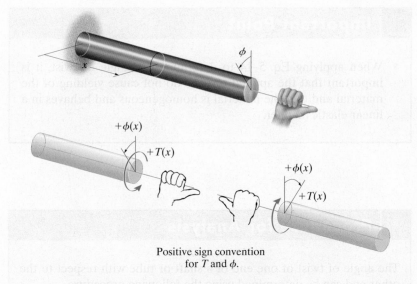

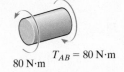

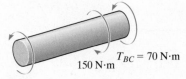

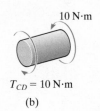

Positive sign convention
for T and ϕ.

Fig. 5–17

internal torque will change at B and at C. Using the method of sections, the *internal torques* are found for each segment, Fig. 5–18b. By the right-hand rule, with positive torques directed away from the *sectioned end* of the shaft, we have $T_{AB} = +80 \text{ N} \cdot \text{m}$, $T_{BC} = -70 \text{ N} \cdot \text{m}$, and $T_{CD} = -10 \text{ N} \cdot \text{m}$. These results are also shown on the *torque diagram* for the shaft, Fig. 5–18c. Applying Eq. 5–16, we have

$$\phi_{A/D} = \frac{(+80 \text{ N} \cdot \text{m})\, L_{AB}}{JG} + \frac{(-70 \text{ N} \cdot \text{m})\, L_{BC}}{JG} + \frac{(-10 \text{ N} \cdot \text{m})\, L_{CD}}{JG}$$

If the other data is substituted and the answer is found as a *positive* quantity, it means that end A will rotate as indicated by the curl of the right-hand fingers when the thumb is directed *away* from the shaft, Fig. 5–18a. The double subscript notation is used to indicate this relative angle of twist ($\phi_{A/D}$); however, if the angle of twist is to be determined relative to a *fixed support*, then only a single subscript will be used. For example, if D is a fixed support, then the angle of twist will be denoted as ϕ_A.

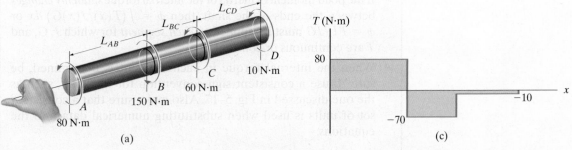

Fig. 5–18

Important Point

- When applying Eq. 5–14 to determine the angle of twist, it is important that the applied torques do not cause yielding of the material and that the material is homogeneous and behaves in a linear elastic manner.

Procedure for Analysis

The angle of twist of one end of a shaft or tube with respect to the other end can be determined using the following procedure.

Internal Torque.

- The *internal torque* is found at a point on the axis of the shaft by using the method of sections and the equation of moment equilibrium, applied along the shaft's axis.
- If the torque varies along the shaft's length, a section should be made at the arbitrary position x along the shaft and the *internal torque* represented as a function of x, i.e., $T(x)$.
- If several constant external torques act on the shaft between its ends, the internal torque in each *segment* of the shaft, between any two external torques, must be determined. The results can be represented graphically as a torque diagram.

Angle of Twist.

- When the circular cross-sectional area of the shaft varies along the shaft's axis, the polar moment of inertia must be expressed as a function of its position x along the axis, $J(x)$.
- If the polar moment of inertia or the internal torque *suddenly changes* between the ends of the shaft, then $\phi = \int (T(x)/J(x)G)\, dx$ or $\phi = TL/JG$ must be applied to *each segment* for which J, G, and T are continuous or constant.
- When the internal torque in each segment is determined, be sure to use a consistent sign convention for the shaft, such as the one discussed in Fig. 5–17. Also make sure that a consistent set of units is used when substituting numerical data into the equations.

EXAMPLE 5.5

The gears attached to the fixed-end steel shaft are subjected to the torques shown in Fig. 5–19a. If the shear modulus of elasticity is 80 GPa and the shaft has a diameter of 14 mm, determine the displacement of the tooth P on gear A. The shaft turns freely within the bearing at B.

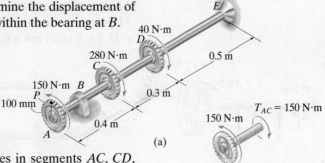

(a)

SOLUTION

Internal Torque. By inspection, the torques in segments AC, CD, and DE are different yet *constant* throughout each segment. Free-body diagrams of appropriate segments of the shaft along with the calculated internal torques are shown in Fig. 5–19b. Using the right-hand rule and the established sign convention that positive torque is directed away from the sectioned end of the shaft, we have

$$T_{AC} = +150 \text{ N} \cdot \text{m} \quad T_{CD} = -130 \text{ N} \cdot \text{m} \quad T_{DE} = -170 \text{ N} \cdot \text{m}$$

These results are also shown on the torque diagram, Fig. 5–19c.

Angle of Twist. The polar moment of inertia for the shaft is

$$J = \frac{\pi}{2}(0.007 \text{ m})^4 = 3.771(10^{-9}) \text{ m}^4$$

Applying Eq. 5–16 to each segment and adding the results algebraically, we have

$$\phi_A = \sum \frac{TL}{JG} = \frac{(+150 \text{ N} \cdot \text{m})(0.4 \text{ m})}{3.771(10^{-9}) \text{ m}^4 \, [80(10^9) \text{ N/m}^2]}$$

$$+ \frac{(-130 \text{ N} \cdot \text{m})(0.3 \text{ m})}{3.771(10^{-9}) \text{ m}^4 \, [80(10^9) \text{ N/m}^2)]}$$

$$+ \frac{(-170 \text{ N} \cdot \text{m})(0.5 \text{ m})}{3.771(10^{-9}) \text{ m}^4 \, [80(10^9) \text{ N/m}^2)]} = -0.2121 \text{ rad}$$

Since the answer is negative, by the right-hand rule the thumb is directed *toward* the end E of the shaft, and therefore gear A will rotate as shown in Fig. 5–19d.

The displacement of tooth P on gear A is

$$s_P = \phi_A r = (0.2121 \text{ rad})(100 \text{ mm}) = 21.2 \text{ mm} \qquad \textit{Ans.}$$

NOTE: Remember that this analysis is valid only if the shear stress does not exceed the proportional limit of the material.

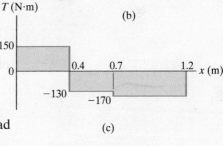

(b)

$$T \text{ (N·m)}$$

(c)

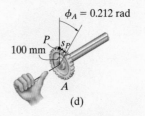

$\phi_A = 0.212$ rad

(d)

Fig. 5–19

EXAMPLE | 5.6

The two solid steel shafts shown in Fig. 5–20a are coupled together using the meshed gears. Determine the angle of twist of end A of shaft AB when the torque $T = 45$ N·m is applied. Take $G = 80$ GPa. Shaft AB is free to rotate within bearings E and F, whereas shaft DC is fixed at D. Each shaft has a diameter of 20 mm.

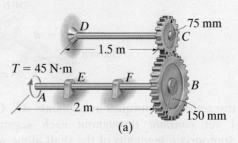

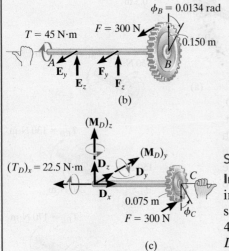

Fig. 5–20

SOLUTION

Internal Torque. Free-body diagrams for each shaft are shown in Fig. 5–20b and 5–20c. Summing moments along the x axis of shaft AB yields the tangential reaction between the gears of $F = 45$ N·m/0.15 m = 300 N. Summing moments about the x axis of shaft DC, this force then creates a torque of $(T_D)_x = 300$ N (0.075 m) = 22.5 N·m on shaft DC.

Angle of Twist. To solve the problem, we will first calculate the rotation of gear C due to the torque of 22.5 N·m in shaft DC, Fig. 5–20c. This angle of twist is

$$\phi_C = \frac{TL_{DC}}{JG} = \frac{(+22.5 \text{ N} \cdot \text{m})(1.5 \text{ m})}{(\pi/2)(0.010 \text{ m})^4[80(10^9) \text{ N/m}^2]} = +0.0269 \text{ rad}$$

Since the gears at the end of the shaft are in mesh, the rotation ϕ_C of gear C causes gear B to rotate ϕ_B, Fig. 5–20b, where

$$\phi_B(0.15 \text{ m}) = (0.0269 \text{ rad})(0.075 \text{ m})$$
$$\phi_B = 0.0134 \text{ rad}$$

We will now determine the angle of twist of end A with respect to end B of shaft AB caused by the 45 N·m torque, Fig. 5–20b. We have

$$\phi_{A/B} = \frac{T_{AB}L_{AB}}{JG} = \frac{(+45 \text{ N} \cdot \text{m})(2 \text{ m})}{(\pi/2)(0.010 \text{ m})^4[80(10^9) \text{ N/m}^2]} = +0.0716 \text{ rad}$$

The rotation of end A is therefore determined by adding ϕ_B and $\phi_{A/B}$, since both angles are in the *same direction*, Fig. 5–20b. We have

$$\phi_A = \phi_B + \phi_{A/B} = 0.0134 \text{ rad} + 0.0716 \text{ rad} = +0.0850 \text{ rad}$$ *Ans.*

EXAMPLE 5.7

The 0.05-m-diameter solid cast-iron post shown in Fig. 5–21a is buried 0.6 m in soil. If a torque is applied to its top using a rigid wrench, determine the maximum shear stress in the post and the angle of twist at its top. Assume that the torque is about to turn the post, and the soil exerts a uniform torsional resistance of t N · mm/mm along its 0.6-m buried length. $G = 40(10^3)$ MPa.

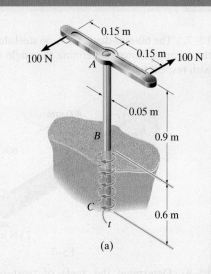

(a)

SOLUTION

Internal Torque. The internal torque in segment AB of the post is constant. From the free-body diagram, Fig. 5–21b, we have

$$\Sigma M_z = 0; \qquad T_{AB} = 100 \text{ N}(0.30 \text{ m}) = 30 \text{ N} \cdot \text{m}$$

The magnitude of the uniform distribution of torque along the buried segment BC can be determined from equilibrium of the entire post, Fig. 5–21c. Here

$$\Sigma M_z = 0 \qquad 100 \text{ N}(0.30 \text{ m}) - t(0.6 \text{ m}) = 0$$

$$t = 50 \text{ N} \cdot \text{m/m}$$

Hence, from a free-body diagram of a section of the post located at the position x, Fig. 5–21d, we have

$$\Sigma M_z = 0; \qquad T_{BC} - 50x = 0$$

$$T_{BC} = 50x$$

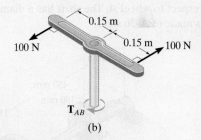

(b)

Maximum Shear Stress. The largest shear stress occurs in region AB, since the torque is largest there and J is constant for the post. Applying the torsion formula, we have

$$\tau_{max} = \frac{T_{AB}c}{J} = \frac{(30 \text{ N} \cdot \text{m})(0.025 \text{ m})}{(\pi/2)(0.025 \text{ m})^4} = 1.22(10^6) \text{ N/m}^2 = 1.22 \text{ MPa} \quad Ans.$$

Angle of Twist. The angle of twist at the top can be determined relative to the bottom of the post, since it is fixed and yet is about to turn. Both segments AB and BC twist, and so in this case we have

$$\phi_A = \frac{T_{AB}L_{AB}}{JG} + \int_0^{L_{BC}} \frac{T_{BC} \, dx}{JG}$$

$$= \frac{(30 \text{ N} \cdot \text{m})0.9 \text{ m}}{JG} + \int_0^{0.6 \text{ m}} \frac{50x \, dx}{JG}$$

$$= \frac{27 \text{ N} \cdot \text{m}^2}{JG} + \frac{50[(0.6)^2/2]\text{N} \cdot \text{m}^2}{JG}$$

$$= \frac{36 \text{ N} \cdot \text{m}^2}{(\pi/2)(0.025 \text{ m})^4 \, 40(10^9) \text{ N/m}^2} = 0.00147 \text{ rad} \qquad Ans.$$

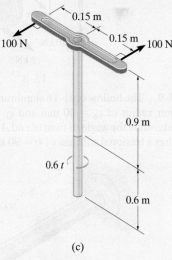

(c)

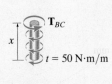

(d)

Fig. 5–21

FUNDAMENTAL PROBLEMS

F5–7. The 60-mm-diameter A-36 steel shaft is subjected to the torques shown. Determine the angle of twist of end A with respect to C.

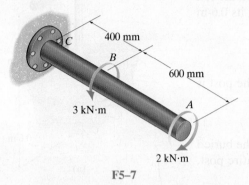

F5–7

F5–8. Determine the angle of twist of wheel B with respect to wheel A. The shaft has a diameter of 40 mm and is made of A-36 steel.

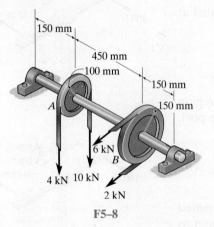

F5–8

F5–9. The hollow 6061-T6 aluminum shaft has an outer and inner radius of $c_o = 40$ mm and $c_i = 30$ mm, respectively. Determine the angle of twist of end A. The flexible support at B has a torsional stiffness of $k = 90$ kN $\cdot$ m/rad.

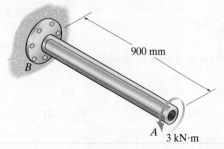

F5–9

F5–10. A series of gears are mounted on the 40-mm-diameter A-36 steel shaft. Determine the angle of twist of gear B relative to gear A.

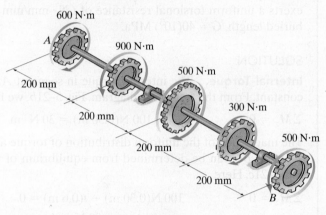

F5–10

F5–11. The 80-mm-diameter shaft is made of A-36 steel. If it is subjected to the uniform distributed torque, determine the angle of twist of end A with respect to B.

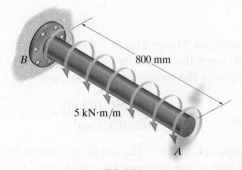

F5–11

F5–12. The 80-mm-diameter shaft is made of A-36 steel. If it is subjected to the triangular distributed load, determine the angle of twist of end A with respect to C.

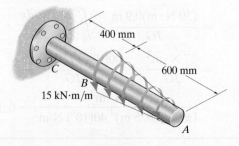

F5–12

PROBLEMS

5-47. The propellers of a ship are connected to a A-36 steel shaft that is 60 m long and has an outer diameter of 340 mm and inner diameter of 260 mm. If the power output is 4.5 MW when the shaft rotates at 20 rad/s, determine the maximum torsional stress in the shaft and its angle of twist.

*5-48. A shaft is subjected to a torque **T**. Compare the effectiveness of using the tube shown in the figure with that of a solid section of radius *c*. To do this, compute the percent increase in torsional stress and angle of twist per unit length for the tube versus the solid section.

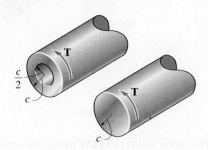

Prob. 5-48

•5-49. The A-36 steel axle is made from tubes *AB* and *CD* and a solid section *BC*. It is supported on smooth bearings that allow it to rotate freely. If the gears, fixed to its ends, are subjected to 85-N · m torques, determine the angle of twist of gear *A* relative to gear *D*. The tubes have an outer diameter of 30 mm and an inner diameter of 20 mm. The solid section has a diameter of 40 mm.

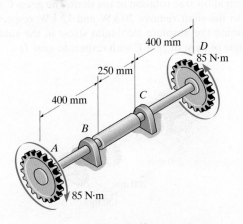

Prob. 5-49

5-50. The hydrofoil boat has an A-36 steel propeller shaft that is 30 m long. It is connected to an in-line diesel engine that delivers a maximum power of 2000 kW and causes the shaft to rotate at 1700 rpm. If the outer diameter of the shaft is 200 mm and the wall thickness is 10 mm, determine the maximum shear stress developed in the shaft. Also, what is the "wind up," or angle of twist in the shaft at full power?

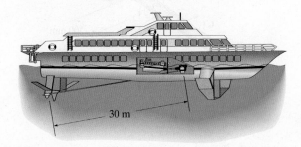

Prob. 5-50

5-51. The engine of the helicopter is delivering 450 kW to the rotor shaft *AB* when the blade is rotating at 1200 rev/min. Determine to the nearest multiples of 5 mm the diameter of the shaft *AB* if the allowable shear stress is $\tau_{\text{allow}} = 56$ MPa and the vibrations limit the angle of twist of the shaft to 0.05 rad. The shaft is 0.6 m long and made from L2 steel.

*5-52. The engine of the helicopter is delivering 450 kW to the rotor shaft *AB* when the blade is rotating at 1200 rev/min. Determine to the nearest multiples of 5 mm the diameter of the shaft *AB* if the allowable shear stress is $\tau_{\text{allow}} = 73.5$ MPa and the vibrations limit the angle of twist of the shaft to 0.05 rad. The shaft is 0.6 m long and made from L2 steel.

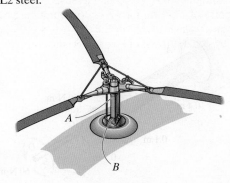

Probs. 5-51/52

•5–53.   The 20-mm-diameter A-36 steel shaft is subjected to the torques shown. Determine the angle of twist of the end B.

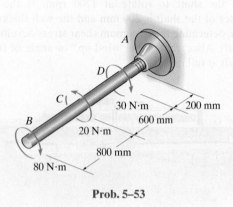

Prob. 5–53

*5–56.   The splined ends and gears attached to the A-36 steel shaft are subjected to the torques shown. Determine the angle of twist of end B with respect to end A. The shaft has a diameter of 40 mm.

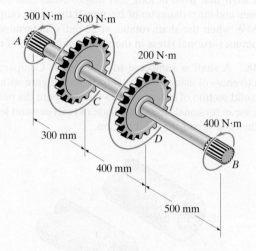

Prob. 5–56

5–54.   The assembly is made of A-36 steel and consists of a solid rod 20 mm in diameter fixed to the inside of a tube using a rigid disk at B. Determine the angle of twist at D. The tube has an outer diameter of 40 mm and wall thickness of 5 mm.

5–55.   The assembly is made of A-36 steel and consists of a solid rod 20 mm in diameter fixed to the inside of a tube using a rigid disk at B. Determine the angle of twist at C. The tube has an outer diameter of 40 mm and wall thickness of 5 mm.

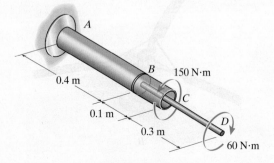

Probs. 5–54/55

•5–57.   The motor delivers 33 kW to the 304 stainless steel shaft while it rotates at 20 Hz. The shaft is supported on smooth bearings at A and B, which allow free rotation of the shaft. The gears C and D fixed to the shaft remove 20 kW and 12 kW, respectively. Determine the diameter of the shaft to the nearest multiples of 5 mm if the allowable shear stress is $\tau_{allow} = 56$ MPa and the allowable angle of twist of C with respect to D is 0.20°.

5–58.   The motor delivers 33 kW to the 304 stainless steel solid shaft while it rotates at 20 Hz. The shaft has a diameter of 37.5 mm and is supported on smooth bearings at A and B, which allow free rotation of the shaft. The gears C and D fixed to the shaft remove 20 kW and 12 kW, respectively. Determine the absolute maximum stress in the shaft and the angle of twist of gear C with respect to gear D.

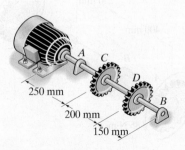

Probs. 5–57/58

5–59. The shaft is made of A-36 steel. It has a diameter of 25 mm and is supported by bearings at *A* and *D*, which allow free rotation. Determine the angle of twist of *B* with respect to *D*.

***5–60.** The shaft is made of A-36 steel. It has a diameter of 25 mm and is supported by bearings at *A* and *D*, which allow free rotation. Determine the angle of twist of gear *C* with respect to *B*.

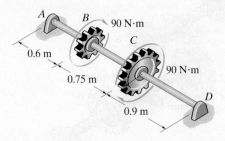

Probs. 5–59/60

•5–61. The two shafts are made of A-36 steel. Each has a diameter of 25 mm, and they are supported by bearings at *A*, *B*, and *C*, which allow free rotation. If the support at *D* is fixed, determine the angle of twist of end *B* when the torques are applied to the assembly as shown.

5–62. The two shafts are made of A-36 steel. Each has a diameter of 25 mm, and they are supported by bearings at *A*, *B*, and *C*, which allow free rotation. If the support at *D* is fixed, determine the angle of twist of end *A* when the torques are applied to the assembly as shown.

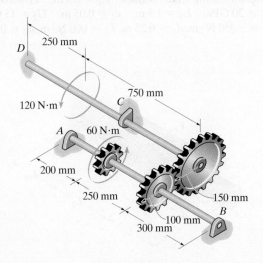

Probs. 5–61/62

5–63. The device serves as a compact torsional spring. It is made of A-36 steel and consists of a solid inner shaft *CB* which is surrounded by and attached to a tube *AB* using a rigid ring at *B*. The ring at *A* can also be assumed rigid and is fixed from rotating. If a torque of $T = 0.25$ kN · m is applied to the shaft, determine the angle of twist at the end *C* and the maximum shear stress in the tube and shaft.

***5–64.** The device serves as a compact torsion spring. It is made of A-36 steel and consists of a solid inner shaft *CB* which is surrounded by and attached to a tube *AB* using a rigid ring at *B*. The ring at *A* can also be assumed rigid and is fixed from rotating. If the allowable shear stress for the material is $\tau_{allow} = 84$ MPa and the angle of twist at *C* is limited to $\phi_{allow} = 3°$, determine the maximum torque *T* that can be applied at the end *C*.

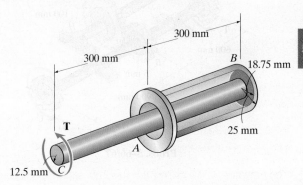

Probs. 5–63/64

•5–65. The A-36 steel assembly consists of a tube having an outer radius of 25 mm and a wall thickness of 3 mm. Using a rigid plate at *B*, it is connected to the solid 25-mm-diameter shaft *AB*. Determine the rotation of the tube's end *C* if a torque of 25 N · m. is applied to the tube at this end. The end *A* of the shaft is fixed supported.

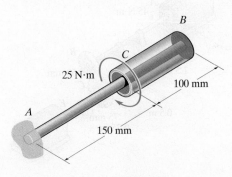

Prob. 5–65

5–66. The 60-mm diameter shaft ABC is supported by two journal bearings, while the 80-mm diameter shaft EH is fixed at E and supported by a journal bearing at H. If $T_1 = 2$ kN $\cdot$ m and $T_2 = 4$ kN $\cdot$ m, determine the angle of twist of gears A and C. The shafts are made of A-36 steel.

5–67. The 60-mm diameter shaft ABC is supported by two journal bearings, while the 80-mm diameter shaft EH is fixed at E and supported by a journal bearing at H. If the angle of twist at gears A and C is required to be 0.04 rad, determine the magnitudes of the torques T_1 and T_2. The shafts are made of A-36 steel.

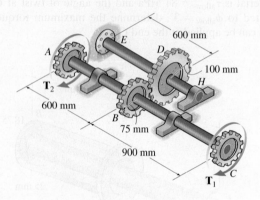

Probs. 5–66/67

*5–68. The 30-mm-diameter shafts are made of L2 tool steel and are supported on journal bearings that allow the shaft to rotate freely. If the motor at A develops a torque of $T = 45$ N $\cdot$ m on the shaft AB, while the turbine at E is fixed from turning, determine the amount of rotation of gears B and C.

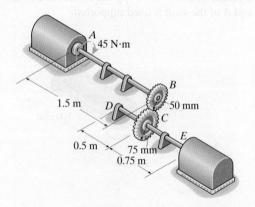

Prob. 5–68

•5–69. The shafts are made of A-36 steel and each has a diameter of 80 mm. Determine the angle of twist at end E.

5–70. The shafts are made of A-36 steel and each has a diameter of 80 mm. Determine the angle of twist of gear D.

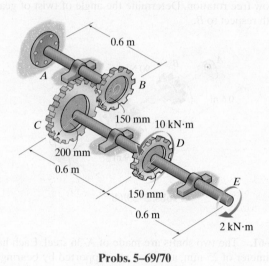

Probs. 5–69/70

5–71. Consider the general problem of a circular shaft made from m segments, each having a radius of c_m and shear modulus G_m. If there are n torques on the shaft as shown, write a computer program that can be used to determine the angle of twist of its end A. Show an application of the program using the values $L_1 = 0.5$ m, $c_1 = 0.02$ m, $G_1 = 30$ GPa, $L_2 = 1.5$ m, $c_2 = 0.05$ m, $G_2 = 15$ GPa, $T_1 = -450$ N $\cdot$ m, $d_1 = 0.25$ m, $T_2 = 600$ N $\cdot$ m, $d_2 = 0.8$ m.

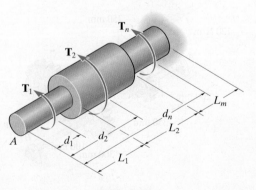

Prob. 5–71

*5–72. The 80-mm diameter shaft is made of 6061-T6 aluminum alloy and subjected to the torsional loading shown. Determine the angle of twist at end A.

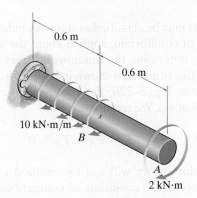

Prob. 5–72

•5–73. The tapered shaft has a length L and a radius r at end A and $2r$ at end B. If it is fixed at end B and is subjected to a torque T, determine the angle of twist of end A. The shear modulus is G.

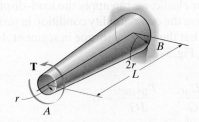

Prob. 5–73

5–74. The rod ABC of radius c is embedded into a medium where the distributed torque reaction varies linearly from zero at C to t_0 at B. If couple forces P are applied to the lever arm, determine the value of t_0 for equilibrium. Also, find the angle of twist of end A. The rod is made from material having a shear modulus of G.

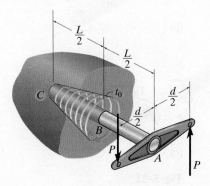

Prob. 5–74

5–75. When drilling a well, the deep end of the drill pipe is assumed to encounter a torsional resistance T_A. Furthermore, soil friction along the sides of the pipe creates a linear distribution of torque per unit length, varying from zero at the surface B to t_0 at A. Determine the necessary torque T_B that must be supplied by the drive unit to turn the pipe. Also, what is the relative angle of twist of one end of the pipe with respect to the other end at the instant the pipe is about to turn? The pipe has an outer radius r_o and an inner radius r_i. The shear modulus is G.

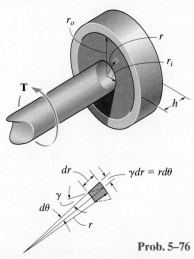

Prob. 5–75

*5–76. A cylindrical spring consists of a rubber annulus bonded to a rigid ring and shaft. If the ring is held fixed and a torque T is applied to the rigid shaft, determine the angle of twist of the shaft. The shear modulus of the rubber is G. *Hint:* As shown in the figure, the deformation of the element at radius r can be determined from $rd\theta = dr\gamma$. Use this expression along with $\tau = T/(2\pi r^2 h)$ from Prob. 5–26, to obtain the result.

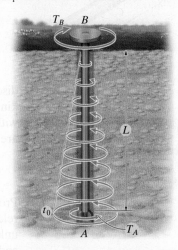

Prob. 5–76

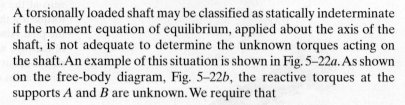

(a)

5.5 Statically Indeterminate Torque-Loaded Members

A torsionally loaded shaft may be classified as statically indeterminate if the moment equation of equilibrium, applied about the axis of the shaft, is not adequate to determine the unknown torques acting on the shaft. An example of this situation is shown in Fig. 5–22a. As shown on the free-body diagram, Fig. 5–22b, the reactive torques at the supports A and B are unknown. We require that

$$\Sigma M_x = 0; \qquad T - T_A - T_B = 0$$

In order to obtain a solution, we will use the method of analysis discussed in Sec. 4.4. The necessary condition of compatibility, or the kinematic condition, requires the angle of twist of one end of the shaft with respect to the other end to be equal to zero, since the end supports are fixed. Therefore,

$$\phi_{A/B} = 0$$

Provided the material is linear elastic, we can apply the load–displacement relation $\phi = TL/JG$ to express the compatibility condition in terms of the unknown torques. Realizing that the internal torque in segment AC is $+T_A$ and in segment CB it is $-T_B$, Fig. 5–22c, we have

$$\frac{T_A L_{AC}}{JG} - \frac{T_B L_{BC}}{JG} = 0$$

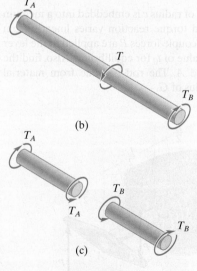

(b)

(c)

Fig. 5–22

Solving the above two equations for the reactions, realizing that $L = L_{AC} + L_{BC}$, we get

$$T_A = T\left(\frac{L_{BC}}{L}\right) \quad \text{and} \quad T_B = T\left(\frac{L_{AC}}{L}\right)$$

Procedure for Analysis

The unknown torques in statically indeterminate shafts are determined by satisfying equilibrium, compatibility, and torque-displacement requirements for the shaft.

Equilibrium.

- Draw a free-body diagram of the shaft in order to identify all the external torques that act on it. Then write the equation of moment equilibrium about the axis of the shaft.

Compatibility.

- Write the compatibility equation between two points along the shaft. Give consideration as to how the supports constrain the shaft when it is twisted.

- Express the angles of twist in the compatibility condition in terms of the torques, using a torque-displacement relation, such as $\phi = TL/JG$.

- Solve the equilibrium and compatibility equations for the unknown reactive torques. If any of the magnitudes have a negative numerical value, it indicates that this torque acts in the opposite sense of direction to that shown on the free-body diagram.

The shaft of this cutting machine is fixed at its ends and subjected to a torque at its center, allowing it to act as a torsional spring.

EXAMPLE | 5.8

The solid steel shaft shown in Fig. 5–23a has a diameter of 20 mm. If it is subjected to the two torques, determine the reactions at the fixed supports A and B.

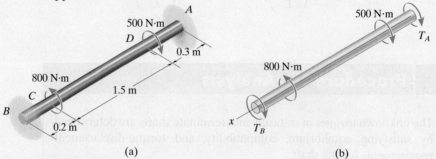

(a)   (b)

SOLUTION

Equilibrium. By inspection of the free-body diagram, Fig. 5–23b, it is seen that the problem is statically indeterminate since there is only *one* available equation of equilibrium and there are two unknowns. We require

$$\Sigma M_x = 0; \qquad -T_B + 800 \text{ N} \cdot \text{m} - 500 \text{ N} \cdot \text{m} - T_A = 0 \qquad (1)$$

Compatibility. Since the ends of the shaft are fixed, the angle of twist of one end of the shaft with respect to the other must be zero. Hence, the compatibility equation becomes

$$\phi_{A/B} = 0$$

This condition can be expressed in terms of the unknown torques by using the load–displacement relationship, $\phi = TL/JG$. Here there are three regions of the shaft where the internal torque is constant. On the free-body diagrams in Fig. 5–23c we have shown the internal torques acting on the left segments of the shaft which are sectioned in each of these regions. This way the internal torque is only a function of T_B. Using the sign convention established in Sec. 5.4, we have

$$\frac{-T_B(0.2 \text{ m})}{JG} + \frac{(800 - T_B)(1.5 \text{ m})}{JG} + \frac{(300 - T_B)(0.3 \text{ m})}{JG} = 0$$

so that

$$T_B = 645 \text{ N} \cdot \text{m} \qquad\qquad Ans.$$

Using Eq. 1,

$$T_A = -345 \text{ N} \cdot \text{m} \qquad\qquad Ans.$$

The negative sign indicates that $\mathbf{T}_A$ acts in the opposite direction of that shown in Fig. 5–23b.

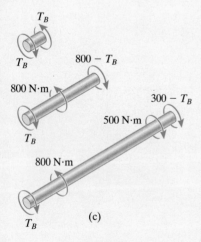

(c)

Fig. 5–23

EXAMPLE 5.9

The shaft shown in Fig. 5–24a is made from a steel tube, which is bonded to a brass core. If a torque of $T = 250$ N · m is applied at its end, plot the shear-stress distribution along a radial line of its cross-sectional area. Take $G_{st} = 80$ GPa, $G_{br} = 36$ GPa.

Refer to the Companion Website for the animation of the concepts that can be found in Example 5.9.

SOLUTION

Equilibrium. A free-body diagram of the shaft is shown in Fig. 5–24b. The reaction at the wall has been represented by the unknown amount of torque resisted by the steel, T_{st}, and by the brass, T_{br}. Working in units of pounds and inches, equilibrium requires

$$-T_{st} - T_{br} + (250 \text{ N} \cdot \text{m}) = 0 \qquad (1)$$

Compatibility. We require the angle of twist of end A to be the same for both the steel and brass since they are bonded together. Thus,

$$\phi = \phi_{st} = \phi_{br}$$

Applying the load–displacement relationship, $\phi = TL/JG$,

$$\frac{T_{st}L}{(\pi/2)[(0.020 \text{ m})^4 - (0.010 \text{ m})^4] \, 80(10^9) \text{ N/m}^2} =$$

$$\frac{T_{br}L}{(\pi/2)(0.010 \text{ m})^4 \, 36(10^9) \text{ N} \cdot \text{m}^2}$$

$$T_{st} = 33.33 T_{br} \qquad (2)$$

Solving Eqs. 1 and 2, we get

$$T_{st} = 242.72 \text{ N} \cdot \text{m}$$
$$T_{br} = 7.28 \text{ N} \cdot \text{m}$$

The shear stress in the brass core varies from zero at its center to a maximum at the interface where it contacts the steel tube. Using the torsion formula,

$$(\tau_{br})_{max} = \frac{(7.28 \text{ N} \cdot \text{m})(0.010 \text{ m})}{(\pi/2)(0.010 \text{ m})^4} = 4.63(10^6) \text{ N/m}^2 = 4.63 \text{ MPa}$$

For the steel, the minimum and maximum shear stresses are

$$(\tau_{st})_{min} = \frac{(242.72 \text{ N} \cdot \text{m})(0.010 \text{ m})}{(\pi/2)[(0.020 \text{ m})^4 - (0.010 \text{ m})^4]} = 10.30(10^6) \text{ N/m}^2 = 10.30 \text{ MPa}$$

$$(\tau_{st})_{max} = \frac{(242.72 \text{ N} \cdot \text{m})(0.020 \text{ m})}{(\pi/2)[(0.020 \text{ m})^4 - (0.010 \text{ m})^4]} = 20.60(10^6) \text{ N/m}^2 = 20.60 \text{ MPa}$$

The results are plotted in Fig. 5–24c. Note the discontinuity of *shear stress* at the brass and steel interface. This is to be expected, since the materials have different moduli of rigidity; i.e., steel is stiffer than brass ($G_{st} > G_{br}$) and thus it carries more shear stress at the interface. Although the shear stress is discontinuous here, the *shear strain* is not. Rather, the shear strain is the *same* for both the brass and the steel.

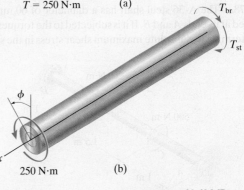

10 mm

20 mm

A

$T = 250$ N·m   (a)

T_{br}

T_{st}

ϕ

x

250 N·m   (b)

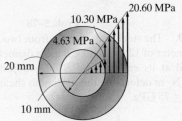

20.60 MPa

10.30 MPa

4.63 MPa

20 mm

10 mm

Shear–stress distribution

(c)

Fig. 5–24

5

PROBLEMS

•5–77. The A-36 steel shaft has a diameter of 50 mm and is fixed at its ends A and B. If it is subjected to the torque, determine the maximum shear stress in regions AC and CB of the shaft.

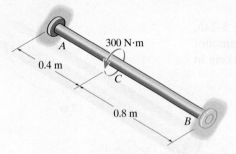

300 N·m

A

0.4 m

C

0.8 m

B

Prob. 5–77

5–78. The A-36 steel shaft has a diameter of 60 mm and is fixed at its ends A and B. If it is subjected to the torques shown, determine the absolute maximum shear stress in the shaft.

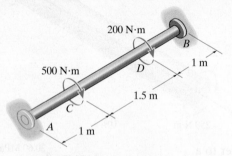

200 N·m

B

1 m

500 N·m

D

1.5 m

C

A

1 m

Prob. 5–78

5–79. The steel shaft is made from two segments: AC has a diameter of 12 mm, and CB has a diameter of 25 mm. If it is fixed at its ends A and B and subjected to a torque of 750 N · m determine the maximum shear stress in the shaft. $G_{st} = 75$ GPa.

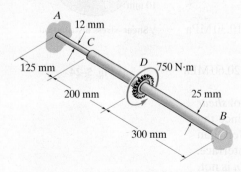

A

12 mm

C

125 mm

200 mm

D   750 N·m

25 mm

300 mm

B

Prob. 5–79

***5–80.** The shaft is made of A-36 steel, has a diameter of 80 mm, and is fixed at B while A is loose and can rotate 0.005 rad before becoming fixed. When the torques are applied to C and D, determine the maximum shear stress in regions AC and CD of the shaft.

•5–81. The shaft is made of A-36 steel and has a diameter of 80 mm. It is fixed at B and the support at A has a torsional stiffness of $k = 0.5$ MN · m/rad. If it is subjected to the gear torques shown, determine the absolute maximum shear stress in the shaft.

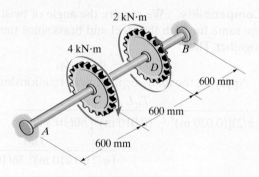

2 kN·m

4 kN·m

B

D

600 mm

C

600 mm

A

600 mm

600 mm

Probs. 5–80/81

5–82. The shaft is made from a solid steel section AB and a tubular portion made of steel and having a brass core. If it is fixed to a rigid support at A, and a torque of $T = 75$ N · m is applied to it at C, determine the angle of twist that occurs at C and compute the maximum shear stress and maximum shear strain in the brass and steel. Take $G_{st} = 80$ GPa, $G_{br} = 40$ GPa.

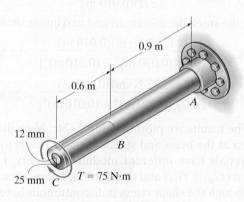

0.9 m

0.6 m

A

12 mm

B

25 mm   C

$T = 75$ N·m

Prob. 5–82

5–83. The motor A develops a torque at gear B of $675 \text{ N} \cdot \text{m}$, which is applied along the axis of the 50-mm-diameter steel shaft CD. This torque is to be transmitted to the pinion gears at E and F. If these gears are temporarily fixed, determine the maximum shear stress in segments CB and BD of the shaft. Also, what is the angle of twist of each of these segments? The bearings at C and D only exert force reactions on the shaft and do not resist torque. $G_{st} = 84 \text{ GPa}$.

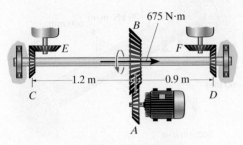

Prob. 5–83

*5–84. A portion of the A-36 steel shaft is subjected to a linearly distributed torsional loading. If the shaft has the dimensions shown, determine the reactions at the fixed supports A and C. Segment AB has a diameter of 30 mm and segment BC has a diameter of 15 mm.

•5–85. Determine the rotation of joint B and the absolute maximum shear stress in the shaft in Prob. 5–84.

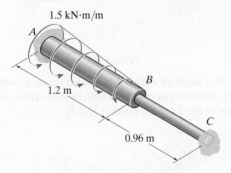

Probs. 5–84/85

5–86. The two shafts are made of A-36 steel. Each has a diameter of 25 mm and they are connected using the gears fixed to their ends. Their other ends are attached to fixed supports at A and B. They are also supported by journal bearings at C and D, which allow free rotation of the shafts along their axes. If a torque of $500 \text{ N} \cdot \text{m}$ is applied to the gear at E as shown, determine the reactions at A and B.

5–87. Determine the rotation of the gear at E in Prob. 5–86.

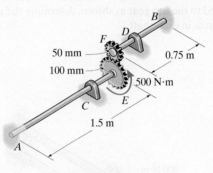

Probs. 5–86/87

*5–88. The shafts are made of A-36 steel and have the same diameter of 100 mm. If a torque of $25 \text{ kN} \cdot \text{m}$ is applied to gear B, determine the absolute maximum shear stress developed in the shaft.

•5–89. The shafts are made of A-36 steel and have the same diameter of 100 mm. If a torque of $25 \text{ kN} \cdot \text{m}$ is applied to gear B, determine the angle of twist of gear B.

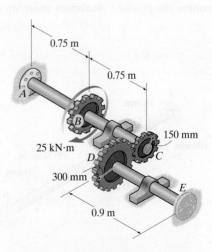

Probs. 5–88/89

5

5–90. The two 1-m-long shafts are made of 2014-T6 aluminum. Each has a diameter of 30 mm and they are connected using the gears fixed to their ends. Their other ends are attached to fixed supports at A and B. They are also supported by bearings at C and D, which allow free rotation of the shafts along their axes. If a torque of 900 N · m is applied to the top gear as shown, determine the maximum shear stress in each shaft.

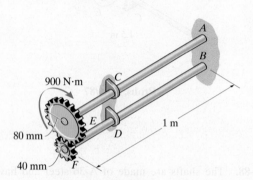

Prob. 5–90

5–91. The A-36 steel shaft is made from two segments: AC has a diameter of 10 mm and CB has a diameter of 20 mm. If the shaft is fixed at its ends A and B and subjected to a uniform distributed torque of 300 N · m/m along segment CB, determine the absolute maximum shear stress in the shaft.

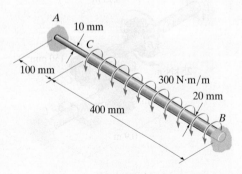

Prob. 5–91

*5–92. If the shaft is subjected to a uniform distributed torque of $t = 20$ kN · m/m, determine the maximum shear stress developed in the shaft. The shaft is made of 2014-T6 aluminum alloy and is fixed at A and C.

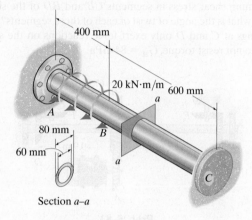

Section a–a

Prob. 5–92

•5–93. The tapered shaft is confined by the fixed supports at A and B. If a torque **T** is applied at its mid-point, determine the reactions at the supports.

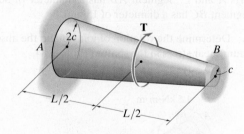

Prob. 5–93

5–94. The shaft of radius c is subjected to a distributed torque t, measured as torque/length of shaft. Determine the reactions at the fixed supports A and B.

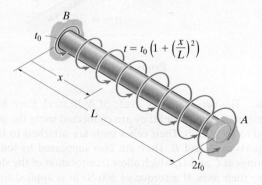

$$t = t_0 \left(1 + \left(\frac{x}{L} \right)^2 \right)$$

Prob. 5–94

*5.6   Solid Noncircular Shafts

It was demonstrated in Sec. 5.1 that when a torque is applied to a shaft having a circular cross section—that is, one that is axisymmetric—the shear strains vary linearly from zero at its center to a maximum at its outer surface. Furthermore, due to the uniformity of the shear strain at all points on the same radius, the cross sections do not deform, but rather remain plane after the shaft has twisted. Shafts that have a noncircular cross section, however, are *not* axisymmetric, and so their cross sections will **bulge** or **warp** when the shaft is twisted. Evidence of this can be seen from the way grid lines deform on a shaft having a square cross section when the shaft is twisted, Fig. 5–25. As a consequence of this deformation the torsional analysis of *noncircular* shafts becomes considerably more complicated and will not be considered in this text.

Using a mathematical analysis based on the theory of elasticity, however, it is possible to determine the shear-stress distribution within a shaft of square cross section. Examples of how this shear stress varies along two radial lines of the shaft are shown in Fig. 5–26a. Because these shear-stress distributions vary in a complex manner, the shear strains they create will *warp* the cross section as shown in Fig. 5–26b. In particular notice that the corner points of the shaft must be subjected to zero shear stress and therefore zero shear strain. The reason for this can be shown by considering an element of material located at one of these points, Fig. 5–26c. One would expect the top face of this element to be subjected to a shear stress in order to aid in resisting the applied torque **T**. This, however, cannot occur since the complementary shear stresses τ and τ', acting on the *outer surface* of the shaft, must be zero.

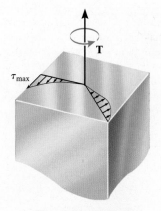

Shear stress distribution
along two radial lines

(a)

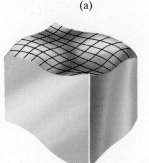

Warping of
cross-sectional area

(b)

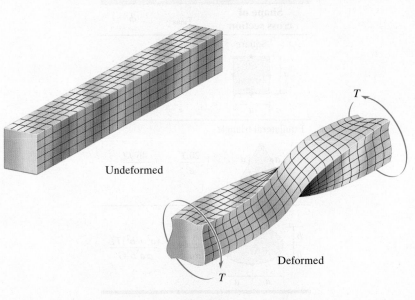

Undeformed

Deformed

Fig. 5–25

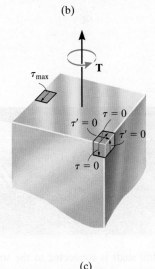

(c)

Fig. 5–26

5

Notice the deformation of the square element when this rubber bar is subjected to a torque.

The results of the analysis for square cross sections, along with other results from the theory of elasticity, for shafts having triangular and elliptical cross sections, are reported in Table 5–1. In all cases the *maximum shear stress* occurs at a point on the edge of the cross section that is *closest to* the center axis of the shaft. In Table 5–1 these points are indicated as "dots" on the cross sections. Also given are formulas for the angle of twist of each shaft. By extending these results to a shaft having an *arbitrary* cross section, it can also be shown that a shaft having a *circular* cross section is most efficient, since it is subjected to both a *smaller* maximum shear stress and a *smaller* angle of twist than a corresponding shaft having a noncircular cross section and subjected to the same torque.

The drill shaft is connected to the soil auger using a shaft having a square cross section.

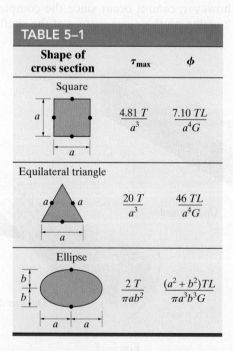

TABLE 5–1		
Shape of cross section	τ_{max}	ϕ
Square	$\dfrac{4.81\,T}{a^3}$	$\dfrac{7.10\,TL}{a^4 G}$
Equilateral triangle	$\dfrac{20\,T}{a^3}$	$\dfrac{46\,TL}{a^4 G}$
Ellipse	$\dfrac{2\,T}{\pi a b^2}$	$\dfrac{(a^2+b^2)TL}{\pi a^3 b^3 G}$

EXAMPLE 5.10

The 6061-T6 aluminum shaft shown in Fig. 5–27 has a cross-sectional area in the shape of an equilateral triangle. Determine the largest torque $\mathbf{T}$ that can be applied to the end of the shaft if the allowable shear stress is $\tau_{allow} = 56$ MPa. and the angle of twist at its end is restricted to $\phi_{allow} = 0.02$ rad. How much torque can be applied to a shaft of circular cross section made from the same amount of material?

SOLUTION

By inspection, the resultant internal torque at any cross section along the shaft's axis is also $\mathbf{T}$. Using the formulas for τ_{max} and ϕ in Table 5–1, we require

$$\tau_{allow} = \frac{20T}{a^3}; \qquad 56(10^6) \text{ N/m}^2 = \frac{20T}{(0.040 \text{ m})^3}$$

$$T = 179.2 \text{ N} \cdot \text{m}$$

Also,

$$\phi_{allow} = \frac{46TL}{a^4 G_{al}}; \qquad 0.02 \text{ rad} = \frac{46T(1.2 \text{ m})}{(0.040 \text{ m})^4[26(10^9) \text{ N/m}^2]}$$

$$T = 24.12 \text{ N} \cdot \text{m} \qquad \qquad Ans.$$

By comparison, the torque is limited due to the angle of twist.

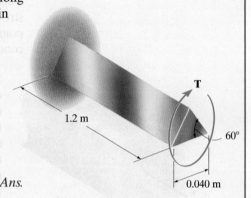

Fig. 5–27

Circular Cross Section. If the same amount of aluminum is to be used in making the same length of shaft having a circular cross section, then the radius of the cross section can be calculated. We have

$$A_{circle} = A_{triangle}; \qquad \pi c^2 = \frac{1}{2}(0.040 \text{ m})(0.040 \text{ m}) \sin 60°$$

$$c = 0.01485 \text{ m}$$

The limitations of stress and angle of twist then require

$$\tau_{allow} = \frac{Tc}{J}; \qquad 56(10^6) \text{ N/m}^2 = \frac{T(0.01485 \text{ m})}{(\pi/2)(0.01485 \text{ m})^4}$$

$$T = 288.06 \text{ N} \cdot \text{m}$$

$$\phi_{allow} = \frac{TL}{JG_{al}}; \qquad 0.02 \text{ rad} = \frac{T(1.2 \text{ m})}{(\pi/2)(0.01485 \text{ m})^4[26(10^9) \text{ N/m}^2]}$$

$$T = 33.10 \text{ N} \cdot \text{m} \qquad \qquad Ans.$$

Again, the angle of twist limits the applied torque.

NOTE: Comparing this result (33.10 N · m) with that given above (24.10 N · m), it is seen that a shaft of circular cross section can support 37% more torque than the one having a triangular cross section.

*5.7   Thin-Walled Tubes Having Closed Cross Sections

Thin-walled tubes of noncircular cross section are often used to construct light-weight frameworks such as those used in aircraft. In some applications, they may be subjected to a torsional loading. In this section we will analyze the effects of applying a torque to a thin-walled tube having a *closed* cross section, that is, a tube that does not have any breaks or slits along its length. Such a tube, having a constant yet arbitrary cross-sectional shape, and variable thickness t, is shown in Fig. 5–28a. Since the walls are thin, we will obtain the average shear stress by assuming that this stress is *uniformly distributed* across the thickness of the tube at any given point. Before we do this, however, we will first discuss some preliminary concepts regarding the action of shear stress over the cross section.

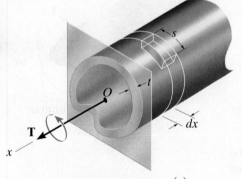

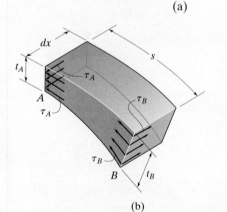

(a)

(b)

Fig. 5–28

Shear Flow.   Shown in Figs. 5–28a and 5–28b is a small element of the tube having a finite length s and differential width dx. At one end the element has a thickness t_A, and at the other end the thickness is t_B. Due to the internal torque $\mathbf{T}$, shear stress is developed on the front face of the element. Specifically, at end A the shear stress is τ_A, and at end B it is τ_B. These stresses can be related by noting that equivalent shear stresses τ_A and τ_B must also act on the longitudinal sides of the element. Since these sides have a *constant* width dx, the forces acting on them are $dF_A = \tau_A(t_A\,dx)$ and $dF_B = \tau_B(t_B\,dx)$. Equilibrium requires these forces to be of equal magnitude but opposite direction, so that

$$\tau_A t_A = \tau_B t_B$$

This important result states that ***the product of the average shear stress times the thickness of the tube is the same at each point on the tube's cross-sectional area***. This product is called ***shear flow***,* q, and in general terms we can express it as

$$\boxed{q = \tau_{\text{avg}}t} \tag{5–17}$$

Since q is constant over the cross section, the *largest* average shear stress must occur where the tube's thickness is the *smallest*.

---

*The terminology "flow" is used since q is analogous to water flowing through a tube of rectangular cross section having a constant depth and variable width w. Although the water's velocity v at each point along the tube will be different (like τ_{avg}), the flow $q = vw$ will be constant.

Now if a differential element having a thickness t, length ds, and width dx is isolated from the tube, Fig. 5–28c, it is seen that the front face over which the average shear stress acts is $dA = t\, ds$. Hence, $dF = \tau_{avg}(t\, ds) = q\, ds$, or $q = dF/ds$. In other words, *the shear flow measures the force per unit length along the tube's cross-sectional area.*

It is important to realize that the shear-stress components shown in Fig. 5–28c are the only ones acting on the tube. Components acting in the other direction, as shown in Fig. 5–28d, cannot exist. This is because the top and bottom faces of the element are at the inner and outer walls of the tube, and these boundaries must be free of stress. Instead, as noted above, the applied torque causes *the shear flow and the average stress to always be directed tangent to the wall of the tube, such that it contributes to the resultant internal torque* **T**.

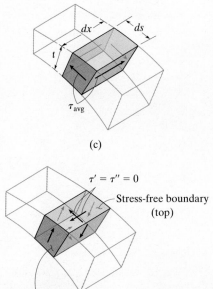

(c)

(d)

Average Shear Stress.

The average shear stress can be related to the torque T by considering the torque produced by this shear stress about a selected point O within the tube's boundary, Fig. 5–28e. As shown, the shear stress develops a force $dF = \tau_{avg}\, dA = \tau_{avg}(t\, ds)$ on an element of the tube. This force acts tangent to the centerline of the tube's wall, and if the moment arm is h, the torque is

$$dT = h(dF) = h(\tau_{avg}\, t\, ds)$$

For the entire cross section, we require

$$T = \oint h\tau_{avg}\, t\, ds$$

Here the "line integral" indicates that integration must be performed *around* the entire boundary of the area. Since the shear flow $q = \tau_{avg} t$ is *constant*, it can be factored out of the integral, so that

$$T = \tau_{avg}\, t \oint h\, ds$$

A graphical simplification can be made for evaluating the integral by noting that the *mean area*, shown by the blue colored triangle in Fig. 5–28e, is $dA_m = (1/2)h\, ds$. Thus,

$$T = 2\tau_{avg}\, t \int dA_m = 2\tau_{avg}\, t A_m$$

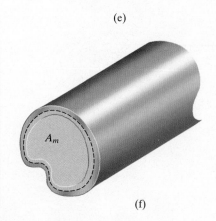

(e)

(f)

Fig. 5–28 (cont.)

Solving for τ_{avg}, we have

$$\tau_{avg} = \frac{T}{2t A_m} \qquad (5\text{--}18)$$

Here

τ_{avg} = the average shear stress acting over a particular thickness of the tube

T = the resultant internal torque at the cross section

t = the thickness of the tube where τ_{avg} is to be determined

A_m = the mean area enclosed within the boundary of the *centerline* of the tube's thickness. A_m is shown shaded in Fig. 5–28f

Since $q = \tau_{avg} t$, then the shear flow throughout the cross section becomes

$$q = \frac{T}{2A_m} \qquad (5\text{--}19)$$

Angle of Twist.   The angle of twist of a thin-walled tube of length L can be determined using energy methods, and the development of the necessary equation is given as a problem later in the text.* If the material behaves in a linear elastic manner and G is the shear modulus, then this angle ϕ, given in radians, can be expressed as

$$\phi = \frac{TL}{4A_m^2 G} \oint \frac{ds}{t} \qquad (5\text{--}20)$$

Here again the integration must be performed around the entire boundary of the tube's cross-sectional area.

Important Points

- Shear flow q is the product of the tube's thickness and the average shear stress. This value is the same at all points along the tube's cross section. As a result, the *largest* average shear stress on the cross section occurs where the thickness is *smallest*.

- Both shear flow and the average shear stress act *tangent* to the wall of the tube at all points and in a direction so as to contribute to the resultant internal torque.

---

*See Prob. 14–12.

EXAMPLE | 5.11

Calculate the average shear stress in a thin-walled tube having a circular cross section of mean radius r_m and thickness t, which is subjected to a torque T, Fig. 5–29a. Also, what is the relative angle of twist if the tube has a length L?

SOLUTION

Average Shear Stress.   The mean area for the tube is $A_m = \pi r_m^2$. Applying Eq. 5–18 gives

$$\tau_{avg} = \frac{T}{2t\,A_m} = \frac{T}{2\pi t r_m^2} \qquad Ans.$$

We can check the validity of this result by applying the torsion formula. In this case, using Eq. 5–9, we have

$$J = \frac{\pi}{2}(r_o^4 - r_i^4)$$

$$= \frac{\pi}{2}(r_o^2 + r_i^2)(r_o^2 - r_i^2)$$

$$= \frac{\pi}{2}(r_o^2 + r_i^2)(r_o + r_i)(r_o - r_i)$$

Since $r_m \approx r_o \approx r_i$ and $t = r_o - r_i$, $J = \dfrac{\pi}{2}(2r_m^2)(2r_m)t = 2\pi r_m^3 t$

so that

$$\tau_{avg} = \frac{T r_m}{J} = \frac{T r_m}{2\pi r_m^3 t} = \frac{T}{2\pi t r_m^2} \qquad Ans.$$

which agrees with the previous result.

The average shear-stress distribution acting throughout the tube's cross section is shown in Fig. 5–29b. Also shown is the shear-stress distribution acting on a radial line as calculated using the torsion formula. Notice how each τ_{avg} acts in a direction such that it contributes to the resultant torque at the section. As the tube's thickness decreases, the shear stress throughout the tube becomes more uniform.

Angle of Twist.   Applying Eq. 5–20, we have

$$\phi = \frac{TL}{4A_m^2 G}\oint \frac{ds}{t} = \frac{TL}{4(\pi r_m^2)^2 Gt}\oint ds$$

The integral represents the length around the centerline boundary, which is $2\pi r_m$. Substituting, the final result is

$$\phi = \frac{TL}{2\pi r_m^3 Gt} \qquad Ans.$$

Show that one obtains this same result using Eq. 5–15.

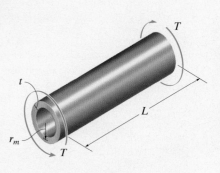

(a)

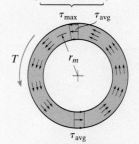

Actual shear-stress
distribution
(torsion formula)

τ_{max} τ_{avg}

r_m

τ_{avg}

Average shear-stress
distribution
(thin-wall approximation)

(b)

Fig. 5–29

EXAMPLE 5.12

The tube is made of C86100 bronze and has a rectangular cross section as shown in Fig. 5–30a. If it is subjected to the two torques, determine the average shear stress in the tube at points A and B. Also, what is the angle of twist of end C? The tube is fixed at E.

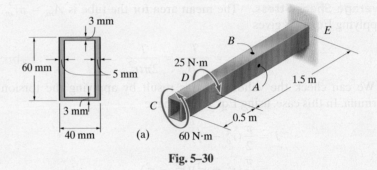

(a)

Fig. 5–30

SOLUTION

Average Shear Stress.   If the tube is sectioned through points A and B, the resulting free-body diagram is shown in Fig. 5–30b. The internal torque is 35 N · m. As shown in Fig. 5–30d, the mean area is

$$A_m = (0.035 \text{ m})(0.057 \text{ m}) = 0.00200 \text{ m}^2$$

Applying Eq. 5–18 for point A, $t_A = 5$ mm, so that

$$\tau_A = \frac{T}{2tA_m} = \frac{35 \text{ N} \cdot \text{m}}{2(0.005 \text{ m})(0.00200 \text{ m}^2)} = 1.75 \text{ MPa} \quad Ans.$$

And for point B, $t_B = 3$ mm, and therefore

$$\tau_B = \frac{T}{2tA_m} = \frac{35 \text{ N} \cdot \text{m}}{2(0.003 \text{ m})(0.00200 \text{ m}^2)} = 2.92 \text{ MPa} \quad Ans.$$

These results are shown on elements of material located at points A and B, Fig. 5–30e. Note carefully how the 35-N · m torque in Fig. 5–30b creates these stresses on the back sides of each element.

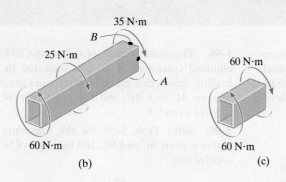

35 N·m

B

25 N·m

A

60 N·m

60 N·m

60 N·m

(b)

(c)

Angle of Twist. From the free-body diagrams in Fig. 5–30*b* and 5–30*c*, the internal torques in regions *DE* and *CD* are 35 N · m and 60 N · m, respectively. Following the sign convention outlined in Sec. 5.4, these torques are both positive. Thus, Eq. 5–20 becomes

$$\phi = \sum \frac{TL}{4A_m^2 G} \oint \frac{ds}{t}$$

$$= \frac{60 \text{ N} \cdot \text{m} \, (0.5 \text{ m})}{4(0.00200 \text{ m}^2)^2(38(10^9) \text{ N/m}^2)} \left[2\left(\frac{57 \text{ mm}}{5 \text{ mm}}\right) + 2\left(\frac{35 \text{ mm}}{3 \text{ mm}}\right) \right]$$

$$+ \frac{35 \text{ N} \cdot \text{m} \, (1.5 \text{ m})}{4(0.00200 \text{ m}^2)^2(38(10^9) \text{ N/m}^2)} \left[2\left(\frac{57 \text{ mm}}{5 \text{ mm}}\right) + 2\left(\frac{35 \text{ mm}}{3 \text{ mm}}\right) \right]$$

$$= 6.29(10^{-3}) \text{ rad} \qquad\qquad\qquad\qquad\qquad \textit{Ans.}$$

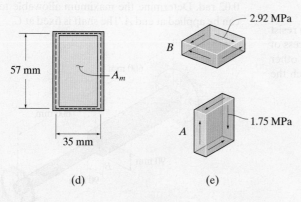

57 mm

A_m

35 mm

B

2.92 MPa

A

1.75 MPa

(d)

(e)

Fig. 5–30

PROBLEMS

5–95.   Compare the values of the maximum elastic shear stress and the angle of twist developed in 304 stainless steel shafts having circular and square cross sections. Each shaft has the same cross-sectional area of 5600 mm², length of 900 mm, and is subjected to a torque of 500 N · m.

5–98.   The shaft is made of red brass C83400 and has an elliptical cross section. If it is subjected to the torsional loading shown, determine the maximum shear stress within regions AC and BC, and the angle of twist ϕ of end B relative to end A.

5–99.   Solve Prob. 5–98 for the maximum shear stress within regions AC and BC, and the angle of twist ϕ of end B relative to C.

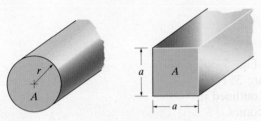

Prob. 5–95

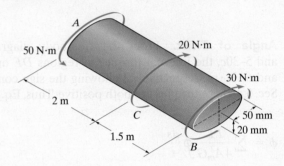

Probs. 5–98/99

*5–96.   If $a = 25$ mm and $b = 15$ mm, determine the maximum shear stress in the circular and elliptical shafts when the applied torque is $T = 80$ N · m. By what percentage is the shaft of circular cross section more efficient at withstanding the torque than the shaft of elliptical cross section?

*5–100.   Segments AB and BC of the shaft have circular and square cross sections, respectively. If end A is subjected to a torque of $T = 2$ kN · m, determine the absolute maximum shear stress developed in the shaft and the angle of twist of end A. The shaft is made from A-36 steel and is fixed at C.

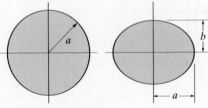

Prob. 5–96

•5–101.   Segments AB and BC of the shaft have circular and square cross sections, respectively. The shaft is made from A-36 steel with an allowable shear stress of $\tau_{\text{allow}} = 75$ MPa, and an angle of twist at end A which is not allowed to exceed 0.02 rad. Determine the maximum allowable torque $\mathbf{T}$ that can be applied at end A. The shaft is fixed at C.

•5–97.   It is intended to manufacture a circular bar to resist torque; however, the bar is made elliptical in the process of manufacturing, with one dimension smaller than the other by a factor k as shown. Determine the factor by which the maximum shear stress is increased.

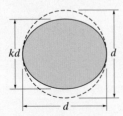

Prob. 5–97

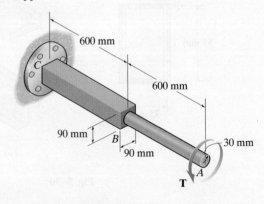

Probs. 5–100/101

5–102. The aluminum strut is fixed between the two walls at A and B. If it has a 50 mm by 50 mm square cross section, and it is subjected to the torque of 120 N · m at C, determine the reactions at the fixed supports. Also, what is the angle of twist at C? $G_{al} = 27$ GPa.

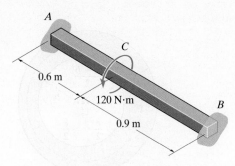

Prob. 5–102

5–103. The square shaft is used at the end of a drive cable in order to register the rotation of the cable on a gauge. If it has the dimensions shown and is subjected to a torque of 8 N · m, determine the shear stress in the shaft at point A. Sketch the shear stress on a volume element located at this point.

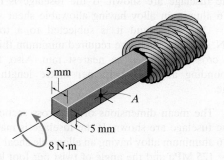

Prob. 5–103

*5–104. The 6061-T6 aluminum bar has a square cross section of 25 mm by 25 mm. If it is 2 m long, determine the maximum shear stress in the bar and the rotation of one end relative to the other end.

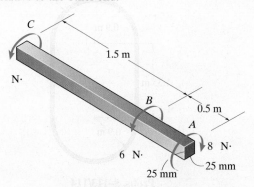

Prob. 5–104

•5–105. The steel shaft is 300 mm long and is screwed into the wall using a wrench. Determine the largest couple forces F that can be applied to the shaft without causing the steel to yield. $\tau_Y = 56$ MPa.

5–106. The steel shaft is 300 mm long and is screwed into the wall using a wrench. Determine the maximum shear stress in the shaft and the amount of displacement that each couple force undergoes if the couple forces have a magnitude of $F = 150$ N, $G_{st} = 75$ GPa.

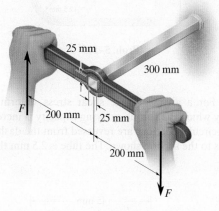

Probs. 5–105/106

5–107. Determine the constant thickness of the rectangular tube if the average shear stress is not to exceed 84 MPa when a torque of $T = 2.5$ kN · m is applied to the tube. Neglect stress concentrations at the corners. The mean dimensions of the tube are shown.

*5–108. Determine the torque T that can be applied to the rectangular tube if the average shear stress is not to exceed 84 MPa. Neglect stress concentrations at the corners. The mean dimensions of the tube are shown and the tube has a thickness of 3 mm.

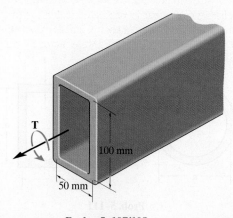

Probs. 5–107/108

•5–109.   For a given maximum shear stress, determine the factor by which the torque carrying capacity is increased if the half-circular section is reversed from the dashed-line position to the section shown. The tube is 2.5 mm thick.

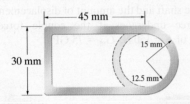

Prob. 5–109

5–110.   For a given average shear stress, determine the factor by which the torque-carrying capacity is increased if the half-circular sections are reversed from the dashed-line positions to the section shown. The tube is 2.5 mm thick.

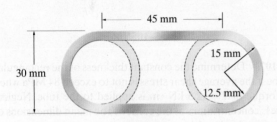

Prob. 5–110

5–111.   A torque **T** is applied to two tubes having the cross sections shown. Compare the shear flow developed in each tube.

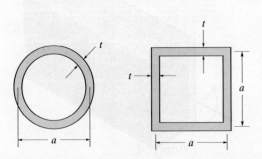

Prob. 5–111

*5–112.   Due to a fabrication error the inner circle of the tube is eccentric with respect to the outer circle. By what percentage is the torsional strength reduced when the eccentricity e is one-fourth of the difference in the radii?

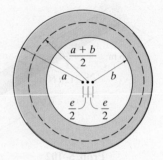

Prob. 5–112

•5–113.   The mean dimensions of the cross section of an airplane fuselage are shown. If the fuselage is made of 2014-T6 aluminum alloy having allowable shear stress of $\tau_{\text{allow}} = 125$ MPa, and it is subjected to a torque of 9000 kN · m, determine the required minimum thickness t of the cross section to the nearest mm. Also, find the corresponding angle of twist per metre length of the fuselage.

5–114.   The mean dimensions of the cross section of an airplane fuselage are shown. If the fuselage is made from 2014-T6 aluminum alloy having an allowable shear stress of $\tau_{\text{allow}} = 125$ MPa and the angle of twist per foot length of fuselage is not allowed to exceed 0.0033 rad/m, determine the maximum allowable torque that can be sustained by the fuselage. The thickness of the wall is $t = 6$ mm.

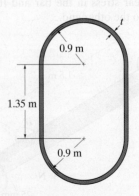

Probs. 5–113/114

5–115. The tube is subjected to a torque of 750 N·m. Determine the average shear stress in the tube at points A and B.

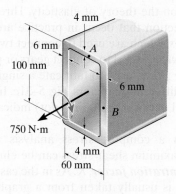

Prob. 5–115

***5–116.** The tube is made of plastic, is 5 mm thick, and has the mean dimensions shown. Determine the average shear stress at points A and B if it is subjected to the torque of $T = 5$ N·m. Show the shear stress on volume elements located at these points.

•**5–117.** The mean dimensions of the cross section of the leading edge and torsion box of an airplane wing can be approximated as shown. If the wing is made of 2014-T6 aluminum alloy having an allowable shear stress of $\tau_{allow} = 125$ MPa and the wall thickness is 10 mm, determine the maximum allowable torque and the corresponding angle of twist per meter length of the wing.

5–118. The mean dimensions of the cross section of the leading edge and torsion box of an airplane wing can be approximated as shown. If the wing is subjected to a torque of 4.5 MN·m and the wall thickness is 10 mm, determine the average shear stress developed in the wing and the angle of twist per meter length of the wing. The wing is made of 2014-T6 aluminum alloy.

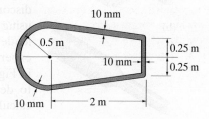

Probs. 5–117/118

5–119. The symmetric tube is made from a high-strength steel, having the mean dimensions shown and a thickness of 5 mm. If it is subjected to a torque of $T = 40$ N·m, determine the average shear stress developed at points A and B. Indicate the shear stress on volume elements located at these points.

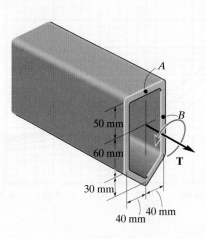

Prob. 5–116

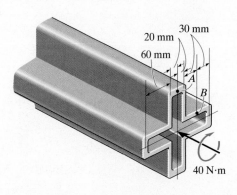

Prob. 5–119

5.8 Stress Concentration

The torsion formula, $\tau_{max} = Tc/J$, cannot be applied to regions of a shaft having a sudden change in the cross section. Here the shear-stress and shear-strain distributions in the shaft become complex and can be obtained only by using experimental methods or possibly by a mathematical analysis based on the theory of elasticity. Three common discontinuities of the cross section that occur in practice are shown in Fig. 5–31. They are at *couplings*, which are used to connect two collinear shafts together, Fig. 5–31a, *keyways*, used to connect gears or pulleys to a shaft, Fig. 5–31b, and *shoulder fillets*, used to fabricate a single collinear shaft from two shafts having different diameters, Fig. 5–31c. In each case the maximum shear stress will occur at the point (dot) indicated on the cross section.

The necessity to perform a complex stress analysis at a shaft discontinuity to obtain the maximum shear stress can be eliminated by using a ***torsional stress-concentration factor***, *K*. As in the case of axially loaded members, Sec. 4.7, *K* is usually taken from a graph based on experimental data. An example, for the shoulder-fillet shaft, is shown in Fig. 5–32. To use this graph, one first finds the geometric ratio *D*/*d* to define the appropriate curve, and then once the abscissa *r*/*d* is calculated, the value of *K* is found along the ordinate.

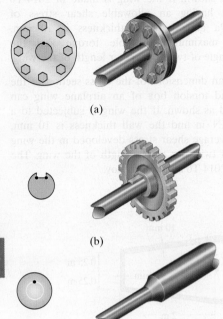

(a)

(b)

(c)

Fig. 5–31

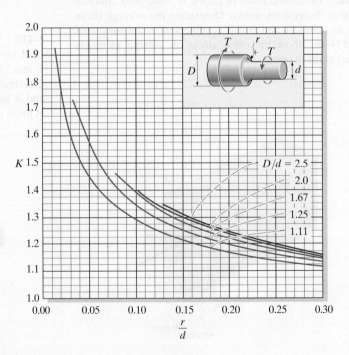

Fig. 5–32

The maximum shear stress is then determined from

$$\tau_{max} = K\frac{Tc}{J} \qquad (5\text{–}21)$$

Here the torsion formula is applied to the *smaller* of the two connected shafts, since τ_{max} occurs at the base of the fillet, Fig. 5–31c.

Note from the graph that an *increase* in fillet radius r causes a *decrease* in K. Hence the maximum shear stress in the shaft can be *reduced* by *increasing* the fillet radius. Also, if the diameter of the larger shaft is reduced, the D/d ratio will be lower and so the value of K and therefore τ_{max} will be lower.

Like the case of axially loaded members, torsional stress concentration factors should *always* be used when designing shafts made from *brittle materials*, or when designing shafts that will be subjected to *fatigue or cyclic torsional loadings*. These conditions give rise to the formation of cracks at the stress concentration, and this can often lead to a sudden fracture. On the other hand, if large *static* torsional loadings are applied to a shaft made from *ductile material*, then *inelastic strains* will develop within the shaft. Yielding of the material will cause the stress distribution to become more *evenly distributed* throughout the shaft, so that the maximum stress will not be limited to the region of stress concentration. This phenomenon will be discussed further in the next section.

Stress concentrations can arise at the coupling of these shafts, and this must be taken into account when the shaft is designed.

5

Important Points

- *Stress concentrations* in shafts occur at points of sudden cross-sectional change, such as couplings, keyways, and at shoulder fillets. The more severe the change in geometry, the larger the stress concentration.

- For design or analysis, it is not necessary to know the exact shear-stress distribution on the cross section. Instead, it is possible to obtain the maximum shear stress using a stress concentration factor, K, that has been determined through experiment, and is only a function of the geometry of the shaft.

- Normally a stress concentration in a ductile shaft subjected to a static torque will *not* have to be considered in design; however, if the material is *brittle*, or subjected to *fatigue* loadings, then stress concentrations become important.

EXAMPLE 5.13

The stepped shaft shown in Fig. 5–33a is supported by bearings at A and B. Determine the maximum stress in the shaft due to the applied torques. The shoulder fillet at the junction of each shaft has a radius of r = 6 mm.

30 N·m

60 N·m

B

30 N·m

40 mm

A

20 mm

(a)

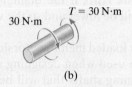

30 N·m   **T = 30 N·m**

(b)

τ_max = 3.10 MPa

Shear-stress distribution predicted by torsion formula

Actual shear-stress distribution caused by stress concentration

(c)

Fig. 5–33

SOLUTION

Internal Torque. By inspection, moment equilibrium about the axis of the shaft is satisfied. Since the maximum shear stress occurs at the rooted ends of the *smaller* diameter shafts, the internal torque (30 N · m) can be found there by applying the method of sections, Fig. 5–33b.

Maximum Shear Stress. The stress-concentration factor can be determined by using Fig. 5–32. From the shaft geometry we have

$$\frac{D}{d} = \frac{2(40 \text{ mm})}{2(20 \text{ mm})} = 2$$

$$\frac{r}{d} = \frac{6 \text{ mm}}{2(20 \text{ mm})} = 0.15$$

Thus, the value of $K = 1.3$ is obtained.
     Applying Eq. 5–21, we have

$$\tau_{max} = K\frac{Tc}{J}; \quad \tau_{max} = 1.3\left[\frac{30 \text{ N} \cdot \text{m} (0.020 \text{ m})}{(\pi/2)(0.020 \text{ m})^4}\right] = 3.10 \text{ MPa} \quad Ans.$$

NOTE: From experimental evidence, the actual stress distribution along a radial line of the cross section at the critical section looks similar to that shown in Fig. 5–33c. Notice how this compares with the linear stress distribution found from the torsion formula.

*5.9   Inelastic Torsion

If the torsional loadings applied to the shaft are excessive, then the material may yield, and, consequently, a "plastic analysis" must be used to determine the shear-stress distribution and the angle of twist. To perform this analysis, then as before, it is necessary to meet the conditions of both deformation and equilibrium for the shaft.

It was shown in Sec. 5.1 that regardless of the material behavior, the shear strains that develop in a circular shaft will vary *linearly*, from zero at the center of the shaft to a maximum at its outer boundary, Fig. 5–34*a*. Also, the resultant internal torque at the section must be equivalent to the torque caused by the entire shear-stress distribution over the cross section. This condition can be expressed mathematically by considering the shear stress τ acting on an element of area dA located a distance ρ from the center of the shaft, Fig. 5–34*b*. The force produced by this stress is $dF = \tau \, dA$, and the torque produced is $dT = \rho \, dF = \rho(\tau \, dA)$. For the entire shaft we require

$$T = \int_A \rho \tau \, dA \qquad (5\text{–}22)$$

If the area dA over which τ acts can be defined as a *differential ring* having an area of $dA = 2\pi\rho \, d\rho$, Fig. 5–34*c*, then the above equation can be written as

$$\boxed{T = 2\pi \int_0^c \tau \rho^2 \, d\rho} \qquad (5\text{–}23)$$

These conditions of geometry and loading will now be used to determine the shear-stress distribution in a shaft when the shaft is subjected to two types of torque.

Elastic-Plastic Torque.

Let us consider the material in the shaft to exhibit an elastic-perfectly plastic behavior. As shown in Fig. 5–35*a*, this is characterized by a shear stress–strain diagram for which the material undergoes an increasing amount of shear strain when the shear stress reaches the yield point τ_Y.

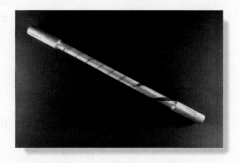

Severe twist of an aluminum specimen caused by the application of a plastic torque.

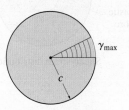

Linear shear–strain distribution

(a)

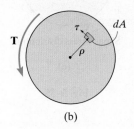

(b)

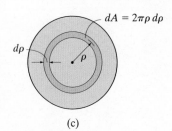

(c)

Fig. 5–34

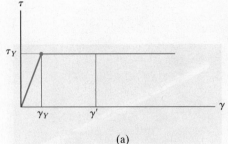

(a)

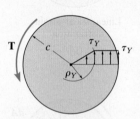

Plastic annulus

Elastic core

Shear–strain distribution

(b)

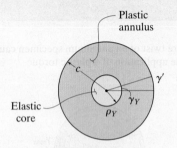

Shear–stress distribution

(c)

Fig. 5–35

If the internal torque produces the maximum *elastic* shear strain γ_Y, at the outer boundary of the shaft, then the maximum elastic torque T_Y that produces this distribution can be found from the torsion formula, $\tau_Y = T_Y c / [(\pi/2)c^4]$, so that

$$T_Y = \frac{\pi}{2} \tau_Y c^3 \qquad (5\text{–}24)$$

Furthermore, the angle of twist can be determined from Eq. 5–13, namely,

$$d\phi = \gamma \frac{dx}{\rho} \qquad (5\text{–}25)$$

If the applied torque increases in magnitude above T_Y, it will begin to cause yielding. First at the outer boundary of the shaft, $\rho = c$, and then, as the maximum shear strain increases to, say, γ' in Fig. 5–35a, the yielding boundary will progress inward toward the shaft's center, Fig. 5–35b. As shown, this produces an *elastic core*, where, by proportion, the radius of the core is $\rho_Y = (\gamma_Y/\gamma')c$. Also, the outer portion of the material forms a *plastic annulus* or ring, since the shear strains γ within this region are greater than γ_Y. The corresponding shear-stress distribution along a radial line of the shaft is shown in Fig. 5–35c. It is established by taking successive points on the shear-strain distribution in Fig. 5–35b and finding the corresponding value of shear stress from the τ–γ diagram, Fig. 5–35a. For example, at $\rho = c$, γ' gives τ_Y, and at $\rho = \rho_Y$, γ_Y also gives τ_Y; etc.

Since τ in Fig. 5–35c can now be expressed as a function of ρ, we can apply Eq. 5–23 to determine the torque. We have

$$T = 2\pi \int_0^c \tau \rho^2 \, d\rho$$

$$= 2\pi \int_0^{\rho_Y} \left(\tau_Y \frac{\rho}{\rho_Y} \right) \rho^2 \, d\rho + 2\pi \int_{\rho_Y}^c \tau_Y \rho^2 \, d\rho$$

$$= \frac{2\pi}{\rho_Y} \tau_Y \int_0^{\rho_Y} \rho^3 \, d\rho + 2\pi \tau_Y \int_{\rho_Y}^c \rho^2 \, d\rho$$

$$= \frac{\pi}{2\rho_Y} \tau_Y \rho_Y^4 + \frac{2\pi}{3} \tau_Y (c^3 - \rho_Y^3)$$

$$= \frac{\pi \tau_Y}{6} (4c^3 - \rho_Y^3) \qquad (5\text{–}26)$$

Plastic Torque. Further increases in T tend to shrink the radius of the elastic core until all the material will yield, i.e., $\rho_Y \rightarrow 0$, Fig. 5–35b. The material of the shaft will then be subjected to *perfectly plastic behavior* and the shear-stress distribution becomes uniform, so that $\tau = \tau_Y$, Fig. 5–35d. We can now apply Eq. 5–23 to determine the **plastic torque** T_p, which represents the largest possible torque the shaft will support.

$$T_p = 2\pi \int_0^c \tau_Y \rho^2 \, d\rho$$

$$= \frac{2\pi}{3} \tau_Y c^3 \qquad (5\text{–}27)$$

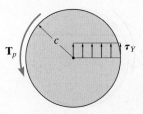

Fully plastic torque

(d)

Fig. 5–35 (cont.)

Compared with the maximum elastic torque T_Y, Eq. 5–24, it can be seen that

$$T_p = \frac{4}{3} T_Y$$

In other words, the plastic torque is 33% greater than the maximum elastic torque.

Unfortunately, the angle of twist ϕ for the shear-stress distribution *cannot* be uniquely defined. This is because $\tau = \tau_Y$ does not correspond to any unique value of shear strain $\gamma \geq \gamma_Y$. As a result, once $\mathbf{T}_p$ is applied, the shaft will continue to deform or twist with no corresponding increase in shear stress.

*5.10 Residual Stress

When a shaft is subjected to plastic shear strains caused by torsion, removal of the torque will cause some shear stress to remain in the shaft. This stress is referred to as **residual stress**, and its distribution can be calculated using superposition and elastic recovery. (See Sec. 4.9.)

For example, if T_p causes the material at the outer boundary of the shaft to be strained to γ_1, shown as point C on the τ–γ curve in Fig. 5–36, the release of T_p will cause a reverse shear stress, such that the material behavior will follow the straight-lined segment CD, creating some *elastic recovery* of the shear strain γ_1. This line is parallel to the initial straight-lined portion AB of the τ–γ diagram, and thus both lines have a slope G as indicated.

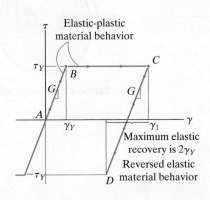

Fig. 5–36

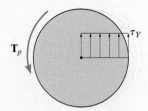

Plastic torque applied
causing plastic shear strains
throughout the shaft
(a)

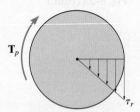

Plastic torque reversed
causing elastic shear strains
throughout the shaft
(b)

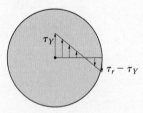

Residual shear–stress
distribution in shaft
(c)

Since elastic recovery occurs, we can superimpose on the plastic torque stress distribution in Fig. 5–37a a *linear stress distribution* caused by applying the plastic torque $\mathbf{T}_p$ in the *opposite* direction, Fig. 5–37b. Here the maximum shear stress τ_r, for this stress distribution, is called the *modulus of rupture* for torsion. It is determined from the torsion formula,* which gives

$$\tau_r = \frac{T_p c}{J} = \frac{T_p c}{(\pi/2)c^4}$$

Using Eq. 5–27,

$$\tau_r = \frac{[(2/3)\pi\tau_Y c^3]c}{(\pi/2)c^4} = \frac{4}{3}\tau_Y$$

Note that reversed application of $\mathbf{T}_p$ using the linear shear-stress distribution in Fig. 5–37b is possible here, since the maximum recovery for the elastic shear strain is $2\gamma_Y$, as noted in Fig. 5–37. This corresponds to a maximum applied shear stress of $2\tau_Y$, which is *greater* than the maximum shear stress of $\frac{4}{3}\tau_Y$ calculated above. Hence, by superimposing the stress distributions involving applications and then removal of the plastic torque, we obtain the residual shear-stress distribution in the shaft as shown in Fig. 5–37c. It should be noted from this diagram that the shear stress at the center of the shaft, shown as τ_Y, must actually be *zero*, since the material along the axis of the shaft is never strained. The reason this is not zero is because we assumed that *all* the material of the shaft was strained beyond the yield point in order to determine the plastic torque, Fig. 5–37a. To be more realistic, an elastic–plastic torque should be considered when modeling the material behavior. Doing so leads to the superposition of the stress distribution shown in Fig. 5–37d.

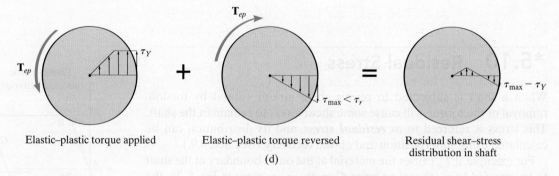

Elastic–plastic torque applied   Elastic–plastic torque reversed   Residual shear–stress
distribution in shaft
(d)

Fig. 5–37

*The torsion formula is valid only when the material behaves in a linear elastic manner; however, the modulus of rupture is so named because it assumes that the material behaves elastically and then suddenly *ruptures* at the proportional limit.

Ultimate Torque.

In the general case, most engineering materials will have a shear stress–strain diagram as shown in Fig. 5–38a. Consequently, if T is increased so that the maximum shear strain in the shaft becomes $\gamma = \gamma_u$, Fig. 5–38b, then, by proportion γ_Y occurs at $\rho_Y = (\gamma_Y/\gamma_u)c$. Likewise, the shear strains at, say, $\rho = \rho_1$ and $\rho = \rho_2$, can be found by proportion, i.e., $\gamma_1 = (\rho_1/c)\gamma_u$ and $\gamma_2 = (\rho_2/c)\gamma_u$. If the corresponding values of τ_1, τ_Y, τ_2, and τ_u are taken from the $\tau-\gamma$ diagram and plotted, we obtain the shear-stress distribution, which acts along a radial line on the cross section, Fig. 5–38c. The torque produced by this stress distribution is called the ***ultimate torque***, T_u.

The magnitude of $\mathbf{T}_u$ can be determined by "graphically" integrating Eq. 5–23. To do this, the cross-sectional area of the shaft is segmented into a finite number of rings, such as the one shown shaded in Fig. 5–38d. The area of this ring, $\Delta A = 2\pi\rho \, \Delta\rho$, is multiplied by the shear stress τ that acts on it, so that the force $\Delta F = \tau \, \Delta A$ can be determined. The torque created by this force is then $\Delta T = \rho \, \Delta F = \rho(\tau \, \Delta A)$. The addition of all the torques for the entire cross section, as determined in this manner, gives the ultimate torque T_u; that is, Eq. 5–23 becomes $T_u \approx 2\pi\Sigma\tau\rho^2 \, \Delta\rho$. If, however, the stress distribution can be expressed as an analytical function, $\tau = f(\rho)$, as in the elastic and plastic torque cases, then the integration of Eq. 5–23 can be carried out directly.

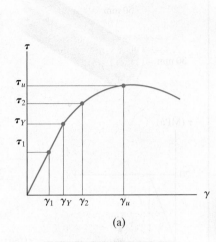

(a)

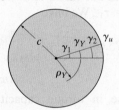

Ultimate shear-strain distribution

(b)

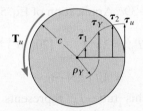

Ultimate shear-stress distribution

(c)

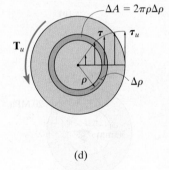

(d)

Fig. 5–38

Important Points

- The *shear-strain distribution* along a radial line on the cross section of a shaft is based on geometric considerations, and it is found to *always* vary linearly along the radial line. Once it is established, the shear-stress distribution can then be determined using the shear stress–strain diagram.

- If the shear-stress distribution for the shaft is established, it produces a torque about the axis of the shaft that is equivalent to the resultant internal torque acting on the cross section.

- *Perfectly plastic behavior* assumes the shear-stress distribution is *constant*. When it occurs, the shaft will continue to twist with no increase in torque. This torque is called the *plastic torque*.

EXAMPLE 5.14

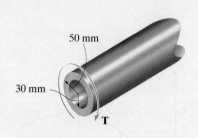

50 mm

30 mm

T

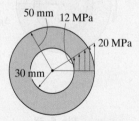

τ (MPa)

20

0.286 (10^{-3})

γ (rad)

(a)

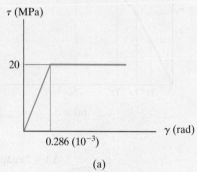

50 mm   12 MPa

20 MPa

30 mm

Elastic shear–stress distribution

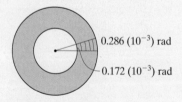

0.286 (10^{-3}) rad

0.172 (10^{-3}) rad

Elastic shear–strain distribution

(b)

Fig. 5–39

The tubular shaft in Fig. 5–39a is made of an aluminum alloy that is assumed to have an elastic-plastic τ–γ diagram as shown. Determine the maximum torque that can be applied to the shaft without causing the material to yield, and the maximum torque or plastic torque that can be applied to the shaft. Also, what should the minimum shear strain at the outer wall be in order to develop a fully plastic torque?

SOLUTION

Maximum Elastic Torque.   We require the shear stress at the outer fiber to be 20 MPa. Using the torsion formula, we have

$$\tau_Y = \frac{T_Y c}{J}; \qquad 20(10^6)\ \text{N/m}^2 = \frac{T_Y(0.05\ \text{m})}{(\pi/2)[(0.05\ \text{m})^4 - (0.03\ \text{m})^4]}$$

$$T_Y = 3.42\ \text{kN} \cdot \text{m} \qquad\qquad Ans.$$

The shear-stress and shear-strain distributions for this case are shown in Fig. 5–39b. The values at the tube's inner wall have been obtained by proportion.

Plastic Torque.   The shear-stress distribution in this case is shown in Fig. 5–39c. Application of Eq. 5–23 requires $\tau = \tau_Y$. We have

$$T_p = 2\pi \int_{0.03\ \text{m}}^{0.05\ \text{m}} [20(10^6)\ \text{N/m}^2]\rho^2\ d\rho = 125.66(10^6)\frac{1}{3}\rho^3 \Big|_{0.03\ \text{m}}^{0.05\ \text{m}}$$

$$= 4.11\ \text{kN} \cdot \text{m} \qquad\qquad Ans.$$

For this tube T_p represents a 20% increase in torque capacity compared with the elastic torque T_Y.

Outer Radius Shear Strain.   The tube becomes fully plastic when the shear strain at the *inner wall* becomes $0.286(10^{-3})$ rad, as shown in Fig. 5–39c. Since the shear strain *remains linear* over the cross section, the plastic strain at the outer fibers of the tube in Fig. 5–39c is determined by proportion.

$$\frac{\gamma_o}{50\ \text{mm}} = \frac{0.286(10^{-3})\ \text{rad}}{30\ \text{mm}}$$

$$\gamma_o = 0.477(10^{-3})\ \text{rad} \qquad\qquad Ans.$$

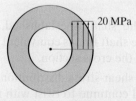

20 MPa

Plastic shear–stress distribution

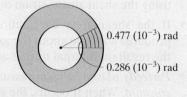

0.477 (10^{-3}) rad

0.286 (10^{-3}) rad

Initial plastic shear–strain distribution

(c)

EXAMPLE | 5.15

A solid circular shaft has a radius of 20 mm and length of 1.5 m. The material has an elastic–plastic $\tau-\gamma$ diagram as shown in Fig. 5–40a. Determine the torque needed to twist the shaft $\phi = 0.6$ rad.

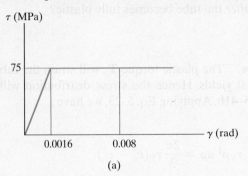

(a)

SOLUTION

We will first obtain the shear-strain distribution, then establish the shear-stress distribution. Once this is known, the applied torque can be determined.

The maximum shear strain occurs at the surface of the shaft, $\rho = c$. Since the angle of twist is $\phi = 0.6$ rad for the entire 1.5-m length of the shaft, then using Eq. 5–25, for the entire length we have

$$\phi = \gamma \frac{L}{\rho}; \qquad 0.6 = \frac{\gamma_{max}(1.5 \text{ m})}{(0.02 \text{ m})}$$

$$\gamma_{max} = 0.008 \text{ rad}$$

The shear-strain distribution is shown in Fig. 5–40b. Note that yielding of the material occurs since $\gamma_{max} > \gamma_Y = 0.0016$ rad in Fig. 5–40a. The radius of the elastic core, ρ_Y, can be obtained by proportion. From Fig. 5–40b,

$$\frac{\rho_Y}{0.0016} = \frac{0.02 \text{ m}}{0.008}$$

$$\rho_Y = 0.004 \text{ m} = 4 \text{ mm}$$

Based on the shear-strain distribution, the shear-stress distribution, plotted over a radial line segment, is shown in Fig. 5–40c. The torque can now be obtained using Eq. 5–26. Substituting in the numerical data yields

$$T = \frac{\pi \tau_Y}{6}(4c^3 - \rho_Y^3)$$

$$= \frac{\pi[75(10^6) \text{ N/m}^2]}{6}[4(0.02 \text{ m})^3 - (0.004 \text{ m})^3]$$

$$= 1.25 \text{ kN} \cdot \text{m} \qquad \qquad Ans.$$

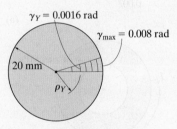

Shear–strain distribution

(b)

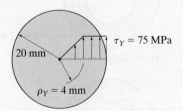

Shear–stress distribution

(c)

Fig. 5–40

EXAMPLE | 5.16

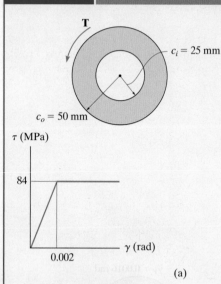

A tube in Fig. 5–41a has a length of 1.5 mm and the material has an elastic-plastic τ–γ diagram, also shown in Fig. 5–41a. Determine the plastic torque T_p. What is the residual shear-stress distribution if $\mathbf{T}_p$ is removed *just after* the tube becomes fully plastic?

SOLUTION

Plastic Torque. The plastic torque $\mathbf{T}_p$ will strain the tube such that all the material yields. Hence the stress distribution will appear as shown in Fig. 5–41b. Applying Eq. 5–23, we have

$$T_p = 2\pi \int_{c_i}^{c_o} \tau_Y \rho^2 \, d\rho = \frac{2\pi}{3} \tau_Y (c_o^3 - c_i^3)$$

$$= \frac{2\pi}{3} (84(10^3) \text{ kN/m}^2)[(0.050 \text{ m})^3 - (0.025 \text{ m})^3] = 19.24 \text{ kN} \cdot \text{m}$$
$$Ans.$$

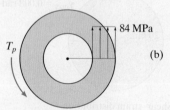

Plastic torque applied

When the tube just becomes fully plastic, yielding has started at the inner wall, i.e., at $c_i = 0.025$ m, $\gamma_Y = 0.002$ rad, Fig. 5–41a. The angle of twist that occurs can be determined from Eq. 5–25, which for the entire tube becomes

$$\phi_p = \gamma_Y \frac{L}{c_i} = \frac{(0.002)(1.5 \text{ m})}{(0.025 \text{ m})} = 0.120 \text{ rad}$$

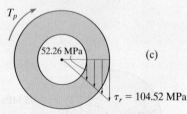

Plastic torque reversed

When $\mathbf{T}_p$ is *removed*, or in effect reapplied in the opposite direction, then the "fictitious" linear shear-stress distribution shown in Fig. 5–41c must be superimposed on the one shown in Fig. 5–41b. In Fig. 5–41c the maximum shear stress or the modulus of rupture is found from the torsion formula

$$\tau_r = \frac{T_p c_o}{J} = \frac{(19.24 \text{ kN} \cdot \text{m})(0.050 \text{ m})}{(\pi/2)[(0.050 \text{ m})^4 - (0.025 \text{ m})^4]} = 104.52(10^3) \text{ kN/m}^2$$

$$= 104.52 \text{ MPa}$$

Also, at the inner wall of the tube the shear stress is

$$\tau_i = 104.52 \text{ MPa} \left(\frac{25 \text{ mm}}{50 \text{ mm}} \right) = 52.26 \text{ MPa} \qquad Ans.$$

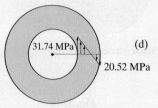

Residual shear–stress distribution

Fig. 5–41

The resultant residual shear-stress distribution is shown in Fig. 5–41d.

PROBLEMS

*5–120. The steel used for the shaft has an allowable shear stress of $\tau_{allow} = 8$ MPa. If the members are connected with a fillet weld of radius $r = 4$ mm, determine the maximum torque T that can be applied.

5–123. The steel shaft is made from two segments: AB and BC, which are connected using a fillet weld having a radius of 2.8 mm. Determine the maximum shear stress developed in the shaft.

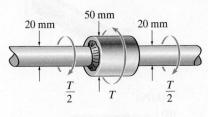

Prob. 5–120

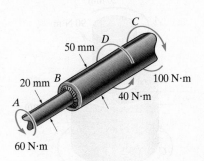

Prob. 5–123

•5–121. The built-up shaft is to be designed to rotate at 720 rpm while transmitting 30 kW of power. Is this possible? The allowable shear stress is $\tau_{allow} = 12$ MPa.

5–122. The built-up shaft is designed to rotate at 540 rpm. If the radius of the fillet weld connecting the shafts is $r = 7.20$ mm, and the allowable shear stress for the material is $\tau_{allow} = 55$ MPa, determine the maximum power the shaft can transmit.

*5–124. The steel used for the shaft has an allowable shear stress of $\tau_{allow} = 8$ MPa. If the members are connected together with a fillet weld of radius $r = 2.25$ mm, determine the maximum torque T that can be applied.

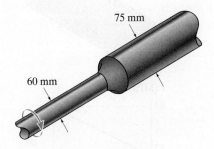

Probs. 5–121/122

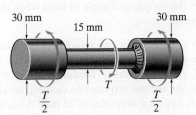

Prob. 5–124

•5–125. The assembly is subjected to a torque of 90 N · m. If the allowable shear stress for the material is $\tau_{allow} = 84$ MPa, determine the radius of the smallest size fillet that can be used to transmit the torque.

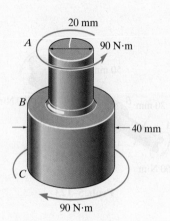

Prob. 5–125

5–126. A solid shaft is subjected to the torque T, which causes the material to yield. If the material is elastic plastic, show that the torque can be expressed in terms of the angle of twist ϕ of the shaft as $T = \frac{4}{3}T_Y(1 - \phi^3_Y/4\phi^3)$, where T_Y and ϕ_Y are the torque and angle of twist when the material begins to yield.

5–127. A solid shaft having a diameter of 50 mm is made of elastic-plastic material having a yield stress of $\tau_Y = 112$ MPa and shear modulus of $G = 84$ GPa. Determine the torque required to develop an elastic core in the shaft having a diameter of 25 mm. Also, what is the plastic torque?

*5–128. Determine the torque needed to twist a short 3-mm-diameter steel wire through several revolutions if it is made from steel assumed to be elastic plastic and having a yield stress of $\tau_Y = 80$ MPa. Assume that the material becomes fully plastic.

•5–129. The solid shaft is made of an elastic-perfectly plastic material as shown. Determine the torque T needed to form an elastic core in the shaft having a radius of $\rho_Y = 20$ mm. If the shaft is 3 m long, through what angle does one end of the shaft twist with respect to the other end? When the torque is removed, determine the residual stress distribution in the shaft and the permanent angle of twist.

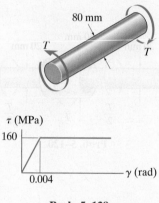

Prob. 5–129

5–130. The shaft is subjected to a maximum shear strain of 0.0048 rad. Determine the torque applied to the shaft if the material has strain hardening as shown by the shear stress–strain diagram.

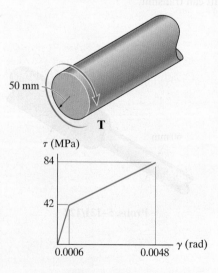

Prob. 5–130

5–131. An 80-mm diameter solid circular shaft is made of an elastic-perfectly plastic material having a yield shear stress of $\tau_Y = 125$ MPa. Determine (a) the maximum elastic torque T_Y; and (b) the plastic torque T_p.

*5–132. The hollow shaft has the cross section shown and is made of an elastic-perfectly plastic material having a yield shear stress of τ_Y. Determine the ratio of the plastic torque T_p to the maximum elastic torque T_Y.

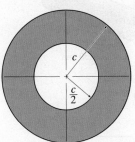

Prob. 5–132

5–133. The shaft consists of two sections that are rigidly connected. If the material is elastic plastic as shown, determine the largest torque T that can be applied to the shaft. Also, draw the shear-stress distribution over a radial line for each section. Neglect the effect of stress concentration.

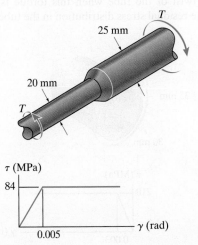

Prob. 5–133

5–134. The hollow shaft is made of an elastic-perfectly plastic material having a shear modulus of G and a yield shear stress of τ_Y. Determine the applied torque $\mathbf{T}_p$ when the material of the inner surface is about to yield (plastic torque). Also, find the corresponding angle of twist and the maximum shear strain. The shaft has a length of L.

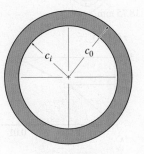

Prob. 5–134

5–135. The hollow shaft has inner and outer diameters of 60 mm and 80 mm, respectively. If it is made of an elastic-perfectly plastic material, which has the $\tau-\gamma$ diagram shown, determine the reactions at the fixed supports A and C.

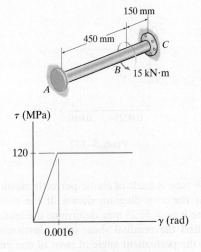

Prob. 5–135

*5–136. The tubular shaft is made of a strain-hardening material having a $\tau-\gamma$ diagram as shown. Determine the torque T that must be applied to the shaft so that the maximum shear strain is 0.01 rad.

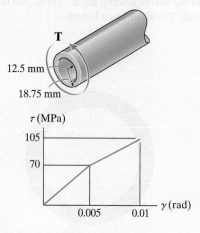

Prob. 5–136

•5–137. The shear stress–strain diagram for a solid 50-mm-diameter shaft can be approximated as shown in the figure. Determine the torque T required to cause a maximum shear stress in the shaft of 125 MPa. If the shaft is 1.5 m long, what is the corresponding angle of twist?

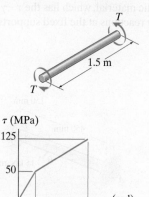

Prob. 5–137

5–138. A tube is made of elastic-perfectly plastic material, which has the $\tau-\gamma$ diagram shown. If the radius of the elastic core is $\rho_Y = 56.25$ mm, determine the applied torque T. Also, find the residual shear-stress distribution in the shaft and the permanent angle of twist of one end relative to the other when the torque is removed.

5–139. The tube is made of elastic-perfectly plastic material, which has the $\tau-\gamma$ diagram shown. Determine the torque T that just causes the inner surface of the shaft to yield. Also, find the residual shear-stress distribution in the shaft when the torque is removed.

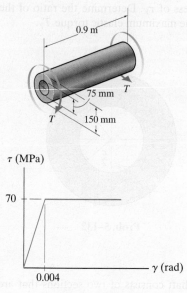

Probs. 5–138/139

*5–140. The 2-m-long tube is made of an elastic-perfectly plastic material as shown. Determine the applied torque T that subjects the material at the tube's outer edge to a shear strain of $\gamma_{max} = 0.006$ rad. What would be the permanent angle of twist of the tube when this torque is removed? Sketch the residual stress distribution in the tube.

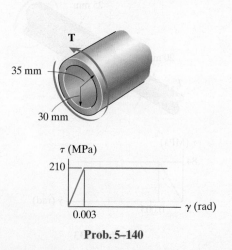

Prob. 5–140

•**5–141.**   A steel alloy core is bonded firmly to the copper alloy tube to form the shaft shown. If the materials have the $\tau-\gamma$ diagrams shown, determine the torque resisted by the core and the tube.

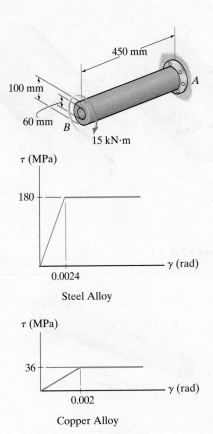

450 mm

100 mm

60 mm

B

A

15 kN·m

τ (MPa)

180

0.0024

γ (rad)

Steel Alloy

τ (MPa)

36

0.002

γ (rad)

Copper Alloy

Prob. 5–141

5–142.   A torque is applied to the shaft of radius r. If the material has a shear stress–strain relation of $\tau = k\gamma^{1/6}$, where k is a constant, determine the maximum shear stress in the shaft.

r

T

Prob. 5–142

CHAPTER REVIEW

Torque causes a shaft having a circular cross section to twist, such that the shear strain in the shaft is proportional to its radial distance from the center of the shaft. Provided the material is homogeneous and linear elastic, then the shear stress is determined from the torsion formula,

$$\tau = \frac{T\rho}{J}$$

The design of a shaft requires finding the geometric parameter,

$$\frac{J}{c} = \frac{T}{\tau_{allow}}$$

Often the power P supplied to a shaft rotating at ω is reported, in which case the torque is determined from $P = T\omega$.

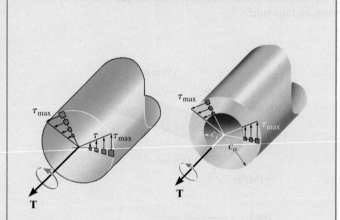

Ref.: Sections 5.2 & 5.3

The angle of twist of a circular shaft is determined from

$$\phi = \int_0^L \frac{T(x)\,dx}{JG}$$

If the internal torque and JG are constant within each segment of the shaft then

$$\phi = \sum \frac{TL}{JG}$$

For application, it is necessary to use a sign convention for the internal torque and to be sure the material remains linear elastic.

Ref.: Section 5.4

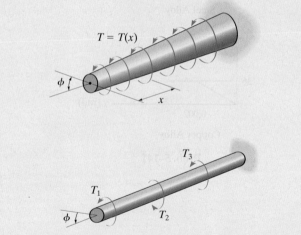

If the shaft is statically indeterminate, then the reactive torques are determined from equilibrium, compatibility of twist, and a torque-twist relationship, such as $\phi = TL/JG$.

Ref.: Section 5.5

Solid non-circular shafts tend to warp out of plane when subjected to a torque. Formulas are available to determine the maximum elastic shear stress and the twist for these cases.

Ref.: Section 5.6

The average shear stress in thin-walled tubes is determined by assuming the shear stress across each thickness t of the tube is constant. Its value is determined from $\tau_{\text{avg}} = \dfrac{T}{2t A_m}$.

Ref.: Section 5.7

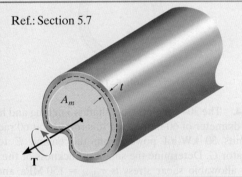

Stress concentrations occur in shafts when the cross section suddenly changes. The maximum shear stress is determined using a stress concentration factor K, which is determined from experiment and represented in graphical form. Once obtained, $\tau_{\max} = K\left(\dfrac{Tc}{J}\right)$.

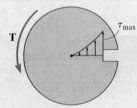

Ref.: Section 5.8

If the applied torque causes the material to exceed the elastic limit, then the stress distribution will not be proportional to the radial distance from the centerline of the shaft. Instead, the internal torque is related to the stress distribution using the shear-stress–shear-strain diagram and equilibrium.

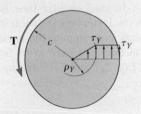

Ref.: Section 5.9

If a shaft is subjected to a plastic torque, which is then released, it will cause the material to respond elastically, thereby causing residual shear stress to be developed in the shaft.

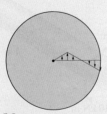

Ref.: Section 5.9

5

REVIEW PROBLEMS

5–143. Consider a thin-walled tube of mean radius r and thickness t. Show that the maximum shear stress in the tube due to an applied torque T approaches the average shear stress computed from Eq. 5–18 as $r/t \rightarrow \infty$.

Prob. 5–143

*5–144. The 304 stainless steel shaft is 3 m long and has an outer diameter of 60 mm. When it is rotating at 60 rad/s, it transmits 30 kW of power from the engine E to the generator G. Determine the smallest thickness of the shaft if the allowable shear stress is $\tau_{\text{allow}} = 150$ MPa and the shaft is restricted not to twist more than 0.08 rad.

Prob. 5–144

•5–145. The A-36 steel circular tube is subjected to a torque of 10 kN · m. Determine the shear stress at the mean radius $\rho = 60$ mm and compute the angle of twist of the tube if it is 4 m long and fixed at its far end. Solve the problem using Eqs. 5–7 and 5–15 and by using Eqs. 5–18 and 5–20.

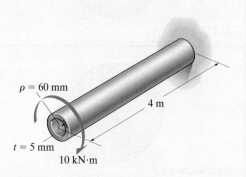

$\rho = 60$ mm

4 m

$t = 5$ mm

10 kN·m

Prob. 5–145

5–146. Rod AB is made of A-36 steel with an allowable shear stress of $(\tau_{\text{allow}})_{\text{st}} = 75$ MPa, and tube BC is made of AM1004-T61 magnesium alloy with an allowable shear stress of $(\tau_{\text{allow}})_{\text{mg}} = 45$ MPa. The angle of twist of end C is not allowed to exceed 0.05 rad. Determine the maximum allowable torque $\mathbf{T}$ that can be applied to the assembly.

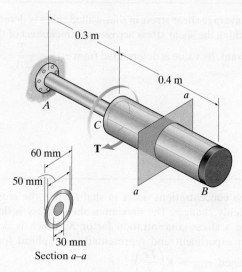

0.3 m

0.4 m

a

A

C

$\mathbf{T}$

60 mm

50 mm

30 mm

a

B

Section a–a

Prob. 5–146

5–147. A shaft has the cross section shown and is made of 2014-T6 aluminum alloy having an allowable shear stress of $\tau_{\text{allow}} = 125$ MPa. If the angle of twist per meter length is not allowed to exceed 0.03 rad, determine the required minimum wall thickness t to the nearest millimeter when the shaft is subjected to a torque of $T = 15$ kN · m.

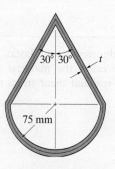

30° 30° t

75 mm

Prob. 5–147

*5–148. The motor A develops a torque at gear B of 750 N · m which is applied along the axis of the 50-mm-diameter A-36 steel shaft CD. This torque is to be transmitted to the pinion gears at E and F. If these gears are temporarily fixed, determine the maximum shear stress in segments CB and BD of the shaft. Also, what is the angle of twist of each of these segments? The bearings at C and D only exert force reactions on the shaft.

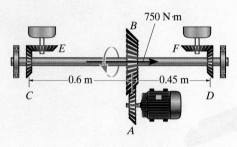

Prob. 5–148

5–149. The coupling consists of two disks fixed to separate shafts, each 25 mm in diameter. The shafts are supported on journal bearings that allow free rotation. In order to limit the torque **T** that can be transmitted, a "shear pin" P is used to connect the disks together. If this pin can sustain an *average* shear force of 550 N before it fails, determine the maximum constant torque T that can be transmitted from one shaft to the other. Also, what is the maximum shear stress in each shaft when the "shear pin" is about to fail?

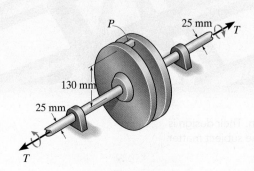

Prob. 5–149

5–150. The rotating flywheel and shaft is brought to a sudden stop at D when the bearing freezes. This causes the flywheel to oscillate clockwise–counterclockwise, so that a point A on the outer edge of the flywheel is displaced through a 10-mm arc in either direction. Determine the maximum shear stress developed in the tubular 304 stainless steel shaft due to this oscillation. The shaft has an inner diameter of 25 mm and an outer diameter of 35 mm. The journal bearings at B and C allow the shaft to rotate freely.

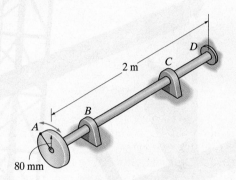

Prob. 5–150

5–151. If the solid shaft AB to which the valve handle is attached is made of C83400 red brass and has a diameter of 10 mm, determine the maximum couple forces F that can be applied to the handle just before the material starts to fail. Take $\tau_{allow} = 40$ MPa. What is the angle of twist of the handle? The shaft is fixed at A.

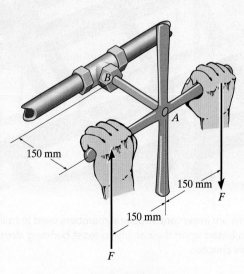

Prob. 5–151

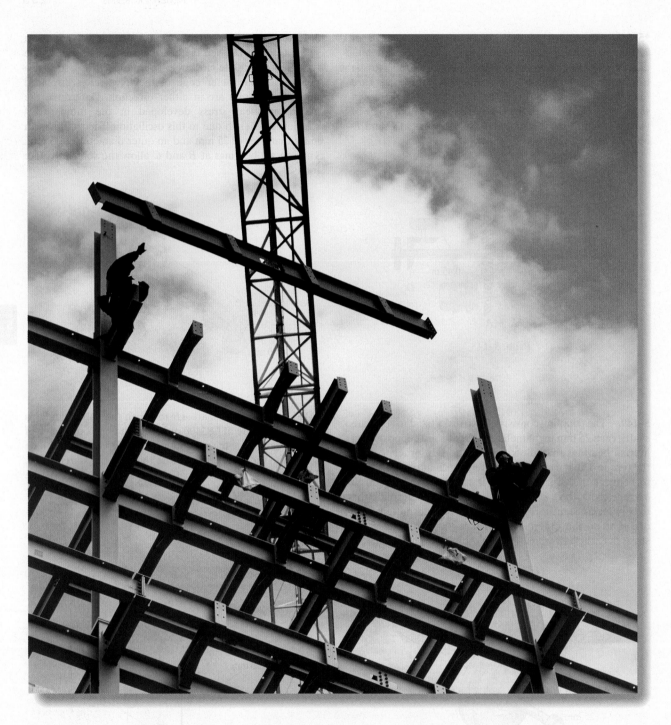

Beams are important structural members used in building construction. Their design is often based upon their ability to resist bending stress, which forms the subject matter of this chapter.

Bending

6

CHAPTER OBJECTIVES

Beams and shafts are important structural and mechanical elements in engineering. In this chapter we will determine the stress in these members caused by bending. The chapter begins with a discussion of how to establish the shear and moment diagrams for a beam or shaft. Like the normal-force and torque diagrams, the shear and moment diagrams provide a useful means for determining the largest shear and moment in a member, and they specify where these maximums occur. Once the internal moment at a section is determined, the bending stress can then be calculated. First we will consider members that are straight, have a symmetric cross section, and are made of homogeneous linear elastic material. Afterward we will discuss special cases involving unsymmetric bending and members made of composite materials. Consideration will also be given to curved members, stress concentrations, inelastic bending, and residual stresses.

Check out the Companion Website for the following video solution(s) that can be found in this chapter.

- Section 1, 2
  Shear and Moment of a T-Beam
- Section 4
  Flexure Formula and Resultant Force
- Section 5
  Unsymmetric Bending
- Section 6
  Sandwich Composite Beams
- Section 7
  Steel-Reinforced Composite Beams
- Section 9
  Stress Concentrations – Bending
- Section 10
  Elastic and Plastic Moment in Bending

6.1 Shear and Moment Diagrams

Members that are slender and support loadings that are applied perpendicular to their longitudinal axis are called *beams*. In general, beams are long, straight bars having a constant cross-sectional area. Often they are classified as to how they are supported. For example, a *simply supported beam* is pinned at one end and roller supported at the other, Fig. 6–1, a *cantilevered beam* is fixed at one end and free at the other, and an *overhanging beam* has one or both of its ends freely extended over the supports. Beams are considered among the most important of all structural elements. They are used to support the floor of a building, the deck of a bridge, or the wing of an aircraft. Also, the axle of an automobile, the boom of a crane, even many of the bones of the body act as beams.

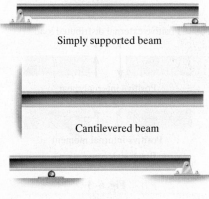

Simply supported beam

Cantilevered beam

Overhanging beam

Fig. 6–1

Because of the applied loadings, beams develop an internal shear force and bending moment that, in general, vary from point to point along the axis of the beam. In order to properly design a beam it therefore becomes necessary to determine the *maximum* shear and moment in the beam. One way to do this is to express V and M as functions of their arbitrary position x along the beam's axis. These *shear and moment functions* can then be plotted and represented by graphs called **shear and moment diagrams**. The maximum values of V and M can then be obtained from these graphs. Also, since the shear and moment diagrams provide detailed information about the *variation* of the shear and moment along the beam's axis, they are often used by engineers to decide where to place reinforcement materials within the beam or how to proportion the size of the beam at various points along its length.

In order to formulate V and M in terms of x we must choose the origin and the positive direction for x. Although the choice is arbitrary, most often the origin is located at the left end of the beam and the positive direction is to the right.

In general, the internal shear and moment functions of x will be *discontinuous*, or their slope will be discontinuous, at points where a distributed load changes or where concentrated forces or couple moments are applied. Because of this, the shear and moment functions must be determined for *each region* of the beam *between* any two discontinuities of loading. For example, coordinates x_1, x_2, and x_3 will have to be used to describe the variation of V and M throughout the length of the beam in Fig. 6–2. These coordinates will be valid *only* within the regions from A to B for x_1, from B to C for x_2, and from C to D for x_3.

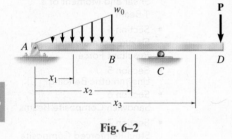

Fig. 6–2

Positive external distributed load

Positive internal shear

Positive internal moment

Beam sign convention

Fig. 6–3

Beam Sign Convention.

Before presenting a method for determining the shear and moment as functions of x and later plotting these functions (shear and moment diagrams), it is first necessary to establish a *sign convention* so as to define "positive" and "negative" values for V and M. Although the choice of a sign convention is arbitrary, here we will use the one often used in engineering practice and shown in Fig. 6–3. The *positive directions* are as follows: the *distributed load* acts *upward* on the beam; the internal *shear force* causes a *clockwise* rotation of the beam segment on which it acts; and the internal *moment* causes *compression* in the *top fibers* of the segment such that it bends the segment so that it holds water. Loadings that are opposite to these are considered negative.

Important Points

- *Beams* are long straight members that are subjected to loads perpendicular to their longitudinal axis. They are classified according to the way they are supported, e.g., simply supported, cantilevered, or overhanging.

- In order to properly design a beam, it is important to know the *variation* of the internal shear and moment along its axis in order to find the points where these values are a maximum.

- Using an established sign convention for positive shear and moment, the shear and moment in the beam can be determined as a function of its position x on the beam, and then these functions can be plotted to form the shear and moment diagrams.

Procedure for Analysis

The shear and moment diagrams for a beam can be constructed using the following procedure.

Support Reactions.

- Determine all the reactive forces and couple moments acting on the beam, and resolve all the forces into components acting perpendicular and parallel to the beam's axis.

Shear and Moment Functions.

- Specify separate coordinates x having an origin at the beam's *left end* and extending to regions of the beam between concentrated forces and/or couple moments, or where there is no discontinuity of distributed loading.

- Section the beam at each distance x, and draw the free-body diagram of one of the segments. Be sure **V** and **M** are shown acting in their positive sense, in accordance with the sign convention given in Fig. 6–3.

- The shear is obtained by summing forces perpendicular to the beam's axis.

- To eliminate V, the moment is obtained directly by summing moments about the sectioned end of the segment.

Shear and Moment Diagrams.

- Plot the shear diagram (V versus x) and the moment diagram (M versus x). If numerical values of the functions describing V and M are *positive*, the values are plotted above the x axis, whereas negative values are plotted below the axis.

- Generally it is convenient to show the shear and moment diagrams below the free-body diagram of the beam.

EXAMPLE | 6.1

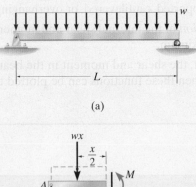

(a)

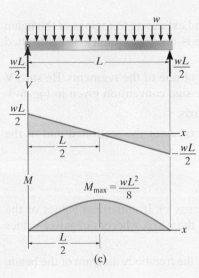

(b)

Draw the shear and moment diagrams for the beam shown in Fig. 6–4a.

SOLUTION

Support Reactions.   The support reactions are shown in Fig. 6–4c.

Shear and Moment Functions.   A free-body diagram of the left segment of the beam is shown in Fig. 6–4b. The distributed loading on this segment, wx, is represented by its resultant force only *after* the segment is isolated as a free-body diagram. This force acts through the centroid of the area comprising the distributed loading, a distance of $x/2$ from the right end. Applying the two equations of equilibrium yields

$$+\uparrow\Sigma F_y = 0; \qquad \frac{wL}{2} - wx - V = 0$$

$$V = w\left(\frac{L}{2} - x\right) \qquad (1)$$

$$\zeta+\Sigma M = 0; \qquad -\left(\frac{wL}{2}\right)x + (wx)\left(\frac{x}{2}\right) + M = 0$$

$$M = \frac{w}{2}(Lx - x^2) \qquad (2)$$

Shear and Moment Diagrams.   The shear and moment diagrams shown in Fig. 6–4c are obtained by plotting Eqs. 1 and 2. The point of *zero shear* can be found from Eq. 1:

$$V = w\left(\frac{L}{2} - x\right) = 0$$

$$x = \frac{L}{2}$$

NOTE: From the moment diagram, this value of x represents the point on the beam where the *maximum moment* occurs, since by Eq. 6–2 (see Sec. 6.2) the *slope* $V = dM/dx = 0$. From Eq. 2, we have

$$M_{max} = \frac{w}{2}\left[L\left(\frac{L}{2}\right) - \left(\frac{L}{2}\right)^2\right]$$

$$= \frac{wL^2}{8}$$

Fig. 6–4

EXAMPLE 6.2

Draw the shear and moment diagrams for the beam shown in Fig. 6–5a.

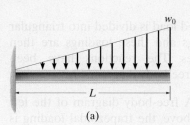

(a)

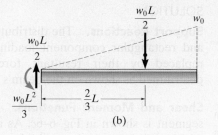

(b)

SOLUTION

Support Reactions. The distributed load is replaced by its resultant force and the reactions have been determined as shown in Fig. 6–5b.

Shear and Moment Functions. A free-body diagram of a beam segment of length x is shown in Fig. 6–5c. Note that the intensity of the triangular load at the section is found by proportion, that is, $w/x = w_0/L$ or $w = w_0 x/L$. With the load intensity known, the resultant of the distributed loading is determined from the area under the diagram. Thus,

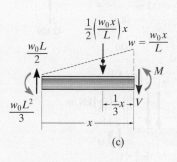

(c)

$$+\uparrow \Sigma F_y = 0; \qquad \frac{w_0 L}{2} - \frac{1}{2}\left(\frac{w_0 x}{L}\right)x - V = 0$$

$$V = \frac{w_0}{2L}(L^2 - x^2) \qquad (1)$$

$$\zeta + \Sigma M = 0; \qquad \frac{w_0 L^2}{3} - \frac{w_0 L}{2}(x) + \frac{1}{2}\left(\frac{w_0 x}{L}\right)x\left(\frac{1}{3}x\right) + M = 0$$

$$M = \frac{w_0}{6L}(-2L^3 + 3L^2 x - x^3) \qquad (2)$$

These results can be checked by applying Eqs. 6–1 and 6–2 of Sec. 6.2, that is,

$$w = \frac{dV}{dx} = \frac{w_0}{2L}(0 - 2x) = -\frac{w_0 x}{L} \qquad \text{OK}$$

$$V = \frac{dM}{dx} = \frac{w_0}{6L}(0 + 3L^2 - 3x^2) = \frac{w_0}{2L}(L^2 - x^2) \qquad \text{OK}$$

Shear and Moment Diagrams. The graphs of Eqs. 1 and 2 are shown in Fig. 6–5d.

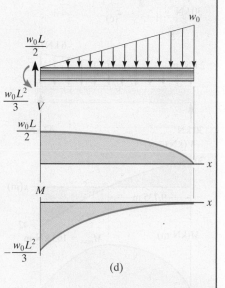

(d)

Fig. 6–5

EXAMPLE | 6.3

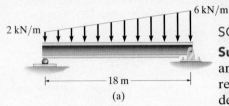

(a)

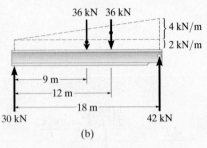

(b)

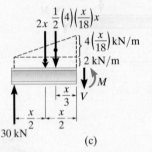

(c)

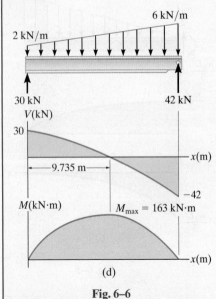

(d)

Fig. 6–6

Draw the shear and moment diagrams for the beam shown in Fig. 6–6a.

SOLUTION

Support Reactions.   The distributed load is divided into triangular and rectangular component loadings and these loadings are then replaced by their resultant forces. The reactions have been determined as shown on the beam's free-body diagram, Fig. 6–6b.

Shear and Moment Functions.   A free-body diagram of the left segment is shown in Fig. 6–6c. As above, the trapezoidal loading is replaced by rectangular and triangular distributions. Note that the intensity of the triangular load at the section is found by proportion. The resultant force and the location of each distributed loading are also shown. Applying the equilibrium equations, we have

$$+\uparrow\Sigma F_y = 0; \quad 30\,\text{kN} - (2\,\text{kN/m})x - \frac{1}{2}(4\,\text{kN/m})\left(\frac{x}{18\,\text{m}}\right)x - V = 0$$

$$V = \left(30 - 2x - \frac{x^2}{9}\right)\text{kN} \tag{1}$$

$$\zeta + \Sigma M = 0;$$

$$-30\,\text{kN}(x) + (2\,\text{kN/m})x\left(\frac{x}{2}\right) + \frac{1}{2}(4\,\text{kN/m})\left(\frac{x}{18\,\text{m}}\right)x\left(\frac{x}{2}\right) + M = 0$$

$$M = \left(30x - x^2 - \frac{x^3}{27}\right)\text{kN}\cdot\text{m} \tag{2}$$

Equation 2 may be checked by noting that $dM/dx = V$, that is, Eq. 1. Also, $w = dV/dx = -2 - \frac{2}{9}x$. This equation checks, since when $x = 0$, $w = -2$ kN/m, and when $x = 18$ m, $w = -6$ kN/m, Fig. 6–6a.

Shear and Moment Diagrams.   Equations 1 and 2 are plotted in Fig. 6–6d. Since the point of maximum moment occurs when $dM/dx = V = 0$ (Eq. 6–2), then, from Eq. 1,

$$V = 0 = 30 - 2x - \frac{x^2}{9}$$

Choosing the positive root,

$$x = 9.735\,\text{m}$$

Thus, from Eq. 2,

$$M_{max} = 30(9.735) - (9.735)^2 - \frac{(9.735)^3}{27}$$

$$= 163\,\text{kN}\cdot\text{m}$$

EXAMPLE 6.4

Draw the shear and moment diagrams for the beam shown in Fig. 6–7a.

 Refer to the Companion Website for the animation of the concepts that can be found in Example 6.4.

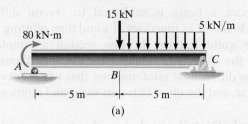

(a)

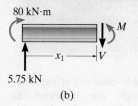

(b)

SOLUTION

Support Reactions. The reactions at the supports have been determined and are shown on the free-body diagram of the beam, Fig. 6–7d.

Shear and Moment Functions. Since there is a discontinuity of distributed load and also a concentrated load at the beam's center, two regions of x must be considered in order to describe the shear and moment functions for the entire beam.

$0 \leq x_1 < 5$ m, Fig. 6–7b:

$$+\uparrow \Sigma F_y = 0; \qquad 5.75 \text{ kN} - V = 0$$
$$V = 5.75 \text{ kN} \qquad (1)$$

$$\zeta + \Sigma M = 0; \qquad -80 \text{ kN} \cdot \text{m} - 5.75 \text{ kN } x_1 + M = 0$$
$$M = (5.75x_1 + 80) \text{ kN} \cdot \text{m} \qquad (2)$$

5 m $< x_2 \leq 10$ m, Fig. 6–7c:

$$+\uparrow \Sigma F_y = 0; \quad 5.75 \text{ kN} - 15 \text{ kN} - 5 \text{ kN/m}(x_2 - 5 \text{ m}) - V = 0$$
$$V = (15.75 - 5x_2) \text{ kN} \qquad (3)$$

$$\zeta + \Sigma M = 0; \quad -80 \text{ kN} \cdot \text{m} - 5.75 \text{ kN } x_2 + 15 \text{ kN}(x_2 - 5 \text{ m})$$
$$+ 5 \text{ kN/m}(x_2 - 5 \text{ m})\left(\frac{x_2 - 5 \text{ m}}{2}\right) + M = 0$$
$$M = (-2.5x_2^2 + 15.75x_2 + 92.5) \text{ kN} \cdot \text{m} \qquad (4)$$

These results can be checked in part by noting that $w = dV/dx$ and $V = dM/dx$. Also, when $x_1 = 0$, Eqs. 1 and 2 give $V = 5.75$ kN and $M = 80$ kN $\cdot$ m; when $x_2 = 10$ m, Eqs. 3 and 4 give $V = -34.25$ kN and $M = 0$. These values check with the support reactions shown on the free-body diagram, Fig. 6–7d.

Shear and Moment Diagrams. Equations 1 through 4 are plotted in Fig. 6–7d.

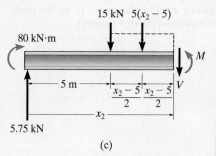

(c)

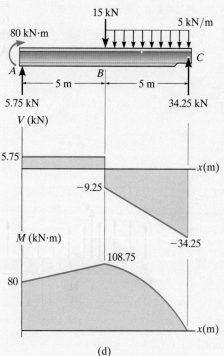

(d)

Fig. 6–7

6

Failure of this table occurred at the brace support on its right side. If drawn, the bending moment diagram for the table loading would indicate this to be the point of maximum internal moment.

6.2   Graphical Method for Constructing Shear and Moment Diagrams

In cases where a beam is subjected to *several* different loadings, determining V and M as functions of x and then plotting these equations can become quite tedious. In this section a simpler method for constructing the shear and moment diagrams is discussed—a method based on two differential relations, one that exists between distributed load and shear, and the other between shear and moment.

Regions of Distributed Load.

For purposes of generality, consider the beam shown in Fig. 6–8a, which is subjected to an arbitrary loading. A free-body diagram for a small segment Δx of the beam is shown in Fig. 6–8b. Since this segment has been chosen at a position x where there is no concentrated force or couple moment, the results to be obtained will *not* apply at these points of concentrated loading.

Notice that all the loadings shown on the segment act in their positive directions according to the established sign convention, Fig. 6–3. Also, both the internal resultant shear and moment, acting on the right face of the segment, must be changed by a small amount in order to keep the segment in equilibrium. The distributed load has been replaced by a resultant force $w(x)\,\Delta x$ that acts at a fractional distance $k(\Delta x)$ from the right side, where $0 < k < 1$ [for example, if $w(x)$ is *uniform*, $k = \frac{1}{2}$]. Applying the equations of equilibrium to the segment, we have

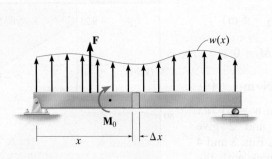

(a)

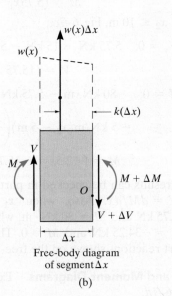

Free-body diagram
of segment Δx

(b)

Fig. 6–8

$+\!\uparrow\!\Sigma F_y = 0;\qquad V + w(x)\,\Delta x - (V + \Delta V) = 0$

$$\Delta V = w(x)\,\Delta x$$

$\zeta\!+\!\Sigma M_O = 0;\quad -V\,\Delta x - M - w(x)\,\Delta x[k(\Delta x)] + (M + \Delta M) = 0$

$$\Delta M = V\,\Delta x + w(x)\,k(\Delta x)^2$$

Dividing by Δx and taking the limit as $\Delta x \to 0$, the above two equations become

$$\frac{dV}{dx} = w(x)$$

$$\begin{array}{ccc} \text{slope of} & & \text{distributed}\\ \text{shear diagram} & = & \text{load intensity}\\ \text{at each point} & & \text{at each point} \end{array} \qquad (6\text{--}1)$$

$$\frac{dM}{dx} = V$$

$$\begin{array}{ccc} \text{slope of} & & \text{shear}\\ \text{moment diagram} & = & \text{at each}\\ \text{at each point} & & \text{point} \end{array} \qquad (6\text{--}2)$$

These two equations provide a convenient means for quickly obtaining the shear and moment diagrams for a beam. Equation 6–1 states that at a point the *slope* of the shear diagram equals the intensity of the distributed loading. For example, consider the beam in Fig. 6–9a. The distributed loading is negative and increases from zero to w_B. Therefore, the shear diagram will be a curve that has a *negative slope*, increasing from zero to $-w_B$. Specific slopes $w_A = 0$, $-w_C$, $-w_D$, and $-w_B$ are shown in Fig. 6–9b.

In a similar manner, Eq. 6–2 states that at a point the *slope* of the moment diagram is equal to the shear. Notice that the shear diagram in Fig. 6–9b starts at $+V_A$, decreases to zero, and then becomes negative and decreases to $-V_B$. The moment diagram will then have an initial slope of $+V_A$ which decreases to zero, then the slope becomes negative and decreases to $-V_B$. Specific slopes V_A, V_C, V_D, 0, and $-V_B$ are shown in Fig. 6–9c.

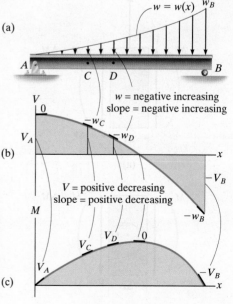

Fig. 6–9

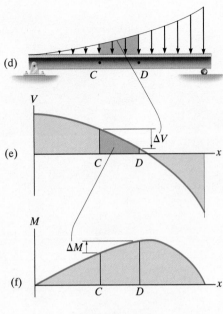

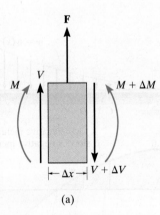

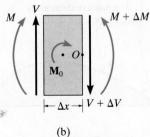

Fig. 6–10

Equations 6–1 and 6–2 may also be rewritten in the form $dV = w(x)\,dx$ and $dM = V\,dx$. Noting that $w(x)\,dx$ and $V\,dx$ represent differential areas under the distributed loading and shear diagram, respectively, we can integrate these areas between any two points C and D on the beam, Fig. 6–9d, and write

$$\Delta V = \int w(x)\,dx$$

$$\begin{array}{ccc} \text{change in} & = & \text{area under} \\ \text{shear} & & \text{distributed loading} \end{array}$$

(6–3)

$$\Delta M = \int V(x)\,dx$$

$$\begin{array}{ccc} \text{change in} & = & \text{area under} \\ \text{moment} & & \text{shear diagram} \end{array}$$

(6–4)

Equation 6–3 states that the *change in shear* between C and D is equal to the *area* under the distributed-loading curve between these two points, Fig. 6–9d. In this case the change is negative since the distributed load acts downward. Similarly, from Eq. 6–4, the change in moment between C and D, Fig. 6–9f, is equal to the area under the shear diagram within the region from C to D. Here the change is positive.

Since the above equations do not apply at points where a concentrated force or couple moment acts, we will now consider each of these cases.

Regions of Concentrated Force and Moment.

A free-body diagram of a small segment of the beam in Fig. 6–10a taken from under the force is shown in Fig. 6–10a. Here it can be seen that force equilibrium requires

$$+\uparrow \Sigma F_y = 0; \qquad V + F - (V + \Delta V) = 0$$

$$\Delta V = F \qquad (6\text{–}5)$$

Thus, when **F** acts *upward* on the beam, ΔV is *positive* so the shear will "jump" *upward*. Likewise, if **F** acts *downward*, the jump (ΔV) will be *downward*.

When the beam segment includes the couple moment M_0, Fig. 6–10b, then moment equilibrium requires the change in moment to be

$$\zeta + \Sigma M_O = 0; \qquad M + \Delta M - M_0 - V\,\Delta x - M = 0$$

Letting $\Delta x \to 0$, we get

$$\Delta M = M_0 \qquad (6\text{–}6)$$

In this case, if **M**$_0$ is applied *clockwise*, ΔM is *positive* so the moment diagram will "jump" *upward*. Likewise, when **M**$_0$ acts *counterclockwise*, the jump (ΔM) will be *downward*.

Procedure for Analysis

The following procedure provides a method for constructing the shear and moment diagrams for a beam based on the relations among distributed load, shear, and moment.

Support Reactions.

- Determine the support reactions and resolve the forces acting on the beam into components that are perpendicular and parallel to the beam's axis.

Shear Diagram.

- Establish the V and x axes and plot the known values of the shear at the two *ends* of the beam.
- Notice how the values of the distributed load vary along the beam, and realize that each of these values indicates the way the shear diagram will slope ($dV/dx = w$). Here w is positive when it acts upward.
- If a numerical value of the shear is to be determined at a point, one can find this value either by using the method of sections and the equation of force equilibrium, or by using $\Delta V = \int w(x)\, dx$, which states that the *change in the shear* between any two points is equal to the *area under the load diagram* between the two points.

Moment Diagram.

- Establish the M and x axes and plot the known values of the moment at the *ends* of the beam.
- Notice how the values of the shear diagram vary along the beam, and realize that each of these values indicates the way the moment diagram will slope ($dM/dx = V$).
- At the point where the shear is zero, $dM/dx = 0$, and therefore this would be a point of maximum or minimum moment.
- If a numerical value of the moment is to be determined at the point, one can find this value either by using the method of sections and the equation of moment equilibrium, or by using $\Delta M = \int V(x)\, dx$, which states that the *change in moment* between any two points is equal to the *area under the shear diagram* between the two points.
- Since $w(x)$ must be *integrated* to obtain ΔV, and $V(x)$ is integrated to obtain $M(x)$, then if $w(x)$ is a curve of degree n, $V(x)$ will be a curve of degree $n + 1$ and $M(x)$ will be a curve of degree $n + 2$. For example, if $w(x)$ is uniform, $V(x)$ will be linear and $M(x)$ will be parabolic.

EXAMPLE | 6.5

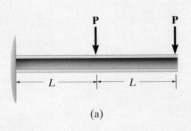

(a)

Draw the shear and moment diagrams for the beam shown in Fig. 6–11a.

SOLUTION

Support Reactions.   The reaction at the fixed support is shown on the free-body diagram, Fig. 6–11b.

Shear Diagram.   The shear at each end of the beam is plotted first, Fig. 6–11c. Since there is no distributed loading on the beam, the slope of the shear diagram is zero as indicated. Note how the force P at the center of the beam causes the shear diagram to jump downward an amount P, since this force acts downward.

Moment Diagram.   The moments at the ends of the beam are plotted, Fig. 6–11d. Here the moment diagram consists of two sloping lines, one with a slope of $+2P$ and the other with a slope of $+P$.

The value of the moment in the center of the beam can be determined by the method of sections, or from the area under the shear diagram. If we choose the left half of the shear diagram,

$$M|_{x=L} = M|_{x=0} + \Delta M$$

$$M|_{x=L} = -3PL + (2P)(L) = -PL$$

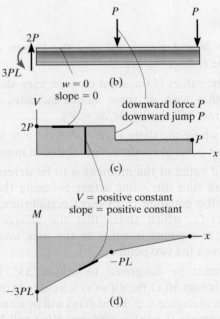

Fig. 6–11

EXAMPLE 6.6

Draw the shear and moment diagrams for the beam shown in Fig. 6–12a.

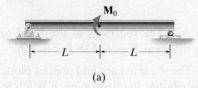

(a)

SOLUTION

Support Reactions.   The reactions are shown on the free-body diagram in Fig. 6–12b.

Shear Diagram.   The shear at each end is plotted first, Fig. 6–12c. Since there is no distributed load on the beam, the shear diagram has zero slope and is therefore a horizontal line.

Moment Diagram.   The moment is zero at each end, Fig. 6–12d. The moment diagram has a constant negative slope of $-M_0/2L$ since this is the shear in the beam at each point. Note that the couple moment M_0 causes a jump in the moment diagram at the beam's center, but it does not affect the shear diagram at this point.

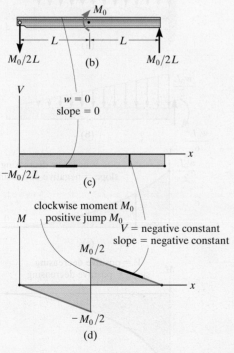

Fig. 6–12

EXAMPLE | 6.7

Draw the shear and moment diagrams for each of the beams shown in Figs. 6–13a and 6–14a.

SOLUTION

Support Reactions. The reactions at the fixed support are shown on each free-body diagram, Figs. 6–13b and 6–14b.

Shear Diagram. The shear at each end point is plotted first, Figs. 6–13c and 6–14c. The distributed loading on each beam indicates the slope of the shear diagram and thus produces the shapes shown.

Moment Diagram. The moment at each end point is plotted first, Figs. 6–13d and 6–14d. Various values of the shear at each point on the beam indicate the slope of the moment diagram at the point. Notice how this variation produces the curves shown.

NOTE: Observe how the degree of the curves from w to V to M increases exponentially due to the integration of $dV = w \, dx$ and $dM = V \, dx$. For example, in Fig. 6–14, the linear distributed load produces a parabolic shear diagram and cubic moment diagram.

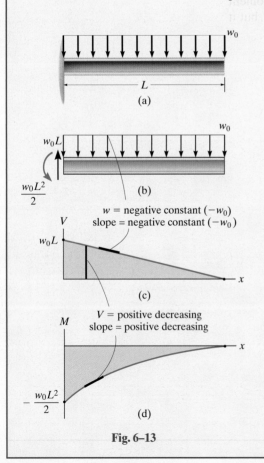

Fig. 6–13

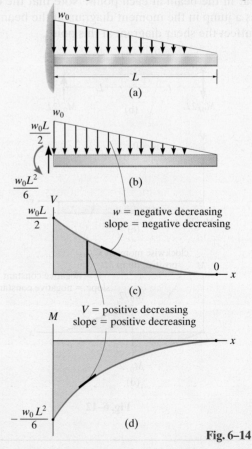

Fig. 6–14

EXAMPLE 6.8

Draw the shear and moment diagrams for the cantilever beam in Fig. 6–15a.

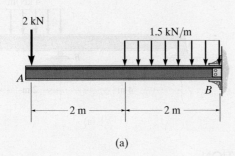

(a)

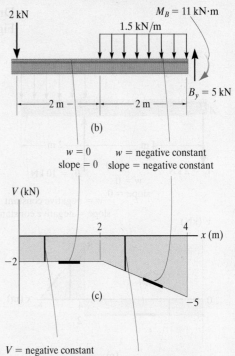

(b)

(c)

(d)

SOLUTION

Support Reactions.   The support reactions at the fixed support B are shown in Fig. 6–15b.

Shear Diagram.   The shear at end A is -2 kN. This value is plotted at $x = 0$, Fig. 6–15c. Notice how the shear diagram is constructed by following the slopes defined by the loading w. The shear at $x = 4$ m is -5 kN, the reaction on the beam. This value can be verified by finding the area under the distributed loading, Eq. 6–3.

$$V|_{x=4\,m} = V|_{x=2\,m} + \Delta V = -2\text{ kN} - (1.5\text{ kN/m})(2\text{ m}) = -5\text{ kN}$$

Moment Diagram.   The moment of zero at $x = 0$ is plotted in Fig. 6–15d. Notice how the moment diagram is constructed based on knowing its slope, which is equal to the shear at each point. The change of moment from $x = 0$ to $x = 2$ m is determined from the area under the shear diagram. Hence, the moment at $x = 2$ m is

$$M|_{x=2\,m} = M|_{x=0} + \Delta M = 0 + [-2\text{ kN}(2\text{ m})] = -4\text{ kN} \cdot \text{m}$$

This same value can be determined from the method of sections, Fig. 6–15e.

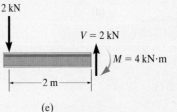

(e)

Fig. 6–15

EXAMPLE 6.9

Draw the shear and moment diagrams for the overhang beam in Fig. 6–16a.

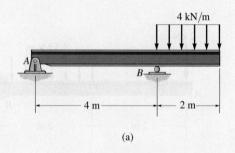

(a)

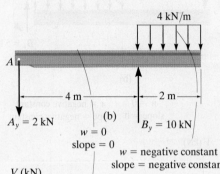

(b)

$A_y = 2$ kN $B_y = 10$ kN

$w = 0$
slope $= 0$ $w =$ negative constant
slope $=$ negative constant

V (kN)

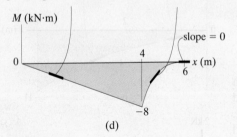

(c)

$V =$ negative constant $V =$ negative decreasing
slope $=$ negative constant   slope $=$ negative decreasing

M (kN·m)

slope $= 0$

(d)

SOLUTION

Support Reactions. The support reactions are shown in Fig. 6–16b.

Shear Diagram. The shear of -2 kN at end A of the beam is plotted at $x = 0$, Fig. 6–16c. The slopes are determined from the loading and from this the shear diagram is constructed, as indicated in the figure. In particular, notice the positive jump of 10 kN at $x = 4$ m due to the force B_y, as indicated in the figure.

Moment Diagram. The moment of zero at $x = 0$ is plotted, Fig. 6–16d. Then following the behavior of the slope found from the shear diagram, the moment diagram is constructed. The moment at $x = 4$ m is found from the area under the shear diagram.

$$M|_{x=4\,m} = M|_{x=0} + \Delta M = 0 + [-2\text{ kN}(4\text{ m})] = -8 \text{ kN} \cdot \text{m}$$

We can also obtain this value by using the method of sections, as shown in Fig. 6–16e.

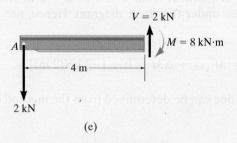

(e)

Fig. 6–16

EXAMPLE 6.10

The shaft in Fig. 6–17a is supported by a thrust bearing at A and a journal bearing at B. Draw the shear and moment diagrams.

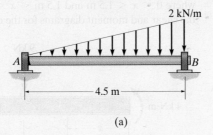

(a)

SOLUTION

Support Reactions. The support reactions are shown in Fig. 6–17b.

Shear Diagram. As shown in Fig. 6–17c, the shear at $x = 0$ is $+1.5$ kN. Following the slope defined by the loading, the shear diagram is constructed, where at B its value is -3 kN. Since the shear changes sign, the point where $V = 0$ must be located. To do this we will use the method of sections. The free-body diagram of the left segment of the shaft, sectioned at an arbitrary position x, is shown in Fig. 6–17e. Notice that the intensity of the distributed load at x is $w = 2(\frac{x}{4.5})$, which has been found by proportional triangles, i.e., $\frac{2}{4.5} = w/x$.

Thus, for $V = 0$,

$$+\uparrow \Sigma F_y = 0; \qquad 1.5 \text{ kN} - \tfrac{1}{2}[2(\tfrac{x}{4.5})]x = 0$$

$$x = 2.60 \text{ m}$$

Moment Diagram. The moment diagram starts at 0 since there is no moment at A; then it is constructed based on the slope as determined from the shear diagram. The maximum moment occurs at $x = 2.60$ m, where the shear is equal to zero, since $dM/dx = V = 0$, Fig. 6–17d,

$$\zeta + \Sigma M = 0;$$

$$M_{\text{max}} + \tfrac{1}{2}[(\tfrac{2}{4.5})(2.60)]\, 2.60(\tfrac{1}{3}(2.60)) - 1.5(2.60) = 0$$

$$M_{\text{max}} = 2.60 \text{ kN} \cdot \text{m}$$

Finally, notice how integration, first of the loading w which is linear, produces a shear diagram which is parabolic, and then a moment diagram which is cubic.

NOTE: Having studied these examples, test yourself by covering over the shear and moment diagrams in Examples 6–1 through 6–4 and see if you can construct them using the concepts discussed here.

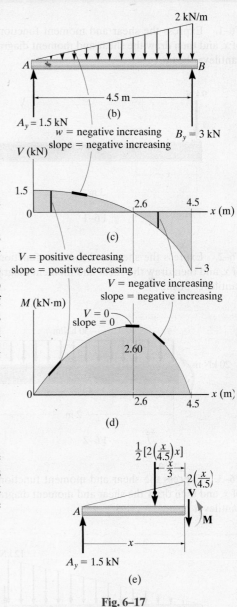

Fig. 6–17

FUNDAMENTAL PROBLEMS

F6–1. Express the shear and moment functions in terms of x, and then draw the shear and moment diagrams for the cantilever beam.

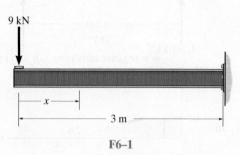

F6–1

F6–2. Express the shear and moment functions in terms of x, and then draw the shear and moment diagrams for the cantilever beam.

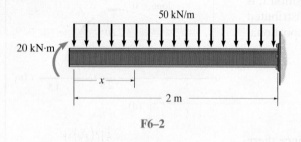

F6–2

F6–3. Express the shear and moment functions in terms of x, and then draw the shear and moment diagrams for the cantilever beam.

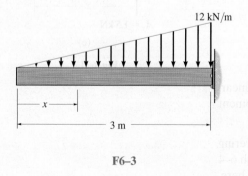

F6–3

F6–4. Express the shear and moment functions in terms of x, where $0 < x < 1.5$ m and 1.5 m $< x < 3$ m, and then draw the shear and moment diagrams for the cantilever beam.

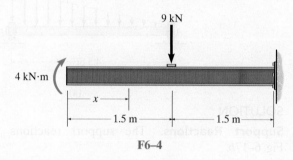

F6–4

F6–5. Express the shear and moment functions in terms of x, and then draw the shear and moment diagrams for the simply supported beam.

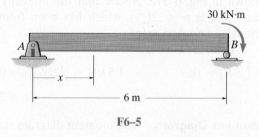

F6–5

F6–6. Express the shear and moment functions in terms of x, and then draw the shear and moment diagrams for the simply supported beam.

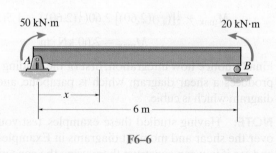

F6–6

F6–7. Draw the shear and moment diagrams for the simply supported beam.

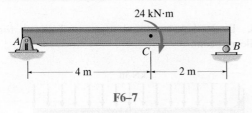

F6–7

F6–8. Draw the shear and moment diagrams for the cantilever beam.

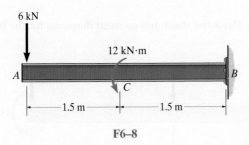

F6–8

F6–9. Draw the shear and moment diagrams for the double overhang beam.

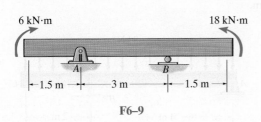

F6–9

F6–10. Draw the shear and moment diagrams for the simply supported beam.

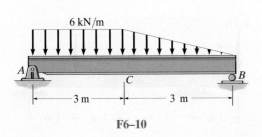

F6–10

F6–11. Draw the shear and moment diagrams for the double-overhang beam.

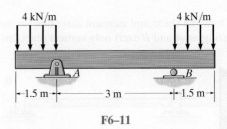

F6–11

F6–12. Draw the shear and moment diagrams for the simply supported beam.

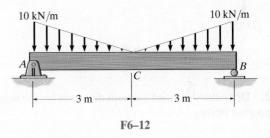

F6–12

F6–13. Draw the shear and moment diagrams for the simply supported beam.

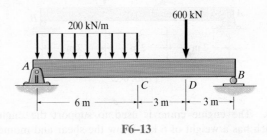

F6–13

F6–14. Draw the shear and moment diagrams for the overhang beam.

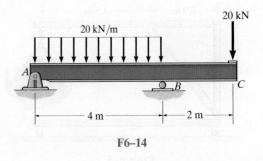

F6–14

6

PROBLEMS

6–1. Draw the shear and moment diagrams for the shaft. The bearings at A and B exert only vertical reactions on the shaft.

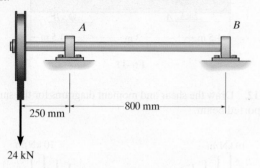

Prob. 6–1

6–2. Draw the shear and moment diagrams for the simply supported beam.

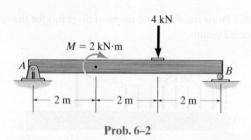

Prob. 6–2

6–3. The engine crane is used to support the engine, which has a weight of 6 kN. Draw the shear and moment diagrams of the boom ABC when it is in the horizontal position shown.

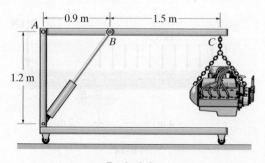

Prob. 6–3

*6–4.** Draw the shear and moment diagrams for the cantilever beam.

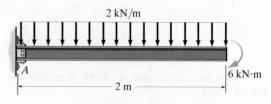

Prob. 6–4

6–5. Draw the shear and moment diagrams for the beam.

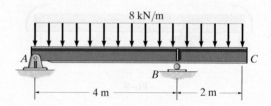

Prob. 6–5

6–6. Draw the shear and moment diagrams for the overhang beam.

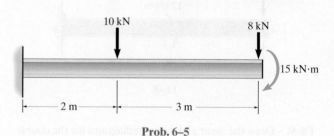

Prob. 6–6

6–7. Draw the shear and moment diagrams for the compound beam which is pin connected at B.

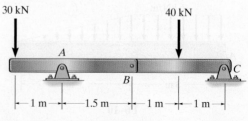

Prob. 6–7

*6–8. Draw the shear and moment diagrams for the simply supported beam.

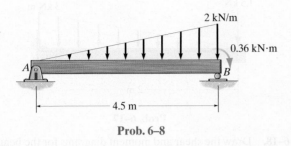

2 kN/m

0.36 kN·m

A

B

4.5 m

Prob. 6–8

6–9. Draw the shear and moment diagrams for the beam. *Hint:* The 100-kN load must be replaced by equivalent loadings at point C on the axis of the beam.

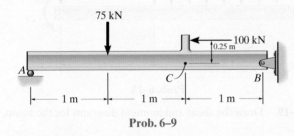

75 kN

100 kN

0.25 m

A

C

B

1 m   1 m   1 m

Prob. 6–9

6–10. Members ABC and BD of the counter chair are rigidly connected at B and the smooth collar at D is allowed to move freely along the vertical slot. Draw the shear and moment diagrams for member ABC.

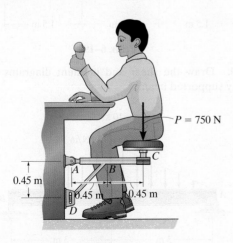

$P = 750$ N

C

0.45 m

A   B

0.45 m   0.45 m

D

Prob. 6–10

6–11. The overhanging beam has been fabricated with a projected arm BD on it. Draw the shear and moment diagrams for the beam ABC if it supports a load of 4 kN. *Hint:* The loading in the supporting strut DE must be replaced by equivalent loads at point B on the axis of the beam.

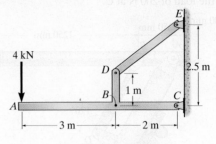

E

4 kN

D

2.5 m

B   1 m

A   C

3 m   2 m

Prob. 6–11

*6–12. A reinforced concrete pier is used to support the stringers for a bridge deck. Draw the shear and moment diagrams for the pier when it is subjected to the stringer loads shown. Assume the columns at A and B exert only vertical reactions on the pier.

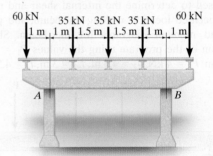

60 kN   35 kN 35 kN 35 kN   60 kN

1 m 1 m 1.5 m 1.5 m 1 m 1 m

A   B

Prob. 6–12

6–13. Draw the shear and moment diagrams for the compound beam. It is supported by a smooth plate at A which slides within the groove and so it cannot support a vertical force, although it can support a moment and axial load.

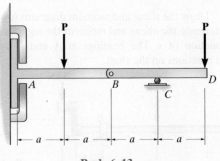

P   P

A   B   C   D

a   a   a   a

Prob. 6–13

6

6–14. The industrial robot is held in the stationary position shown. Draw the shear and moment diagrams of the arm *ABC* if it is pin connected at *A* and connected to a hydraulic cylinder (two-force member) *BD*. Assume the arm and grip have a uniform weight of 0.3 N/mm and support the load of 200 N at *C*.

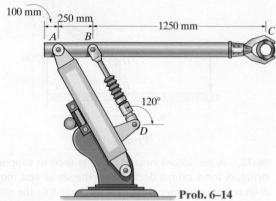

Prob. 6–14

6–15. Consider the general problem of the beam subjected to *n* concentrated loads. Write a computer program that can be used to determine the internal shear and moment at any specified location *x* along the beam, and plot the shear and moment diagrams for the beam. Show an application of the program using the values $P_1 = 2.5$ kN, $d_1 = 1.5$ m, $P_2 = 4$ kN, $d_2 = 4.5$ m, $L_1 = 3$ m, $L = 4.5$ m.

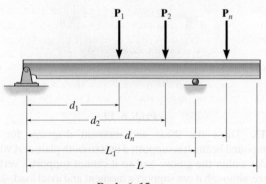

Prob. 6–15

*6–16. Draw the shear and moment diagrams for the shaft and determine the shear and moment throughout the shaft as a function of *x*. The bearings at *A* and *B* exert only vertical reactions on the shaft.

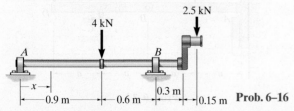

Prob. 6–16

•6–17. Draw the shear and moment diagrams for the cantilevered beam.

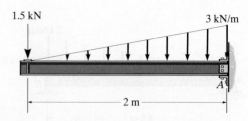

Prob. 6–17

6–18. Draw the shear and moment diagrams for the beam, and determine the shear and moment throughout the beam as functions of *x*.

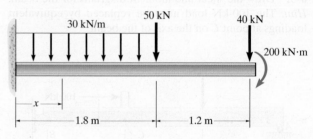

Prob. 6–18

6–19. Draw the shear and moment diagrams for the beam.

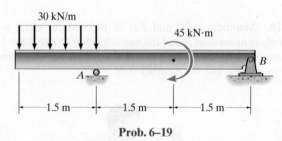

Prob. 6–19

*6–20. Draw the shear and moment diagrams for the simply supported beam.

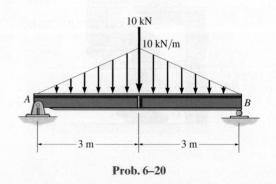

Prob. 6–20

•6–21. The beam is subjected to the uniform distributed load shown. Draw the shear and moment diagrams for the beam.

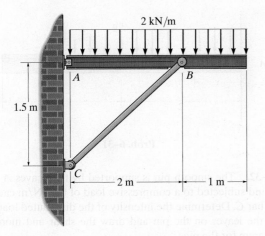

Prob. 6–21

6–22. Draw the shear and moment diagrams for the overhang beam.

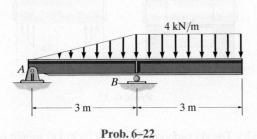

Prob. 6–22

6–23. Draw the shear and moment diagrams for the beam. It is supported by a smooth plate at *A* which slides within the groove and so it cannot support a vertical force, although it can support a moment and axial load.

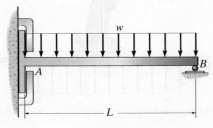

Prob. 6–23

*6–24. Determine the placement distance *a* of the roller support so that the largest absolute value of the moment is a minimum. Draw the shear and moment diagrams for this condition.

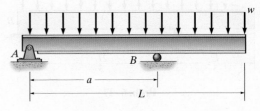

Prob. 6–24

6–25. The beam is subjected to the uniformly distributed moment *m* (moment/length). Draw the shear and moment diagrams for the beam.

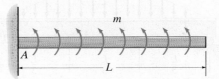

Prob. 6–25

6–26. Consider the general problem of a cantilevered beam subjected to *n* concentrated loads and a constant distributed loading *w*. Write a computer program that can be used to determine the internal shear and moment at any specified location *x* along the beam, and plot the shear and moment diagrams for the beam. Show an application of the program using the values $P_1 = 4$ kN, $d_1 = 2$ m, $w = 800$ N/m, $a_1 = 2$ m, $a_2 = 4$ m, $L = 4$ m.

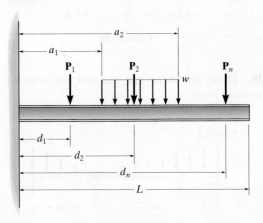

Prob. 6–26

6–27. Draw the shear and moment diagrams for the beam.

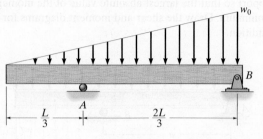

Prob. 6–27

*****6–28.** Draw the shear and moment diagrams for the beam.

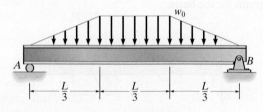

Prob. 6–28

•**6–29.** Draw the shear and moment diagrams for the beam.

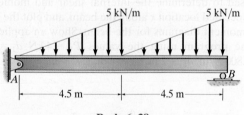

Prob. 6–29

6–30. Draw the shear and moment diagrams for the compound beam.

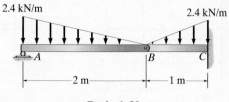

Prob. 6–30

6–31. Draw the shear and moment diagrams for the beam and determine the shear and moment in the beam as functions of x.

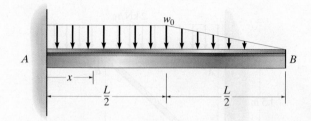

Prob. 6–31

*****6–32.** The smooth pin is supported by two leaves A and B and subjected to a compressive load of 0.4 kN/m caused by bar C. Determine the intensity of the distributed load w_0 of the leaves on the pin and draw the shear and moment diagram for the pin.

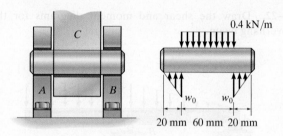

Prob. 6–32

•**6–33.** The ski supports the 900-N ($\approx$90-kg) weight of the man. If the snow loading on its bottom surface is trapezoidal as shown, determine the intensity w, and then draw the shear and moment diagrams for the ski.

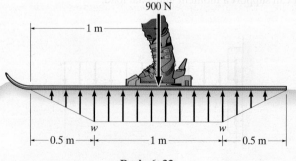

Prob. 6–33

6–34. Draw the shear and moment diagrams for the compound beam.

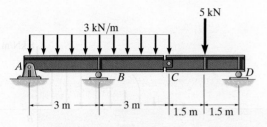

Prob. 6–34

6–35. Draw the shear and moment diagrams for the beam and determine the shear and moment as functions of x.

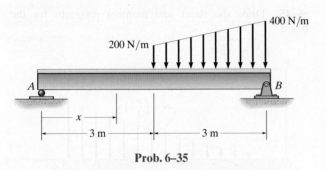

Prob. 6–35

*6–36. Draw the shear and moment diagrams for the overhang beam.

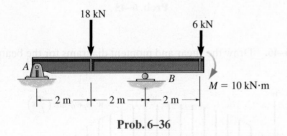

Prob. 6–36

6–37. Draw the shear and moment diagrams for the beam.

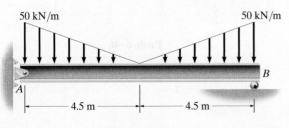

Prob. 6–37

6–38. The dead-weight loading along the centerline of the airplane wing is shown. If the wing is fixed to the fuselage at A, determine the reactions at A, and then draw the shear and moment diagram for the wing.

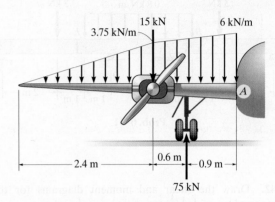

Prob. 6–38

6–39. The compound beam consists of two segments that are pinned together at B. Draw the shear and moment diagrams if it supports the distributed loading shown.

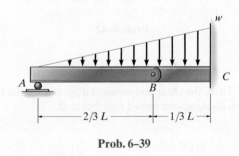

Prob. 6–39

*6–40. Draw the shear and moment diagrams for the simply supported beam.

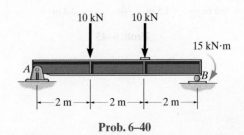

Prob. 6–40

6

6–41. Draw the shear and moment diagrams for the compound beam. The three segments are connected by pins at B and E.

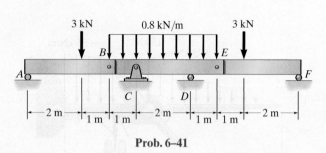

Prob. 6–41

6–42. Draw the shear and moment diagrams for the compound beam.

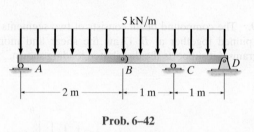

Prob. 6–42

6–43. Draw the shear and moment diagrams for the beam. The two segments are joined together at B.

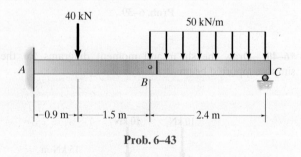

Prob. 6–43

*6–44. Draw the shear and moment diagrams for the beam.

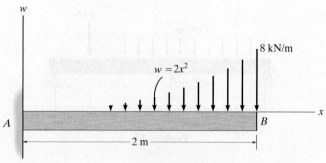

Prob. 6–44

•6–45. Draw the shear and moment diagrams for the beam.

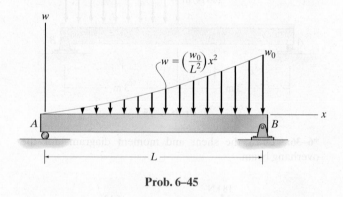

Prob. 6–45

6–46. Draw the shear and moment diagrams for the beam.

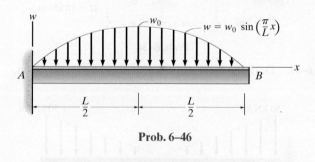

Prob. 6–46

6.3 Bending Deformation of a Straight Member

In this section, we will discuss the deformations that occur when a straight prismatic beam, made of homogeneous material, is subjected to bending. The discussion will be limited to beams having a cross-sectional area that is symmetrical with respect to an axis, and the bending moment is applied about an axis perpendicular to this axis of symmetry as shown in Fig. 6–18. The behavior of members that have unsymmetrical cross sections, or are made from several different materials, is based on similar observations and will be discussed separately in later sections of this chapter.

By using a highly deformable material such as rubber, we can illustrate what happens when a straight prismatic member is subjected to a bending moment. Consider, for example, the undeformed bar in Fig. 6–19a, which has a square cross section and is marked with longitudinal and transverse grid lines. When a bending moment is applied, it tends to distort these lines into the pattern shown in Fig. 6–19b. Notice that the longitudinal lines become *curved* and the vertical transverse lines *remain straight* and yet undergo a *rotation*.

The bending moment causes the material within the *bottom* portion of the bar to *stretch* and the material within the *top* portion to *compress*. Consequently, between these two regions there must be a surface, called the *neutral surface*, in which longitudinal fibers of the material will not undergo a change in length, Fig. 6–18.

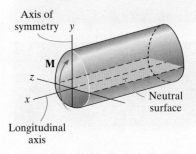

Fig. 6–18

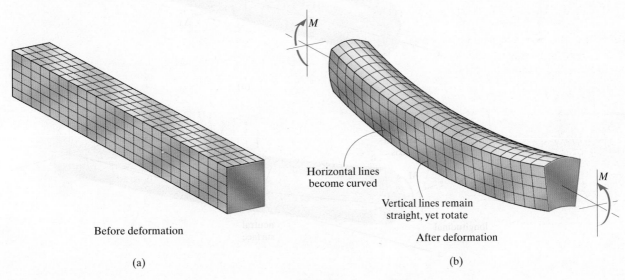

Before deformation

(a)

Horizontal lines become curved

Vertical lines remain straight, yet rotate

After deformation

(b)

Fig. 6–19

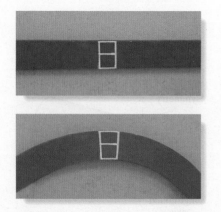

Note the distortion of the lines due to bending of this rubber bar. The top line stretches, the bottom line compresses, and the center line remains the same length. Furthermore the vertical lines rotate and yet remain straight.

From these observations we will make the following three assumptions regarding the way the stress deforms the material. First, the *longitudinal axis x*, which lies within the neutral surface, Fig. 6–20a, does *not* experience any *change in length*. Rather the moment will tend to deform the beam so that this line *becomes a curve* that lies in the *x–y* plane of symmetry, Fig. 6–20b. Second, all **cross sections** of the beam **remain plane** and perpendicular to the longitudinal axis during the deformation. And third, *any deformation* of the *cross section* within its own plane, as noticed in Fig. 6–19b, will be *neglected*. In particular, the *z* axis, lying in the plane of the cross section and about which the cross section rotates, is called the *neutral axis*, Fig. 6–20b.

In order to show how this distortion will strain the material, we will isolate a small segment of the beam located a distance *x* along the beam's length and having an undeformed thickness Δx, Fig. 6–20a. This element, taken from the beam, is shown in profile view in the undeformed and

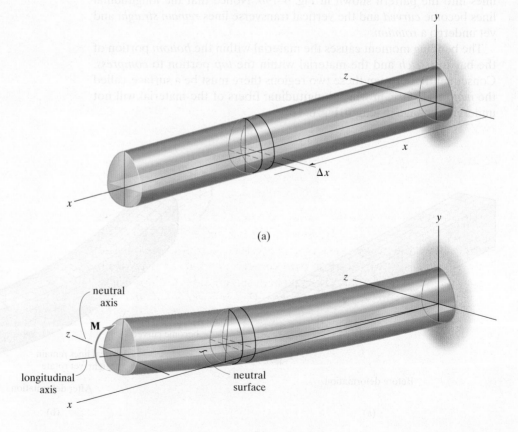

(a)

(b)

Fig. 6–20

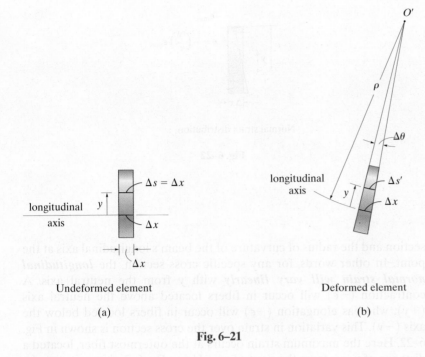

Undeformed element

(a)

Deformed element

(b)

Fig. 6–21

deformed positions in Fig. 6–21. Notice that any line segment Δx, located on the neutral surface, does not change its length, whereas any line segment Δs, located at the arbitrary distance y above the neutral surface, will contract and become $\Delta s'$ after deformation. By definition, the normal strain along Δs is determined from Eq. 2–2, namely,

$$\epsilon = \lim_{\Delta s \to 0} \frac{\Delta s' - \Delta s}{\Delta s}$$

We will now represent this strain in terms of the location y of the segment and the radius of curvature ρ of the longitudinal axis of the element. Before deformation, $\Delta s = \Delta x$, Fig. 6–21a. After deformation Δx has a radius of curvature ρ, with center of curvature at point O', Fig. 6–21b. Since $\Delta\theta$ defines the angle between the sides of the element, $\Delta x = \Delta s = \rho\Delta\theta$. In the same manner, the deformed length of Δs becomes $\Delta s' = (\rho - y)\Delta\theta$. Substituting into the above equation, we get

$$\epsilon = \lim_{\Delta\theta \to 0} \frac{(\rho - y)\Delta\theta - \rho\Delta\theta}{\rho\Delta\theta}$$

or

$$\epsilon = -\frac{y}{\rho} \qquad (6\text{–}7)$$

This important result indicates that the longitudinal normal strain of any element within the beam depends on its location y on the cross

$$-\epsilon_{max}$$

$$\epsilon = -\left(\frac{y}{c}\right)\epsilon_{max}$$

Normal strain distribution

Fig. 6–22

section and the radius of curvature of the beam's longitudinal axis at the point. In other words, for any specific cross section, the ***longitudinal normal strain will vary linearly*** with y from the neutral axis. A contraction $(-\epsilon)$ will occur in fibers located above the neutral axis $(+y)$, whereas elongation $(+\epsilon)$ will occur in fibers located below the axis $(-y)$. This variation in strain over the cross section is shown in Fig. 6–22. Here the maximum strain occurs at the outermost fiber, located a distance of $y = c$ from the neutral axis. Using Eq. 6–7, since $\epsilon_{max} = c/\rho$, then by division,

$$\frac{\epsilon}{\epsilon_{max}} = -\left(\frac{y/\rho}{c/\rho}\right)$$

So that

$$\epsilon = -\left(\frac{y}{c}\right)\epsilon_{max} \qquad (6\text{–}8)$$

This normal strain depends only on the assumptions made with regards to the *deformation*. When a moment is applied to the beam, therefore, it will only cause a *normal stress* in the longitudinal or x direction. All the other components of normal and shear stress will be zero. It is this uniaxial state of stress that causes the material to have the longitudinal normal strain component ϵ_x, defined by Eq. 6–8. Furthermore, by Poisson's ratio, there must *also* be associated strain components $\epsilon_y = -\nu\epsilon_x$ and $\epsilon_z = -\nu\epsilon_x$, which deform the plane of the cross-sectional area, although here we have neglected these deformations. Such deformations will, however, cause the *cross-sectional dimensions* to become smaller below the neutral axis and larger above the neutral axis. For example, if the beam has a square cross section, it will actually deform as shown in Fig. 6–23.

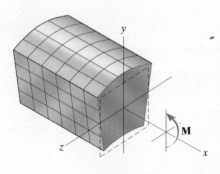

Fig. 6–23

6.4   The Flexure Formula

In this section, we will develop an equation that relates the stress distribution in a beam to the internal resultant bending moment acting on the beam's cross section. To do this we will assume that the material behaves in a linear-elastic manner and therefore a *linear variation of normal strain*, Fig. 6–24a, must then be the result of a *linear variation in normal stress*, Fig. 6–24b. Hence, like the normal strain variation, σ will vary from zero at the member's neutral axis to a maximum value, σ_{max}, a distance c farthest from the neutral axis. Because of the proportionality of triangles, Fig. 6–23b, or by using Hooke's law, $\sigma = E\epsilon$, and Eq. 6–8, we can write

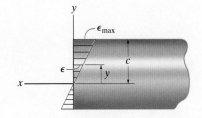

Normal strain variation
(profile view)

(a)

$$\sigma = -\left(\frac{y}{c}\right)\sigma_{max} \qquad (6\text{–}9)$$

This equation describes the stress distribution over the cross-sectional area. The sign convention established here is significant. For positive **M**, which acts in the $+z$ direction, positive values of y give negative values for σ, that is, a compressive stress since it acts in the negative x direction. Similarly, negative y values will give positive or tensile values for σ. If a volume element of material is selected at a specific point on the cross section, only these tensile or compressive normal stresses will act on it. For example, the element located at $+y$ is shown in Fig. 6–24c.

We can locate the position of the neutral axis on the cross section by satisfying the condition that the *resultant force* produced by the stress distribution over the cross-sectional area must be equal to *zero*. Noting that the force $dF = \sigma \, dA$ acts on the arbitrary element dA in Fig. 6–24c, we require

Bending stress variation
(profile view)

(b)

Fig. 6–24

$$F_R = \Sigma F_x; \qquad 0 = \int_A dF = \int_A \sigma \, dA$$

$$= \int_A -\left(\frac{y}{c}\right)\sigma_{max} \, dA$$

$$= \frac{-\sigma_{max}}{c} \int_A y \, dA$$

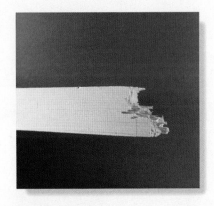

This wood specimen failed in bending due to its fibers being crushed at its top and torn apart at its bottom.

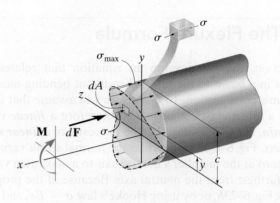

Bending stress variation

(c)

Fig. 6–24 (cont.)

Since σ_{max}/c is not equal to zero, then

$$\int_A y \, dA = 0 \qquad (6\text{–}10)$$

In other words, the first moment of the member's cross-sectional area about the neutral axis must be zero. This condition can only be satisfied if the *neutral axis* is also the horizontal *centroidal axis* for the cross section.* Consequently, once the centroid for the member's cross-sectional area is determined, the location of the neutral axis is known

We can determine the stress in the beam from the requirement that the resultant internal moment M must be equal to the moment produced by the stress distribution about the neutral axis. The moment of $d\mathbf{F}$ in Fig. 6–24c about the neutral axis is $dM = y \, dF$. Since $dF = \sigma \, dA$, using Eq. 6–9, we have for the entire cross section,

$$(M_R)_z = \Sigma M_z; \quad M = \int_A y \, dF = \int_A y(\sigma \, dA) = \int_A y\left(\frac{y}{c}\sigma_{max}\right) dA$$

or

$$M = \frac{\sigma_{max}}{c}\int_A y^2 \, dA \qquad (6\text{–}11)$$

---

*Recall that the location $\bar{y}$ for the centroid of the cross-sectional area is defined from the equation $\bar{y} = \int y \, dA / \int dA$. If $\int y \, dA = 0$, then $\bar{y} = 0$, and so the centroid lies on the reference (neutral) axis. See Appendix A.

The integral represents the *moment of inertia* of the cross-sectional area about the neutral axis. We will symbolize its value as *I*. Hence, Eq. 6–11 can be solved for σ_{max} and written as

$$\sigma_{max} = \frac{Mc}{I} \qquad (6\text{–}12)$$

Here

σ_{max} = the maximum normal stress in the member, which occurs at a point on the cross-sectional area *farthest away* from the neutral axis

M = the resultant internal moment, determined from the method of sections and the equations of equilibrium, and calculated about the neutral axis of the cross section

c = the perpendicular distance from the neutral axis to a point farthest away from the neutral axis. This is where σ_{max} acts

I = the moment of inertia of the cross-sectional area about the neutral axis

Since $\sigma_{max}/c = -\sigma/y$, Eq. 6–9, the normal stress at the intermediate distance y can be determined from an equation similar to Eq. 6–12. We have

$$\sigma = -\frac{My}{I} \qquad (6\text{–}13)$$

Note that the negative sign is necessary since it agrees with the established x, y, z axes. By the right-hand rule, M is positive along the $+z$ axis, y is positive upward, and σ therefore must be negative (compressive) since it acts in the negative x direction, Fig. 6–24c.

Either of the above two equations is often referred to as the **flexure formula**. It is used to determine the normal stress in a straight member, having a cross section that is symmetrical with respect to an axis, and the moment is applied perpendicular to this axis. Although we have assumed that the member is prismatic, we can in most cases of engineering design also use the flexure formula to determine the normal stress in members that have a *slight taper*. For example, using a mathematical analysis based on the theory of elasticity, a member having a rectangular cross section and a length that is tapered 15° will have an actual maximum normal stress that is about 5.4% *less* than that calculated using the flexure formula.

Important Points

- The cross section of a straight beam *remains plane* when the beam deforms due to bending. This causes tensile stress on one portion of the cross section and compressive stress on the other portion. In between these portions, there exists the *neutral axis* which is subjected to *zero stress*.
- Due to the deformation, the *longitudinal strain* varies *linearly* from zero at the neutral axis to a maximum at the outer fibers of the beam. Provided the material is homogeneous and linear elastic, then the *stress* also varies in a *linear* fashion over the cross section.
- The neutral axis passes through the *centroid* of the cross-sectional area. This result is based on the fact that the resultant normal force acting on the cross section must be zero.
- The flexure formula is based on the requirement that the resultant internal moment on the cross section is equal to the moment produced by the normal stress distribution about the neutral axis.

Procedure for Analysis

In order to apply the flexure formula, the following procedure is suggested.

Internal Moment.

- Section the member at the point where the bending or normal stress is to be determined, and obtain the internal moment M at the section. The centroidal or neutral axis for the cross section must be known, since M *must* be calculated about this axis.
- If the absolute maximum bending stress is to be determined, then draw the moment diagram in order to determine the maximum moment in the member.

Section Property.

- Determine the moment of inertia of the cross-sectional area about the neutral axis. Methods used for its calculation are discussed in Appendix A, and a table listing values of I for several common shapes is given on the inside front cover.

Normal Stress.

- Specify the distance y, measured perpendicular to the neutral axis to the point where the normal stress is to be determined. Then apply the equation $\sigma = -My/I$, or if the maximum bending stress is to be calculated, use $\sigma_{max} = Mc/I$. When substituting the data, make sure the units are consistent.
- The stress acts in a direction such that the force it creates at the point contributes a moment about the neutral axis that is in the same direction as the internal moment **M**, Fig. 6–24c. In this manner the stress distribution acting over the entire cross section can be sketched, or a volume element of the material can be isolated and used to graphically represent the normal stress acting at the point.

EXAMPLE 6.11

A beam has a rectangular cross section and is subjected to the stress distribution shown in Fig. 6–25a. Determine the internal moment **M** at the section caused by the stress distribution (a) using the flexure formula, (b) by finding the resultant of the stress distribution using basic principles.

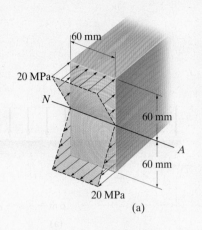

(a)

SOLUTION

Part (a). The flexure formula is $\sigma_{max} = Mc/I$. From Fig. 6–25a, $c = 60$ mm and $\sigma_{max} = 20$ MPa. The neutral axis is defined as line NA, because the stress is zero along this line. Since the cross section has a rectangular shape, the moment of inertia for the area about NA is determined from the formula for a rectangle given on the inside front cover; i.e.,

$$I = \frac{1}{12}bh^3 = \frac{1}{12}(60 \text{ mm})(120 \text{ mm})^3 = 840(10^4) \text{ mm}^4$$

Therefore,

$$\sigma_{max} = \frac{Mc}{I}; \qquad 20 \text{ N/mm}^2 = \frac{M(60 \text{ mm})}{840(10^4) \text{ mm}^4}$$

$$M = 288(10^4) \text{ N} \cdot \text{mm} = 2.88 \text{ kN} \cdot \text{m} \qquad \textit{Ans.}$$

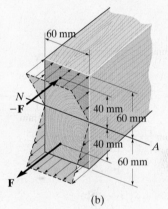

(b)

Fig. 6–25

Part (b). The resultant force for each of the two *triangular* stress distributions in Fig. 6–25b is graphically equivalent to the *volume* contained within each stress distribution. Thus, each volume is

$$F = \frac{1}{2}(60 \text{ mm})[20(10^3) \text{ kN/m}^2](60 \text{ mm}) = 36 \text{ kN}$$

These forces, which form a couple, act in the same direction as the stresses within each distribution, Fig. 6–25b. Furthermore, they act through the *centroid* of each volume, i.e., $\frac{2}{3}(60 \text{ mm}) = 40$ mm, from the neutral axis of the beam. Hence the distance between them is 80 mm as shown. The moment of the couple is therefore

$$M = 36 \text{ kN}(80 \text{ mm}) = 2880 \text{ kN} \cdot \text{m} = 2.88 \text{ kN} \cdot \text{m} \qquad \textit{Ans.}$$

NOTE: This result can also be obtained by choosing a horizontal strip of area $dA = (60 \text{ mm}) \, dy$ and using integration by applying Eq. 6–11.

EXAMPLE | 6.12

The simply supported beam in Fig. 6–26a has the cross-sectional area shown in Fig. 6–26b. Determine the absolute maximum bending stress in the beam and draw the stress distribution over the cross section at this location.

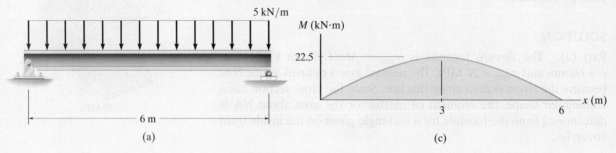

5 kN/m

(a)

(c)

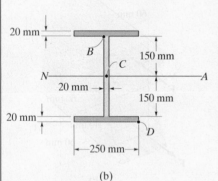

(b)

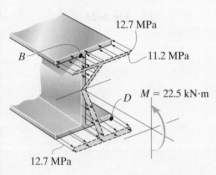

(d)

Fig. 6–26

SOLUTION

Maximum Internal Moment. The maximum internal moment in the beam, $M = 22.5$ kN · m, occurs at the center.

Section Property. By reasons of symmetry, the neutral axis passes through the centroid C at the midheight of the beam, Fig. 6–26b. The area is subdivided into the three parts shown, and the moment of inertia of each part is calculated about the neutral axis using the parallel-axis theorem. (See Eq. A–5 of Appendix A.) Choosing to work in meters, we have

$$I = \Sigma(\bar{I} + Ad^2)$$

$$= 2\left[\frac{1}{12}(0.25 \text{ m})(0.020 \text{ m})^3 + (0.25 \text{ m})(0.020 \text{ m})(0.160 \text{ m})^2\right]$$

$$+ \left[\frac{1}{12}(0.020 \text{ m})(0.300 \text{ m})^3\right]$$

$$= 301.3(10^{-6}) \text{ m}^4$$

$$\sigma_{max} = \frac{Mc}{I}; \qquad \sigma_{max} = \frac{22.5(10^3) \text{ N} \cdot \text{m}(0.170 \text{ m})}{301.3(10^{-6}) \text{ m}^4} = 12.7 \text{ MPa} \quad Ans.$$

A three-dimensional view of the stress distribution is shown in Fig. 6–26d. Notice how the stress at points B and D on the cross section develops a force that contributes a moment about the neutral axis that has the same direction as **M**. Specifically, at point B, $y_B = 150$ mm, and so

$$\sigma_B = -\frac{My_B}{I}; \qquad \sigma_B = -\frac{22.5(10^3) \text{ N} \cdot \text{m}(0.150 \text{ m})}{301.3(10^{-6}) \text{ m}^4} = -11.2 \text{ MPa}$$

EXAMPLE | 6.13

The beam shown in Fig. 6–27a has a cross-sectional area in the shape of a channel, Fig. 6–27b. Determine the maximum bending stress that occurs in the beam at section a–a.

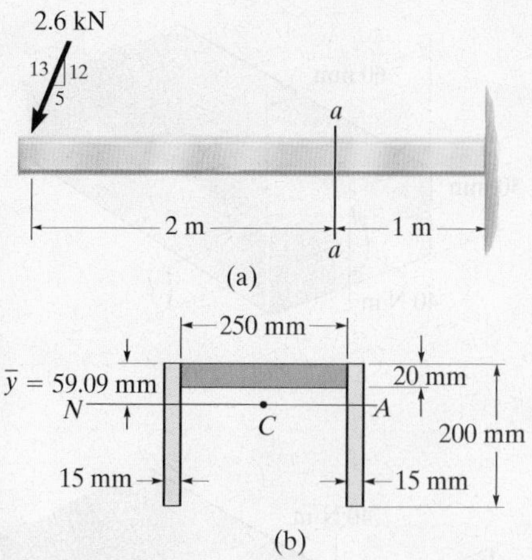

SOLUTION

Internal Moment.   Here the beam's support reactions do not have to be determined. Instead, by the method of sections, the segment to the left of section a–a can be used, Fig. 6–27c. In particular, note that the resultant internal axial force **N** passes through the centroid of the cross section. Also, realize that *the resultant internal moment must be calculated about the beam's neutral axis* at section a–a.

To find the location of the neutral axis, the cross-sectional area is subdivided into three composite parts as shown in Fig. 6–27b. Using Eq. A–2 of Appendix A, we have

$$\bar{y} = \frac{\Sigma \bar{y}A}{\Sigma A} = \frac{2[0.100 \text{ m}](0.200 \text{ m})(0.015 \text{ m}) + [0.010 \text{ m}](0.02 \text{ m})(0.250 \text{ m})}{2(0.200 \text{ m})(0.015 \text{ m}) + 0.020 \text{ m}(0.250 \text{ m})}$$

$$= 0.05909 \text{ m} = 59.09 \text{ mm}$$

This dimension is shown in Fig. 6–27c.

Applying the moment equation of equilibrium about the neutral axis, we have

$$\zeta + \Sigma M_{NA} = 0; \quad 2.4 \text{ kN}(2 \text{ m}) + 1.0 \text{ kN}(0.05909 \text{ m}) - M = 0$$

$$M = 4.859 \text{ kN} \cdot \text{m}$$

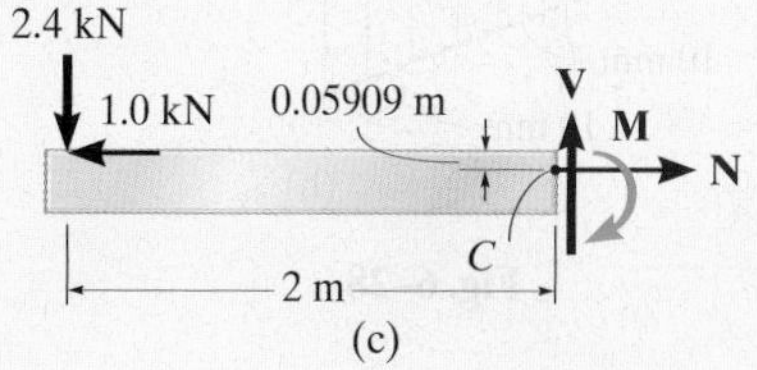

Section Property.   The moment of inertia about the neutral axis is determined using the parallel-axis theorem applied to each of the three composite parts of the cross-sectional area. Working in meters, we have

$$I = \left[\frac{1}{12}(0.250 \text{ m})(0.020 \text{ m})^3 + (0.250 \text{ m})(0.020 \text{ m})(0.05909 \text{ m} - 0.010 \text{ m})^2 \right]$$

$$+ 2\left[\frac{1}{12}(0.015 \text{ m})(0.200 \text{ m})^3 + (0.015 \text{ m})(0.200 \text{ m})(0.100 \text{ m} - 0.05909 \text{ m})^2 \right]$$

$$= 42.26(10^{-6}) \text{ m}^4$$

Maximum Bending Stress.   The maximum bending stress occurs at points farthest away from the neutral axis. This is at the bottom of the beam, c = 0.200 m − 0.05909 m = 0.1409 m. Thus,

$$\sigma_{max} = \frac{Mc}{I} = \frac{4.859(10^3) \text{ N} \cdot \text{m}(0.1409 \text{ m})}{42.26(10^{-6}) \text{ m}^4} = 16.2 \text{ MPa} \quad Ans.$$

Show that at the top of the beam the bending stress is $\sigma' = 6.79$ MPa.

NOTE: The normal force of $N = 1$ kN and shear force $V = 2.4$ kN will also contribute additional stress on the cross section. The superposition of all these effects will be discussed in Chapter 8.

2.6 kN

13 / 12 / 5

a
a

2 m — 1 m

(a)

250 mm

$\bar{y} = 59.09$ mm
N
C
A
20 mm
200 mm
15 mm — 15 mm

(b)

2.4 kN
1.0 kN   0.05909 m
V
M
N
C
2 m

(c)

Fig. 6–27

EXAMPLE | 6.14

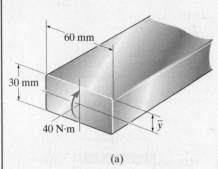

(a)

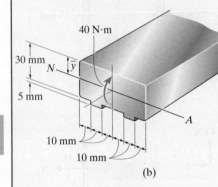

(b)

Fig. 6–28

The member having a rectangular cross section, Fig. 6–28a, is designed to resist a moment of 40 N · m. In order to increase its strength and rigidity, it is proposed that two small ribs be added at its bottom, Fig. 6–28b. Determine the maximum normal stress in the member for both cases.

SOLUTION

Without Ribs. Clearly the neutral axis is at the center of the cross section, Fig. 6–28a, so $\bar{y} = c = 15$ mm $= 0.015$ m. Thus,

$$I = \frac{1}{12}bh^3 = \frac{1}{12}(0.060 \text{ m})(0.030 \text{ m})^3 = 0.135(10^{-6}) \text{ m}^4$$

Therefore the maximum normal stress is

$$\sigma_{max} = \frac{Mc}{I} = \frac{(40 \text{ N} \cdot \text{m})(0.015 \text{ m})}{0.135(10^{-6}) \text{ m}^4} = 4.44 \text{ MPa} \qquad Ans.$$

With Ribs. From Fig. 6–28b, segmenting the area into the large main rectangle and the bottom two rectangles (ribs), the location $\bar{y}$ of the centroid and the neutral axis is determined as follows:

$$\bar{y} = \frac{\Sigma \bar{y}A}{\Sigma A}$$

$$= \frac{[0.015 \text{ m}](0.030 \text{ m})(0.060 \text{ m}) + 2[0.0325 \text{ m}](0.005 \text{ m})(0.010 \text{ m})}{(0.03 \text{ m})(0.060 \text{ m}) + 2(0.005 \text{ m})(0.010 \text{ m})}$$

$$= 0.01592 \text{ m}$$

This value does not represent c. Instead

$$c = 0.035 \text{ m} - 0.01592 \text{ m} = 0.01908 \text{ m}$$

Using the parallel-axis theorem, the moment of inertia about the neutral axis is

$$I = \left[\frac{1}{12}(0.060 \text{ m})(0.030 \text{ m})^3 + (0.060 \text{ m})(0.030 \text{ m})(0.01592 \text{ m} - 0.015 \text{ m})^2 \right]$$

$$+ 2\left[\frac{1}{12}(0.010 \text{ m})(0.005 \text{ m})^3 + (0.010 \text{ m})(0.005 \text{ m})(0.0325 \text{ m} - 0.01592 \text{ m})^2 \right]$$

$$= 0.1642(10^{-6}) \text{ m}^4$$

Therefore, the maximum normal stress is

$$\sigma_{max} = \frac{Mc}{I} = \frac{40 \text{ N} \cdot \text{m}(0.01908 \text{ m})}{0.1642(10^{-6}) \text{ m}^4} = 4.65 \text{ MPa} \qquad Ans.$$

NOTE: This surprising result indicates that the addition of the ribs to the cross section will *increase* the normal stress rather than decrease it, and for this reason they should be omitted.

FUNDAMENTAL PROBLEMS

F6–15. If the beam is subjected to a bending moment of $M = 20$ kN·m, determine the maximum bending stress in the beam.

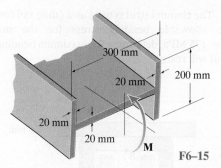

F6–15

F6–16. If the beam is subjected to a bending moment of $M = 50$ kN·m, sketch the bending stress distribution over the beam's cross section.

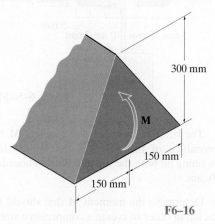

F6–16

F6–17. If the beam is subjected to a bending moment of $M = 50$ kN·m, determine the maximum bending stress in the beam.

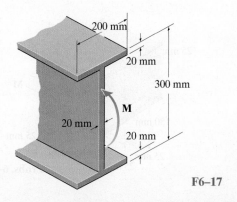

F6–17

F6–18. If the beam is subjected to a bending moment of $M = 10$ kN·m, determine the maximum bending stress in the beam.

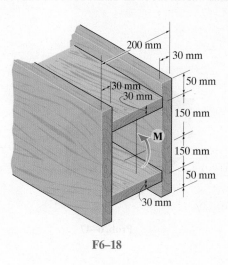

F6–18

F6–19. If the beam is subjected to a bending moment of $M = 5$ kN·m, determine the bending stress developed at point A.

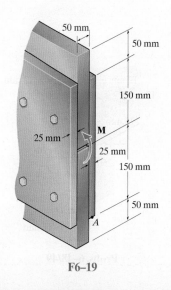

F6–19

6

PROBLEMS

6–47. A member having the dimensions shown is used to resist an internal bending moment of $M = 90$ kN·m. Determine the maximum stress in the member if the moment is applied (a) about the z axis (as shown) (b) about the y axis. Sketch the stress distribution for each case.

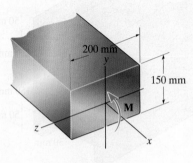

Prob. 6–47

*6–48. Determine the moment M that will produce a maximum stress of 70 MPa on the cross section.

•6–49. Determine the maximum tensile and compressive bending stress in the beam if it is subjected to a moment of $M = 6$ kN·m.

6–50. The channel strut is used as a guide rail for a trolley. If the maximum moment in the strut is $M = 30$ N·m, determine the bending stress at points A, B, and C.

6–51. The channel strut is used as a guide rail for a trolley. If the allowable bending stress for the material is $\sigma_{allow} = 175$ MPa, determine the maximum bending moment the strut will resist.

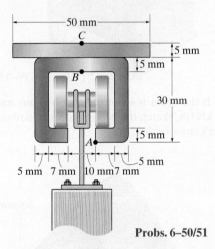

Probs. 6–50/51

*6–52. The beam is subjected to a moment **M**. Determine the percentage of this moment that is resisted by the stresses acting on both the top and bottom boards, A and B, of the beam.

•6–53. Determine the moment **M** that should be applied to the beam in order to create a compressive stress at point D of $\sigma_D = 30$ MPa. Also sketch the stress distribution acting over the cross section and compute the maximum stress developed in the beam.

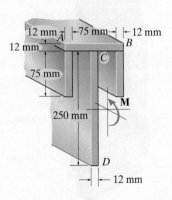

Probs. 6–48/49

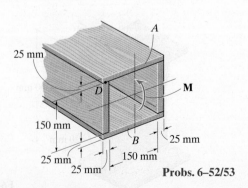

Probs. 6–52/53

6–54.   The beam is made from three boards nailed together as shown. If the moment acting on the cross section is $M = 600$ N · m, determine the maximum bending stress in the beam. Sketch a three-dimensional view of the stress distribution acting over the cross section.

6–55.   The beam is made from three boards nailed together as shown. If the moment acting on the cross section is $M = 600$ N · m, determine the resultant force the bending stress produces on the top board.

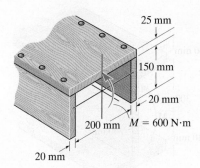

25 mm
150 mm
20 mm
200 mm $M = 600$ N·m
20 mm

Probs. 6–54/55

*6–56.   The aluminum strut has a cross-sectional area in the form of a cross. If it is subjected to the moment $M = 8$ kN · m, determine the bending stress acting at points A and B, and show the results acting on volume elements located at these points.

•6–57.   The aluminum strut has a cross-sectional area in the form of a cross. If it is subjected to the moment $M = 8$ kN · m, determine the maximum bending stress in the beam, and sketch a three-dimensional view of the stress distribution acting over the entire cross-sectional area.

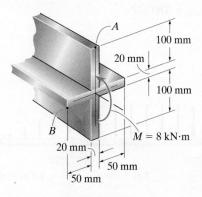

A
100 mm
20 mm
100 mm
B
20 mm $M = 8$ kN·m
50 mm
50 mm

Probs. 6–56/57

6–58.   If the beam is subjected to an internal moment of $M = 150$ kN · m. determine the maximum tensile and compressive bending stress in the beam.

6–59.   If the beam is made of material having an allowable tensile and compressive stress of $(\sigma_{allow})_t = 168$ MPa and $(\sigma_{allow})_c = 154$ MPa, respectively, determine the maximum allowable internal moment **M** that can be applied to the beam.

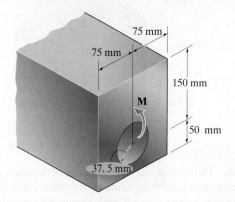

75 mm
75 mm
150 mm
M
50 mm
37. 5 mm

Probs. 6–58/59

*6–60.   The beam is constructed from four boards as shown. If it is subjected to a moment of $M_z = 24$ kN · m, determine the stress at points A and B. Sketch a three-dimensional view of the stress distribution.

•6–61.   The beam is constructed from four boards as shown. If it is subjected to a moment of $M_z = 24$ kN · m, determine the resultant force the stress produces on the top board C.

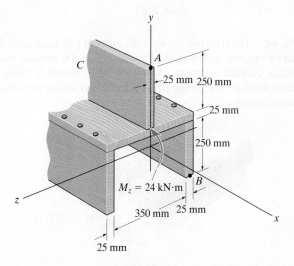

y
C
A
25 mm   250 mm
25 mm
250 mm
$M_z = 24$ kN·m B
z
350 mm   25 mm x
25 mm

Probs. 6–60/61

6–62. A box beam is constructed from four pieces of wood, glued together as shown. If the moment acting on the cross section is $10 \text{ kN} \cdot \text{m}$, determine the stress at points A and B and show the results acting on volume elements located at these points.

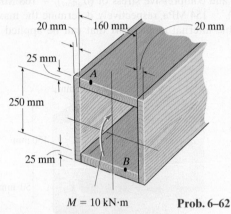

$M = 10 \text{ kN·m}$ **Prob. 6–62**

6–63. Determine the dimension a of a beam having a square cross section in terms of the radius r of a beam with a circular cross section if both beams are subjected to the same internal moment which results in the same maximum bending stress.

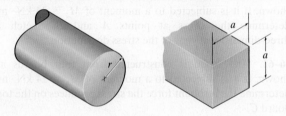

Prob. 6–63

*6–64. The steel rod having a diameter of 1 in. is subjected to an internal moment of $M = 450 \text{ kN} \cdot \text{m}$. Determine the stress created at points A and B. Also, sketch a three-dimensional view of the stress distribution acting over the cross section.

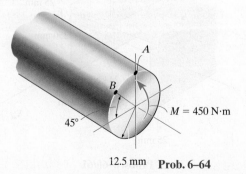

$M = 450 \text{ N·m}$

$45°$

12.5 mm **Prob. 6–64**

•6–65. If the moment acting on the cross section of the beam is $M = 6 \text{ kN} \cdot \text{m}$, determine the maximum bending stress in the beam. Sketch a three-dimensional view of the stress distribution acting over the cross section.

6–66. If $M = 6 \text{ kN} \cdot \text{m}$, determine the resultant force the bending stress produces on the top board A of the beam.

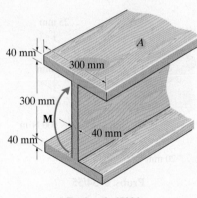

Probs. 6–65/66

6–67. The rod is supported by smooth journal bearings at A and B that only exert vertical reactions on the shaft. If $d = 90 \text{ mm}$, determine the absolute maximum bending stress in the beam, and sketch the stress distribution acting over the cross section.

*6–68. The rod is supported by smooth journal bearings at A and B that only exert vertical reactions on the shaft. Determine its smallest diameter d if the allowable bending stress is $\sigma_{\text{allow}} = 180 \text{ MPa}$.

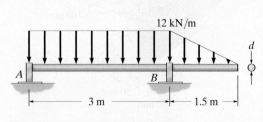

Probs. 6–67/68

•6–69.   Two designs for a beam are to be considered. Determine which one will support a moment of $M = 150 \text{ kN} \cdot \text{m}$ with the least amount of bending stress. What is that stress?

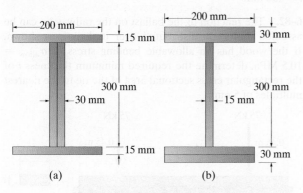

(a)   (b)

Prob. 6–69

6–70.   The simply supported truss is subjected to the central distributed load. Neglect the effect of the diagonal lacing and determine the absolute maximum bending stress in the truss. The top member is a pipe having an outer diameter of 25 mm and thickness of 5 mm and the bottom member is a solid rod having a diameter of 12 mm.

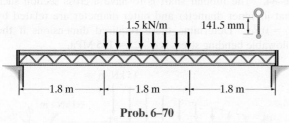

Prob. 6–70

6–71.   The axle of the freight car is subjected to wheel loadings of 100 kN. If it is supported by two journal bearings at C and D, determine the maximum bending stress developed at the center of the axle, where the diameter is 137.5 mm.

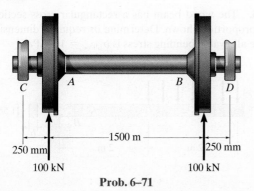

Prob. 6–71

*6–72.   The steel beam has the cross-sectional area shown. Determine the largest intensity of distributed load w_0 that it can support so that the maximum bending stress in the beam does not exceed $\sigma_{max} = 150$ MPa.

•6–73.   The steel beam has the cross-sectional area shown. If $w_0 = 10$ kN/m, determine the maximum bending stress in the beam.

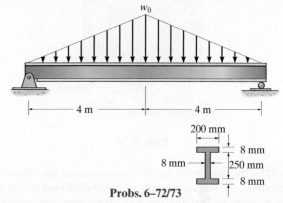

Probs. 6–72/73

6–74.   The boat has a weight of 11.5 kN and a center of gravity at G. If it rests on the trailer at the smooth contact A and can be considered pinned at B, determine the absolute maximum bending stress developed in the main strut of the trailer. Consider the strut to be a box-beam having the dimensions shown and pinned at C.

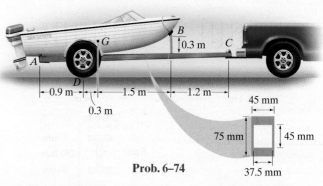

Prob. 6–74

6–75.   The shaft is supported by a smooth thrust bearing at A and smooth journal bearing at D. If the shaft has the cross section shown, determine the absolute maximum bending stress in the shaft.

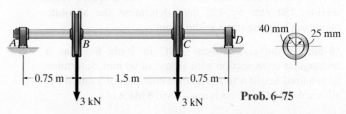

Prob. 6–75

*6–76. Determine the moment **M** that must be applied to the beam in order to create a maximum stress of 80 MPa. Also sketch the stress distribution acting over the cross section.

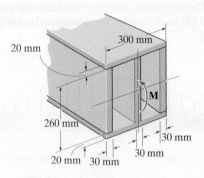

Prob. 6–76

•6–77. The steel beam has the cross-sectional area shown. Determine the largest intensity of distributed load w that it can support so that the bending stress does not exceed $\sigma_{max} = 150$ MPa.

6–78. The steel beam has the cross-sectional area shown. If $w = 75$ kN/m, determine the absolute maximum bending stress in the beam.

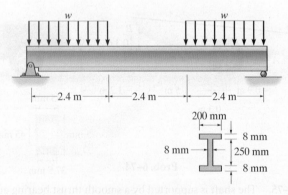

Probs. 6–77/78

6–79. If the beam ACB in Prob. 6–9 has a square cross section, 150 mm by 150 mm, determine the absolute maximum bending stress in the beam.

*6–80. If the crane boom ABC in Prob. 6–3 has a rectangular cross section with a base of 60 mm, determine its required height h to the nearest multiples of 5 mm if the allowable bending stress is $\sigma_{allow} = 170$ MPa.

•6–81. If the reaction of the ballast on the railway tie can be assumed uniformly distributed over its length as shown, determine the maximum bending stress developed in the tie. The tie has the rectangular cross section with thickness $t = 150$ mm.

6–82. The reaction of the ballast on the railway tie can be assumed uniformly distributed over its length as shown. If the wood has an allowable bending stress of $\sigma_{allow} = 10.5$ MPa, determine the required minimum thickness t of the rectangular cross sectional area of the tie to the nearest multiples of 5 mm.

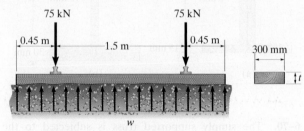

Probs. 6–81/82

6–83. Determine the absolute maximum bending stress in the tubular shaft if $d_i = 160$ mm and $d_o = 200$ mm.

*6–84. The tubular shaft is to have a cross section such that its inner diameter and outer diameter are related by $d_i = 0.8d_o$. Determine these required dimensions if the allowable bending stress is $\sigma_{allow} = 155$ MPa.

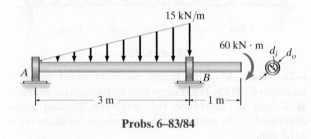

Probs. 6–83/84

6–85. The wood beam has a rectangular cross section in the proportion shown. Determine its required dimension b if the allowable bending stress is $\sigma_{allow} = 10$ MPa.

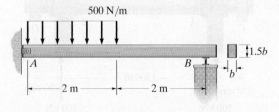

Prob. 6–85

6–86. Determine the absolute maximum bending stress in the 50-mm-diameter shaft which is subjected to the concentrated forces. The journal bearings at A and B only support vertical forces.

6–87. Determine the smallest allowable diameter of the shaft which is subjected to the concentrated forces. The journal bearings at A and B only support vertical forces. The allowable bending stress is $\sigma_{allow} = 150$ MPa.

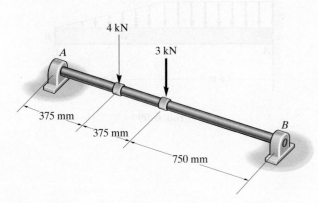

Probs. 6–86/87

*6–88. If the beam has a square cross section of 225 mm on each side, determine the absolute maximum bending stress in the beam.

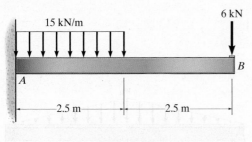

Prob. 6–88

•6–89. If the compound beam in Prob. 6–42 has a square cross section, determine its dimension a if the allowable bending stress is $\sigma_{allow} = 150$ MPa.

6–90. If the beam in Prob. 6–28 has a rectangular cross section with a width b and a height h, determine the absolute maximum bending stress in the beam.

6–91. Determine the absolute maximum bending stress in the 80-mm-diameter shaft which is subjected to the concentrated forces. The journal bearings at A and B only support vertical forces.

*6–92. Determine the smallest allowable diameter of the shaft which is subjected to the concentrated forces. The journal bearings at A and B only support vertical forces. The allowable bending stress is $\sigma_{allow} = 150$ MPa.

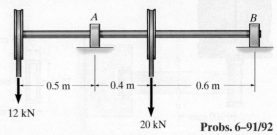

Probs. 6–91/92

•6–93. The man has a mass of 78 kg and stands motionless at the end of the diving board. If the board has the cross section shown, determine the maximum normal strain developed in the board. The modulus of elasticity for the material is $E = 125$ GPa. Assume A is a pin and B is a roller.

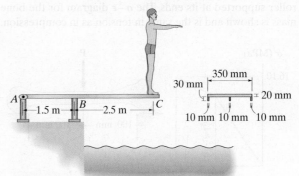

Prob. 6–93

6–94. The two solid steel rods are bolted together along their length and support the loading shown. Assume the support at A is a pin and B is a roller. Determine the required diameter d of each of the rods if the allowable bending stress is $\sigma_{allow} = 130$ MPa.

6–95. Solve Prob. 6–94 if the rods are rotated 90° so that both rods rest on the supports at A (pin) and B (roller).

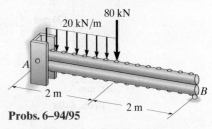

Probs. 6–94/95

*6–96. The chair is supported by an arm that is hinged so it rotates about the vertical axis at A. If the load on the chair is 900 N and the arm is a hollow tube section having the dimensions shown, determine the maximum bending stress at section a–a.

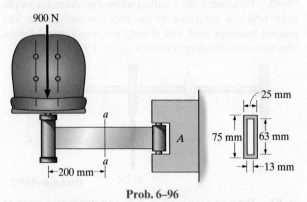

900 N

25 mm

75 mm   63 mm

13 mm

200 mm

a

a

A

Prob. 6–96

•6–97. A portion of the femur can be modeled as a tube having an inner diameter of 9.5 mm and an outer diameter of 32 mm. Determine the maximum elastic static force P that can be applied to its center. Assume the bone to be roller supported at its ends. The σ–ϵ diagram for the bone mass is shown and is the same in tension as in compression.

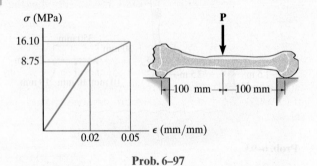

σ (MPa)

16.10

8.75

0.02   0.05

ϵ (mm/mm)

P

100 mm   100 mm

Prob. 6–97

6–98. If the beam in Prob. 6–18 has a rectangular cross section with a width of 200 mm and a height of 400 mm, determine the absolute maximum bending stress in the beam.

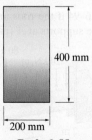

400 mm

200 mm

Prob. 6–98

6–99. If the beam has a square cross section of 150 mm on each side, determine the absolute maximum bending stress in the beam.

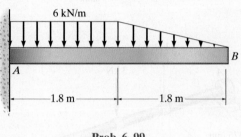

6 kN/m

A

B

1.8 m

1.8 m

Prob. 6–99

*6–100. The steel beam has the cross-sectional area shown. Determine the largest intensity of the distributed load w_0 that it can support so that the maximum bending stress in the beam does not exceed $\sigma_{allow} = 160$ MPa.

•6–101. The steel beam has the cross-sectional area shown. If $w_0 = 30$ kN/m, determine the maximum bending stress in the beam.

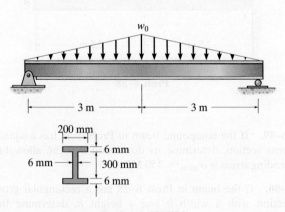

w_0

3 m   3 m

200 mm

6 mm

6 mm   300 mm

6 mm

Probs. 6–100/101

6–102. The bolster or main supporting girder of a truck body is subjected to the uniform distributed load. Determine the bending stress at points A and B.

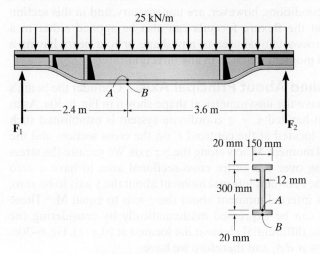

25 kN/m

2.4 m   3.6 m

F_1 F_2

20 mm 150 mm

12 mm

300 mm

A

B

20 mm

Prob. 6–102

6–103. Determine the largest uniform distributed load w that can be supported so that the bending stress in the beam does not exceed $\sigma_{allow} = 5$ MPa.

*6–104. If $w = 10$ kN/m, determine the maximum bending stress in the beam. Sketch the stress distribution acting over the cross section.

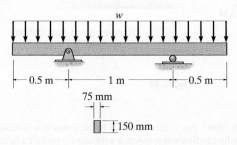

w

0.5 m   1 m   0.5 m

75 mm

150 mm

Probs. 6–103/104

•6–105. If the allowable bending stress for the wood beam is $\sigma_{allow} = 1$ MPa, determine the required dimension b to the nearest multiples of 5 mm of its cross section. Assume the support at A is a pin and B is a roller.

6–106. The wood beam has a rectangular cross section in the proportion shown. If $b = 200$ mm, determine the absolute maximum bending stress in the beam.

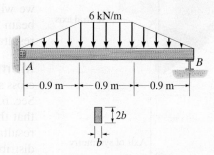

6 kN/m

A B

0.9 m   0.9 m   0.9 m

$2b$

b

Probs. 6–105/106

6–107. A beam is made of a material that has a modulus of elasticity in compression different from that given for tension. Determine the location c of the neutral axis, and derive an expression for the maximum tensile stress in the beam having the dimensions shown if it is subjected to the bending moment **M**.

*6–108. The beam has a rectangular cross section and is subjected to a bending moment **M**. If the material from which it is made has a different modulus of elasticity for tension and compression as shown, determine the location c of the neutral axis and the maximum compressive stress in the beam.

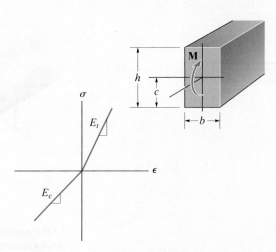

M

h

c

b

σ

E_t

ϵ

E_c

Probs. 6–107/108

6

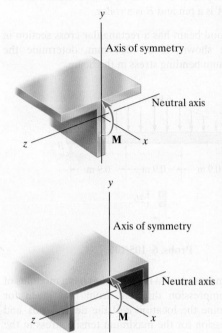

Axis of symmetry

Neutral axis

M

Fig. 6–29

Axis of symmetry

Neutral axis

M

6.5 Unsymmetric Bending

When developing the flexure formula we imposed a condition that the cross-sectional area be *symmetric* about an axis perpendicular to the neutral axis; furthermore, the resultant internal moment **M** acts along the neutral axis. Such is the case for the "T" or channel sections shown in Fig. 6–29. These conditions, however, are unnecessary, and in this section we will show that the flexure formula can also be applied either to a beam having a cross-sectional area of any shape or to a beam having a resultant internal moment that acts in any direction.

Moment Applied About Principal Axis.

Consider the beam's cross section to have the unsymmetrical shape shown in Fig. 6–30a. As in Sec. 6.4, the right-handed x, y, z coordinate system is established such that the origin is located at the centroid C on the cross section, and the resultant internal moment **M** acts along the $+z$ axis. We require the stress distribution acting over the entire cross-sectional area to have a zero force resultant, the resultant internal moment about the y axis to be zero, and the resultant internal moment about the z axis to equal **M**.* These three conditions can be expressed mathematically by considering the force acting on the differential element dA located at $(0, y, z)$, Fig. 6–30a. This force is $dF = \sigma\, dA$, and therefore we have

$$F_R = \Sigma F_x; \qquad\qquad 0 = -\int_A \sigma\, dA \qquad\qquad (6\text{–}14)$$

$$(M_R)_y = \Sigma M_y; \qquad\qquad 0 = -\int_A z\sigma\, dA \qquad\qquad (6\text{–}15)$$

$$(M_R)_z = \Sigma M_z; \qquad\qquad M = \int_A y\sigma\, dA \qquad\qquad (6\text{–}16)$$

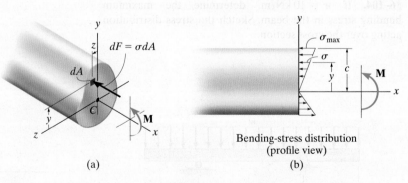

$dF = \sigma dA$

dA

C

M

(a)

σ_{max}

σ

c

y

M

Bending-stress distribution
(profile view)

(b)

Fig. 6–30

*The condition that moments about the y axis be equal to zero was not considered in Sec. 6.4, since the bending-stress distribution was *symmetric* with respect to the y axis and such a distribution of stress automatically produces zero moment about the y axis. See Fig. 6–24c.

As shown in Sec. 6.4, Eq. 6–14 is satisfied since the z axis passes through the *centroid* of the area. Also, since the z axis represents the *neutral axis* for the cross section, the normal stress will vary linearly from zero at the neutral axis, to a maximum at $y = c$, Fig. 6–30b. Hence the stress distribution is defined by $\sigma = -(y/c)\sigma_{max}$. When this equation is substituted into Eq. 6–16 and integrated, it leads to the flexure formula $\sigma_{max} = Mc/I$. When it is substituted into Eq. 6–15, we get

$$0 = \frac{-\sigma_{max}}{c} \int_A yz \, dA$$

which requires

$$\int_A yz \, dA = 0$$

This integral is called the **product of inertia** for the area. As indicated in Appendix A, it will indeed be zero provided the y and z axes are chosen as **principal axes of inertia** for the area. For an arbitrarily shaped area, the orientation of the principal axes can always be determined, using either the inertia transformation equations or Mohr's circle of inertia as explained in Appendix A, Secs. A.4 and A.5. If the area has an axis of symmetry, however, the **principal axes** can easily be established *since they will always be oriented along the axis of symmetry and perpendicular to it*.

For example, consider the members shown in Fig. 6–31. In each of these cases, y and z must define the principal axes of inertia for the cross section in order to satisfy Eqs. 6–14 through 6–16. In Fig. 6–31a the principal axes are located by symmetry, and in Figs. 6–31b and 6–31c their orientation is determined using the methods of Appendix A. Since **M** is applied about one of the principal axes (z axis), the stress distribution is determined from the flexure formula, $\sigma = -My/I_z$, and is shown for each case.

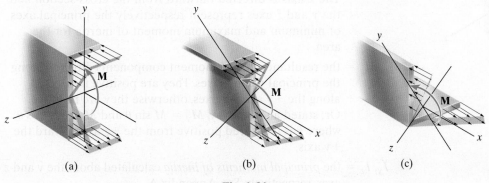

(a)   (b)   (c)

Fig. 6–31

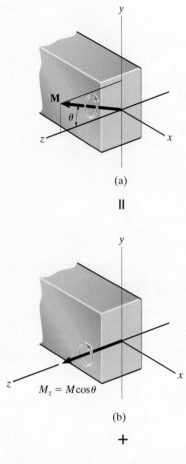

(a)

$\parallel$

$M_z = M\cos\theta$

(b)

$+$

$M_y = M\sin\theta$

(c)

Fig. 6–32

Moment Arbitrarily Applied. Sometimes a member may be loaded such that **M** does not act about one of the principal axes of the cross section. When this occurs, the moment should first be resolved into components directed along the principal axes, then the flexure formula can be used to determine the normal stress caused by *each* moment component. Finally, using the principle of superposition, the resultant normal stress at the point can be determined.

To show this, consider the beam to have a rectangular cross section and to be subjected to the moment **M**, Fig. 6–32a. Here **M** makes an angle θ with the *principal z* axis. We will assume θ is positive when it is directed from the $+z$ axis toward the $+y$ axis, as shown. Resolving **M** into components along the z and y axes, we have $M_z = M\cos\theta$ and $M_y = M\sin\theta$, as shown in Figs. 6–32b and 6–32c. The normal-stress distributions that produce **M** and its components **M$_z$** and **M$_y$** are shown in Figs. 6–32d, 6–32e, and 6–32f, where it is assumed that $(\sigma_x)_{max} > (\sigma'_x)_{max}$. By inspection, the maximum tensile and compressive stresses $[(\sigma_x)_{max} + (\sigma'_x)_{max}]$ occur at two opposite corners of the cross section, Fig. 6–32d.

Applying the flexure formula to each moment component in Figs. 6–32b and 6–32c, and adding the results algebraically, the resultant normal stress at any point on the cross section, Fig. 6–32d, is then

$$\sigma = -\frac{M_z y}{I_z} + \frac{M_y z}{I_y} \qquad (6\text{–}17)$$

Here,

σ = the normal stress at the point

y, z = the coordinates of the point measured from x, y, z axes having their origin at the centroid of the cross-sectional area and forming a right-handed coordinate system
The x axis is directed outward from the cross-section and the y and z axes represent respectively the principal axes of minimum and maximum moment of inertia for the area

M_y, M_z = the resultant internal moment components directed along the principal y and z axes. They are positive if directed along the $+y$ and $+z$ axes, otherwise they are negative. Or, stated another way, $M_y = M\sin\theta$ and $M_z = M\cos\theta$, where θ is measured positive from the $+z$ axis toward the $+y$ axis

I_y, I_z = the *principal moments of inertia* calculated about the y and z axes, respectively. See Appendix A

The x, y, z axes form a right-handed system, and the proper algebraic signs must be assigned to the moment components and the coordinates when applying this equation. When this is the case, the resulting stress will be *tensile* if it is *positive* and *compressive* if it is *negative*.

Orientation of the Neutral Axis.

The angle α of the neutral axis in Fig. 6–32d can be determined by applying Eq. 6–17 with $\sigma = 0$, since by definition no normal stress acts on the neutral axis. We have

$$y = \frac{M_y I_z}{M_z I_y} z$$

Since $M_z = M \cos \theta$ and $M_y = M \sin \theta$, then

$$y = \left(\frac{I_z}{I_y} \tan \theta \right) z$$

(6–18)

This equation defines the neutral axis for the cross section. Since the slope of this line is $\tan \alpha = y/z$, then

$$\boxed{\tan \alpha = \frac{I_z}{I_y} \tan \theta} \qquad (6-19)$$

Here it can be seen that unless $I_z = I_y$ the angle θ, defining the direction of the moment **M**, Fig. 6–32a, will *not equal* α, the angle defining the inclination of the neutral axis, Fig. 6–32d.

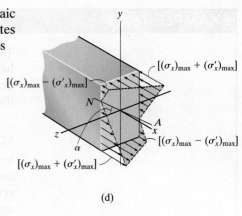
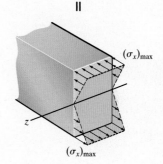

(d)

||

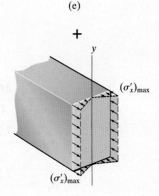

$(\sigma_x)_{max}$

$(\sigma_x)_{max}$

(e)

+

$(\sigma'_x)_{max}$

$(\sigma'_x)_{max}$

(f)

Fig. 6–32 (cont.)

6

Important Points

- The flexure formula can be applied only when bending occurs about axes that represent the *principal axes of inertia* for the cross section. These axes have their origin at the centroid and are oriented along an axis of symmetry, if there is one, and perpendicular to it.

- If the moment is applied about some arbitrary axis, then the moment must be resolved into components along each of the principal axes, and the stress at a point is determined by superposition of the stress caused by each of the moment components.

EXAMPLE | 6.15

The rectangular cross section shown in Fig. 6–33a is subjected to a bending moment of $M = 12 \text{ kN} \cdot \text{m}$. Determine the normal stress developed at each corner of the section, and specify the orientation of the neutral axis.

SOLUTION

Internal Moment Components. By inspection it is seen that the y and z axes represent the principal axes of inertia since they are axes of symmetry for the cross section. As required we have established the z axis as the principal axis for *maximum* moment of inertia. The moment is resolved into its y and z components, where

$$M_y = -\frac{4}{5}(12 \text{ kN} \cdot \text{m}) = -9.60 \text{ kN} \cdot \text{m}$$

$$M_z = \frac{3}{5}(12 \text{ kN} \cdot \text{m}) = 7.20 \text{ kN} \cdot \text{m}$$

Section Properties. The moments of inertia about the y and z axes are

$$I_y = \frac{1}{12}(0.4 \text{ m})(0.2 \text{ m})^3 = 0.2667(10^{-3}) \text{ m}^4$$

$$I_z = \frac{1}{12}(0.2 \text{ m})(0.4 \text{ m})^3 = 1.067(10^{-3}) \text{ m}^4$$

Bending Stress. Thus,

$$\sigma = -\frac{M_z y}{I_z} + \frac{M_y z}{I_y}$$

$$\sigma_B = -\frac{7.20(10^3) \text{ N} \cdot \text{m}(0.2 \text{ m})}{1.067(10^{-3}) \text{ m}^4} + \frac{-9.60(10^3) \text{ N} \cdot \text{m}(-0.1 \text{ m})}{0.2667(10^{-3}) \text{ m}^4} = 2.25 \text{ MPa} \qquad Ans.$$

$$\sigma_C = -\frac{7.20(10^3) \text{ N} \cdot \text{m}(0.2 \text{ m})}{1.067(10^{-3}) \text{ m}^4} + \frac{-9.60(10^3) \text{ N} \cdot \text{m}(0.1 \text{ m})}{0.2667(10^{-3}) \text{ m}^4} = -4.95 \text{ MPa} \qquad Ans.$$

$$\sigma_D = -\frac{7.20(10^3) \text{ N} \cdot \text{m}(-0.2 \text{ m})}{1.067(10^{-3}) \text{ m}^4} + \frac{-9.60(10^3) \text{ N} \cdot \text{m}(0.1 \text{ m})}{0.2667(10^{-3}) \text{ m}^4} = -2.25 \text{ MPa} \qquad Ans.$$

$$\sigma_E = -\frac{7.20(10^3) \text{ N} \cdot \text{m}(-0.2 \text{ m})}{1.067(10^{-3}) \text{ m}^4} + \frac{-9.60(10^3) \text{ N} \cdot \text{m}(-0.1 \text{ m})}{0.2667(10^{-3}) \text{ m}^4} = 4.95 \text{ MPa} \qquad Ans.$$

The resultant normal-stress distribution has been sketched using these values, Fig. 6–33b. Since superposition applies, the distribution is linear as shown.

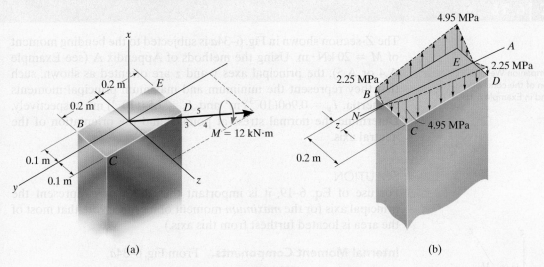

(a)

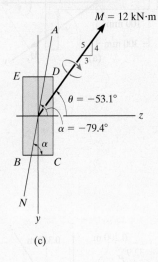

(b)

Fig. 6–33

Orientation of Neutral Axis. The location z of the neutral axis (*NA*), Fig. 6–33*b*, can be established by proportion. Along the edge *BC*, we require

$$\frac{2.25 \text{ MPa}}{z} = \frac{4.95 \text{ MPa}}{(0.2 \text{ m} - z)}$$

$$0.450 - 2.25z = 4.95z$$

$$z = 0.0625 \text{ m}$$

In the same manner this is also the distance from *D* to the neutral axis in Fig. 6–33*b*.

We can also establish the orientation of the *NA* using Eq. 6–19, which is used to specify the angle α that the axis makes with the z or *maximum* principal axis. According to our sign convention, θ must be measured from the $+z$ axis toward the $+y$ axis. By comparison, in Fig. 6–33*c*, $\theta = -\tan^{-1}\frac{4}{3} = -53.1°$ (or $\theta = +306.9°$). Thus,

$$\tan \alpha = \frac{I_z}{I_y} \tan \theta$$

$$\tan \alpha = \frac{1.067(10^{-3}) \text{ m}^4}{0.2667(10^{-3}) \text{ m}^4} \tan(-53.1°)$$

$$\alpha = -79.4° \hspace{3cm} Ans.$$

This result is shown in Fig. 6–33*c*. Using the value of z calculated above, verify, using the geometry of the cross section, that one obtains the same answer.

EXAMPLE | 6.16

Refer to the Companion Website for the animation of the concepts that can be found in Example 6.16

The Z-section shown in Fig. 6–34a is subjected to the bending moment of $M = 20$ kN·m. Using the methods of Appendix A (see Example A.4 or A.5), the principal axes y and z are oriented as shown, such that they represent the minimum and maximum principal moments of inertia, $I_y = 0.960(10^{-3})$ m^4 and $I_z = 7.54(10^{-3})$ m^4, respectively. Determine the normal stress at point P and the orientation of the neutral axis.

SOLUTION

For use of Eq. 6–19, it is important that the z axis represent the principal axis for the *maximum* moment of inertia. (Note that most of the area is located furthest from this axis.)

Internal Moment Components. From Fig. 6–34a,

$$M_y = 20 \text{ kN} \cdot \text{m} \sin 57.1° = 16.79 \text{ kN} \cdot \text{m}$$

$$M_z = 20 \text{ kN} \cdot \text{m} \cos 57.1° = 10.86 \text{ kN} \cdot \text{m}$$

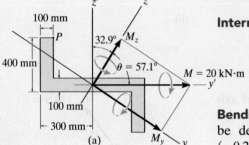

(a)

Bending Stress. The y and z coordinates of point P must be determined first. Note that the y', z' coordinates of P are $(-0.2 \text{ m}, 0.35 \text{ m})$. Using the colored triangles from the construction shown in Fig. 6–34b, we have

$$y_P = -0.35 \sin 32.9° - 0.2 \cos 32.9° = -0.3580 \text{ m}$$

$$z_P = 0.35 \cos 32.9° - 0.2 \sin 32.9° = 0.1852 \text{ m}$$

Applying Eq. 6–17,

$$\sigma_P = -\frac{M_z y_P}{I_z} + \frac{M_y z_P}{I_y}$$

$$= -\frac{(10.86(10^3) \text{ N} \cdot \text{m})(-0.3580 \text{ m})}{7.54(10^{-3}) \text{ m}^4} + \frac{(16.79(10^3) \text{ N} \cdot \text{m})(0.1852 \text{ m})}{0.960(10^{-3}) \text{ m}^4}$$

$$= 3.76 \text{ MPa} \qquad \qquad \qquad Ans.$$

Orientation of Neutral Axis. The angle $\theta = 57.1°$ is shown in Fig. 6–34a. Thus,

$$\tan \alpha = \left[\frac{7.54(10^{-3}) \text{ m}^4}{0.960(10^{-3}) \text{ m}^4}\right] \tan 57.1°$$

$$\alpha = 85.3° \qquad \qquad \qquad Ans.$$

The neutral axis is oriented as shown in Fig. 6–34b.

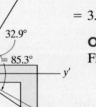

(b)

Fig. 6–34

FUNDAMENTAL PROBLEMS

F6–20. Determine the bending stress developed at corners A and B. What is the orientation of the neutral axis?

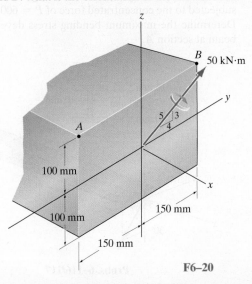

F6–20

F6–21. Determine the maximum stress in the beam's cross section.

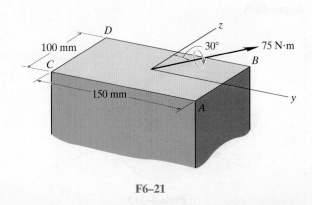

F6–21

PROBLEMS

•6–109. The beam is subjected to a bending moment of $M = 30$ kN · m directed as shown. Determine the maximum bending stress in the beam and the orientation of the neutral axis.

6–110. Determine the maximum magnitude of the bending moment **M** that can be applied to the beam so that the bending stress in the member does not exceed 100 MPa.

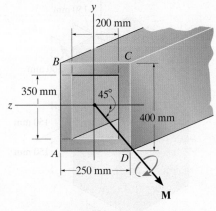

Probs. 6–109/110

6–111. If the resultant internal moment acting on the cross section of the aluminum strut has a magnitude of $M = 520$ N · m and is directed as shown, determine the bending stress at points A and B. The location $\overline{y}$ of the centroid C of the strut's cross-sectional area must be determined. Also, specify the orientation of the neutral axis.

*6–112. The resultant internal moment acting on the cross section of the aluminum strut has a magnitude of $M = 520$ N · m and is directed as shown. Determine maximum bending stress in the strut. The location $\overline{y}$ of the centroid C of the strut's cross-sectional area must be determined. Also, specify the orientation of the neutral axis.

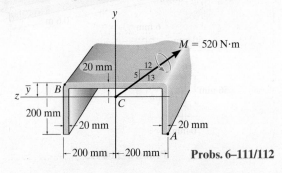

Probs. 6–111/112

6–113. Consider the general case of a prismatic beam subjected to bending-moment components M_y and M_z, as shown, when the x, y, z axes pass through the centroid of the cross section. If the material is linear-elastic, the normal stress in the beam is a linear function of position such that $\sigma = a + by + cz$. Using the equilibrium conditions $0 = \int_A \sigma \, dA$, $M_y = \int_A z\sigma \, dA$, $M_z = \int_A - y\sigma \, dA$, determine the constants a, b, and c, and show that the normal stress can be determined from the equation $\sigma = [-(M_z I_y + M_y I_{yz})y + (M_y I_z + M_z I_{yz})z]/(I_y I_z - I_{yz}^2)$, where the moments and products of inertia are defined in Appendix A.

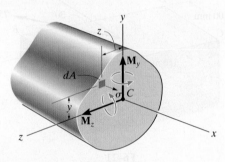

Prob. 6–113

6–114. The cantilevered beam is made from the Z-section having the cross-section shown. If it supports the two loadings, determine the bending stress at the wall in the beam at point A. Use the result of Prob. 6–113.

6–115. The cantilevered beam is made from the Z-section having the cross-section shown. If it supports the two loadings, determine the bending stress at the wall in the beam at point B. Use the result of Prob. 6–113.

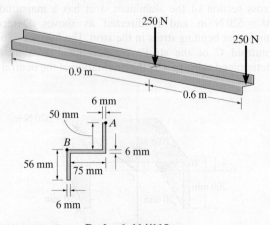

Probs. 6–114/115

*6–116.** The cantilevered wide-flange steel beam is subjected to the concentrated force **P** at its end. Determine the largest magnitude of this force so that the bending stress developed at A does not exceed $\sigma_{\text{allow}} = 180$ MPa.

•6–117.** The cantilevered wide-flange steel beam is subjected to the concentrated force of $P = 600$ N at its end. Determine the maximum bending stress developed in the beam at section A.

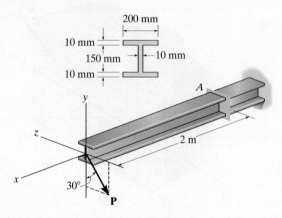

Probs. 6–116/117

6–118. If the beam is subjected to the internal moment of $M = 1200$ kN · m, determine the maximum bending stress acting on the beam and the orientation of the neutral axis.

6–119. If the beam is made from a material having an allowable tensile and compressive stress of $(\sigma_{\text{allow}})_t = 125$ MPa and $(\sigma_{\text{allow}})_c = 150$ MPa, respectively, determine the maximum allowable internal moment **M** that can be applied to the beam.

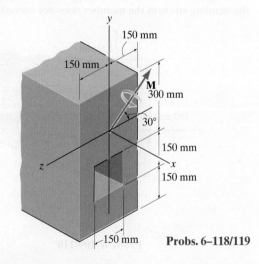

Probs. 6–118/119

*6–120. The shaft is supported on two journal bearings at A and B which offer no resistance to axial loading. Determine the required diameter d of the shaft if the allowable bending stress for the material is $\sigma_{allow} = 150$ MPa.

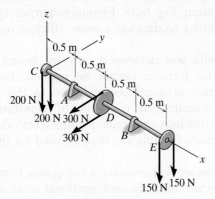

200 N

200 N 300 N

300 N

150 N 150 N

Prob. 6–120

•6–121. The 30-mm-diameter shaft is subjected to the vertical and horizontal loadings of two pulleys as shown. It is supported on two journal bearings at A and B which offer no resistance to axial loading. Furthermore, the coupling to the motor at C can be assumed not to offer any support to the shaft. Determine the maximum bending stress developed in the shaft.

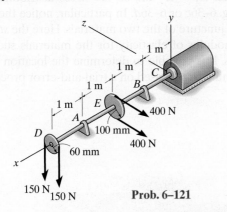

400 N

400 N

150 N 150 N

Prob. 6–121

6–122. Using the techniques outlined in Appendix A, Example A.5 or A.6, the Z section has principal moments of inertia of $I_y = 0.060(10^{-3})$ m^4 and $I_z = 0.471(10^{-3})$ m^4, computed about the principal axes of inertia y and z, respectively. If the section is subjected to an internal moment of $M = 250$ N·m directed horizontally as shown, determine the stress produced at point A. Solve the problem using Eq. 6–17.

6–123. Solve Prob. 6–122 using the equation developed in Prob. 6–113.

*6–124. Using the techniques outlined in Appendix A, Example A.5 or A.6, the Z section has principal moments of inertia of $I_y = 0.060(10^{-3})$ m^4 and $I_z = 0.471(10^{-3})$ m^4, computed about the principal axes of inertia y and z, respectively. If the section is subjected to an internal moment of $M = 250$ N·m directed horizontally as shown, determine the stress produced at point B. Solve the problem using Eq. 6–17.

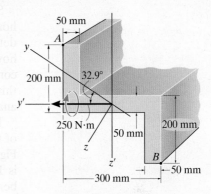

Probs. 6–122/123/124

•6–125. Determine the bending stress at point A of the beam, and the orientation of the neutral axis. Using the method in Appendix A, the principal moments of inertia of the cross section are $I'_z = 3.48(10^6)$ mm^4 and $I'_y = 0.92(10^6)$ mm^4, where z' and y' are the principal axes. Solve the problem using Eq. 6–17.

6–126. Determine the bending stress at point A of the beam using the result obtained in Prob. 6–113. The moments of inertia of the cross sectional area about the z and y axes are $I_z = I_y = 2.20(10^6)$ mm^4, and the product of inertia of the cross sectional area with respect to the z and y axes is $I_{yz} = -1.28(10^6)$ mm^4. (See Appendix A)

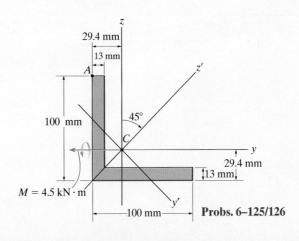

Probs. 6–125/126

*6.6 Composite Beams

Beams constructed of two or more different materials are referred to as *composite beams*. For example, a beam can be made of wood with straps of steel at its top and bottom, Fig. 6–35. Engineers purposely design beams in this manner in order to develop a more efficient means for supporting loads.

Since the flexure formula was developed only for beams having homogeneous material, this formula cannot be applied directly to determine the normal stress in a composite beam. In this section, however, we will develop a method for modifying or "transforming" a composite beam's cross section into one made of a single material. Once this has been done, the flexure formula can then be used for the stress analysis.

To explain how to do this we will consider a composite beam made of two materials, 1 and 2, which have the cross-sectional areas shown in Fig. 6–36a. If a bending moment is applied to this beam, then, like one that is homogeneous, the total cross-sectional area will *remain plane* after bending, and hence the normal strains will vary linearly from zero at the neutral axis to a maximum in the material located farthest from this axis, Fig. 6–36b. Provided the material is linear elastic, then at any point the normal stress in material 1 is determined from $\sigma = E_1\epsilon$, and for material 2 the stress distribution is found from $\sigma = E_2\epsilon$. Assuming material 1 is stiffer than material 2, then $E_1 > E_2$ and so the stress distribution will look like that shown in Fig. 6–36c or 6–36d. In particular, notice the jump in stress that occurs at the juncture of the two materials. Here the *strain* is the *same*, but since the modulus of elasticity for the materials suddenly changes, so does the stress. It is possible to determine the location of the neutral axis and the maximum stress based on a trial-and-error procedure.

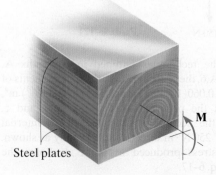

Steel plates

Fig. 6–35

This requires satisfying the conditions that the stress distribution produces a zero resultant force on the cross section and the moment of the stress distribution about the neutral axis must be equal to **M**.

A simpler way to satisfy these two conditions is to use the *transformed section method*, which transforms the beam into one made of a *single material*. For example, if the beam is thought to consist entirely of the less stiff material 2, then the cross section will look like that shown in Fig. 6–36e. Here the height h of the beam remains the *same*, since the strain distribution in Fig. 6–36b must be preserved. However, the upper portion of the beam must be widened in order to carry a load *equivalent* to that carried by the stiffer material 1 in Fig. 6–36d. The necessary width can be determined by considering the force $d\mathbf{F}$ acting on an area $dA = dz\, dy$ of the beam in Fig. 6–36a. It is $dF = \sigma\, dA = (E_1\epsilon)\, dz\, dy$. Assuming the width of a *corresponding element* of height dy in Fig. 6–36e is $n\, dz$, then $dF' = \sigma'dA' = (E_2\epsilon)n\, dz\, dy$. Equating these forces, so that they produce the same moment about the z (neutral) axis, we have

$$E_1\epsilon\, dz\, dy = E_2\epsilon n\, dz\, dy$$

or

$$n = \frac{E_1}{E_2} \qquad (6\text{–}20)$$

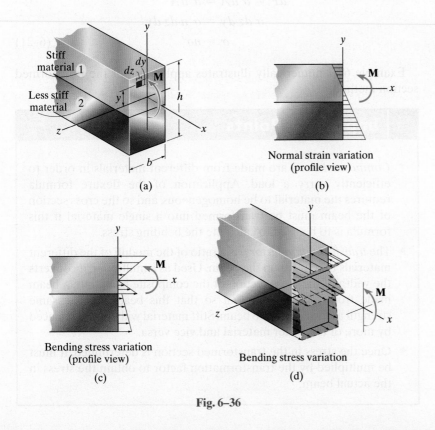

Stiff material 1

Less stiff material 2

(a)

Normal strain variation (profile view)

(b)

Bending stress variation (profile view)

(c)

Bending stress variation

(d)

Fig. 6–36

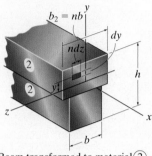

Beam transformed to material ②
(e)

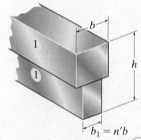

Beam transformed to material ①
(f)

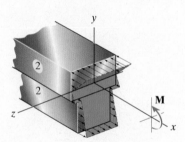

Bending-stress variation for
beam transformed to material ②
(g)

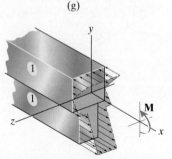

Bending-stress variation for
beam transformed to material ①
(h)

Fig. 6–36 (cont.)

This dimensionless number n is called the ***transformation factor***. It indicates that the cross section, having a width b on the original beam, Fig. 6–36a, must be increased in width to $b_2 = nb$ in the region where material 1 is being transformed into material 2, Fig. 6–36e.

In a similar manner, if the less stiff material 2 is transformed into the stiffer material 1, the cross section will look like that shown in Fig. 6–36f. Here the width of material 2 has been changed to $b_1 = n'b$, where $n' = E_2/E_1$. In this case the transformation factor n' will be *less than one* since $E_1 > E_2$. In other words, we need less of the stiffer material to support the moment.

Once the beam has been transformed into one having a *single material*, the normal-stress distribution over the transformed cross section will be linear as shown in Fig. 6–36g or 6–36h. Consequently, the centroid (neutral axis) and moment of inertia for the transformed area can be determined and the flexure formula applied in the usual manner to determine the stress at each point on the transformed beam. The stress in the transformed beam will be equivalent to the stress in the *same material* of the actual beam; however, the stress found on the transformed material has to be multiplied by the transformation factor n (or n'), since the area of the transformed material, $dA' = n\, dz\, dy$, is n times the area of actual material $dA = dz\, dy$. That is,

$$dF = \sigma\, dA = \sigma'\, dA'$$
$$\sigma\, dz\, dy = \sigma' n\, dz\, dy$$
$$\sigma = n\sigma' \tag{6–21}$$

Example 6.17 numerically illustrates application of the transformed section method.

Important Points

- *Composite beams* are made from different materials in order to efficiently carry a load. Application of the flexure formula requires the material to be homogeneous, and so the cross section of the beam must be transformed into a single material if this formula is to be used to calculate the bending stress.

- The *transformation factor n* is a ratio of the moduli of the different materials that make up the beam. Used as a multiplier, it converts the width of the cross section of the composite beam into a beam made from a single material so that this beam has the same strength as the composite beam. Stiff material will thus be replaced by more of the softer material and vice versa.

- Once the stress in the transformed section is determined, it must be multiplied by the transformation factor to obtain the stress in the actual beam.

*6.7 Reinforced Concrete Beams

All beams subjected to pure bending must resist both tensile and compressive stresses. Concrete, however, is very susceptible to cracking when it is in tension, and therefore by itself it will not be suitable for resisting a bending moment.* In order to circumvent this shortcoming, engineers place steel reinforcing rods within a concrete beam at a location where the concrete is in tension, Fig. 6–37a. To be most effective, these rods are located farthest from the beam's neutral axis, so that the moment created by the forces developed in them is greatest about the neutral axis. Furthermore, the rods are required to have some concrete coverage to protect them from corrosion or loss of strength in the event of a fire. Codes used for actual reinforced concrete design assume the ability of concrete will not support any tensile loading, since the possible cracking of concrete is unpredictable. As a result, the normal stress distribution acting on the cross-sectional area of a reinforced concrete beam is assumed to look like that shown in Fig. 6–37b.

The stress analysis requires locating the neutral axis and determining the maximum stress in the steel and concrete. To do this, the area of steel A_{st} is first transformed into an equivalent area of concrete using the transformation factor $n = E_{st}/E_{conc}$. This ratio, which gives $n > 1$, requires a "greater" amount of concrete to replace the steel. The transformed area is nA_{st} and the transformed section looks like that shown in Fig. 6–37c. Here d represents the distance from the top of the beam to the (transformed) steel, b is the beam's width, and h' is the yet unknown distance from the top of the beam to the neutral axis. To obtain h', we require the centroid C of the cross-sectional area of the transformed section to lie on the neutral axis, Fig. 6–37c. With reference to the neutral axis, therefore, the moment of the two areas, $\Sigma \tilde{y} A$, must be zero, since $\bar{y} = \Sigma \tilde{y} A / \Sigma A = 0$. Thus,

$$bh'\left(\frac{h'}{2}\right) - nA_{st}(d - h') = 0$$

$$\frac{b}{2}h'^2 + nA_{st}h' - nA_{st}d = 0$$

Once h' is obtained from this quadratic equation, the solution proceeds in the usual manner for obtaining the stress in the beam. Example 6.18 numerically illustrates application of this method.

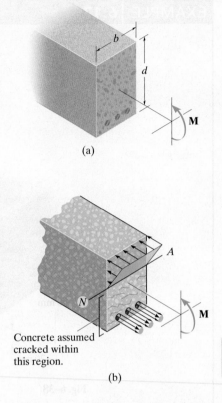

(a)

(b)

Concrete assumed cracked within this region.

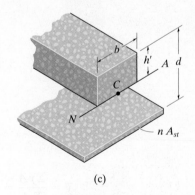

(c)

Fig. 6–37

*Inspection of its particular stress–strain diagram in Fig. 3–11 reveals that concrete can be 12.5 times stronger in compression than in tension.

EXAMPLE 6.17

A composite beam is made of wood and reinforced with a steel strap located on its bottom side. It has the cross-sectional area shown in Fig. 6–38a. If the beam is subjected to a bending moment of $M = 2$ kN·m, determine the normal stress at points B and C. Take $E_w = 12$ GPa and $E_{st} = 200$ GPa.

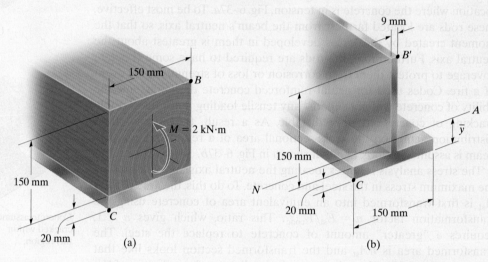

Fig. 6–38

SOLUTION

Section Properties. Although the choice is arbitrary, here we will transform the section into one made entirely of steel. Since steel has a greater stiffness than wood ($E_{st} > E_w$), the width of the wood must be *reduced* to an equivalent width for steel. Hence n must be less than one. For this to be the case, $n = E_w/E_{st}$, so that

$$b_{st} = nb_w = \frac{12 \text{ GPa}}{200 \text{ GPa}}(150 \text{ mm}) = 9 \text{ mm}$$

The transformed section is shown in Fig. 6–38b.

The location of the centroid (neutral axis), calculated from a reference axis located at the *bottom* of the section, is

$$\bar{y} = \frac{\Sigma \bar{y} A}{\Sigma A} = \frac{[0.01 \text{ m}](0.02 \text{ m})(0.150 \text{ m}) + [0.095 \text{ m}](0.009 \text{ m})(0.150 \text{ m})}{0.02 \text{ m}(0.150 \text{ m}) + 0.009 \text{ m}(0.150 \text{ m})} = 0.03638 \text{ m}$$

The moment of inertia about the neutral axis is therefore

$$I_{NA} = \left[\frac{1}{12}(0.150 \text{ m})(0.02 \text{ m})^3 + (0.150 \text{ m})(0.02 \text{ m})(0.03638 \text{ m} - 0.01 \text{ m})^2 \right]$$

$$+ \left[\frac{1}{12}(0.009 \text{ m})(0.150 \text{ m})^3 + (0.009 \text{ m})(0.150 \text{ m})(0.095 \text{ m} - 0.03638 \text{ m})^2 \right]$$

$$= 9.358(10^{-6}) \text{ m}^4$$

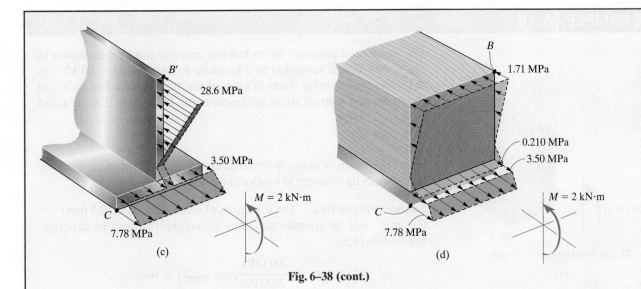

Fig. 6–38 (cont.)

Normal Stress. Applying the flexure formula, the normal stress at B' and C is

$$\sigma_{B'} = \frac{2(10^3)\ \text{N} \cdot \text{m}(0.170\ \text{m} - 0.03638\ \text{m})}{9.358(10^{-6})\ \text{m}^4} = 28.6\ \text{MPa}$$

$$\sigma_C = \frac{2(10^3)\ \text{N} \cdot \text{m}(0.03638\ \text{m})}{9.358(10^{-6})\ \text{m}^4} = 7.78\ \text{MPa} \qquad Ans.$$

The normal-stress distribution on the transformed (all steel) section is shown in Fig. 6–38c.

The normal stress in the wood at B in Fig. 6–38a, is determined from Eq. 6–21; that is,

$$\sigma_B = n\sigma_{B'} = \frac{12\ \text{GPa}}{200\ \text{GPa}}(28.56\ \text{MPa}) = 1.71\ \text{MPa} \qquad Ans.$$

Using these concepts, show that the normal stress in the steel and the wood at the point where they are in contact is $\sigma_{st} = 3.50$ MPa and $\sigma_w = 0.210$ MPa, respectively. The normal-stress distribution in the actual beam is shown in Fig. 6–38d.

EXAMPLE 6.18

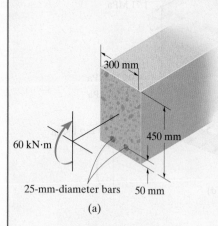

(a)

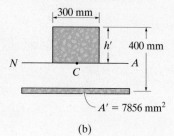

(b)

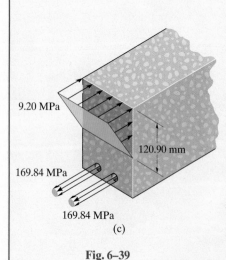

(c)

Fig. 6–39

The reinforced concrete beam has the cross-sectional area shown in Fig. 6–39a. If it is subjected to a bending moment of $M = 60$ kN·m, determine the normal stress in each of the steel reinforcing rods and the maximum normal stress in the concrete. Take $E_{st} = 200$ GPa and $E_{conc} = 25$ GPa.

SOLUTION

Since the beam is made from concrete, in the following analysis we will neglect its strength in supporting a tensile stress.

Section Properties.   The total area of steel, $A_{st} = 2[\pi(12.5 \text{ mm})^2] = 982 \text{ mm}^2$ will be transformed into an equivalent area of concrete, Fig. 6–39b. Here

$$A' = nA_{st} = \frac{200 \text{ GPa}}{25 \text{ GPa}}(982 \text{ mm}^2) = 7856 \text{ mm}^2$$

We require the centroid to lie on the neutral axis. Thus $\Sigma \widetilde{y} A = 0$, or

$$300 \text{ mm }(h')\frac{h'}{2} - 7856 \text{ mm}^2(400 \text{ mm} - h') = 0$$

$$h'^2 + 52.37 - 20949.33 = 0$$

Solving for the positive root,

$$h' = 120.90 \text{ mm}$$

Using this value for h', the moment of inertia of the transformed section about the neutral axis is

$$I = \left[\frac{1}{12}(300 \text{ mm})(120.90 \text{ mm})^3 + 300 \text{ mm }(120.90 \text{ mm})\left(\frac{120.90 \text{ mm}}{2}\right)^2\right]$$
$$+ 7856 \text{ mm}^2(400 \text{ mm} - 120.90 \text{ mm})^2 = 788.67(10^6) \text{ mm}^4$$

Normal Stress.   Applying the flexure formula to the transformed section, the maximum normal stress in the concrete is

$$(\sigma_{conc})_{max} = \frac{[60(10^3) \text{ N·m}(1000 \text{ mm/m})](120.90 \text{ mm})}{788.67(10^6) \text{ mm}^4} = 9.20 \text{ N/mm}^2 = 9.20 \text{ MPa} \qquad Ans.$$

The normal stress resisted by the "concrete" strip, which replaced the steel, is

$$\sigma'_{conc} = \frac{[60(10^3) \text{ N·m}(1000 \text{ mm/m})](400 \text{ mm} - 120.90 \text{ mm})}{788.67(10^6) \text{ mm}^4} = 21.23 \text{ MPa}$$

The normal stress in each of the two reinforcing rods is therefore

$$\sigma_{st} = n\sigma'_{conc} = \left(\frac{200 \text{ GPa}}{25 \text{ GPa}}\right)21.23 \text{ MPa} = 169.84 \text{ MPa} \qquad Ans.$$

The normal-stress distribution is shown graphically in Fig. 6–39c.

*6.8   Curved Beams

The flexure formula applies to a straight member, since it was shown that the normal strain within it varies linearly from the neutral axis. If the member is *curved*, however, this assumption becomes inaccurate, and so we must develop another method to describe the stress distribution. In this section we will consider the analysis of a *curved beam*, that is, a member that has a curved axis and is subjected to bending. Typical examples include hooks and chain links. In all cases, the members are not slender, but rather have a sharp curve, and their cross-sectional dimensions are large compared with their radius of curvature.

The following analysis assumes that the cross section is constant and has an axis of symmetry that is perpendicular to the direction of the applied moment **M**, Fig. 6–40a. Also, the material is homogeneous and isotropic, and it behaves in a linear-elastic manner when the load is applied. Like the case of a straight beam, we will also assume that the *cross sections* of the member *remain plane* after the moment is applied. Furthermore, any distortion of the cross section within its own plane will be neglected.

To perform the analysis, three radii, extending from the center of curvature O' of the member, are identified in Fig. 6–40a. Here $\bar{r}$ references the known location of the *centroid* for the cross-sectional area, R references the yet unspecified location of the *neutral axis*, and r locates the *arbitrary point* or area element dA on the cross section.

This crane hook represents a typical example of a curved beam.

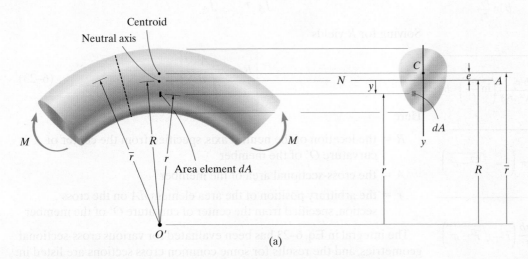

(a)

Fig. 6–40

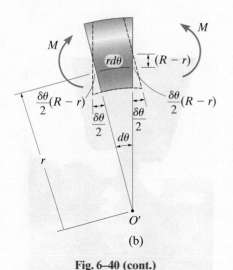

(b)

Fig. 6–40 (cont.)

If we isolate a differential segment of the beam, Fig. 6–40b, the stress tends to deform the material such that each cross section will rotate through an angle $\delta\theta/2$. The normal strain ϵ in the strip (or line) of material located at r will now be determined. This strip has an original length $r\,d\theta$, Fig. 6–40b. However, due to the rotations $\delta\theta/2$ the strip's total change in length is $\delta\theta(R-r)$. Consequently, $\epsilon = \delta\theta(R-r)/r\,d\theta$. If we let $k = \delta\theta/d\theta$, which is the same for any particular strip, we have $\epsilon = k(R-r)/r$. Unlike the case of straight beams, here it can be seen that the ***normal strain*** is a nonlinear function of r, in fact it varies in a ***hyperbolic fashion***. This occurs even though the cross section of the beam remains plane after deformation. If the material remains linearly elastic then $\sigma = E\epsilon$ and so

$$\sigma = Ek\left(\frac{R-r}{r}\right) \qquad (6\text{–}22)$$

This variation is also hyperbolic, and since it has now been established, we can determine the location of the neutral axis and relate the stress distribution to the resultant internal moment M.

To obtain the location R of the neutral axis, we require the resultant internal force caused by the stress distribution acting over the cross section to be equal to zero; i.e.,

$$F_R = \Sigma F_x; \qquad \int_A \sigma\,dA = 0$$

$$\int_A Ek\left(\frac{R-r}{r}\right)dA = 0$$

Since Ek and R are constants, we have

$$R\int_A \frac{dA}{r} - \int_A dA = 0$$

Solving for R yields

$$\boxed{R = \frac{A}{\displaystyle\int_A \frac{dA}{r}}} \qquad (6\text{–}23)$$

Here

   R = the location of the neutral axis, specified from the center of curvature O' of the member

   A = the cross-sectional area of the member

   r = the arbitrary position of the area element dA on the cross section, specified from the center of curvature O' of the member

The integral in Eq. 6–23 has been evaluated for various cross-sectional geometries, and the results for some common cross sections are listed in Table 6–1.

TABLE 6–1

Shape	$\int_A \dfrac{dA}{r}$
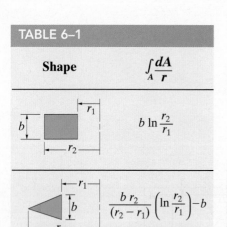	$b\ln\dfrac{r_2}{r_1}$
	$\dfrac{b\,r_2}{(r_2-r_1)}\left(\ln\dfrac{r_2}{r_1}\right)-b$
	$2\pi\left(\bar{r}-\sqrt{\bar{r}^2-c^2}\right)$
	$\dfrac{2\pi b}{a}\left(\bar{r}-\sqrt{\bar{r}^2-a^2}\right)$

In order to relate the stress distribution to the resultant bending moment, we require the resultant internal moment to be equal to the moment of the stress distribution calculated about the neutral axis. From Fig. 6–40a, the stress σ, acting on the area element dA and located a distance y from the neutral axis, creates a moment about the neutral axis of $dM = y(\sigma\, dA)$. For the entire cross section, we require $M = \int y\sigma\, dA$. Since $y = R - r$, and σ is defined by Eq. 6–22, we have

$$M = \int_A (R - r)Ek\left(\frac{R - r}{r}\right) dA$$

Expanding, realizing that Ek and R are constants, then

$$M = Ek\left(R^2 \int_A \frac{dA}{r} - 2R \int_A dA + \int_A r\, dA\right)$$

The first integral is equivalent to A/R as determined from Eq. 6–23, and the second integral is simply the cross-sectional area A. Realizing that the location of the centroid of the cross section is determined from $\bar{r} = \int r\, dA/A$, the third integral can be replaced by $\bar{r}A$. Thus,

$$M = EkA(\bar{r} - R)$$

Finally, solving for Ek in Eq. 6–22, substituting into the above equation, and solving for σ, we have

$$\sigma = \frac{M(R - r)}{Ar(\bar{r} - R)} \qquad (6\text{–}24)$$

Here

σ = the normal stress in the member

M = the internal moment, determined from the method of sections and the equations of equilibrium and calculated about the neutral axis for the cross section. This moment is *positive* if it tends to increase the member's radius of curvature, i.e., it tends to straighten out the member

A = the cross-sectional area of the member

R = the distance measured from the center of curvature to the neutral axis, determined from Eq. 6–23

$\bar{r}$ = the distance measured from the center of curvature to the centroid of the cross section

r = the distance measured from the center of curvature to the point where the stress σ is to be determined

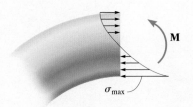

Bending stress variation
(profile view)

(c)

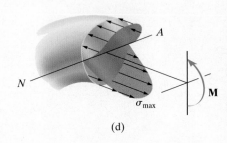

(d)

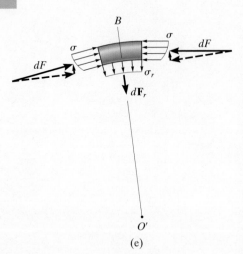

(e)

Fig. 6–40 (cont.)

From Fig. 6–40a, $r = R - y$. Also, the constant and usually very small distance between the neutral axis and the centroid is $e = \bar{r} - R$. When these results are substituted into Eq. 6–24, we can also write

$$\sigma = \frac{My}{Ae(R - y)} \qquad (6\text{–}25)$$

These two equations represent two forms of the so-called *curved-beam formula*, which like the flexure formula can be used to determine the normal-stress distribution in a curved member. This distribution is, as previously stated, hyperbolic; an example is shown in Fig. 6–40c and 6–40d. Since the stress acts along the circumference of the beam, it is sometimes called **circumferential stress**. Note that due to the curvature of the beam, the circumferential stress will create a corresponding component of **radial stress**, so called since this component acts in the radial direction. To show how it is developed, consider the free-body diagram of the segment shown in Fig. 6–40e. Here the radial stress σ_r is necessary since it creates the force dF_r that is required to balance the two components of circumferential forces dF which act along the line $O'B$.

Sometimes the radial stresses within curved members may become significant, especially if the member is constructed from thin plates and has, for example, the shape of an I-section. In this case the radial stress can become as large as the circumferential stress, and consequently the member must be designed to resist both stresses. For most cases, however, these stresses can be neglected, especially if the member has a *solid section*. Here the curved-beam formula gives results that are in very close agreement with those determined either by experiment or by a mathematical analysis based on the theory of elasticity.

The curved-beam formula is normally used when the curvature of the member is very pronounced, as in the case of hooks or rings. However, if the radius of curvature is greater than five times the depth of the member, the *flexure formula* can normally be used to determine the stress. For example, for rectangular sections for which this ratio equals 5, the maximum normal stress, when determined by the flexure formula, will be about 7% *less* than its value when determined by the curved-beam formula. This error is further reduced when the radius of curvature-to-depth ratio is more than 5.*

*See, for example, Boresi, A. P., et al., *Advanced Mechanics of Materials*, 3rd ed., p. 333, 1978, John Wiley & Sons, New York.

Important Points

- The *curved-beam formula* should be used to determine the circumferential stress in a beam when the radius of curvature is less than five times the depth of the beam.

- Due to the curvature of the beam, the normal strain in the beam *does not* vary linearly with depth as in the case of a straight beam. As a result, the neutral axis does not pass through the centroid of the cross section.

- The radial stress component caused by bending can generally be neglected, especially if the cross section is a solid section and not made from thin plates.

Procedure for Analysis

In order to apply the curved-beam formula the following procedure is suggested.

Section Properties.

- Determine the cross-sectional area A and the location of the centroid, $\bar{r}$, measured from the center of curvature.

- Find the location of the neutral axis, R, using Eq. 6–23 or Table 6–1. If the cross-sectional area consists of n "composite" parts, determine $\int dA/r$ for *each part*. Then, from Eq. 6–23, for the entire section, $R = \Sigma A / \Sigma(\int dA/r)$. In all cases, $R < \bar{r}$.

Normal Stress.

- The normal stress located at a point r away from the center of curvature is determined from Eq. 6–24. If the distance y to the point is measured from the neutral axis, then find $e = \bar{r} - R$ and use Eq. 6–25.

- Since $\bar{r} - R$ generally produces a very *small number*, it is best to calculate $\bar{r}$ and R with sufficient accuracy so that the subtraction leads to a number e having at least four significant figures.

- If the stress is positive it will be tensile, whereas if it is negative it will be compressive.

- The stress distribution over the entire cross section can be graphed, or a volume element of the material can be isolated and used to represent the stress acting at the point on the cross section where it has been calculated.

EXAMPLE 6.19

The curved bar has a cross-sectional area shown in Fig. 6–41a. If it is subjected to bending moments of 4 kN·m, determine the maximum normal stress developed in the bar.

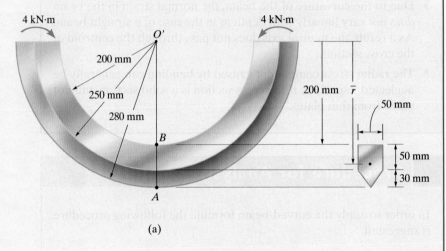

(a)

Fig. 6–41

SOLUTION

Internal Moment. Each section of the bar is subjected to the same resultant internal moment of 4 kN·m. Since this moment tends to decrease the bar's radius of curvature, it is negative. Thus, $M = -4$ kN·m.

Section Properties. Here we will consider the cross section to be composed of a rectangle and triangle. The total cross-sectional area is

$$\Sigma A = (0.05 \text{ m})^2 + \frac{1}{2}(0.05 \text{ m})(0.03 \text{ m}) = 3.250(10^{-3}) \text{ m}^2$$

The location of the centroid is determined with reference to the center of curvature, point O', Fig. 6–41a.

$$\bar{r} = \frac{\Sigma \tilde{r} A}{\Sigma A}$$

$$= \frac{[0.225 \text{ m}](0.05 \text{ m})(0.05 \text{ m}) + [0.260 \text{ m}]\frac{1}{2}(0.050 \text{ m})(0.030 \text{ m})}{3.250(10^{-3}) \text{ m}^2}$$

$$= 0.23308 \text{ m}$$

We can find $\int_A dA/r$ for each part using Table 6–1. For the rectangle,

$$\int_A \frac{dA}{r} = 0.05 \text{ m}\left(\ln\frac{0.250 \text{ m}}{0.200 \text{ m}} \right) = 0.011157 \text{ m}$$

And for the triangle,

$$\int_A \frac{dA}{r} = \frac{(0.05 \text{ m})(0.280 \text{ m})}{(0.280 \text{ m} - 0.250 \text{ m})}\left(\ln\frac{0.280 \text{ m}}{0.250 \text{ m}} \right) - 0.05 \text{ m} = 0.0028867 \text{ m}$$

Thus the location of the neutral axis is determined from

$$R = \frac{\Sigma A}{\Sigma \int_A dA/r} = \frac{3.250(10^{-3}) \text{ m}^2}{0.011157 \text{ m} + 0.0028867 \text{ m}} = 0.23142 \text{ m}$$

Note that $R < \bar{r}$ as expected. Also, the calculations were performed with sufficient accuracy so that $(\bar{r} - R) = 0.23308 \text{ m} - 0.23142 \text{ m} = 0.00166 \text{ m}$ is now accurate to three significant figures.

Normal Stress. The maximum normal stress occurs either at A or B. Applying the curved-beam formula to calculate the normal stress at B, $r_B = 0.200 \text{ m}$, we have

$$\sigma_B = \frac{M(R - r_B)}{Ar_B(\bar{r} - R)} = \frac{(-4 \text{ kN} \cdot \text{m})(0.23142 \text{ m} - 0.200 \text{ m})}{3.250(10^{-3}) \text{ m}^2(0.200 \text{ m})(0.00166 \text{ m})}$$

$$= -116 \text{ MPa}$$

At point A, $r_A = 0.280 \text{ m}$ and the normal stress is

$$\sigma_A = \frac{M(R - r_A)}{Ar_A(\bar{r} - R)} = \frac{(-4 \text{ kN} \cdot \text{m})(0.23142 \text{ m} - 0.280 \text{ m})}{3.250(10^{-3}) \text{ m}^2(0.280 \text{ m})(0.00166 \text{ m})}$$

$$= 129 \text{ MPa} \qquad Ans.$$

By comparison, the maximum normal stress is at A. A two-dimensional representation of the stress distribution is shown in Fig. 6–41b.

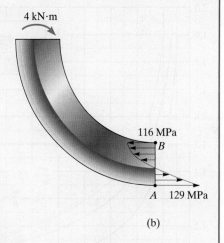

Fig. 6–41 (cont.)

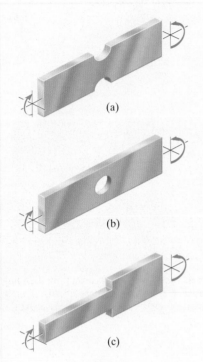

(a)

(b)

(c)

Fig. 6–42

6.9 Stress Concentrations

The flexure formula cannot be used to determine the stress distribution within regions of a member where the cross-sectional area suddenly changes, since the normal-stress and strain distributions at the section become *nonlinear*. The results can only be obtained through experiment or, in some cases, by using the theory of elasticity. Common discontinuities include members having notches on their surfaces, Fig. 6–42a, holes for passage of fasteners or other items, Fig. 6–42b, or abrupt changes in the outer dimensions of the member's cross section, Fig. 6–42c. The *maximum* normal stress at each of these discontinuities occurs at the section taken through the *smallest* cross-sectional area.

For design, it is generally important to only know the maximum normal stress developed at these sections, not the actual stress distribution. As in the previous cases of axially loaded bars and torsionally loaded shafts, we can obtain the maximum normal stress due to bending using a stress-concentration factor K. For example, Fig. 6–43 gives values of K for a flat bar that has a change in cross section using shoulder fillets. To use this graph simply find the geometric ratios w/h and r/h and then find the corresponding value of K for a particular

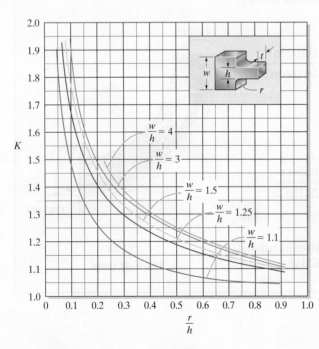

Fig. 6–43

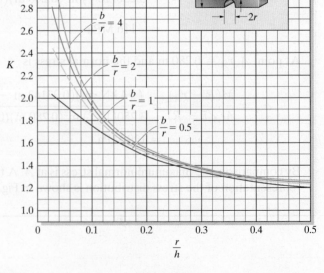

Fig. 6–44

geometry. Once K is obtained, the maximum bending stress shown in Fig. 6–45 is determined using

$$\sigma_{max} = K\frac{Mc}{I}$$ (6–26)

Fig. 6–45

In the same manner, Fig. 6–44 can be used if the discontinuity consists of circular grooves or notches.

Like axial load and torsion, stress concentration for bending should always be considered when designing members made of brittle materials or those that are subjected to fatigue or cyclic loadings. Also, realize that stress-concentration factors apply *only* when the material is subjected to *elastic behavior.* If the applied moment causes yielding of the material, as is the case with ductile materials, the stress becomes redistributed throughout the member, and the maximum stress that results will be *lower* than that determined using stress-concentration factors. This phenomenon is discussed further in the next section.

Stress concentrations caused by bending occur at the sharp corners of this window lintel and are responsible for the crack at the corner.

Important Points

- Stress concentrations occur at points of sudden cross-sectional change, caused by notches and holes, because here the stress and strain become nonlinear. The more severe the change, the larger the stress concentration.

- For design or analysis, the maximum normal stress occurs on the *smallest* cross-sectional area. This stress can be obtained by using a stress concentration factor, K, that has been determined through experiment and is only a function of the geometry of the member.

- Normally, the stress concentration in a ductile material subjected to a static moment will not have to be considered in design; however, if the material is *brittle*, or subjected to *fatigue* loading, then stress concentrations become important.

EXAMPLE | **6.20**

The transition in the cross-sectional area of the steel bar is achieved using shoulder fillets as shown in Fig. 6–46a. If the bar is subjected to a bending moment of 5 kN · m, determine the maximum normal stress developed in the steel. The yield stress is $\sigma_Y = 500$ MPa.

SOLUTION

The moment creates the largest stress in the bar at the base of the fillet, where the cross-sectional area is smallest. The stress-concentration factor can be determined by using Fig. 6–43. From the geometry of the bar, we have $r = 16$ mm, $h = 80$ mm, $w = 120$ mm. Thus,

$$\frac{r}{h} = \frac{16 \text{ mm}}{80 \text{ mm}} = 0.2 \qquad \frac{w}{h} = \frac{120 \text{ mm}}{80 \text{ mm}} = 1.5$$

These values give $K = 1.45$. Applying Eq. 6–26, we have

$$\sigma_{max} = K\frac{Mc}{I} = (1.45)\frac{(5(10^3) \text{ N} \cdot \text{m})(0.04 \text{ m})}{\left[\frac{1}{12}(0.020 \text{ m})(0.08 \text{ m})^3\right]} = 340 \text{ MPa} \qquad Ans.$$

This result indicates that the steel remains elastic since the stress is below the yield stress (500 MPa).

NOTE: The normal-stress distribution is nonlinear and is shown in Fig. 6–46b. Realize, however, that by Saint-Venant's principle, Sec. 4.1, these localized stresses smooth out and become linear when one moves (approximately) a distance of 80 mm or more to the right of the transition. In this case, the flexure formula gives $\sigma_{max} = 234$ MPa, Fig. 6–46c. Also note that the choice of a larger-radius fillet will significantly reduce σ_{max}, since as r increases in Fig. 6–43, K will decrease.

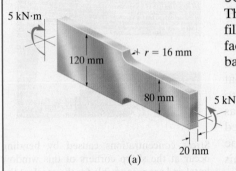

5 kN·m

120 mm

$r = 16$ mm

80 mm

5 kN·m

20 mm

(a)

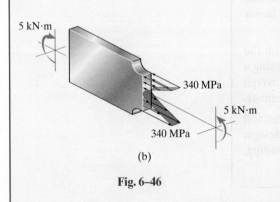

5 kN·m

340 MPa

5 kN·m

340 MPa

(b)

Fig. 6–46

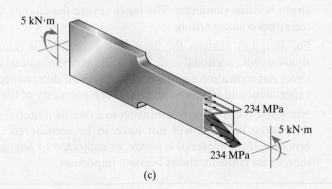

5 kN·m

234 MPa

5 kN·m

234 MPa

(c)

PROBLEMS

6–127. The composite beam is made of 6061-T6 aluminum (A) and C83400 red brass (B). Determine the dimension h of the brass strip so that the neutral axis of the beam is located at the seam of the two metals. What maximum moment will this beam support if the allowable bending stress for the aluminum is $(\sigma_{allow})_{al} = 128$ MPa and for the brass $(\sigma_{allow})_{br} = 35$ MPa?

***6–128.** The composite beam is made of 6061-T6 aluminum (A) and C83400 red brass (B). If the height $h = 40$ mm, determine the maximum moment that can be applied to the beam if the allowable bending stress for the aluminum is $(\sigma_{allow})_{al} = 128$ MPa and for the brass $(\sigma_{allow})_{br} = 35$ MPa.

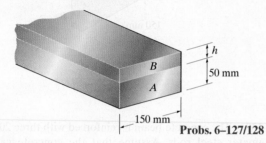

Probs. 6–127/128

•6–129. Segment A of the composite beam is made from 2014-T6 aluminum alloy and segment B is A-36 steel. If $w = 15$ kN/m, determine the absolute maximum bending stress developed in the aluminum and steel. Sketch the stress distribution on the cross section.

6–130. Segment A of the composite beam is made from 2014-T6 aluminum alloy and segment B is A-36 steel. If the allowable bending stress for the aluminum and steel are $(\sigma_{allow})_{al} = 100$ MPa and $(\sigma_{allow})_{st} = 150$ MPa, determine the maximum allowable intensity w of the uniform distributed load.

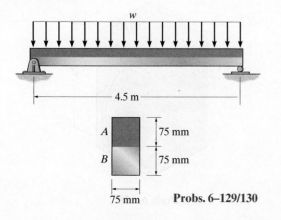

Probs. 6–129/130

6–131. The Douglas fir beam is reinforced with A-36 straps at its center and sides. Determine the maximum stress developed in the wood and steel if the beam is subjected to a bending moment of $M_z = 10$ kN · m. Sketch the stress distribution acting over the cross section.

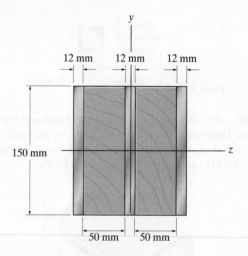

Prob. 6–131

***6–132.** The top plate is made of 2014-T6 aluminum and is used to reinforce a Kevlar 49 plastic beam. Determine the maximum stress in the aluminum and in the Kevlar if the beam is subjected to a moment of $M = 1.35$ kN · m.

•6–133. The top plate made of 2014-T6 aluminum is used to reinforce a Kevlar 49 plastic beam. If the allowable bending stress for the aluminum is $(\sigma_{allow})_{al} = 280$ MPa and for the Kevlar $(\sigma_{allow})_k = 56$ MPa, determine the maximum moment M that can be applied to the beam.

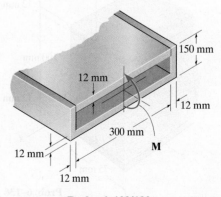

Probs. 6–132/133

6

6–134. The member has a brass core bonded to a steel casing. If a couple moment of 8 kN·m is applied at its end, determine the maximum bending stress in the member. $E_{br} = 100$ GPa, $E_{st} = 200$ GPa.

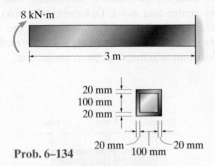

20 mm
100 mm
20 mm

20 mm   100 mm   20 mm

Prob. 6–134

6–135. The steel channel is used to reinforce the wood beam. Determine the maximum stress in the steel and in the wood if the beam is subjected to a moment of $M = 1.2$ kN · m. $E_{st} = 200$ GPa, $E_w = 12$ GPa.

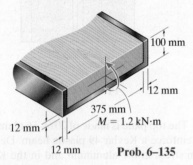

100 mm

12 mm

375 mm

$M = 1.2$ kN·m

12 mm

12 mm   **Prob. 6–135**

*6–136. A white spruce beam is reinforced with A-36 steel straps at its top and bottom as shown. Determine the bending moment M it can support if $(\sigma_{allow})_{st} = 150$ MPa and $(\sigma_{allow})_w = 14$ MPa.

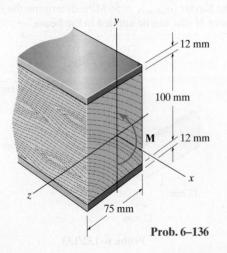

12 mm

100 mm

M   12 mm

75 mm

Prob. 6–136

•6–137. If the beam is subjected to an internal moment of $M = 45$ kN·m, determine the maximum bending stress developed in the A-36 steel section A and the 2014-T6 aluminum alloy section B.

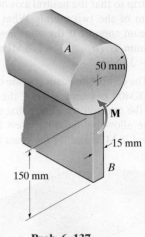

A

50 mm

M

15 mm

B

150 mm

Prob. 6–137

6–138. The concrete beam is reinforced with three 20-mm diameter steel rods. Assume that the concrete cannot support tensile stress. If the allowable compressive stress for concrete is $(\sigma_{allow})_{con} = 12.5$ MPa and the allowable tensile stress for steel is $(\sigma_{allow})_{st} = 220$ MPa, determine the required dimension d so that both the concrete and steel achieve their allowable stress simultaneously. This condition is said to be 'balanced'. Also, compute the corresponding maximum allowable internal moment **M** that can be applied to the beam. The moduli of elasticity for concrete and steel are $E_{con} = 25$ GPa and $E_{st} = 200$ GPa, respectively.

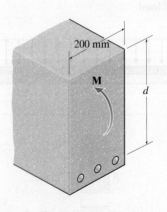

200 mm

M

d

Prob. 6–138

6–139. The beam is made from three types of plastic that are identified and have the moduli of elasticity shown in the figure. Determine the maximum bending stress in the PVC.

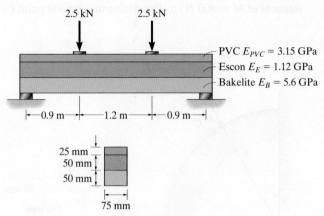

PVC E_{PVC} = 3.15 GPa
Escon E_E = 1.12 GPa
Bakelite E_B = 5.6 GPa

Prob. 6–139

*6–140. The low strength concrete floor slab is integrated with a wide-flange A-36 steel beam using shear studs (not shown) to form the composite beam. If the allowable bending stress for the concrete is $(\sigma_{allow})_{con}$ = 10 MPa, and allowable bending stress for steel is $(\sigma_{allow})_{st}$ = 165 MPa, determine the maximum allowable internal moment **M** that can be applied to the beam.

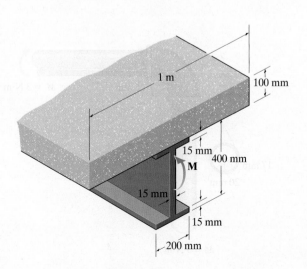

Prob. 6–140

•6–141. The reinforced concrete beam is used to support the loading shown. Determine the absolute maximum normal stress in each of the A-36 steel reinforcing rods and the absolute maximum compressive stress in the concrete. Assume the concrete has a high strength in compression and yet neglect its strength in supporting tension.

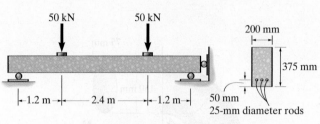

Prob. 6–141

6–142. The reinforced concrete beam is made using two steel reinforcing rods. If the allowable tensile stress for the steel is $(\sigma_{st})_{allow}$ = 280 MPa and the allowable compressive stress for the concrete is $(\sigma_{conc})_{allow}$ = 21 MPa, determine the maximum moment M that can be applied to the section. Assume the concrete cannot support a tensile stress. E_{st} = 200 GPa, E_{conc} = 26.5 GPa.

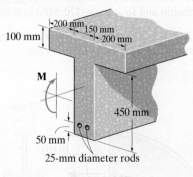

Prob. 6–142

6–143. For the curved beam in Fig. 6–40a, show that when the radius of curvature approaches infinity, the curved-beam formula, Eq. 6–24, reduces to the flexure formula, Eq. 6–13.

*6–144. The member has an elliptical cross section. If it is subjected to a moment of $M = 50$ N·m, determine the stress at points A and B. Is the stress at point A', which is located on the member near the wall, the same as that at A? Explain.

•6–145. The member has an elliptical cross section. If the allowable bending stress is $\sigma_{allow} = 125$ MPa, determine the maximum moment M that can be applied to the member.

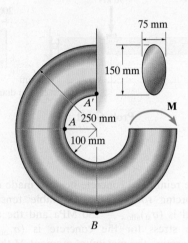

Probs. 6–144/145

6–146. Determine the greatest magnitude of the applied forces P if the allowable bending stress is $(\sigma_{allow})_c = 50$ MPa in compression and $(\sigma_{allow})_t = 120$ MPa in tension.

6–147. If $P = 6$ kN, determine the maximum tensile and compressive bending stresses in the beam.

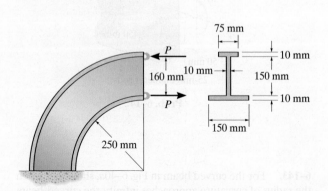

Probs. 6–146/147

*6–148. The curved beam is subjected to a bending moment of $M = 900$ N·m as shown. Determine the stress at points A and B, and show the stress on a volume element located at each of these points.

•6–149. The curved beam is subjected to a bending moment of $M = 900$ N·m. Determine the stress at point C.

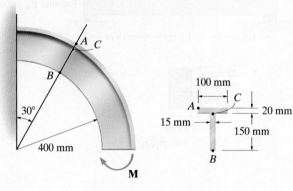

Probs. 6–148/149

6–150. The elbow of the pipe has an outer radius of 20 mm and an inner radius of 17 mm. If the assembly is subjected to the moments of $M = 3$ N · m, determine the maximum stress developed at section $a–a$.

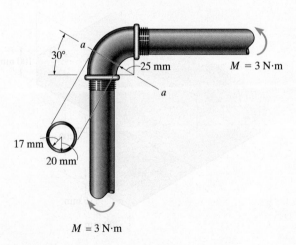

Prob. 6–150

6–151. The curved member is symmetric and is subjected to a moment of $M = 900$ N · m. Determine the bending stress in the member at points A and B. Show the stress acting on volume elements located at these points.

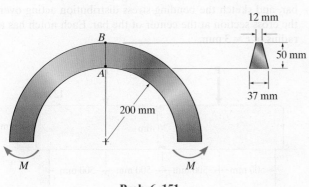

Prob. 6–151

•6–153. The ceiling-suspended C-arm is used to support the X-ray camera used in medical diagnoses. If the camera has a mass of 150 kg, with center of mass at G, determine the maximum bending stress at section A.

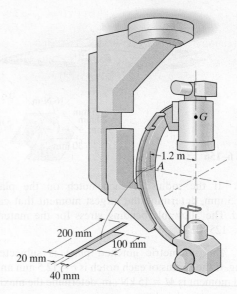

Prob. 6–153

*6–152. The curved bar used on a machine has a rectangular cross section. If the bar is subjected to a couple as shown, determine the maximum tensile and compressive stress acting at section a–a. Sketch the stress distribution on the section in three dimensions.

Prob. 6–152

6–154. The circular spring clamp produces a compressive force of 3 N on the plates. Determine the maximum bending stress produced in the spring at A. The spring has a rectangular cross section as shown.

6–155. Determine the maximum compressive force the spring clamp can exert on the plates if the allowable bending stress for the clamp is $\sigma_{allow} = 4$ MPa.

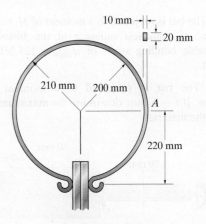

Probs. 6–154/155

*6–156. While in flight, the curved rib on the jet plane is subjected to an anticipated moment of $M = 16$ N·m at the section. Determine the maximum bending stress in the rib at this section, and sketch a two-dimensional view of the stress distribution.

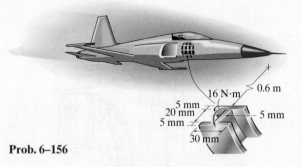

Prob. 6–156

•6–157. If the radius of each notch on the plate is $r = 12.5$ mm, determine the largest moment that can be applied. The allowable bending stress for the material is $\sigma_{allow} = 125$ MPa.

6–158. The symmetric notched plate is subjected to bending. If the radius of each notch is $r = 12.5$ mm and the applied moment is $M = 15$ kN · m, determine the maximum bending stress in the plate.

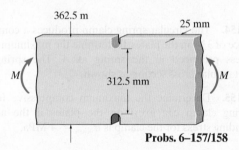

Probs. 6–157/158

6–159. The bar is subjected to a moment of $M = 40$ N·m. Determine the smallest radius r of the fillets so that an allowable bending stress of $\sigma_{allow} = 124$ MPa is not exceeded.

*6–160. The bar is subjected to a moment of $M = 17.5$ N·m. If $r = 5$ mm, determine the maximum bending stress in the material.

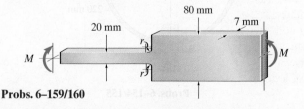

Probs. 6–159/160

•6–161. The simply supported notched bar is subjected to two forces **P**. Determine the largest magnitude of **P** that can be applied without causing the material to yield. The material is A-36 steel. Each notch has a radius of $r = 3$ mm.

6–162. The simply supported notched bar is subjected to the two loads, each having a magnitude of $P = 500$ N. Determine the maximum bending stress developed in the bar, and sketch the bending-stress distribution acting over the cross section at the center of the bar. Each notch has a radius of $r = 3$ mm.

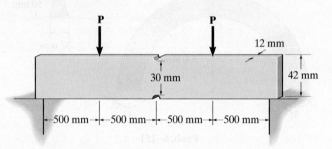

Probs. 6–161/162

6–163. Determine the length L of the center portion of the bar so that the maximum bending stress at A, B, and C is the same. The bar has a thickness of 10 mm.

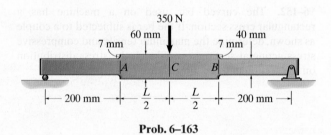

Prob. 6–163

*6–164. The stepped bar has a thickness of 15 mm. Determine the maximum moment that can be applied to its ends if it is made of a material having an allowable bending stress of $\sigma_{allow} = 200$ MPa.

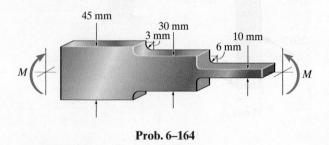

Prob. 6–164

*6.10 Inelastic Bending

The equations for determining the normal stress due to bending that have previously been developed are valid only if the material behaves in a linear-elastic manner. If the applied moment causes the material to *yield*, a plastic analysis must then be used to determine the stress distribution. For bending of straight members three conditions must be met.

Linear Normal-Strain Distribution. Based only on geometric considerations, it was shown in Sec. 6.3 that the normal strains always vary *linearly* from zero at the neutral axis of the cross section to a maximum at the farthest point from the neutral axis.

Resultant Force Equals Zero. Since there is only a resultant internal moment acting on the cross section, the resultant force caused by the stress distribution must be equal to zero. Since σ creates a force on the area dA of Fig. 6–47, then for the entire cross-sectional area A, we have

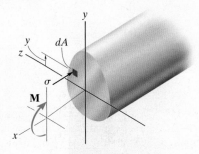

Fig. 6–47

$$F_R = \Sigma F_x; \qquad \int_A \sigma \, dA = 0 \qquad (6\text{–}27)$$

This equation provides a means for obtaining the *location of the neutral axis.*

Resultant Moment. The resultant moment at the section must be equivalent to the moment caused by the stress distribution about the neutral axis. Since the moment of the force $dF = \sigma \, dA$ about the neutral axis is $dM = y(\sigma \, dA)$, Fig 6–47, then summing the results over the entire cross section, we have,

$$(M_R)_z = \Sigma M_z; \qquad M = \int_A y(\sigma \, dA) \qquad (6\text{–}28)$$

These conditions of geometry and loading will now be used to show how to determine the stress distribution in a beam when it is subjected to a resultant internal moment that causes yielding of the material. Throughout the discussion we will assume that the material has a stress-strain diagram that is the *same* in tension as it is in compression. For simplicity, we will begin by considering the beam to have a cross-sectional area with two axes of symmetry; in this case, a rectangle of height h and width b, as shown in Fig. 6–48a. Two cases of loading that are of special interest will be considered.

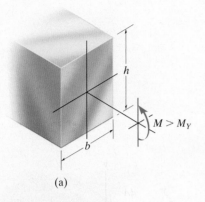

(a)

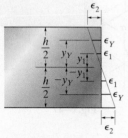

Strain distribution
(profile view)

(b)

Plastic Moment.

Some materials, such as steel, tend to exhibit elastic-perfectly plastic behavior when the stress in the material reaches σ_Y. If the applied moment $M = M_Y$ is just sufficient to produce yielding in the top and bottom fibers of the beam as shown in Fig. 6–48b, then we can determine M_Y using the flexure formula $\sigma_Y = M_Y(h/2)/[bh^3/12]$ or

$$M_Y = \frac{1}{6}bh^2\sigma_Y \qquad (6-29)$$

If the internal moment $M > M_Y$, the material at the top and bottom of the beam will begin to yield, causing a redistribution of stress over the cross section until the required internal moment M is developed. If this causes normal-strain distribution as shown in Fig. 6–48b, then the corresponding normal-stress distribution is determined from the stress–strain diagram shown in Fig. 6–48c. Here the strains ϵ_1, ϵ_Y, ϵ_2, correspond to stresses σ_1, σ_Y, σ_Y, respectively. When these and other stresses like them are plotted on the cross section, we obtain the stress distribution shown in Fig. 6–48d or 6–48e. Here the tension and compression stress "blocks" each consist of component rectangular and triangular blocks. The resultant forces they produce are equivalent to their volumes.

$$T_1 = C_1 = \frac{1}{2}y_Y\sigma_Yb$$

$$T_2 = C_2 = \left(\frac{h}{2} - y_Y\right)\sigma_Yb$$

Because of the symmetry, Eq. 6–27 is satisfied and the neutral axis passes through the centroid of the cross section as shown. The applied moment M can be related to the yield stress σ_Y using Eq. 6–28. From Fig. 6–48e, we require

$$M = T_1\left(\frac{2}{3}y_Y\right) + C_1\left(\frac{2}{3}y_Y\right) + T_2\left[y_Y + \frac{1}{2}\left(\frac{h}{2} - y_Y\right)\right]$$
$$+ C_2\left[y_Y + \frac{1}{2}\left(\frac{h}{2} - y_Y\right)\right]$$
$$= 2\left(\frac{1}{2}y_Y\sigma_Yb\right)\left(\frac{2}{3}y_Y\right) + 2\left[\left(\frac{h}{2} - y_Y\right)\sigma_Yb\right]\left[\frac{1}{2}\left(\frac{h}{2} + y_Y\right)\right]$$
$$= \frac{1}{4}bh^2\sigma_Y\left(1 - \frac{4}{3}\frac{y_Y^2}{h^2}\right)$$

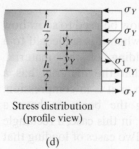

Stress distribution
(profile view)

(d)

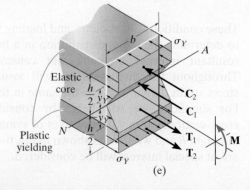

(e)

Fig. 6–48

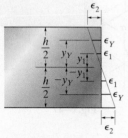

σ

σ_Y

σ_1

ϵ_1 ϵ_Y ϵ_2 ϵ

Stress–strain diagram
(elastic-plastic region)

(c)

Or using Eq. 6–29,

$$M = \frac{3}{2}M_Y\left(1 - \frac{4}{3}\frac{y_Y^2}{h^2}\right) \qquad (6\text{–}30)$$

As noted in Fig. 6–48e, **M** produces two zones of plastic yielding and an elastic core in the member. The boundary between them is located a distance $\pm y_Y$ from the neutral axis. As **M** increases in magnitude, y_Y approaches zero. This would render the material entirely plastic and the stress distribution will then look like that shown in Fig. 6–48f. From Eq. 6–30 with $y_Y = 0$, or by finding the moments of the stress "blocks" around the neutral axis, we can write this limiting value as

$$M_p = \frac{1}{4}bh^2\sigma_Y \qquad (6\text{–}31)$$

Using Eq. 6–29, or Eq. 6–30 with $y = 0$, we have

$$M_p = \frac{3}{2}M_Y \qquad (6\text{–}32)$$

This moment is referred to as the ***plastic moment***. Its value applies only for a rectangular section, since the analysis depends on the geometry of the cross section.

Beams used in steel buildings are sometimes designed to resist a plastic moment. When this is the case, codes usually list a design property for a beam called the shape factor. The ***shape factor*** is defined as the ratio

$$k = \frac{M_p}{M_Y} \qquad (6\text{–}33)$$

This value specifies the additional moment capacity that a beam can support beyond its maximum elastic moment. For example, from Eq. 6–32, a beam having a rectangular cross section has a shape factor of $k = 1.5$. Therefore this section will support 50% more bending moment than its maximum elastic moment when it becomes fully plastic.

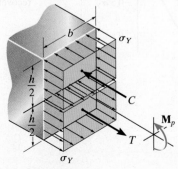

Plastic moment
(f)

Fig. 6–48 (cont.)

Residual Stress.

When the plastic moment is removed from the beam then it will cause **residual stress** to be developed in the beam. These stresses are often important when considering fatigue and other types of mechanical behavior, and so we will discuss a method used for their computation. To explain how this is done, we will assume that M_p causes the material at the top and bottom of the beam to be strained to $\epsilon_1\ (\gg \epsilon_Y)$, as shown by point B on the $\sigma-\epsilon$ curve in Fig. 6–49a. A release of this moment will cause this material to recover some of this strain elastically by following the dashed path BC. Since this recovery is elastic, we can superimpose on the stress distribution in Fig. 6–49b a linear stress distribution caused by applying the plastic moment in the opposite direction, Fig. 6–49c. Here the maximum stress, which is called the **modulus of rupture** for bending, σ_r, can be determined from the flexure formula when the beam is loaded with the plastic moment. We have

$$\sigma_{max} = \frac{Mc}{I} = \frac{M_p\left(\tfrac{1}{2}h\right)}{\left(\tfrac{1}{12}bh^3\right)} = \frac{\left(\tfrac{1}{4}bh^2\sigma_Y\right)\left(\tfrac{1}{2}h\right)}{\left(\tfrac{1}{12}bh^3\right)} = 1.5\sigma_Y$$

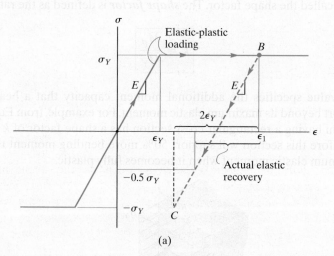

(a)

Fig. 6–49

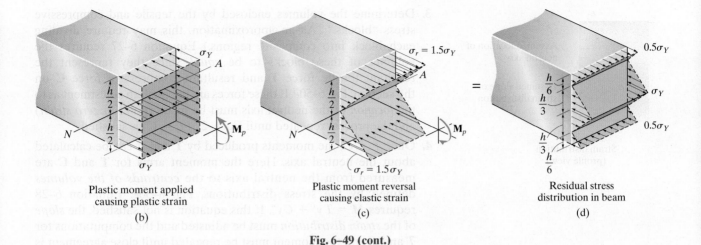

Plastic moment applied causing plastic strain
(b)

Plastic moment reversal causing elastic strain
(c)

Residual stress distribution in beam
(d)

Fig. 6–49 (cont.)

This reversed application of the plastic moment is possible here, since maximum elastic recovery strain at the top and bottom of the beam is $2\epsilon_Y$ as shown in Fig. 6–49a. This corresponds to a maximum stress of $2\sigma_Y$ which is greater than the *required* stress of $1.5\sigma_Y$ as calculated above, Fig. 6–49c.

The superposition of the plastic moment, Fig. 6–49b, and its removal, Fig. 6–49c, gives the residual-stress distribution shown in Fig. 6–49d. As an exercise, use the component triangular "blocks" that represent this stress distribution and show that it results in a zero-force and zero-moment resultant on the member as required.

Ultimate Moment.

Consider now the more general case of a beam having a cross section that is symmetrical only with respect to the vertical axis, while the moment is applied about the horizontal axis, Fig. 6–50a. We will assume that the material exhibits strain hardening and that its stress–strain diagrams for tension and compression are different, Fig. 6–50b.

If the moment **M** produces yielding of the beam, difficulty arises in finding *both* the location of the neutral axis and the maximum strain that is produced in the beam. This is because the cross section is unsymmetrical about the horizontal axis and the stress–strain behavior of the material is not the same in tension and compression. To solve this problem, a trial-and-error procedure requires the following steps:

1. For a given moment **M**, *assume* the location of the neutral axis and the slope of the "linear" strain distribution, Fig. 6–50c.

2. Graphically establish the stress distribution on the member's cross section using the $\sigma-\epsilon$ curve to plot values of stress corresponding to values of strain. The resulting stress distribution, Fig. 6–50d, will then have the same shape as the $\sigma-\epsilon$ curve.

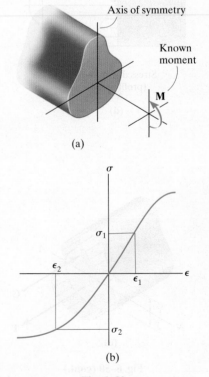

(a)

(b)

Fig. 6–50

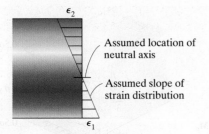

Assumed location of
neutral axis

Assumed slope of
strain distribution

Strain distribution
(profile view)

(c)

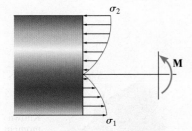

M

Stress distribution
(profile view)

(d)

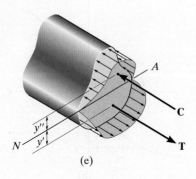

(e)

Fig. 6–50 (cont.)

3. Determine the volumes enclosed by the tensile and compressive stress "blocks." (As an approximation, this may require dividing each block into composite regions.) Equation 6–27 requires the volumes of these blocks to be *equal*, since they represent the resultant tensile force **T** and resultant compressive force **C** on the section, Fig. 6–50*e*. If these forces are unequal, an adjustment as to the *location* of the neutral axis must be made (point of *zero strain*) and the process repeated until Eq. 6–27 ($T = C$) is satisfied.

4. Once $T = C$, the moments produced by **T** and **C** can be calculated about the neutral axis. Here the moment arms for **T** and **C** are measured from the neutral axis to the *centroids of the volumes* defined by the stress distributions, Fig. 6–50*e*. Equation 6–28 requires $M = Ty' + Cy''$. If this equation is not satisfied, the *slope* of the *strain distribution* must be adjusted and the computations for T and C and the moment must be repeated until close agreement is obtained.

This trial-and-error procedure is obviously very tedious, and fortunately it does not occur very often in engineering practice. Most beams are symmetric about two axes, and they are constructed from materials that are assumed to have similar tension-and-compression stress–strain diagrams. Whenever this occurs, the neutral axis will pass through the centroid of the cross section, and the process of relating the stress distribution to the resultant moment is thereby simplified.

Important Points

- The *normal strain distribution* over the cross section of a beam is based only on geometric considerations and has been found to always remain *linear*, regardless of the applied load. The normal stress distribution, however, must be determined from the material behavior, or stress–strain diagram once the strain distribution is established.

- The *location of the neutral axis* is determined from the condition that the *resultant force* on the cross section is *zero*.

- The resultant internal moment on the cross section must be equal to the moment of the stress distribution about the neutral axis.

- Perfectly plastic behavior assumes the normal stress distribution is *constant* over the cross section, and the beam will continue to bend, with no increase in moment. This moment is called the *plastic moment*.

EXAMPLE | 6.21

The steel wide-flange beam has the dimensions shown in Fig. 6–51a. If it is made of an elastic perfectly plastic material having a tensile and compressive yield stress of $\sigma_Y = 250$ MPa determine the shape factor for the beam.

SOLUTION

In order to determine the shape factor, it is first necessary to calculate the maximum elastic moment M_Y and the plastic moment M_p.

Maximum Elastic Moment.   The normal-stress distribution for the maximum elastic moment is shown in Fig. 6–51b. The moment of inertia about the neutral axis is

$$I = \left[\frac{1}{12}(12.5 \text{ mm})(225 \text{ mm})^3\right] + 2\left[\frac{1}{12}(200 \text{ mm})(12.5 \text{ mm})^3 + 200 \text{ mm}(12.5 \text{ mm})(118.75 \text{ mm})^2\right] = 82.44(10^6)\text{mm}^4$$

Applying the flexure formula, we have

$$\sigma_{\max} = \frac{Mc}{I}; \qquad 250 \text{ N/m}^2 = \frac{M_Y(125 \text{ mm})}{82.44(10^6) \text{ mm}^4} \qquad M_Y = 164.88 \text{ kN} \cdot \text{m}$$

Plastic Moment.   The plastic moment causes the steel over the entire cross section of the beam to yield, so that the normal-stress distribution looks like that shown in Fig. 6–51c. Due to symmetry of the cross-sectional area and since the tension and compression stress–strain diagrams are the same, the neutral axis passes through the centroid of the cross section. In order to determine the plastic moment, the stress distribution is divided into four composite rectangular "blocks," and the force produced by each "block" is equal to the volume of the block. Therefore, we have

$$C_1 = T_1 = 250 \text{ N/mm}^2(12.5 \text{ mm})(112.5 \text{ mm}) = 351.56 \text{ kN}$$
$$C_2 = T_2 = 250 \text{ N/mm}^2(12.5 \text{ mm})(200 \text{ mm}) = 625 \text{ kN}$$

These forces act through the *centroid* of the volume for each block. Calculating the moments of these forces about the neutral axis, we obtain the plastic moment:

$$M_p = 2[(56.25 \text{ mm})(351.56 \text{ kN})] + 2[(118.75 \text{ mm})(625 \text{ kN})] = 188 \text{ kN} \cdot \text{m}$$

Shape Factor.   Applying Eq. 6–33 gives

$$k = \frac{M_p}{M_Y} = \frac{188 \text{ kN} \cdot \text{m}}{164.88 \text{ kN} \cdot \text{m}} = 1.14 \qquad\qquad Ans.$$

NOTE: This value indicates that a wide-flange beam provides a very efficient section for resisting an *elastic moment*. Most of the moment is developed in the flanges, i.e., in the top and bottom segments, whereas the web or vertical segment contributes very little. In this particular case, only 14% additional moment can be supported by the beam beyond that which can be supported elastically.

12.5 mm
12.5 mm
22.5 mm
12.5 mm
200 mm

(a)

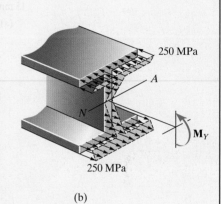

250 MPa

A

N

$\mathbf{M}_Y$

250 MPa

(b)

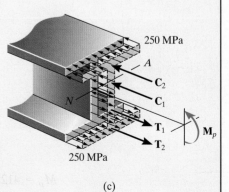

250 MPa

A

C_2
C_1

N

T_1
T_2

$\mathbf{M}_p$

250 MPa

(c)

Fig. 6–51

EXAMPLE 6.22

A T-beam has the dimensions shown in Fig. 6–52a. If it is made of an elastic perfectly plastic material having a tensile and compressive yield stress of $\sigma_Y = 250$ MPa, determine the plastic moment that can be resisted by the beam.

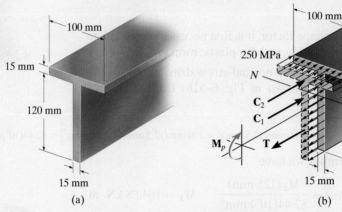

Fig. 6–52

SOLUTION

The "plastic" stress distribution acting over the beam's cross-sectional area is shown in Fig. 6–52b. In this case the cross section is not symmetric with respect to a horizontal axis, and consequently, the neutral axis will *not* pass through the centroid of the cross section. To determine the *location* of the neutral axis, d, we require the stress distribution to produce a zero resultant force on the cross section. Assuming that $d \leq 120$ mm, we have

$$\int_A \sigma\, dA = 0; \qquad T - C_1 - C_2 = 0$$

$$250 \text{ MPa } (0.015 \text{ m})(d) - 250 \text{ MPa } (0.015 \text{ m})(0.120 \text{ m} - d)$$
$$- 250 \text{ MPa } (0.015 \text{ m})(0.100 \text{ m}) = 0$$
$$d = 0.110 \text{ m} < 0.120 \text{ m} \qquad \text{OK}$$

Using this result, the forces acting on each segment are

$$T = 250 \text{ MN/m}^2 \,(0.015 \text{ m})(0.110 \text{ m}) = 412.5 \text{ kN}$$
$$C_1 = 250 \text{ MN/m}^2 \,(0.015 \text{ m})(0.010 \text{ m}) = 37.5 \text{ kN}$$
$$C_2 = 250 \text{ MN/m}^2 \,(0.015 \text{ m})(0.100 \text{ m}) = 375 \text{ kN}$$

Hence the resultant plastic moment about the neutral axis is

$$M_p = 412.5 \text{ kN}\left(\frac{0.110 \text{ m}}{2}\right) + 37.5 \text{ kN}\left(\frac{0.01 \text{ m}}{2}\right) + 375 \text{ kN}\left(0.01 \text{ m} + \frac{0.015 \text{ m}}{2}\right)$$
$$M_p = 29.4 \text{ kN} \cdot \text{m} \qquad\qquad Ans.$$

EXAMPLE | 6.23

The steel wide-flange beam shown in Fig. 6–53a is subjected to a fully plastic moment of $\mathbf{M}_p$. If this moment is removed, determine the residual-stress distribution in the beam. The material is elastic perfectly plastic and has a yield stress of $\sigma_Y = 250$ MPa.

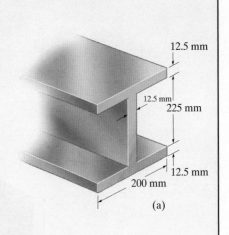

(a)

SOLUTION

The normal-stress distribution in the beam caused by $\mathbf{M}_p$ is shown in Fig. 6–53b. When $\mathbf{M}_p$ is removed, the material responds elastically. Removal of $\mathbf{M}_p$ requires applying $\mathbf{M}_p$ in its reverse direction and therefore leads to an assumed elastic stress distribution as shown in Fig. 6–53c. The modulus of rupture σ_r is computed from the flexure formula. Using $M_p = 188$ kN · m and $I = 82.44(10^6)$ mm^4 from Example 6.21, we have

$$\sigma_{\max} = \frac{Mc}{I};$$

$$\sigma_r = \frac{188(10^6) \text{ N} \cdot \text{mm}}{82.44(10^6) \text{ mm}^4} (125 \text{ mm}) = 285.1 \text{ N/mm}^2 = 285.1 \text{ MPa}$$

As expected, $\sigma_r < 2\sigma_Y$.

Superposition of the stresses gives the residual-stress distribution shown in Fig. 6–53d. Note that the point of zero normal stress was determined by proportion; i.e., from Fig. 6–53b and 6–53c, we require that

$$\frac{285.1 \text{ MPa}}{125 \text{ mm}} = \frac{250 \text{ MPa}}{y}$$

$$y = 109.6 \text{ mm}$$

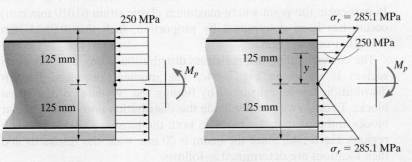

Plastic moment applied
(profile view)

(b)

Plastic moment reversed
(profile view)

(c)

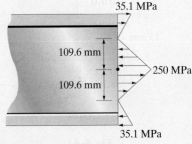

Residual stress distribution

(d)

Fig. 6–53

EXAMPLE | **6.24**

The beam in Fig. 6–54a is made of an alloy of titanium that has a stress–strain diagram that can in part be approximated by two straight lines. If the material behavior is the *same* in both tension and compression, determine the bending moment that can be applied to the beam that will cause the material at the top and bottom of the beam to be subjected to a strain of 0.050 mm/mm.

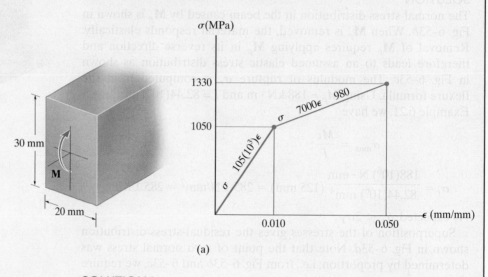

(a)

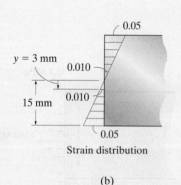

Strain distribution

(b)

Fig. 6–54

SOLUTION I

By inspection of the stress–strain diagram, the material is said to exhibit "elastic-plastic behavior with strain hardening." Since the cross section is symmetric and the tension–compression σ–ϵ diagrams are the same, the neutral axis must pass through the centroid of the cross section. The strain distribution, which is always linear, is shown in Fig. 6–54b. In particular, the point where maximum elastic strain (0.010 mm/mm) occurs has been determined by proportion, such that 0.05/15 mm = 0.010/y or y = 3 mm.

The corresponding normal-stress distribution acting over the cross section is shown in Fig. 6–54c. The moment produced by this distribution can be calculated by finding the "volume" of the stress blocks. To do so we will subdivide this distribution into two triangular blocks and a rectangular block in both the tension and compression regions, Fig. 6–54d. Since the beam is 20 mm wide, the resultants and their locations are determined as follows:

$$T_1 = C_1 = \frac{1}{2}(12 \text{ mm})(280 \text{ N/mm}^2)(20 \text{ mm}) = 33600 \text{ N} = 33.6 \text{ kN}$$

$$y_1 = 3 \text{ mm} + \frac{2}{3}(12 \text{ mm}) = 11 \text{ mm}$$

$$T_2 = C_2 = (12 \text{ mm})(1050 \text{ N/mm}^2)(20 \text{ mm}) = 252000 \text{ N} = 252 \text{ kN}$$

$$y_2 = 3 \text{ mm} + \frac{1}{2}(12 \text{ mm}) = 9 \text{ mm}$$

$$T_3 = C_3 = \frac{2}{3}(3 \text{ mm})(1050 \text{ N/mm}^2)(20 \text{ mm}) = 31500 \text{ N} = 31.5 \text{ kN}$$

$$y_3 = \frac{2}{3}(3 \text{ mm}) = 2 \text{ mm}$$

The moment produced by this normal-stress distribution about the neutral axis is therefore

$$M = 2[33.6 \text{ kN}(110 \text{ mm}) + 252 \text{ kN}(9\text{mm}) + 31.5 \text{ kN}(2 \text{ mm})]$$

$$= 5401.2 \text{ kN} \cdot \text{mm} = 5.40 \text{ kN} \cdot \text{m} \qquad\qquad \textit{Ans.}$$

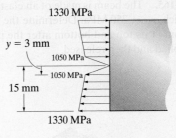

Stress distribution

(c)

SOLUTION II

Rather than using the above semigraphical technique, it is also possible to find the moment analytically. To do this we must express the stress distribution in Fig. 6–54c as a function of position y along the beam. Note that $\sigma = f(\epsilon)$ has been given in Fig. 6–54a. Also, from Fig. 6–54b, the normal strain can be determined as a function of position y by proportional triangles; i.e.,

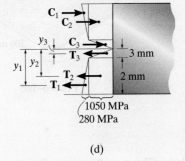

(d)

$$\epsilon = \frac{0.05}{15}y \qquad 0 \le y \le 15 \text{ mm}$$

Substituting this into the $\sigma-\epsilon$ functions shown in Fig. 6–54a gives

$$\sigma = 350y \qquad\qquad 0 \le y \le 3 \text{ mm} \qquad (1)$$

$$\sigma = 23.33y + 980 \qquad 3 \text{ mm} \le y \le 15 \text{ mm} \qquad (2)$$

From Fig. 6–54e, the moment caused by σ acting on the area strip $dA = 20 \, dy$ is

$$dM = y(\sigma \, dA) = y\sigma \, (20 \, dy)$$

Using Eqs. 1 and 2, the moment for the entire cross section is thus

$$M = 2\left[20\int_0^{3 \text{ mm}} 350y^2 \, dy + 20\int_{3 \text{ mm}}^{15 \text{ mm}} (23.33y^2 + 980y) \, dy \right]$$

$$= 5401(10^3) \text{ N} \cdot \text{mm} = 5.40 \text{ kN} \cdot \text{m} \qquad\qquad \textit{Ans.}$$

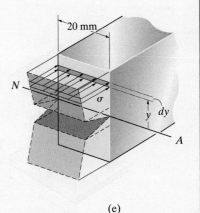

(e)

Fig. 6–54 (cont.)

PROBLEMS

•6–165.   The beam is made of an elastic plastic material for which $\sigma_Y = 250$ MPa. Determine the residual stress in the beam at its top and bottom after the plastic moment $\mathbf{M}_p$ is applied and then released.

6–167.   Determine the shape factor for the cross section.

*6–168.   The beam is made of elastic perfectly plastic material. Determine the maximum elastic moment and the plastic moment that can be applied to the cross section. Take $a = 50$ mm and $\sigma_Y = 250$ MPa.

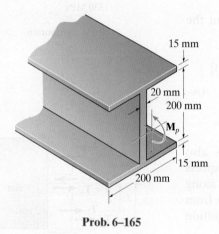

Prob. 6–165

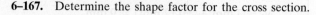

Prob. 6–167/168

6–166.   The wide-flange member is made from an elastic-plastic material. Determine the shape factor.

•6–169.   The box beam is made of an elastic perfectly plastic material for which $\sigma_Y = 250$ MPa. Determine the residual stress in the top and bottom of the beam after the plastic moment $\mathbf{M}_p$ is applied and then released.

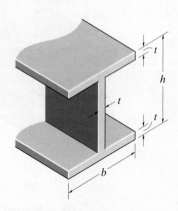

Prob. 6–166

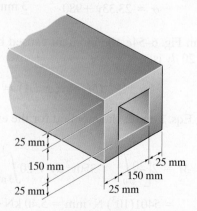

Prob. 6–169

6–170. Determine the shape factor for the wide-flange beam.

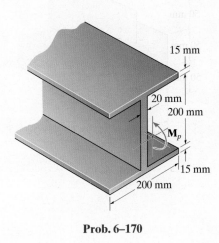

Prob. 6–170

6–171. Determine the shape factor of the beam's cross section.

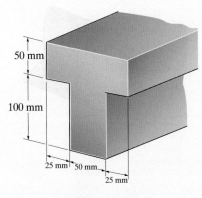

Prob. 6–171

***6–172.** The beam is made of elastic-perfectly plastic material. Determine the maximum elastic moment and the plastic moment that can be applied to the cross section. Take $\sigma_Y = 250$ MPa.

Prob. 6–172

•6–173. Determine the shape factor for the cross section of the H-beam.

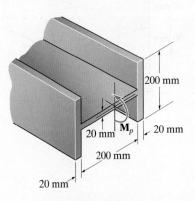

Prob. 6–173

6–174. The H-beam is made of an elastic-plastic material for which $\sigma_Y = 250$ MPa. Determine the residual stress in the top and bottom of the beam after the plastic moment $\mathbf{M}_p$ is applied and then released.

*6–176. The beam is made of elastic-perfectly plastic material. Determine the maximum elastic moment and the plastic moment that can be applied to the cross section. Take $\sigma_Y = 250$ MPa.

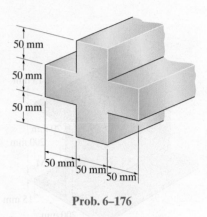

Prob. 6–176

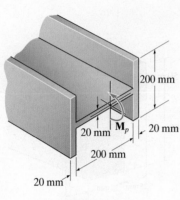

Prob. 6–174

6–175. Determine the shape factor of the cross section.

•6–177. Determine the shape factor of the cross section for the tube.

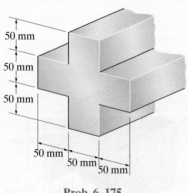

Prob. 6–175

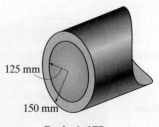

Prob. 6–177

6–178. The beam is made from elastic-perfectly plastic material. Determine the shape factor for the thick-walled tube.

Prob. 6–178

6–179. Determine the shape factor for the member.

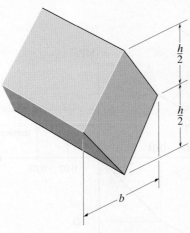

Prob. 6–179

*6–180. The member is made from an elastic-plastic material. Determine the maximum elastic moment and the plastic moment that can be applied to the cross section. Take $b = 100$ mm, $h = 150$ mm, $\sigma_Y = 250$ MPa.

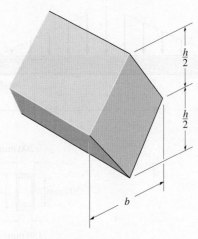

Prob. 6–180

•6–181. The beam is made of a material that can be assumed perfectly plastic in tension and elastic perfectly plastic in compression. Determine the maximum bending moment M that can be supported by the beam so that the compressive material at the outer edge starts to yield.

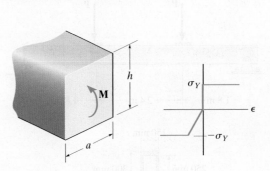

Prob. 6–181

6–182. The box beam is made from an elastic-plastic material for which $\sigma_Y = 175$ MPa. Determine the intensity of the distributed load w_0 that will cause the moment to be (a) the largest elastic moment and (b) the largest plastic moment.

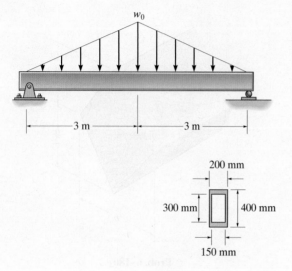

Prob. 6–182

6–183. The box beam is made from an elastic-plastic material for which $\sigma_Y = 250$ MPa. Determine the magnitude of each concentrated force **P** that will cause the moment to be (a) the largest elastic moment and (b) the largest plastic moment.

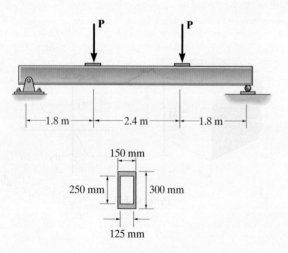

Prob. 6–183

*6–184.** The beam is made of a polyester that has the stress–strain curve shown. If the curve can be represented by the equation $\sigma = [140 \tan^{-1}(15\epsilon)]$ MPa where $\tan^{-1}(15\epsilon)$ is in radians, determine the magnitude of the force **P** that can be applied to the beam without causing the maximum strain in its fibers at the critical section to exceed $\epsilon_{max} = 0.003$ mm/mm.

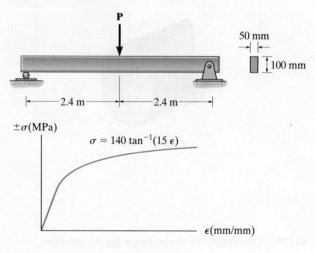

Prob. 6–184

•6–185.** The plexiglass bar has a stress–strain curve that can be approximated by the straight-line segments shown. Determine the largest moment M that can be applied to the bar before it fails.

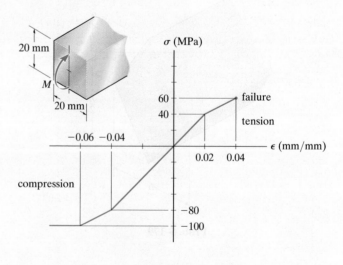

Prob. 6–185

6–186. The stress–strain diagram for a titanium alloy can be approximated by the two straight lines. If a strut made of this material is subjected to bending, determine the moment resisted by the strut if the maximum stress reaches a value of (a) σ_A and (b) σ_B.

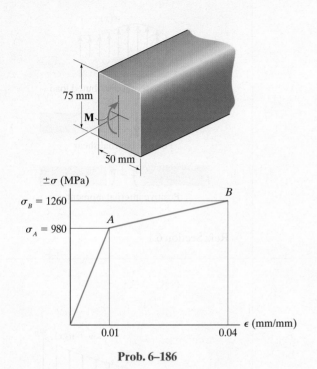

Prob. 6–186

6–187. A beam is made from polypropylene plastic and has a stress–strain diagram that can be approximated by the curve shown. If the beam is subjected to a maximum tensile and compressive strain of $\epsilon = 0.02$ mm/mm, determine the maximum moment M.

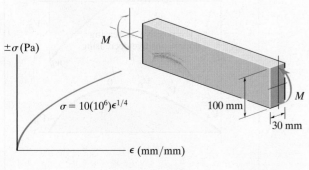

Prob. 6–187

*6–188. The beam has a rectangular cross section and is made of an elastic-plastic material having a stress–strain diagram as shown. Determine the magnitude of the moment **M** that must be applied to the beam in order to create a maximum strain in its outer fibers of $\epsilon_{max} = 0.008$.

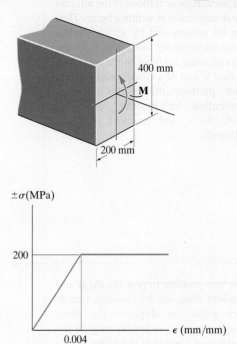

Prob. 6–188

•6–189. The bar is made of an aluminum alloy having a stress–strain diagram that can be approximated by the straight line segments shown. Assuming that this diagram is the same for both tension and compression, determine the moment the bar will support if the maximum strain at the top and bottom fibers of the beam is $\epsilon_{max} = 0.03$.

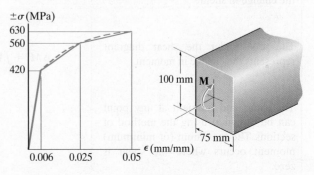

Prob. 6–189

CHAPTER REVIEW

Shear and moment diagrams are graphical representations of the internal shear and moment within a beam. They can be constructed by sectioning the beam an arbitrary distance x from the left end, using the equilibrium equations to find V and M as functions of x, and then plotting the results. A sign convention for positive distributed load, shear, and moment must be followed.

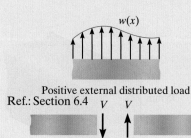

Positive external distributed load
Ref.: Section 6.4

Positive internal shear

Positive internal moment

Ref.: Section 6.1

It is also possible to plot the shear and moment diagrams by realizing that at each point the slope of the shear diagram is equal to the intensity of the distributed loading at the point.

$$w = \frac{dV}{dx}$$

Likewise, the slope of the moment diagram is equal to the shear at the point.

$$V = \frac{dM}{dx}$$

The area under the distributed loading diagram between the points represents the change in shear.

$$\Delta V = \int w \, dx$$

The area under the shear diagram represents the change in moment.

$$\Delta M = \int V \, dx$$

The shear and moment at any point can be obtained using the method of sections. The maximum (or minimum) moment occurs where the shear is zero.

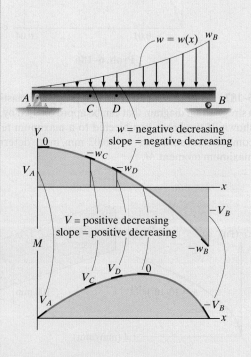

Ref.: Section 6.2

A bending moment tends to produce a linear variation of normal strain within a straight beam. Provided the material is homogeneous, and linear elastic, then equilibrium can be used to relate the internal moment in the beam to the stress distribution. The result is the flexure formula,

$$\sigma_{max} = \frac{Mc}{I}$$

where I and c are determined from the neutral axis that passes through the centroid of the cross section.

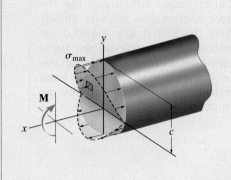

Ref.: Section 6.4

If the cross-sectional area of the beam is not symmetric about an axis that is perpendicular to the neutral axis, then unsymmetrical bending will occur. The maximum stress can be determined from formulas, or the problem can be solved by considering the superposition of bending caused by the moment components $\mathbf{M}_y$ and $\mathbf{M}_z$ about the principal axes of inertia for the area.

$$\sigma = -\frac{M_z y}{I_z} + \frac{M_y z}{I_y}$$

Ref.: Section 6.5

Beams made from composite materials can be "transformed" so their cross section is considered as if it were made from a single material. To do this, the transformation factor n, which is a ratio of the moduli of elasticity of the materials, is used to change the width b of the beam.

Once the cross section is transformed, then the stress in the beam can be determined in the usual manner using the flexure formula.

$$n = \frac{E_1}{E_2}$$

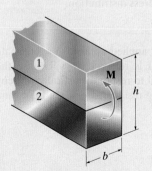

Ref.: Section 6.6

Curved beams deform such that the normal strain does not vary linearly from the neutral axis. Provided the material is homogeneous and linear elastic and the cross section has an axis of symmetry, then the curved beam formula can be used to determine the bending stress.	$$\sigma = \frac{M(R - r)}{Ar(\bar{r} - R)}$$ or $$\sigma = \frac{My}{Ae(R - y)}$$	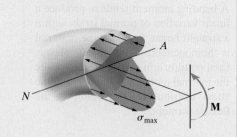 Ref.: Section 6.8
Stress concentrations occur in members having a sudden change in their cross section, caused, for example, by holes and notches. The maximum bending stress at these locations is determined using a stress concentration factor K that is found from graphs determined from experiment.	$$\sigma_{max} = K\frac{Mc}{I}$$	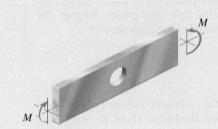 Ref.: Section 6.9
If the bending moment causes the stress in the material to exceed its elastic limit, then the normal strain will remain linear; however, the stress distribution will vary in accordance with the stress–strain diagram. The plastic and ultimate moments supported by the beam can be determined by requiring the resultant force to be zero and the resultant moment to be equivalent to the moment of the stress distribution.		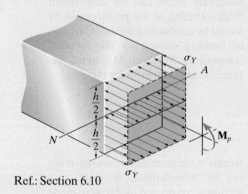 Ref.: Section 6.10
If an applied plastic or ultimate moment is released, it will cause the material to respond elastically, thereby inducing residual stresses in the beam.		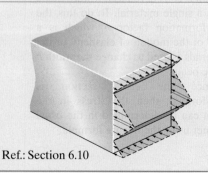 Ref.: Section 6.10

CONCEPTUAL PROBLEMS

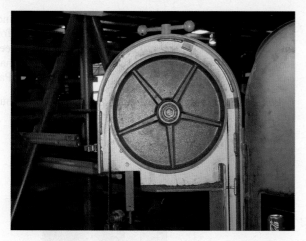

P6–1

P6–1. The steel saw blade passes over the drive wheel of the band saw. Using appropriate measurements and data, explain how to determine the bending stress in the blade.

P6–2

P6–2. This crane boom on a ship has a moment of inertia that varies along its length. Draw the moment diagram for the boom to explain why the boom tapers as shown.

P6–3

P6–3. Hurricane winds caused failure of this highway sign by bending the supporting pipes at their connections with the column. Assuming the pipes are made of A-36 steel, use reasonable dimensions for the sign and pipes, and try and estimate the smallest uniform wind pressure acting on the face of the sign that caused the pipes to yield.

(a)

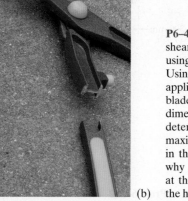

(b)

P6–4

P6–4. These garden shears were manufactured using an inferior material. Using a loading of 250 N applied normal to the blades, and appropriate dimensions for the shears, determine the absolute maximum bending stress in the material and show why the failure occurred at the critical location on the handle.

REVIEW PROBLEMS

6–190. The beam is made from three boards nailed together as shown. If the moment acting on the cross section is $M = 650$ N·m, determine the resultant force the bending stress produces on the top board.

6–191. The beam is made from three boards nailed together as shown. Determine the maximum tensile and compressive stresses in the beam.

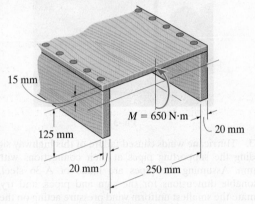

Probs. 6–190/191

*6–192. Determine the bending stress distribution in the beam at section *a–a*. Sketch the distribution in three dimensions acting over the cross section.

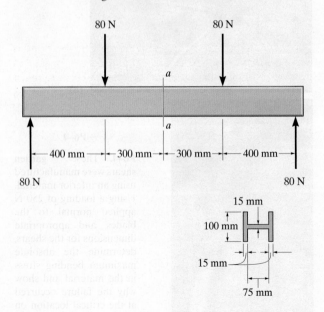

Prob. 6–192

•6–193. The composite beam consists of a wood core and two plates of steel. If the allowable bending stress for the wood is $(\sigma_{allow})_w = 20$ MPa, and for the steel $(\sigma_{allow})_{st} = 130$ MPa, determine the maximum moment that can be applied to the beam. $E_w = 11$ GPa, $E_{st} = 200$ GPa.

6–194. Solve Prob. 6–193 if the moment is applied about the *y* axis instead of the *z* axis as shown.

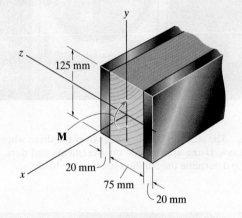

Probs. 6–193/194

6–195. A shaft is made of a polymer having a parabolic cross section. If it resists an internal moment of $M = 125$ N·m, determine the maximum bending stress developed in the material (a) using the flexure formula and (b) using integration. Sketch a three-dimensional view of the stress distribution acting over the cross-sectional area. *Hint:* The moment of inertia is determined using Eq. A–3 of Appendix A.

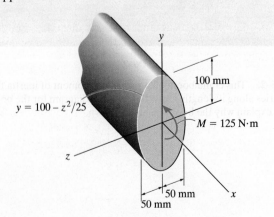

Prob. 6–195

*6–196. Determine the maximum bending stress in the handle of the cable cutter at section *a–a*. A force of 225 N is applied to the handles. The cross-sectional area is shown in the figure.

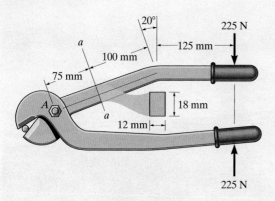

Prob. 6–196

•6–197. The curved beam is subjected to a bending moment of $M = 85$ N·m as shown. Determine the stress at points A and B and show the stress on a volume element located at these points.

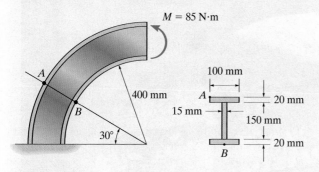

Prob. 6–197

6–198. Draw the shear and moment diagrams for the beam and determine the shear and moment in the beam as functions of x, where $0 " x < 1.8$ m.

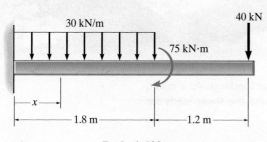

Prob. 6–198

6–199. Draw the shear and moment diagrams for the shaft if it is subjected to the vertical loadings of the belt, gear, and flywheel. The bearings at A and B exert only vertical reactions on the shaft.

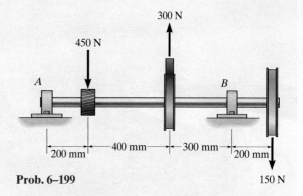

Prob. 6–199

*6–200. A member has the triangular cross section shown. Determine the largest internal moment M that can be applied to the cross section without exceeding allowable tensile and compressive stresses of $(\sigma_{allow})_t = 150$ MPa and $(\sigma_{allow})_c = 100$ MPa, respectively.

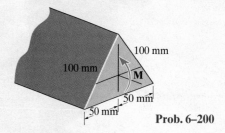

Prob. 6–200

•6–201. The strut has a square cross section a by a and is subjected to the bending moment $\mathbf{M}$ applied at an angle θ as shown. Determine the maximum bending stress in terms of a, M, and θ. What angle θ will give the largest bending stress in the strut? Specify the orientation of the neutral axis for this case.

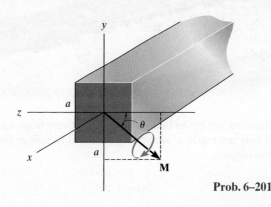

Prob. 6–201

Railroad ties act as beams that support very large transverse shear loadings. As a result, if they are made of wood they will tend to split at their ends, where the shear loads are the largest.

Transverse Shear

7

CHAPTER OBJECTIVES

In this chapter, we will develop a method for finding the shear stress in a beam having a prismatic cross section and made from homogeneous material that behaves in a linear-elastic manner. The method of analysis to be developed will be somewhat limited to special cases of cross-sectional geometry. Although this is the case, it has many wide-range applications in engineering design and analysis. The concept of shear flow, along with shear stress, will be discussed for beams and thin-walled members. The chapter ends with a discussion of the shear center.

7.1 Shear in Straight Members

In general, a beam will support both shear and moment. The shear **V** is the result of a transverse shear-stress distribution that acts over the beam's cross section. Due to the complementary property of shear, however, this stress will create corresponding longitudinal shear stresses which will act along longitudinal planes of the beam as shown in Fig. 7–1.

Check out the Companion Website for the following video solution(s) that can be found in this chapter.

- Section 1,2
  Transverse Shear and Shear Stress
- Section 1,2
  Transverse Shear and Force Resultants
- Section 3
  Maximum Shear Stress in a Structure
- Section 4
  Shear Flow in a Nailed Structure
- Section 4
  Shear Flow in a Box Beam
- Section 5
  Shear Flow in a Thin Walled Structure

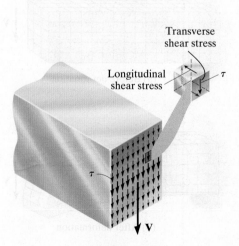

Transverse shear stress

Longitudinal shear stress

τ

τ

V

Fig. 7–1

359

Boards not bonded together

(a)

Boards bonded together

(b)

Fig. 7–2

Shear connectors are "tack welded" to this corrugated metal floor liner so that when the concrete floor is poured, the connectors will prevent the concrete slab from slipping on the liner surface. The two materials will thus act as a composite slab.

To illustrate this effect, consider the beam to be made from three boards, Fig. 7–2a. If the top and bottom surfaces of each board are smooth, and the boards are *not* bonded together, then application of the load **P** will cause the boards to *slide* relative to one another when the beam deflects. However, if the boards are bonded together, then the longitudinal shear stresses acting between the boards will prevent their relative sliding, and consequently the beam will act as a single unit, Fig. 7–2b.

As a result of the shear stress, shear strains will be developed and these will tend to distort the cross section in a rather complex manner. For example, consider the short bar in Fig. 7–3a made of a highly deformable material and marked with horizontal and vertical grid lines. When a shear **V** is applied, it tends to deform these lines into the pattern shown in Fig. 7–3b. This nonuniform shear-strain distribution will cause the cross section to *warp*.

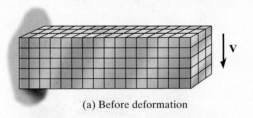

(a) Before deformation

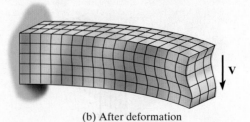

(b) After deformation

Fig. 7–3

As a result, when a beam is subjected to *both* bending and shear, the cross section will not remain plane as assumed in the development of the flexure formula. Although this is the case, we can generally assume the cross-sectional warping due to shear is small enough so that it can be *neglected*. This assumption is particularly true for the most common case of a *slender beam*; that is, one that has a small depth compared with its length.

7.2 The Shear Formula

Because the strain distribution for shear is not easily defined, as in the case of axial load, torsion, and bending, we will develop the shear formula in an indirect manner. To do this we will consider the *horizontal force equilibrium* of a portion of the element taken from the beam in Fig. 7–4a. A free-body diagram of this *element* is shown in Fig. 7–4b. This distribution is caused by the bending moments M and $M + dM$. We have excluded the effects of $V, V + dV$, and $w(x)$ on the free-body diagram since these loadings are vertical and will therefore not be involved in a horizontal force summation. The element in Fig. 7–4b will indeed satisfy $\Sigma F_x = 0$ since the stress distribution on each side of the element forms only a couple moment and therefore a zero force resultant.

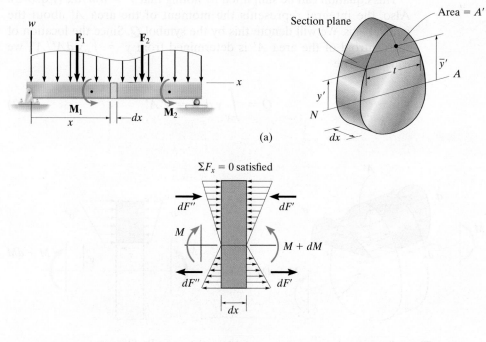

(a)

(b)

Fig. 7–4

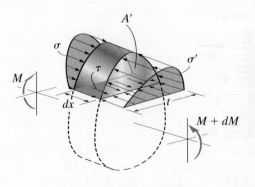

(a)

Now consider the shaded top *portion* of the element that has been sectioned at y' from the neutral axis, Fig. 7–4a. This segment has a width t at the section, and the two cross-sectional sides each have an area A'. Because the resultant moments on each side of the element differ by dM, it can be seen in Fig. 7–4c that $\Sigma F_x = 0$ will not be satisfied *unless* a longitudinal shear stress τ acts over the bottom face of the segment. We will assume this shear stress is *constant* across the width t of the bottom face. It acts on the area $t\,dx$. Applying the equation of horizontal force equilibrium, and using the flexure formula, Eq. 6–13, we have

$$\pm\!\!\rightarrow \Sigma F_x = 0; \qquad \int_{A'} \sigma'\,dA' - \int_{A'} \sigma\,dA' - \tau(t\,dx) = 0$$

$$\int_{A'}\left(\frac{M + dM}{I}\right)y\,dA' - \int_{A'}\left(\frac{M}{I}\right)y\,dA' - \tau(t\,dx) = 0$$

$$\left(\frac{dM}{I}\right)\int_{A'} y\,dA' = \tau(t\,dx) \qquad (7\text{–}1)$$

Solving for τ, we get

$$\tau = \frac{1}{It}\left(\frac{dM}{dx}\right)\int_{A'} y\,dA'$$

This equation can be simplified by noting that $V = dM/dx$ (Eq. 6–2). Also, the integral represents the moment of the area A' about the neutral axis. We will denote this by the symbol Q. Since the location of the centroid of the area A' is determined from $\bar{y}' = \int_{A'} y\,dA'/A'$, we can also write

$$Q = \int_{A'} y\,dA' = \bar{y}'A' \qquad (7\text{–}2)$$

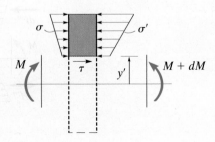

Three-dimensional view   (c)   Profile view

Fig. 7–4 (cont.)

The final result is therefore

$$\tau = \frac{VQ}{It} \qquad (7\text{--}3)$$

Here, as shown in Fig. 7–5,

τ = the shear stress in the member at the point located a distance y' from the neutral axis. This stress is assumed to be constant and therefore *averaged* across the width t of the member

V = the internal resultant shear force, determined from the method of sections and the equations of equilibrium

I = the moment of inertia of the *entire* cross-sectional area calculated about the neutral axis

t = the width of the member's cross-sectional area, measured at the point where τ is to be determined

$Q = \bar{y}' A'$, where A' is the area of the top (or bottom) portion of the member's cross-sectional area, above (or below) the section plane where t is measured, and $\bar{y}'$ is the distance from the neutral axis to the centroid of A'

The above equation is referred to as the *shear formula*. Although in the derivation we considered only the shear stresses acting on the beam's longitudinal plane, the formula applies as well for finding the transverse shear stress on the beam's cross-section. Recall that these stresses are complementary and numerically equal.

Also, because the flexure formula was used in the derivation, it is necessary that the material behave in a linear elastic manner and have a modulus of elasticity that is the *same* in tension as it is in compression.

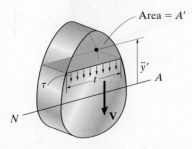

Fig. 7–5

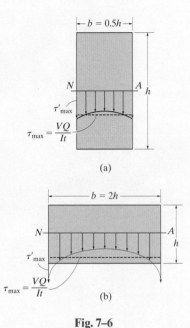

(a)

(b)

Fig. 7–6

Limitations on the Use of the Shear Formula.

One of the major assumptions used in the development of the shear formula is that the shear stress is *uniformly* distributed over the *width t* at the section. In other words, the *average shear stress* is calculated across the width. We can test the accuracy of this assumption by comparing it with a more exact mathematical analysis based on the theory of elasticity. For example, if the beam's cross section is rectangular, the shear-stress distribution across the neutral axis as calculated from the theory of elasticity varies as shown in Fig. 7–6. The maximum value, τ'_{max}, occurs at the *sides* of the cross section, and its magnitude depends on the ratio b/h (width/depth). For sections having a $b/h = 0.5$, τ'_{max} is only about 3% greater than the shear stress calculated from the shear formula, Fig. 7–6a. However, for *flat sections*, say $b/h = 2$, τ'_{max} is about 40% greater than τ_{max}, Fig. 7–6b. The error becomes even greater as the section becomes flatter, or as the b/h ratio increases. Errors of this magnitude are certainly intolerable if one uses the shear formula to determine the shear stress in the *flange* of the wide-flange beam shown in Fig. 7–7.

It should also be pointed out that the shear formula will not give accurate results when used to determine the shear stress at the flange–web junction of a wide-flange beam, since this is a point of sudden cross-sectional change and therefore a *stress concentration* occurs here. Fortunately, these limitations for applying the shear formula to the flanges of a wide-flange beam are not important in engineering practice. Most often engineers must only calculate the *average maximum shear stress* in the beam, which occurs at the neutral axis, where the b/h (width/depth) ratio for the web is *very small*, and therefore the calculated result is very close to the *actual* maximum shear stress as explained above.

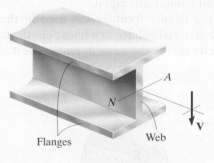

Fig. 7–7

Another important limitation on the use of the shear formula can be illustrated with reference to Fig. 7–8a, which shows a member having a cross section with an irregular or nonrectangular boundary. If we apply the shear formula to determine the (average) shear stress τ along the line AB, it will be directed downward as shown in Fig. 7–8b. However, consider an element of material taken from the boundary point B, Fig. 7–8c. Here τ on the front face of the element is resolved into components, τ' and τ'' acting perpendicular and parallel to the boundary. By inspection, τ' must be equal to zero since its corresponding longitudinal component τ', on the stress-free boundary surface, must be zero. To satisfy this boundary condition, therefore, the shear stress acting on this element must *actually be directed tangent to the boundary*. As a result, the shear-stress distribution across line AB is directed as shown in Fig. 7–8d. Here specific values for the shear stress must be obtained using the theory of elasticity. Note, however, that we can apply the shear formula to obtain the shear stress acting across each of the colored lines in Fig. 7–8a. These lines intersect the tangents to the boundary at *right angles*, and as shown in Fig. 7–8e, the transverse shear stress is vertical and constant along each line.

To summarize the above points, the shear formula does not give accurate results when applied to members having cross sections that are *short or flat*, or at points where the cross section suddenly changes. Nor should it be applied across a section that intersects the boundary of the member at an angle other than 90°. Instead, for these cases the shear stress should be determined using more advanced methods based on the theory of elasticity.

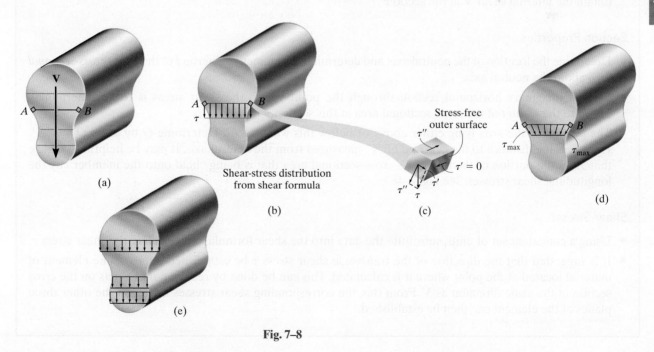

(a)

Shear-stress distribution
from shear formula

(b)

τ'' τ' $\tau'' \quad \tau$ Stress-free outer surface $\tau' = 0$

(c)

τ_{max} τ_{max}

(d)

(e)

Fig. 7–8

Important Points

- Shear forces in beams cause *nonlinear shear-strain* distributions over the cross section, causing it to *warp*.
- Due to the complementary property of shear stress, the shear stress developed in a beam acts over the cross section of the beam and along its longitudinal planes.
- The *shear formula* was derived by considering horizontal force equilibrium of the longitudinal shear-stress and bending-stress distributions acting on a portion of a differential segment of the beam.
- The shear formula is to be used on straight prismatic members made of homogeneous material that has linear elastic behavior. Also, the internal resultant shear force must be directed along an axis of symmetry for the cross-sectional area.
- The shear formula should not be used to determine the shear stress on cross sections that are short or flat, at points of sudden cross-sectional changes, or at a point on an inclined boundary.

Procedure for Analysis

In order to apply the shear formula, the following procedure is suggested.

Internal Shear.

- Section the member perpendicular to its axis at the point where the shear stress is to be determined, and obtain the internal shear **V** at the section.

Section Properties.

- Determine the location of the neutral axis, and determine the moment of inertia I of the *entire cross-sectional area* about the neutral axis.
- Pass an imaginary horizontal section through the point where the shear stress is to be determined. Measure the width t of the cross-sectional area at this section.
- The portion of the area lying either above or below this width is A'. Determine Q by using $Q = \bar{y}'A'$. Here $\bar{y}'$ is the distance to the centroid of A', measured from the neutral axis. It may be helpful to realize that A' is the portion of the member's cross-sectional area that is being "held onto the member" by the longitudinal shear stresses. See Fig. 7–4c.

Shear Stress.

- Using a consistent set of units, substitute the data into the shear formula and calculate the shear stress τ.
- It is suggested that the direction of the transverse shear stress τ be established on a volume element of material located at the point where it is calculated. This can be done by realizing that τ acts on the cross section in the same direction as **V**. From this, the corresponding shear stresses acting on the other three planes of the element can then be established.

EXAMPLE 7.1

The solid shaft and tube shown in Fig. 7–9a are subjected to the shear force of 4 kN. Determine the shear stress acting over the diameter of each cross section.

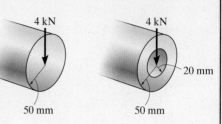

(a)

SOLUTION

Section Properties. Using the table on the inside front cover, the moment of inertia of each section, calculated about its diameter (or neutral axis), is

$$I_{\text{solid}} = \frac{1}{4}\pi c^4 = \frac{1}{4}\pi (0.05 \text{ m})^4 = 4.909(10^{-6}) \text{ m}^4$$

$$I_{\text{tube}} = \frac{1}{4}\pi (c_o^4 - c_i^4) = \frac{1}{4}\pi [(0.05 \text{ m})^4 - (0.02 \text{ m})^4] = 4.783(10^{-6}) \text{ m}^4$$

The semicircular area shown shaded in Fig. 7–9b, above (or below) each diameter, represents Q, because this area is "held onto the member" by the longitudinal shear stress along the diameter.

$$Q_{\text{solid}} = \bar{y}'A' = \frac{4c}{3\pi}\left(\frac{\pi c^2}{2}\right) = \frac{4(0.05 \text{ m})}{3\pi}\left(\frac{\pi(0.05 \text{ m})^2}{2}\right) = 83.33(10^{-6}) \text{ m}^3$$

$$Q_{\text{tube}} = \Sigma\bar{y}'A' = \frac{4c_o}{3\pi}\left(\frac{\pi c_o^2}{2}\right) - \frac{4c_i}{3\pi}\left(\frac{\pi c_i^2}{2}\right)$$

$$= \frac{4(0.05 \text{ m})}{3\pi}\left(\frac{\pi(0.05 \text{ m})^2}{2}\right) - \frac{4(0.02 \text{ m})}{3\pi}\left(\frac{\pi(0.02 \text{ m})^2}{2}\right)$$

$$= 78.0(10^{-6}) \text{ m}^3$$

Shear Stress. Applying the shear formula where $t = 0.1$ m for the solid section, and $t = 2(0.03 \text{ m}) = 0.06$ m for the tube, we have

$$\tau_{\text{solid}} = \frac{VQ}{It} = \frac{4(10^3) \text{ N}(83.33(10^{-6}) \text{ m}^3)}{4.909(10^{-6}) \text{ m}^4(0.1 \text{ m})} = 679 \text{ kPa} \qquad Ans.$$

$$\tau_{\text{tube}} = \frac{VQ}{It} = \frac{4(10^3) \text{ N}(78.0(10^{-6}) \text{ m}^3)}{4.783(10^{-6}) \text{ m}^4(0.06 \text{ m})} = 1.09 \text{ MPa} \qquad Ans.$$

NOTE: As discussed in the limitations for the shear formula, the calculations performed here are valid since the shear stress along the diameter is vertical and therefore tangent to the boundary of the cross section. An element of material on the diameter is subjected to "pure shear" as shown in Fig. 7–9b.

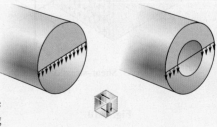

(b)

Fig. 7–9

EXAMPLE | 7.2

Determine the distribution of the shear stress over the cross section of the beam shown in Fig. 7–10a.

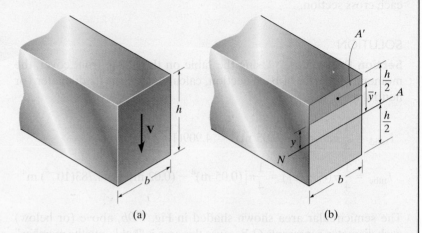

(a)   (b)

SOLUTION

The distribution can be determined by finding the shear stress at an *arbitrary height y* from the neutral axis, Fig. 7–10b, and then plotting this function. Here, the dark colored area A' will be used for Q.* Hence

$$Q = \bar{y}'A' = \left[y + \frac{1}{2}\left(\frac{h}{2} - y \right) \right]\left(\frac{h}{2} - y \right)b = \frac{1}{2}\left(\frac{h^2}{4} - y^2 \right)b$$

Applying the shear formula, we have

$$\tau = \frac{VQ}{It} = \frac{V\left(\frac{1}{2}\right)[(h^2/4) - y^2]b}{\left(\frac{1}{12}bh^3\right)b} = \frac{6V}{bh^3}\left(\frac{h^2}{4} - y^2 \right) \qquad (1)$$

This result indicates that the shear-stress distribution over the cross section is **parabolic**. As shown in Fig. 7–10c, the intensity varies from zero at the top and bottom, $y = \pm h/2$, to a maximum value at the neutral axis, $y = 0$. Specifically, since the area of the cross section is $A = bh$, then at $y = 0$ we have

$$\tau_{max} = 1.5\frac{V}{A} \qquad (2)$$

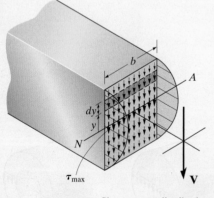

Shear–stress distribution

(c)

Fig. 7–10

*The area below y can also be used [$A' = b(h/2 + y)$], but doing so involves a bit more algebraic manipulation.

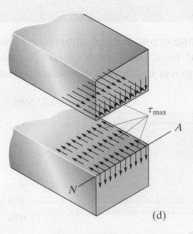

Typical shear failure of this wooden beam occurred at the support and through the approximate center of its cross section.

(d)

Fig. 7–10 (cont.)

This same value for τ_{max} can be obtained directly from the shear formula, $\tau = VQ/It$, by realizing that τ_{max} occurs where Q is *largest*, since V, I, and t are *constant*. By inspection, Q will be a maximum when the entire area above (or below) the neutral axis is considered; that is, $A' = bh/2$ and $\overline{y}' = h/4$. Thus,

$$\tau_{max} = \frac{VQ}{It} = \frac{V(h/4)(bh/2)}{\left[\frac{1}{12}bh^3\right]b} = 1.5\frac{V}{A}$$

By comparison, τ_{max} is 50% greater than the *average* shear stress determined from Eq. 1–7; that is, $\tau_{avg} = V/A$.

It is important to realize that τ_{max} also acts in the longitudinal direction of the beam, Fig. 7–10d. It is this stress that can cause a timber beam to fail as shown Fig. 7–10e. Here horizontal splitting of the wood starts to occur through the neutral axis at the beam's ends, since there the vertical reactions subject the beam to large shear stress and wood has a low resistance to shear along its grains, which are oriented in the longitudinal direction.

It is instructive to show that when the shear-stress distribution, Eq. 1, is integrated over the cross section it yields the resultant shear V. To do this, a differential strip of area $dA = b\,dy$ is chosen, Fig. 7–10c, and since τ acts uniformly over this strip, we have

$$\int_A \tau\,dA = \int_{-h/2}^{h/2}\frac{6V}{bh^3}\left(\frac{h^2}{4} - y^2\right)b\,dy$$

$$= \frac{6V}{h^3}\left[\frac{h^2}{4}y - \frac{1}{3}y^3\right]_{-h/2}^{h/2}$$

$$= \frac{6V}{h^3}\left[\frac{h^2}{4}\left(\frac{h}{2} + \frac{h}{2}\right) - \frac{1}{3}\left(\frac{h^3}{8} + \frac{h^3}{8}\right)\right] = V$$

(e)

EXAMPLE 7.3

A steel wide-flange beam has the dimensions shown in Fig. 7–11a. If it is subjected to a shear of $V = 80\ kN$, plot the shear-stress distribution acting over the beam's cross-sectional area.

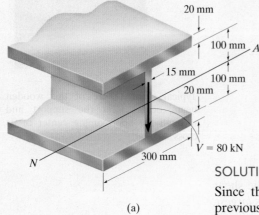

20 mm

100 mm A

15 mm

100 mm

20 mm

300 mm

$V = 80\ kN$

N

(a)

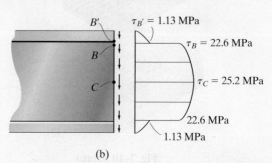

$\tau_{B'} = 1.13\ MPa$

$\tau_B = 22.6\ MPa$

$\tau_C = 25.2\ MPa$

22.6 MPa

1.13 MPa

(b)

SOLUTION

Since the flange and web are rectangular elements, then like the previous example, the shear-stress distribution will be parabolic and in this case it will vary in the manner shown in Fig. 7–11b. Due to symmetry, only the shear stresses at points B', B, and C have to be determined. To show how these values are obtained, we must first determine the moment of inertia of the cross-sectional area about the neutral axis. Working in meters, we have

$$I = \left[\frac{1}{12}(0.015\ m)(0.200\ m)^3\right]$$

$$+ 2\left[\frac{1}{12}(0.300\ m)(0.02\ m)^3 + (0.300\ m)(0.02\ m)(0.110\ m)^2\right]$$

$$= 155.6(10^{-6})\ m^4$$

For point B', $t_{B'} = 0.300\ m$, and A' is the dark shaded area shown in Fig. 7–11c. Thus,

$$Q_{B'} = \overline{y}'A' = [0.110\ m](0.300\ m)(0.02\ m) = 0.660(10^{-3})\ m^3$$

so that

$$\tau_{B'} = \frac{VQ_{B'}}{It_{B'}} = \frac{80(10^3)\ N(0.660(10^{-3})\ m^3)}{155.6(10^{-6})\ m^4(0.300\ m)} = 1.13\ MPa$$

For point B, $t_B = 0.015\ m$ and $Q_B = Q_{B'}$, Fig. 7–11c. Hence

$$\tau_B = \frac{VQ_B}{It_B} = \frac{80(10^3)\ N(0.660(10^{-3})\ m^3)}{155.6(10^{-6})\ m^4(0.015\ m)} = 22.6\ MPa$$

Note from the discussion of "Limitations on the Use of the Shear Formula" that the calculated value for both $\tau_{B'}$ and τ_B will actually be very misleading. Why?

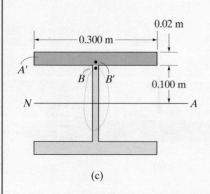

0.02 m

0.300 m

A'

B   B'

0.100 m

N ———————————————— A

(c)

Fig. 7–11

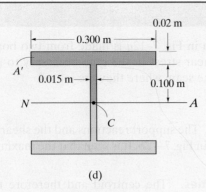

(d)

Fig. 7–11 (cont.)

For point C, $t_C = 0.015$ m and A' is the dark shaded area shown in Fig. 7–11d. Considering this area to be composed of two rectangles, we have

$$Q_C = \Sigma\bar{y}'A' = [0.110 \text{ m}](0.300 \text{ m})(0.02 \text{ m})$$

$$+ [0.05 \text{ m}](0.015 \text{ m})(0.100 \text{ m})$$

$$= 0.735(10^{-3}) \text{ m}^3$$

Thus,

$$\tau_C = \tau_{max} = \frac{VQ_C}{It_C} = \frac{80(10^3) \text{ N}[0.735(10^{-3}) \text{ m}^3]}{155.6(10^{-6}) \text{ m}^4(0.015 \text{ m})} = 25.2 \text{ MPa}$$

NOTE: From Fig. 7–11b, note that most of the shear stress occurs in the web and is almost uniform throughout its depth, varying from 22.6 MPa to 25.2 MPa. It is for this reason that for design, some codes permit the use of calculating the average shear stress on the cross section of the web rather than using the shear formula. This will be discussed further in Chapter 11.

EXAMPLE 7.4

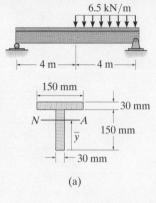

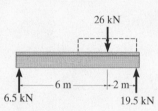

(a)

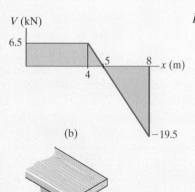

(b)

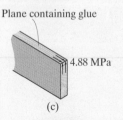

Plane containing glue

4.88 MPa

(c)

Fig. 7–12

The beam shown in Fig. 7–12a is made from two boards. Determine the maximum shear stress in the glue necessary to hold the boards together along the seam where they are joined.

SOLUTION

Internal Shear.   The support reactions and the shear diagram for the beam are shown in Fig. 7–12b. It is seen that the maximum shear in the beam is 19.5 kN.

Section Properties.   The centroid and therefore the neutral axis will be determined from the reference axis placed at the bottom of the cross-sectional area, Fig. 7–12a. Working in units of meters, we have

$$\bar{y} = \frac{\Sigma \tilde{y} A}{\Sigma A}$$

$$= \frac{[0.075\text{ m}](0.150\text{ m})(0.030\text{ m}) + [0.165\text{ m}](0.030\text{ m})(0.150\text{ m})}{(0.150\text{ m})(0.030\text{ m}) + (0.030\text{ m})(0.150\text{ m})} = 0.120\text{ m}$$

The moment of inertia, about the neutral axis, Fig. 7–12a, is therefore

$$I = \left[\frac{1}{12}(0.030\text{ m})(0.150\text{ m})^3 + (0.150\text{ m})(0.030\text{ m})(0.120\text{ m} - 0.075\text{ m})^2\right]$$

$$+ \left[\frac{1}{12}(0.150\text{ m})(0.030\text{ m})^3 + (0.030\text{ m})(0.150\text{ m})(0.165\text{ m} - 0.120\text{ m})^2\right]$$

$$= 27.0(10^{-6})\text{ m}^4$$

The top board (flange) is being held onto the bottom board (web) by the glue, which is applied over the thickness $t = 0.03$ m. Consequently A' is defined as the area of the top board, Fig. 7–12a. We have

$$Q = \bar{y}'A' = [0.180\text{ m} - 0.015\text{ m} - 0.120\text{ m}](0.03\text{ m})(0.150\text{ m})$$

$$= 0.2025(10^{-3})\text{ m}^3$$

Shear Stress.   Using the above data and applying the shear formula yields

$$\tau_{max} = \frac{VQ}{It} = \frac{19.5(10^3)\text{ N}(0.2025(10^{-3})\text{ m}^3)}{27.0(10^{-6})\text{ m}^4(0.030\text{ m})} = 4.88\text{ MPa} \qquad Ans.$$

The shear stress acting at the top of the bottom board is shown in Fig. 7–12c.

NOTE: It is the glue's resistance to this longitudinal *shear* stress that holds the boards from slipping at the right-hand support.

FUNDAMENTAL PROBLEMS

F7–1. If the beam is subjected to a shear force of $V = 100$ kN, determine the shear stress developed at point A. Represent the state of stress at A on a volume element.

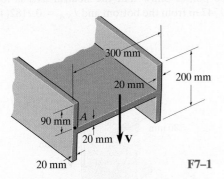

F7–1

F7–2. Determine the shear stress at points A and B on the beam if it is subjected to a shear force of $V = 600$ kN.

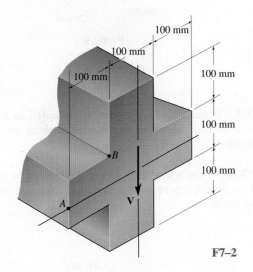

F7–2

F7–3. Determine the absolute maximum shear stress developed in the beam.

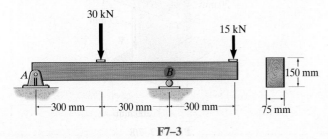

F7–3

F7–4. If the beam is subjected to a shear force of $V = 20$ kN, determine the maximum shear stress developed in the beam.

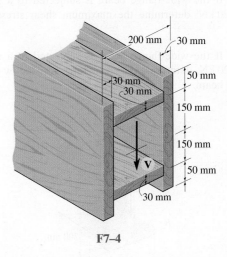

F7–4

F7–5. If the beam is made from four plates and subjected to a shear force of $V = 20$ kN, determine the maximum shear stress developed in the beam.

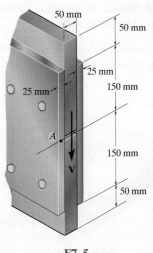

F7–5

7

PROBLEMS

•7–1. If the wide-flange beam is subjected to a shear of $V = 20$ kN, determine the shear stress on the web at A. Indicate the shear-stress components on a volume element located at this point.

7–2. If the wide-flange beam is subjected to a shear of $V = 20$ kN, determine the maximum shear stress in the beam.

7–3. If the wide-flange beam is subjected to a shear of $V = 20$ kN, determine the shear force resisted by the web of the beam.

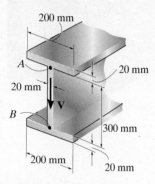

Probs. 7–1/2/3

*7–4. If the T-beam is subjected to a vertical shear of $V = 60$ kN, determine the maximum shear stress in the beam. Also, compute the shear-stress jump at the flange-web junction AB. Sketch the variation of the shear-stress intensity over the entire cross section.

•7–5. If the T-beam is subjected to a vertical shear of $V = 60$ kN, determine the vertical shear force resisted by the flange.

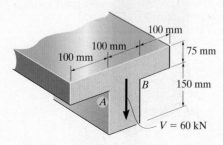

Probs. 7–4/5

7–6. If the beam is subjected to a shear of $V = 15$ kN, determine the web's shear stress at A and B. Indicate the shear-stress components on a volume element located at these points. Show that the neutral axis is located at $\bar{y} = 0.1747$ m from the bottom and $I_{NA} = 0.2182(10^{-3})$ m^4.

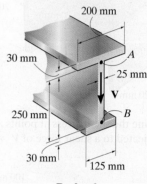

Probs. 6

7–7. If the wide-flange beam is subjected to a shear of $V = 30$ kN, determine the maximum shear stress in the beam.

*7–8. If the wide-flange beam is subjected to a shear of $V = 30$ kN, determine the shear force resisted by the web of the beam.

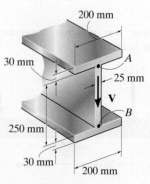

Probs. 7–7/8

•7–9. Determine the largest shear force V that the member can sustain if the allowable shear stress is $\tau_{allow} = 56$ MPa.

7–10. If the applied shear force $V = 90$ kN, determine the maximum shear stress in the member.

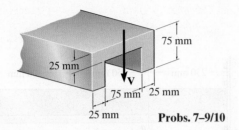

Probs. 7–9/10

7–11. The wood beam has an allowable shear stress of $\tau_{allow} = 7$ MPa. Determine the maximum shear force V that can be applied to the cross section.

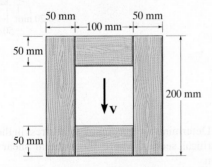

Prob. 7–11

*7–12. The beam has a rectangular cross section and is made of wood having an allowable shear stress of $\tau_{allow} = 1.4$ MPa. Determine the maximum shear force V that can be developed in the cross section of the beam. Also, plot the shear-stress variation over the cross section.

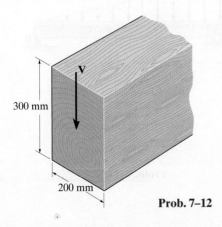

Prob. 7–12

7–13. Determine the maximum shear stress in the strut if it is subjected to a shear force of $V = 20$ kN.

7–14. Determine the maximum shear force V that the strut can support if the allowable shear stress for the material is $\tau_{allow} = 40$ MPa.

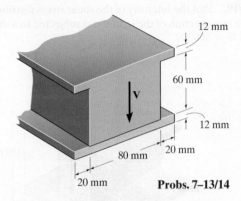

Probs. 7–13/14

7–15. Plot the shear-stress distribution over the cross section of a rod that has a radius c. By what factor is the maximum shear stress greater than the average shear stress acting over the cross section?

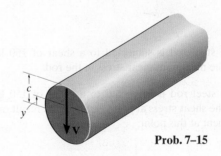

Prob. 7–15

*7–16. A member has a cross section in the form of an equilateral triangle. If it is subjected to a shear force $\mathbf{V}$, determine the maximum average shear stress in the member using the shear formula. Should the shear formula actually be used to predict this value? Explain.

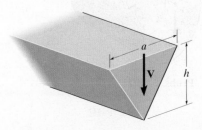

Prob. 7–16

7

•**7–17.** Determine the maximum shear stress in the strut if it is subjected to a shear force of $V = 600$ kN.

7–18. Determine the maximum shear force V that the strut can support if the allowable shear stress for the material is $\tau_{allow} = 45$ MPa.

7–19. Plot the intensity of the shear stress distributed over the cross section of the strut if it is subjected to a shear force of $V = 600$ kN.

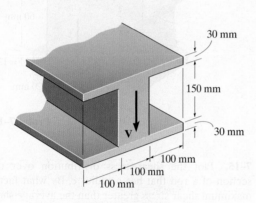

30 mm

150 mm

V

30 mm

100 mm

100 mm

100 mm

Probs. 7–17/18/19

*7–20. The steel rod is subjected to a shear of 150 kN. Determine the maximum shear stress in the rod.

•**7–21.** The steel rod is subjected to a shear of 150 kN. Determine the shear stress at point A. Show the result on a volume element at this point.

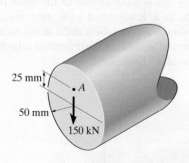

25 mm

• A

50 mm

150 kN

Probs. 7–20/21

7–22. Determine the shear stress at point B on the web of the cantilevered strut at section a–a.

7–23. Determine the maximum shear stress acting at section a–a of the cantilevered strut.

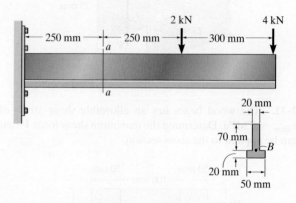

2 kN   4 kN

250 mm   250 mm   300 mm

a

a

20 mm

70 mm

B

20 mm

50 mm

Probs. 7–22/23

*7–24. Determine the maximum shear stress in the T-beam at the critical section where the internal shear force is maximum.

•**7–25.** Determine the maximum shear stress in the T-beam at point C. Show the result on a volume element at this point.

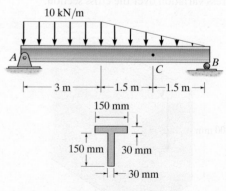

10 kN/m

A

C

B

3 m   1.5 m   1.5 m

150 mm

150 mm   30 mm

30 mm

Probs. 7–24/25

7–26. Determine the maximum shear stress acting in the fiberglass beam at the section where the internal shear force is maximum.

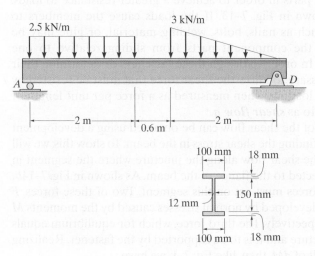

Prob. 7–26

7–27. Determine the shear stress at points C and D located on the web of the beam.

*7–28. Determine the maximum shear stress acting in the beam at the critical section where the internal shear force is maximum.

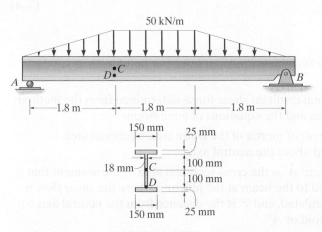

Probs. 7–27/28

7–29. Write a computer program that can be used to determine the maximum shear stress in the beam that has the cross section shown, and is subjected to a specified constant distributed load w and concentrated force P. Show an application of the program using the values $L = 4$ m, $a = 2$ m, $P = 1.5$ kN, $d_1 = 0$, $d_2 = 2$ m, $w = 400$ N/m, $t_1 = 15$ mm, $t_2 = 20$ mm, $b = 50$ mm, and $h = 150$ mm.

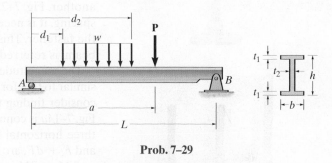

Prob. 7–29

7–30. The beam has a rectangular cross section and is subjected to a load P that is just large enough to develop a fully plastic moment $M_p = PL$ at the fixed support. If the material is elastic-plastic, then at a distance $x < L$ the moment $M = Px$ creates a region of plastic yielding with an associated elastic core having a height $2y'$. This situation has been described by Eq. 6–30 and the moment $\mathbf{M}$ is distributed over the cross section as shown in Fig. 6–48e. Prove that the maximum shear stress developed in the beam is given by $\tau_{max} = \frac{3}{2}(P/A')$, where $A' = 2y'b$, the cross-sectional area of the elastic core.

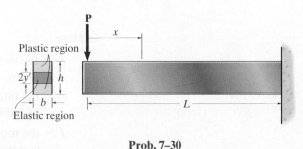

Prob. 7–30

7–31. The beam in Fig. 6–48f is subjected to a fully plastic moment $\mathbf{M}_p$. Prove that the longitudinal and transverse shear stresses in the beam are zero. *Hint:* Consider an element of the beam as shown in Fig. 7–4c.

7

Fig. 7–13

7.3   Shear Flow in Built-Up Members

Occasionally in engineering practice, members are "built up" from several composite parts in order to achieve a greater resistance to loads. Examples are shown in Fig. 7–13. If the loads cause the members to bend, fasteners such as nails, bolts, welding material, or glue may be needed to keep the component parts from sliding relative to one another, Fig. 7–2. In order to design these fasteners or determine their spacing, it is necessary to know the shear force that must be resisted by the fastener. This loading, when measured as a force per unit length of beam, is referred to as **shear flow q**.*

The magnitude of the shear flow can be obtained using a development similar to that for finding the shear stress in the beam. To show this, we will consider finding the shear flow along the juncture where the segment in Fig. 7–14a is connected to the flange of the beam. As shown in Fig. 7–14b, three horizontal forces must act on this segment. Two of these forces, F and $F + dF$, are developed by normal stresses caused by the moments M and $M + dM$, respectively. The third force, which for equilibrium equals dF, acts at the juncture and it is to be supported by the fastener. Realizing that dF is the result of dM, then, like Eq. 7–1, we have

$$dF = \frac{dM}{I}\int_{A'} y\, dA'$$

The integral represents Q, that is, the moment of the segment's area A' in Fig. 7–14b about the neutral axis for the entire cross section. Since the segment has a length dx, the shear flow, or force per unit length along the beam, is $q = dF/dx$. Hence dividing both sides by dx and noting that $V = dM/dx$, Eq. 6–2, we can write

$$\boxed{q = \frac{VQ}{I}} \tag{7–4}$$

Here

q = the shear flow, measured as a force per unit length along the beam

V = the internal resultant shear force, determined from the method of sections and the equations of equilibrium

I = the moment of inertia of the *entire* cross-sectional area computed about the neutral axis

$Q = \bar{y}'A'$, where A' is the cross-sectional area of the segment that is connected to the beam at the juncture where the shear flow is to be calculated, and $\bar{y}'$ is the distance from the neutral axis to the centroid of A'

---

*The use of the word "flow" in this terminology will become meaningful as it pertains to the discussion in Sec. 7.5.

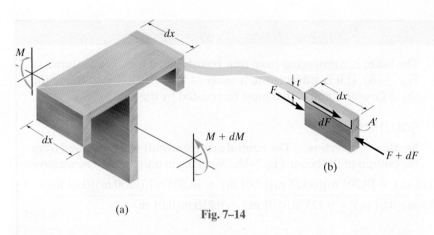

(a)

(b)

Fig. 7–14

Application of this equation follows the same "procedure for analysis" as outlined in Sec. 7.2 for the shear formula. It is very important to identify Q correctly when determining the shear flow at a particular junction on the cross section. A few examples should serve to illustrate how this is done. Consider the beam cross sections shown in Fig. 7–15. The shaded segments are connected to the beam by fasteners and at the planes of connection, identified by the thick black lines, the shear flow q is determined by using a value of Q calculated from A' and $\bar{y}'$ indicated in each figure. This value of q will be resisted by a *single* fastener in Fig. 7–15a, by *two* fasteners in Fig. 7–15b, and by *three* fasteners in Fig. 7–15c. In other words, the fastener in Fig. 7–15a supports the calculated value of q, and in Figs. 7–15b and 7–15c each fastener supports $q/2$ and $q/3$, respectively.

Important Point

- *Shear flow* is a measure of the force per unit length along the axis of a beam. This value is found from the shear formula and is used to determine the shear force developed in fasteners and glue that holds the various segments of a composite beam together.

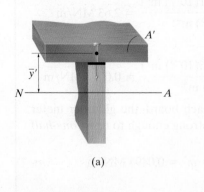

(a)

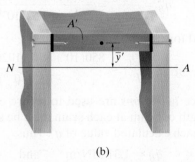

(b)

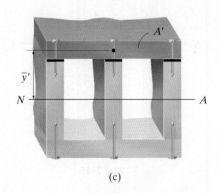

(c)

Fig. 7–15

EXAMPLE 7.5

The beam is constructed from four boards glued together as shown in Fig. 7–16a. If it is subjected to a shear of $V = 850$ kN, determine the shear flow at B and C that must be resisted by the glue.

SOLUTION

Section Properties. The neutral axis (centroid) will be located from the bottom of the beam, Fig. 7–16a. Working in units of meters, we have

$$\bar{y} = \frac{\Sigma \bar{y} A}{\Sigma A} = \frac{2[0.15 \text{ m}](0.3 \text{ m})(0.01 \text{ m}) + [0.205 \text{ m}](0.125 \text{ m})(0.01 \text{ m}) + [0.305 \text{ m}](0.250 \text{ m})(0.01 \text{ m})}{2(0.3 \text{ m})(0.01 \text{ m}) + 0.125 \text{ m}(0.01 \text{ m}) + 0.250 \text{ m}(0.01 \text{ m})}$$

$$= 0.1968 \text{ m}$$

The moment of inertia about the neutral axis is thus

$$I = 2\left[\frac{1}{12}(0.01 \text{ m})(0.3 \text{ m})^3 + (0.01 \text{ m})(0.3 \text{ m})(0.1968 \text{ m} - 0.150 \text{ m})^2\right]$$

$$+ \left[\frac{1}{12}(0.125 \text{ m})(0.01 \text{ m})^3 + (0.125 \text{ m})(0.01 \text{ m})(0.205 \text{ m} - 0.1968 \text{ m})^2\right]$$

$$+ \left[\frac{1}{12}(0.250 \text{ m})(0.01 \text{ m})^3 + (0.250 \text{ m})(0.01 \text{ m})(0.305 \text{ m} - 0.1968 \text{ m})^2\right]$$

$$= 87.52(10^{-6}) \text{ m}^4$$

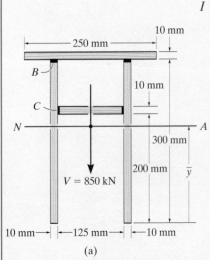

Since the glue at B and B' in Fig. 7–16b "holds" the top board to the beam, we have

$$Q_B = \bar{y}'_B A'_B = [0.305 \text{ m} - 0.1968 \text{ m}](0.250 \text{ m})(0.01 \text{ m})$$

$$= 0.271(10^{-3}) \text{ m}^3$$

Likewise, the glue at C and C' "holds" the inner board to the beam, Fig. 7–16b, and so

$$Q_C = \bar{y}'_C A'_C = [0.205 \text{ m} - 0.1968 \text{ m}](0.125 \text{ m})(0.01 \text{ m})$$

$$= 0.01026(10^{-3}) \text{ m}^3$$

Shear Flow. For B and B' we have

$$q'_B = \frac{V Q_B}{I} = \frac{850(10^3) \text{ N}(0.271(10^{-3}) \text{ m}^3)}{87.52(10^{-6}) \text{ m}^4} = 2.63 \text{ MN/m}$$

And for C and C',

$$q'_C = \frac{V Q_C}{I} = \frac{850(10^3) \text{ N}(0.01026(10^{-3}) \text{ m}^3)}{87.52(10^{-6}) \text{ m}^4} = 0.0996 \text{ MN/m}$$

Since *two seams* are used to secure each board, the glue per meter length of beam at each seam must be strong enough to resist *one-half* of each calculated value of q'. Thus,

$$q_B = 1.31 \text{ MN/m} \quad \text{and} \quad q_C = 0.0498 \text{ MN/m} \quad \textit{Ans.}$$

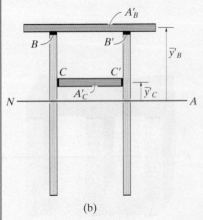

(a)

(b)

Fig. 7–16

EXAMPLE | 7.6

A box beam is constructed from four boards nailed together as shown in Fig. 7–17a. If each nail can support a shear force of 30 N, determine the maximum spacing s of the nails at B and at C so that the beam will support the force of 80 N.

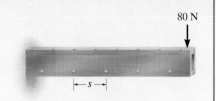

(a)

SOLUTION

Internal Shear. If the beam is sectioned at an *arbitrary point* along its length, the internal shear required for equilibrium is always $V = 80$ N and so the shear diagram is shown in Fig. 7–17b.

Section Properties. The moment of inertia of the cross-sectional area about the neutral axis can be determined by considering a 75-mm × 75-mm square minus a 45-mm × 45-mm square.

$$I = \frac{1}{12}(75 \text{ mm})(75 \text{ mm})^3 - \frac{1}{12}(45 \text{ mm})(45 \text{ mm})^3 = 2295(10^3) \text{ mm}^4$$

The shear flow at B is determined using Q_B found from the darker shaded area shown in Fig. 7–17c. It is this "symmetric" portion of the beam that is to be "held" onto the rest of the beam by nails on the left side and by the fibers of the board on the right side. Thus,

$$Q_B = \bar{y}'A' = [30 \text{ mm}](75 \text{ mm})(15 \text{ mm}) = 33750 \text{ mm}^3$$

Likewise, the shear flow at C can be determined using the "symmetric" shaded area shown in Fig. 7–17d. We have

$$Q_C = \bar{y}'A' = [30 \text{ mm}](45 \text{ mm})(15 \text{ mm}) = 20250 \text{ mm}^3$$

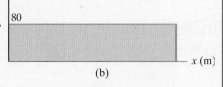

(b)

Shear Flow.

$$q_B = \frac{VQ_B}{I} = \frac{80 \text{ N}(33750 \text{ mm}^3)}{2295(10)^3 \text{ mm}^4} = 1.176 \text{ N/mm}$$

$$q_C = \frac{VQ_C}{I} = \frac{80 \text{ N}(20250 \text{ mm}^3)}{2295(10)^3 \text{ mm}^4} = 0.7059 \text{ N/mm}$$

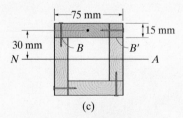

(c)

These values represent the shear force per unit length of the beam that must be resisted by the nails at B and the fibers at B', Fig. 7–17c, and the nails at C and the fibers at C', Fig. 7–17d, respectively. Since in each case the shear flow is resisted at *two* surfaces and each nail can resist 30 N, for B the spacing is

$$s_B = \frac{30 \text{ N}}{(1.176/2) \text{ N/mm}} = 51.0 \text{ mm} \qquad \text{Use } s_B = 55 \text{ mm} \quad \textit{Ans.}$$

And for C,

$$s_C = \frac{30 \text{ N}}{(0.7059/2) \text{ N/mm}} = 85.0 \text{ mm} \qquad \text{Use } s_C = 85 \text{ mm} \quad \textit{Ans.}$$

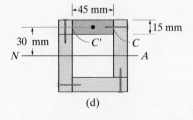

(d)

Fig. 7–17

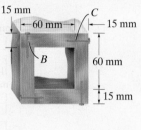

EXAMPLE | 7.7

Nails having a total shear strength of 40 N are used in a beam that can be constructed either as in Case I or as in Case II, Fig. 7–18. If the nails are spaced at 90 mm, determine the largest vertical shear that can be supported in each case so that the fasteners will not fail.

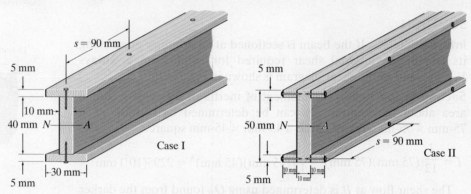

Fig. 7–18

SOLUTION

Since the cross section is the same in both cases, the moment of inertia about the neutral axis is

$$I = \frac{1}{12}(30 \text{ mm})(50 \text{ mm})^3 - 2\left[\frac{1}{12}(10 \text{ mm})(40 \text{ mm})^3\right] = 205\ 833 \text{ mm}^4$$

Case I. For this design a single row of nails holds the top or bottom flange onto the web. For one of these flanges,

$$Q = \bar{y}'A' = [22.5 \text{ mm}](30 \text{ mm}(5 \text{ mm})) = 3375 \text{ mm}^3$$

so that

$$q = \frac{VQ}{I}$$

$$\frac{40 \text{ N}}{90 \text{ mm}} = \frac{V(3375 \text{ mm}^3)}{205\ 833 \text{ mm}^4}$$

$$V = 27.1 \text{ N} \qquad\qquad Ans.$$

Case II. Here a single row of nails holds one of the side boards onto the web. Thus,

$$Q = \bar{y}'A' = [22.5 \text{ mm}](10 \text{ mm}(5 \text{ mm})) = 1125 \text{ mm}^3$$

$$q = \frac{VQ}{I}$$

$$\frac{40 \text{ N}}{90 \text{ mm}} = \frac{V(1125 \text{ mm}^3)}{205\ 833 \text{ mm}^4}$$

$$V = 81.3 \text{ N} \qquad\qquad Ans.$$

FUNDAMENTAL PROBLEMS

F7–6. The two identical boards are bolted together to form the beam. Determine the maximum allowable spacing s of the bolts to the nearest mm if each bolt has a shear strength of 15 kN. The beam is subjected to a shear force of $V = 50$ kN.

F7–7. The two identical boards are bolted together to form the beam. If the spacing of the bolts is $s = 100$ mm and each bolt has a shear strength of 15 kN, determine the maximum shear force **V** the beam can resist.

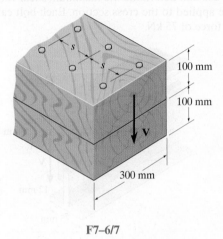

F7–6/7

F7–8. Two identical 20-mm thick plates are bolted to the top and bottom flange to form the built-up beam. If the beam is subjected to a shear force of $V = 300$ kN, determine the allowable maximum spacing s of the bolts to the nearest mm. Each bolt has a shear strength of 30 kN.

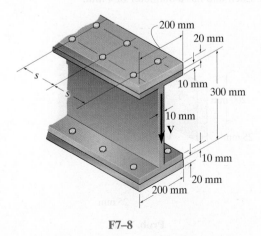

F7–8

F7–9. The boards are bolted together to form the built-up beam. If the beam is subjected to a shear force of $V = 20$ kN, determine the allowable maximum spacing of the bolts to the nearest mm. Each bolt has a shear strength of 8 kN.

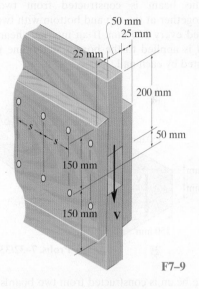

F7–9

F7–10. The boards are bolted together to form the built-up beam. If the beam is subjected to a shear force of $V = 75$ kN, determine the allowable maximum spacing of the bolts to the nearest multiples of 5 mm. Each bolt has a shear strength of 30 kN.

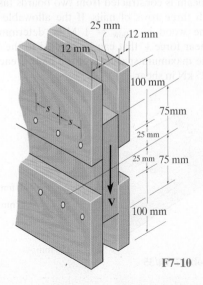

F7–10

PROBLEMS

*7–32. The beam is constructed from two boards fastened together at the top and bottom with two rows of nails spaced every 150 mm. If each nail can support a 2.5-kN shear force, determine the maximum shear force V that can be applied to the beam.

•7–33. The beam is constructed from two boards fastened together at the top and bottom with two rows of nails spaced every 150 mm. If an internal shear force of $V = 3$ kN is applied to the boards, determine the shear force resisted by each nail.

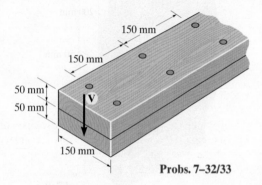

Probs. 7–32/33

7–34. The beam is constructed from two boards fastened together with three rows of nails spaced $s = 50$ mm apart. If each nail can support a 2.25-kN shear force, determine the maximum shear force V that can be applied to the beam. The allowable shear stress for the wood is $\tau_{allow} = 2.1$ MPa.

7–35. The beam is constructed from two boards fastened together with three rows of nails. If the allowable shear stress for the wood is $\tau_{allow} = 1$ MPa. determine the maximum shear force V that can be applied to the beam. Also, find the maximum spacing s of the nails if each nail can resist 3.25 kN in shear.

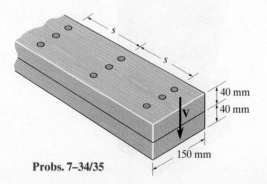

Probs. 7–34/35

*7–36. The beam is fabricated from two equivalent structural tees and two plates. Each plate has a height of 150 mm and a thickness of 12 mm. If a shear of $V = 250$ kN is applied to the cross section, determine the maximum spacing of the bolts. Each bolt can resist a shear force of 75 kN.

•7–37. The beam is fabricated from two equivalent structural tees and two plates. Each plate has a height of 150 mm and a thickness of 12 mm. If the bolts are spaced at $s = 200$ mm determine the maximum shear force V that can be applied to the cross section. Each bolt can resist a shear force of 75 kN.

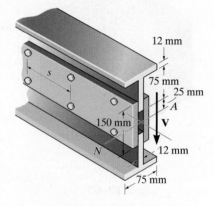

Probs. 7–36/37

7–38. The beam is subjected to a shear of $V = 2$ kN. Determine the average shear stress developed in each nail if the nails are spaced 75 mm apart on each side of the beam. Each nail has a diameter of 4 mm.

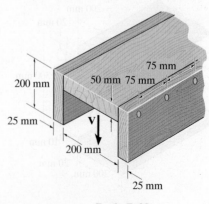

Prob. 7–38

7–39. A beam is constructed from three boards bolted together as shown. Determine the shear force developed in each bolt if the bolts are spaced $s = 250$ mm apart and the applied shear is $V = 35$ kN.

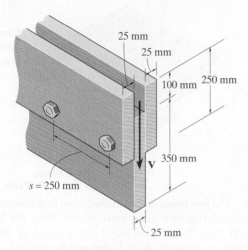

Prob. 7–39

*7–40. The double-web girder is constructed from two plywood sheets that are secured to wood members at its top and bottom. If each fastener can support 3 kN in single shear, determine the required spacing s of the fasteners needed to support the loading $P = 15$ kN. Assume A is pinned and B is a roller.

•7–41. The double-web girder is constructed from two plywood sheets that are secured to wood members at its top and bottom. The allowable bending stress for the wood is $\sigma_{allow} = 56$ MPa and the allowable shear stress is $\tau_{allow} = 21$ MPa. If the fasteners are spaced $s = 150$ mm and each fastener can support 3 kN in single shear, determine the maximum load P that can be applied to the beam.

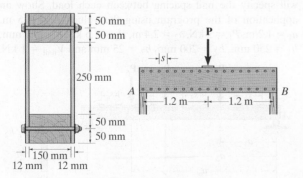

Probs. 7–40/41

7–42. The T-beam is nailed together as shown. If the nails can each support a shear force of 4.5 kN, determine the maximum shear force V that the beam can support and the corresponding maximum nail spacing s to the nearest multiples of 5 mm. The allowable shear stress for the wood is $\tau_{allow} = 3$ MPa.

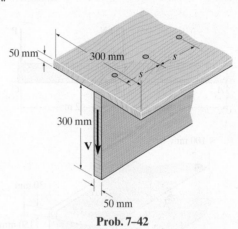

Prob. 7–42

7–43. Determine the average shear stress developed in the nails within region AB of the beam. The nails are located on each side of the beam and are spaced 100 mm apart. Each nail has a diameter of 4 mm. Take $P = 2$ kN.

*7–44. The nails are on both sides of the beam and each can resist a shear of 2 kN. In addition to the distributed loading, determine the maximum load P that can be applied to the end of the beam. The nails are spaced 100 mm apart and the allowable shear stress for the wood is $\tau_{allow} = 3$ MPa.

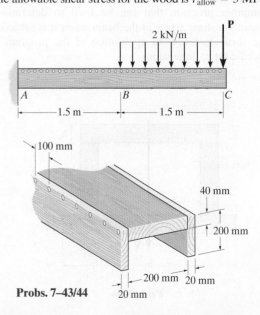

Probs. 7–43/44

7

•7–45. The beam is constructed from four boards which are nailed together. If the nails are on both sides of the beam and each can resist a shear of 3 kN, determine the maximum load P that can be applied to the end of the beam.

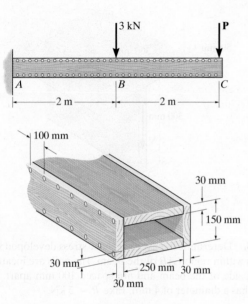

Prob. 7–45

7–46. A built-up timber beam is made from the four boards, each having a rectangular cross section. Write a computer program that can be used to determine the maximum shear stress in the beam when it is subjected to the shear V. Show an application of the program for a specific set of dimensions.

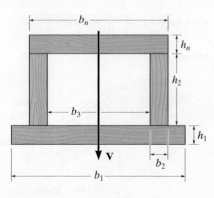

Prob. 7–46

7–47. The beam is made from four boards nailed together as shown. If the nails can each support a shear force of 500 N, determine their required spacing s' and s if the beam is subjected to a shear of $V = 3.5$ kN.

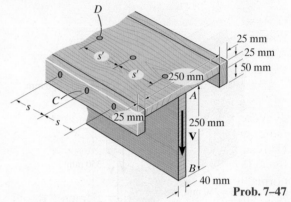

Prob. 7–47

*7–48. The box beam is constructed from four boards that are fastened together using nails spaced along the beam every 50 mm. If each nail can resist a shear of 250 N, determine the greatest shear V that can be applied to the beam without causing failure of the nails.

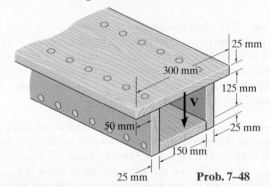

Prob. 7–48

7–49. The timber T-beam is subjected to a load consisting of n concentrated forces, P_n. If the allowable shear V_{nail} for each of the nails is known, write a computer program that will specify the nail spacing between each load. Show an application of the program using the values $L = 4.5$ m, $a_1 = 1.2$ m, $P_1 = 3$ kN, $a_2 = 2.4$ m, $P_2 = 7.5$ kN, $b_1 = 40$ mm, $h_1 = 250$ mm, $b_2 = 200$ mm, $h_2 = 25$ mm and $V_{nail} = 1$ kN.

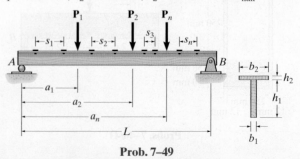

Prob. 7–49

7.4 Shear Flow in Thin-Walled Members

In this section we will show how to apply the shear flow equation $q = VQ/I$ to find the shear-flow *distribution* throughout a member's cross-sectional area. We will assume that the member has *thin walls*, that is, the wall thickness is small compared to its height or width. As will be shown in the next section, this analysis has important applications in structural and mechanical design.

Like the shear stress, the shear flow acts on both the longitudinal *and* transverse planes of the member. To show how to establish its direction on the cross section, consider the segment dx of the wide-flange beam in Fig. 7–19a. Free-body diagrams of two segments B and C taken from the top flange are shown in Figs. 7–19b and 7–19c. The force dF must act on the longitudinal section in order to balance the normal forces F and $F + dF$ created by the moments M and $M + dM$, respectively. Now, if the corner elements B and C of each segment are removed, then the transverse components q act on the cross section as shown in Figs. 7–19b and 7–19c. Using this method, show that the shear flow at the corresponding points B' and C' on the bottom flange in Fig. 7–19d is directed as shown.

Although it is also true that $V + dV$ will create *vertical* shear-flow components on this element, here we will neglect its effects. This is because this component, like the shear stress, is approximately zero throughout the thickness of the element. Here, the flange is thin and the top and bottom surfaces of the element are free of stress, Fig. 7–19e. To summarize then, only the shear flow component that acts *parallel* to the sides of the flange will be considered.

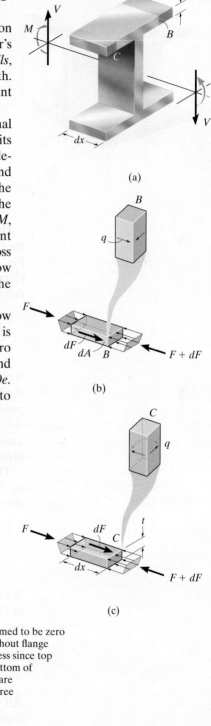

(a)

(b)

(c)

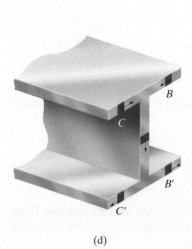

(d)

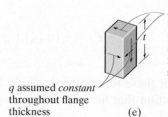

q' assumed to be zero throughout flange thickness since top and bottom of flange are stress free

q assumed *constant* throughout flange thickness

(e)

Fig. 7–19

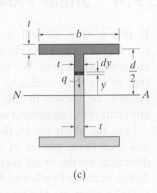
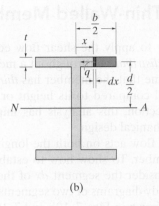

Fig. 7–20

Having determined the direction of the shear flow in each flange, we can now find its distribution along the top right flange of the beam in Fig. 7–20a. To do this, consider the shear flow q, acting on the colored element dx, located an arbitrary distance x from the centerline of the cross section, Fig. 7–20b. Here $Q = \bar{y}'A' = [d/2](b/2 - x)t$, so that

$$q = \frac{VQ}{I} = \frac{V[d/2](b/2 - x)t}{I} = \frac{Vtd}{2I}\left(\frac{b}{2} - x\right) \tag{7-5}$$

By inspection, this distribution varies in a *linear* manner from $q = 0$ at $x = b/2$ to $(q_{max})_f = Vt\,db/4I$ at $x = 0$. (The limitation of $x = 0$ is possible here since the member is assumed to have "thin walls" and so the thickness of the web is neglected.) Due to symmetry, a similar analysis yields the same distribution of shear flow for the other flange segments, so that the results are as shown in Fig. 7–20d.

The total force developed in each flange segment can be determined by integration. Since the force on the element dx in Fig. 7–20b is $dF = q\,dx$, then

$$F_f = \int q\,dx = \int_0^{b/2} \frac{Vtd}{2I}\left(\frac{b}{2} - x\right)dx = \frac{Vtdb^2}{16I}$$

We can also determine this result by finding the area under the triangle in Fig. 7–20d. Hence,

$$F_f = \frac{1}{2}(q_{max})_f\left(\frac{b}{2}\right) = \frac{Vtdb^2}{16I}$$

All four of these forces are shown in Fig. 7–20e, and we can see from their direction that horizontal force equilibrium of the cross section is maintained.

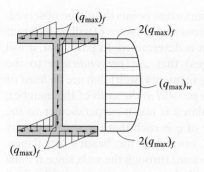

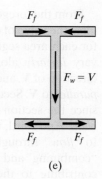

(e)

Shear-flow distribution

(d)

Fig. 7–20 (cont.)

A similar analysis can be performed for the web, Fig. 7–20c. Here q must act downward, and at element dy we have $Q = \Sigma\bar{y}'A' = [d/2](bt) + [y + (1/2)(d/2 - y)]t(d/2 - y) = bt\,d/2 + (t/2)(d^2/4 - y^2)$, so that

$$q = \frac{VQ}{I} = \frac{Vt}{I}\left[\frac{db}{2} + \frac{1}{2}\left(\frac{d^2}{4} - y^2\right)\right] \qquad (7\text{–}6)$$

For the web, the shear flow varies in a *parabolic manner*, from $q = 2(q_{max})_f = Vt\,db/2I$ at $y = d/2$ to $(q_{max})_w = (Vt\,d/I)(b/2 + d/8)$ at $y = 0$, Fig. 7–20d.

Integrating to determine the force in the web, F_w, we have,

$$F_w = \int q\,dy = \int_{-d/2}^{d/2}\frac{Vt}{I}\left[\frac{db}{2} + \frac{1}{2}\left(\frac{d^2}{4} - y^2\right)\right]dy$$

$$= \frac{Vt}{I}\left[\frac{db}{2}y + \frac{1}{2}\left(\frac{d^2}{4}y - \frac{1}{3}y^3\right)\right]\Bigg|_{-d/2}^{d/2}$$

$$= \frac{Vt\,d^2}{4I}\left(2b + \frac{1}{3}d\right)$$

Simplification is possible by noting that the moment of inertia for the cross-sectional area is

$$I = 2\left[\frac{1}{12}bt^3 + bt\left(\frac{d}{2}\right)^2\right] + \frac{1}{12}td^3$$

Neglecting the first term, since the thickness of each flange is small, then

$$I = \frac{td^2}{4}\left(2b + \frac{1}{3}d\right)$$

Substituting this into the above equation, we see that $F_w = V$, which is to be expected, Fig. 7–20e.

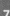

From the foregoing analysis, three important points should be observed. First, the value of q changes over the cross section, since Q will be different for each area segment A' for which it is determined. In particular, q will vary *linearly* along segments (flanges) that are *perpendicular* to the direction of $\mathbf{V}$, and *parabolically* along segments (web) that are *inclined or parallel* to $\mathbf{V}$. Second, q will *always act parallel to the walls* of the member, since the section on which q is calculated is taken perpendicular to the walls. And third, the *directional sense* of q is such that the shear appears to "*flow*" through the cross section, *inward* at the beam's top flange, "combining" and then "flowing" *downward* through the web, since it must contribute to the shear force $\mathbf{V}$, and then separating and "flowing" *outward* at the bottom flange. If one is able to "visualize" this "flow" it will provide an easy means for establishing not only the direction of q, but *also* the corresponding direction of τ. Other examples of how q is directed along the segments of thin-walled members are shown in Fig. 7–21. In all cases, symmetry prevails about an axis that is collinear with $\mathbf{V}$. As a result, q "flows" in a direction such that it will provide the vertical force $\mathbf{V}$ and yet also satisfy horizontal force equilibrium for the cross section.

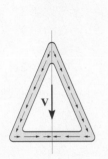

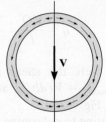

Shear flow q

Fig. 7–21

Important Points

- The shear flow formula $q = VQ/I$ can be used to determine the distribution of the shear flow throughout a thin-walled member provided the shear $\mathbf{V}$ acts along an axis of symmetry or principal centroidal axis of inertia for the cross section.

- If a member is made from segments having thin walls, only the shear flow *parallel* to the walls of the member is important.

- The shear flow varies *linearly* along segments that are *perpendicular* to the direction of the shear $\mathbf{V}$.

- The shear flow varies *parabolically* along segments that are *inclined* or *parallel* to the direction of the shear $\mathbf{V}$.

- On the cross section, the shear "flows" along the segments so that it results in the vertical shear force $\mathbf{V}$ and yet satisfies horizontal force equilibrium.

EXAMPLE 7.8

The thin-walled box beam in Fig. 7–22a is subjected to a shear of 10 kN. Determine the variation of the shear flow throughout the cross section.

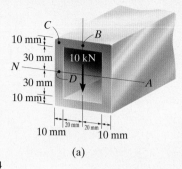

(a)

SOLUTION

By symmetry, the neutral axis passes through the center of the cross section. For thin-walled members we use centerline dimensions for calculating the moment of inertia.

$$I = \frac{1}{12}(20 \text{ mm})(70 \text{ mm})^3 + 2[(50 \text{ mm})(10 \text{ mm})(35 \text{ mm})^2] = 1.797(10^6)\text{mm}^4$$

Only the shear flow at points $B, C,$ and D has to be determined. For point B, the area $A' \approx 0$, Fig. 7–22b, since it can be thought of as being located entirely at point B. Alternatively, A' can also represent the *entire* cross-sectional area, in which case $Q_B = \bar{y}'A' = 0$ since $\bar{y}' = 0$. Because $Q_B = 0$, then

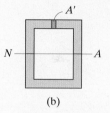

(b)

$$q_B = 0$$

For point C, the area A' is shown dark shaded in Fig. 7–22c. Here, we have used the mean dimensions since point C is on the centerline of each segment. We have

$$Q_C = \bar{y}'A' = (35 \text{ mm})(50 \text{ mm})(10 \text{ mm}) = 17\,500 \text{ mm}^3$$

Since there are two points of attachment,

$$q_C = \frac{1}{2}\left(\frac{VQ_C}{I}\right) = \frac{1}{2}\left(\frac{10 \text{ kN}(17500 \text{ mm}^3)}{1.797(10^6) \text{ mm}^4}\right) = 0.0487 \text{ kN/mm} = 48.7 \text{ kN/m}$$

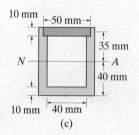

(c)

The shear flow at D is determined using the three dark-shaded rectangles shown in Fig. 7–22d. Again, using centerline dimensions

$$Q_D = \Sigma\bar{y}'A' = 2\left[\frac{35 \text{ mm}}{2}\right](10 \text{ mm})(35 \text{ mm}) + [35 \text{ mm}](50 \text{ mm})(10 \text{ mm}) = 29\,750 \text{ mm}^3$$

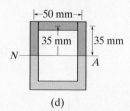

(d)

Because there are two points of attachment,

$$q_D = \frac{1}{2}\left(\frac{VQ_D}{I}\right) = \frac{1}{2}\left(\frac{10 \text{ kN}(29\,750 \text{ mm}^3)}{1.797(10^6) \text{ mm}^4}\right) = 0.0828 \text{ kN/mm} = 82.8 \text{ kN/m}$$

Using these results, and the symmetry of the cross section, the shear-flow distribution is plotted in Fig. 7–22e. The distribution is linear along the horizontal segments (perpendicular to **V**) and parabolic along the vertical segments (parallel to **V**).

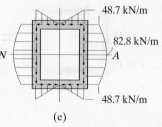

(e)

Fig. 7–22

*7.5 Shear Center for Open Thin-Walled Members

In the previous section, it was assumed that the internal shear V was applied along a principal centroidal axis of inertia that *also* represents an *axis of symmetry* for the cross section. In this section we will consider the effect of applying the shear along a principal centroidal axis that is *not* an axis of symmetry. As before, only open thin-walled members will be analyzed, so the dimensions to the centerline of the walls of the members will be used. A typical example of this case is the channel section shown in Fig. 7–23a. Here it is cantilevered from a fixed support and is subjected to the force **P**. If this force is applied along the once vertical, unsymmetrical axis that passes through the *centroid C* of the cross section, the channel will not only bend downward, *it will also twist* clockwise as shown.

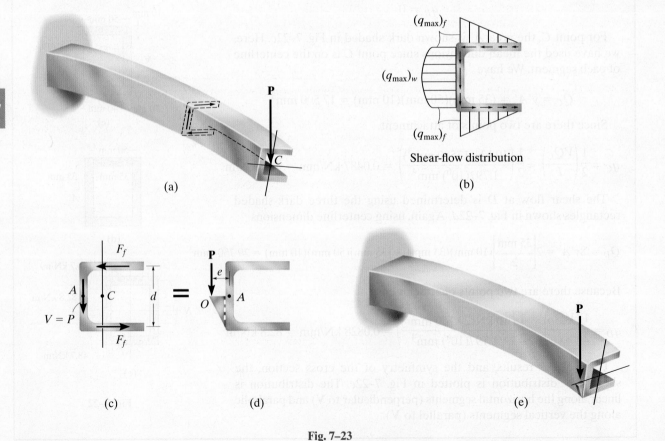

(a)

(b)

Shear-flow distribution

(c)   (d)   (e)

Fig. 7–23

To understand why the member twists, it is necessary to show the shear-flow distribution along the channel's flanges and web, Fig. 7–23b. When this distribution is integrated over the flange and web areas, it will give resultant forces of F_f in each flange and a force of $V = P$ in the web, Fig. 7–23c. If the moments of these forces are summed about point A, it can be seen that the couple or torque created by the flange forces is responsible for twisting the member. The actual twist is clockwise when viewed from the front of the beam as shown in Fig. 7–23a, since *reactive* internal "equilibrium" forces F_f cause the twisting. In order to *prevent* this twisting it is therefore necessary to apply **P** at a point O located an eccentric distance e from the web of the channel, Fig. 7–23d. We require $\Sigma M_A = F_f d = Pe$, or

$$e = \frac{F_f d}{P}$$

Using the method discussed in the previous section, F_f can be evaluated in terms of $P \, (= V)$ and the dimensions of the flanges and web. Once this is done, then P will cancel upon substitution into the above equation, and it becomes possible to express e simply as a function of the cross-sectional geometry (see Example 7.9). The point O so located is called the **shear center** or **flexural center**. When **P** is applied at the shear center, the **beam will bend without twisting** as shown in Fig. 7–23e. Design handbooks often list the location of this point for a variety of beams having thin-walled cross sections that are commonly used in practice.

From this analysis, it should be noted that **the shear center will always lie on an axis of symmetry** of a member's cross-sectional area. For example, if the channel is rotated 90° and **P** is applied at A, Fig. 7–24a, no twisting will occur since the shear flow in the web and flanges for this case is *symmetrical*, and therefore the force resultants in these elements will create zero moments about A, Fig. 7–24b. Obviously, if a member has a cross section with *two* axes of symmetry, as in the case of a wide-flange beam, the shear center will then coincide with the intersection of these axes (the centroid).

Demonstration of how a cantilever beam deflects when loaded through the centroid (above) and through the shear center (below).

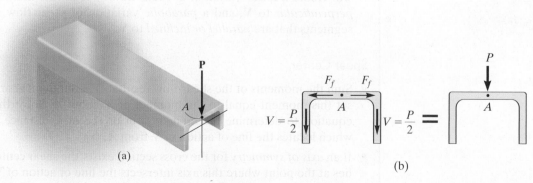

(a)

(b)

Fig. 7–24

Important Points

- The *shear center* is the point through which a force can be applied which will cause a beam to bend and yet not twist.
- The shear center will always lie on an axis of symmetry of the cross section.
- The location of the shear center is only a function of the geometry of the cross section, and does not depend upon the applied loading.

Procedure for Analysis

The location of the shear center for an open thin-walled member for which the internal shear is in the *same direction* as a principal centroidal axis for the cross section may be determined by using the following procedure.

Shear-Flow Resultants.

- By observation, determine the direction of the shear flow through the various segments of the cross section, and sketch the force resultants on each segment of the cross section. (For example, see Fig. 7–23*c*.) Since the shear center is determined by taking the moments of these force resultants about a point (*A*), choose this point at a location that eliminates the moments of as many force resultants as possible.
- The magnitudes of the force resultants that create a moment about *A* must be calculated. For any segment this is done by determining the shear flow *q* at an arbitrary point on the segment and then integrating *q* along the segment's length. Realize that **V** will create a *linear* variation of shear flow in segments that are *perpendicular* to **V**, and a *parabolic* variation of shear flow in segments that are *parallel or inclined* to **V**.

Shear Center.

- Sum the moments of the shear-flow resultants about point *A* and set this moment equal to the moment of **V** about *A*. Solve this equation to determine the moment-arm or eccentric distance *e*, which locates the line of action of **V** from *A*.
- If an *axis of symmetry* for the cross section exists, the shear center lies at the point where this axis intersects the line of action of **V**.

EXAMPLE 7.9

Determine the location of the shear center for the thin-walled channel section having the dimensions shown in Fig. 7–25a.

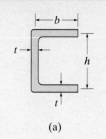

(a)

SOLUTION

Shear-Flow Resultants.   A vertical downward shear $\mathbf{V}$ applied to the section causes the shear to flow through the flanges and web as shown in Fig. 7–25b. This causes force resultants F_f and V in the flanges and web as shown in Fig. 7–25c. We will take moments about point A so that only the force F_f on the lower flange has to be determined.

The cross-sectional area can be divided into three component rectangles—a web and two flanges. Since each component is assumed to be thin, the moment of inertia of the area about the neutral axis is

$$I = \frac{1}{12}th^3 + 2\left[bt\left(\frac{h}{2}\right)^2\right] = \frac{th^2}{2}\left(\frac{h}{6} + b\right)$$

From Fig. 7–25d, q at the arbitrary position x is

$$q = \frac{VQ}{I} = \frac{V(h/2)[b - x]t}{(th^2/2)[(h/6) + b]} = \frac{V(b - x)}{h[(h/6) + b]}$$

Hence, the force F_f is

$$F_f = \int_0^b q \, dx = \frac{V}{h[(h/6) + b]}\int_0^b (b - x) \, dx = \frac{Vb^2}{2h[(h/6) + b]}$$

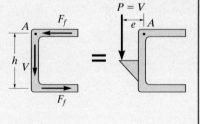

(c)

This same result can also be determined by first finding $(q_{max})_f$, Fig. 7–25b, then determining the triangular area $\frac{1}{2}b(q_{max})_f = F_f$.

Shear Center.   Summing moments about point A, Fig. 7–25c, we require

$$Ve = F_f h = \frac{Vb^2h}{2h[(h/6) + b]}$$

Thus,

$$e = \frac{b^2}{[(h/3) + 2b]} \qquad Ans.$$

As stated previously, e depends only on the geometry of the cross section.

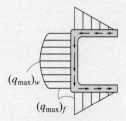

(b)

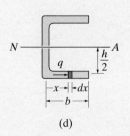

(d)

Fig. 7–25

EXAMPLE | 7.10

Determine the location of the shear center for the angle having equal legs, Fig. 7–26a. Also, find the internal shear force resultant in each leg.

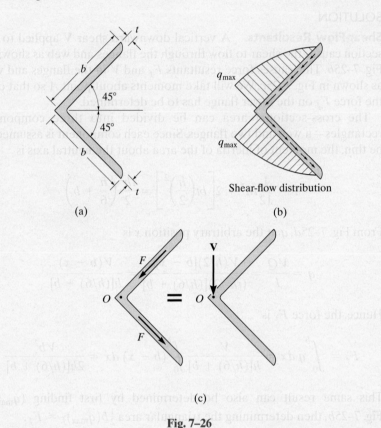

(a)

q_{max}

q_{max}

Shear-flow distribution

(b)

(c)

Fig. 7–26

SOLUTION

When a vertical downward shear **V** is applied at the section, the shear flow and shear-flow resultants are directed as shown in Figs. 7–26b and 7–26c, respectively. Note that the force F in each leg must be equal, since for equilibrium the sum of their horizontal components must be equal to zero. Also, the lines of action of both forces intersect point O; therefore, this point *must be the shear center* since the sum of the moments of these forces and **V** about O is zero, Fig. 7–26c.

The magnitude of **F** can be determined by first finding the shear flow at the arbitrary location s along the top leg, Fig. 7–26d. Here

$$Q = \bar{y}'A' = \frac{1}{\sqrt{2}}\left((b-s) + \frac{s}{2}\right)ts = \frac{1}{\sqrt{2}}\left(b - \frac{s}{2}\right)st$$

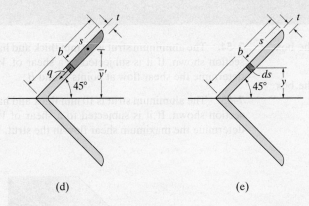

(d)   (e)

Fig. 7–26 (cont.)

The moment of inertia of the angle, about the neutral axis, must be determined from "first principles," since the legs are inclined with respect to the neutral axis. For the area element $dA = t\,ds$, Fig. 7–26e, we have

$$I = \int_A y^2\, dA = 2\int_0^b \left[\frac{1}{\sqrt{2}}(b - s)\right]^2 t\,ds = t\left(b^2 s - bs^2 + \frac{1}{3}s^3\right)\Bigg|_0^b = \frac{tb^3}{3}$$

Thus, the shear flow is

$$q = \frac{VQ}{I} = \frac{V}{(tb^3/3)}\left[\frac{1}{\sqrt{2}}\left(b - \frac{s}{2}\right)st\right]$$

$$= \frac{3V}{\sqrt{2}\,b^3}s\left(b - \frac{s}{2}\right)$$

The variation of q is parabolic, and it reaches a maximum value when $s = b$ as shown in Fig. 7–26b. The force F is therefore

$$F = \int_0^b q\,ds = \frac{3V}{\sqrt{2}\,b^3}\int_0^b s\left(b - \frac{s}{2}\right) ds$$

$$= \frac{3V}{\sqrt{2}\,b^3}\left(b\frac{s^2}{2} - \frac{1}{6}s^3\right)\Bigg|_0^b$$

$$= \frac{1}{\sqrt{2}}V \qquad\qquad Ans.$$

NOTE: This result can be easily verified since the sum of the vertical components of the force F in each leg must equal V and, as stated previously, the sum of the horizontal components equals zero.

PROBLEMS

7–50. A shear force of $V = 300$ kN is applied to the box girder. Determine the shear flow at points A and B.

7–51. A shear force of $V = 450$ kN is applied to the box girder. Determine the shear flow at points C and D.

7–54. The aluminum strut is 10 mm thick and has the cross section shown. If it is subjected to a shear of, $V = 150$ N, determine the shear flow at points A and B.

7–55. The aluminum strut is 10 mm thick and has the cross section shown. If it is subjected to a shear of $V = 150$ N, determine the maximum shear flow in the strut.

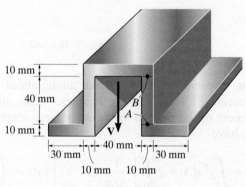

Probs. 7–54/55

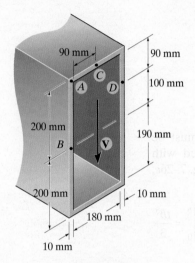

Probs. 7–50/51

*7–52. A shear force of $V = 18$ kN is applied to the symmetric box girder. Determine the shear flow at A and B.

•7–53. A shear force of $V = 18$ kN is applied to the box girder. Determine the shear flow at C.

*7–56. The beam is subjected to a shear force of $V = 25$ kN. Determine the shear flow at points A and B.

•7–57. The beam is constructed from four plates and is subjected to a shear force of $V = 25$ kN. Determine the maximum shear flow in the cross section.

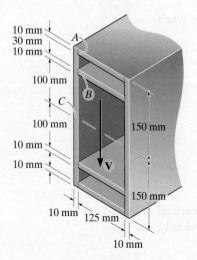

Probs. 7–52/53

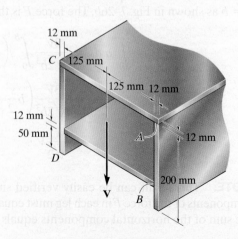

Probs. 7–56/57

7–58. The channel is subjected to a shear of $V = 75$ kN. Determine the shear flow developed at point A.

7–59. The channel is subjected to a shear of $V = 75$ kN. Determine the maximum shear flow in the channel.

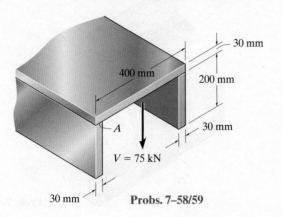

Probs. 7–58/59

*7–60. The angle is subjected to a shear of $V = 10$ kN. Sketch the distribution of shear flow along the leg AB. Indicate numerical values at all peaks.

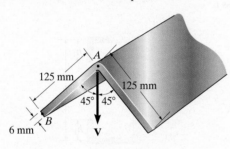

Prob. 7–60

•7–61. The assembly is subjected to a vertical shear of $V = 35$ kN. Determine the shear flow at points A and B and the maximum shear flow in the cross section.

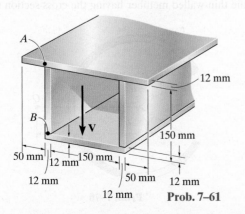

Prob. 7–61

7–62. Determine the shear-stress variation over the cross section of the thin-walled tube as a function of elevation y and show that $\tau_{\max} = 2V/A$, where $A = 2\pi rt$. *Hint:* Choose a differential area element $dA = Rt\, d\theta$. Using $dQ = y\, dA$, formulate Q for a circular section from θ to $(\pi - \theta)$ and show that $Q = 2R^2 t \cos\theta$, where $\cos\theta = \sqrt{R^2 - y^2}/R$.

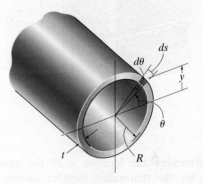

Prob. 7–62

7–63. Determine the location e of the shear center, point O, for the thin-walled member having the cross section shown where $b_2 > b_1$. The member segments have the same thickness t.

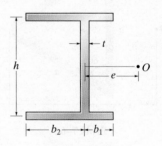

Prob. 7–63

*7–64. Determine the location e of the shear center, point O, for the thin-walled member having the cross section shown. The member segments have the same thickness t.

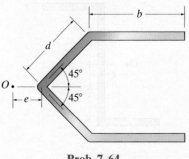

Prob. 7–64

•**7–65.** Determine the location e of the shear center, point O, for the thin-walled member having a slit along its side. Each element has a constant thickness t.

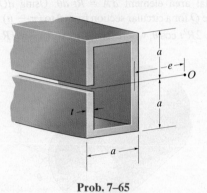

Prob. 7–65

7–66. Determine the location e of the shear center, point O, for the thin-walled member having the cross section shown.

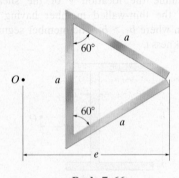

Prob. 7–66

7–67. Determine the location e of the shear center, point O, for the thin-walled member having the cross section shown. The member segments have the same thickness t.

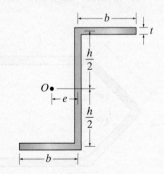

Prob. 7–67

7–68. Determine the location e of the shear center, point O, for the beam having the cross section shown. The thickness is t.

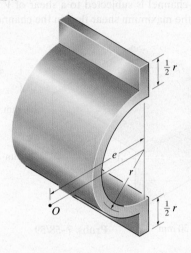

Prob. 7–68

•**7–69.** Determine the location e of the shear center, point O, for the thin-walled member having the cross section shown. The member segments have the same thickness t.

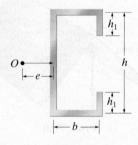

Prob. 7–69

7–70. Determine the location e of the shear center, point O, for the thin-walled member having the cross section shown.

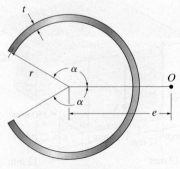

Prob. 7–70

CHAPTER REVIEW

Transverse shear stress in beams is determined indirectly by using the flexure formula and the relationship between moment and shear ($V = dM/dx$). The result is the shear formula

$$\tau = \frac{VQ}{It}$$

In particular, the value for Q is the moment of the area A' about the neutral axis, $Q = \bar{y}' A'$. This area is the portion of the cross-sectional area that is "held on" to the beam above (or below) the thickness t where τ is to be determined.

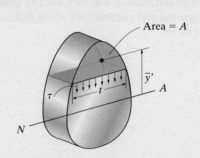

Ref.: Section 7.2

If the beam has a *rectangular* cross section, then the shear-stress distribution will be parabolic, having a maximum value at the neutral axis. The maximum shear stress can be determined using $\tau = 1.5\dfrac{V}{A}$.

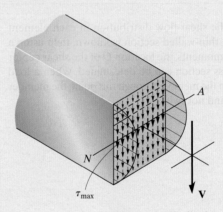

Ref.: Section 7.1

Shear–stress distribution

Fasteners, such as nails, bolts, glue, or welds, are used to connect the composite parts of a "built-up" section. The shear force resisted by these fasteners is determined from the shear flow, q, or force per unit length, that must be carried by the beam. The shear flow is

$$q = \frac{VQ}{I}$$

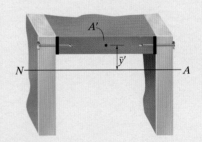

Ref.: Section 7.3

If the beam is made from thin-walled segments, then the shear-flow distribution along each segment can be determined. This distribution will vary linearly along horizontal segments, and parabolically along inclined or vertical segments.

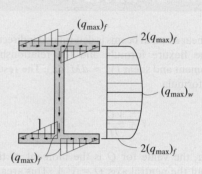

Shear-flow distribution

Ref.: Section 7.4

Provided the shear-flow distribution in each element of an open thin-walled section is known, then using a balance of moments, the location O of the shear center for the cross section can be determined. When a load is applied to the member through this point, the member will bend, and not twist.

Ref.: Section 7.5

7

REVIEW PROBLEMS

7–71. Sketch the intensity of the shear-stress distribution acting over the beam's cross-sectional area, and determine the resultant shear force acting on the segment AB. The shear acting at the section is $V = 175$ kN. Show that $I_{NA} = 340.82(10^6)$ mm^4.

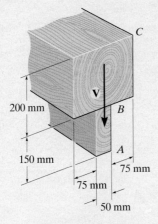

Prob. 7–71

*7–72. The beam is fabricated from four boards nailed together as shown. Determine the shear force each nail along the sides C and the top D must resist if the nails are uniformly spaced at $s = 75$ mm. The beam is subjected to a shear of $V = 22.5$ kN.

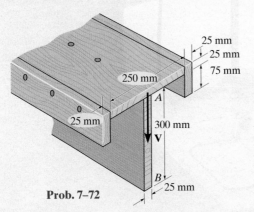

Prob. 7–72

•7–73. The member is subjected to a shear force of $V = 2$ kN. Determine the shear flow at points A, B, and C. The thickness of each thin-walled segment is 15 mm.

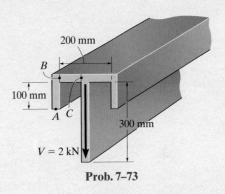

Prob. 7–73

7–74. The beam is constructed from four boards glued together at their seams. If the glue can withstand 15 kN/m, what is the maximum vertical shear V that the beam can support?

7–75. Solve Prob. 7–74 if the beam is rotated 90° from the position shown.

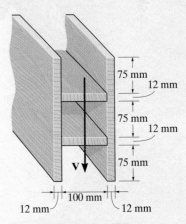

Probs. 7–74/75

7

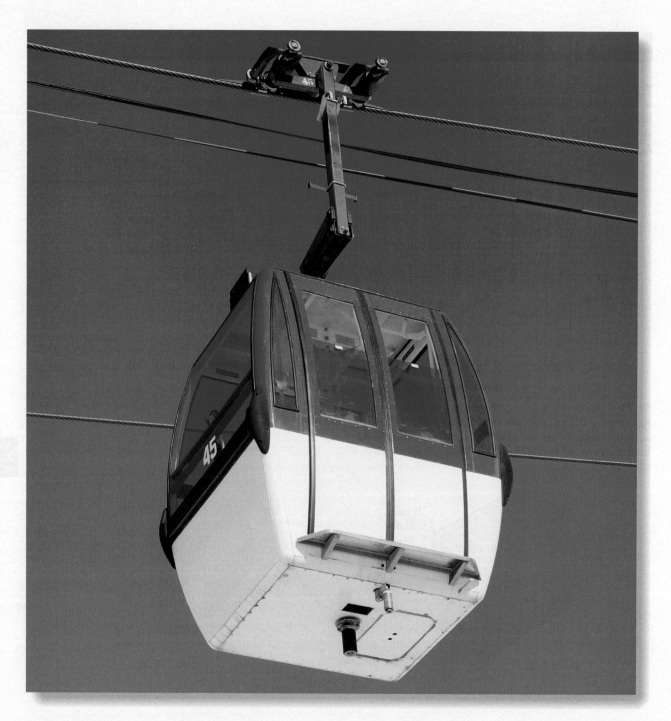

The offset hanger supporting this ski gondola is subjected to the combined loadings of axial force and bending moment.

Combined Loadings

Check out the Companion Website for the following video solution(s) that can be found in this chapter.

- Section 1
  Pressure Vessels and Thermal Strain
- Section 2
  Combined Loading – Multiple Forces
- Section 2
  Combined Loading – Hook Geometry
- Section 2
  Combined Loading – Curved Geometry

CHAPTER OBJECTIVES

This chapter serves as a review of the stress analysis that has been developed in the previous chapters regarding axial load, torsion, bending, and shear. We will discuss the solution of problems where several of these internal loads occur simultaneously on a member's cross section. Before doing this, however, the chapter begins with an analysis of stress developed in thin-walled pressure vessels.

8.1 Thin-Walled Pressure Vessels

Cylindrical or spherical vessels are commonly used in industry to serve as boilers or tanks. When under pressure, the material of which they are made is subjected to a loading from all directions. Although this is the case, the vessel can be analyzed in a simple manner provided it has a thin wall. In general, "***thin wall***" refers to a vessel having an inner-radius-to-wall-thickness ratio of 10 or more ($r/t \geq 10$). Specifically, when $r/t = 10$ the results of a thin-wall analysis will predict a stress that is approximately 4% *less* than the actual maximum stress in the vessel. For larger r/t ratios this error will be even smaller.

Provided the vessel wall is "thin," the stress distribution throughout its thickness will not vary significantly, and so we will assume that it is *uniform* or *constant*. Using this assumption, we will now analyze the state of stress in thin-walled cylindrical and spherical pressure vessels. In both cases, the pressure in the vessel is understood to be the *gauge pressure*, that is, it measures the pressure *above* atmospheric pressure, since atmospheric pressure is assumed to exist both inside and outside the vessel's wall before the vessel is pressurized.

Cylindrical pressure vessels, such as this gas tank, have semi-spherical end caps rather than flat ones in order to reduce the stress in the tank.

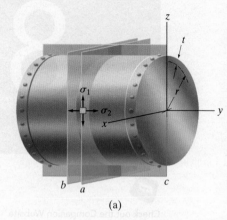

(a)

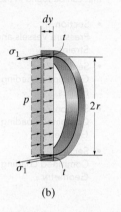

(b)

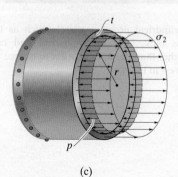

(c)

Fig. 8–1

Cylindrical Vessels.   Consider the cylindrical vessel in Fig. 8–1a, having a wall thickness t, inner radius r, and subjected to a gauge pressure p that developed within the vessel by a contained gas. Due to this loading, a small element of the vessel that is sufficiently removed from the ends and oriented as shown in Fig. 8–1a, is subjected to normal stresses σ_1 in the *circumferential or hoop direction* and σ_2 in the *longitudinal or axial direction*.

The hoop stress can be determined by considering the vessel to be sectioned by planes a, b, and c. A free-body diagram of the back segment along with the contained gas is shown in Fig. 8–1b. Here only the loadings in the x direction are shown. These loadings are developed by the uniform hoop stress σ_1, acting on the vessel's wall, and the pressure acting on the vertical face of the gas. For equilibrium in the x direction, we require

$$\Sigma F_x = 0; \qquad 2[\sigma_1(t\,dy)] - p(2r\,dy) = 0$$

$$\boxed{\sigma_1 = \frac{pr}{t}} \qquad (8\text{–}1)$$

The longitudinal stress can be determined by considering the left portion of section b of the cylinder, Fig. 8–1a. As shown in Fig. 8–1c, σ_2 acts uniformly throughout the wall, and p acts on the section of the contained gas. Since the mean radius is approximately equal to the vessel's inner radius, equilibrium in the y direction requires

$$\Sigma F_y = 0; \qquad \sigma_2(2\pi rt) - p(\pi r^2) = 0$$

$$\boxed{\sigma_2 = \frac{pr}{2t}} \qquad (8\text{–}2)$$

In the above equations,

$\sigma_1, \sigma_2 =$ the normal stress in the hoop and longitudinal directions, respectively. Each is assumed to be *constant* throughout the wall of the cylinder, and each subjects the material to tension.

$p =$ the internal gauge pressure developed by the contained gas

$r =$ the inner radius of the cylinder

$t =$ the thickness of the wall ($r/t \geq 10$)

By comparison, note that the hoop or circumferential stress is twice as large as the longitudinal or axial stress. Consequently, when fabricating cylindrical pressure vessels from rolled-formed plates, the longitudinal joints must be designed to carry twice as much stress as the circumferential joints.

Spherical Vessels.

We can analyze a spherical pressure vessel in a similar manner. To do this, consider the vessel to have a wall thickness t, inner radius r, and subjected to an internal gauge pressure p, Fig. 8–2a. If the vessel is sectioned in half, the resulting free-body diagram is shown in Fig. 8–2b. Like the cylinder, equilibrium in the y direction requires

$$\Sigma F_y = 0; \qquad \sigma_2(2\pi rt) - p(\pi r^2) = 0$$

Shown is the barrel of a shotgun which was clogged with debris just before firing. Gas pressure from the charge increased the circumferential stress within the barrel enough to cause the rupture.

$$\boxed{\sigma_2 = \frac{pr}{2t}} \qquad (8\text{–}3)$$

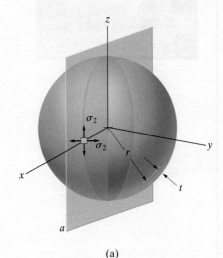

(a)

This is the *same result* as that obtained for the longitudinal stress in the cylindrical pressure vessel. Furthermore, from the analysis, this stress will be the same *regardless* of the orientation of the hemispheric free-body diagram. Consequently, a small element of the material is subjected to the state of stress shown in Fig. 8–2a.

The above analysis indicates that an element of material taken from either a cylindrical or a spherical pressure vessel is subjected to **biaxial stress**, i.e., normal stress existing in only two directions. Actually, the pressure also subjects the material to a **radial stress**, σ_3, which acts along a radial line. This stress has a maximum value equal to the pressure p at the interior wall and it decreases through the wall to zero at the exterior surface of the vessel, since the gauge pressure there is zero. For thin-walled vessels, however, we will *ignore* this radial-stress component, since our limiting assumption of $r/t = 10$ results in σ_2 and σ_1 being, respectively, 5 and 10 times *higher* than the maximum radial stress, $(\sigma_3)_{\max} = p$. Finally, if the vessel is subjected to an *external pressure*, the compressive stress developed within the thin wall may cause the vessel to become unstable, and collapse may occur by buckling rather than causing the material to fracture.

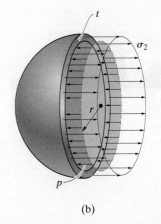

(b)

Fig. 8–2

EXAMPLE   8.1

A cylindrical pressure vessel has an inner diameter of 1.2 m and a thickness of 12 mm. Determine the maximum internal pressure it can sustain so that neither its circumferential nor its longitudinal stress component exceeds 140 MPa. Under the same conditions, what is the maximum internal pressure that a similar-size spherical vessel can sustain?

SOLUTION

Cylindrical Pressure Vessel. The maximum stress occurs in the circumferential direction. From Eq. 8–1 we have

$$\sigma_1 = \frac{pr}{t}; \qquad 140 \text{ N/mm}^2 = \frac{p(600 \text{ mm})}{12 \text{ mm}}$$

$$p = 2.8 \text{ N/mm}^2 = 2.8 \text{ MPa} \qquad\qquad Ans.$$

Note that when this pressure is reached, from Eq. 8–2, the stress in the longitudinal direction will be $\sigma_2 = \frac{1}{2}(140 \text{ MPa}) = 70 \text{ MPa}$. Furthermore, the *maximum stress* in the *radial direction* occurs on the material at the inner wall of the vessel and is $(\sigma_3)_{max} = p = 2.8 \text{ MPa}$. This value is 48 times smaller than the circumferential stress (140 MPa), and as stated earlier, its effects will be neglected.

Spherical Vessel. Here the maximum stress occurs in any two perpendicular directions on an element of the vessel, Fig. 8–2a. From Eq. 8–3, we have

$$\sigma_2 = \frac{pr}{2t}; \qquad 140 \text{ N/mm}^2 = \frac{p(600 \text{ mm})}{2(12 \text{ mm})}$$

$$p = 5.6 \text{ N/mm}^2 = 5.6 \text{ MPa} \qquad\qquad Ans.$$

NOTE: Although it is more difficult to fabricate, the spherical pressure vessel will carry twice as much internal pressure as a cylindrical vessel.

PROBLEMS

8–1. A spherical gas tank has an inner radius of $r = 1.5$ m. If it is subjected to an internal pressure of $p = 300$ kPa, determine its required thickness if the maximum normal stress is not to exceed 12 MPa.

8–2. A pressurized spherical tank is to be made of 12-mm-thick steel. If it is subjected to an internal pressure of $p = 1.4$ MPa, determine its outer radius if the maximum normal stress is not to exceed 105 MPa.

8–3. The thin-walled cylinder can be supported in one of two ways as shown. Determine the state of stress in the wall of the cylinder for both cases if the piston P causes the internal pressure to be 0.5 MPa. The wall has a thickness of 6 mm and the inner diameter of the cylinder is 200 mm.

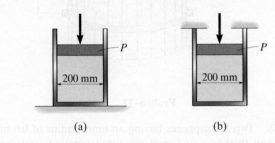

(a)   (b)

Prob. 8–3

*8–4. The tank of the air compressor is subjected to an internal pressure of 0.63 MPa. If the internal diameter of the tank is 550 mm, and the wall thickness is 6 mm, determine the stress components acting at point A. Draw a volume element of the material at this point, and show the results on the element.

Prob. 8–4

•8–5. The spherical gas tank is fabricated by bolting together two hemispherical thin shells of thickness 30 mm. If the gas contained in the tank is under a gauge pressure of 2 MPa, determine the normal stress developed in the wall of the tank and in each of the bolts. The tank has an inner diameter of 8 m and is sealed with 900 bolts each 25 mm in diameter.

8–6. The spherical gas tank is fabricated by bolting together two hemispherical thin shells. If the 8-m inner diameter tank is to be designed to withstand a gauge pressure of 2 MPa, determine the minimum wall thickness of the tank and the minimum number of 25-mm diameter bolts that must be used to seal it. The tank and the bolts are made from material having an allowable normal stress of 150 MPa and 250 MPa, respectively.

Probs. 8–5/6

8–7. A boiler is constructed of 8-mm thick steel plates that are fastened together at their ends using a butt joint consisting of two 8-mm cover plates and rivets having a diameter of 10 mm and spaced 50 mm apart as shown. If the steam pressure in the boiler is 1.35 MPa, determine (a) the circumferential stress in the boiler's plate apart from the seam, (b) the circumferential stress in the outer cover plate along the rivet line a–a, and (c) the shear stress in the rivets.

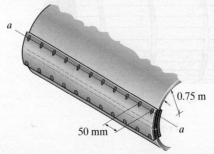

0.75 m

50 mm

Prob. 8–7

8

*8–8. The gas storage tank is fabricated by bolting together two half cylindrical thin shells and two hemispherical shells as shown. If the tank is designed to withstand a pressure of 3 MPa, determine the required minimum thickness of the cylindrical and hemispherical shells and the minimum required number of longitudinal bolts per meter length at each side of the cylindrical shell. The tank and the 25 mm diameter bolts are made from material having an allowable normal stress of 150 MPa and 250 MPa, respectively. The tank has an inner diameter of 4 m.

•8–9. The gas storage tank is fabricated by bolting together two half cylindrical thin shells and two hemispherical shells as shown. If the tank is designed to withstand a pressure of 3 MPa, determine the required minimum thickness of the cylindrical and hemispherical shells and the minimum required number of bolts for each hemispherical cap. The tank and the 25 mm diameter bolts are made from material having an allowable normal stress of 150 MPa and 250 MPa, respectively. The tank has an inner diameter of 4 m.

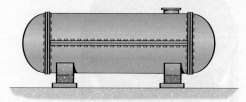

Probs. 8–8/9

8–10. A wood pipe having an inner diameter of 0.9 m is bound together using steel hoops each having a cross-sectional area of 125 mm². If the allowable stress for the hoops is $\sigma_{allow} = 84$ MPa, determine their maximum spacing s along the section of pipe so that the pipe can resist an internal gauge pressure of 28 kPa. Assume each hoop supports the pressure loading acting along the length s of the pipe.

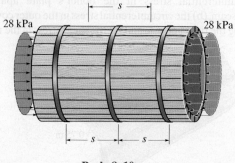

Prob. 8–10

8–11. The staves or vertical members of the wooden tank are held together using semicircular hoops having a thickness of 12 mm and a width of 50 mm. Determine the normal stress in hoop AB if the tank is subjected to an internal gauge pressure of 14 kPa and this loading is transmitted directly to the hoops. Also, if 6-mm-diameter bolts are used to connect each hoop together, determine the tensile stress in each bolt at A and B. Assume hoop AB supports the pressure loading within a 300-mm length of the tank as shown.

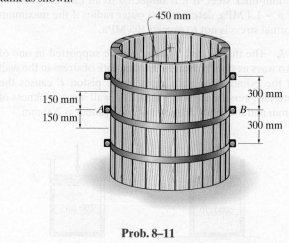

Prob. 8–11

*8–12. Two hemispheres having an inner radius of 0.6 m and wall thickness of 6 mm are fitted together, and the inside gauge pressure is reduced to −70 kPa. If the coefficient of static friction is $\mu_s = 0.5$ between the hemispheres, determine (a) the torque T needed to initiate the rotation of the top hemisphere relative to the bottom one, (b) the vertical force needed to pull the top hemisphere off the bottom one, and (c) the horizontal force needed to slide the top hemisphere off the bottom one.

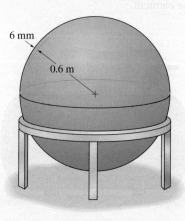

Prob. 8–12

•8–13. The 304 stainless steel band initially fits snugly around the smooth rigid cylinder. If the band is then subjected to a nonlinear temperature drop of $\Delta T = 12 \sin^2 \theta$ °C, where θ is in radians, determine the circumferential stress in the band.

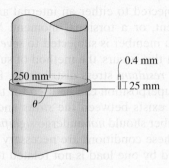

Prob. 8–13

8–14. The ring, having the dimensions shown, is placed over a flexible membrane which is pumped up with a pressure p. Determine the change in the internal radius of the ring after this pressure is applied. The modulus of elasticity for the ring is E.

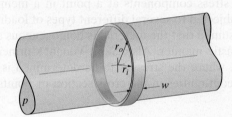

Prob. 8–14

8–15. The inner ring A has an inner radius r_1 and outer radius r_2. Before heating, the outer ring B has an inner radius r_3 and an outer radius r_4, and $r_2 > r_3$. If the outer ring is heated and then fitted over the inner ring, determine the pressure between the two rings when ring B reaches the temperature of the inner ring. The material has a modulus of elasticity of E and a coefficient of thermal expansion of α.

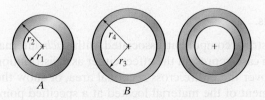

Prob. 8–15

***8–16.** The cylindrical tank is fabricated by welding a strip of thin plate helically, making an angle θ with the longitudinal axis of the tank. If the strip has a width w and thickness t, and the gas within the tank of diameter d is pressured to p, show that the normal stress developed along the strip is given by $\sigma_\theta = (pd/8t)(3 - \cos 2\theta)$.

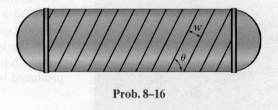

Prob. 8–16

8–17. In order to increase the strength of the pressure vessel, filament winding of the same material is wrapped around the circumference of the vessel as shown. If the pretension in the filament is T and the vessel is subjected to an internal pressure p, determine the hoop stresses in the filament and in the wall of the vessel. Use the free-body diagram shown, and assume the filament winding has a thickness t' and width w for a corresponding length of the vessel.

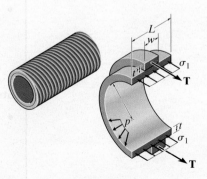

Prob. 8–17

This chimney is subjected to the combined loading of wind and weight. It is important to investigate the tensile stress in the chimney since masonry is weak in tension.

8.2   State of Stress Caused by Combined Loadings

In previous chapters we developed methods for determining the stress distributions in a member subjected to either an internal axial force, a shear force, a bending moment, or a torsional moment. Most often, however, the cross section of a member is subjected to *several* of these loadings *simultaneously*. When this occurs, the method of superposition can be used to determine the *resultant* stress distribution. Recall from Sec. 4.3 that the principle of superposition can be used for this purpose provided a *linear relationship* exists between the *stress* and the *loads*. Also, the geometry of the member should *not* undergo *significant change* when the loads are applied. These conditions are necessary in order to ensure that the stress produced by one load is not related to the stress produced by any other load.

Procedure for Analysis

The following procedure provides a general means for establishing the normal and shear stress components at a point in a member when the member is subjected to several different types of loadings simultaneously. It is assumed that the material is homogeneous and behaves in a linear elastic manner. Also, Saint-Venant's principle requires that the point where the stress is to be determined is far removed from any discontinuities in the cross section or points of applied load.

Internal Loading.

• Section the member perpendicular to its axis at the point where the stress is to be determined and obtain the resultant internal normal and shear force components and the bending and torsional moment components.

• The force components should act through the *centroid* of the cross section, and the moment components should be computed about *centroidal axes*, which represent the principal axes of inertia for the cross section.

Stress Components.

• Determine the stress component associated with *each* internal loading. For each case, represent the effect either as a distribution of stress acting over the entire cross-sectional area, or show the stress on an element of the material located at a specified point on the cross section.

Normal Force.

- The internal normal force is developed by a uniform normal-stress distribution determined from $\sigma = P/A$.

Shear Force.

- The internal shear force in a member is developed by a shear-stress distribution determined from the shear formula, $\tau = VQ/It$. Special care, however, must be exercised when applying this equation, as noted in Sec. 7.2.

Bending Moment.

- For *straight members* the internal bending moment is developed by a normal-stress distribution that varies linearly from zero at the neutral axis to a maximum at the outer boundary of the member. This stress distribution is determined from the flexure formula, $\sigma = My/I$. If the member is *curved*, the stress distribution is nonlinear and is determined from $\sigma = My/[Ae(R - y)]$.

Torsional Moment.

- For circular shafts and tubes the internal torsional moment is developed by a shear-stress distribution that varies linearly from the central axis of the shaft to a maximum at the shaft's outer boundary. This stress distribution is determined from the torsional formula, $\tau = T\rho/J$.

Thin-Walled Pressure Vessels.

- If the vessel is a thin-walled cylinder, the internal pressure p will cause a biaxial state of stress in the material such that the hoop or circumferential stress component is $\sigma_1 = pr/t$ and the longitudinal stress component is $\sigma_2 = pr/2t$. If the vessel is a thin-walled sphere, then the biaxial state of stress is represented by two equivalent components, each having a magnitude of $\sigma_2 = pr/2t$.

Superposition.

- Once the normal and shear stress components for each loading have been calculated, use the principle of superposition and determine the resultant normal and shear stress components.
- Represent the results on an element of material located at the point, or show the results as a distribution of stress acting over the member's cross-sectional area.

Problems in this section, which involve combined loadings, serve as a basic *review* of the application of the stress equations mentioned above. A thorough understanding of how these equations are applied, as indicated in the previous chapters, is necessary if one is to successfully solve the problems at the end of this section. The following examples should be carefully studied before proceeding to solve the problems.

EXAMPLE 8.2

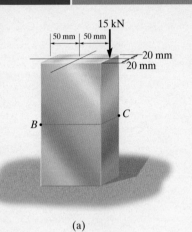

15 kN

50 mm | 50 mm

20 mm
20 mm

B •

• C

(a)

Fig. 8–3

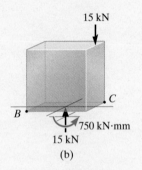

15 kN

• C

B •

750 kN·mm

15 kN

(b)

A force of 15 kN is applied to the edge of the member shown in Fig. 8–3a. Neglect the weight of the member and determine the state of stress at points B and C.

SOLUTION

Internal Loadings. The member is sectioned through B and C. For equilibrium at the section there must be an axial force of 15 kN acting through the *centroid* and a bending moment of 750 kN · mm about the centroidal or principal axis, Fig. 8–3b.

Stress Components.

Normal Force. The uniform normal-stress distribution due to the normal force is shown in Fig. 8–3c. Here

$$\sigma = \frac{P}{A} = \frac{15(10^3)\ \text{N}}{(100\ \text{mm})(40\ \text{mm})} = 3.75\ \text{N/mm}^2 = 3.75\ \text{MPa}$$

Bending Moment. The normal-stress distribution due to the bending moment is shown in Fig. 8–3d. The maximum stress is

$$\sigma_{\max} = \frac{Mc}{I} = \frac{750(10^3)\ \text{N} \cdot \text{mm}(50\ \text{mm})}{\frac{1}{12}(40\ \text{mm})(100\ \text{mm})^3} = 11.25\ \text{N/mm}^2 = 11.25\ \text{MPa}$$

Superposition. If the above normal-stress distributions are added algebraically, the resultant stress distribution is shown in Fig. 8–3e. Although it is not needed here, the location of the line of zero stress can be determined by proportional triangles; i.e.,

$$\frac{7.5\ \text{MPa}}{x} = \frac{15\ \text{MPa}}{(100\ \text{mm} - x)}$$

$$x = 33.3\ \text{mm}$$

Elements of material at B and C are subjected only to normal or *uniaxial stress* as shown in Figs. 8–3f and 8–3g. Hence,

$$\sigma_B = 7.5\ \text{MPa (tension)} \qquad\qquad Ans.$$
$$\sigma_C = 15\ \text{MPa (compression)} \qquad Ans.$$

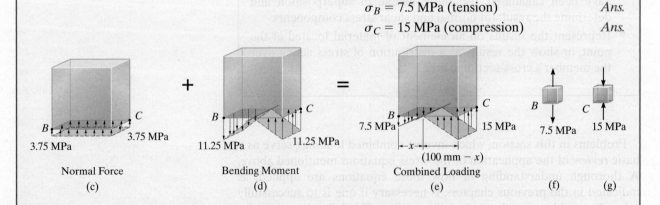

B •

3.75 MPa

• C

3.75 MPa

Normal Force
(c)

+

B •

11.25 MPa

• C

11.25 MPa

Bending Moment
(d)

=

B •
7.5 MPa

• C

15 MPa

x

(100 mm − x)

Combined Loading
(e)

B

7.5 MPa

(f)

C

15 MPa

(g)

EXAMPLE | 8.3

The tank in Fig. 8–4a has an inner radius of 600 mm and a thickness of 12 mm. It is filled to the top with water having a specific weight of $\gamma_w = 10$ kN/m³. If it is made of steel having a specific weight of $\gamma_{st} = 78$ kN/m³ determine the state of stress at point A. The tank is open at the top.

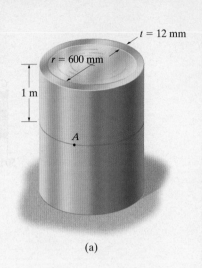

(a)

SOLUTION

Internal Loadings. The free-body diagram of the section of both the tank and the water above point A is shown in Fig. 8–4b. Notice that the weight of the water is supported by the water surface just *below* the section, *not* by the walls of the tank. In the vertical direction, the walls simply hold up the weight of the tank. This weight is

$$W_{st} = \gamma_{st} V_{st} = (78 \text{ kN/m}^3)\left[\pi\left(\frac{612}{1000}\text{m}\right)^2 - \pi\left(\frac{600}{1000}\text{m}\right)^2\right](1\text{ m})$$

$$= 3.56 \text{ kN}$$

The stress in the circumferential direction is developed by the water pressure at level A. To obtain this pressure we must use *Pascal's law*, which states that the pressure at a point located a depth z in the water is $p = \gamma_w z$. Consequently, the pressure on the tank at level A is

$$P = \gamma_w z = (10 \text{ kN/m}^3)(1\text{ m}) = 10 \text{ kN/m}^2 = 10 \text{ kPa}$$

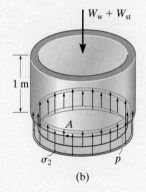

(b)

Stress Components.

Circumferential Stress. Since $r/t = 600$ mm/12 mm $= 50 > 10$, the tank is a thin-walled vessel. Applying Eq. 8–1, using the inner radius $r = 600$ mm, we have

$$\sigma_1 = \frac{pr}{t} = \frac{10 \text{ kPa (600 mm)}}{12 \text{ mm}} = 500 \text{ kPa} \qquad Ans.$$

Longitudinal Stress. Since the weight of the tank is supported uniformly by the walls, we have

$$\sigma_2 = \frac{W_{st}}{A_{st}} = \frac{3.56 \text{ kN}}{\pi[(0.615 \text{ m})^2 - (0.6 \text{ m})^2]} = 77.9 \text{ kN/m}^2 = 77.9 \text{ kPa} \qquad Ans.$$

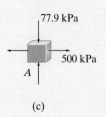

(c)

NOTE: Equation 8–2, $\sigma_2 = pr/2t$, does *not apply* here, since the tank is open at the top and therefore, as stated previously, the water cannot develop a loading on the walls in the longitudinal direction.

Point A is therefore subjected to the biaxial stress shown in Fig. 8–4c.

Fig. 8–4

EXAMPLE 8.4

The member shown in Fig. 8–5a has a rectangular cross section. Determine the state of stress that the loading produces at point C.

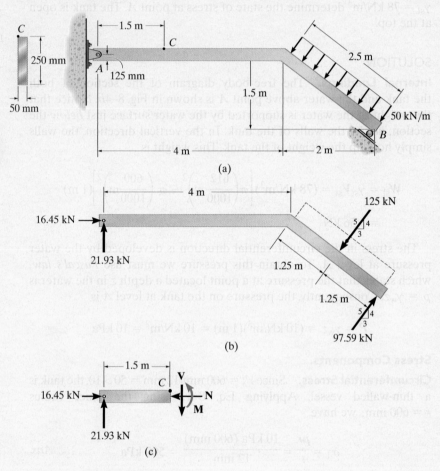

(a)

(b)

(c)

Fig. 8–5

SOLUTION

Internal Loadings. The support reactions on the member have been determined and are shown in Fig. 8–5b. If the left segment AC of the member is considered, Fig. 8–5c, the resultant internal loadings at the section consist of a normal force, a shear force, and a bending moment. Solving,

$$N = 16.45 \text{ kN} \qquad V = 21.93 \text{ kN} \qquad M = 32.89 \text{ kN} \cdot \text{m}$$

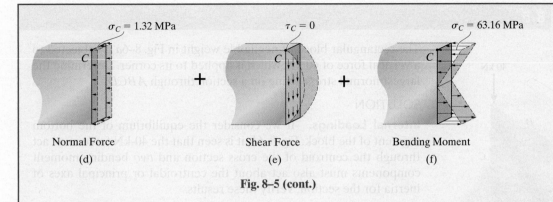

$\sigma_C = 1.32$ MPa $\tau_C = 0$ $\sigma_C = 63.16$ MPa

Normal Force   Shear Force   Bending Moment
   (d)   (e)   (f)

Fig. 8–5 (cont.)

Stress Components.

Normal Force.   The uniform normal-stress distribution acting over the cross section is produced by the normal force, Fig. 8–5d. At point C,

$$\sigma_C = \frac{P}{A} = \frac{16.45(10^3)\text{ N}}{(0.050\text{ m})(0.250\text{ m})} = 1.32 \text{ MPa}$$

Shear Force.   Here the area $A' = 0$, since point C is located at the top of the member. Thus $Q = \bar{y}'A' = 0$ and for C, Fig. 8–5e, the shear stress

$$\tau_C = 0$$

Bending Moment.   Point C is located at $y = c = 0.125$ m from the neutral axis, so the normal stress at C, Fig. 8–5f, is

$$\sigma_C = \frac{Mc}{I} = \frac{(32.89(10^3)\text{ N} \cdot \text{m})(0.125\text{ m})}{\left[\frac{1}{12}(0.050\text{ m})(0.250\text{ m})^3\right]} = 63.16 \text{ MPa}$$

Superposition.   The shear stress is zero. Adding the normal stresses determined above gives a compressive stress at C having a value of

$$\sigma_C = 1.32 \text{ MPa} + 63.16 \text{ MPa} = 64.5 \text{ MPa} \qquad Ans.$$

This result, acting on an element at C, is shown in Fig. 8–5g.

64.5 MPa

(g)

EXAMPLE | 8.5

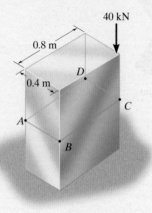

40 kN

0.8 m

D

0.4 m

C

A

B

(a)

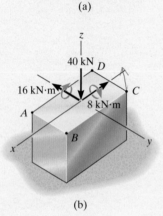

z

40 kN D

16 kN·m

C

8 kN·m

A

B

x

y

(b)

Fig. 8–6

The rectangular block of negligible weight in Fig. 8–6a is subjected to a vertical force of 40 kN, which is applied to its corner. Determine the largest normal stress acting on a section through *ABCD*.

SOLUTION

Internal Loadings. If we consider the equilibrium of the bottom segment of the block, Fig. 8–6b, it is seen that the 40-kN force must act through the centroid of the cross section and *two* bending-moment components must also act about the centroidal or principal axes of inertia for the section. Verify these results.

Stress Components.

Normal Force. The uniform normal-stress distribution is shown in Fig. 8–6c. We have

$$\sigma = \frac{P}{A} = \frac{40(10^3) \text{ N}}{(0.8 \text{ m})(0.4 \text{ m})} = 125 \text{ kPa}$$

Bending Moments. The normal-stress distribution for the 8-kN·m moment is shown in Fig. 8–6d. The maximum stress is

$$\sigma_{max} = \frac{M_x c_y}{I_x} = \frac{8(10^3) \text{ N·m}(0.2 \text{ m})}{\left[\frac{1}{12}(0.8 \text{ m})(0.4 \text{ m})^3\right]} = 375 \text{ kPa}$$

Likewise, for the 16-kN·m moment, Fig. 8–6e, the maximum normal stress is

$$\sigma_{max} = \frac{M_y c_x}{I_y} = \frac{16(10^3) \text{ N·m}(0.4 \text{ m})}{\left[\frac{1}{12}(0.4 \text{ m})(0.8 \text{ m})^3\right]} = 375 \text{ kPa}$$

Superposition. By inspection the normal stress at point *C* is the largest since each loading creates a compressive stress there. Therefore

$$\sigma_C = -125 \text{ kPa} - 375 \text{ kPa} - 375 \text{ kPa} = -875 \text{ kPa} \quad Ans.$$

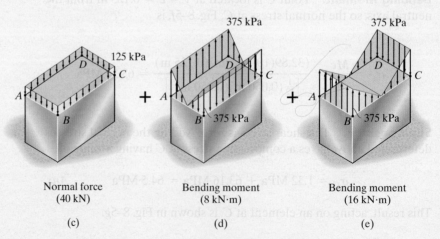

375 kPa   375 kPa

125 kPa

D   C   +   D   C   +   D   C

A   A   A

B   B 375 kPa   B 375 kPa

Normal force   Bending moment   Bending moment
(40 kN)   (8 kN·m)   (16 kN·m)

(c)   (d)   (e)

EXAMPLE | 8.6

A rectangular block has a negligible weight and is subjected to a vertical force **P**, Fig. 8–7a. (a) Determine the range of values for the eccentricity e_y of the load along the y axis so that it does not cause any tensile stress in the block. (b) Specify the region on the cross section where **P** may be applied without causing a tensile stress in the block.

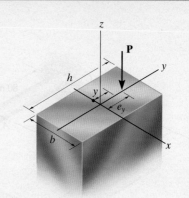

(a)

SOLUTION

Part (a). When **P** is moved to the centroid of the cross section, Fig. 8–7b, it is necessary to add a couple moment $M_x = Pe_y$ in order to maintain a statically equivalent loading. The combined normal stress at any coordinate location y on the cross section caused by these two loadings is

$$\sigma = -\frac{P}{A} - \frac{(Pe_y)y}{I_x} = -\frac{P}{A}\left(1 + \frac{Ae_y y}{I_x}\right)$$

Here the negative sign indicates compressive stress. For positive e_y, Fig. 8–7a, the *smallest* compressive stress will occur along edge AB, where $y = -h/2$, Fig. 8–7b. (By inspection, **P** causes compression there, but $\mathbf{M}_x$ causes tension.) Hence,

$$\sigma_{\min} = -\frac{P}{A}\left(1 - \frac{Ae_y h}{2I_x}\right)$$

This stress will remain negative, i.e., compressive, provided the term in parentheses is positive; i.e.,

$$1 > \frac{Ae_y h}{2I_x}$$

Since $A = bh$ and $I_x = \frac{1}{12}bh^3$, then

$$1 > \frac{6e_y}{h} \quad \text{or} \quad e_y < \frac{1}{6}h \qquad \textit{Ans.}$$

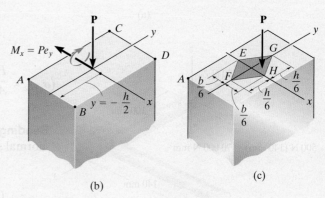

(b)

(c)

Fig. 8–7

In other words, if $-\frac{1}{6}h \le e_y \le \frac{1}{6}h$, the stress in the block along edge AB or CD will be zero or remain *compressive*.

NOTE: This is sometimes referred to as the "*middle-third rule.*" It is very important to keep this rule in mind when loading columns or arches having a rectangular cross section and made of material such as stone or concrete, which can support little or no tensile stress. We can extend this analysis in the same way by placing **P** along the x axis in Fig. 8–7. The result will produce a shaded parallelogram shown in Fig. 8–7c. This region is referred to as the core or kern of the section. When **P** is applied within the kern, the normal stress at the corners of the cross section will be compressive.

Here is an example of where combined axial and bending stress can occur.

8

EXAMPLE 8.7

The solid rod shown in Fig. 8–8a has a radius of 7.5 mm. If it is subjected to the force of 500 N, determine the state of stress at point A.

SOLUTION

Internal Loadings. The rod is sectioned through point A. Using the free-body diagram of segment AB, Fig. 8–8b, the resultant internal loadings are determined from the equations of equilibrium. Verify these results. In order to better "visualize" the stress distributions due to these loadings, we can consider the *equal but opposite resultants* acting on segment AC, Fig. 8–8c.

Stress Components.

Normal Force. The normal-stress distribution is shown in Fig. 8–8d. For point A, we have

$$(\sigma_A)_y = \frac{P}{A} = \frac{500 \text{ N}}{\pi(7.5 \text{ mm})^2} = 2.83 \text{ N/mm}^2 = 2.83 \text{ MPa}$$

Bending Moment. For the moment, c = 7.5 mm, so the normal stress at point A, Fig. 8–8e, is

$$(\sigma_A)_y = \frac{Mc}{I} = \frac{70\ 000 \text{ N} \cdot \text{mm}(7.5 \text{ mm})}{[\frac{1}{4}\pi(7.5 \text{ mm})^4]}$$

$$= 211.3 \text{ N/mm}^2 = 211.3 \text{ MPa}$$

Superposition. When the above results are superimposed, it is seen that an element of material at A is subjected to the normal stress

$$(\sigma_A)_y = 2.83 \text{ Mpa} = 211.3 \text{ MPa} = 214.1 \text{ MPa}$$

Ans.

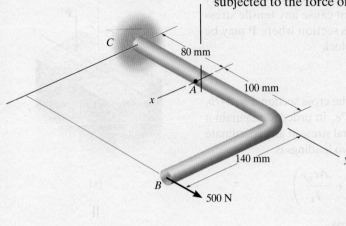

(a)

500 N

x

500 N (140 mm) = 70 000 N·mm

z

100 mm

140 mm

500 N

y

(b)

70 000 N·mm

=

500 N

A

2.83 MPa

Normal force
(500 N)

+

A

211.3 MPa

Bending moment
(70 000 N·mm)

(c)   (d)   (e)

Fig. 8–8

EXAMPLE 8.8

The solid rod shown in Fig. 8–9a has a radius of 7.5 mm. If it is subjected to the force of 800 N, determine the state of stress at point A.

SOLUTION

Internal Loadings.   The rod is sectioned through point A. Using the free-body diagram of segment AB, Fig. 8–9b, the resultant internal loadings are determined from the six equations of equilibrium. Verify these results. The *equal but opposite resultants* are shown acting on segment AC, Fig. 8–9c.

Stress Components.

Shear Force.   The shear-stress distribution is shown in Fig. 8–9d. For point A, Q is determined from the shaded *semicircular* area. Using the table on the inside front cover, we have

$$Q = \bar{y}'A' = \frac{4(7.5 \text{ mm})}{3\pi}\left[\frac{1}{2}\pi(0.75 \text{ mm})^2\right] = 281.25 \text{ mm}^3$$

so that

$$(\tau_{yz})_A = \frac{VQ}{It} = \frac{800 \text{ N}(281.25 \text{ mm}^3)}{[\frac{1}{4}\pi(7.5 \text{ mm})^4]2(7.5 \text{ mm})}$$

$$= 6.04 \text{ N/mm}^2 = 6.04 \text{ MPa}$$

Bending Moment.   Since point A lies on the neutral axis, Fig. 8–9e, the normal stress is

$$\sigma_A = 0$$

Torque.   At point A, $\rho_A = c = 7.5$ mm, Fig. 8–9f. Thus the shear stress is

$$(\tau_{yz})_A = \frac{Tc}{J} = \frac{112\,000 \text{ N} \cdot \text{mm} (7.5 \text{ mm})}{[\frac{1}{2}\pi(7.5 \text{ mm})^4]} = 169.0 \text{ N/mm}^2 = 169.0 \text{ MPa}$$

Superposition.   Here the element of material at A is subjected only to a shear stress component, where

$$(\tau_{yz})_A = 6.04 \text{ MPa} + 169.0 \text{ MPa} = 175.0 \text{ MPa} \qquad Ans.$$

(a)

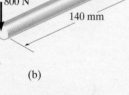

800 N (140 mm) = 112 000 N·mm

800 N (100 mm) = 80 000 N·mm.

(b)

Fig. 8–9

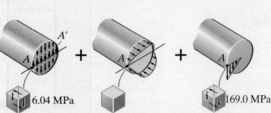

(c)   (d)   (e)   (f)

Shear force   Bending moment   Torsional moment
(800 N)   (80 000 N·mm)   (112 000 N·mm)

6.04 MPa   169.0 MPa

FUNDAMENTAL PROBLEMS

F8–1. Determine the normal stress developed at corners A and B of the column.

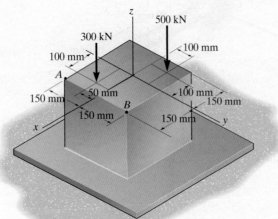

F8–1

F8–3. Determine the state of stress at point A on the cross section of the beam at section a–a.

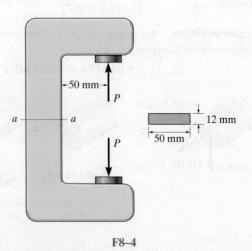

F8–3

F8–2. Determine the state of stress at point A on the cross section at section a–a of the cantilever beam.

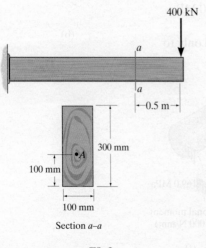

Section a–a

F8–2

F8–4. Determine the magnitude of the load P that will cause a maximum normal stress of $\sigma_{max} = 210$ MPa in the link along section a–a.

F8–4

F8–5. The beam has a rectangular cross section and is subjected to the loading shown. Determine the components of stress σ_x, σ_y, and τ_{xy} at point B.

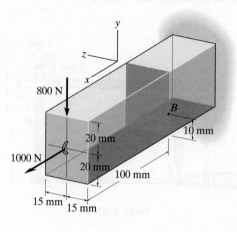

F8–5

F8–6. Determine the state of stress at point A on the cross section of the pipe assembly at section a–a.

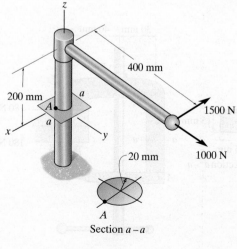

Section a–a

F8–6

F8–7. Determine the state of stress at point A on the cross section of the pipe at section a–a.

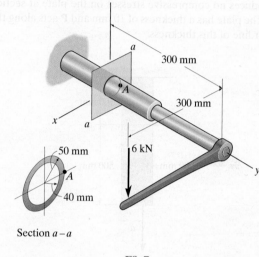

Section a–a

F8–7

F8–8. Determine the state of stress at point A on the cross section of the shaft at section a–a.

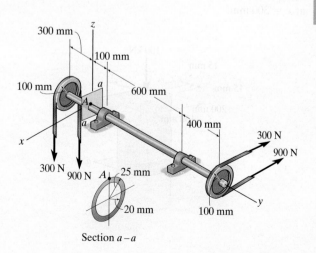

Section a–a

F8–8

PROBLEMS

8–18. The vertical force **P** acts on the bottom of the plate having a negligible weight. Determine the shortest distance d to the edge of the plate at which it can be applied so that it produces no compressive stresses on the plate at section a–a. The plate has a thickness of 10 mm and **P** acts along the center line of this thickness.

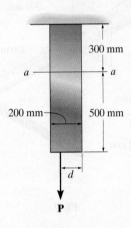

Prob. 8–18

8–19. Determine the maximum and minimum normal stress in the bracket at section a–a when the load is applied at $x = 0$.

***8–20.** Determine the maximum and minimum normal stress in the bracket at section a–a when the load is applied at $x = 300$ mm.

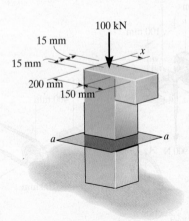

Probs. 8–19/20

•8–21. The coping saw has an adjustable blade that is tightened with a tension of 40 N. Determine the state of stress in the frame at points A and B.

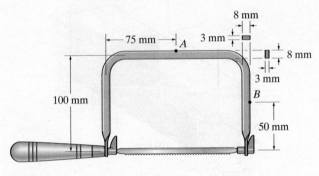

Prob. 8–21

8–22. The clamp is made from members AB and AC, which are pin connected at A. If it exerts a compressive force at C and B of 180 N, determine the maximum compressive stress in the clamp at section a–a. The screw EF is subjected only to a tensile force along its axis.

8–23. The clamp is made from members AB and AC, which are pin connected at A. If it exerts a compressive force at C and B of 180 N, sketch the stress distribution acting over section a–a. The screw EF is subjected only to a tensile force along its axis.

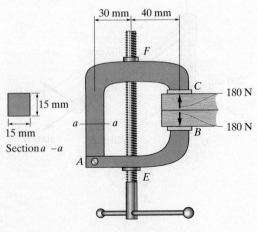

Probs. 8–22/23

*8–24. The bearing pin supports the load of 3.5 kN. Determine the stress components in the support member at point A. The support is 12 mm thick.

•8–25. The bearing pin supports the load of 3.5 kN. Determine the stress components in the support member at point B. The support is 12 mm thick.

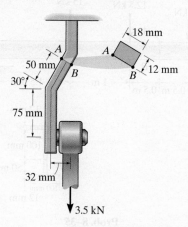

Probs. 8–24/25

8–26. The offset link supports the loading of $P = 30$ kN. Determine its required width w if the allowable normal stress is $\sigma_{\text{allow}} = 73$ MPa. The link has a thickness of 40 mm.

8–27. The offset link has a width of $w = 200$ mm and a thickness of 40 mm. If the allowable normal stress is $\sigma_{\text{allow}} = 75$ MPa, determine the maximum load P that can be applied to the cables.

*8–28. The joint is subjected to a force of $P = 0.4$ kN and $F = 0$. Sketch the normal-stress distribution acting over section a–a if the member has a rectangular cross-sectional area of width 50 mm and thickness 12 mm.

•8–29. The joint is subjected to a force of $P = 1$ kN and $F = 0.75$ kN. Determine the state of stress at points A and B and sketch the results on differential elements located at these points. The member has a rectangular cross-sectional area of width 18 mm and thickness 12 mm.

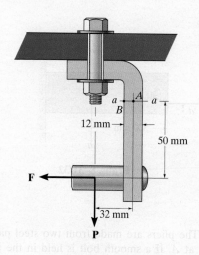

Probs. 8–28/29

8–30. If the 75-kg man stands in the position shown, determine the state of stress at point A on the cross section of the plank at section a–a. The center of gravity of the man is at G. Assume that the contact point at C is smooth.

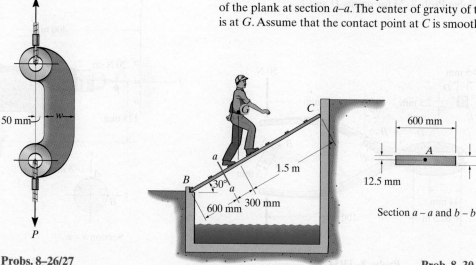

Probs. 8–26/27

Prob. 8–30

8–31. Determine the smallest distance d to the edge of the plate at which the force **P** can be applied so that it produces no compressive stresses in the plate at section a–a. The plate has a thickness of 20 mm and **P** acts along the centerline of this thickness.

*8–32. The horizontal force of $P = 80$ kN acts at the end of the plate. The plate has a thickness of 10 mm and **P** acts along the centerline of this thickness such that $d = 50$ mm. Plot the distribution of normal stress acting along section a–a.

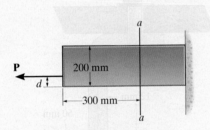

Probs. 8–31/32

•8–33. The pliers are made from two steel parts pinned together at A. If a smooth bolt is held in the jaws and a gripping force of 50 N is applied at the handles, determine the state of stress developed in the pliers at points B and C. Here the cross section is rectangular, having the dimensions shown in the figure.

8–34. Solve Prob. 8–33 for points D and E.

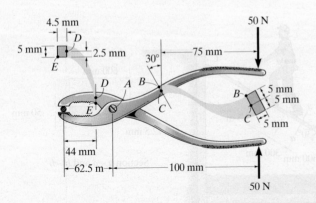

Probs. 8–33/34

8–35. The wide-flange beam is subjected to the loading shown. Determine the stress components at points A and B and show the results on a volume element at each of these points. Use the shear formula to compute the shear stress.

Prob. 8–35

*8–36. The drill is jammed in the wall and is subjected to the torque and force shown. Determine the state of stress at point A on the cross section of drill bit at section a–a.

•8–37. The drill is jammed in the wall and is subjected to the torque and force shown. Determine the state of stress at point B on the cross section of drill bit at section a–a.

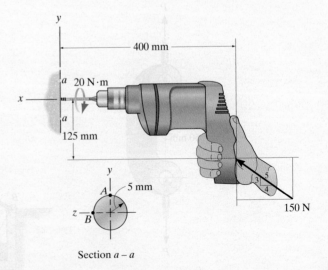

Section a – a

Probs. 8–36/37

8–38. Since concrete can support little or no tension, this problem can be avoided by using wires or rods to *prestress* the concrete once it is formed. Consider the simply supported beam shown, which has a rectangular cross section of 450 mm by 300 mm. If concrete has a specific weight of 24 kN/m³ determine the required tension in rod *AB*, which runs through the beam so that no tensile stress is developed in the concrete at its center section *a–a*. Neglect the size of the rod and any deflection of the beam.

8–39. Solve Prob. 8–38 if the rod has a diameter of 12 mm. Use the transformed area method discussed in Sec. 6.6. $E_{st} = 200$ GPa, $E_c = 25$ GPa.

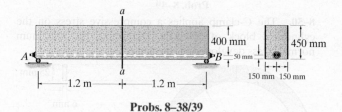

Probs. 8–38/39

*8–40. Determine the state of stress at point *A* when the beam is subjected to the cable force of 4 kN. Indicate the result as a differential volume element.

•8–41. Determine the state of stress at point *B* when the beam is subjected to the cable force of 4 kN. Indicate the result as a differential volume element.

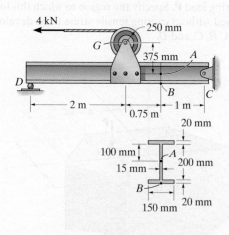

Probs. 8–40/41

8–42. The bar has a diameter of 80 mm. Determine the stress components that act at point *A* and show the results on a volume element located at this point.

8–43. The bar has a diameter of 80 mm. Determine the stress components that act at point *B* and show the results on a volume element located at this point.

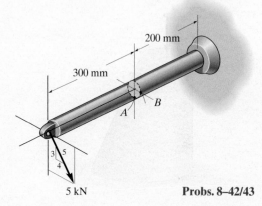

Probs. 8–42/43

*8–44. Determine the normal stress developed at points *A* and *B*. Neglect the weight of the block.

•8–45. Sketch the normal stress distribution acting over the cross section at section *a–a*. Neglect the weight of the block.

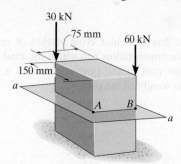

Probs. 8–44/45

8–46. The support is subjected to the compressive load **P**. Determine the absolute maximum and minimum normal stress acting in the material.

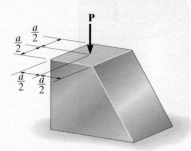

Prob. 8–46

8–47. The support is subjected to the compressive load **P**. Determine the maximum and minimum normal stress acting in the material. All horizontal cross sections are circular.

Prob. 8–47

*****8–48.** The post has a circular cross section of radius *c*. Determine the maximum radius *e* at which the load can be applied so that no part of the post experiences a tensile stress. Neglect the weight of the post.

Prob. 8–48

•**8–49.** If the baby has a mass of 5 kg and his center of mass is at *G*, determine the normal stress at points *A* and *B* on the cross section of the rod at section *a–a*. There are two rods, one on each side of the cradle.

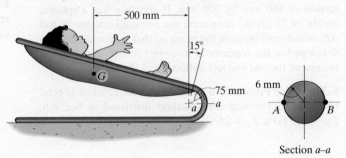

Section *a–a*

Prob. 8–49

8–50. The C-clamp applies a compressive stress on the cylindrical block of 560 kPa. Determine the maximum normal stress developed in the clamp.

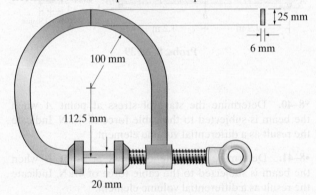

Prob. 8–50

8–51. A post having the dimensions shown is subjected to the bearing load **P**. Specify the region to which this load can be applied without causing tensile stress to be developed at points *A*, *B*, *C*, and *D*.

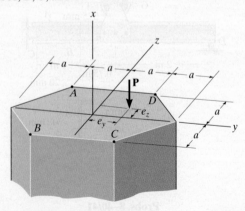

Prob. 8–51

*8–52. The hook is used to lift the force of 3 kN. Determine the maximum tensile and compressive stresses at section *a–a*. The cross section is circular and has a diameter of 25 mm. Use the curved-beam formula to compute the bending stress.

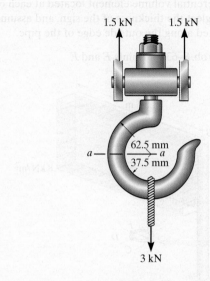

Prob. 8–52

•8–53. The masonry pier is subjected to the 800-kN load. Determine the equation of the line $y = f(x)$ along which the load can be placed without causing a tensile stress in the pier. Neglect the weight of the pier.

8–54. The masonry pier is subjected to the 800-kN load. If $x = 0.25$ m and $y = 0.5$ m, determine the normal stress at each corner A, B, C, D (not shown) and plot the stress distribution over the cross section. Neglect the weight of the pier.

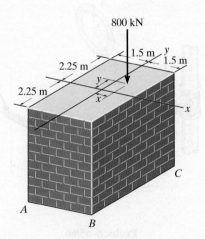

Probs. 8–53/54

8–55. The bar has a diameter of 40 mm. If it is subjected to the two force components at its end as shown, determine the state of stress at point A and show the results on a differential volume element located at this point.

*8–56. Solve Prob. 8–55 for point B.

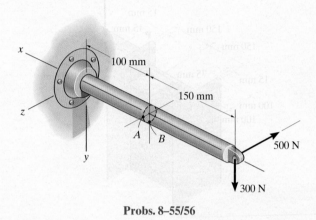

Probs. 8–55/56

•8–57. The 50-mm-diameter rod is subjected to the loads shown. Determine the state of stress at point A, and show the results on a differential element located at this point.

8–58. The 50-mm-diameter rod is subjected to the loads shown. Determine the state of stress at point B, and show the results on a differential element located at this point.

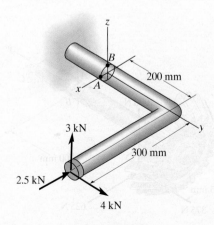

Probs. 8–57/58

8

8–59. If $P = 60$ kN, determine the maximum normal stress developed on the cross section of the column.

***8–60.** Determine the maximum allowable force **P**, if the column is made from material having an allowable normal stress of $\sigma_{allow} = 100$ MPa.

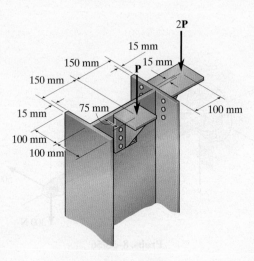

Probs. 8–59/60

•8–61. The beveled gear is subjected to the loads shown. Determine the stress components acting on the shaft at point A, and show the results on a volume element located at this point. The shaft has a diameter of 25 mm and is fixed to the wall at C.

8–62. The beveled gear is subjected to the loads shown. Determine the stress components acting on the shaft at point B, and show the results on a volume element located at this point. The shaft has a diameter of 25 mm and is fixed to the wall at C.

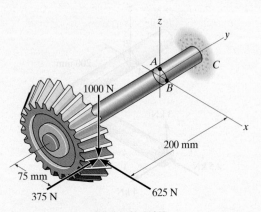

Probs. 8–61/62

8–63. The uniform sign has a weight of 7.5 kN and is supported by the pipe AB, which has an inner radius of 68 mm and an outer radius of 75 mm. If the face of the sign is subjected to a uniform wind pressure of $p = 8$ kN/m^2 determine the state of stress at points C and D. Show the results on a differential volume element located at each of these points. Neglect the thickness of the sign, and assume that it is supported along the outside edge of the pipe.

***8–64.** Solve Prob. 8–63 for points E and F.

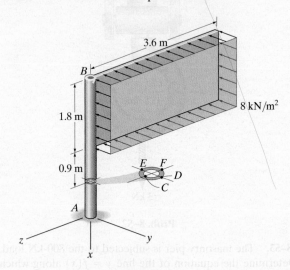

Probs. 8–63/64

•8–65. Determine the state of stress at point A on the cross section of the pipe at section a–a.

8–66. Determine the state of stress at point B on the cross section of the pipe at section a–a.

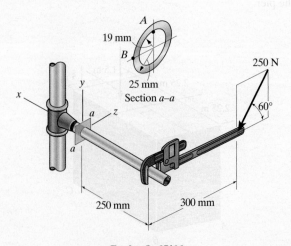

Probs. 8–65/66

•8–67. The eccentric force **P** is applied at a distance e_y from the centroid on the concrete support shown. Determine the range along the y axis where **P** can be applied on the cross section so that no tensile stress is developed in the material.

8–70. The 18-mm-diameter shaft is subjected to the loading shown. Determine the stress components at point A. Sketch the results on a volume element located at this point. The journal bearing at C can exert only force components C_y and C_z on the shaft, and the thrust bearing at D can exert force components D_x, D_y, and D_z on the shaft.

8–71. Solve Prob. 8–70 for the stress components at point B.

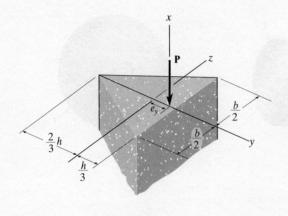

Prob. 8–67

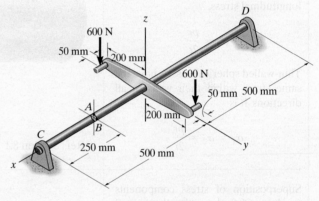

Probs. 8–70/71

*8–68. The bar has a diameter of 40 mm. If it is subjected to a force of 800 N as shown, determine the stress components that act at point A and show the results on a volume element located at this point.

•8–69. Solve Prob. 8–68 for point B.

*8–72. The hook is subjected to the force of 400 N. Determine the state of stress at point A at section a–a. The cross section is circular and has a diameter of 12 mm. Use the curved-beam formula to compute the bending stress.

•8–73. The hook is subjected to the force of 400 N. Determine the state of stress at point B at section a–a. The cross section has a diameter of 12 mm. Use the curved-beam formula to compute the bending stress.

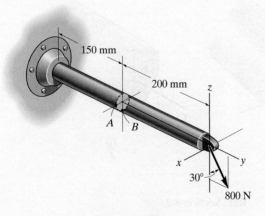

Probs. 8–68/69

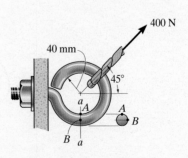

Probs. 8–72/73

CHAPTER REVIEW

A pressure vessel is considered to have a thin wall provided $r/t \geq 10$. For a thin-walled cylindrical vessel, the circumferential or hoop stress is

$$\sigma_1 = \frac{pr}{t}$$

This stress is twice as great as the longitudinal stress,

$$\sigma_2 = \frac{pr}{2t}$$

Thin-walled spherical vessels have the same stress within their walls in all directions. It is

$$\sigma_1 = \sigma_2 = \frac{pr}{2t}$$

Ref.: Section 8.1

Superposition of stress components can be used to determine the normal and shear stress at a point in a member subjected to a combined loading. To do this, it is first necessary to determine the resultant axial and shear forces and the internal resultant torsional and bending moments at the section where the point is located. Then the normal and shear stress resultant components at the point are determined by algebraically adding the normal and shear stress components of each loading.

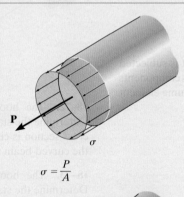

$$\sigma = \frac{P}{A}$$

$$\sigma = \frac{My}{I}$$

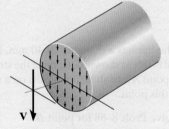

$$\tau = \frac{VQ}{It}$$

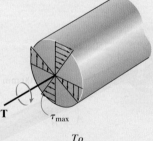

$$\tau = \frac{T\rho}{J}$$

Ref.: Section 8.2

CONCEPTUAL PROBLEMS

P8–1

P8–3

P8–1. Explain why failure of this garden hose occurred near its end and why the tear occurred along its length. Use numerical values to explain your result. Assume the water pressure is 0.2 MPa.

P8–3. Unlike the turnbuckle at B, which is connected along the axis of the rod, the one at A has been welded to the edges of the rod, and so it will be subjected to additional stress. Use the same numerical values for the tensile load in each rod and the rod's diameter, and compare the stress in each rod.

P8–2

P8–4

P8–2. This open-ended silo contains granular material. It is constructed from wood slats and held together with steel bands. Explain, using numerical values, why the bands are not spaced evenly along the height of the cylinder. Also, how would you find this spacing if each band is to be subjected to the same stress?

P8–4. A constant wind blowing against the side of this chimney has caused creeping strains in the mortar joints, such that the chimney has a noticeable deformation. Explain how to obtain the stress distribution over a section at the base of the chimney, and sketch this distribution over the section.

8

REVIEW PROBLEMS

8–74. The block is subjected to the three axial loads shown. Determine the normal stress developed at points A and B. Neglect the weight of the block.

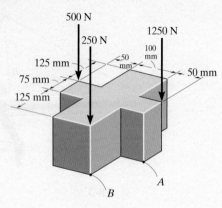

Prob. 8–74

8–75. The 20-kg drum is suspended from the hook mounted on the wooden frame. Determine the state of stress at point E on the cross section of the frame at section a–a. Indicate the results on an element.

*8–76. The 20-kg drum is suspended from the hook mounted on the wooden frame. Determine the state of stress at point F on the cross section of the frame at section b–b. Indicate the results on an element.

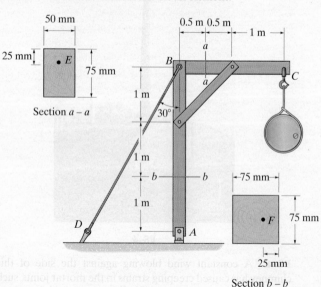

Probs. 8–75/76

•8–77. The eye is subjected to the force of 250 N. Determine the maximum tensile and compressive stresses at section a–a. The cross section is circular and has a diameter of 6 mm. Use the curved-beam formula to compute the bending stress.

8–78. Solve Prob. 8–77 if the cross section is square, having dimensions of 6 mm by 6 mm.

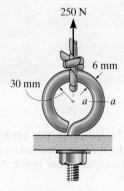

Probs. 8–77/78

8–79. If the cross section of the femur at section a–a can be approximated as a circular tube as shown, determine the maximum normal stress developed on the cross section at section a–a due to the load of 375 N.

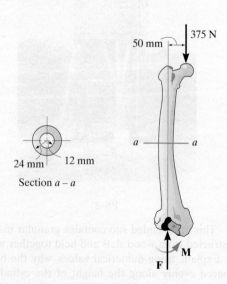

Prob. 8–79

*8–80. The hydraulic cylinder is required to support a force of $P = 100$ kN. If the cylinder has an inner diameter of 100 mm and is made from a material having an allowable normal stress of $\sigma_{\text{allow}} = 150$ MPa, determine the required minimum thickness t of the wall of the cylinder.

•8–81. The hydraulic cylinder has an inner diameter of 100 mm and wall thickness of $t = 4$ mm. If it is made from a material having an allowable normal stress of $\sigma_{\text{allow}} = 150$ MPa, determine the maximum allowable force **P**.

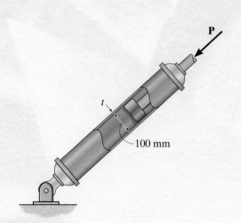

Probs. 8–80/81

8–82. The screw of the clamp exerts a compressive force of 2.5 kN on the wood blocks. Determine the maximum normal stress developed along section a–a. The cross section there is rectangular, 18 mm by 12 mm.

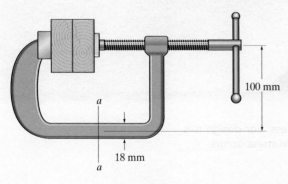

Prob. 8–82

8–83. Air pressure in the cylinder is increased by exerting forces $P = 2$ kN on the two pistons, each having a radius of 45 mm. If the cylinder has a wall thickness of 2 mm, determine the state of stress in the wall of the cylinder.

*8–84. Determine the maximum force P that can be exerted on each of the two pistons so that the circumferential stress component in the cylinder does not exceed 3 MPa. Each piston has a radius of 45 mm and the cylinder has a wall thickness of 2 mm.

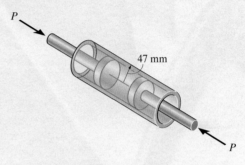

Probs. 8–83/84

•8–85. The cap on the cylindrical tank is bolted to the tank along the flanges. The tank has an inner diameter of 1.5 m and a wall thickness of 18 mm. If the largest normal stress is not to exceed 150 MPa, determine the maximum pressure the tank can sustain. Also, compute the number of bolts required to attach the cap to the tank if each bolt has a diameter of 20 mm. The allowable stress for the bolts is $(\sigma_{\text{allow}})_b = 180$ MPa.

8–86. The cap on the cylindrical tank is bolted to the tank along the flanges. The tank has an inner diameter of 1.5 m and a wall thickness of 18 mm. If the pressure in the tank is $p = 1.20$ MPa, determine the force in each of the 16 bolts that are used to attach the cap to the tank. Also, specify the state of stress in the wall of the tank.

Probs. 8–85/86

8

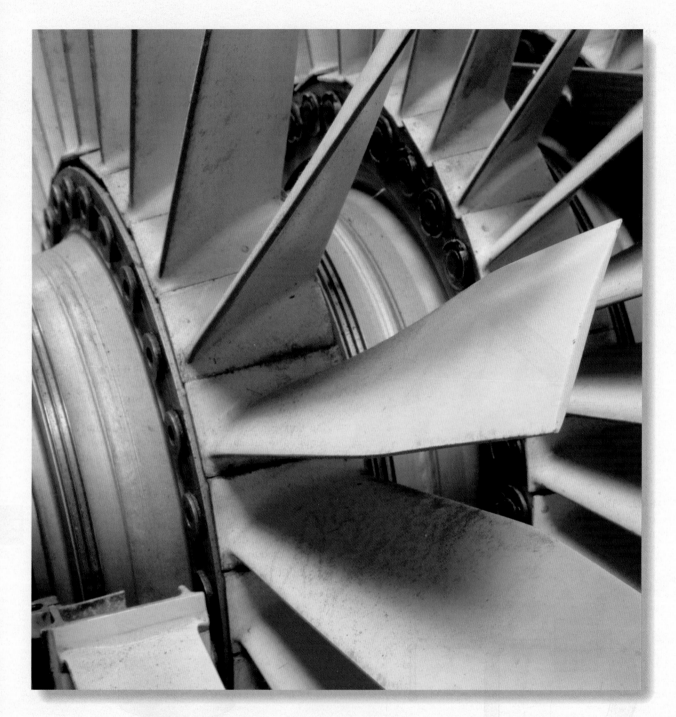

These turbine blades are subjected to a complex pattern of stress. For design it is necessary to determine where and in what direction the maximum stress occurs.

Stress Transformation

9

CHAPTER OBJECTIVES

In this chapter, we will show how to transform the stress components that are associated with a particular coordinate system into components associated with a coordinate system having a different orientation. Once the necessary transformation equations are established, we will then be able to obtain the maximum normal and maximum shear stress at a point and find the orientation of elements upon which they act. Plane-stress transformation will be discussed in the first part of the chapter, since this condition is most common in engineering practice. At the end of the chapter we will discuss a method for finding the absolute maximum shear stress at a point when the material is subjected to both plane and three-dimensional states of stress.

Check out the Companion Website for the following video solution(s) that can be found in this chapter.

- Section 1, 2
  Stress Transformations –
  Transform Equations
- Section 1, 2
  Stress Transformations =
  Stress Summation
- Section 4
  Principal Stresses and Max.
  In-Plane Shear
- Section 4
  Principal Stresses in a Dam
- Section 4
  Principal Stresses – 3-D
  Loading
- Section 7
  Absolute Maximum Shear
  Stress

9.1 Plane-Stress Transformation

It was shown in Sec. 1.3 that the general state of stress at a point is characterized by _six_ independent normal and shear stress components, which act on the faces of an element of material located at the point, Fig. 9–1a. This state of stress, however, is not often encountered in engineering practice. Instead, engineers frequently make approximations or simplifications of the loadings on a body in order that the stress produced in a structural member or mechanical element can be analyzed in a _single plane_. When this is the case, the material is said to be subjected to _plane stress_, Fig. 9–1b. For example, if there is no load on the surface of a body, then the normal and shear stress components will be zero on the face of an element that lies on this surface. Consequently, the corresponding stress components on the opposite face will also be zero, and so the material at the point will be subjected to plane stress. This case occurred throughout the previous chapter.

437

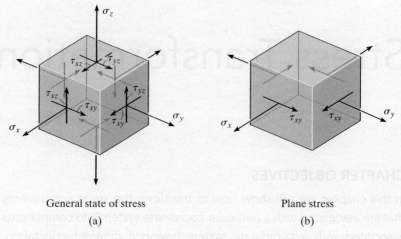

General state of stress

(a)

Plane stress

(b)

Fig. 9–1

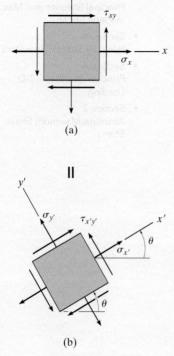

(a)

‖

(b)

Fig. 9–2

The general state of ***plane stress*** at a point is therefore represented by a combination of two normal-stress components, σ_x, σ_y, and one shear-stress component, τ_{xy}, which act on four faces of the element. For convenience, in this text we will view this state of stress in the x–y plane, Fig. 9–2a. If this state of stress is defined on an element having a *different orientation* as in Fig. 9–2b, then it will be subjected to three *different* stress components defined as $\sigma_{x'}$, $\sigma_{y'}$, $\tau_{x'y'}$. In other words, ***the state of plane stress at the point is uniquely represented by two normal stress components and one shear stress component acting on an element that has a specific orientation at the point***.

In this section, we will show how to *transform* the stress components from the orientation of an element in Fig. 9–2a to the orientation of the element in Fig. 9–2b. This is like knowing two force components, say, $\mathbf{F}_x$ and $\mathbf{F}_y$, directed along the x, y axes, that produce a resultant force $\mathbf{F}_R$, and then trying to find the force components $\mathbf{F}_{x'}$ and $\mathbf{F}_{y'}$, directed along the x', y' axes, so they produce the *same* resultant. The transformation for force must only account for the force component's magnitude and direction. The transformation of stress components, however, is more difficult since the transformation must account for the magnitude and direction of each stress component *and* the orientation of the area upon which each component acts.

Procedure for Analysis

If the state of stress at a point is known for a given orientation of an element of material, Fig. 9–3a, then the state of stress in an element having some other orientation, θ, Fig. 9–3b, can be determined using the following procedure.

- To determine the normal and shear stress components $\sigma_{x'}$, $\tau_{x'y'}$ acting on the $+x'$ face of the element, Fig. 9–3b, section the element in Fig. 9–3a as shown in Fig. 9–3c. If the sectioned area is ΔA, then the adjacent areas of the segment will be $\Delta A \sin \theta$ and $\Delta A \cos \theta$.

- Draw the free-body diagram of the segment, which requires showing the *forces* that act on the segment, Fig. 9–3d. This is done by multiplying the stress components on each face by the area upon which they act.

- Apply the force equations of equilibrium in the x' and y' directions. The area ΔA will cancel from the equations and so the two unknown stress components $\sigma_{x'}$ and $\tau_{x'y'}$ can be determined.

- If $\sigma_{y'}$, acting on the $+y'$ face of the element in Fig. 9–3b, is to be determined, then it is necessary to consider a segment of the element as shown in Fig. 9–3e and follow the same procedure just described. Here, however, the shear stress $\tau_{x'y'}$ does not have to be determined if it was previously calculated, since it is complementary, that is, it must have the same magnitude on each of the four faces of the element, Fig. 9–3b.

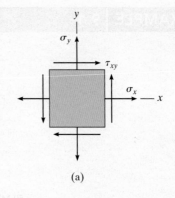

(a)

||

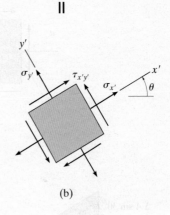

(b)

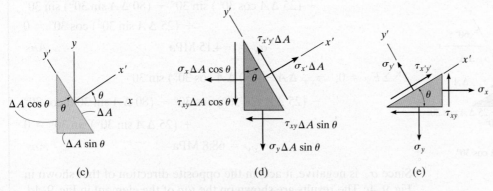

(c)   (d)   (e)

Fig. 9–3

EXAMPLE | 9.1

The state of plane stress at a point on the surface of the airplane fuselage is represented on the element oriented as shown in Fig. 9–4a. Represent the state of stress at the point on an element that is oriented 30° clockwise from the position shown.

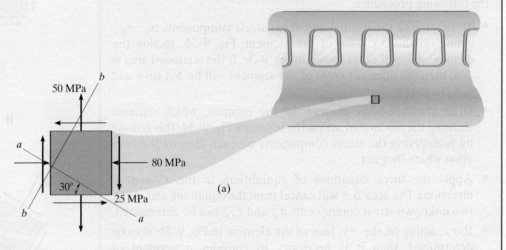

(a)

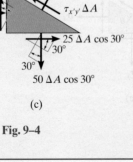

(b)

SOLUTION

The rotated element is shown in Fig. 9–4d. To obtain the stress component on this element we will first section the element in Fig. 9–4a by the line a–a. The bottom segment is removed, and assuming the sectioned (inclined) plane has an area ΔA, the horizontal and vertical planes have the areas shown in Fig. 9–4b. The free-body diagram of this segment is shown in Fig. 9–4c. Applying the equations of force equilibrium in the x' and y' directions to avoid a simultaneous solution for the two unknowns $\sigma_{x'}$ and $\tau_{x'y'}$, we have

$$+\nearrow \Sigma F_{x'} = 0; \quad \sigma_{x'} \Delta A - (50 \, \Delta A \cos 30°) \cos 30°$$
$$+ (25 \, \Delta A \cos 30°) \sin 30° + (80 \, \Delta A \sin 30°) \sin 30°$$
$$+ (25 \, \Delta A \sin 30°) \cos 30° = 0$$
$$\sigma_{x'} = -4.15 \text{ MPa} \qquad Ans.$$

$$+\nwarrow \Sigma F_{y'} = 0; \quad \tau_{x'y'} \Delta A - (50 \, \Delta A \cos 30°) \sin 30°$$
$$- (25 \, \Delta A \cos 30°) \cos 30° - (80 \, \Delta A \sin 30°) \cos 30°$$
$$+ (25 \, \Delta A \sin 30°) \sin 30° = 0$$
$$\tau_{x'y'} = 68.8 \text{ MPa} \qquad Ans.$$

Since $\sigma_{x'}$ is negative, it acts in the opposite direction of that shown in Fig. 9–4c. The results are shown on the *top* of the element in Fig. 9–4d, since this surface is the one considered in Fig. 9–4c.

(c)

Fig. 9–4

We must now repeat the procedure to obtain the stress on the *perpendicular* plane *b–b*. Sectioning the element in Fig. 9–4a along *b–b* results in a segment having sides with areas shown in Fig. 9–4e. Orienting the $+x'$ axis outward, perpendicular to the sectioned face, the associated free-body diagram is shown in Fig. 9–4f. Thus,

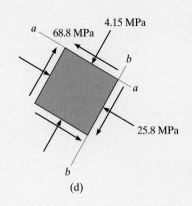

(d)

$$+\searrow \Sigma F_{x'} = 0; \quad \sigma_{x'} \, \Delta A - (25 \, \Delta A \cos 30°) \sin 30°$$
$$+ (80 \, \Delta A \cos 30°) \cos 30° - (25 \, \Delta A \sin 30°) \cos 30°$$
$$- (50 \, \Delta A \sin 30°) \sin 30° = 0$$
$$\sigma_{x'} = -25.8 \text{ MPa} \qquad\qquad Ans.$$

$$+\nearrow \Sigma F_{y'} = 0; \quad -\tau_{x'y'} \, \Delta A + (25 \, \Delta A \cos 30°) \cos 30°$$
$$+ (80 \, \Delta A \cos 30°) \sin 30° - (25 \, \Delta A \sin 30°) \sin 30°$$
$$+ (50 \, \Delta A \sin 30°) \cos 30° = 0$$
$$\tau_{x'y'} = 68.8 \text{ MPa} \qquad\qquad Ans.$$

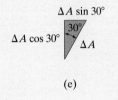

(e)

Since $\sigma_{x'}$ is a negative quantity, it acts opposite to its direction shown in Fig. 9–4f. The stress components are shown acting on the *right side* of the element in Fig. 9–4d.

From this analysis we may therefore conclude that the state of stress at the point can be represented by choosing an element oriented as shown in Fig. 9–4a, or by choosing one oriented as shown in Fig. 9–4d. In other words, these states of stress are equivalent.

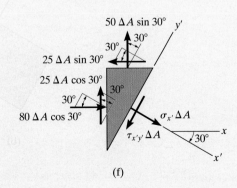

(f)

9.2   General Equations of Plane-Stress Transformation

The method of transforming the normal and shear stress components from the x, y to the x', y' coordinate axes, as discussed in the previous section, can be developed in a general manner and expressed as a set of stress-transformation equations.

Sign Convention.   First we must establish a sign convention for the stress components. To do this the $+x$ and $+x'$ axes are used to define the outward normal from a side of the element. Then σ_x and $\sigma_{x'}$ are positive when they act in the positive x and x' directions, and τ_{xy} and $\tau_{x'y'}$ are positive when they act in the positive y and y' directions, Fig. 9–5.

The orientation of the plane on which the normal and shear stress components are to be determined will be defined by the angle θ, which is measured from the $+x$ axis to the $+x'$ axis, Fig. 9–5b. Notice that the unprimed and primed sets of axes in this figure both form right-handed coordinate systems; that is, the positive z (or z') axis is established by the right-hand rule. Curling the fingers from x (or x') toward y (or y') gives the direction for the positive z (or z') axis that points outward, along the thumb. The *angle* θ will be *positive* provided it follows the curl of the right-hand fingers, i.e., counterclockwise as shown in Fig. 9–5b.

Normal   and   Shear   Stress   Components.   Using the established sign convention, the element in Fig. 9–6a is sectioned along the inclined plane and the segment shown in Fig. 9–6b is isolated. Assuming the sectioned area is ΔA, then the horizontal and vertical faces of the segment have an area of $\Delta A \sin \theta$ and $\Delta A \cos \theta$, respectively.

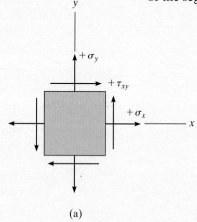

(a)

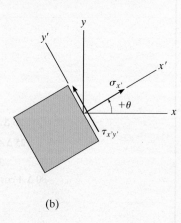

(b)

Positive Sign Convention

Fig. 9–5

The resulting *free-body diagram* of the segment is shown in Fig. 9–6c. Applying the equations of equilibrium to determine the unknown normal and shear stress components $\sigma_{x'}$ and $\tau_{x'y'}$, we have

$$+\nearrow \Sigma F_{x'} = 0; \quad \sigma_{x'}\,\Delta A - (\tau_{xy}\,\Delta A \sin\theta)\cos\theta - (\sigma_y\,\Delta A \sin\theta)\sin\theta$$
$$- (\tau_{xy}\,\Delta A \cos\theta)\sin\theta - (\sigma_x\,\Delta A \cos\theta)\cos\theta = 0$$
$$\sigma_{x'} = \sigma_x\cos^2\theta + \sigma_y\sin^2\theta + \tau_{xy}(2\sin\theta\cos\theta)$$

$$+\nwarrow \Sigma F_{y'} = 0; \quad \tau_{x'y'}\,\Delta A + (\tau_{xy}\,\Delta A \sin\theta)\sin\theta - (\sigma_y\,\Delta A \sin\theta)\cos\theta$$
$$- (\tau_{xy}\,\Delta A \cos\theta)\cos\theta + (\sigma_x\,\Delta A \cos\theta)\sin\theta = 0$$
$$\tau_{x'y'} = (\sigma_y - \sigma_x)\sin\theta\cos\theta + \tau_{xy}(\cos^2\theta - \sin^2\theta)$$

These two equations may be simplified by using the trigonometric identities $\sin 2\theta = 2\sin\theta\cos\theta$, $\sin^2\theta = (1 - \cos 2\theta)/2$, and $\cos^2\theta = (1 + \cos 2\theta)/2$, in which case,

$$\boxed{\sigma_{x'} = \frac{\sigma_x + \sigma_y}{2} + \frac{\sigma_x - \sigma_y}{2}\cos 2\theta + \tau_{xy}\sin 2\theta} \qquad (9\text{–}1)$$

$$\boxed{\tau_{x'y'} = -\frac{\sigma_x - \sigma_y}{2}\sin 2\theta + \tau_{xy}\cos 2\theta} \qquad (9\text{–}2)$$

If the normal stress acting in the y' direction is needed, it can be obtained by simply substituting $(\theta = \theta + 90°)$ for θ into Eq. 9–1, Fig. 9–6d. This yields

$$\sigma_{y'} = \frac{\sigma_x + \sigma_y}{2} - \frac{\sigma_x - \sigma_y}{2}\cos 2\theta - \tau_{xy}\sin 2\theta \qquad (9\text{–}3)$$

If $\sigma_{y'}$ is calculated as a positive quantity, it indicates that it acts in the positive y' direction as shown in Fig. 9–6d.

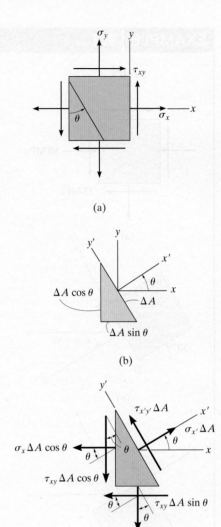

(a)

(b)

(c)

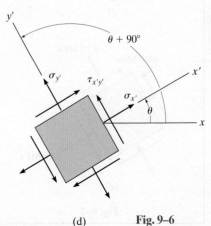

(d)   **Fig. 9–6**

Procedure for Analysis

To apply the stress transformation Eqs. 9–1 and 9–2, it is simply necessary to substitute in the known data for σ_x, σ_y, τ_{xy}, and θ in accordance with the established sign convention, Fig. 9–5. If $\sigma_{x'}$ and $\tau_{x'y'}$ are calculated as positive quantities, then these stresses act in the positive direction of the x' and y' axes.

For convenience, these equations can easily be programmed on a pocket calculator.

EXAMPLE │ 9.2

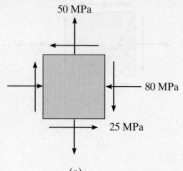

(a)

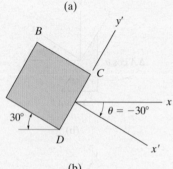

(b)

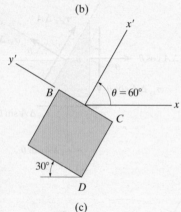

(c)

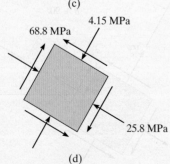

(d)

Fig. 9–7

The state of plane stress at a point is represented by the element shown in Fig. 9–7a. Determine the state of stress at the point on another element oriented 30° clockwise from the position shown.

SOLUTION

This problem was solved in Example 9.1 using basic principles. Here we will apply Eqs. 9–1 and 9–2. From the established sign convention, Fig. 9–5, it is seen that

$$\sigma_x = -80 \text{ MPa} \qquad \sigma_y = 50 \text{ MPa} \qquad \tau_{xy} = -25 \text{ MPa}$$

Plane CD. To obtain the stress components on plane CD, Fig. 9–7b, the positive x' axis is directed outward, perpendicular to CD, and the associated y' axis is directed along CD. The angle measured from the x to the x' axis is $\theta = -30°$ (clockwise). Applying Eqs. 9–1 and 9–2 yields

$$\sigma_{x'} = \frac{\sigma_x + \sigma_y}{2} + \frac{\sigma_x - \sigma_y}{2} \cos 2\theta + \tau_{xy} \sin 2\theta$$

$$= \frac{-80 + 50}{2} + \frac{-80 - 50}{2} \cos 2(-30°) + (-25) \sin 2(-30°)$$

$$= -25.8 \text{ MPa} \qquad\qquad Ans.$$

$$\tau_{x'y'} = -\frac{\sigma_x - \sigma_y}{2} \sin 2\theta + \tau_{xy} \cos 2\theta$$

$$= -\frac{-80 - 50}{2} \sin 2(-30°) + (-25) \cos 2(-30°)$$

$$= -68.8 \text{ MPa} \qquad\qquad Ans.$$

The negative signs indicate that $\sigma_{x'}$ and $\tau_{x'y'}$ act in the negative x' and y' directions, respectively. The results are shown acting on the element in Fig. 9–7d.

Plane BC. In a similar manner, the stress components acting on face BC, Fig. 9–7c, are obtained using $\theta = 60°$. Applying Eqs. 9–1 and 9–2,* we get

$$\sigma_{x'} = \frac{-80 + 50}{2} + \frac{-80 - 50}{2} \cos 2(60°) + (-25) \sin 2(60°)$$

$$= -4.15 \text{ MPa} \qquad\qquad Ans.$$

$$\tau_{x'y'} = -\frac{-80 - 50}{2} \sin 2(60°) + (-25) \cos 2(60°)$$

$$= 68.8 \text{ MPa} \qquad\qquad Ans.$$

Here $\tau_{x'y'}$ has been calculated twice in order to provide a check. The negative sign for $\sigma_{x'}$ indicates that this stress acts in the negative x' direction, Fig. 9–7c. The results are shown on the element in Fig. 9–7d.

*Alternatively, we could apply Eq. 9–3 with $\theta = -30°$ rather than Eq. 9–1.

9.3 Principal Stresses and Maximum In-Plane Shear Stress

From Eqs. 9–1 and 9–2, it can be seen that the magnitudes of $\sigma_{x'}$ and $\tau_{x'y'}$ depend on the angle of inclination θ of the planes on which these stresses act. In engineering practice it is often important to determine the orientation of the element that causes the normal stress to be a maximum and a minimum and the orientation that causes the shear stress to be a maximum. In this section each of these problems will be considered.

In-Plane Principal Stresses. To determine the maximum and minimum *normal stress*, we must differentiate Eq. 9–1 with respect to θ and set the result equal to zero. This gives

$$\frac{d\sigma_{x'}}{d\theta} = -\frac{\sigma_x - \sigma_y}{2}(2 \sin 2\theta) + 2\tau_{xy} \cos 2\theta = 0$$

Solving this equation we obtain the orientation $\theta = \theta_p$ of the planes of maximum and minimum normal stress.

$$\tan 2\theta_p = \frac{\tau_{xy}}{(\sigma_x - \sigma_y)/2} \tag{9–4}$$

The solution has two roots, θ_{p_1}, and θ_{p_2}. Specifically, the values of $2\theta_{p_1}$ and $2\theta_{p_2}$ are 180° apart, so θ_{p_1} and θ_{p_2} will be 90° apart.

The values of θ_{p_1} and θ_{p_2} must be substituted into Eq. 9–1 if we are to obtain the required normal stresses. To do this we can obtain the necessary sine and cosine of $2\theta_{p_1}$ and $2\theta_{p_2}$ from the shaded triangles shown in Fig. 9–8. The construction of these triangles is based on Eq. 9–4, assuming that τ_{xy} and $(\sigma_x - \sigma_y)$ are both positive or both negative quantities.

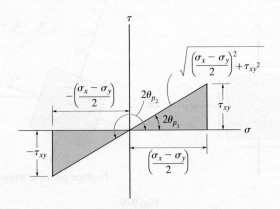

Fig. 9–8

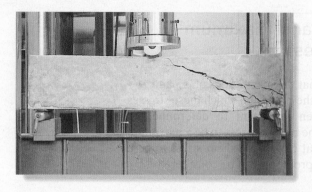

The cracks in this concrete beam were caused by tension stress, even though the beam was subjected to both an internal moment and shear. The stress-transformation equations can be used to predict the direction of the cracks, and the principal normal stresses that caused them.

Substituting these trigonometric values into Eq. 9–1 and simplifying, we obtain

$$\sigma_{1,2} = \frac{\sigma_x + \sigma_y}{2} \pm \sqrt{\left(\frac{\sigma_x - \sigma_y}{2}\right)^2 + \tau_{xy}^2} \qquad (9\text{–}5)$$

Depending upon the sign chosen, this result gives the maximum or minimum in-plane normal stress acting at a point, where $\sigma_1 \geq \sigma_2$. This particular set of values are called the in-plane ***principal stresses***, and the corresponding planes on which they act are called the ***principal planes*** of stress, Fig. 9–9. Furthermore, if the trigonometric relations for θ_{p_1} or θ_{p_2} are substituted into Eq. 9–2, it can be seen that $\tau_{x'y'} = 0$; in other words, ***no shear stress acts on the principal planes***.

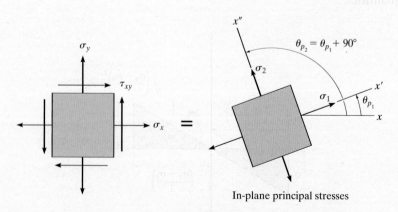

In-plane principal stresses

Fig. 9–9

Maximum In-Plane Shear Stress.

The orientation of an element that is subjected to maximum shear stress on its sides can be determined by taking the derivative of Eq. 9–2 with respect to θ and setting the result equal to zero. This gives

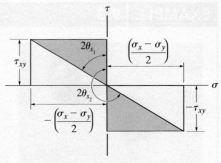

Fig. 9–10

$$\tan 2\theta_s = \frac{-(\sigma_x - \sigma_y)/2}{\tau_{xy}} \qquad (9\text{–}6)$$

The two roots of this equation, θ_{s_1} and θ_{s_2}, can be determined from the shaded triangles shown in Fig. 9–10. By comparison with Eq. 9–4, $\tan 2\theta_s$ is the negative reciprocal of $\tan 2\theta_p$ and so each root $2\theta_s$ is 90° from $2\theta_p$, and the roots θ_s and θ_p are 45° apart. Therefore, an element subjected to *maximum shear stress will be 45° from the position of an element that is subjected to the principal stress*.

Using either one of the roots θ_{s_1} or θ_{s_2}, the maximum shear stress can be found by taking the trigonometric values of $\sin 2\theta_s$ and $\cos 2\theta_s$ from Fig. 9–10 and substituting them into Eq. 9–2. The result is

$$\tau_{\substack{\max \\ \text{in-plane}}} = \sqrt{\left(\frac{\sigma_x - \sigma_y}{2}\right)^2 + \tau_{xy}{}^2} \qquad (9\text{–}7)$$

The value of $\tau_{\substack{\max \\ \text{in-plane}}}$ as calculated from this equation is referred to as the *maximum in-plane shear stress* because it acts on the element in the *x–y* plane.

Substituting the values for $\sin 2\theta_s$ and $\cos 2\theta_s$ into Eq. 9–1, we see that there is *also* an average normal stress on the planes of maximum in-plane shear stress. We get

$$\sigma_{\text{avg}} = \frac{\sigma_x + \sigma_y}{2} \qquad (9\text{–}8)$$

Like the stress-transformation equations, it may be convenient to program Eqs. 9–4 through 9–8 for use on a pocket calculator.

Important Points

- The *principal stresses* represent the maximum and minimum normal stress at the point.

- When the state of stress is represented by the principal stresses, *no shear stress* will act on the element.

- The state of stress at the point can also be represented in terms of the *maximum in-plane shear stress*. In this case an *average normal stress* will also act on the element.

- The element representing the maximum in-plane shear stress with the associated average normal stresses is *oriented 45° from the element representing the principal stresses*.

9

EXAMPLE 9.3

Notice how the failure plane is at an angle (23.7°) due to tearing of the material, Fig. 9–11c.

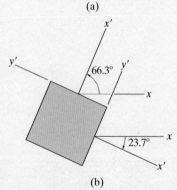

(a)

(b)

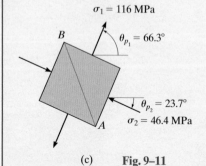

(c)   **Fig. 9–11**

The state of plane stress at a failure point on the shaft is shown on the element in Fig. 9–11a. Represent this stress state in terms of the principal stresses.

SOLUTION

From the established sign convention, we have

$$\sigma_x = -20 \text{ MPa} \qquad \sigma_y = 90 \text{ MPa} \qquad \tau_{xy} = 60 \text{ MPa}$$

Orientation of Element.   Applying Eq. 9–4,

$$\tan 2\theta_p = \frac{\tau_{xy}}{(\sigma_x - \sigma_y)/2} = \frac{60}{(-20 - 90)/2}$$

Solving, and referring to this root as θ_{p_2}, as will be shown below, yields

$$2\theta_{p_2} = -47.49° \qquad \theta_{p_2} = -23.7°$$

Since the difference between $2\theta_{p_1}$ and $2\theta_{p_2}$ is 180°, we have

$$2\theta_{p_1} = 180° + 2\theta_{p_2} = 132.51° \qquad \theta_{p_1} = 66.3°$$

Recall that θ is measured positive *counterclockwise* from the x axis to the outward normal (x' axis) on the face of the element, and so the results are shown in Fig. 9–11b.

Principal Stress.   We have

$$\sigma_{1,2} = \frac{\sigma_x + \sigma_y}{2} \pm \sqrt{\left(\frac{\sigma_x - \sigma_y}{2}\right)^2 + \tau_{xy}^2}$$

$$= \frac{-20 + 90}{2} \pm \sqrt{\left(\frac{-20 - 90}{2}\right)^2 + (60)^2}$$

$$= 35.0 \pm 81.4$$

$$\sigma_1 = 116 \text{ MPa} \qquad\qquad\qquad\qquad\qquad Ans.$$

$$\sigma_2 = -46.4 \text{ MPa} \qquad\qquad\qquad\qquad\qquad Ans.$$

The principal plane on which each normal stress acts can be determined by applying Eq. 9–1 with, say, $\theta = \theta_{p_2} = -23.7°$. We have

$$\sigma_{x'} = \frac{\sigma_x + \sigma_y}{2} + \frac{\sigma_x - \sigma_y}{2} \cos 2\theta + \tau_{xy} \sin 2\theta$$

$$= \frac{-20 + 90}{2} + \frac{-20 - 90}{2} \cos 2(-23.7°) + 60 \sin 2(-23.7°)$$

$$= -46.4 \text{ MPa}$$

Hence, $\sigma_2 = -46.4$ MPa acts on the plane defined by $\theta_{p_2} = -23.7°$, whereas $\sigma_1 = 116$ MPa acts on the plane defined by $\theta_{p_1} = 66.3°$. The results are shown on the element in Fig. 9–11c. Recall that no shear stress acts on this element.

EXAMPLE | 9.4

The state of plane stress at a point on a body is represented on the element shown in Fig. 9–12a. Represent this stress state in terms of the maximum in-plane shear stress and associated average normal stress.

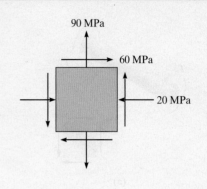

(a)

SOLUTION

Orientation of Element. Since $\sigma_x = -20$ MPa, $\sigma_y = 90$ MPa, and $\tau_{xy} = 60$ MPa, applying Eq. 9–6, we have

$$\tan 2\theta_s = \frac{-(\sigma_x - \sigma_y)/2}{\tau_{xy}} = \frac{-(-20 - 90)/2}{60}$$

$$2\theta_{s_2} = 42.5° \qquad \theta_{s_2} = 21.3°$$

$$2\theta_{s_1} = 180° + 2\theta_{s_2} \qquad \theta_{s_1} = 111.3°$$

Note that these angles shown in Fig. 9–12b are 45° away from the principal planes of stress, which were determined in Example 9.3.

Maximum In-Plane Shear Stress. Applying Eq. 9–7,

$$\tau_{\substack{\max \\ \text{in-plane}}} = \sqrt{\left(\frac{\sigma_x - \sigma_y}{2}\right)^2 + \tau_{xy}^2} = \sqrt{\left(\frac{-20 - 90}{2}\right)^2 + (60)^2}$$

$$= \pm 81.4 \text{ MPa} \qquad\qquad\qquad Ans.$$

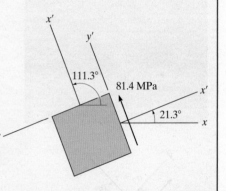

(b)

The proper direction of $\tau_{\substack{\max \\ \text{in-plane}}}$ on the element can be determined by substituting $\theta = \theta_{s_2} = 21.3°$ into Eq. 9–2. We have

$$\tau_{x'y'} = -\left(\frac{\sigma_x - \sigma_y}{2}\right) \sin 2\theta + \tau_{xy} \cos 2\theta$$

$$= -\left(\frac{-20 - 90}{2}\right) \sin 2(21.3°) + 60 \cos 2(21.3°)$$

$$= 81.4 \text{ MPa}$$

This positive result indicates that $\tau_{\substack{\max \\ \text{in-plane}}} = \tau_{x'y'}$ acts in the *positive y'* direction on this face ($\theta = 21.3°$) Fig. 9–12b. The shear stresses on the other three faces are directed as shown in Fig. 9–12c.

Average Normal Stress. Besides the maximum shear stress, as calculated above, the element is also subjected to an average normal stress determined from Eq. 9–8; that is,

$$\sigma_{\text{avg}} = \frac{\sigma_x + \sigma_y}{2} = \frac{-20 + 90}{2} = 35 \text{ MPa} \qquad Ans.$$

This is a tensile stress. The results are shown in Fig. 9–12c.

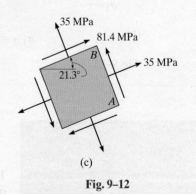

(c)

Fig. 9–12

9

EXAMPLE | **9.5**

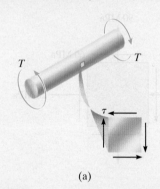

(a)

When the torsional loading T is applied to the bar in Fig. 9–13a, it produces a state of pure shear stress in the material. Determine (a) the maximum in-plane shear stress and the associated average normal stress, and (b) the principal stress.

SOLUTION

From the established sign convention,

$$\sigma_x = 0 \qquad \sigma_y = 0 \qquad \tau_{xy} = -\tau$$

Maximum In-Plane Shear Stress.   Applying Eqs. 9–7 and 9–8, we have

$$\tau_{\substack{max \\ in\text{-}plane}} = \sqrt{\left(\frac{\sigma_x - \sigma_y}{2}\right)^2 + \tau_{xy}^2} = \sqrt{(0)^2 + (-\tau)^2} = \pm\tau \quad Ans.$$

$$\sigma_{avg} = \frac{\sigma_x + \sigma_y}{2} = \frac{0+0}{2} = 0 \qquad\qquad Ans.$$

Thus, as expected, the maximum in-plane shear stress is represented by the element in Fig. 9–13a.

NOTE: Through experiment it has been found that materials that are *ductile* will *fail* due to *shear stress*. As a result, if the bar in Fig. 9–13a is made of mild steel, the maximum in-plane shear stress will cause it to fail as shown in the adjacent photo.

Principal Stress.   Applying Eqs. 9–4 and 9–5 yields

$$\tan 2\theta_p = \frac{\tau_{xy}}{(\sigma_x - \sigma_y)/2} = \frac{-\tau}{(0-0)/2}, \quad \theta_{p_2} = 45°, \theta_{p_1} = -45°$$

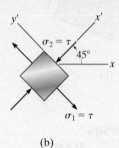

(b)

Fig. 9–13

$$\sigma_{1,2} = \frac{\sigma_x + \sigma_y}{2} \pm \sqrt{\left(\frac{\sigma_x - \sigma_y}{2}\right)^2 + \tau_{xy}^2} = 0 \pm \sqrt{(0)^2 + \tau^2} = \pm\tau \qquad Ans.$$

If we now apply Eq. 9–1 with $\theta_{p_2} = 45°$, then

$$\sigma_{x'} = \frac{\sigma_x + \sigma_y}{2} + \frac{\sigma_x - \sigma_y}{2}\cos 2\theta + \tau_{xy}\sin 2\theta$$

$$= 0 + 0 + (-\tau)\sin 90° = -\tau$$

Thus, $\sigma_2 = -\tau$ acts at $\theta_{p_2} = 45°$ as shown in Fig. 9–13b, and $\sigma_1 = \tau$ acts on the other face, $\theta_{p_1} = -45°$.

NOTE: Materials that are *brittle* fail due to *normal stress*. Therefore, if the bar in Fig. 9–13a is made of cast iron it will fail in tension at a 45° inclination as seen in the adjacent photo.

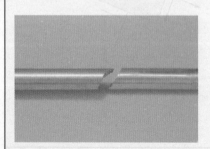

EXAMPLE | 9.6

When the axial loading P is applied to the bar in Fig. 9–14a, it produces a tensile stress in the material. Determine (a) the principal stress and (b) the maximum in-plane shear stress and associated average normal stress.

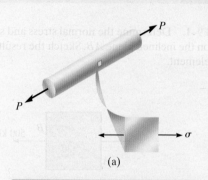

(a)

SOLUTION
From the established sign convention,

$$\sigma_x = \sigma \qquad \sigma_y = 0 \qquad \tau_{xy} = 0$$

Principal Stress. By observation, the element oriented as shown in Fig. 9–14a illustrates a condition of principal stress since no shear stress acts on this element. This can also be shown by direct substitution of the above values into Eqs. 9–4 and 9–5. Thus,

$$\sigma_1 = \sigma \qquad \sigma_2 = 0 \qquad\qquad Ans.$$

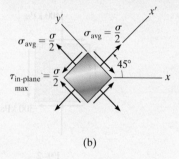

NOTE: Experiments have shown that *brittle materials* will fail due to normal stress. Thus if the bar in Fig. 9–14a is made of cast iron, it will cause failure as shown in the adjacent photo.

Maximum In-Plane Shear Stress. Applying Eqs. 9–6, 9–7, and 9–8, we have

$$\tan 2\theta_s = \frac{-(\sigma_x - \sigma_y)/2}{\tau_{xy}} = \frac{-(\sigma - 0)/2}{0}; \quad \theta_{s_1} = 45°, \theta_{s_2} = -45°$$

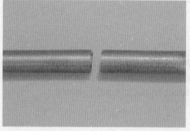

$$\tau_{\substack{max \\ in\text{-}plane}} = \sqrt{\left(\frac{\sigma_x - \sigma_y}{2}\right)^2 + \tau_{xy}^2} = \sqrt{\left(\frac{\sigma - 0}{2}\right)^2 + (0)^2} = \pm\frac{\sigma}{2} \quad Ans.$$

$$\sigma_{avg} = \frac{\sigma_x + \sigma_y}{2} = \frac{\sigma + 0}{2} = \frac{\sigma}{2} \qquad\qquad Ans.$$

(b)

Fig. 9–14

To determine the proper orientation of the element, apply Eq. 9–2.

$$\tau_{x'y'} = -\frac{\sigma_x - \sigma_y}{2}\sin 2\theta + \tau_{xy}\cos 2\theta = -\frac{\sigma - 0}{2}\sin 90° + 0 = -\frac{\sigma}{2}$$

This negative shear stress acts on the x' face, in the negative y' direction as shown in Fig. 9–14b.

NOTE: If the bar in Fig. 9–14a is made of a *ductile material* such as mild steel then *shear stress* will cause it to fail. This can be noted in the adjacent photo, where within the region of necking, shear stress has caused "slipping" along the steel's crystalline boundaries, resulting in a plane of failure that has formed a *cone* around the bar oriented at approximately 45° as calculated above.

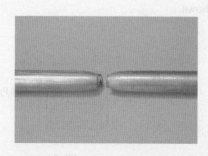

9

FUNDAMENTAL PROBLEMS

F9–1. Determine the normal stress and shear stress acting on the inclined plane *AB*. Sketch the result on the sectioned element.

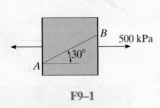

F9–1

F9–2. Determine the equivalent state of stress on an element at the same point oriented 45° clockwise with respect to the element shown.

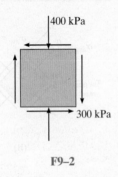

F9–2

F9–3. Determine the equivalent state of stress on an element at the same point that represents the principal stresses at the point. Also, find the corresponding orientation of the element with respect to the element shown.

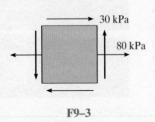

F9–3

F9–4. Determine the equivalent state of stress on an element at the same point that represents the maximum in-plane shear stress at the point.

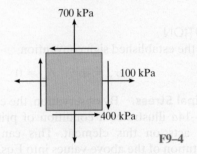

F9–4

F9–5. The beam is subjected to the load at its end. Determine the maximum principal stress at point *B*.

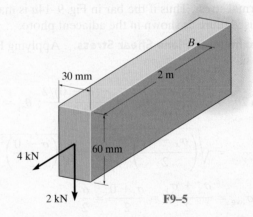

F9–5

F9–6. The beam is subjected to the loading shown. Determine the principal stress at point *C*.

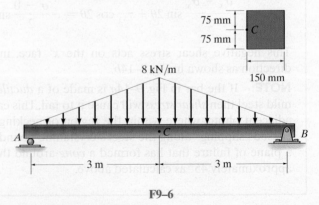

F9–6

PROBLEMS

9–1. Prove that the sum of the normal stresses $\sigma_x + \sigma_y = \sigma_{x'} + \sigma_{y'}$ is constant. See Figs. 9–2a and 9–2b.

9–2. The state of stress at a point in a member is shown on the element. Determine the stress components acting on the inclined plane AB. Solve the problem using the method of equilibrium described in Sec. 9.1.

*9–4. The state of stress at a point in a member is shown on the element. Determine the stress components acting on the inclined plane AB. Solve the problem using the method of equilibrium described in Sec. 9.1.

•9–5. Solve Prob. 9–4 using the stress-transformation equations developed in Sec. 9.2.

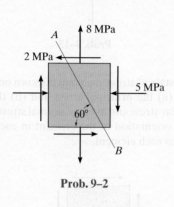

Prob. 9–2

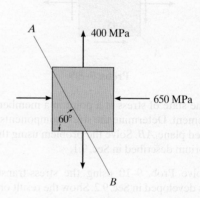

Probs. 9–4/5

9–3. The state of stress at a point in a member is shown on the element. Determine the stress components acting on the inclined plane AB. Solve the problem using the method of equilibrium described in Sec. 9.1.

9–6. The state of stress at a point in a member is shown on the element. Determine the stress components acting on the inclined plane AB. Solve the problem using the method of equilibrium described in Sec. 9.1.

9–7. Solve Prob. 9–6 using the stress-transformation equations developed in Sec. 9.2. Show the result on a sketch.

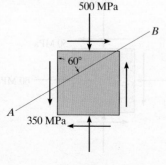

Prob. 9–3

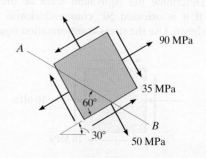

Probs. 9–6/7

*9–8. Determine the normal stress and shear stress acting on the inclined plane AB. Solve the problem using the method of equilibrium described in Sec. 9.1.

•9–9. Determine the normal stress and shear stress acting on the inclined plane AB. Solve the problem using the stress transformation equations. Show the result on the sectioned element.

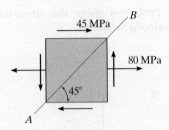

Probs. 9–8/9

9–10. The state of stress at a point in a member is shown on the element. Determine the stress components acting on the inclined plane AB. Solve the problem using the method of equilibrium described in Sec. 9.1.

9–11. Solve Prob. 9–10 using the stress-transformation equations developed in Sec. 9.2. Show the result on a sketch.

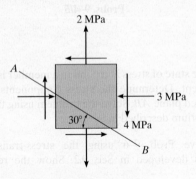

Probs. 9–10/11

*9–12. Determine the equivalent state of stress on an element if it is oriented 50° counterclockwise from the element shown. Use the stress-transformation equations.

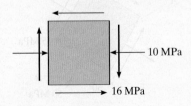

Prob. 9–12

•9–13. Determine the equivalent state of stress on an element if the element is oriented 60° clockwise from the element shown. Show the result on a sketch.

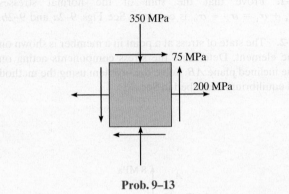

Prob. 9–13

9–14. The state of stress at a point is shown on the element. Determine (a) the principal stress and (b) the maximum in-plane shear stress and average normal stress at the point. Specify the orientation of the element in each case. Show the results on each element.

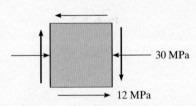

Prob. 9–14

9–15. The state of stress at a point is shown on the element. Determine (a) the principal stress and (b) the maximum in-plane shear stress and average normal stress at the point. Specify the orientation of the element in each case. Show the results on each element.

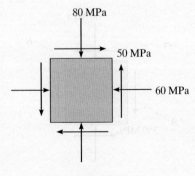

Prob. 9–15

*9–16. The state of stress at a point is shown on the element. Determine (a) the principal stress and (b) the maximum in-plane shear stress and average normal stress at the point. Specify the orientation of the element in each case. Sketch the results on each element.

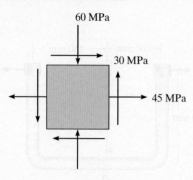

60 MPa

30 MPa

45 MPa

Prob. 9–16

•9–17. Determine the equivalent state of stress on an element at the same point which represents (a) the principal stress, and (b) the maximum in-plane shear stress and the associated average normal stress. Also, for each case, determine the corresponding orientation of the element with respect to the element shown. Sketch the results on each element.

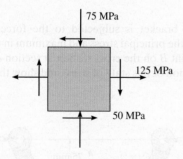

75 MPa

125 MPa

50 MPa

Prob. 9–17

9–18. A point on a thin plate is subjected to the two successive states of stress shown. Determine the resultant state of stress represented on the element oriented as shown on the right.

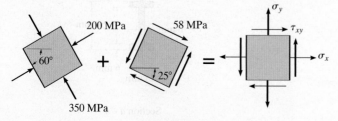

200 MPa

60°

350 MPa

58 MPa

25°

σ_y

τ_{xy}

σ_x

Prob. 9–18

9–19. The state of stress at a point is shown on the element. Determine (a) the principal stress and (b) the maximum in-plane shear stress and average normal stress at the point. Specify the orientation of the element in each case. Sketch the results on each element.

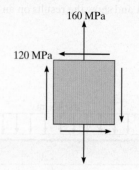

160 MPa

120 MPa

Prob. 9–19

*9–20. The stress acting on two planes at a point is indicated. Determine the normal stress σ_b and the principal stresses at the point.

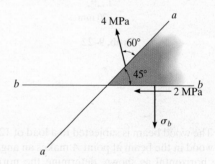

4 MPa

60°

45°

2 MPa

σ_b

a

b

b

a

Prob. 9–20

•9–21. The stress acting on two planes at a point is indicated. Determine the shear stress on plane a–a and the principal stresses at the point.

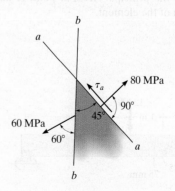

b

a

τ_a

80 MPa

90°

45°

60 MPa

60°

a

b

Prob. 9–21

9

The following problems involve material covered in Chapter 8.

9–22. The T-beam is subjected to the distributed loading that is applied along its centerline. Determine the principal stress at point A and show the results on an element located at this point.

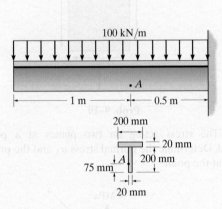

Prob. 9–22

•9–23. The wood beam is subjected to a load of 12 kN. If a grain of wood in the beam at point A makes an angle of 25° with the horizontal as shown, determine the normal and shear stress that act perpendicular and parallel to the grain due to the loading.

*9–24. The wood beam is subjected to a load of 12 kN. Determine the principal stress at point A and specify the orientation of the element.

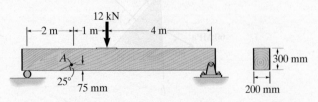

Probs. 9–23/24

•9–25. The bent rod has a diameter of 20 mm and is subjected to the force of 400 N. Determine the principal stress and the maximum in-plane shear stress that is developed at point A. Show the results on a properly oriented element located at this point.

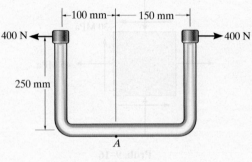

Prob. 9–25

9–26. The bracket is subjected to the force of 15 kN. Determine the principal stress and maximum in-plane shear stress at point A on the cross section at section a–a. Specify the orientation of this state of stress and show the results on elements.

9–27. The bracket is subjected to the force of 15 kN. Determine the principal stress and maximum in-plane shear stress at point B on the cross section at section a–a. Specify the orientation of this state of stress and show the results on elements.

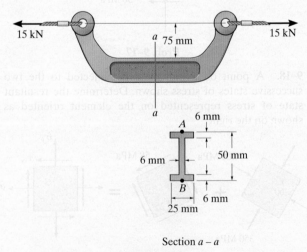

Section a – a

Probs. 9–26/27

*9–28. The wide-flange beam is subjected to the loading shown. Determine the principal stress in the beam at point A and at point B. These points are located at the top and bottom of the web, respectively. Although it is not very accurate, use the shear formula to determine the shear stress.

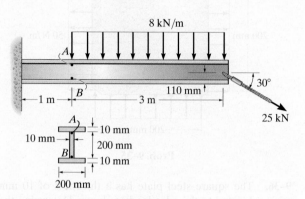

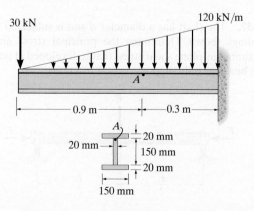

Prob. 9–28

•9–29. The wide-flange beam is subjected to the loading shown. Determine the principal stress in the beam at point A, which is located at the top of the web. Although it is not very accurate, use the shear formula to determine the shear stress. Show the result on an element located at this point.

Prob. 9–29

9–30. The cantilevered rectangular bar is subjected to the force of 20 kN. Determine the principal stress at points A and B.

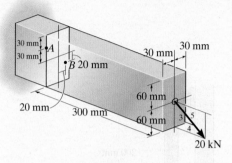

Prob. 9–30

9–31. Determine the principal stress at point A on the cross section of the arm at section a–a. Specify the orientation of this state of stress and indicate the results on an element at the point.

*9–32. Determine the maximum in-plane shear stress developed at point A on the cross section of the arm at section a–a. Specify the orientation of this state of stress and indicate the results on an element at the point.

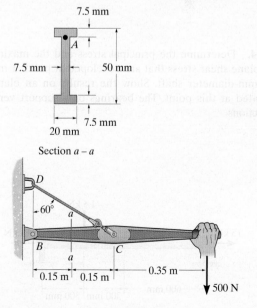

Probs. 9–31/32

•9–33. The clamp bears down on the smooth surface at E by tightening the bolt. If the tensile force in the bolt is 40 kN, determine the principal stress at points A and B and show the results on elements located at each of these points. The cross-sectional area at A and B is shown in the adjacent figure.

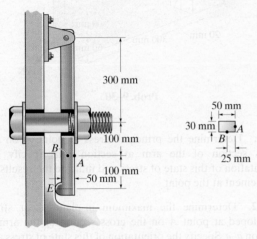

Prob. 9–33

9–34. Determine the principal stress and the maximum in-plane shear stress that are developed at point A in the 50-mm-diameter shaft. Show the results on an element located at this point. The bearings only support vertical reactions.

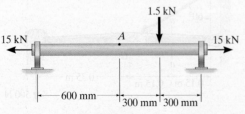

Prob. 9–34

9–35. The square steel plate has a thickness of 10 mm and is subjected to the edge loading shown. Determine the maximum in-plane shear stress and the average normal stress developed in the steel.

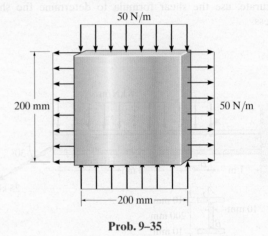

Prob. 9–35

*9–36. The square steel plate has a thickness of 10 mm and is subjected to the edge loading shown. Determine the principal stresses developed in the steel.

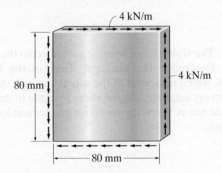

Prob. 9–36

•9–37. The shaft has a diameter d and is subjected to the loadings shown. Determine the principal stress and the maximum in-plane shear stress that is developed at point A. The bearings only support vertical reactions.

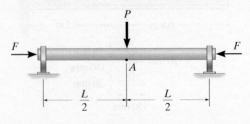

Prob. 9–37

9–38. A paper tube is formed by rolling a paper strip in a spiral and then gluing the edges together as shown. Determine the shear stress acting along the seam, which is at 30° from the vertical, when the tube is subjected to an axial force of 10 N. The paper is 1 mm thick and the tube has an outer diameter of 30 mm.

9–39. Solve Prob. 9–38 for the normal stress acting perpendicular to the seam.

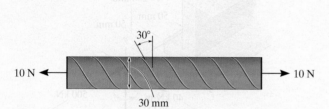

Probs. 9–38/39

*9–40. Determine the principal stresses acting at point A of the supporting frame. Show the results on a properly oriented element located at this point.

•9–41. Determine the principal stress acting at point B, which is located just on the web, below the horizontal segment on the cross section. Show the results on a properly oriented element located at this point. Although it is not very accurate, use the shear formula to calculate the shear stress.

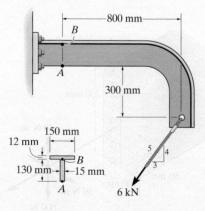

Probs. 9–40/41

9–42. The drill pipe has an outer diameter of 75 mm, a wall thickness of 6 mm, and a weight of 0.8 kN/m. If it is subjected to a torque and axial load as shown, determine (a) the principal stress and (b) the maximum in-plane shear stress at a point on its surface at section a.

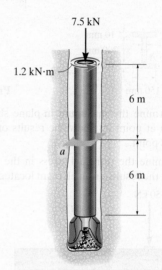

Prob. 9–42

9–43. Determine the principal stress in the beam at point A.

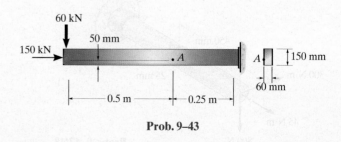

Prob. 9–43

*9–44. Determine the principal stress at point A which is located at the bottom of the web. Show the results on an element located at this point.

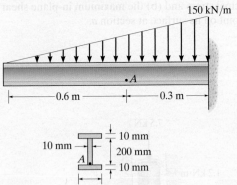

Prob. 9–44

•9–45. Determine the maximum in-plane shear stress in the box beam at point A. Show the results on an element located at this point.

9–46. Determine the principal stress in the box beam at point B. Show the results on an element located at this point.

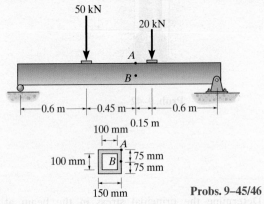

Probs. 9–45/46

9–47. The solid shaft is subjected to a torque, bending moment, and shear force as shown. Determine the principal stresses acting at point A.

*9–48. Solve Prob. 9–47 for point B.

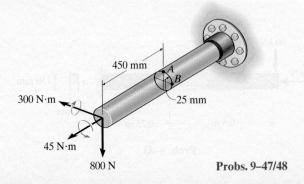

Probs. 9–47/48

•9–49. The internal loadings at a section of the beam are shown. Determine the principal stress at point A. Also compute the maximum in-plane shear stress at this point.

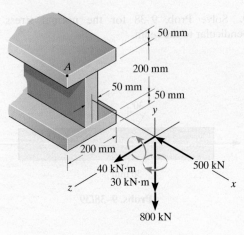

Prob. 9–49

9–50. The internal loadings at a section of the beam consist of an axial force of 500 N, a shear force of 800 N, and two moment components of 30 N · m and 40 N · m. Determine the principal stress at point A. Also calculate the maximum in-plane shear stress at this point.

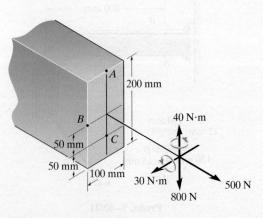

Prob. 9–50

9.4 Mohr's Circle—Plane Stress

In this section, we will show how to apply the equations for plane stress transformation using a graphical solution that is often convenient to use and easy to remember. Furthermore, this approach will allow us to "visualize" how the normal and shear stress components $\sigma_{x'}$ and $\tau_{x'y'}$ vary as the plane on which they act is oriented in different directions, Fig. 9–15a.

If we write Eqs. 9–1 and 9–2 in the form

$$\sigma_{x'} - \left(\frac{\sigma_x + \sigma_y}{2}\right) = \left(\frac{\sigma_x - \sigma_y}{2}\right)\cos 2\theta + \tau_{xy}\sin 2\theta \qquad (9\text{–}9)$$

$$\tau_{x'y'} = -\left(\frac{\sigma_x - \sigma_y}{2}\right)\sin 2\theta + \tau_{xy}\cos 2\theta \qquad (9\text{–}10)$$

then the parameter θ can be *eliminated* by squaring each equation and adding the equations together. The result is

$$\left[\sigma_{x'} - \left(\frac{\sigma_x + \sigma_y}{2}\right)\right]^2 + \tau_{x'y'}^2 = \left(\frac{\sigma_x - \sigma_y}{2}\right)^2 + \tau_{xy}^2$$

For a specific problem, σ_x, σ_y, τ_{xy} are *known constants*. Thus the above equation can be written in a more compact form as

$$(\sigma_{x'} - \sigma_{avg})^2 + \tau_{x'y'}^2 = R^2 \qquad (9\text{–}11)$$

where

$$\sigma_{avg} = \frac{\sigma_x + \sigma_y}{2}$$

$$R = \sqrt{\left(\frac{\sigma_x - \sigma_y}{2}\right)^2 + \tau_{xy}^2} \qquad (9\text{–}12)$$

If we establish coordinate axes, σ *positive to the right* and τ *positive downward*, and then plot Eq. 9–11, it will be seen that this equation represents a *circle* having a radius R and center on the σ axis at point $C(\sigma_{avg}, 0)$, Fig. 9–15b. This circle is called *Mohr's circle*, because it was developed by the German engineer Otto Mohr.

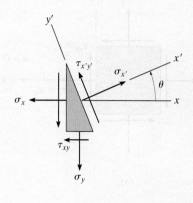

(a)

Fig. 9–15

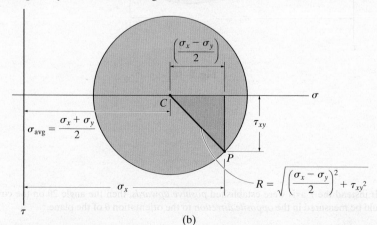

(b)

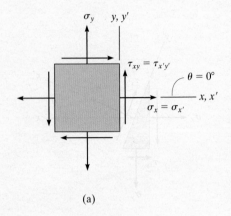

(a)

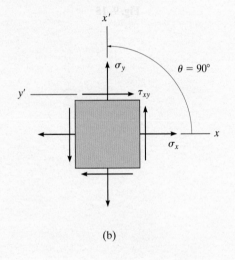

(b)

Each point on Mohr's circle represents the two stress components $\sigma_{x'}$ and $\tau_{x'y'}$ acting on the side of the element defined by the x' axis, when the axis is in a specific direction θ. For example, when x' is coincident with the x axis as shown in Fig. 9–16a, then $\theta = 0°$ and $\sigma_{x'} = \sigma_x$, $\tau_{x'y'} = \tau_{xy}$. We will refer to this as the "reference point" A and plot its coordinates $A(\sigma_x, \tau_{xy})$, Fig. 9–16c.

Now consider rotating the x' axis 90° counterclockwise, Fig. 9–16b. Then $\sigma_{x'} = \sigma_y$, $\tau_{x'y'} = -\tau_{xy}$. These values are the coordinates of point $G(\sigma_y, -\tau_{xy})$ on the circle, Fig. 9–16c. Hence, the radial line CG is 180° counterclockwise from the "reference line" CA. In other words, a rotation θ of the x' axis on the element will correspond to a rotation 2θ on the circle in the *same direction*.*

Once constructed, Mohr's circle can be used to determine the principal stresses, the maximum in-plane shear stress and associated average normal stress, or the stress on any arbitrary plane.

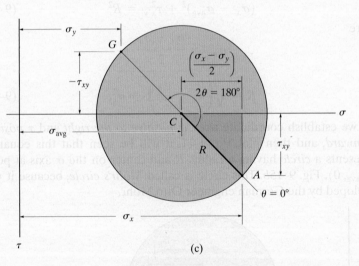

(c)

Fig. 9–16

*If instead the τ axis were established *positive upwards*, then the angle 2θ on the circle would be measured in the *opposite direction* to the orientation θ of the plane.

Procedure for Analysis

The following steps are required to draw and use Mohr's circle.

Construction of the Circle.

- Establish a coordinate system such that the horizontal axis represents the normal stress σ, with *positive to the right*, and the vertical axis represents the shear stress τ, with *positive downwards*, Fig. 9–17a.*

- Using the positive sign convention for σ_x, σ_y, τ_{xy}, as shown in Fig. 9–17b, plot the center of the circle C, which is located on the σ axis at a distance $\sigma_{avg} = (\sigma_x + \sigma_y)/2$ from the origin, Fig. 9–17a.

- Plot the "reference point" A having coordinates $A(\sigma_x, \tau_{xy})$. This point represents the normal and shear stress components on the element's right-hand vertical face, and since the x' axis coincides with the x axis, this represents $\theta = 0°$, Fig. 9–17a.

- Connect point A with the center C of the circle and determine CA by trigonometry. This distance represents the radius R of the circle, Fig. 9–17a.

- Once R has been determined, sketch the circle.

Principal Stress.

- The principal stresses σ_1 and σ_2 ($\sigma_1 \geq \sigma_2$) are the coordinates of points B and D where the circle intersects the σ axis, i.e., where $\tau = 0$, Fig. 9–17a.

- These stresses act on planes defined by angles θ_{p_1} and θ_{p_2}, Fig. 9–17c. They are represented on the circle by angles $2\theta_{p_1}$ (shown) and $2\theta_{p_2}$ (not shown) and are measured *from* the radial reference line CA *to* lines CB and CD, respectively.

- Using trigonometry, only one of these angles needs to be calculated from the circle, since θ_{p_1} and θ_{p_2} are 90° apart. Remember that the direction of rotation $2\theta_p$ on the circle (here it happens to be counterclockwise) represents the *same* direction of rotation θ_p from the reference axis $(+x)$ to the principal plane $(+x')$, Fig. 9–17c.*

Maximum In-Plane Shear Stress.

- The average normal stress and maximum in-plane shear stress components are determined from the circle as the coordinates of either point E or F, Fig. 9–17a.

- In this case the angles θ_{s_1} and θ_{s_2} give the orientation of the planes that contain these components, Fig. 9–17d. The angle $2\theta_{s_1}$ is shown in Fig. 9–17a and can be determined using trigonometry. Here the rotation happens to be clockwise, from CA to CE, and so θ_{s_1} must be clockwise on the element, Fig. 9–17d.*

Procedure for Analysis (continued)

Stresses on Arbitrary Plane.

- The normal and shear stress components $\sigma_{x'}$ and $\tau_{x'y'}$ acting on a specified plane or x' axis, defined by the angle θ, Fig. 9–17e, can be obtained from the circle using trigonometry to determine the coordinates of point P, Fig. 9–17a.

- To locate P, the known angle θ (in this case counterclockwise), Fig. 9–17e, must be measured on the circle in the *same direction* 2θ (counterclockwise), *from* the radial reference line CA to the radial line CP, Fig. 9–17a.*

*If the τ axis were constructed *positive upwards*, then the angle 2θ on the circle would be measured in the *opposite direction* to the orientation θ of the x' axis.

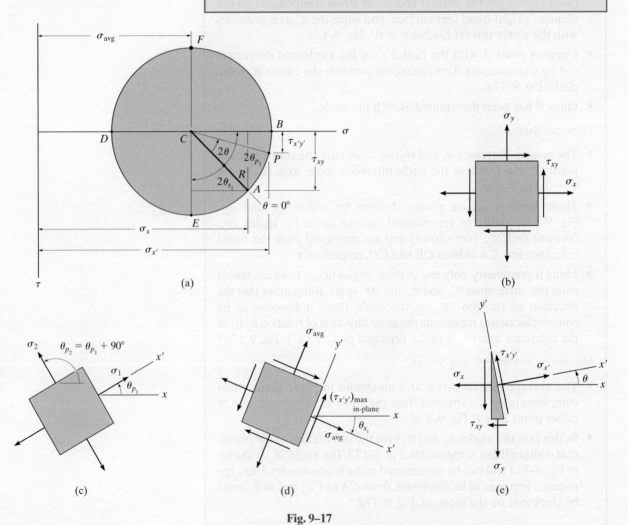

Fig. 9–17

EXAMPLE 9.7

Due to the applied loading, the element at point A on the solid shaft in Fig. 9–18a is subjected to the state of stress shown. Determine the principal stresses acting at this point.

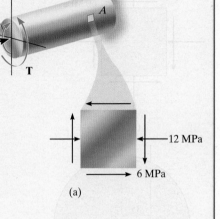

(a)

SOLUTION

Construction of the Circle. From Fig. 9–18a,

$$\sigma_x = -12 \text{ MPa} \qquad \sigma_y = 0 \qquad \tau_{xy} = -6 \text{ MPa}$$

The center of the circle is at

$$\sigma_{\text{avg}} = \frac{-12 + 0}{2} = -6 \text{ MPa}$$

The reference point $A(-12, -6)$ and the center $C(-6, 0)$ are plotted in Fig. 9–18b. The circle is constructed having a radius of

$$R = \sqrt{(12 - 6)^2 + (6)^2} = 8.49 \text{ MPa}$$

Principal Stress. The principal stresses are indicated by the coordinates of points B and D. We have, for $\sigma_1 > \sigma_2$,

$$\sigma_1 = 8.49 - 6 = 2.49 \text{ MPa} \qquad\qquad Ans.$$

$$\sigma_2 = -6 - 8.49 = -14.5 \text{ MPa} \qquad\qquad Ans.$$

The orientation of the element can be determined by calculating the angle $2\theta_{p_2}$ in Fig. 9–18b, which here is measured *counterclockwise* from CA to CD. It defines the direction θ_{p_2} of σ_2 and its associated principal plane. We have

$$2\theta_{p_2} = \tan^{-1} \frac{6}{12 - 6} = 45.0°$$

$$\theta_{p_2} = 22.5°$$

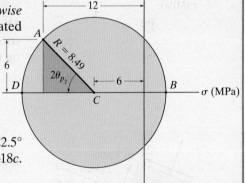

(b)

The element is oriented such that the x' axis or σ_2 is directed 22.5° *counterclockwise* from the horizontal (x axis) as shown in Fig. 9–18c.

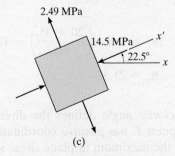

(c)

Fig. 9–18

Refer to the Companion Website for the animation of the concepts that can be found in Example 9.7.

EXAMPLE | 9.8

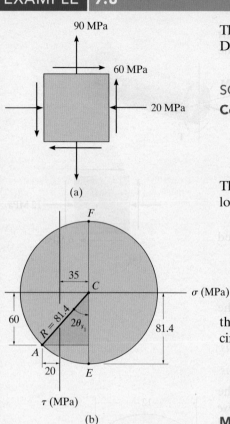

(a)

(b)

(c)

Fig. 9–19

The state of plane stress at a point is shown on the element in Fig. 9–19a. Determine the maximum in-plane shear stress at this point.

SOLUTION

Construction of the Circle.   From the problem data,

$$\sigma_x = -20 \text{ MPa} \qquad \sigma_y = 90 \text{ MPa} \qquad \tau_{xy} = 60 \text{ MPa}$$

The σ, τ axes are established in Fig.9–19b. The center of the circle C is located on the σ axis, at the point

$$\sigma_{avg} = \frac{-20 + 90}{2} = 35 \text{ MPa}$$

Point C and the reference point $A(-20, 60)$ are plotted. Applying the Pythagorean theorem to the shaded triangle to determine the circle's radius CA, we have

$$R = \sqrt{(60)^2 + (55)^2} = 81.4 \text{ MPa}$$

Maximum In-Plane Shear Stress.   The maximum in-plane shear stress and the average normal stress are identified by point E (or F) on the circle. The coordinates of point $E(35, 81.4)$ give

$$\tau_{\substack{max \\ in\text{-}plane}} = 81.4 \text{ MPa} \qquad\qquad Ans.$$

$$\sigma_{avg} = 35 \text{ MPa} \qquad\qquad Ans.$$

The angle θ_{s_1}, measured *counterclockwise* from CA to CE, can be found from the circle, identified as $2\theta_{s_1}$. We have

$$2\theta_{s_1} = \tan^{-1}\left(\frac{20 + 35}{60}\right) = 42.5°$$

$$\theta_{s_1} = 21.3° \qquad\qquad Ans.$$

This *counterclockwise* angle defines the direction of the x' axis, Fig. 9–19c. Since point E has *positive* coordinates, then the average normal stress and the maximum in-plane shear stress both act in the *positive* x' and y' directions as shown.

EXAMPLE 9.9

The state of plane stress at a point is shown on the element in Fig. 9–20a. Represent this state of stress on an element oriented 30° counterclockwise from the position shown.

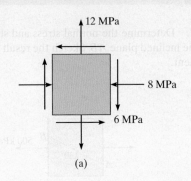

(a)

SOLUTION

Construction of the Circle. From the problem data,

$$\sigma_x = -8 \text{ MPa} \qquad \sigma_y = 12 \text{ MPa} \qquad \tau_{xy} = -6 \text{ MPa}$$

The σ and τ axes are established in Fig. 9–20b. The center of the circle C is on the σ axis at

$$\sigma_{\text{avg}} = \frac{-8 + 12}{2} = 2 \text{ MPa}$$

The reference point for $\theta = 0°$ has coordinates $A(-8, -6)$. Hence from the shaded triangle the radius CA is

$$R = \sqrt{(10)^2 + (6)^2} = 11.66$$

Stresses on 30° Element. Since the element is to be rotated 30° *counterclockwise*, we must construct a radial line CP, $2(30°) = 60°$ *counterclockwise*, measured from CA ($\theta = 0°$), Fig. 9–20b. The coordinates of point $P(\sigma_{x'}, \tau_{x'y'})$ must now be obtained. From the geometry of the circle,

$$\phi = \tan^{-1}\frac{6}{10} = 30.96° \qquad \psi = 60° - 30.96° = 29.04°$$

$$\sigma_{x'} = 2 - 11.66 \cos 29.04° = -8.20 \text{ MPa} \qquad Ans.$$

$$\tau_{x'y'} = 11.66 \sin 29.04° = 5.66 \text{ MPa} \qquad Ans.$$

These two stress components act on face BD of the element shown in Fig. 9–20c since the x' axis for this face is oriented 30° *counterclockwise* from the x axis.

The stress components acting on the adjacent face DE of the element, which is 60° *clockwise* from the positive x axis, Fig. 9–20c, are represented by the coordinates of point Q on the circle. This point lies on the radial line CQ, which is 180° from CP. The coordinates of point Q are

$$\sigma_{x'} = 2 + 11.66 \cos 29.04° = 12.2 \text{ MPa} \qquad Ans.$$

$$\tau_{x'y'} = -(11.66 \sin 29.04) = -5.66 \text{ MPa (check)} \qquad Ans.$$

NOTE: Here $\tau_{x'y'}$ acts in the $-y'$ direction.

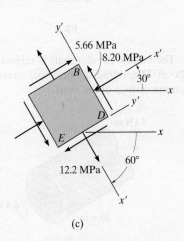

(b)

(c)

Fig. 9–20

FUNDAMENTAL PROBLEMS

F9–7. Determine the normal stress and shear stress acting on the inclined plane *AB*. Sketch the result on the sectioned element.

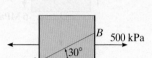

F9–7

F9–8. Determine the equivalent state of stress on an element at the same point that represents the principal stresses at the point. Also, find the corresponding orientation of the element with respect to the element shown. Sketch the results on the element.

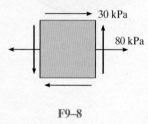

F9–8

F9–9. The hollow circular shaft is subjected to the torque of 4 kN · m. Determine the principal stress developed at a point on the surface of the shaft.

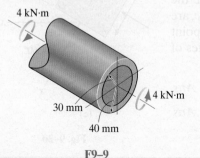

F9–9

F9–10. Determine the principal stress developed at point *A* on the cross section of the beam at section *a–a*.

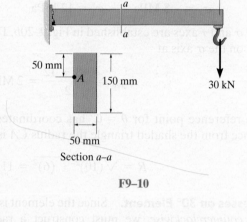

F9–10

F9–11. Determine the maximum in-plane shear stress developed at point *A* on the cross section of the beam at section *a–a*, which is located just to the left of the 60-kN force. Point *A* is just below the flange.

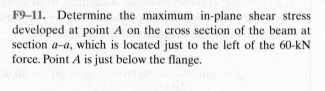

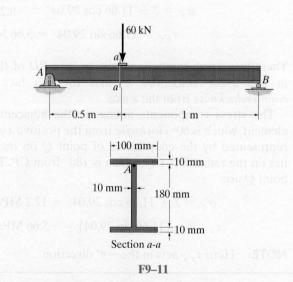

F9–11

PROBLEMS

9–51. Solve Prob. 9–4 using Mohr's circle.

*9–52. Solve Prob. 9–6 using Mohr's circle.

•9–53. Solve Prob. 9–14 using Mohr's circle.

9–54. Solve Prob. 9–16 using Mohr's circle.

9–55. Solve Prob. 9–12 using Mohr's circle.

*9–56. Solve Prob. 9–11 using Mohr's circle.

9–57. Mohr's circle for the state of stress in Fig. 9–15a is shown in Fig. 9–15b. Show that finding the coordinates of point $P(\sigma_{x'}, \tau_{x'y'})$ on the circle gives the same value as the stress-transformation Eqs. 9–1 and 9–2.

9–58. Determine the equivalent state of stress if an element is oriented 25° counterclockwise from the element shown.

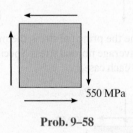

Prob. 9–58

9–59. Determine the equivalent state of stress if an element is oriented 20° clockwise from the element shown.

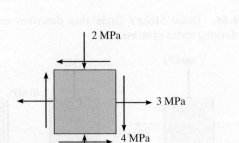

Prob. 9–59

*9–60. Determine the equivalent state of stress if an element is oriented 30° clockwise from the element shown. Show the result on the element.

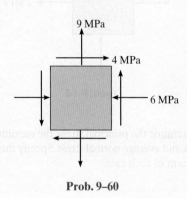

Prob. 9–60

•9–61. Determine the equivalent state of stress for an element oriented 60° counterclockwise from the element shown. Show the result on the element.

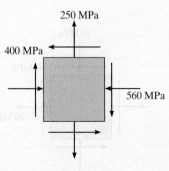

Prob. 9–61

9–62. Determine the equivalent state of stress for an element oriented 30° clockwise from the element shown. Show the result on the element.

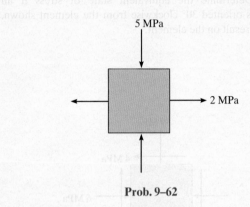

Prob. 9–62

9–63. Determine the principal stress, the maximum in-plane shear stress, and average normal stress. Specify the orientation of the element in each case.

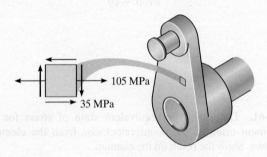

Prob. 9–63

*9–64. Determine the principal stress, the maximum in-plane shear stress, and average normal stress. Specify the orientation of the element in each case.

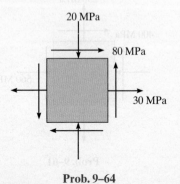

Prob. 9–64

•9–65. Determine the principal stress, the maximum in-plane shear stress, and average normal stress. Specify the orientation of the element in each case.

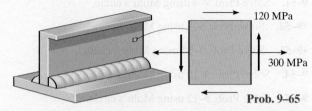

Prob. 9–65

9–66. Determine the principal stress, the maximum in-plane shear stress, and average normal stress. Specify the orientation of the element in each case.

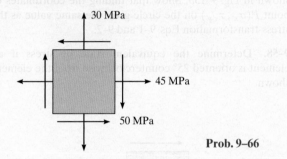

Prob. 9–66

9–67. Determine the principal stress, the maximum in-plane shear stress, and average normal stress. Specify the orientation of the element in each case.

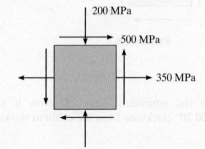

Prob. 9–67

*9–68. Draw Mohr's circle that describes each of the following states of stress.

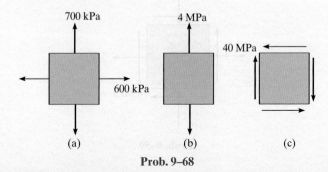

Prob. 9–68

The following problems involve material covered in Chapter 8.

9–69. The frame supports the distributed loading of 200 N/m. Determine the normal and shear stresses at point D that act perpendicular and parallel, respectively, to the grain. The grain at this point makes an angle of 30° with the horizontal as shown.

9–70. The frame supports the distributed loading of 200 N/m. Determine the normal and shear stresses at point E that act perpendicular and parallel, respectively, to the grain. The grain at this point makes an angle of 60° with the horizontal as shown.

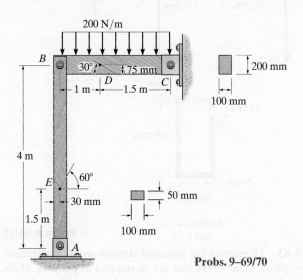

Probs. 9–69/70

9–71. The stair tread of the escalator is supported on two of its sides by the moving pin at A and the roller at B. If a man having a weight of 1500 N ($\approx$ 150 kg) stands in the center of the tread, determine the principal stresses developed in the supporting truck on the cross section at point C. The stairs move at constant velocity.

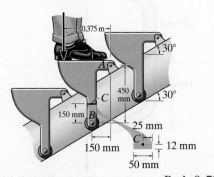

Prob. 9–71

*9–72. The thin-walled pipe has an inner diameter of 12 mm and a thickness of 0.6 mm. If it is subjected to an internal pressure of 3.5 MPa and the axial tension and torsional loadings shown, determine the principal stress at a point on the surface of the pipe.

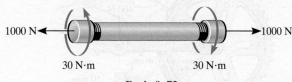

Prob. 9–72

•9–73. The cantilevered rectangular bar is subjected to the force of 20 kN. Determine the principal stress at point A.

9–74. Solve Prob. 9–73 for the principal stress at point B.

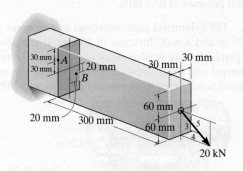

Probs. 9–73/74

9–75. The 50-mm-diameter drive shaft AB on the helicopter is subjected to an axial tension of 50 kN and a torque of 0.45 kN · m. Determine the principal stress and the maximum in-plane shear stress that act at a point on the surface of the shaft.

Prob. 9–75

9

*9–76.   The pedal crank for a bicycle has the cross section shown. If it is fixed to the gear at B and does not rotate while subjected to a force of 400 N, determine the principal stress in the material on the cross section at point C.

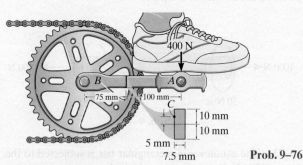

Prob. 9–76

•9–77.   A spherical pressure vessel has an inner radius of 1.5 m and a wall thickness of 12 mm. Draw Mohr's circle for the state of stress at a point on the vessel and explain the significance of the result. The vessel is subjected to an internal pressure of 0.56 MPa.

9–78.   The cylindrical pressure vessel has an inner radius of 1.25 m and a wall thickness of 15 mm. It is made from steel plates that are welded along the 45° seam. Determine the normal and shear stress components along this seam if the vessel is subjected to an internal pressure of 8 MPa.

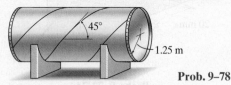

Prob. 9–78

•9–79.   Determine the normal and shear stresses at point D that act perpendicular and parallel, respectively, to the grains. The grains at this point make an angle of 30° with the horizontal as shown. Point D is located just to the left of the 10-kN force.

*9–80.   Determine the principal stress at point D, which is located just to the left of the 10-kN force.

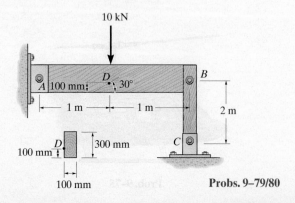

Probs. 9–79/80

•9–81.   Determine the principal stress at point A on the cross section of the hanger at section a–a. Specify the orientation of this state of stress and indicate the result on an element at the point.

9–82.   Determine the principal stress at point A on the cross section of the hanger at section b–b. Specify the orientation of the state of stress and indicate the results on an element at the point.

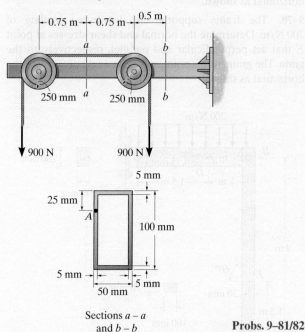

Sections $a - a$
and $b - b$ **Probs. 9–81/82**

9–83.   Determine the principal stresses and the maximum in-plane shear stress that are developed at point A. Show the results on an element located at this point. The rod has a diameter of 40 mm.

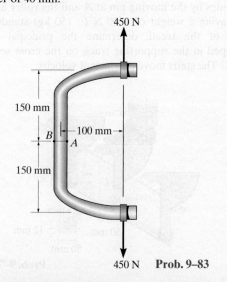

450 N   **Prob. 9–83**

9.5 Absolute Maximum Shear Stress

When a point in a body is subjected to a general three-dimensional state of stress, an element of material has a normal-stress and two shear-stress components acting on each of its faces, Fig. 9–21a. Like the case of plane stress, it is possible to develop stress-transformation equations that can be used to determine the normal and shear stress components σ and τ acting on *any* skewed plane of the element, Fig. 9–21b. Furthermore, at the point it is also possible to determine the unique orientation of an element having only principal stresses acting on its faces. As shown in Fig. 9–21c, in general these principal stresses will have magnitudes of maximum, intermediate, and minimum intensity, i.e., $\sigma_{max} \geq \sigma_{int} \geq \sigma_{min}$. This is a condition known as ***triaxial stress***.

A discussion of the transformation of stress in three dimensions is beyond the scope of this text; however, it is discussed in books related to the theory of elasticity. For our purposes, we will confine our attention only to the case of plane stress. For example, consider the

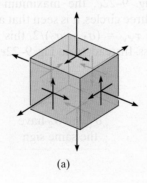

(a)

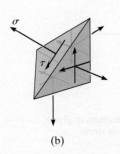

(b)

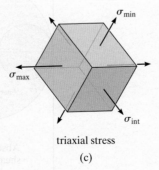

triaxial stress

(c)

Fig. 9–21

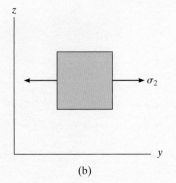

x–y plane stress

(a)

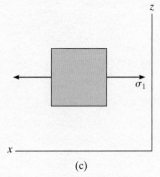

(b)

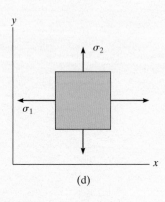

(c)

material to be subjected to the in-plane principal stresses σ_1 and σ_2 shown in Fig. 9–22a, where *both* of these stresses are tensile. If we view the element in two dimensions, that is, in the y–z, x–z, and x–y planes, Fig. 9–22b, 9–22c, and 9–22d, then we can use Mohr's circle to determine the maximum in-plane shear for each case and from this, determine the *absolute maximum shear stress* in the material. For example, the diameter of Mohr's circle extends between 0 and σ_2 for the case shown in Fig. 9–22b. From this circle, Fig. 9–22e, the maximum in-plane shear stress is $\tau_{y'z'} = \sigma_2/2$. For all three circles, it is seen that although the maximum in-plane shear stress is $\tau_{x'y'} = (\sigma_1 - \sigma_2)/2$, this value is *not* the absolute maximum shear stress. Instead, from Fig. 9–22e,

$$\tau_{\substack{abs \\ max}} = \frac{\sigma_1}{2}$$

σ_1 and σ_2 have the same sign

(9–13)

(d)

Fig. 9–22

(e)

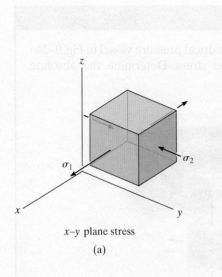

x–y plane stress

(a)

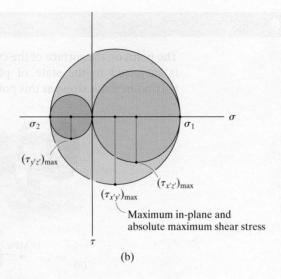

$(\tau_{y'z'})_{max}$

$(\tau_{x'z'})_{max}$

$(\tau_{x'y'})_{max}$

Maximum in-plane and
absolute maximum shear stress

(b)

Fig. 9–23

If one of the in-plane principal stresses has the *opposite sign* of that of the other, Fig. 9–23*a*, then the three Mohr's circles that describe the state of stress for element orientations about each coordinate axis are shown in Fig. 9–23*b*. Clearly, in this case

$$\tau_{max}^{abs} = \frac{\sigma_1 - \sigma_2}{2}$$

σ_1 and σ_2 have opposite signs

(9–14)

Calculation of the absolute maximum shear stress as indicated here is important when designing members made of a ductile material, since the strength of the material depends on its ability to resist shear stress. This situation will be discussed further in Sec. 10.7.

Important Points

- The general three-dimensional state of stress at a point can be represented by an element oriented so that only three principal stresses σ_{max}, σ_{int}, σ_{min} act on it.

- In the case of plane stress, if the in-plane principal stresses both have the *same sign*, the *absolute maximum shear stress* will occur *out of the plane* and has a value of $\tau_{max}^{abs} = \sigma_{max}/2$. This value is greater than the in-plane shear stress.

- If the in-plane principal stresses are of *opposite signs*, then the *absolute maximum shear stress will equal the maximum in-plane shear stress*; that is, $\tau_{max}^{abs} = (\sigma_{max} - \sigma_{min})/2$.

9

EXAMPLE | 9.10

The point on the surface of the cylindrical pressure vessel in Fig. 9–24a is subjected to the state of plane stress. Determine the absolute maximum shear stress at this point.

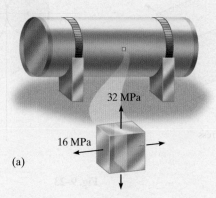

32 MPa

16 MPa

(a)

SOLUTION

The principal stresses are $\sigma_1 = 32$ MPa, $\sigma_2 = 16$ MPa. If these stresses are plotted along the σ axis, the three Mohr's circles can be constructed that describe the stress state viewed in each of the three perpendicular planes, Fig. 9–24b. The largest circle has a radius of 16 MPa and describes the state of stress in the plane only containing $\sigma_1 = 32$ M Pa, shown shaded in Fig. 9–24a. An orientation of an element 45° within this plane yields the state of absolute maximum shear stress and the associated average normal stress, namely,

$$\tau_{\text{max}}^{\text{abs}} = 16 \text{ MPa} \qquad\qquad Ans.$$
$$\sigma_{\text{avg}} = 16 \text{ MPa}$$

This same result for $\tau_{\text{max}}^{\text{abs}}$ can be obtained from direct application of Eq. 9–13.

$$\tau_{\text{max}}^{\text{abs}} = \frac{\sigma_1}{2} = \frac{32}{2} = 16 \text{ MPa} \qquad\qquad Ans.$$

$$\sigma_{\text{avg}} = \frac{32 + 0}{2} = 16 \text{ MPa}$$

By comparison, the maximum in-plane shear stress can be determined from the Mohr's circle drawn between $\sigma_1 = 32$ MPa and $\sigma_2 = 16$ MPa, Fig. 9–24b. This gives a value of

$$\tau_{\text{max}}^{\text{in-plane}} = \frac{32 - 16}{2} = 8 \text{ MPa}$$

$$\sigma_{\text{avg}} = \frac{32 + 16}{2} = 24 \text{ MPa}$$

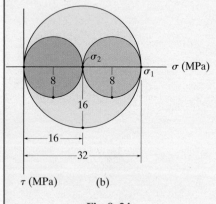

Fig. 9–24

EXAMPLE | 9.11

Due to an applied loading, an element at the point on a machine shaft is subjected to the state of plane stress shown in Fig. 9–25a. Determine the principal stresses and the absolute maximum shear stress at the point.

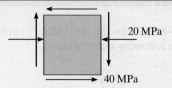

(a)

SOLUTION

Principal Stresses. The in-plane principal stresses can be determined from Mohr's circle. The center of the circle is on the σ axis at $\sigma_{avg} = (-20 + 0)/2 = -10$ MPa. Plotting the reference point $A(-20, -40)$, the radius CA is established and the circle is drawn as shown in Fig. 9–25b. The radius is

$$R = \sqrt{(20 - 10)^2 + (40)^2} = 41.2 \text{ MPa}$$

The principal stresses are at the points where the circle intersects the σ axis; i.e.,

$$\sigma_1 = -10 + 41.2 = 31.2 \text{ MPa}$$
$$\sigma_2 = -10 - 41.2 = -51.2 \text{ MPa}$$

From the circle, the *counterclockwise* angle 2θ, measured from CA to the $-\sigma$ axis, is

$$2\theta = \tan^{-1}\left(\frac{40}{20 - 10}\right) = 76.0°$$

Thus,

$$\theta = 38.0°$$

This *counterclockwise* rotation defines the direction of the x' axis and σ_2 and its associated principal plane, Fig. 9–25c. We have

$$\sigma_1 = 31.2 \text{ MPa} \qquad \sigma_2 = -51.2 \text{ MPa} \qquad \textit{Ans.}$$

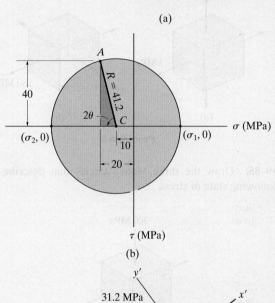

(b)

(c)

Absolute Maximum Shear Stress. Since these stresses have opposite signs, applying Eq. 9–14 we have

$$\tau_{\max}^{abs} = \frac{\sigma_1 - \sigma_2}{2} = \frac{31.2 - (-51.2)}{2} = 41.2 \text{ MPa} \quad \textit{Ans.}$$

$$\sigma_{avg} = \frac{31.2 - 51.2}{2} = -10 \text{ MPa}$$

NOTE: These same results can also be obtained by drawing Mohr's circle for each orientation of an element about the x, y, and z axes, Fig. 9–25d. Since σ_1 and σ_2 are of *opposite signs*, then the absolute maximum shear stress equals the maximum in-plane shear stress.

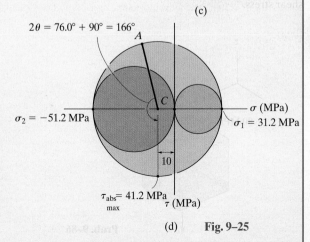

(d)   **Fig. 9–25**

PROBLEMS

*9–84. Draw the three Mohr's circles that describe each of the following states of stress.

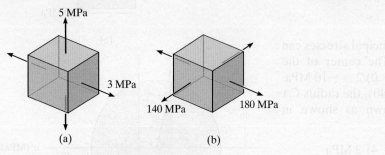

(a)   (b)

Prob. 9–84

•9–85. Draw the three Mohr's circles that describe the following state of stress.

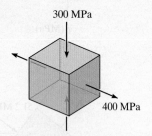

Prob. 9–85

9–86. The stress at a point is shown on the element. Determine the principal stress and the absolute maximum shear stress.

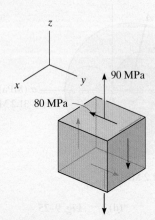

Prob. 9–86

9–87. The stress at a point is shown on the element. Determine the principal stress and the absolute maximum shear stress.

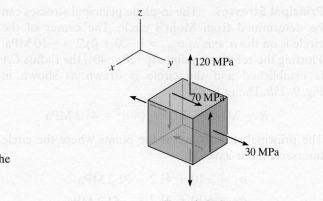

Prob. 9–87

*9–88. The stress at a point is shown on the element. Determine the principal stress and the absolute maximum shear stress.

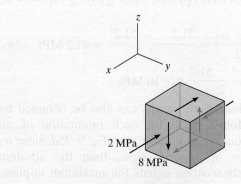

Prob. 9–88

•9–89. The stress at a point is shown on the element. Determine the principal stress and the absolute maximum shear stress.

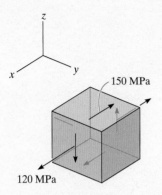

150 MPa

120 MPa

Prob. 9–89

9–90. The state of stress at a point is shown on the element. Determine the principal stress and the absolute maximum shear stress.

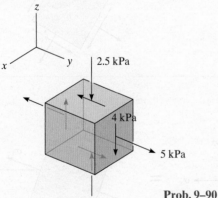

2.5 kPa

4 kPa

5 kPa

Prob. 9–90

9–91. Consider the general case of plane stress as shown. Write a computer program that will show a plot of the three Mohr's circles for the element, and will also calculate the maximum in-plane shear stress and the absolute maximum shear stress.

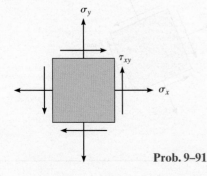

σ_y

τ_{xy}

σ_x

Prob. 9–91

*9–92. The solid shaft is subjected to a torque, bending moment, and shear force as shown. Determine the principal stress acting at points A and B and the absolute maximum shear stress.

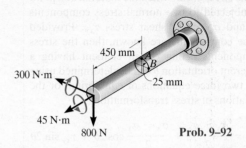

450 mm

300 N·m

25 mm

45 N·m

800 N

Prob. 9–92

•9–93. The propane gas tank has an inner diameter of 1500 mm and wall thickness of 15 mm. If the tank is pressurized to 2 MPa, determine the absolute maximum shear stress in the wall of the tank.

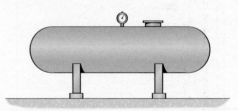

Prob. 9–93

9–94. Determine the principal stress and absolute maximum shear stress developed at point A on the cross section of the bracket at section a–a.

9–95. Determine the principal stress and absolute maximum shear stress developed at point B on the cross section of the bracket at section a–a.

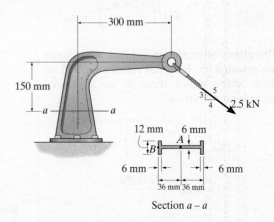

300 mm

150 mm

a   a

3

5

4

2.5 kN

12 mm   6 mm

A

B

6 mm

6 mm   6 mm

36 mm 36 mm

Section a – a

Probs. 9–94/95

9

CHAPTER REVIEW

Plane stress occurs when the material at a point is subjected to two normal stress components σ_x and σ_y and a shear stress τ_{xy}. Provided these components are known, then the stress components acting on an element having a different orientation θ can be determined using the two force equations of equilibrium or the equations of stress transformation.

$$\sigma_{x'} = \frac{\sigma_x + \sigma_y}{2} + \frac{\sigma_x - \sigma_y}{2}\cos 2\theta + \tau_{xy}\sin 2\theta$$

$$\tau_{x'y'} = -\frac{\sigma_x - \sigma_y}{2}\sin 2\theta + \tau_{xy}\cos 2\theta$$

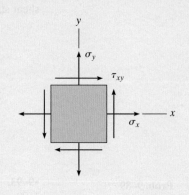

Ref.: Section 9.2

For design, it is important to determine the orientation of the element that produces the maximum principal normal stresses and the maximum in-plane shear stress. Using the stress transformation equations, it is found that no shear stress acts on the planes of principal stress. The principal stresses are

$$\sigma_{1,2} = \frac{\sigma_x + \sigma_y}{2} \pm \sqrt{\left(\frac{\sigma_x - \sigma_y}{2}\right)^2 + \tau_{xy}^2}$$

The planes of maximum in-plane shear stress are oriented 45° from this orientation, and on these shear planes there is an associated average normal stress.

$$\tau_{\substack{max \\ in\text{-}plane}} = \sqrt{\left(\frac{\sigma_x - \sigma_y}{2}\right)^2 + \tau_{xy}^2}$$

$$\sigma_{avg} = \frac{\sigma_x + \sigma_y}{2}$$

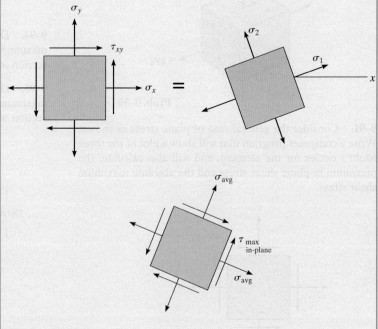

Ref.: Section 9.3

9

Mohr's circle provides a semi-graphical method for finding the stress on any plane, the principal normal stresses, and the maximum in-plane shear stress. To draw the circle, the σ and τ axes are established, the center of the circle $C[(\sigma_x + \sigma_y)/2, 0]$ and the reference point $A(\sigma_x, \tau_{xy})$ are plotted. The radius R of the circle extends between these two points and is determined from trigonometry.

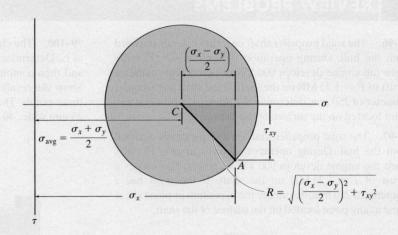

Ref.: Section 9.4

If σ_1 and σ_2 are of the same sign, then the absolute maximum shear stress will lie out of plane.

$$\tau_{\max}^{abs} = \frac{\sigma_1}{2}$$

In the case of plane stress, the absolute maximum shear stress will be equal to the maximum in-plane shear stress provided the principal stresses σ_1 and σ_2 have the opposite sign.

$$\tau_{\max}^{abs} = \frac{\sigma_1 - \sigma_2}{2}$$

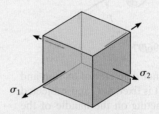

x–y plane stress

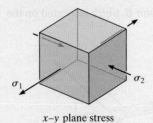

x–y plane stress

Ref.: Section 9.5

9

REVIEW PROBLEMS

*9–96.   The solid propeller shaft on a ship extends outward from the hull. During operation it turns at $\omega = 15$ rad/s when the engine develops 900 kW of power. This causes a thrust of $F = 1.23$ MN on the shaft. If the shaft has an outer diameter of 250 mm, determine the principal stresses at any point located on the surface of the shaft.

•9–97.   The solid propeller shaft on a ship extends outward from the hull. During operation it turns at $\omega = 15$ rad/s when the engine develops 900 kW of power. This causes a thrust of $F = 1.23$ MN on the shaft. If the shaft has a diameter of 250 mm, determine the maximum in-plane shear stress at any point located on the surface of the shaft.

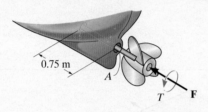

Probs. 9–96/97

9–98.   The steel pipe has an inner diameter of 68 mm and an outer diameter of 75 mm. If it is fixed at C and subjected to the horizontal 100-N force acting on the handle of the pipe wrench at its end, determine the principal stresses in the pipe at point A, which is located on the surface of the pipe.

9–99.   Solve Prob. 9–98 for point B, which is located on the surface of the pipe.

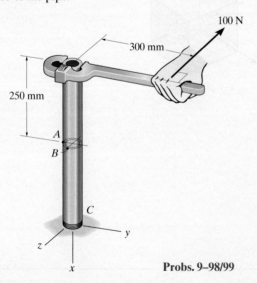

Probs. 9–98/99

*9–100.   The clamp exerts a force of 0.75 kN on the boards at G. Determine the axial force in each screw, AB and CD, and then compute the principal stresses at points E and F. Show the results on properly oriented elements located at these points. The section through EF is rectangular and is 25 mm wide., 40 mm deep.

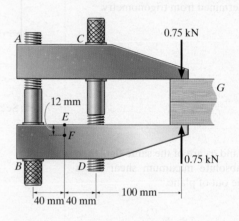

Prob. 9–100

9–101.   The shaft has a diameter d and is subjected to the loadings shown. Determine the principal stress and the maximum in-plane shear stress that is developed anywhere on the surface of the shaft.

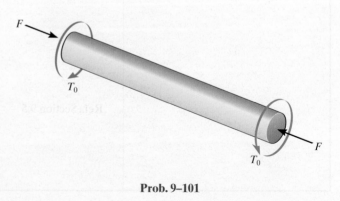

Prob. 9–101

9–102. The state of stress at a point in a member is shown on the element. Determine the stress components acting on the plane AB.

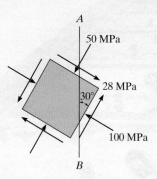

Prob. 9–102

9–103. The propeller shaft of the tugboat is subjected to the compressive force and torque shown. If the shaft has an inner diameter of 100 mm and an outer diameter of 150 mm, determine the principal stress at a point A located on the outer surface.

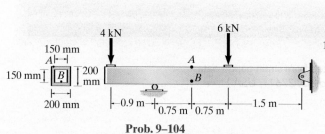

Prob. 9–103

*9–104. The box beam is subjected to the loading shown. Determine the principal stress in the beam at points A and B.

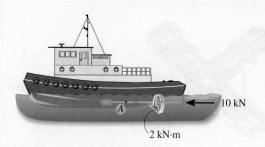

Prob. 9–104

•9–105. The wooden strut is subjected to the loading shown. Determine the principal stresses that act at point C and specify the orientation of the element at this point. The strut is supported by a bolt (pin) at B and smooth support at A.

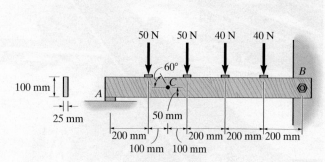

Prob. 9–105

9–106. The wooden strut is subjected to the loading shown. If grains of wood in the strut at point C make an angle of $60°$ with the horizontal as shown, determine the normal and shear stresses that act perpendicular and parallel to the grains, respectively, due to the loading. The strut is supported by a bolt (pin) at B and smooth support at A.

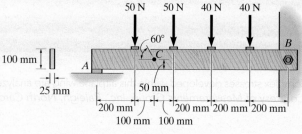

Prob. 9–106

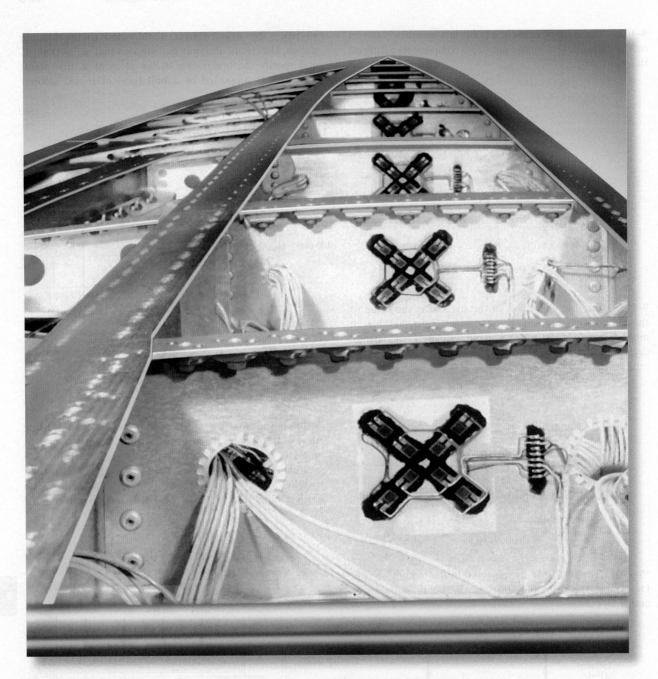

Complex stresses developed within this airplane wing are analyzed from strain gauge data.
(Courtesy of Measurements Group, Inc., Raleigh, North Carolina, 27611, USA.)

Strain Transformation

10

CHAPTER OBJECTIVES

The transformation of strain at a point is similar to the transformation of stress, and as a result the methods of Chapter 9 will be applied in this chapter. Here we will also discuss various ways for measuring strain and develop some important material-property relationships, including a generalized form of Hooke's law. At the end of the chapter, a few of the theories used to predict the failure of a material will be discussed.

Check out the Companion Website for the following video solution(s) that can be found in this chapter.

- Section 7
  Failure Theories – Ductile Failure
- Section 7
  Failure Theories – Factor of Safety
- Section 7
  Failure Theories – Brittle Failure
- Section 7
  Failure Theories – F.S. Estimates

10.1 Plane Strain

As outlined in Sec. 2.2, the general state of strain at a point in a body is represented by a combination of three components of normal strain, ϵ_x, ϵ_y, ϵ_z, and three components of shear strain γ_{xy}, γ_{xz}, γ_{yz}. These six components tend to deform each face of an element of the material, and like stress, the normal and shear strain *components* at the point will vary according to the orientation of the element. The *strains* at a point are often determined by using strain gauges, which measure normal strain in *specified directions*. For both analysis and design, however, engineers must sometimes transform this data in order to obtain the strain in other directions.

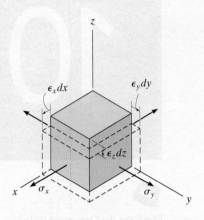

Plane stress, σ_x, σ_y, does not cause plane strain in the x–y plane since $\epsilon_z \neq 0$.

Fig. 10–1

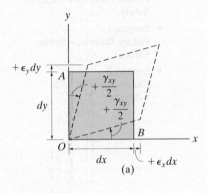

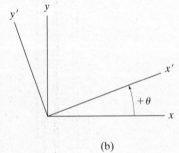

Positive sign convention

Fig. 10–2

To understand how this is done, we will first confine our attention to a study of **plane strain**. Specifically, we will not consider the effects of the components ϵ_z, γ_{xz}, and γ_{yz}. In general, then, a plane-strained element is subjected to two components of normal strain, ϵ_x, ϵ_y, and one component of shear strain, γ_{xy}. Although plane strain and plane stress each have three components lying in the same plane, realize that plane stress *does not* necessarily cause plane strain or vice versa. The reason for this has to do with the Poisson effect discussed in Sec. 3.6. For example, if the element in Fig. 10–1 is subjected to plane stress σ_x and σ_y, not only are normal strains ϵ_x and ϵ_y produced, but there is *also* an associated normal strain, ϵ_z. This is obviously *not* a case of plane strain. In general, then, unless $\nu = 0$, the Poisson effect will *prevent* the simultaneous occurrence of plane strain and plane stress.

10.2   General Equations of Plane-Strain Transformation

It is important in plane-strain analysis to establish transformation equations that can be used to determine the x', y' components of normal and shear strain at a point, provided the x, y components of strain are known. Essentially this problem is one of geometry and requires relating the deformations and rotations of line segments, which represent the sides of differential elements that are parallel to each set of axes.

Sign Convention.   Before the strain-transformation equations can be developed, we must first establish a sign convention for the strains. With reference to the differential element shown in Fig. 10–2a, *normal strains* ϵ_x and ϵ_y are *positive* if they cause *elongation* along the x and y axes, respectively, and the *shear strain* γ_{xy} is *positive* if the interior angle *AOB becomes smaller* than 90°. This sign convention also follows the corresponding one used for plane stress, Fig. 9–5a, that is, positive σ_x, σ_y, τ_{xy} will cause the element to *deform* in the positive ϵ_x, ϵ_y, γ_{xy} directions, respectively.

The problem here will be to determine at a point the normal and shear strains $\epsilon_{x'}$, $\epsilon_{y'}$, $\gamma_{x'y'}$, measured relative to the x', y' axes, if we know ϵ_x, ϵ_y, γ_{xy}, measured relative to the x, y axes. If the angle between the x and x' axes is θ, then, like the case of plane stress, θ will be *positive* provided it follows the curl of the right-hand fingers, i.e., counterclockwise, as shown in Fig. 10–2b.

Normal and Shear Strains. In order to develop the strain-transformation equation for $\epsilon_{x'}$, we must determine the elongation of a line segment dx' that lies along the x' axis and is subjected to strain components ϵ_x, ϵ_y, γ_{xy}. As shown in Fig. 10–3a, the components of the line dx' along the x and y axes are

$$dx = dx' \cos \theta$$
$$dy = dx' \sin \theta \qquad (10\text{–}1)$$

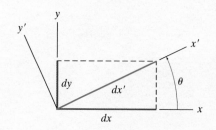

Before deformation

(a)

When the positive normal strain ϵ_x occurs, the line dx is elongated $\epsilon_x \, dx$, Fig. 10–3b, which causes line dx' to elongate $\epsilon_x \, dx \cos \theta$. Likewise, when ϵ_y occurs, line dy elongates $\epsilon_y \, dy$, Fig. 10–3c, which causes line dx' to elongate $\epsilon_y \, dy \sin \theta$. Finally, assuming that dx remains fixed in position, the shear strain γ_{xy}, which is the change in angle between dx and dy, causes the top of line dy to be displaced $\gamma_{xy} \, dy$ to the right, as shown in Fig. 10–3d. This causes dx' to elongate $\gamma_{xy} \, dy \cos \theta$. If all three of these elongations are added together, the resultant elongation of dx' is then

$$\delta x' = \epsilon_x \, dx \cos \theta + \epsilon_y \, dy \sin \theta + \gamma_{xy} \, dy \cos \theta$$

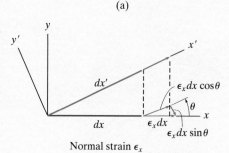

Normal strain ϵ_x

(b)

From Eq. 2–2, the normal strain along the line dx' is $\epsilon_{x'} = \delta x' / dx'$. Using Eq. 10–1, we therefore have

$$\epsilon_{x'} = \epsilon_x \cos^2 \theta + \epsilon_y \sin^2 \theta + \gamma_{xy} \sin \theta \cos \theta \qquad (10\text{–}2)$$

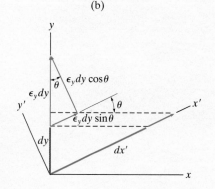

Normal strain ϵ_y

(c)

The rubber specimen is constrained between the two fixed supports, and so it will undergo plane strain when loads are applied to it in the horizontal plane.

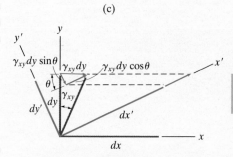

Shear strain γ_{xy}

(d)

Fig. 10–3

10

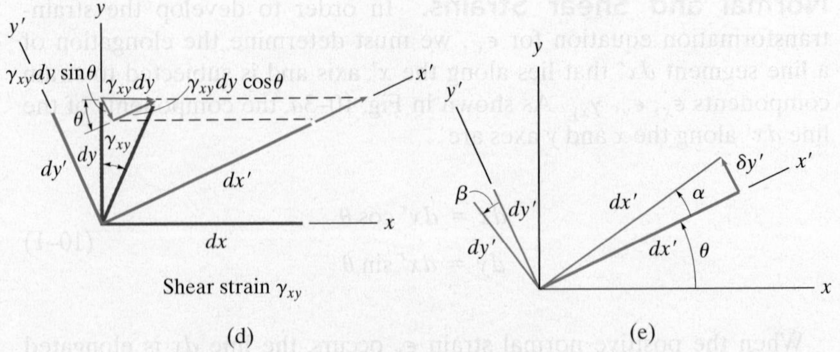

Shear strain γ_{xy}

(d)

(e)

Fig. 10–3 (cont.)

The strain-transformation equation for $\gamma_{x'y'}$ can be developed by considering the amount of rotation each of the line segments dx' and dy' undergo when subjected to the strain components ϵ_x, ϵ_y, γ_{xy}. First we will consider the rotation of dx', which is defined by the counterclockwise angle α shown in Fig. 10–3e. It can be determined by the displacement caused by $\delta y'$ using $\alpha = \delta y'/dx'$. To obtain $\delta y'$, consider the following three displacement components acting in the y' direction: one from ϵ_x, giving $-\epsilon_x\, dx \sin \theta$, Fig. 10–3b; another from ϵ_y, giving $\epsilon_y\, dy \cos \theta$, Fig. 10–3c; and the last from γ_{xy}, giving $-\gamma_{xy}\, dy \sin \theta$, Fig. 10–3d. Thus, $\delta y'$, as caused by all three strain components, is

$$\delta y' = -\epsilon_x\, dx \sin \theta + \epsilon_y\, dy \cos \theta - \gamma_{xy}\, dy \sin \theta$$

Dividing each term by dx' and using Eq. 10–1, with $\alpha = \delta y'/dx'$, we have

$$\alpha = (-\epsilon_x + \epsilon_y) \sin \theta \cos \theta - \gamma_{xy} \sin^2 \theta \qquad (10\text{–}3)$$

As shown in Fig. 10–3e, the line dy' rotates by an amount β. We can determine this angle by a similar analysis, or by simply substituting $\theta + 90°$ for θ into Eq. 10–3. Using the identities $\sin(\theta + 90°) = \cos \theta$, $\cos(\theta + 90°) = -\sin \theta$, we have

$$\beta = (-\epsilon_x + \epsilon_y) \sin(\theta + 90°) \cos(\theta + 90°) - \gamma_{xy} \sin^2(\theta + 90°)$$
$$= -(-\epsilon_x + \epsilon_y) \cos \theta \sin \theta - \gamma_{xy} \cos^2 \theta$$

Since α and β represent the rotation of the sides dx' and dy' of a differential element whose sides were originally oriented along the x' and y' axes, Fig. 10–3e, the element is then subjected to a shear strain of

$$\gamma_{x'y'} = \alpha - \beta = -2(\epsilon_x - \epsilon_y) \sin \theta \cos \theta + \gamma_{xy}(\cos^2 \theta - \sin^2 \theta) \qquad (10\text{–}4)$$

10

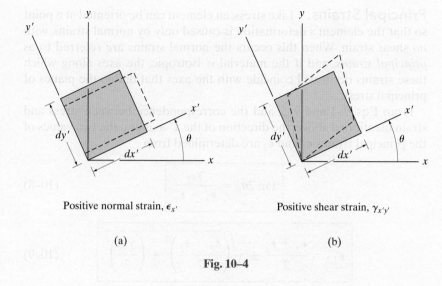

Positive normal strain, $\epsilon_{x'}$ Positive shear strain, $\gamma_{x'y'}$

(a)   (b)

Fig. 10–4

Using the trigonometric identities $\sin 2\theta = 2 \sin \theta \cos \theta$, $\cos^2 \theta = (1 + \cos 2\theta)/2$, and $\sin^2 \theta + \cos^2 \theta = 1$, we can rewrite Eqs. 10–2 and 10–4 in the final form

$$\epsilon_{x'} = \frac{\epsilon_x + \epsilon_y}{2} + \frac{\epsilon_x - \epsilon_y}{2} \cos 2\theta + \frac{\gamma_{xy}}{2} \sin 2\theta \qquad (10\text{–}5)$$

$$\frac{\gamma_{x'y'}}{2} = -\left(\frac{\epsilon_x - \epsilon_y}{2} \right) \sin 2\theta + \frac{\gamma_{xy}}{2} \cos 2\theta \qquad (10\text{–}6)$$

These strain-transformation equations give the normal strain $\epsilon_{x'}$ in the x' direction and the shear strain $\gamma_{x'y'}$ of an element oriented at an angle θ, as shown in Fig. 10–4. According to the established sign convention, if $\epsilon_{x'}$ is *positive*, the element *elongates* in the positive x' direction, Fig. 10–4a, and if $\gamma_{x'y'}$ is positive, the element deforms as shown in Fig. 10–4b.

If the normal strain in the y' direction is required, it can be obtained from Eq. 10–5 by simply substituting $(\theta + 90°)$ for θ. The result is

$$\epsilon_{y'} = \frac{\epsilon_x + \epsilon_y}{2} - \frac{\epsilon_x - \epsilon_y}{2} \cos 2\theta - \frac{\gamma_{xy}}{2} \sin 2\theta \qquad (10\text{–}7)$$

The similarity between the above three equations and those for plane-stress transformation, Eqs. 9–1, 9–2, and 9–3, should be noted. By comparison, σ_x, σ_y, $\sigma_{x'}$, $\sigma_{y'}$ correspond to ϵ_x, ϵ_y, $\epsilon_{x'}$, $\epsilon_{y'}$; and τ_{xy}, $\tau_{x'y'}$ correspond to $\gamma_{xy}/2$, $\gamma_{x'y'}/2$.

10

Principal Strains.

Like stress, an element can be oriented at a point so that the element's deformation is caused only by normal strains, with *no* shear strain. When this occurs the normal strains are referred to as *principal strains*, and if the material is isotropic, the axes along which these strains occur will coincide with the axes that define the planes of principal stress.

From Eqs. 9–4 and 9–5, and the correspondence between stress and strain mentioned above, the direction of the x' axis and the two values of the principal strains ϵ_1 and ϵ_2 are determined from

$$\tan 2\theta_p = \frac{\gamma_{xy}}{\epsilon_x - \epsilon_y} \qquad (10\text{–}8)$$

$$\epsilon_{1,2} = \frac{\epsilon_x + \epsilon_y}{2} \pm \sqrt{\left(\frac{\epsilon_x - \epsilon_y}{2}\right)^2 + \left(\frac{\gamma_{xy}}{2}\right)^2} \qquad (10\text{–}9)$$

Complex stresses are often developed at the joints where the cylindrical and hemispherical vessels are joined together. The stresses are determined by making measurements of strain.

Maximum In-Plane Shear Strain.

Using Eqs. 9–6, 9–7, and 9–8, the direction of the x' axis, and the maximum in-plane shear strain and associated average normal strain are determined from the following equations:

$$\tan 2\theta_s = -\left(\frac{\epsilon_x - \epsilon_y}{\gamma_{xy}}\right) \qquad (10\text{–}10)$$

$$\frac{\gamma_{\substack{max \\ in\text{-}plane}}}{2} = \sqrt{\left(\frac{\epsilon_x - \epsilon_y}{2}\right)^2 + \left(\frac{\gamma_{xy}}{2}\right)^2} \qquad (10\text{–}11)$$

$$\epsilon_{avg} = \frac{\epsilon_x + \epsilon_y}{2} \qquad (10\text{–}12)$$

Important Points

- In the case of plane stress, plane-strain analysis may be used within the plane of the stresses to analyze the data from strain gauges. Remember, though, there will be a normal strain that is perpendicular to the gauges due to the Poisson effect.
- When the state of strain is represented by the principal strains, no shear strain will act on the element.
- The state of strain at a point can also be represented in terms of the maximum in-plane shear strain. In this case an average normal strain will also act on the element.
- The element representing the maximum in-plane shear strain and its associated average normal strains is 45° from the orientation of an element representing the principal strains.

EXAMPLE 10.1

A differential element of material at a point is subjected to a state of plane strain $\epsilon_x = 500(10^{-6})$, $\epsilon_y = -300(10^{-6})$, $\gamma_{xy} = 200(10^{-6})$, which tends to distort the element as shown in Fig. 10–5a. Determine the equivalent strains acting on an element of the material oriented at the point, *clockwise* 30° from the original position.

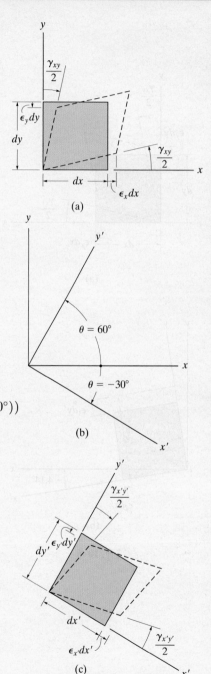

(a)

(b)

(c)

Fig. 10–5

SOLUTION
The strain-transformation Eqs. 10–5 and 10–6 will be used to solve the problem. Since θ is *positive counterclockwise*, then for this problem $\theta = -30°$. Thus,

$$\epsilon_{x'} = \frac{\epsilon_x + \epsilon_y}{2} + \frac{\epsilon_x - \epsilon_y}{2}\cos 2\theta + \frac{\gamma_{xy}}{2}\sin 2\theta$$

$$= \left[\frac{500 + (-300)}{2}\right](10^{-6}) + \left[\frac{500 - (-300)}{2}\right](10^{-6})\cos(2(-30°))$$

$$+ \left[\frac{200(10^{-6})}{2}\right]\sin(2(-30°))$$

$$\epsilon_{x'} = 213(10^{-6}) \qquad \qquad Ans.$$

$$\frac{\gamma_{x'y'}}{2} = -\left(\frac{\epsilon_x - \epsilon_y}{2}\right)\sin 2\theta + \frac{\gamma_{xy}}{2}\cos 2\theta$$

$$= -\left[\frac{500 - (-300)}{2}\right](10^{-6})\sin(2(-30°)) + \frac{200(10^{-6})}{2}\cos(2(-30°))$$

$$\gamma_{x'y'} = 793(10^{-6}) \qquad \qquad Ans.$$

The strain in the y' direction can be obtained from Eq. 10–7 with $\theta = -30°$. However, we can also obtain $\epsilon_{y'}$ using Eq. 10–5 with $\theta = 60°(\theta = -30° + 90°)$, Fig. 10–5b. We have with $\epsilon_{y'}$ replacing $\epsilon_{x'}$,

$$\epsilon_{y'} = \frac{\epsilon_x + \epsilon_y}{2} + \frac{\epsilon_x - \epsilon_y}{2}\cos 2\theta + \frac{\gamma_{xy}}{2}\sin 2\theta$$

$$= \left[\frac{500 + (-300)}{2}\right](10^{-6}) + \left[\frac{500 - (-300)}{2}\right](10^{-6})\cos(2(60°))$$

$$+ \frac{200(10^{-6})}{2}\sin(2(60°))$$

$$\epsilon_{y'} = -13.4(10^{-6}) \qquad \qquad Ans.$$

These results tend to distort the element as shown in Fig. 10–5c.

EXAMPLE | 10.2

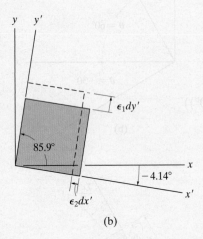

(a)

y   y'

85.9°

$\epsilon_1 dy'$

$\epsilon_2 dx'$

-4.14°

x

x'

(b)

Fig. 10–6

A differential element of material at a point is subjected to a state of plane strain defined by $\epsilon_x = -350(10^{-6})$, $\epsilon_y = 200(10^{-6})$, $\gamma_{xy} = 80(10^{-6})$, which tends to distort the element as shown in Fig. 10–6a. Determine the principal strains at the point and the associated orientation of the element.

SOLUTION

Orientation of the Element. From Eq. 10–8 we have

$$\tan 2\theta_p = \frac{\gamma_{xy}}{\epsilon_x - \epsilon_y}$$

$$= \frac{80(10^{-6})}{(-350 - 200)(10^{-6})}$$

Thus, $2\theta_p = -8.28°$ and $-8.28° + 180° = 171.72°$, so that

$$\theta_p = -4.14° \text{ and } 85.9° \qquad Ans.$$

Each of these angles is measured *positive counterclockwise*, from the x axis to the outward normals on each face of the element, Fig. 10–6b.

Principal Strains. The principal strains are determined from Eq. 10–9. We have

$$\epsilon_{1,2} = \frac{\epsilon_x + \epsilon_y}{2} \pm \sqrt{\left(\frac{\epsilon_x - \epsilon_y}{2}\right)^2 + \left(\frac{\gamma_{xy}}{2}\right)^2}$$

$$= \frac{(-350 + 200)(10^{-6})}{2} \pm \left[\sqrt{\left(\frac{-350 - 200}{2}\right)^2 + \left(\frac{80}{2}\right)^2}\right](10^{-6})$$

$$= -75.0(10^{-6}) \pm 277.9(10^{-6})$$

$$\epsilon_1 = 203(10^{-6}) \qquad \epsilon_2 = -353(10^{-6}) \qquad Ans.$$

We can determine which of these two strains deforms the element in the x' direction by applying Eq. 10–5 with $\theta = -4.14°$. Thus,

$$\epsilon_{x'} = \frac{\epsilon_x + \epsilon_y}{2} + \frac{\epsilon_x - \epsilon_y}{2}\cos 2\theta + \frac{\gamma_{xy}}{2}\sin 2\theta$$

$$= \left(\frac{-350 + 200}{2}\right)(10^{-6}) + \left(\frac{-350 - 200}{2}\right)(10^{-6})\cos 2(-4.14°)$$

$$+ \frac{80(10^{-6})}{2}\sin 2(-4.14°)$$

$$\epsilon_{x'} = -353(10^{-6})$$

Hence $\epsilon_{x'} = \epsilon_2$. When subjected to the principal strains, the element is distorted as shown in Fig. 10–6b.

EXAMPLE 10.3

A differential element of material at a point is subjected to a state of plane strain defined by $\epsilon_x = -350(10^{-6})$, $\epsilon_y = 200(10^{-6})$, $\gamma_{xy} = 80(10^{-6})$, which tends to distort the element as shown in Fig. 10–7a. Determine the maximum in-plane shear strain at the point and the associated orientation of the element.

SOLUTION

Orientation of the Element. From Eq. 10–10 we have

$$\tan 2\theta_s = -\left(\frac{\epsilon_x - \epsilon_y}{\gamma_{xy}}\right) = -\frac{(-350 - 200)(10^{-6})}{80(10^{-6})}$$

Thus, $2\theta_s = 81.72°$ and $81.72° + 180° = 261.72°$, so that

$$\theta_s = 40.9° \text{ and } 131°$$

Note that this orientation is 45° from that shown in Fig. 10–6b in Example 10.2 as expected.

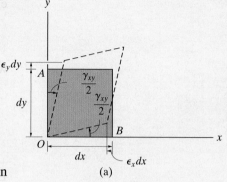

(a)

Maximum In-Plane Shear Strain. Applying Eq. 10–11 gives

$$\frac{\gamma_{\text{in-plane}}^{\text{max}}}{2} = \sqrt{\left(\frac{\epsilon_x - \epsilon_y}{2}\right)^2 + \left(\frac{\gamma_{xy}}{2}\right)^2}$$

$$= \left[\sqrt{\left(\frac{-350 - 200}{2}\right)^2 + \left(\frac{80}{2}\right)^2}\right](10^{-6})$$

$$\gamma_{\text{in-plane}}^{\text{max}} = 556(10^{-6}) \qquad \qquad Ans.$$

Due to the square root, the proper sign of $\gamma_{\text{in-plane}}^{\text{max}}$ can be obtained by applying Eq. 10–6 with $\theta_s = 40.9°$. We have

$$\frac{\gamma_{x'y'}}{2} = -\frac{\epsilon_x - \epsilon_y}{2}\sin 2\theta + \frac{\gamma_{xy}}{2}\cos 2\theta$$

$$= -\left(\frac{-350 - 200}{2}\right)(10^{-6})\sin 2(40.9°) + \frac{80(10^{-6})}{2}\cos 2(40.9°)$$

$$\gamma_{x'y'} = 556(10^{-6})$$

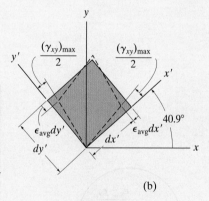

Fig. 10–7

This result is positive and so $\gamma_{\text{in-plane}}^{\text{max}}$ tends to distort the element so that the right angle between dx' and dy' is *decreased* (positive sign convention), Fig. 10–7b.

Also, there are associated average normal strains imposed on the element that are determined from Eq. 10–12:

$$\epsilon_{\text{avg}} = \frac{\epsilon_x + \epsilon_y}{2} = \frac{-350 + 200}{2}(10^{-6}) = -75(10^{-6})$$

These strains tend to cause the element to contract, Fig. 10–7b.

*10.3 Mohr's Circle—Plane Strain

Since the equations of plane-strain transformation are mathematically similar to the equations of plane-stress transformation, we can also solve problems involving the transformation of strain using Mohr's circle.

Like the case for stress, the parameter θ in Eqs. 10–5 and 10–6 can be eliminated and the result rewritten in the form

$$(\epsilon_{x'} - \epsilon_{avg})^2 + \left(\frac{\gamma_{x'y'}}{2}\right)^2 = R^2 \tag{10–13}$$

where

$$\epsilon_{avg} = \frac{\epsilon_x + \epsilon_y}{2}$$

$$R = \sqrt{\left(\frac{\epsilon_x - \epsilon_y}{2}\right)^2 + \left(\frac{\gamma_{xy}}{2}\right)^2}$$

Equation 10–13 represents the equation of Mohr's circle for strain. It has a center on the ϵ axis at point $C(\epsilon_{avg}, 0)$ and a radius R.

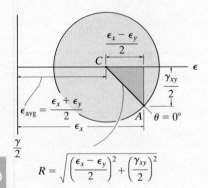

$$R = \sqrt{\left(\frac{\epsilon_x - \epsilon_y}{2}\right)^2 + \left(\frac{\gamma_{xy}}{2}\right)^2}$$

Fig. 10–8

Procedure for Analysis

The procedure for drawing Mohr's circle for strain follows the same one established for stress.

Construction of the Circle.

- Establish a coordinate system such that the abscissa represents the normal strain ϵ, with *positive to the right*, and the ordinate represents *half* the value of the shear strain, $\gamma/2$, with *positive downward*, Fig. 10–8.

- Using the positive sign convention for ϵ_x, ϵ_y, γ_{xy}, as shown in Fig. 10–2, determine the center of the circle C, which is located on the ϵ axis at a distance $\epsilon_{avg} = (\epsilon_x + \epsilon_y)/2$ from the origin, Fig. 10–8.

- Plot the reference point A having coordinates $A(\epsilon_x, \gamma_{xy}/2)$. This point represents the case for which the x' axis coincides with the x axis. Hence $\theta = 0°$, Fig. 10–8.

- Connect point A with the center C of the circle and from the shaded triangle determine the radius R of the circle, Fig. 10–8.

- Once R has been determined, sketch the circle.

Principal Strains.

- The principal strains ϵ_1 and ϵ_2 are determined from the circle as the coordinates of points B and D, that is where $\gamma/2 = 0$, Fig. 10–9a.

- The orientation of the plane on which ϵ_1 acts can be determined from the circle by calculating $2\theta_{p_1}$ using trigonometry. Here this angle happens to be counterclockwise *from* the radial reference line *CA to* line *CB*, Fig. 10–9a. Remember that the *rotation* of θ_{p_1} must be in this *same direction*, from the element's reference axis x to the x' axis, Fig. 10–9b.*

- When ϵ_1 and ϵ_2 are indicated as being positive as in Fig. 10–9a, the element in Fig. 10–9b will elongate in the x' and y' directions as shown by the dashed outline.

Maximum In-Plane Shear Strain.

- The average normal strain and half the maximum in-plane shear strain are determined from the circle as the coordinates of point E or F, Fig. 10–9a.

- The orientation of the plane on which $\gamma^{\max}_{\text{in-plane}}$ and ϵ_{avg} act can be determined from the circle by calculating $2\theta_{s_1}$ using trigonometry. Here this angle happens to be clockwise *from* the radial reference line *CA to* line *CE*, Fig. 10–9a. Remember that the *rotation* of θ_{s_1} must be in this *same direction*, from the element's reference axis x to the x' axis, Fig. 10–9c.*

Strains on Arbitrary Plane.

- The normal and shear strain components $\epsilon_{x'}$ and $\gamma_{x'y'}$ for a plane oriented at an angle θ, Fig. 10–9d, can be obtained from the circle using trigonometry to determine the coordinates of point P, Fig. 10–9a.

- To locate P, the known angle θ of the x' axis is measured on the circle as 2θ. This measurement is made *from* the radial reference line *CA to* the radial line *CP*. Remember that measurements for 2θ on the circle must be in the same direction as θ for the x' axis.*

- If the value of $\epsilon_{y'}$ is required, it can be determined by calculating the ϵ coordinate of point Q in Fig. 10–9a. The line CQ lies $180°$ away from CP and thus represents a $90°$ rotation of the x' axis.

*If the $\gamma/2$ axis were constructed *positive upwards*, then the angle 2θ on the circle would be measured in the *opposite direction* to the orientation θ of the plane.

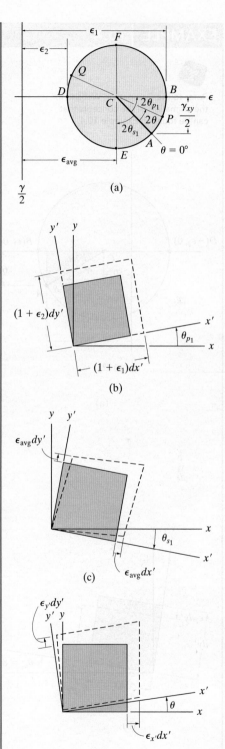

(a)

(b)

(c)

(d)

Fig. 10–9

10

EXAMPLE   10.4

Refer to the Companion Website for the animation of the concepts that can be found in Example 10.4.

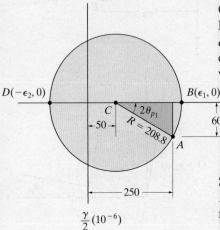

$D(-\epsilon_2, 0)$

C

$2\theta_{p_1}$

$R = 208.8$

A

$B(\epsilon_1, 0)$

$\epsilon \, (10^{-6})$

$\frac{\gamma}{2} (10^{-6})$

(a)

y' y

$\epsilon_2 dy'$

dy'

$\theta_{p_1} = 8.35°$

x'
x

dx'

$\epsilon_1 dx'$

(b)

Fig. 10–10

The state of plane strain at a point is represented by the components $\epsilon_x = 250(10^{-6})$, $\epsilon_y = -150(10^{-6})$, and $\gamma_{xy} = 120(10^{-6})$. Determine the principal strains and the orientation of the element.

SOLUTION

Construction of the Circle. The ϵ and $\gamma/2$ axes are established in Fig. 10–10a. Remember that the *positive* $\gamma/2$ axis must be directed *downward* so that *counterclockwise* rotations of the element correspond to *counterclockwise* rotation around the circle, and vice versa. The center of the circle C is located on the ϵ axis at

$$\epsilon_{avg} = \frac{250 + (-150)}{2}(10^{-6}) = 50(10^{-6})$$

Since $\gamma_{xy}/2 = 60(10^{-6})$, the reference point A ($\theta = 0°$) has coordinates $A(250(10^{-6}), 60(10^{-6}))$. From the shaded triangle in Fig. 10–10a, the radius of the circle is CA; that is,

$$R = \left[\sqrt{(250 - 50)^2 + (60)^2}\right](10^{-6}) = 208.8(10^{-6})$$

Principal Strains. The ϵ coordinates of points B and D represent the principal strains. They are

$$\epsilon_1 = (50 + 208.8)(10^{-6}) = 259(10^{-6}) \qquad Ans.$$
$$\epsilon_2 = (50 - 208.8)(10^{-6}) = -159(10^{-6}) \qquad Ans.$$

The direction of the positive principal strain ϵ_1 is defined by the *counterclockwise* angle $2\theta_{p_1}$, measured from the radial reference line CA ($\theta = 0°$) to the line CB. We have

$$\tan 2\theta_{p_1} = \frac{60}{(250 - 50)}$$

$$\theta_{p_1} = 8.35° \qquad Ans.$$

Hence, the side dx' of the element is oriented *counterclockwise* 8.35° as shown in Fig. 10–10b. This also defines the direction of ϵ_1. The deformation of the element is also shown in the figure.

EXAMPLE | 10.5

The state of plane strain at a point is represented by the components $\epsilon_x = 250(10^{-6})$, $\epsilon_y = -150(10^{-6})$, and $\gamma_{xy} = 120(10^{-6})$. Determine the maximum in-plane shear strains and the orientation of an element.

SOLUTION
The circle has been established in the previous example and is shown in Fig. 10–11a.

Maximum In-Plane Shear Strain.   Half the maximum in-plane shear strain and average normal strain are represented by the coordinates of point E or F on the circle. From the coordinates of point E,

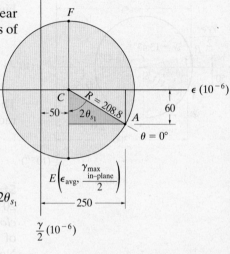

$$\frac{(\gamma_{x'y'})_{\text{in-plane}}^{\max}}{2} = 208.8(10^{-6})$$

$$(\gamma_{x'y'})_{\text{in-plane}}^{\max} = 418(10^{-6}) \qquad Ans.$$

$$\epsilon_{\text{avg}} = 50(10^{-6})$$

To orient the element, we can determine the clockwise angle $2\theta_{s_1}$ measured from CA ($\theta = 0°$) to CE.

$$2\theta_{s_1} = 90° - 2(8.35°)$$

$$\theta_{s_1} = 36.7° \qquad Ans.$$

Fig. 10–11

This angle is shown in Fig. 10–11b. Since the shear strain defined from point E on the circle has a positive value and the average normal strain is also positive, these strains deform the element into the dashed shape shown in the figure.

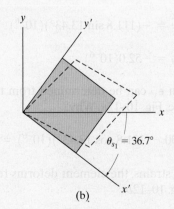

(b)

EXAMPLE 10.6

The state of plane strain at a point is represented on an element having components $\epsilon_x = -300(10^{-6})$, $\epsilon_y = -100(10^{-6})$, and $\gamma_{xy} = 100(10^{-6})$. Determine the state of strain on an element oriented 20° clockwise from this reported position.

SOLUTION

Construction of the Circle.   The ϵ and $\gamma/2$ axes are established in Fig. 10–12a. The center of the circle is on the ϵ axis at

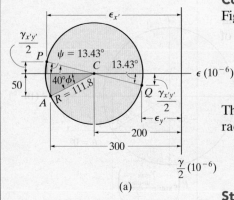

$$\epsilon_{avg} = \left(\frac{-300 - 100}{2}\right)(10^{-6}) = -200(10^{-6})$$

The reference point A has coordinates $A(-300(10^{-6}), 50(10^{-6}))$. The radius CA determined from the shaded triangle is therefore

$$R = \left[\sqrt{(300 - 200)^2 + (50)^2}\right](10^{-6}) = 111.8(10^{-6})$$

Strains on Inclined Element.   Since the element is to be oriented 20° *clockwise*, we must establish a radial line CP, $2(20°) = 40°$ *clockwise*, measured from CA ($\theta = 0°$), Fig. 10–12a. The coordinates of point P ($\epsilon_{x'}, \gamma_{x'y'}/2$) are obtained from the geometry of the circle. Note that

$$\phi = \tan^{-1}\left(\frac{50}{(300 - 200)}\right) = 26.57°, \qquad \psi = 40° - 26.57° = 13.43°$$

Thus,

$$\epsilon_{x'} = -(200 + 111.8 \cos 13.43°)(10^{-6})$$
$$= -309(10^{-6}) \qquad\qquad Ans.$$

$$\frac{\gamma_{x'y'}}{2} = -(111.8 \sin 13.43°)(10^{-6})$$

$$\gamma_{x'y'} = -52.0(10^{-6}) \qquad\qquad Ans.$$

The normal strain $\epsilon_{y'}$ can be determined from the ϵ coordinate of point Q on the circle, Fig. 10–12a. Why?

$$\epsilon_{y'} = -(200 - 111.8 \cos 13.43°)(10^{-6}) = -91.3(10^{-6}) \;\; Ans.$$

As a result of these strains, the element deforms relative to the x', y' axes as shown in Fig. 10–12b.

(a)

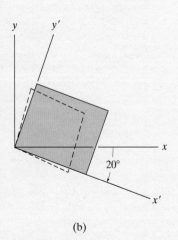

(b)

Fig. 10–12

PROBLEMS

10–1. Prove that the sum of the normal strains in perpendicular directions is constant.

10–2. The state of strain at the point has components of $\epsilon_x = 200(10^{-6})$, $\epsilon_y = -300(10^{-6})$, and $\gamma_{xy} = 400(10^{-6})$. Use the strain-transformation equations to determine the equivalent in-plane strains on an element oriented at an angle of $30°$ counterclockwise from the original position. Sketch the deformed element due to these strains within the x–y plane.

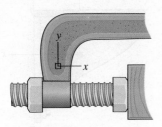

Prob. 10–2

10–3. A strain gauge is mounted on the 25-mm-diameter A-36 steel shaft in the manner shown. When the shaft is rotating with an angular velocity of $\omega = 1760$ rev/min, the reading on the strain gauge is $\epsilon = 800(10^{-6})$. Determine the power output of the motor. Assume the shaft is only subjected to a torque.

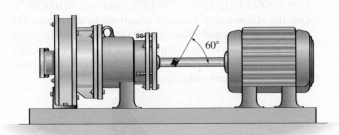

Prob. 10–3

*10–4. The state of strain at a point on a wrench has components $\epsilon_x = 120(10^{-6})$, $\epsilon_y = -180(10^{-6})$, $\gamma_{xy} = 150(10^{-6})$. Use the strain-transformation equations to determine (a) the in-plane principal strains and (b) the maximum in-plane shear strain and average normal strain. In each case specify the orientation of the element and show how the strains deform the element within the x–y plane.

10–5. The state of strain at the point on the arm has components $\epsilon_x = 250(10^{-6})$, $\epsilon_y = -450(10^{-6})$, $\gamma_{xy} = -825(10^{-6})$. Use the strain-transformation equations to determine (a) the in-plane principal strains and (b) the maximum in-plane shear strain and average normal strain. In each case specify the orientation of the element and show how the strains deform the element within the x–y plane.

Prob. 10–5

10–6. The state of strain at the point has components of $\epsilon_x = -100(10^{-6})$, $\epsilon_y = 400(10^{-6})$, and $\gamma_{xy} = -300(10^{-6})$. Use the strain-transformation equations to determine the equivalent in-plane strains on an element oriented at an angle of $60°$ counterclockwise from the original position. Sketch the deformed element due to these strains within the x–y plane.

10–7. The state of strain at the point has components of $\epsilon_x = 100(10^{-6})$, $\epsilon_y = 300(10^{-6})$, and $\gamma_{xy} = -150(10^{-6})$. Use the strain-transformation equations to determine the equivalent in-plane strains on an element oriented $\theta = 30°$ clockwise. Sketch the deformed element due to these strains within the x–y plane.

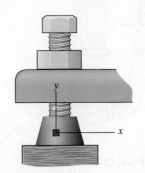

Probs. 10–6/7

10

*10–8. The state of strain at the point on the bracket has components $\epsilon_x = -200(10^{-6})$, $\epsilon_y = -650(10^{-6})$, $\gamma_{xy} = -175(10^{-6})$. Use the strain-transformation equations to determine the equivalent in-plane strains on an element oriented at an angle of $\theta = 20°$ counterclockwise from the original position. Sketch the deformed element due to these strains within the x–y plane.

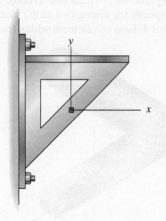

Prob. 10–8

10–9. The state of strain at the point has components of $\epsilon_x = 180(10^{-6})$, $\epsilon_y = -120(10^{-6})$, and $\gamma_{xy} = -100(10^{-6})$. Use the strain-transformation equations to determine (a) the in-plane principal strains and (b) the maximum in-plane shear strain and average normal strain. In each case specify the orientation of the element and show how the strains deform the element within the x–y plane.

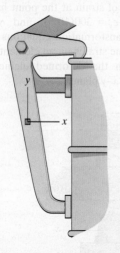

Prob. 10–9

10–10. The state of strain at the point on the bracket has components $\epsilon_x = 400(10^{-6})$, $\epsilon_y = -250(10^{-6})$, $\gamma_{xy} = 310(10^{-6})$. Use the strain-transformation equations to determine the equivalent in-plane strains on an element oriented at an angle of $\theta = 30°$ clockwise from the original position. Sketch the deformed element due to these strains within the x–y plane.

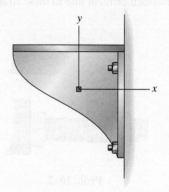

Prob. 10–10

10–11. The state of strain at the point has components of $\epsilon_x = -100(10^{-6})$, $\epsilon_y = -200(10^{-6})$, and $\gamma_{xy} = 100(10^{-6})$. Use the strain-transformation equations to determine (a) the in-plane principal strains and (b) the maximum in-plane shear strain and average normal strain. In each case specify the orientation of the element and show how the strains deform the element within the x–y plane.

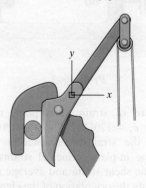

Prob. 10–11

*10–12. The state of plane strain on an element is given by $\epsilon_x = 500(10^{-6})$, $\epsilon_y = 300(10^{-6})$, and $\gamma_{xy} = -200(10^{-6})$. Determine the equivalent state of strain on an element at the same point oriented 45° clockwise with respect to the original element.

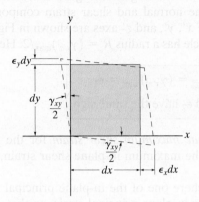

Prob. 10–12

10–13. The state of plane strain on an element is $\epsilon_x = -300(10^{-6})$, $\epsilon_y = 0$, and $\gamma_{xy} = 150(10^{-6})$. Determine the equivalent state of strain which represents (a) the principal strains, and (b) the maximum in-plane shear strain and the associated average normal strain. Specify the orientation of the corresponding elements for these states of strain with respect to the original element.

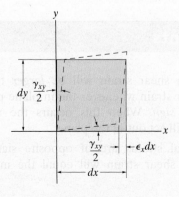

Prob. 10–13

10–14. The state of strain at the point on a boom of an hydraulic engine crane has components of $\epsilon_x = 250(10^{-6})$, $\epsilon_y = 300(10^{-6})$, and $\gamma_{xy} = -180(10^{-6})$. Use the strain-transformation equations to determine (a) the in-plane principal strains and (b) the maximum in-plane shear strain and average normal strain. In each case, specify the orientation of the element and show how the strains deform the element within the x–y plane.

Prob. 10–14

■10–15. Consider the general case of plane strain where ϵ_x, ϵ_y, and γ_{xy} are known. Write a computer program that can be used to determine the normal and shear strain, $\epsilon_{x'}$ and $\gamma_{x'y'}$, on the plane of an element oriented θ from the horizontal. Also, include the principal strains and the element's orientation, and the maximum in-plane shear strain, the average normal strain, and the element's orientation.

*10–16. The state of strain at a point on a support has components of $\epsilon_x = 350(10^{-6})$, $\epsilon_y = 400(10^{-6})$, $\gamma_{xy} = -675(10^{-6})$. Use the strain-transformation equations to determine (a) the in-plane principal strains and (b) the maximum in-plane shear strain and average normal strain. In each case specify the orientation of the element and show how the strains deform the element within the x–y plane.

•10–17. Solve part (a) of Prob. 10–4 using Mohr's circle.

10–18. Solve part (b) of Prob. 10–4 using Mohr's circle.

10–19. Solve Prob. 10–8 using Mohr's circle.

*10–20. Solve Prob. 10–10 using Mohr's circle.

•10–21. Solve Prob. 10–14 using Mohr's circle.

10

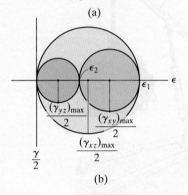

x–y plane strain

(a)

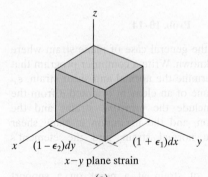

(b)

Fig. 10–13

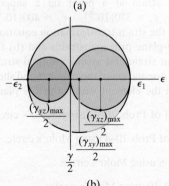

x–y plane strain

(a)

(b)

Fig. 10–14

*10.4 Absolute Maximum Shear Strain

In Sec. 9.5 it was pointed out that in the case of plane stress, the absolute maximum shear stress in an element of material will occur out of the plane when the principal stresses have the same sign, i.e., both are tensile or both are compressive. A similar result occurs for plane strain. For example, if the principal in-plane strains cause elongations, Fig. 10–13a, then the three Mohr's circles describing the normal and shear strain components for elements oriented about the x', y', and z' axes are shown in Fig. 10–13b. By inspection, the largest circle has a radius $R = (\gamma_{x'z'})_{max}/2$. Hence,

$$\gamma_{max}^{abs} = (\gamma_{x'z'})_{max} = \epsilon_1$$

ϵ_1 and ϵ_2 have the same sign

(10–14)

This value gives the *absolute maximum shear strain* for the material. Note that it is *larger* than the maximum in-plane shear strain, which is $(\gamma_{x'y'})_{max} = \epsilon_1 - \epsilon_2$.

Now consider the case where one of the in-plane principal strains is of *opposite sign* to the other in-plane principal strain, so that ϵ_1 causes elongation and ϵ_2 causes contraction, Fig. 10–14a. Mohr's circles, which describe the strains on each element's orientation about the x', y', z' axes, are shown in Fig. 10–14b. Here

$$\gamma_{max}^{abs} = (\gamma_{x'y'})_{max}^{\text{in-plane}} = \epsilon_1 - \epsilon_2$$

ϵ_1 and ϵ_2 have opposite signs

(10–15)

We may therefore summarize the above two cases as follows. If the in-plane principal strains both have the *same sign*, the *absolute maximum shear strain* will occur *out of plane* and has a value of $\gamma_{max}^{abs} = \epsilon_{max}$. However, if the in-plane principal strains are of *opposite signs*, then the absolute maximum shear strain *equals* the maximum in-plane shear strain.

Important Points

- The absolute maximum shear strain will be *larger* than the maximum in-plane shear strain whenever the in-plane principal strains have the *same sign*. When this occurs the absolute maximum shear strain will act out of the plane.

- If the in-plane principal strains are of opposite signs, then the absolute maximum shear strain will equal the maximum in-plane shear strain.

EXAMPLE 10.7

The state of plane strain at a point is represented by the strain components $\epsilon_x = -400(10^{-6})$, $\epsilon_y = 200(10^{-6})$, $\gamma_{xy} = 150(10^{-6})$. Determine the maximum in-plane shear strain and the absolute maximum shear strain.

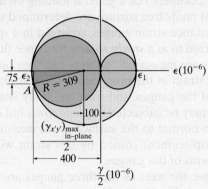

Fig. 10–15

SOLUTION

Maximum In-Plane Shear Strain.   We will solve this problem using Mohr's circle. From the strain components, the center of the circle is on the ϵ axis at

$$\epsilon_{\text{avg}} = \frac{-400 + 200}{2}(10^{-6}) = -100(10^{-6})$$

Since $\gamma_{xy}/2 = 75(10^{-6})$, the reference point A has coordinates $(-400(10^{-6}), 75(10^{-6}))$. As shown in Fig. 10–15, the radius of the circle is therefore

$$R = \left[\sqrt{(400 - 100)^2 + (75)^2}\right](10^{-6}) = 309(10^{-6})$$

Calculating the in-plane principal strains from the circle, we have

$$\epsilon_1 = (-100 + 309)(10^{-6}) = 209(10^{-6})$$

$$\epsilon_2 = (-100 - 309)(10^{-6}) = -409(10^{-6})$$

Also, the maximum in-plane shear strain is

$$\gamma_{\substack{\text{max} \\ \text{in-plane}}} = \epsilon_1 - \epsilon_2 = [209 - (-409)](10^{-6}) = 618(10^{-6}) \quad Ans.$$

Absolute Maximum Shear Strain.   From the above results, we have $\epsilon_1 = 209(10^{-6})$, $\epsilon_2 = -409(10^{-6})$. The three Mohr's circles, plotted for element orientations about each of the x, y, z axes, are also shown in Fig. 10–15. It is seen that since the *principal in-plane strains have opposite signs*, the maximum in-plane shear strain is *also* the absolute maximum shear strain; i.e.,

$$\gamma_{\substack{\text{abs} \\ \text{max}}} = 618(10^{-6}) \quad Ans.$$

10

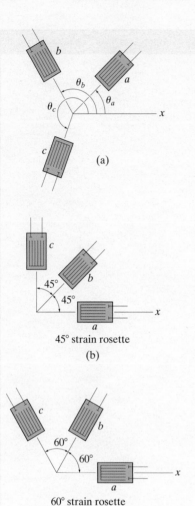

(a)

45° strain rosette

(b)

60° strain rosette

(c)

Fig. 10–16

Typical electrical resistance 45° strain rosette.

10.5 Strain Rosettes

When performing a tension test on a specimen as discussed in Sec. 3.1, the normal strain in the material is measured using an **electrical-resistance strain gauge**, which consists of a wire grid or piece of metal foil bonded to the specimen. For a general loading on a body, however, the strains at a point on its free surface are determined using a cluster of three electrical-resistance strain gauges, arranged in a specified pattern. This pattern is referred to as a **strain rosette**, and once the normal strains on the three gauges are measured, the data can then be transformed to specify the state of strain at the point. Since these strains are measured *only* in the plane of the gauges, and since the body is stress-free on its surface, the gauges may be subjected to *plane stress* but *not* plane strain. Although the strain normal to the surface is not measured, realize that the out-of-plane displacement caused by this strain will *not* affect the in-plane measurements of the gauges.

In the general case, the axes of the three gauges are arranged at the angles $\theta_a, \theta_b, \theta_c$ shown in Fig. 10–16a. If the readings $\epsilon_a, \epsilon_b, \epsilon_c$ are taken, we can determine the strain components $\epsilon_x, \epsilon_y, \gamma_{xy}$ at the point by applying the strain-transformation equation, Eq. 10–2, for each gauge. We have

$$\epsilon_a = \epsilon_x \cos^2 \theta_a + \epsilon_y \sin^2 \theta_a + \gamma_{xy} \sin \theta_a \cos \theta_a$$
$$\epsilon_b = \epsilon_x \cos^2 \theta_b + \epsilon_y \sin^2 \theta_b + \gamma_{xy} \sin \theta_b \cos \theta_b \qquad (10\text{–}16)$$
$$\epsilon_c = \epsilon_x \cos^2 \theta_c + \epsilon_y \sin^2 \theta_c + \gamma_{xy} \sin \theta_c \cos \theta_c$$

The values of $\epsilon_x, \epsilon_y, \gamma_{xy}$ are determined by solving these three equations simultaneously.

Strain rosettes are often arranged in 45° or 60° patterns. In the case of the 45° or "rectangular" strain rosette shown in Fig. 10–16b, $\theta_a = 0°$, $\theta_b = 45°$, $\theta_c = 90°$, so that Eq. 10–16 gives

$$\epsilon_x = \epsilon_a$$
$$\epsilon_y = \epsilon_c$$
$$\gamma_{xy} = 2\epsilon_b - (\epsilon_a + \epsilon_c)$$

And for the 60° strain rosette in Fig. 10–16c, $\theta_a = 0°$, $\theta_b = 60°$, $\theta_c = 120°$. Here Eq. 10–16 gives

$$\epsilon_x = \epsilon_a$$
$$\epsilon_y = \frac{1}{3}(2\epsilon_b + 2\epsilon_c - \epsilon_a) \qquad (10\text{–}17)$$
$$\gamma_{xy} = \frac{2}{\sqrt{3}}(\epsilon_b - \epsilon_c)$$

Once $\epsilon_x, \epsilon_y, \gamma_{xy}$ are determined, the transformation equations of Sec. 10.2 or Mohr's circle can then be used to determine the principal in-plane strains and the maximum in-plane shear strain at the point.

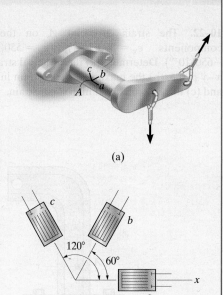

(a)

EXAMPLE | 10.8

The state of strain at point A on the bracket in Fig. 10–17a is measured using the strain rosette shown in Fig. 10–17b. Due to the loadings, the readings from the gauges give $\epsilon_a = 60(10^{-6})$, $\epsilon_b = 135(10^{-6})$, and $\epsilon_c = 264(10^{-6})$. Determine the in-plane principal strains at the point and the directions in which they act.

SOLUTION

We will use Eqs. 10–16 for the solution. Establishing an x axis as shown in Fig. 10–17b and measuring the angles counterclockwise from the $+x$ axis to the centerlines of each gauge, we have $\theta_a = 0°$, $\theta_b = 60°$, and $\theta_c = 120°$. Substituting these results, along with the problem data, into the equations gives

$$60(10^{-6}) = \epsilon_x \cos^2 0° + \epsilon_y \sin^2 0° + \gamma_{xy} \sin 0° \cos 0°$$
$$= \epsilon_x \qquad (1)$$
$$135(10^{-6}) = \epsilon_x \cos^2 60° + \epsilon_y \sin^2 60° + \gamma_{xy} \sin 60° \cos 60°$$
$$= 0.25\epsilon_x + 0.75\epsilon_y + 0.433\gamma_{xy} \qquad (2)$$
$$264(10^{-6}) = \epsilon_x \cos^2 120° + \epsilon_y \sin^2 120° + \gamma_{xy} \sin 120° \cos 120°$$
$$= 0.25\epsilon_x + 0.75\epsilon_y - 0.433\gamma_{xy} \qquad (3)$$

Using Eq. 1 and solving Eqs. 2 and 3 simultaneously, we get

$$\epsilon_x = 60(10^{-6}) \qquad \epsilon_y = 246(10^{-6}) \qquad \gamma_{xy} = -149(10^{-6})$$

These same results can also be obtained in a more direct manner from Eq. 10–17.

The in-plane principal strains can be determined using Mohr's circle. The reference point on the circle is at A $[60(10^{-6}), -74.5(10^{-6})]$ and the center of the circle, C, is on the ϵ axis at $\epsilon_{avg} = 153(10^{-6})$, Fig. 10–17c. From the shaded triangle, the radius is

$$R = \left[\sqrt{(153 - 60)^2 + (74.5)^2}\right](10^{-6}) = 119.1(10^{-6})$$

The in-plane principal strains are thus

$$\epsilon_1 = 153(10^{-6}) + 119.1(10^{-6}) = 272(10^{-6}) \qquad Ans.$$
$$\epsilon_2 = 153(10^{-6}) - 119.1(10^{-6}) = 33.9(10^{-6}) \qquad Ans.$$
$$2\theta_{p2} = \tan^{-1}\frac{74.5}{(153 - 60)} = 38.7°$$
$$\theta_{p2} = 19.3° \qquad Ans.$$

NOTE: The deformed element is shown in the dashed position in Fig. 10–17d. Realize that, due to the Poisson effect, the element is *also* subjected to an out-of-plane strain, i.e., in the z direction, although this value will not influence the calculated results.

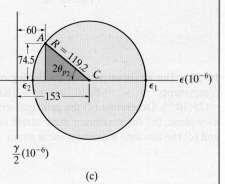

(b)

(c)

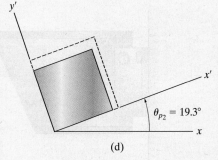

(d)

Fig. 10–17

10

PROBLEMS

10–22. The strain at point A on the bracket has components $\epsilon_x = 300(10^{-6})$, $\epsilon_y = 550(10^{-6})$, $\gamma_{xy} = -650(10^{-6})$. Determine (a) the principal strains at A in the x–y plane, (b) the maximum shear strain in the x–y plane, and (c) the absolute maximum shear strain.

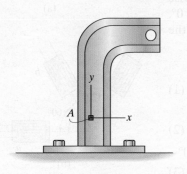

Prob. 10–22

10–23. The strain at point A on the leg of the angle has components $\epsilon_x = -140(10^{-6})$, $\epsilon_y = 180(10^{-6})$, $\gamma_{xy} = -125(10^{-6})$. Determine (a) the principal strains at A in the x–y plane, (b) the maximum shear strain in the x–y plane, and (c) the absolute maximum shear strain.

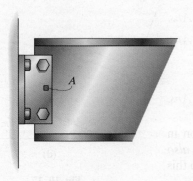

Prob. 10–23

*10–24. The strain at point A on the pressure-vessel wall has components $\epsilon_x = 480(10^{-6})$, $\epsilon_y = 720(10^{-6})$, $\gamma_{xy} = 650(10^{-6})$. Determine (a) the principal strains at A, in the x–y plane, (b) the maximum shear strain in the x–y plane, and (c) the absolute maximum shear strain.

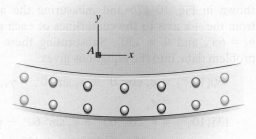

Prob. 10–24

•10–25. The 60° strain rosette is mounted on the bracket. The following readings are obtained for each gauge: $\epsilon_a = -100(10^{-6})$, $\epsilon_b = 250(10^{-6})$, and $\epsilon_c = 150(10^{-6})$. Determine (a) the principal strains and (b) the maximum in-plane shear strain and associated average normal strain. In each case show the deformed element due to these strains.

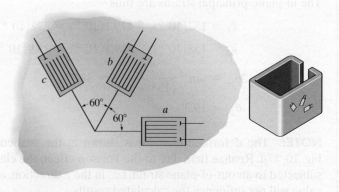

Prob. 10–25

10–26. The 60° strain rosette is mounted on a beam. The following readings are obtained for each gauge: $\epsilon_a = 200(10^{-6})$, $\epsilon_b = -450(10^{-6})$, and $\epsilon_c = 250(10^{-6})$. Determine (a) the in-plane principal strains and (b) the maximum in-plane shear strain and average normal strain. In each case show the deformed element due to these strains.

*10–28. The 45° strain rosette is mounted on the link of the backhoe. The following readings are obtained from each gauge: $\epsilon_a = 650(10^{-6})$, $\epsilon_b = -300(10^{-6})$, $\epsilon_c = 480(10^{-6})$. Determine (a) the in-plane principal strains and (b) the maximum in-plane shear strain and associated average normal strain.

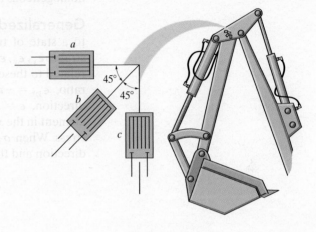

Prob. 10–28

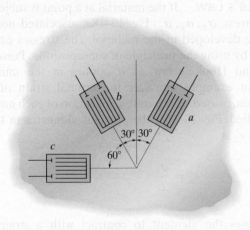

Prob. 10–26

10–27. The 45° strain rosette is mounted on a steel shaft. The following readings are obtained from each gauge: $\epsilon_a = 300(10^{-6})$, $\epsilon_b = -250(10^{-6})$, and $\epsilon_c = -450(10^{-6})$. Determine (a) the in-plane principal strains and (b) the maximum in-plane shear strain and average normal strain. In each case show the deformed element due to these strains.

10–29. Consider the general orientation of three strain gauges at a point as shown. Write a computer program that can be used to determine the principal in-plane strains and the maximum in-plane shear strain at the point. Show an application of the program using the values $\theta_a = 40°$, $\epsilon_a = 160(10^{-6})$, $\theta_b = 125°$, $\epsilon_b = 100(10^{-6})$, $\theta_c = 220°$, $\epsilon_c = 80(10^{-6})$.

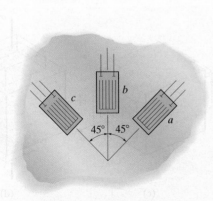

Prob. 10–27

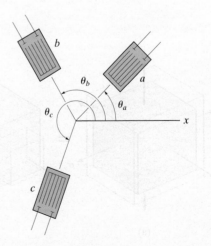

Prob. 10–29

10

10.6 Material-Property Relationships

In this section we will present some important relationships involving a material's properties that are used when the material is subjected to multiaxial stress and strain. To do so we will assume that the material is homogeneous and isotropic and behaves in a linear-elastic manner.

Generalized Hooke's Law.

If the material at a point is subjected to a state of triaxial stress, $\sigma_x, \sigma_y, \sigma_z$, Fig. 10–18a, associated normal strains $\epsilon_x, \epsilon_y, \epsilon_z$ will be developed in the material. The stresses can be related to these strains by using the principle of superposition, Poisson's ratio, $\epsilon_{\text{lat}} = -\nu\epsilon_{\text{long}}$, and Hooke's law, as it applies in the uniaxial direction, $\epsilon = \sigma/E$. For example, consider the normal strain of the element in the x direction, caused by separate application of each normal stress. When σ_x is applied, Fig. 10–18b, the element elongates in the x direction and the strain ϵ_x' is

$$\epsilon_x' = \frac{\sigma_x}{E}$$

Application of σ_y causes the element to contract with a strain ϵ_x'', Fig. 10–18c. Here

$$\epsilon_x'' = -\nu\frac{\sigma_y}{E}$$

Likewise, application of σ_z, Fig. 10–18d, causes a contraction such that

$$\epsilon_x''' = -\nu\frac{\sigma_z}{E}$$

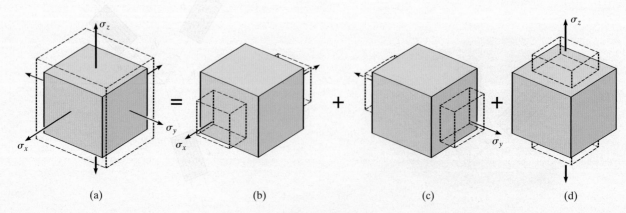

(a)   (b)   (c)   (d)

Fig. 10–18

When these three normal strains are superimposed, the normal strain ϵ_x is determined for the state of stress in Fig. 10–18a. Similar equations can be developed for the normal strains in the y and z directions. The final results can be written as

$$\boxed{\begin{aligned}\epsilon_x &= \frac{1}{E}[\sigma_x - \nu(\sigma_y + \sigma_z)] \\[4pt] \epsilon_y &= \frac{1}{E}[\sigma_y - \nu(\sigma_x + \sigma_z)] \\[4pt] \epsilon_z &= \frac{1}{E}[\sigma_z - \nu(\sigma_x + \sigma_y)]\end{aligned}} \qquad (10\text{--}18)$$

These three equations express Hooke's law in a general form for a triaxial state of stress. For application tensile stresses are considered positive quantities, and compressive stresses are negative. If a resulting normal strain is *positive*, it indicates that the material *elongates*, whereas a *negative* normal strain indicates the material *contracts*.

If we now apply a shear stress τ_{xy} to the element, Fig. 10–19a, experimental observations indicate that the material will deform *only* due to a shear strain γ_{xy}; that is, τ_{xy} will not cause other strains in the material. Likewise, τ_{yz} and τ_{xz} will only cause shear strains γ_{yz} and γ_{xz}, Figs. 10–19b and 10–19c, and so Hooke's law for shear stress and shear strain can be written as

$$\boxed{\gamma_{xy} = \frac{1}{G}\tau_{xy} \qquad \gamma_{yz} = \frac{1}{G}\tau_{yz} \qquad \gamma_{xz} = \frac{1}{G}\tau_{xz}} \qquad (10\text{--}19)$$

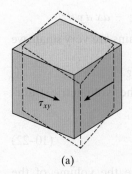

(a)

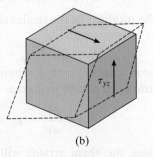

(b)

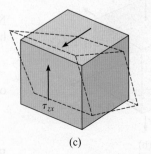

(c)

Fig. 10–19

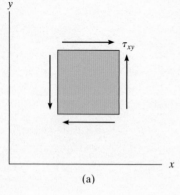

(a)

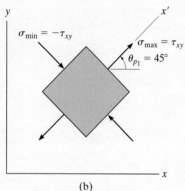

(b)

Fig. 10–20

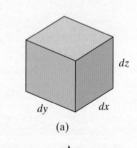

(a)

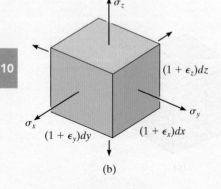

(b)

Fig. 10–21

Relationship Involving E, ν, and G.

In Sec. 3.7 it was stated that the modulus of elasticity E is related to the shear modulus G by Eq. 3–11, namely,

$$G = \frac{E}{2(1 + \nu)} \tag{10–20}$$

One way to derive this relationship is to consider an element of the material to be subjected to pure shear ($\sigma_x = \sigma_y = \sigma_z = 0$), Fig. 10–20a. Applying Eq. 9–5 to obtain the principal stresses yields $\sigma_{max} = \tau_{xy}$ and $\sigma_{min} = -\tau_{xy}$. This element must be oriented $\theta_{p_1} = 45°$ counterclockwise from the x axis as shown in Fig. 10–20b. If the three principal stresses $\sigma_{max} = \tau_{xy}$, $\sigma_{int} = 0$, and $\sigma_{min} = -\tau_{xy}$ are substituted into the first of Eqs. 10–18, the principal strain ϵ_{max} can be related to the shear stress τ_{xy}. The result is

$$\epsilon_{max} = \frac{\tau_{xy}}{E}(1 + \nu) \tag{10–21}$$

This strain, which deforms the element along the x' axis, can also be related to the shear strain γ_{xy}. To do this, first note that since $\sigma_x = \sigma_y = \sigma_z = 0$, then from the first and second Eqs. 10–18, $\epsilon_x = \epsilon_y = 0$. Substituting these results into the strain transformation Eq. 10–9, we get

$$\epsilon_1 = \epsilon_{max} = \frac{\gamma_{xy}}{2}$$

By Hooke's law, $\gamma_{xy} = \tau_{xy}/G$, so that $\epsilon_{max} = \tau_{xy}/2G$. Substituting into Eq. 10–21 and rearranging terms gives the final result, namely, Eq. 10–20.

Dilatation and Bulk Modulus.

When an elastic material is subjected to normal stress, its volume will change. For example, consider a volume element which is subjected to the principal stresses σ_x, σ_y, σ_z. If the sides of the element are originally dx, dy, dz, Fig. 10–21a, then after application of the stress they become $(1 + \epsilon_x)\, dx$, $(1 + \epsilon_y)\, dy$, $(1 + \epsilon_z)\, dz$, Fig. 10–21b. The change in volume of the element is therefore

$$\delta V = (1 + \epsilon_x)(1 + \epsilon_y)(1 + \epsilon_z)\, dx\, dy\, dz - dx\, dy\, dz$$

Neglecting the products of the strains since the strains are very small, we have

$$\delta V = (\epsilon_x + \epsilon_y + \epsilon_z)\, dx\, dy\, dz$$

The change in volume per unit volume is called the "volumetric strain" or the **dilatation** e. It can be written as

$$e = \frac{\delta V}{dV} = \epsilon_x + \epsilon_y + \epsilon_z \tag{10–22}$$

By comparison, the shear strains will *not* change the volume of the element, rather they will only change its rectangular shape.

Also, if we use Hooke's law, as defined by Eq. 10–18, we can write the dilatation in terms of the applied stress. We have

$$e = \frac{1 - 2\nu}{E}(\sigma_x + \sigma_y + \sigma_z) \tag{10–23}$$

When a volume element of material is subjected to the uniform pressure p of a liquid, the pressure on the body is the same in all directions and is always normal to any surface on which it acts. Shear stresses are *not present*, since the shear resistance of a liquid is zero. This state of "hydrostatic" loading requires the normal stresses to be equal in any and all directions, and therefore an element of the body is subjected to principal stresses $\sigma_x = \sigma_y = \sigma_z = -p$, Fig. 10–22. Substituting into Eq. 10–23 and rearranging terms yields

$$\frac{p}{e} = -\frac{E}{3(1 - 2\nu)} \tag{10–24}$$

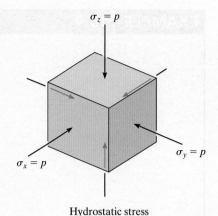

Hydrostatic stress

Fig. 10–22

Since this ratio is *similar* to the ratio of linear elastic stress to strain, which defines E, i.e., $\sigma/\epsilon = E$, the term on the right is called the *volume modulus of elasticity* or the **bulk modulus**. It has the same units as stress and will be symbolized by the letter k; that is,

$$\boxed{k = \frac{E}{3(1 - 2\nu)}} \tag{10–25}$$

Note that for most metals $\nu \approx \frac{1}{3}$ so $k \approx E$. If a material existed that did not change its volume then $\delta V = e = 0$, and k would have to be infinite. From Eq. 10–25 the theoretical *maximum* value for Poisson's ratio is therefore $\nu = 0.5$. During yielding, no actual volume change of the material is observed, and so $\nu = 0.5$ is used when plastic yielding occurs.

Important Points

- When a homogeneous isotropic material is subjected to a state of triaxial stress, the strain in each direction is influenced by the strains produced by *all* the stresses. This is the result of the Poisson effect, and results in the form of a generalized Hooke's law.

- Unlike normal stress, a shear stress applied to homogeneous isotropic material will only produce shear strain in the same plane.

- The material constants E, G, and ν are related mathematically.

- *Dilatation*, or *volumetric strain*, is caused only by normal strain, not shear strain.

- The *bulk modulus* is a measure of the stiffness of a volume of material. This material property provides an upper limit to Poisson's ratio of $\nu = 0.5$, which remains at this value while plastic yielding occurs.

10

EXAMPLE | 10.9

The bracket in Example 10–8, Fig. 10–23a, is made of steel for which $E_{st} = 200$ GPa, $\nu_{st} = 0.3$. Determine the principal stresses at point A.

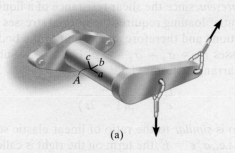

(a)

Fig. 10–23

SOLUTION I

From Example 10.8 the principal strains have been determined as

$$\epsilon_1 = 272(10^{-6})$$

$$\epsilon_2 = 33.9(10^{-6})$$

Since point A is on the *surface* of the bracket for which there is no loading, the stress on the surface is zero, and so point A is subjected to plane stress. Applying Hooke's law with $\sigma_3 = 0$, we have

$$\epsilon_1 = \frac{\sigma_1}{E} - \frac{\nu}{E}\sigma_2; \quad 272(10^{-6}) = \frac{\sigma_1}{200(10^9)} - \frac{0.3}{200(10^9)}\sigma_2$$

$$54.4(10^6) = \sigma_1 - 0.3\sigma_2 \tag{1}$$

$$\epsilon_2 = \frac{\sigma_2}{E} - \frac{\nu}{E}\sigma_1; \quad 33.9(10^{-6}) = \frac{\sigma_2}{200(10^9)} - \frac{0.3}{200(10^9)}\sigma_1$$

$$6.78(10^6) = \sigma_2 - 0.3\sigma_1 \tag{2}$$

Solving Eqs. 1 and 2 simultaneously yields

$$\sigma_1 = 62.0 \text{ MPa} \hspace{3cm} Ans.$$

$$\sigma_2 = 25.4 \text{ MPa} \hspace{3cm} Ans.$$

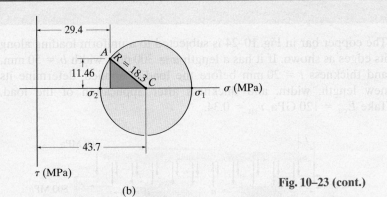

Fig. 10–23 (cont.)

(b)

SOLUTION II

It is also possible to solve the problem using the given state of strain,

$$\epsilon_x = 60(10^{-6}) \qquad \epsilon_y = 246(10^{-6}) \qquad \gamma_{xy} = -149(10^{-6})$$

as specified in Example 10.8. Applying Hooke's law in the x–y plane, we have

$$\epsilon_x = \frac{\sigma_x}{E} - \frac{\nu}{E}\sigma_y; \qquad 60(10^{-6}) = \frac{\sigma_x}{200(10^9)\ \text{Pa}} - \frac{0.3\sigma_y}{200(10^9)\ \text{Pa}}$$

$$\epsilon_y = \frac{\sigma_y}{E} - \frac{\nu}{E}\sigma_x; \qquad 246(10^{-6}) = \frac{\sigma_y}{200(10^9)\ \text{Pa}} - \frac{0.3\sigma_x}{200(10^9)\ \text{Pa}}$$

$$\sigma_x = 29.4\ \text{MPa} \qquad \sigma_y = 58.0\ \text{MPa}$$

The shear stress is determined using Hooke's law for shear. First, however, we must calculate G.

$$G = \frac{E}{2(1 + \nu)} = \frac{200\ \text{GPa}}{2(1 + 0.3)} = 76.9\ \text{GPa}$$

Thus,

$$\tau_{xy} = G\gamma_{xy}; \qquad \tau_{xy} = 76.9(10^9)[-149(10^{-6})] = -11.46\ \text{MPa}$$

The Mohr's circle for this state of plane stress has a reference point $A(29.4\ \text{MPa}, -11.46\ \text{MPa})$ and center at $\sigma_{\text{avg}} = 43.7\ \text{MPa}$, Fig. 10–23$b$. The radius is determined from the shaded triangle.

$$R = \sqrt{(43.7 - 29.4)^2 + (11.46)^2} = 18.3\ \text{MPa}$$

Therefore,

$$\sigma_1 = 43.7\ \text{MPa} + 18.3\ \text{MPa} = 62.0\ \text{MPa} \qquad Ans.$$

$$\sigma_2 = 43.7\ \text{MPa} - 18.3\ \text{MPa} = 25.4\ \text{MPa} \qquad Ans.$$

NOTE: Each of these solutions is valid provided the material is both linear elastic and isotropic, since then the principal planes of stress and strain coincide.

10

EXAMPLE | 10.10

The copper bar in Fig. 10–24 is subjected to a uniform loading along its edges as shown. If it has a length $a = 300$ mm, width $b = 50$ mm, and thickness $t = 20$ mm before the load is applied, determine its new length, width, and thickness after application of the load. Take $E_{cu} = 120$ GPa, $\nu_{cu} = 0.34$.

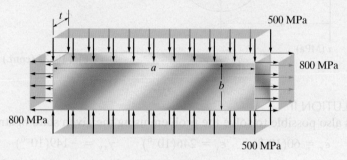

Fig. 10–24

SOLUTION

By inspection, the bar is subjected to a state of plane stress. From the loading we have

$$\sigma_x = 800 \text{ MPa} \qquad \sigma_y = -500 \text{ MPa} \qquad \tau_{xy} = 0 \qquad \sigma_z = 0$$

The associated normal strains are determined from the generalized Hooke's law, Eq. 10–18; that is,

$$\epsilon_x = \frac{\sigma_x}{E} - \frac{\nu}{E}(\sigma_y + \sigma_z)$$

$$= \frac{800 \text{ MPa}}{120(10^3) \text{ MPa}} - \frac{0.34}{120(10^3) \text{ MPa}}(-500 \text{ MPa} + 0) = 0.00808$$

$$\epsilon_y = \frac{\sigma_y}{E} - \frac{\nu}{E}(\sigma_x + \sigma_z)$$

$$= \frac{-500 \text{ MPa}}{120(10^3) \text{ MPa}} - \frac{0.34}{120(10^3) \text{ MPa}}(800 \text{ MPa} + 0) = -0.00643$$

$$\epsilon_z = \frac{\sigma_z}{E} - \frac{\nu}{E}(\sigma_x + \sigma_y)$$

$$= 0 - \frac{0.34}{120(10^3) \text{ MPa}}(800 \text{ MPa} - 500 \text{ MPa}) = -0.000850$$

The new bar length, width, and thickness are therefore

$$a' = 300 \text{ mm} + 0.00808(300 \text{ mm}) = 302.4 \text{ mm} \qquad Ans.$$

$$b' = 50 \text{ mm} + (-0.00643)(50 \text{ mm}) = 49.68 \text{ mm} \qquad Ans.$$

$$t' = 20 \text{ mm} + (-0.000850)(20 \text{ mm}) = 19.98 \text{ mm} \qquad Ans.$$

EXAMPLE 10.11

If the rectangular block shown in Fig. 10–25 is subjected to a uniform pressure of $p = 0.2$ MPa, determine the dilatation and the change in length of each side. Take $E = 6$ MPa, $v = 0.45$.

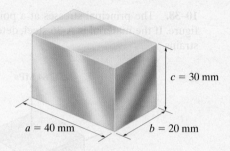

$c = 30$ mm

$a = 40$ mm

$b = 20$ mm

Fig. 10–25

SOLUTION

Dilatation. The dilatation can be determined using Eq. 10–23 with $\sigma x = \sigma y = \sigma z = -0.2$ MPa. We have

$$e = \frac{1 - 2v}{E}(\sigma_x + \sigma_y + \sigma_z)$$

$$= \frac{1 - 2(0.45)}{6 \text{ MPa}}[3(-0.2 \text{ MPa})]$$

$$= -0.01 \text{ m}^3/\text{m}^3 \qquad\qquad Ans.$$

Change in Length. The normal strain on each side can be determined from Hooke's law, Eq. 10–18; that is,

$$\epsilon = \frac{1}{E}[\sigma_x - v(\sigma_y + \sigma_z)]$$

$$= \frac{1}{6 \text{ MPa}}[-0.2 \text{ MPa} - (0.45)(-0.2 \text{ MPa} - 0.2 \text{ MPa})] = -0.00333 \text{ m/m}$$

Thus, the change in length of each side is

$$\delta a = -0.00333(40 \text{ mm}) = -0.133 \text{ mm} \qquad Ans.$$

$$\delta b = -0.00333(20 \text{ mm}) = -0.0667 \text{ mm} \qquad Ans.$$

$$\delta c = -0.00333(30 \text{ mm}) = -0.100 \text{ mm} \qquad Ans.$$

The negative signs indicate that each dimension is decreased.

10

PROBLEMS

10–30. For the case of plane stress, show that Hooke's law can be written as

$$\sigma_x = \frac{E}{(1-\nu^2)}(\epsilon_x + \nu\epsilon_y), \quad \sigma_y = \frac{E}{(1-\nu^2)}(\epsilon_y + \nu\epsilon_x)$$

10–31. Use Hooke's law, Eq. 10–18, to develop the strain-transformation equations, Eqs. 10–5 and 10–6, from the stress-transformation equations, Eqs. 9–1 and 9–2.

*10–32. A bar of copper alloy is loaded in a tension machine and it is determined that $\epsilon_x = 940(10^{-6})$ and $\sigma_x = 100$ MPa, $\sigma_y = 0, \sigma_z = 0$. Determine the modulus of elasticity, E_{cu}, and the dilatation, e_{cu}, of the copper. $\nu_{cu} = 0.35$.

•10–33. The principal strains at a point on the aluminum fuselage of a jet aircraft are $\epsilon_1 = 780(10^{-6})$ and $\epsilon_2 = 400(10^{-6})$. Determine the associated principal stresses at the point in the same plane. $E_{al} = 70$ GPa. *Hint:* See Prob. 10–30.

10–34. The rod is made of aluminum 2014-T6. If it is subjected to the tensile load of 700 N and has a diameter of 20 mm, determine the absolute maximum shear strain in the rod at a point on its surface.

10–35. The rod is made of aluminum 2014-T6. If it is subjected to the tensile load of 700 N and has a diameter of 20 mm, determine the principal strains at a point on the surface of the rod.

Probs. 10–34/35

*10–36. The steel shaft has a radius of 15 mm. Determine the torque T in the shaft if the two strain gauges, attached to the surface of the shaft, report strains of $\epsilon_{x'} = -80(10^{-6})$ and $\epsilon_{y'} = 80(10^{-6})$. Also, compute the strains acting in the x and y directions. $E_{st} = 200$ GPa, $\nu_{st} = 0.3$.

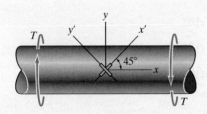

Prob. 10–36

10–37. Determine the bulk modulus for each of the following materials: (a) rubber, $E_r = 2.8$ MPa, $\nu_r = 0.48$, and (b) glass, $E_g = 56$ GPa, $\nu_g = 0.24$.

10–38. The principal stresses at a point are shown in the figure. If the material is A-36 steel, determine the principal strains.

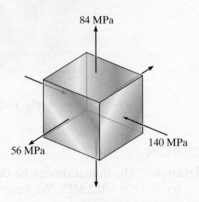

Prob. 10–38

10–39. The spherical pressure vessel has an inner diameter of 2 m and a thickness of 10 mm. A strain gauge having a length of 20 mm is attached to it, and it is observed to increase in length by 0.012 mm when the vessel is pressurized. Determine the pressure causing this deformation, and find the maximum in-plane shear stress, and the absolute maximum shear stress at a point on the outer surface of the vessel. The material is steel, for which $E_{st} = 200$ GPa and $\nu_{st} = 0.3$.

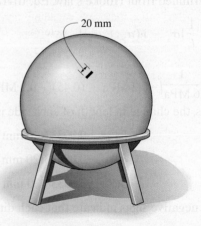

Prob. 10–39

*10–40. The strain in the x direction at point A on the steel beam is measured and found to be $\epsilon_x = -100(10^{-6})$. Determine the applied load P. What is the shear strain γ_{xy} at point A? $E_{st} = 200$ GPa, $\nu_{st} = 0.3$.

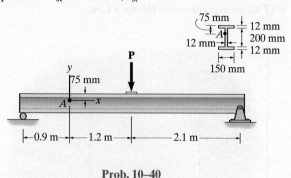

Prob. 10–40

•10–41. The cross section of the rectangular beam is subjected to the bending moment **M**. Determine an expression for the increase in length of lines AB and CD. The material has a modulus of elasticity E and Poisson's ratio is ν.

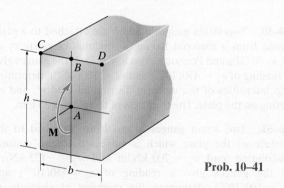

Prob. 10–41

10–42. The principal stresses at a point are shown in the figure. If the material is aluminum for which $E_{al} = 70$ GPa and $\nu_{al} = 0.33$, determine the principal strains.

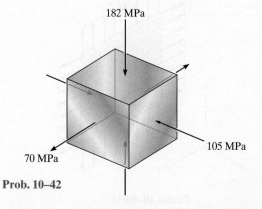

Prob. 10–42

10–43. A single strain gauge, placed on the outer surface and at an angle of 30° to the axis of the pipe, gives a reading at point A of $\epsilon_a = -200(10^{-6})$. Determine the horizontal force P if the pipe has an outer diameter of 50 mm and an inner diameter of 25 mm. The pipe is made of A-36 steel.

*10–44. A single strain gauge, placed in the vertical plane on the outer surface and at an angle of 30° to the axis of the pipe, gives a reading at point A of $\epsilon_a = -200(10^{-6})$. Determine the principal strains in the pipe at point A. The pipe has an outer diameter of 50 mm and an inner diameter of 25 mm and is made of A-36 steel.

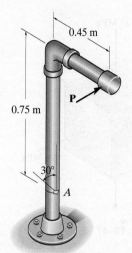

Probs. 10–43/44

10–45. The cylindrical pressure vessel is fabricated using hemispherical end caps in order to reduce the bending stress that would occur if flat ends were used. The bending stresses at the seam where the caps are attached can be eliminated by proper choice of the thickness t_h and t_c of the caps and cylinder, respectively. This requires the radial expansion to be the same for both the hemispheres and cylinder. Show that this ratio is $t_c/t_h = (2 - \nu)/(1 - \nu)$. Assume that the vessel is made of the same material and both the cylinder and hemispheres have the same inner radius. If the cylinder is to have a thickness of 12 mm, what is the required thickness of the hemispheres? Take $\nu = 0.3$.

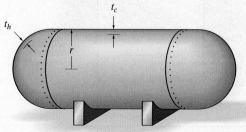

Prob. 10–45

10

10–46. The principal strains in a plane, measured experimentally at a point on the aluminum fuselage of a jet aircraft, are $\epsilon_1 = 630(10^{-6})$ and $\epsilon_2 = 350(10^{-6})$. If this is a case of plane stress, determine the associated principal stresses at the point in the same plane. $E_{al} = 70$ GPa and $\nu_{al} = 0.33$.

10–47. The principal stresses at a point are shown in the figure. If the material is aluminum for which $E_{al} = 70$ GPa and $\nu_{al} = 0.33$, determine the principal strains.

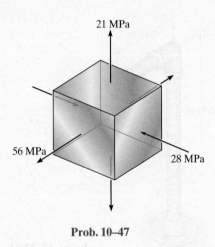

21 MPa

56 MPa

28 MPa

Prob. 10–47

***10–48.** The 6061-T6 aluminum alloy plate fits snugly into the rigid constraint. Determine the normal stresses σ_x and σ_y developed in the plate if the temperature is increased by $\Delta T = 50°C$. To solve, add the thermal strain $\alpha \Delta T$ to the equations for Hooke's Law.

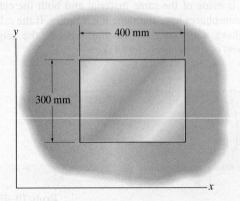

y

400 mm

300 mm

x

Prob. 10–48

•10–49. Initially, gaps between the A-36 steel plate and the rigid constraint are as shown. Determine the normal stresses σ_x and σ_y developed in the plate if the temperature is increased by $\Delta T = 55°C$. To solve, add the thermal strain $\alpha \Delta T$ to the equations for Hooke's Law.

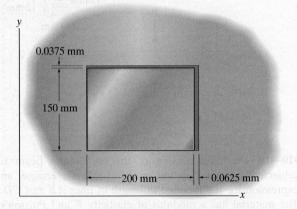

y

0.0375 mm

150 mm

200 mm   0.0625 mm

x

Prob. 10–49

10–50. Two strain gauges a and b are attached to a plate made from a material having a modulus of elasticity of $E = 70$ GPa and Poisson's ratio $\nu = 0.35$. If the gauges give a reading of $\epsilon_a = 450(10^{-6})$ and $\epsilon_b = 100(10^{-6})$, determine the intensities of the uniform distributed load w_x and w_y acting on the plate. The thickness of the plate is 25 mm.

10–51. Two strain gauges a and b are attached to the surface of the plate which is subjected to the uniform distributed load $w_x = 700$ kN/m and $w_y = -175$ kN/m. If the gauges give a reading of $\epsilon_a = 450(10^{-6})$ and $\epsilon_b = 100(10^{-6})$, determine the modulus of elasticity E, shear modulus G, and Poisson's ratio ν for the material.

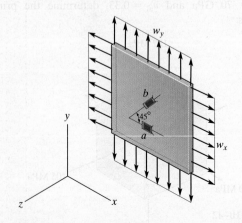

w_y

b

45°

a

y

w_x

z   x

Probs. 10–50/51

*10–52. The block is fitted between the fixed supports. If the glued joint can resist a maximum shear stress of $\tau_{allow} = 14$ MPa. determine the temperature rise that will cause the joint to fail. Take $E = 70$ GPa, $\nu = 0.2$, and *Hint:* Use Eq. 10–18 with an additional strain term of $\alpha \Delta T$ (Eq. 4–4).

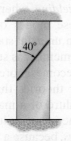

40°

Prob. 10–52

•10–53. The smooth rigid-body cavity is filled with liquid 6061-T6 aluminum. When cooled it is 0.3 mm from the top of the cavity. If the top of the cavity is covered and the temperature is increased by 110°C, determine the stress components σ_x, σ_y, and σ_z in the aluminum. *Hint:* Use Eqs. 10–18 with an additional strain term of $\alpha \Delta T$ (Eq. 4–4).

10–54. The smooth rigid-body cavity is filled with liquid 6061-T6 aluminum. When cooled it is 0.3 mm from the top of the cavity. If the top of the cavity is not covered and the temperature is increased by 110°C, determine the strain components ϵ_x, ϵ_y, and ϵ_z in the aluminum. *Hint:* Use Eqs. 10–18 with an additional strain term of $\alpha \Delta T$ (Eq. 4–4).

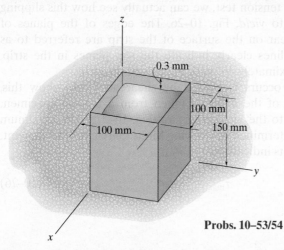

0.3 mm

100 mm

100 mm

150 mm

z

y

x

Probs. 10–53/54

10–55. A thin-walled spherical pressure vessel having an inner radius r and thickness t is subjected to an internal pressure p. Show that the increase in the volume within the vessel is $\Delta V = (2p\pi r^4/Et)(1 - \nu)$. Use a small-strain analysis.

*10–56. A thin-walled cylindrical pressure vessel has an inner radius r, thickness t, and length L. If it is subjected to an internal pressure p, show that the increase in its inner radius is $dr = r\epsilon_1 = pr^2(1 - \frac{1}{2}\nu)/Et$ and the increase in its length is $\Delta L = pLr(\frac{1}{2} - \nu)/Et$. Using these results, show that the change in internal volume becomes $dV = \pi r^2(1 + \epsilon_1)^2(1 + \epsilon_2)L - \pi r^2 L$. Since ϵ_1 and ϵ_2 are small quantities, show further that the change in volume per unit volume, called *volumetric strain*, can be written as $dV/V = pr(2.5 - 2\nu)/Et$.

10–57. The rubber block is confined in the U-shape smooth rigid block. If the rubber has a modulus of elasticity E and Poisson's ratio ν, determine the effective modulus of elasticity of the rubber under the confined condition.

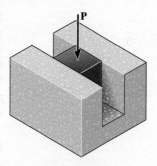

P

Prob. 10–57

10–58. A soft material is placed within the confines of a rigid cylinder which rests on a rigid support. Assuming that $\epsilon_x = 0$ and $\epsilon_y = 0$, determine the factor by which the modulus of elasticity will be increased when a load is applied if $\nu = 0.3$ for the material.

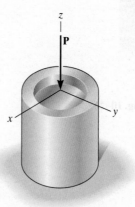

z

P

x

y

Prob. 10–58

10

*10.7 Theories of Failure

When an engineer is faced with the problem of design using a specific material, it becomes important to place an upper *limit* on the state of stress that defines the material's failure. If the material is *ductile*, failure is usually specified by the initiation of *yielding*, whereas if the material is *brittle*, it is specified by *fracture*. These modes of failure are readily defined if the member is subjected to a uniaxial state of stress, as in the case of simple tension; however, if the member is subjected to biaxial or triaxial stress, the criterion for failure becomes more difficult to establish.

In this section we will discuss four theories that are often used in engineering practice to predict the failure of a material subjected to a *multiaxial* state of stress. No single theory of failure, however, can be applied to a specific material at *all times*, because a material may behave in either a ductile or brittle manner depending on the temperature, rate of loading, chemical environment, or the way the material is shaped or formed. When using a particular theory of failure, it is first necessary to calculate the normal and shear stress at points where they are the largest in the member. Once this state of stress is established, the *principal stresses* at these critical points are then determined, since each of the following theories is based on knowing the principal stress.

Ductile Materials

Maximum-Shear-Stress Theory.   The most common type of *yielding of a ductile material* such as steel is caused by *slipping*, which occurs along the contact planes of randomly ordered crystals that make up the material. If we make a specimen into a highly polished thin strip and subject it to a simple tension test, we can actually see how this slipping causes the material to *yield*, Fig. 10–26. The edges of the planes of slipping as they appear on the surface of the strip are referred to as *Lüder's lines*. These lines clearly indicate the slip planes in the strip, which occur at approximately 45° with the axis of the strip.

The slipping that occurs is caused by shear stress. To show this, consider an element of the material taken from a tension specimen, when it is subjected to the yield stress σ_Y, Fig. 10–27a. The maximum shear stress can be determined by drawing Mohr's circle for the element, Fig. 10–27b. The results indicate that

$$\tau_{max} = \frac{\sigma_Y}{2} \qquad (10\text{–}26)$$

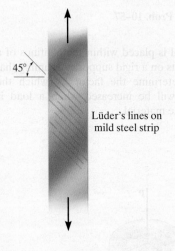

45°

Lüder's lines on mild steel strip

Fig. 10–26

10

Furthermore, this shear stress acts on planes that are 45° from the planes of principal stress, Fig. 10–27c, and these planes *coincide* with the direction of the Lüder lines shown on the specimen, indicating that indeed failure occurs by shear.

Using this idea, that ductile materials fail by shear, in 1868 Henri Tresca proposed the *maximum-shear-stress theory* or *Tresca yield criterion*. This theory can be used to predict the failure stress of a ductile material subjected to any type of loading. The theory states that yielding of the material begins when the absolute maximum shear stress in the material reaches the shear stress that causes the same material to yield when it is subjected *only* to axial tension. Therefore, to avoid failure, it is required that $\tau_{\max}^{\text{abs}}$ in the material must be less than or equal to $\sigma_Y/2$, where σ_Y is determined from a simple tension test.

For application we will express the absolute maximum shear stress in terms of the *principal stresses*. The procedure for doing this was discussed in Sec. 9.5 with reference to a condition of *plane stress*, that is, where the out-of-plane principal stress is zero. If the two in-plane principal stresses have the *same sign*, i.e., they are both tensile or both compressive, then failure will occur *out of the plane*, and from Eq. 9–13,

$$\tau_{\max}^{\text{abs}} = \frac{\sigma_1}{2}$$

If instead the in-plane principal stresses are of *opposite signs*, then failure occurs in the plane, and from Eq. 9–14,

$$\tau_{\max}^{\text{abs}} = \frac{\sigma_1 - \sigma_2}{2}$$

Using these equations and Eq. 10–26, the maximum-shear-stress theory for *plane stress* can be expressed for any two in-plane principal stresses σ_1 and σ_2 by the following criteria:

$$\left.\begin{array}{c} |\sigma_1| = \sigma_Y \\ |\sigma_2| = \sigma_Y \end{array}\right\} \quad \sigma_1, \sigma_2 \text{ have same signs}$$

$$|\sigma_1 - \sigma_2| = \sigma_Y\} \quad \sigma_1, \sigma_2 \text{ have opposite signs} \qquad (10\text{–}27)$$

A graph of these equations is given in Fig. 10–28. Clearly, if any point of the material is subjected to plane stress, and its in-plane principal stresses are represented by a coordinate (σ_1, σ_2) plotted *on the boundary* or *outside* the shaded hexagonal area shown in this figure, the material will yield at the point and failure is said to occur.

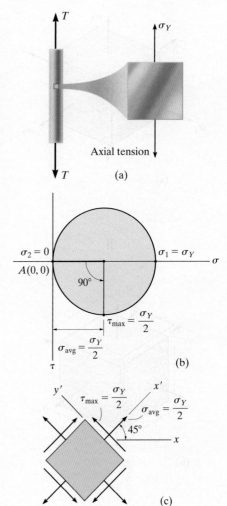

Axial tension

(a)

(b)

(c)

Fig. 10–27

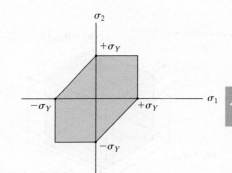

Maximum-shear-stress theory

Fig. 10–28

10

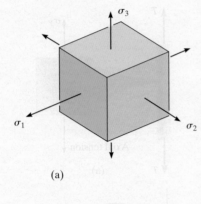

(a)

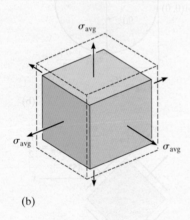

(b)

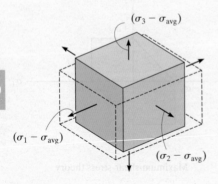

(c)

Fig. 10–29

Maximum-Distortion-Energy Theory.

It was stated in Sec. 3.5 that an external loading will deform a material, causing it to store energy *internally* throughout its volume. The energy per unit volume of material is called the *strain-energy density*, and if the material is subjected to a uniaxial stress the strain-energy density, defined by Eq. 3–6, becomes

$$u = \frac{1}{2}\sigma\epsilon \qquad (10\text{--}28)$$

If the material is subjected to triaxial stress, Fig. 10–29a, then each principal stress contributes a portion of the total strain-energy density, so that

$$u = \frac{1}{2}\sigma_1\epsilon_1 + \frac{1}{2}\sigma_2\epsilon_2 + \frac{1}{2}\sigma_3\epsilon_3$$

Furthermore, if the material behaves in a linear-elastic manner, then Hooke's law applies. Therefore, substituting Eq. 10–18 into the above equation and simplifying, we get

$$u = \frac{1}{2E}\Big[\sigma_1^2 + \sigma_2^2 + \sigma_3^2 - 2\nu(\sigma_1\sigma_2 + \sigma_1\sigma_3 + \sigma_3\sigma_2)\Big] \qquad (10\text{--}29)$$

This strain-energy density can be considered as the sum of two parts, one part representing the energy needed to cause a *volume change* of the element with no change in shape, and the other part representing the energy needed to *distort* the element. Specifically, the energy stored in the element as a result of its volume being changed is caused by application of the average principal stress, $\sigma_{\text{avg}} = (\sigma_1 + \sigma_2 + \sigma_3)/3$, since this stress causes equal principal strains in the material, Fig. 10–29b. The remaining portion of the stress, $(\sigma_1 - \sigma_{\text{avg}})$, $(\sigma_2 - \sigma_{\text{avg}})$, $(\sigma_3 - \sigma_{\text{avg}})$, causes the energy of distortion, Fig. 10–29c.

Experimental evidence has shown that materials do not yield when subjected to a uniform (hydrostatic) stress, such as σ_{avg} discussed above. As a result, in 1904, M. Huber proposed that yielding in a ductile material occurs when the *distortion energy* per unit volume of the material equals or exceeds the distortion energy per unit volume of the same material when it is subjected to yielding in a simple tension test. This theory is called the *maximum-distortion-energy theory*, and since it was later redefined independently by R. von Mises and H. Hencky, it sometimes also bears their names.

To obtain the distortion energy per unit volume, we will substitute the stresses $(\sigma_1 - \sigma_{\text{avg}})$, $(\sigma_2 - \sigma_{\text{avg}})$, and $(\sigma_3 - \sigma_{\text{avg}})$ for σ_1, σ_2, and σ_3, respectively, into Eq. 10–29, realizing that $\sigma_{\text{avg}} = (\sigma_1 + \sigma_2 + \sigma_3)/3$. Expanding and simplifying, we obtain

$$u_d = \frac{1+\nu}{6E}\Big[(\sigma_1 - \sigma_2)^2 + (\sigma_2 - \sigma_3)^2 + (\sigma_3 - \sigma_1)^2\Big]$$

In the case of *plane stress*, $\sigma_3 = 0$, and this equation reduces to

$$u_d = \frac{1 + \nu}{3E}\left(\sigma_1^2 - \sigma_1\sigma_2 + \sigma_2^2\right)$$

For a *uniaxial* tension test, $\sigma_1 = \sigma_Y$, $\sigma_2 = \sigma_3 = 0$, and so

$$(u_d)_Y = \frac{1 + \nu}{3E}\sigma_Y^2$$

Since the maximum-distortion-energy theory requires $u_d = (u_d)_Y$, then for the case of plane or biaxial stress, we have

$$\boxed{\sigma_1^2 - \sigma_1\sigma_2 + \sigma_2^2 = \sigma_Y^2} \qquad (10\text{--}30)$$

This is the equation of an ellipse, Fig. 10–30. Thus, if a point in the material is stressed such that (σ_1, σ_2) is plotted on the boundary or outside the shaded area, the material is said to fail.

A comparison of the above two failure criteria is shown in Fig. 10–31. Note that both theories give the same results when the principal stresses are equal, i.e., $\sigma_1 = \sigma_2 = \sigma_Y$, or when one of the principal stresses is zero and the other has a magnitude of σ_Y. If the material is subjected to pure shear, τ, then the theories have the largest discrepancy in predicting failure. The stress coordinates of these points on the curves can be determined by considering the element shown in Fig. 10–32a. From the associated Mohr's circle for this state of stress, Fig. 10–32b, we obtain principal stresses $\sigma_1 = \tau$ and $\sigma_2 = -\tau$. Thus, with $\sigma_1 = -\sigma_2$, then from Eq. 10–27, the maximum-shear-stress theory gives $(\sigma_Y/2, -\sigma_Y/2)$, and from Eq. 10–30, the maximum-distortion-energy theory gives $(\sigma_Y/\sqrt{3}, -\sigma_Y/\sqrt{3})$, Fig. 10–31.

Actual torsion tests, used to develop a condition of pure shear in a ductile specimen, have shown that the maximum-distortion-energy theory gives more accurate results for pure-shear failure than the maximum-shear-stress theory. In fact, since $(\sigma_Y/\sqrt{3})/(\sigma_Y/2) = 1.15$, the shear stress for yielding of the material, as given by the maximum-distortion-energy theory, is 15% more accurate than that given by the maximum-shear-stress theory.

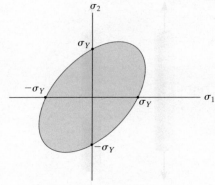

Maximum-distortion-energy theory

Fig. 10–30

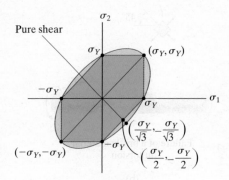

Fig. 10–31

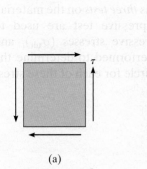

(a)

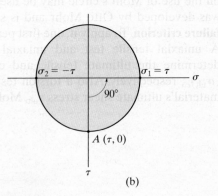

(b)

Fig. 10–32

10

Failure of a brittle material
in tension

(a)

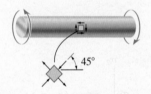

45°

45°

Failure of a brittle material
in torsion

(b)

Fig. 10–33

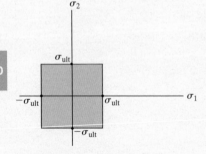

Maximum-normal-stress theory

Fig. 10–34

Brittle Materials

Maximum-Normal-Stress Theory. It was previously stated that brittle materials, such as gray cast iron, tend to fail suddenly by *fracture* with no apparent yielding. In a *tension test*, the fracture occurs when the normal stress reaches the ultimate stress σ_{ult}, Fig. 10–33a. Also, brittle fracture occurs in a torsion test due to tension since the plane of fracture for an element is at 45° to the shear direction, Fig. 10–33b. The fracture surface is therefore helical as shown.* Experiments have further shown that during torsion the material's strength is somewhat *unaffected* by the presence of the associated principal compressive stress being at right angles to the principal tensile stress. Consequently, the tensile stress needed to fracture a specimen during a torsion test is approximately the same as that needed to fracture a specimen in simple tension. Because of this, the ***maximum-normal-stress theory*** states that a brittle material will fail when the maximum tensile stress, σ_1, in the material reaches a value that is equal to the ultimate normal stress the material can sustain when it is subjected to simple tension.

If the material is subjected to *plane stress*, we require that

$$\begin{aligned} |\sigma_1| &= \sigma_{ult} \\ |\sigma_2| &= \sigma_{ult} \end{aligned} \qquad (10\text{--}31)$$

These equations are shown graphically in Fig. 10–34. Therefore, if the stress coordinates (σ_1, σ_2) at a point in the material fall on the boundary or outside the shaded area, the material is said to fracture. This theory is generally credited to W. Rankine, who proposed it in the mid-1800s. Experimentally it has been found to be in close agreement with the behavior of brittle materials that have stress–strain diagrams that are *similar* in both tension and compression.

Mohr's Failure Criterion.

In some brittle materials tension and compression properties are *different*. When this occurs a criterion based on the use of Mohr's circle may be used to predict failure. This method was developed by Otto Mohr and is sometimes referred to as **Mohr's failure criterion**. To apply it, one first performs *three tests* on the material. A uniaxial tensile test and uniaxial compressive test are used to determine the ultimate tensile and compressive stresses $(\sigma_{ult})_t$ and $(\sigma_{ult})_c$, respectively. Also a torsion test is performed to determine the material's ultimate shear stress τ_{ult}. Mohr's circle for each of these stress

*A stick of blackboard chalk fails in this way when its ends are twisted with the fingers.

conditions is then plotted as shown in Fig. 10–35. These three circles are contained in a "failure envelope" indicated by the extrapolated colored curve that is drawn tangent to all three circles. If a plane-stress condition at a point is represented by a circle that has a point of tangency with the envelope, or if it extends beyond the envelope's boundary, then failure is said to occur.

We may also represent this criterion on a graph of principal stresses σ_1 and σ_2. This is shown in Fig. 10–36. Here failure occurs when the absolute value of either one of the principal stresses reaches a value equal to or greater than $(\sigma_{\text{ult}})_t$ or $(\sigma_{\text{ult}})_c$ or in general, if the state of stress at a point defined by the stress coordinates (σ_1, σ_2) is plotted on the boundary or outside the shaded area.

Either the maximum-normal-stress theory or Mohr's failure criterion can be used in practice to predict the failure of a brittle material. However, it should be realized that their usefulness is quite limited. A tensile fracture occurs very suddenly, and its initiation generally depends on stress concentrations developed at microscopic imperfections of the material such as inclusions or voids, surface indentations, and small cracks. Since each of these irregularities varies from specimen to specimen, it becomes difficult to specify fracture on the basis of a single test.

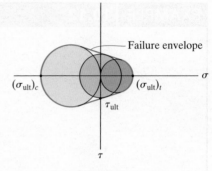

Fig. 10–35

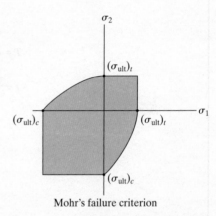

Mohr's failure criterion

Fig. 10–36

Important Points

- If a material is *ductile*, failure is specified by the initiation of *yielding*, whereas if it is *brittle*, it is specified by *fracture*.

- *Ductile failure* can be defined when *slipping* occurs between the crystals that compose the material. This slipping is due to *shear stress* and the *maximum-shear-stress theory* is based on this idea.

- *Strain energy* is stored in a material when it is subjected to normal stress. The *maximum-distortion-energy theory* depends on the *strain energy* that *distorts* the material, and not the part that increases its volume.

- The fracture of a *brittle material* is caused only by the *maximum tensile stress* in the material, and not the compressive stress. This is the basis of the *maximum-normal-stress theory*, and it is applicable if the stress–strain diagram is *similar* in tension and compression.

- If a *brittle material* has a stress–strain diagram that is *different* in tension and compression, then *Mohr's failure criterion* may be used to predict failure.

- Due to material imperfections, *tensile fracture* of a brittle material is *difficult to predict*, and so theories of failure for brittle materials should be used with caution.

10

EXAMPLE | 10.12

The solid cast-iron shaft shown in Fig. 10–37a is subjected to a torque of $T = 400$ N·m. Determine its smallest radius so that it does not fail according to the maximum-normal-stress theory. A specimen of cast iron, tested in tension, has an ultimate stress of $(\sigma_{ult})_t = 150$ MPa.

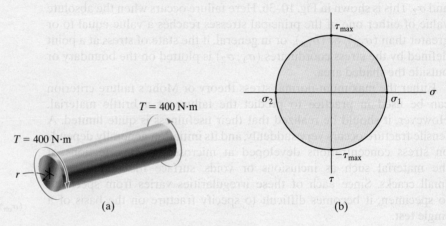

Fig. 10–37

SOLUTION

The maximum or critical stress occurs at a point located on the surface of the shaft. Assuming the shaft to have a radius r, the shear stress is

$$\tau_{max} = \frac{Tc}{J} = \frac{(400 \text{ N} \cdot \text{m})}{(\pi/2)r^4} = \frac{254.65 \text{ N} \cdot \text{m}}{r^3}$$

Mohr's circle for this state of stress (pure shear) is shown in Fig. 10–37b. Since $R = \tau_{max}$, then

$$\sigma_1 = -\sigma_2 = \tau_{max} = \frac{254.65 \text{ N} \cdot \text{m}}{r^3}$$

The maximum-normal-stress theory, Eq. 10–31, requires

$$|\sigma_1| \leq \sigma_{ult}$$

$$\frac{254.65 \text{ N} \cdot \text{m}}{r^3} \leq 150(10^6) \text{ N/m}^2$$

Thus, the smallest radius of the shaft is determined from

$$\frac{254.65 \text{ N} \cdot \text{m}}{r^3} = 150(10^6) \text{ N/m}^2$$

$$r = 0.01193 \text{ m} = 11.93 \text{ mm} \qquad \textit{Ans.}$$

10

EXAMPLE 10.13

The solid shaft shown in Fig. 10–38a has a radius of 5 mm and is made of steel having a yield stress of $\sigma_Y = 360$ MPa. Determine if the loadings cause the shaft to fail according to the maximum-shear-stress theory and the maximum-distortion-energy theory.

SOLUTION
The state of stress in the shaft is caused by both the axial force and the torque. Since maximum shear stress caused by the torque occurs in the material at the outer surface, we have

$$\sigma_x = \frac{P}{A} = \frac{-15 \text{ kN}}{\pi (5 \text{ mm})^2} = -0.191 \text{ kN/mm}^2$$

$$\tau_{xy} = \frac{Tc}{A} = \frac{32.5 \text{ N}\cdot\text{m } (0.005 \text{ m})}{\frac{\pi}{2}(0.005 \text{ m})^4} = 165.5(10^6) \text{ N/mm}^2 = 165.5 \text{ MPa}$$

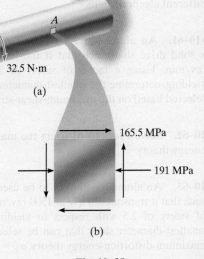

(a)

The stress components are shown acting on an element of material at point A in Fig. 10–38b. Rather than using Mohr's circle, the principal stresses can also be obtained using the stress-transformation Eq. 9–5.

$$\sigma_{1,2} = \frac{\sigma_x + \sigma_y}{2} \pm \sqrt{\left(\frac{\sigma_x - \sigma_y}{2}\right)^2 + \tau_{xy}^2}$$

$$= \frac{-191 + 0}{2} \pm \sqrt{\left(\frac{-191 + 0}{2}\right)^2 - (165.5)^2}$$

$$= -95.5 \pm 191.1$$

$$\sigma_1 = 95.6 \text{ MPa}$$

$$\sigma_2 = -286.6 \text{ MPa}$$

(b)

Fig. 10–38

Maximum-Shear-Stress Theory. Since the principal stresses have *opposite signs*, then from Sec. 9.5, the absolute maximum shear stress will occur in the plane, and therefore, applying the second of Eqs. 10–27, we have

$$|\sigma_1 - \sigma_2| \le \sigma_Y$$

$$|95.6 - (-286.6)| \overset{?}{\le} 360$$

$$382.2 > 360$$

Thus, shear failure of the material will occur according to this theory.

Maximum-Distortion-Energy Theory. Applying Eq. 10–30, we have

$$(\sigma_1^2 - \sigma_1\sigma_2 + \sigma_2^2) \le \sigma_Y^2$$

$$[(95.6)^2 - (95.6)(-286.6) - (-286.6)^2] \overset{?}{\le} (360)^2$$

$$118\,677.9 \le 129\,600$$

Using this theory, failure will not occur.

10

PROBLEMS

10–59. A material is subjected to plane stress. Express the distortion-energy theory of failure in terms of σ_x, σ_y, and τ_{xy}.

*10–60. A material is subjected to plane stress. Express the maximum-shear-stress theory of failure in terms of σ_x, σ_y, and τ_{xy}. Assume that the principal stresses are of different algebraic signs.

•10–61. An aluminum alloy 6061-T6 is to be used for a solid drive shaft such that it transmits 33 kW at 2400 rev/min. Using a factor of safety of 2 with respect to yielding, determine the smallest-diameter shaft that can be selected based on the maximum-shear-stress theory.

10–62. Solve Prob. 10–61 using the maximum-distortion-energy theory.

10–63. An aluminum alloy is to be used for a drive shaft such that it transmits 20 kW at 1500 rev/min. Using a factor of safety of 2.5 with respect to yielding, determine the smallest-diameter shaft that can be selected based on the maximum-distortion-energy theory. $\sigma_Y = 25$ MPa.

*10–64. A bar with a square cross-sectional area is made of a material having a yield stress of $\sigma_Y = 840$ MPa. If the bar is subjected to a bending moment of 10 kN $\cdot$ m, determine the required size of the bar according to the maximum-distortion-energy theory. Use a factor of safety of 1.5 with respect to yielding.

•10–65. Solve Prob. 10–64 using the maximum-shear-stress theory.

10–66. Derive an expression for an equivalent torque T_e that, if applied alone to a solid bar with a circular cross section, would cause the same energy of distortion as the combination of an applied bending moment M and torque T.

10–67. Derive an expression for an equivalent bending moment M_e that, if applied alone to a solid bar with a circular cross section, would cause the same energy of distortion as the combination of an applied bending moment M and torque T.

*10–68. The short concrete cylinder having a diameter of 50 mm is subjected to a torque of 500 N $\cdot$ m and an axial compressive force of 2 kN. Determine if it fails according to the maximum-normal-stress theory. The ultimate stress of the concrete is $\sigma_{ult} = 28$ MPa.

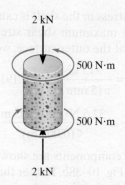

Prob. 10–68

•10–69. Cast iron when tested in tension and compression has an ultimate strength of $(\sigma_{ult})_t = 280$ MPa and $(\sigma_{ult})_c = 420$ MPa, respectively. Also, when subjected to pure torsion it can sustain an ultimate shear stress of $\tau_{ult} = 168$ MPa. Plot the Mohr's circles for each case and establish the failure envelope. If a part made of this material is subjected to the state of plane stress shown, determine if it fails according to Mohr's failure criterion.

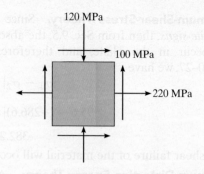

Prob. 10–69

10–70. Derive an expression for an equivalent bending moment M_e that, if applied alone to a solid bar with a circular cross section, would cause the same maximum shear stress as the combination of an applied moment M and torque T. Assume that the principal stresses are of opposite algebraic signs.

10–71. The components of plane stress at a critical point on an A-36 steel shell are shown. Determine if failure (yielding) has occurred on the basis of the maximum-shear-stress theory.

*10–72. The components of plane stress at a critical point on an A-36 steel shell are shown. Determine if failure (yielding) has occurred on the basis of the maximum-distortion-energy theory.

10–75. If the A-36 steel pipe has outer and inner diameters of 30 mm and 20 mm, respectively, determine the factor of safety against yielding of the material at point A according to the maximum-shear-stress theory.

*10–76. If the A-36 steel pipe has an outer and inner diameter of 30 mm and 20 mm, respectively, determine the factor of safety against yielding of the material at point A according to the maximum-distortion-energy theory.

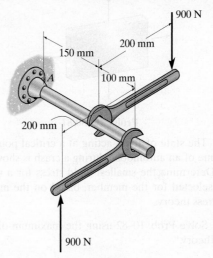

Probs. 10–75/76

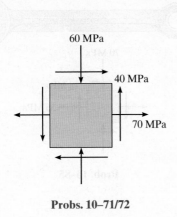

Probs. 10–71/72

•10–73. If the 50-mm diameter shaft is made from brittle material having an ultimate strength of $\sigma_{ult} = 350$ MPa for both tension and compression, determine if the shaft fails according to the maximum-normal-stress theory. Use a factor of safety of 1.5 against rupture.

10–74. If the 50-mm diameter shaft is made from cast iron having tensile and compressive ultimate strengths of $(\sigma_{ult})_t = 350$ MPa and $(\sigma_{ult})_c = 525$ MPa, respectively, determine if the shaft fails in accordance with Mohr's failure criterion.

•10–77. The element is subjected to the stresses shown. If $\sigma_Y = 250$ MPa, determine the factor of safety for the loading based on the maximum-shear-stress theory.

10–78. Solve Prob. 10–77 using the maximum-distortion-energy theory.

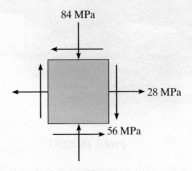

Probs. 10–77/78

10–79. The yield stress for heat-treated beryllium copper is $\sigma_Y = 900$ MPa. If this material is subjected to plane stress and elastic failure occurs when one principal stress is 1000 MPa, what is the smallest magnitude of the other principal stress? Use the maximum-distortion-energy theory.

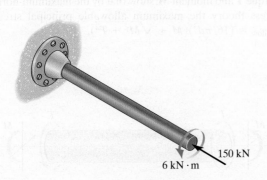

Probs. 10–73/74

*10–80. The plate is made of hard copper, which yields at $\sigma_Y = 735$ MPa. Using the maximum-shear-stress theory, determine the tensile stress σ_x that can be applied to the plate if a tensile stress $\sigma_y = 0.5\sigma_x$ is also applied.

•10–81. Solve Prob. 10–80 using the maximum-distortion-energy theory.

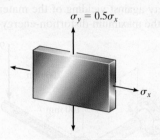

$\sigma_y = 0.5\sigma_x$

σ_x

Probs. 10–80/81

10–82. The state of stress acting at a critical point on the seat frame of an automobile during a crash is shown in the figure. Determine the smallest yield stress for a steel that can be selected for the member, based on the maximum-shear-stress theory.

10–83. Solve Prob. 10–82 using the maximum-distortion-energy theory.

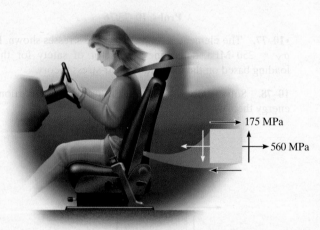

175 MPa

560 MPa

Probs. 10–82/83

*10–84. A bar with a circular cross-sectional area is made of SAE 1045 carbon steel having a yield stress of $\sigma_Y = 1000$ MPa. If the bar is subjected to a torque of 3.75 kN · m and a bending moment of 7 kN · m, determine the required diameter of the bar according to the maximum-distortion-energy theory. Use a factor of safety of 2 with respect to yielding.

•10–85. The state of stress acting at a critical point on a machine element is shown in the figure. Determine the smallest yield stress for a steel that might be selected for the part, based on the maximum-shear-stress theory.

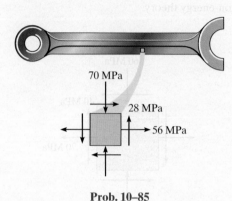

70 MPa

28 MPa

56 MPa

Prob. 10–85

10–86. The principal stresses acting at a point on a thin-walled cylindrical pressure vessel are $\sigma_1 = pr/t$, $\sigma_2 = pr/2t$, and $\sigma_3 = 0$. If the yield stress is σ_Y, determine the maximum value of p based on (a) the maximum-shear-stress theory and (b) the maximum-distortion-energy theory.

10–87. If a solid shaft having a diameter d is subjected to a torque $\mathbf{T}$ and moment $\mathbf{M}$, show that by the maximum-shear-stress theory the maximum allowable shear stress is $\tau_{\text{allow}} = (16/\pi d^3)\sqrt{M^2 + T^2}$. Assume the principal stresses to be of opposite algebraic signs.

*10–88. If a solid shaft having a diameter d is subjected to a torque $\mathbf{T}$ and moment $\mathbf{M}$, show that by the maximum-normal-stress theory the maximum allowable principal stress is $\sigma_{\text{allow}} = (16/\pi d^3)(M + \sqrt{M^2 + T^2})$.

M T T M

Probs. 10–87/88

•10–89. The shaft consists of a solid segment AB and a hollow segment BC, which are rigidly joined by the coupling at B. If the shaft is made from A-36 steel, determine the maximum torque T that can be applied according to the maximum-shear-stress theory. Use a factor of safety of 1.5 against yielding.

10–90. The shaft consists of a solid segment AB and a hollow segment BC, which are rigidly joined by the coupling at B. If the shaft is made from A-36 steel, determine the maximum torque T that can be applied according to the maximum-distortion-energy theory. Use a factor of safety of 1.5 against yielding.

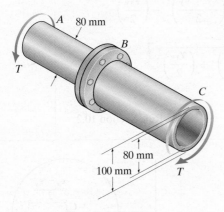

Probs. 10–89/90

10–91. The internal loadings at a critical section along the steel drive shaft of a ship are calculated to be a torque of 3.45 kN · m a bending moment of 2.25 kN · m and an axial thrust of 12.5 kN. If the yield points for tension and shear are $\sigma_Y = 700$ MPa and $\tau_Y = 350$ MPa, respectively, determine the required diameter of the shaft using the maximum-shear-stress theory.

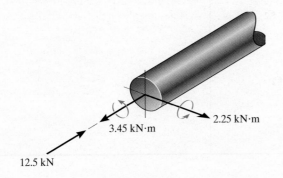

Prob. 10–91

*10–92. The gas tank has an inner diameter of 1.50 m and a wall thickness of 25 mm. If it is made from A-36 steel and the tank is pressured to 5 MPa, determine the factor of safety against yielding using (a) the maximum-shear-stress theory, and (b) the maximum-distortion-energy theory.

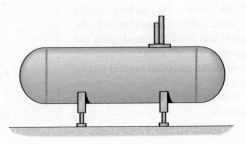

Prob. 10–92

•10–93. The gas tank is made from A-36 steel and has an inner diameter of 1.50 m. If the tank is designed to withstand a pressure of 5 MPa, determine the required minimum wall thickness to the nearest millimeter using (a) the maximum-shear-stress theory, and (b) maximum-distortion-energy theory. Apply a factor of safety of 1.5 against yielding.

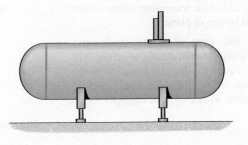

Prob. 10–93

10

CHAPTER REVIEW

When an element of material is subjected to deformations that only occur in a single plane, then it undergoes plane strain. If the strain components ϵ_x, ϵ_y, and γ_{xy} are known for a specified orientation of the element, then the strains acting for some other orientation of the element can be determined using the plane-strain transformation equations. Likewise, the principal normal strains and maximum in-plane shear strain can be determined using transformation equations.

$$\epsilon_{x'} = \frac{\epsilon_x + \epsilon_y}{2} + \frac{\epsilon_x - \epsilon_y}{2}\cos 2\theta + \frac{\gamma_{xy}}{2}\sin 2\theta$$

$$\epsilon_{y'} = \frac{\epsilon_x + \epsilon_y}{2} - \frac{\epsilon_x - \epsilon_y}{2}\cos 2\theta - \frac{\gamma_{xy}}{2}\sin 2\theta$$

$$\frac{\gamma_{x'y'}}{2} = -\left(\frac{\epsilon_x - \epsilon_y}{2}\right)\sin 2\theta + \frac{\gamma_{xy}}{2}\cos 2\theta$$

$$\epsilon_{1,2} = \frac{\epsilon_x + \epsilon_y}{2} \pm \sqrt{\left(\frac{\epsilon_x - \epsilon_y}{2}\right)^2 + \left(\frac{\gamma_{xy}}{2}\right)^2}$$

$$\frac{\gamma_{\substack{max \\ \text{in-plane}}}}{2} = \sqrt{\left(\frac{\epsilon_x - \epsilon_y}{2}\right)^2 + \left(\frac{\gamma_{xy}}{2}\right)^2}$$

$$\epsilon_{avg} = \frac{\epsilon_x + \epsilon_y}{2}$$

Ref.: Section 10.2

Strain transformation problems can also be solved in a semi-graphical manner using Mohr's circle. To draw the circle, the ϵ and $\gamma/2$ axes are established and the center of the circle $C\,[(\epsilon_x + \epsilon_y)/2, 0]$ and the "reference point" A $(\epsilon_x, \gamma_{xy}/2)$ are plotted. The radius of the circle extends between these two points and is determined from trigonometry.

Ref.: Section 10.3

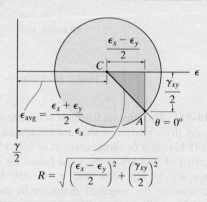

$$R = \sqrt{\left(\frac{\epsilon_x - \epsilon_y}{2}\right)^2 + \left(\frac{\gamma_{xy}}{2}\right)^2}$$

If ϵ_1 and ϵ_2 have the same sign then the absolute maximum shear strain will be out of plane.

$$\gamma_{max}^{abs} = \epsilon_1$$

$$\gamma_{max}^{in\text{-}plane} = \epsilon_1 - \epsilon_2$$

In the case of plane strain, the absolute maximum shear strain will be equal to the maximum in-plane shear strain provided the principal strains ϵ_1 and ϵ_2 have opposite signs.

$$\gamma_{max}^{abs} = \epsilon_1 - \epsilon_2$$

Ref.: Section 10.4

If the material is subjected to triaxial stress, then the strain in each direction is influenced by the strain produced by all three stresses. Hooke's law then involves the material properties E and ν.	$$\epsilon_x = \frac{1}{E}[\sigma_x - \nu(\sigma_y + \sigma_z)]$$ $$\epsilon_y = \frac{1}{E}[\sigma_y - \nu(\sigma_x + \sigma_z)]$$ $$\epsilon_z = \frac{1}{E}[\sigma_z - \nu(\sigma_x + \sigma_y)]$$ Ref.: Section 10.5	
If E and ν are known, then G can be determined.	$$G = \frac{E}{2(1 + \nu)}$$ Ref.: Section 10.6	
The dilatation is a measure of volumetric strain. The bulk modulus is used to measure the stiffness of a volume of material.	$$e = \frac{1 - 2\nu}{E}(\sigma_x + \sigma_y + \sigma_z)$$ $$k = \frac{E}{3(1 - 2\nu)}$$ Ref.: Section 10.6	
If the principal stresses at a critical point in the material are known, then a theory of failure can be used as a basis for design. *Ductile materials fail in shear*, and here the maximum-shear-stress theory or the maximum-distortion-energy theory can be used to predict failure. Both of these theories make comparison to the yield stress of a specimen subjected to a uniaxial tensile stress. *Brittle materials fail in tension*, and so the maximum-normal-stress theory or Mohr's failure criterion can be used to predict failure. Here comparisons are made with the ultimate tensile stress developed in a specimen.	Ref.: Section 10.7	

10

REVIEW PROBLEMS

10–94. A thin-walled spherical pressure vessel has an inner radius r, thickness t, and is subjected to an internal pressure p. If the material constants are E and ν, determine the strain in the circumferential direction in terms of the stated parameters.

10–95. The strain at point A on the shell has components $\epsilon_x = 250(10^{-6})$, $\epsilon_y = 400(10^{-6})$, $\gamma_{xy} = 275(10^{-6})$, $\epsilon_z = 0$. Determine (a) the principal strains at A, (b) the maximum shear strain in the x–y plane, and (c) the absolute maximum shear strain.

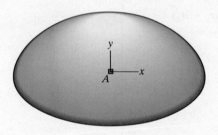

Prob. 10–95

*10–96. The principal plane stresses acting at a point are shown in the figure. If the material is machine steel having a yield stress of $\sigma_Y = 500$ MPa, determine the factor of safety with respect to yielding if the maximum-shear-stress theory is considered.

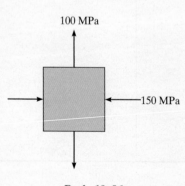

Prob. 10–96

•10–97. The components of plane stress at a critical point on a thin steel shell are shown. Determine if failure (yielding) has occurred on the basis of the maximum-distortion-energy theory. The yield stress for the steel is $\sigma_Y = 650$ MPa.

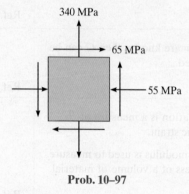

Prob. 10–97

10–98. The 60° strain rosette is mounted on a beam. The following readings are obtained for each gauge: $\epsilon_a = 600(10^{-6})$, $\epsilon_b = -700(10^{-6})$, and $\epsilon_c = 350(10^{-6})$. Determine (a) the in-plane principal strains and (b) the maximum in-plane shear strain and average normal strain. In each case show the deformed element due to these strains.

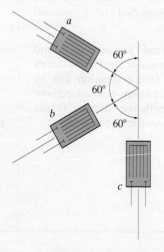

Prob. 10–98

10–99. A strain gauge forms an angle of 45° with the axis of the 50-mm diameter shaft. If it gives a reading of $\epsilon = -200(10^{-6})$ when the torque **T** is applied to the shaft, determine the magnitude of **T**. The shaft is made from A-36 steel.

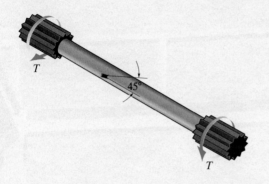

Prob. 10–99

*10–100. The A-36 steel post is subjected to the forces shown. If the strain gauges a and b at point A give readings of $\epsilon_a = 300(10^{-6})$ and $\epsilon_b = 175(10^{-6})$, determine the magnitudes of $\mathbf{P}_1$ and $\mathbf{P}_2$.

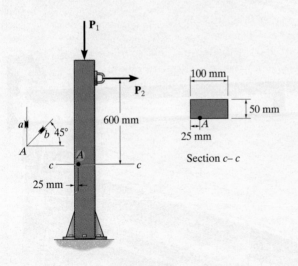

Prob. 10–100

10–101. A differential element is subjected to plane strain that has the following components: $\epsilon_x = 950(10^{-6})$, $\epsilon_y = 420(10^{-6})$, $\gamma_{xy} = -325(10^{-6})$. Use the strain-transformation equations and determine (a) the principal strains and (b) the maximum in-plane shear strain and the associated average strain. In each case specify the orientation of the element and show how the strains deform the element.

10–102. The state of plane strain on an element is $\epsilon_x = 400(10^{-6})$, $\epsilon_y = 200(10^{-6})$, and $\gamma_{xy} = -300(10^{-6})$. Determine the equivalent state of strain on an element at the same point oriented 30° clockwise with respect to the original element. Sketch the results on the element.

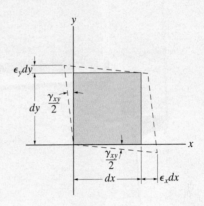

Prob. 10–102

10–103. The state of plane strain on an element is $\epsilon_x = 400(10^{-6})$, $\epsilon_y = 200(10^{-6})$, and $\gamma_{xy} = -300(10^{-6})$. Determine the equivalent state of strain, which represents (a) the principal strains, and (b) the maximum in-plane shear strain and the associated average normal strain. Specify the orientation of the corresponding element at the point with respect to the original element. Sketch the results on the element.

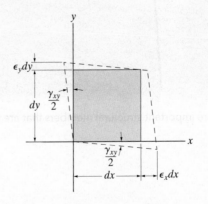

Prob. 10–103

10

Beams are important structural members that are used to support roof and floor loadings.

10–101.   A differential element is subjected to plane strain that has the following components; $\epsilon_x = 950(10^{-6})$, $\epsilon_y = 420(10^{-6})$, $\gamma_{xy} = -325(10^{-6})$. Use the strain-transformation equations and determine (a) the principal strains and (b) the maximum in-plane shear strain and the associated average strain. In each case specify the orientation of the element and show how the strains deform the element.

Design of Beams
and Shafts

11

CHAPTER OBJECTIVES

In this chapter, we will discuss how to design a beam so that it is able to resist both bending and shear loads. Specifically, methods used for designing prismatic beams and determining the shape of fully stressed beams will be developed. At the end of the chapter, we will consider the design of shafts based on the resistance of both bending and torsional moments.

Check out the Companion Website for the following video solution(s) that can be found in this chapter.

- Section 1, 2
  Beam Design – Timber Beams
- Section 1, 2
  Beam Design – Wide-Flange Beams
- Section 1, 2
  Beam Design – Prismatic Beams

11.1 Basis for Beam Design

Beams are said to be designed on the basis of strength so that they can resist the internal shear and moment developed along their length. To design a beam in this way requires application of the shear and flexure formulas provided the material is homogeneous and has linear elastic behavior. Although some beams may also be subjected to an axial force, the effects of this force are often neglected in design since the axial stress is generally much smaller than the stress developed by shear and bending.

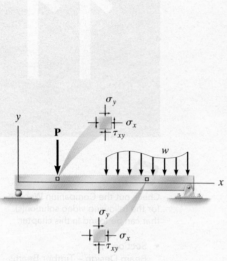

Fig. 11–1

As shown in Fig. 11–1, the external loadings on a beam will create additional stresses in the beam *directly under the load.* Notably, a compressive stress σ_y will be developed, in addition to the bending stress σ_x and shear stress τ_{xy} discussed previously. Using advanced methods of analysis, as treated in the theory of elasticity, it can be shown that σ_y diminishes rapidly throughout the beam's depth, and for *most* beam span-to-depth ratios used in engineering practice, the maximum value of σ_y generally represents only a small percentage compared to the bending stress σ_x, that is, $\sigma_x \gg \sigma_y$. Furthermore, the direct application of concentrated loads is generally avoided in beam design. Instead, *bearing plates* are used to spread these loads more evenly onto the surface of the beam.

Although beams are designed mainly for strength, they must also be braced properly along their sides so that they do not buckle or suddenly become unstable. Furthermore, in some cases beams must be designed to resist a limited amount of *deflection*, as when they support ceilings made of brittle materials such as plaster. Methods for finding beam deflections will be discussed in Chapter 12, and limitations placed on beam buckling are often discussed in codes related to structural or mechanical design.

Since the shear and flexure formulas are used for beam design, we will discuss the general results obtained when these equations are applied to various points in a cantilevered beam that has a rectangular cross section and supports a load **P** at its end, Fig. 11–2a.

In general, at an arbitrary section a–a along the beam's axis, Fig. 11–2b, the internal shear **V** and moment **M** are developed from a *parabolic* shear-stress distribution, and a *linear* normal-stress distribution, Fig. 11–2c. As a result, the stresses acting on elements located at points 1 through 5 along the section will be as shown in Fig. 11–2d. Note that elements 1 and 5 are subjected only to the maximum normal stress, whereas element 3, which is on the neutral axis, is subjected only to the maximum shear stress. The intermediate elements 2 and 4 resist *both* normal and shear stress.

In each case the state of stress can be transformed into *principal stresses*, using either the stress-transformation equations or Mohr's circle. The results are shown in Fig. 11–2e. Here each successive element, 1 through 5, undergoes a counterclockwise orientation. Specifically, relative to element 1, considered to be at the 0° position, element 3 is oriented at 45° and element 5 is oriented at 90°.

Whenever large shear loads occur on a beam it is important to use stiffeners such as at *A*, in order to prevent any localized failure such as crimping of the beam flanges.

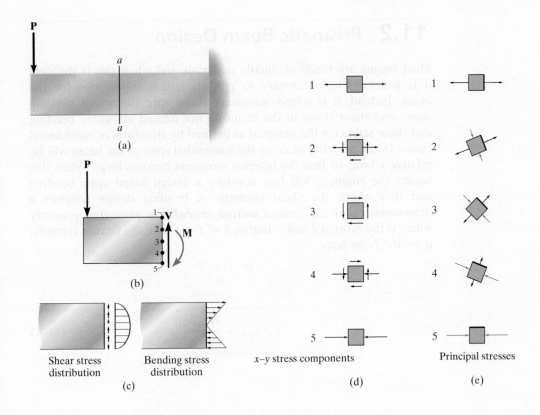

(a)

(b)

Shear stress
distribution

Bending stress
distribution

(c)

x–y stress components

(d)

Principal stresses

(e)

Fig. 11–2

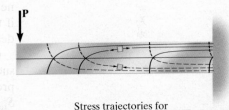

Stress trajectories for
cantilevered beam

Fig. 11–3

If this analysis is extended to many vertical sections along the beam other than a–a, a profile of the results can be represented by curves called *stress trajectories*. Each of these curves indicated the *direction* of a principal stress having a constant magnitude. Some of these trajectories are shown for the cantilevered beam in Fig. 11–3. Here the solid lines represent the direction of the tensile principal stresses and the dashed lines represent the direction of the compressive principal stresses. As expected, the lines intersect the neutral axis at 45° angles (like element 3) and the solid and dashed lines will intersect at 90° because the principal stresses are always 90° apart. Knowing the direction of these lines can help engineers decide where to reinforce a beam if it is made of brittle material so that it does not crack or become unstable.

11

11.2 Prismatic Beam Design

Most beams are made of ductile materials and when this is the case it is generally not necessary to plot the stress trajectories for the beam. Instead, it is simply necessary to be sure the actual bending stress and shear stress in the beam do not exceed allowable bending and shear stress for the material as defined by structural or mechanical codes. In the majority of cases the suspended span of the beam will be relatively long, so that the internal moments become large. When this occurs the engineer will first consider a design based upon bending and then check the shear strength. A bending design requires a determination of the beam's *section modulus*, a geometric property which is the ratio of I and c, that is, $S = I/c$. Using the flexure formula, $\sigma = Mc/I$, we have

$$S_{\text{req'd}} = \frac{M_{\text{max}}}{\sigma_{\text{allow}}} \qquad (11\text{–}1)$$

Here M is determined from the beam's moment diagram, and the allowable bending stress, σ_{allow}, is specified in a design code. In many cases the beam's as yet unknown weight will be small and can be neglected in comparison with the loads the beam must carry. However, if the additional moment caused by the weight is to be included in the design, a selection for S is made so that it slightly *exceeds* $S_{\text{req'd}}$.

Once $S_{\text{req'd}}$ is known, if the beam has a simple cross-sectional shape, such as a square, a circle, or a rectangle of known width-to-height proportions, its *dimensions* can be determined directly from $S_{\text{req'd}}$, since $S_{\text{req'd}} = I/c$. However, if the cross section is made from several elements, such as a wide-flange section, then an infinite number of web and flange dimensions can be determined that satisfy the value of $S_{\text{req'd}}$. In practice, however, engineers choose a particular beam meeting the requirement that $S > S_{\text{req'd}}$ from a handbook that lists the standard shapes available from manufacturers. Often several beams that have the same section modulus can be selected from these tables. If deflections are not restricted, usually the beam having the smallest cross-sectional area is chosen, since it is made of less material and is therefore both lighter and more economical than the others.

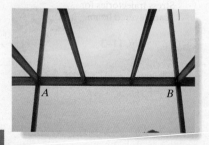

The two floor beams are connected to the beam AB, which transmits the load to the columns of this building frame. For design, all the connections can be considered to act as pins.

Once the beam has been selected, the shear formula $\tau_{\text{allow}} \geq VQ/It$ can then be used to check that the allowable shear stress is not exceeded. Often this requirement will not present a problem. However, if the beam is "short" and supports large concentrated loads, the shear-stress limitation may dictate the size of the beam. This limitation is particularly important in the design of wood beams, because wood tends to split along its grain due to shear (see Fig. 7–10e).

Fabricated Beams.   Since beams are often made of steel or wood, we will now discuss some of the tabulated properties of beams made from these materials.

Steel Sections.   Most manufactured steel beams are produced by rolling a hot ingot of steel until the desired shape is formed. These so-called *rolled shapes* have properties that are tabulated in the American Institute of Steel Construction (AISC) manual. A representative listing for wide-flange beams taken from this manual is given in Appendix B. As noted in this appendix, the wide-flange shapes are designated by their depth and weight per unit length; for example, W460 × 68 indicates a wide-flange cross section (W) having a depth of 460 mm and a weight of 0.68 kN/m, Fig. 11–4. For any given section, the weight per length, dimensions, cross-sectional area, moment of inertia, and section modulus are reported. Also included is the radius of gyration r, which is a geometric property related to the section's buckling strength. This will be discussed in Chapter 13. Appendix B and the *AISC Manual* also list data for other members such as channels and angles.

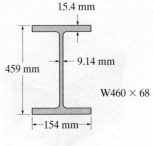

Fig. 11–4

Typical profile view of a steel wide-flange beam.

Wood Sections.   Most beams made of wood have rectangular cross sections because such beams are easy to manufacture and handle. Manuals, such as that of the National Forest Products Association, list the dimensions of lumber often used in the design of wood beams. Often, both the nominal and actual dimensions are reported. Lumber is identified by its *nominal* dimensions, such as 50 × 100 (50 mm by 100 mm); however, its actual or "dressed" dimensions are smaller, being 37.5 mm by 87.5 mm. The reduction in the dimensions occurs in order to obtain a smooth surface from lumber that is rough sawn. Obviously, the *actual dimensions* must be used whenever stress calculations are performed on wood beams.

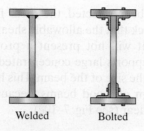

Welded   Bolted

Steel plate girders

Fig. 11–5

Wooden box beam

(a)

Glulam beam

(b)

Fig. 11–6

Built-up Sections. A *built-up section* is constructed from two or more parts joined together to form a single unit. Since $S_{req'd} = M/\sigma_{allow}$, the capacity of the beam to resist a moment will vary directly with its section modulus, and since $S_{req'd} = I/c$, then $S_{req'd}$ is *increased* if I is *increased*. In order to increase I, *most of the material* should be placed as far *away* from the neutral axis as practical. This, of course, is what makes a deep wide-flange beam so efficient in resisting a moment. For a very large load, however, an available rolled-steel section may not have a section modulus great enough to support the load. Rather than using several available beams, instead engineers will usually "build up" a beam made from plates and angles. A deep I-shaped section having this form is called a *plate girder*. For example, the steel plate girder in Fig. 11–5 has two flange plates that are either welded or, using angles, bolted to the web plate.

Wood beams are also "built up," usually in the form of a box beam section, Fig. 11–6a. They may be made having plywood webs and larger boards for the flanges. For very large spans, *glulam beams* are used. These members are made from several boards glue-laminated together to form a single unit, Fig. 11–6b.

Just as in the case of rolled sections or beams made from a single piece, the design of built-up sections requires that the bending and shear stresses be checked. In addition, the shear stress in the fasteners, such as weld, glue, nails, etc., must be checked to be certain the beam acts as a single unit. The principles for doing this were outlined in Sec. 7.4.

Important Points

- Beams support loadings that are applied perpendicular to their axes. If they are designed on the basis of strength, they must resist allowable shear and bending stresses.

- The maximum bending stress in the beam is assumed to be much greater than the localized stresses caused by the application of loadings on the surface of the beam.

Procedure for Analysis

Based on the previous discussion, the following procedure provides a rational method for the design of a beam on the basis of strength.

Shear and Moment Diagrams.

- Determine the maximum shear and moment in the beam. Often this is done by constructing the beam's shear and moment diagrams.

- For built-up beams, shear and moment diagrams are useful for identifying *regions* where the shear and moment are excessively large and may require additional structural reinforcement or fasteners.

Bending Stress.

- If the beam is relatively long, it is designed by finding its section modulus using the flexure formula, $S_{req'd} = M_{max}/\sigma_{allow}$.

- Once $S_{req'd}$ is determined, the cross-sectional dimensions for simple shapes can then be computed, since $S_{req'd} = I/c$.

- If rolled-steel sections are to be used, several possible values of S may be selected from the tables in Appendix B. Of these, choose the one having the smallest cross-sectional area, since this beam has the least weight and is therefore the most economical.

- Make sure that the selected section modulus, S, is *slightly greater* than $S_{req'd}$, so that the additional moment created by the beam's weight is considered.

Shear Stress.

- Normally beams that are short and carry large loads, especially those made of wood, are first designed to resist shear and then later checked against the allowable-bending-stress requirements.

- Using the shear formula, check to see that the allowable shear stress is not exceeded; that is, use $\tau_{allow} \geq V_{max}Q/It$.

- If the beam has a solid *rectangular* cross section, the shear formula becomes $\tau_{allow} \geq 1.5(V_{max}/A)$ (See Eq. 2 of Example 7.2.), and if the cross section is a *wide flange*, it is generally appropriate to assume that the shear stress is *constant* over the cross-sectional area of the beam's web so that $\tau_{allow} \geq V_{max}/A_{web}$, where A_{web} is determined from the product of the beam's depth and the web's thickness. (See the note at the end of Example 7.3.)

Adequacy of Fasteners.

- The adequacy of fasteners used on built-up beams depends upon the shear stress the fasteners can resist. Specifically, the required spacing of nails or bolts of a particular size is determined from the allowable shear flow, $q_{allow} = VQ/I$, calculated at points on the cross section where the fasteners are located. (See Sec. 7.3.)

EXAMPLE 11.1

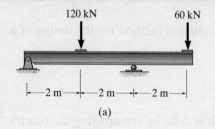

120 kN   60 kN

(a)

120 kN   60 kN

2 m — 2 m — 2 m

30 kN   150 kN

V (kN)

30   60 x (m)

−90

M (kN·m)   60

2.667 m x (m)

−120

(b)

Fig. 11–7

A beam is to be made of steel that has an allowable bending stress of $\sigma_{allow} = 170$ MPa and an allowable shear stress of $\tau_{allow} = 100$ MPa. Select an appropriate W shape that will carry the loading shown in Fig. 11–7a.

SOLUTION

Shear and Moment Diagrams. The support reactions have been calculated, and the shear and moment diagrams are shown in Fig. 11–7b. From these diagrams, $V_{max} = 90$ kN and $M_{max} = 120$ kN·m.

Bending Stress. The required section modulus for the beam is determined from the flexure formula,

$$S_{req'd} = \frac{M_{max}}{\sigma_{allow}} = \frac{120 \text{ kN} \cdot \text{m}}{170(10^3) \text{ kN/m}^2} = 0.706(10^{-3}) \text{ m}^3 = 706(10^3) \text{ mm}^3$$

Using the table in Appendix B, the following beams are adequate:

W460 × 60	$S = 1120(10^3) \text{ mm}^3$
W410 × 67	$S = 1200(10^3) \text{ mm}^3$
W360 × 64	$S = 1030(10^3) \text{ mm}^3$
W310 × 74	$S = 1060(10^3) \text{ mm}^3$
W250 × 80	$S = 984(10^3) \text{ mm}^3$
W200 × 100	$S = 987(10^3) \text{ mm}^3$

The beam having the least weight per foot is chosen, i.e.,

$$\text{W460} \times 60$$

The *actual* maximum moment M_{max}, which includes the weight of the beam, can be calculated and the adequacy of the selected beam can be checked. In comparison with the applied loads, however, the beam's weight, $(60.35 \text{ kg/m})(9.81 \text{ N/kg})(6 \text{ m}) = 3552.2 \text{ N} = 3.55 \text{ kN}$ will only *slightly increase* $S_{req'd}$. In spite of this,

$$S_{req'd} = 706(10^3) \text{ mm}^3 < 1120(10^3) \text{ mm}^3 \qquad \text{OK}$$

Shear Stress. Since the beam is a *wide-flange section*, the *average shear stress* within the web will be considered. (See Example 7.3.) Here the web is assumed to extend from the very top to the very bottom of the beam. From Appendix B, for a W460 × 60, $d = 455$ mm $t_w = 8$ mm. Thus,

$$\tau_{avg} = \frac{V_{max}}{A_w} = \frac{90(10^3) \text{ N}}{(455 \text{ mm})(8 \text{ mm})} = 24.7 \text{ MPa} < 100 \text{ MPa} \qquad \text{OK}$$

Use a W460 × 60. *Ans.*

11

EXAMPLE | 11.2

The laminated wooden beam shown in Fig. 11–8a supports a uniform distributed loading of 12 kN/m. If the beam is to have a height-to-width ratio of 1.5, determine its smallest width. The allowable bending stress is $\sigma_{allow} = 9$ MPa and the allowable shear stress is $\tau_{allow} = 0.6$ MPa. Neglect the weight of the beam.

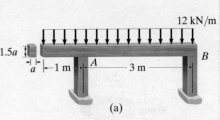

(a)

SOLUTION

Shear and Moment Diagrams. The support reactions at A and B have been calculated and the shear and moment diagrams are shown in Fig. 11–8b. Here $V_{max} = 20$ kN, $M_{max} = 10.67$ kN · m.

Bending Stress. Applying the flexure formula,

$$S_{req'd} = \frac{M_{max}}{\sigma_{allow}} = \frac{10.67(10^3) \text{ N} \cdot \text{m}}{9(10^6) \text{ N/m}^2} = 0.00119 \text{ m}^3$$

Assuming that the width is a, then the height is $1.5a$, Fig. 11–8a. Thus,

$$S_{req'd} = \frac{I}{c} = 0.00119 \text{ m}^3 = \frac{\frac{1}{12}(a)(1.5a)^3}{(0.75a)}$$

$$a^3 = 0.003160 \text{ m}^3$$

$$a = 0.147 \text{ m}$$

Shear Stress. Applying the shear formula for rectangular sections (which is a special case of $\tau_{max} = VQ/It$, Example 7.2), we have

$$\tau_{max} = 1.5\frac{V_{max}}{A} = (1.5)\frac{20(10^3) \text{ N}}{(0.147 \text{ m})(1.5)(0.147 \text{ m})}$$

$$= 0.929 \text{ MPa} > 0.6 \text{ MPa}$$

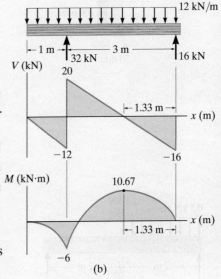

(b)

Fig. 11–8

EQUATION
Since the design fails the shear criterion, the beam must be redesigned on the basis of shear.

$$\tau_{allow} = 1.5\frac{V_{max}}{A}$$

$$600 \text{ kN/m}^2 = 1.5\frac{20(10^3) \text{ N}}{(a)(1.5a)}$$

$$a = 0.183 \text{ m} = 183 \text{ mm} \qquad \qquad Ans.$$

This larger section will also adequately resist the normal stress.

EXAMPLE 11.3

The wooden T-beam shown in Fig. 11–9a is made from two 200 mm × 30 mm boards. If the allowable bending stress is $\sigma_{allow} = 12$ MPa and the allowable shear stress is $\tau_{allow} = 0.8$ MPa, determine if the beam can safely support the loading shown. Also, specify the maximum spacing of nails needed to hold the two boards together if each nail can safely resist 1.50 kN in shear.

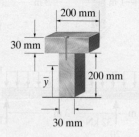

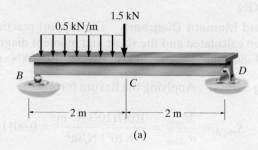

(a)

SOLUTION

Shear and Moment Diagrams. The reactions on the beam are shown, and the shear and moment diagrams are drawn in Fig. 11–9b. Here $V_{max} = 1.5$ kN, $M_{max} = 2$ kN·m.

Bending Stress. The neutral axis (centroid) will be located from the bottom of the beam. Working in units of meters, we have

$$\bar{y} = \frac{\Sigma \bar{y} A}{\Sigma A}$$

$$= \frac{(0.1 \text{ m})(0.03 \text{ m})(0.2 \text{ m}) + 0.215 \text{ m}(0.03 \text{ m})(0.2 \text{ m})}{0.03 \text{ m}(0.2 \text{ m}) + 0.03 \text{ m}(0.2 \text{ m})} = 0.1575 \text{ m}$$

Thus,

$$I = \left[\frac{1}{12}(0.03 \text{ m})(0.2 \text{ m})^3 + (0.03 \text{ m})(0.2 \text{ m})(0.1575 \text{ m} - 0.1 \text{ m})^2\right]$$

$$+ \left[\frac{1}{12}(0.2 \text{ m})(0.03 \text{ m})^3 + (0.03 \text{ m})(0.2 \text{ m})(0.215 \text{ m} - 0.1575 \text{ m})^2\right]$$

$$= 60.125(10^{-6}) \text{ m}^4$$

Since $c = 0.1575$ m (not 0.230 m − 0.1575 m = 0.0725 m), we require

$$\sigma_{allow} \geq \frac{M_{max}c}{I}$$

$$12(10^6) \text{ Pa} \geq \frac{2(10^3) \text{ N} \cdot \text{m}(0.1575 \text{ m})}{60.125(10^{-6}) \text{ m}^4} = 5.24(10^6) \text{ Pa}   \text{OK}$$

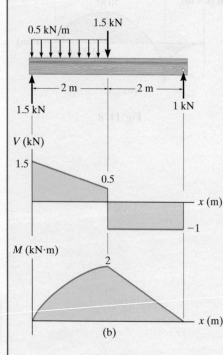

(b)

Fig. 11–9

11

Shear Stress.   Maximum shear stress in the beam depends upon the magnitude of Q and t. It occurs at the neutral axis, since Q is a maximum there and the neutral axis is in the web, where the thickness $t = 0.03$ m is smallest for the cross section. For simplicity, we will use the rectangular area below the neutral axis to calculate Q, rather than a two-part composite area above this axis, Fig. 11–9c. We have

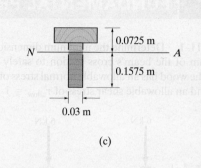

(c)

$$Q = \bar{y}'A' = \left(\frac{0.1575 \text{ m}}{2}\right)[(0.1575 \text{ m})(0.03 \text{ m})] = 0.372(10^{-3}) \text{ m}^3$$

so that

$$\tau_{allow} \geq \frac{V_{max}Q}{It}$$

$$800(10^3) \text{ Pa} \geq \frac{1.5(10^3) \text{ N}[0.372(10^{-3})] \text{ m}^3}{60.125(10^{-6}) \text{ m}^4 (0.03 \text{ m})} = 309(10^3) \text{ Pa} \quad \text{OK}$$

Nail Spacing.   From the shear diagram it is seen that the shear varies over the entire span. Since the nail spacing depends on the magnitude of shear in the beam, for simplicity (and to be conservative), we will design the spacing on the basis of $V = 1.5$ kN for region BC and $V = 1$ kN for region CD. Since the nails join the flange to the web, Fig. 11–9d, we have

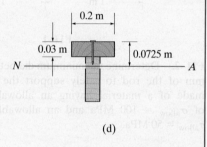

(d)

Fig. 11–9 (cont.)

$$Q = \bar{y}'A' = (0.0725 \text{ m} - 0.015 \text{ m})[(0.2 \text{ m})(0.03 \text{ m})] = 0.345(10^{-3}) \text{ m}^3$$

The shear flow for each region is therefore

$$q_{BC} = \frac{V_{BC}Q}{I} = \frac{1.5(10^3) \text{ N}[0.345(10^{-3}) \text{ m}^3]}{60.125(10^{-6}) \text{ m}^4} = 8.61 \text{ kN/m}$$

$$q_{CD} = \frac{V_{CD}Q}{I} = \frac{1(10^3) \text{ N}[0.345(10^{-3}) \text{ m}^3]}{60.125(10^{-6}) \text{ m}^4} = 5.74 \text{ kN/m}$$

One nail can resist 1.50 kN in shear, so the maximum spacing becomes

$$s_{BC} = \frac{1.50 \text{ kN}}{8.61 \text{ kN/m}} = 0.174 \text{ m}$$

$$s_{CD} = \frac{1.50 \text{ kN}}{5.74 \text{ kN/m}} = 0.261 \text{ m}$$

For ease of measuring, use

$$s_{BC} = 150 \text{ mm} \qquad \qquad Ans.$$

$$s_{CD} = 250 \text{ mm} \qquad \qquad Ans.$$

11

FUNDAMENTAL PROBLEMS

F11–1. Determine the minimum dimension a to the nearest mm of the beam's cross section to safely support the load. The wood has an allowable normal stress of $\sigma_{allow} = 10$ MPa and an allowable shear stress of $\tau_{allow} = 1$ MPa.

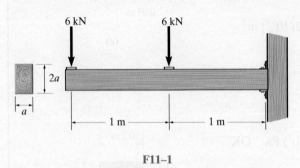

F11–1

F11–2. Determine the minimum diameter d to the nearest mm of the rod to safely support the load. The rod is made of a material having an allowable normal stress of $\sigma_{allow} = 100$ MPa and an allowable shear stress of $\tau_{allow} = 50$ MPa.

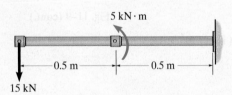

F11–2

F11–3. Determine the minimum dimension a to the nearest mm of the beam's cross section to safely support the load. The wood has an allowable normal stress of $\sigma_{allow} = 12$ MPa and an allowable shear stress of $\tau_{allow} = 1.5$ MPa.

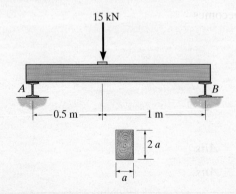

F11–3

F11–4. Determine the minimum dimension h to the nearest mm of the beam's cross section to safely support the load. The wood has an allowable normal stress of $\sigma_{allow} = 15$ MPa and an allowable shear stress of $\tau_{allow} = 1.5$ MPa.

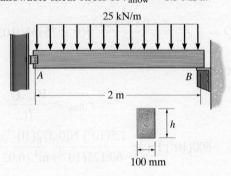

F11–4

F11–5. Determine the minimum dimension b to the nearest mm of the beam's cross section to safely support the load. The wood has an allowable normal stress of $\sigma_{allow} = 12$ MPa and an allowable shear stress of $\tau_{allow} = 1.5$ MPa.

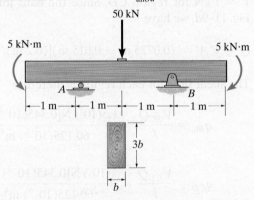

F11–5

F11–6. Select the lightest W410-shaped section that can safely support the load. The beam is made of steel having an allowable normal stress of $\sigma_{allow} = 150$ MPa and an allowable shear stress of $\tau_{allow} = 75$ MPa.

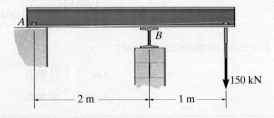

F11–6

PROBLEMS

11–1. The simply supported beam is made of timber that has an allowable bending stress of $\sigma_{allow} = 6.5$ MPa and an allowable shear stress of $\tau_{allow} = 500$ kPa. Determine its dimensions if it is to be rectangular and have a height-to-width ratio of 1.25.

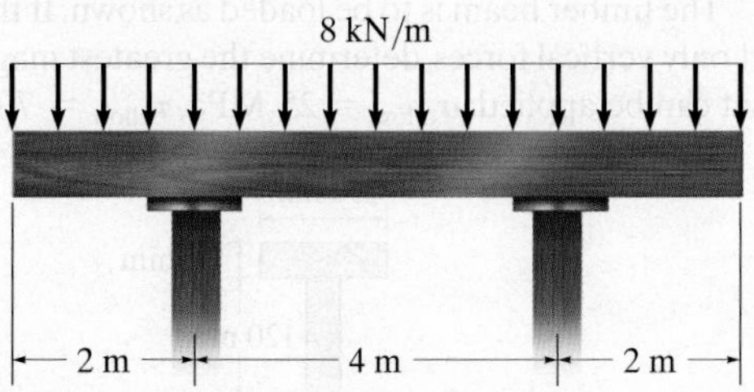

Prob. 11–1

11–2. The brick wall exerts a uniform distributed load of 20 kN/m on the beam. If the allowable bending stress is $\sigma_{allow} = 154$ MPa and the allowable shear stress is $\tau_{allow} = 84$ MPa, select the lightest wide-flange section with the shortest depth from Appendix B that will safely support the load.

Prob. 11–2

11–3. The brick wall exerts a uniform distributed load of 20 kN/m on the beam. If the allowable bending stress is $\sigma_{allow} = 154$ MPa, determine the required width b of the flange to the nearest multiples of 5 mm.

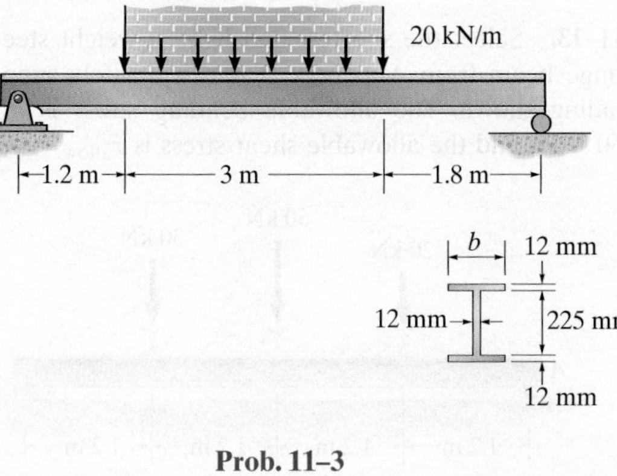

Prob. 11–3

***11–4.** Draw the shear and moment diagrams for the shaft, and determine its required diameter to the nearest multiples of 5 mm if $\sigma_{allow} = 50$ MPa and $\tau_{allow} = 20$ MPa. The bearings at A and D exert only vertical reactions on the shaft. The loading is applied to the pulleys at B, C, and E.

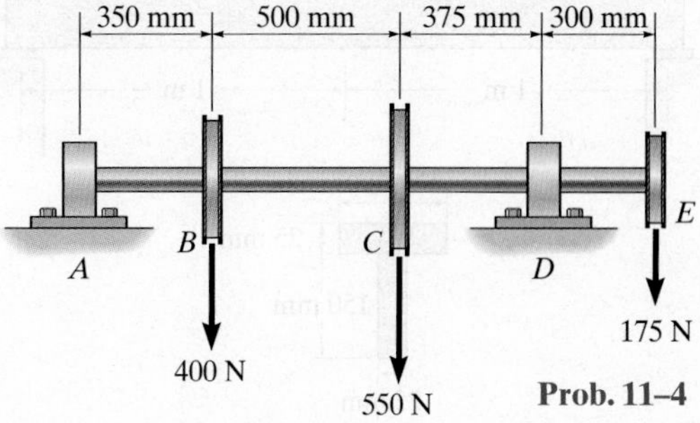

Prob. 11–4

•**11–5.** Select the lightest-weight steel wide-flange beam from Appendix B that will safely support the machine loading shown. The allowable bending stress is $\sigma_{allow} = 168$ MPa and the allowable shear stress is $\tau_{allow} = 100$ MPa.

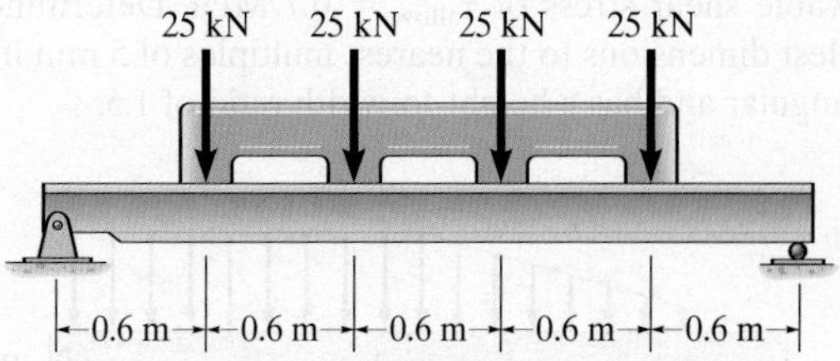

Prob. 11–5

11–6. The compound beam is made from two sections, which are pinned together at B. Use Appendix B and select the lightest-weight wide-flange beam that would be safe for each section if the allowable bending stress is $\sigma_{allow} = 168$ MPa and the allowable shear stress is $\tau_{allow} = 100$ MPa. The beam supports a pipe loading of 6 kN and 9 kN as shown.

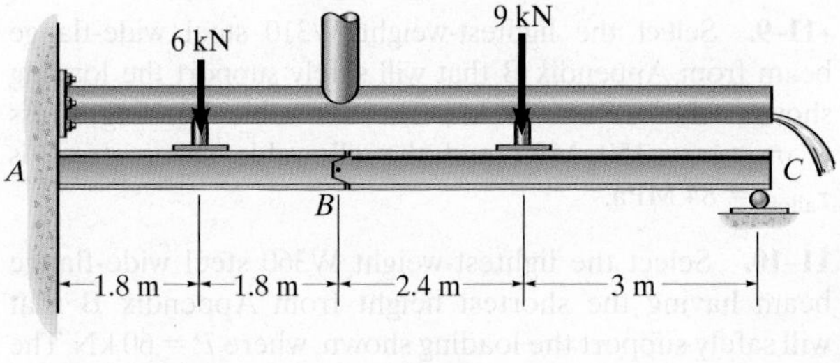

Prob. 11–6

11–7. If the bearing pads at A and B support only vertical forces, determine the greatest magnitude of the uniform distributed loading w that can be applied to the beam. $\sigma_{allow} = 15$ MPa, $\tau_{allow} = 1.5$ MPa.

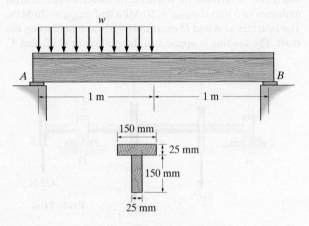

Prob. 11–7

*11–8. The simply supported beam is made of timber that has an allowable bending stress of $\sigma_{allow} = 8.4$ MPa and an allowable shear stress of $\tau_{allow} = 0.7$ MPa. Determine its smallest dimensions to the nearest multiples of 5 mm if it is rectangular and has a height-to-width ratio of 1.5.

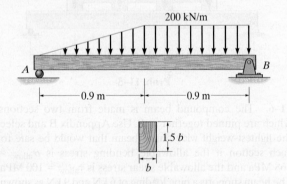

Prob. 11–8

•11–9. Select the lightest-weight W310 steel wide-flange beam from Appendix B that will safely support the loading shown, where $P = 30$ kN. The allowable bending stress is $\sigma_{allow} = 150$ MPa and the allowable shear stress is $\tau_{allow} = 84$ MPa.

11–10. Select the lightest-weight W360 steel wide-flange beam having the shortest height from Appendix B that will safely support the loading shown, where $P = 60$ kN. The allowable bending stress is $\sigma_{allow} = 150$ MPa and the allowable shear stress is $\tau_{allow} = 84$ MPa.

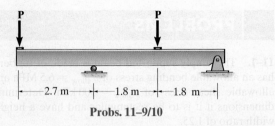

Probs. 11–9/10

11–11. The timber beam is to be loaded as shown. If the ends support only vertical forces, determine the greatest magnitude of **P** that can be applied. $\sigma_{allow} = 25$ MPa, $\tau_{allow} = 700$ kPa.

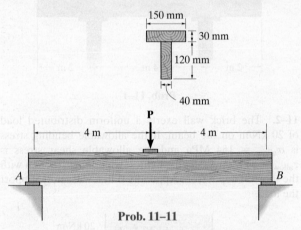

Prob. 11–11

*11–12. Determine the minimum width of the beam to the nearest multiples of 5 mm that will safely support the loading of $P = 40$ kN. The allowable bending stress is $\sigma_{allow} = 168$ MPa and the allowable shear stress is $\tau_{allow} = 100$ MPa.

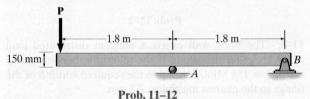

Prob. 11–12

•11–13. Select the shortest and lightest-weight steel wide-flange beam from Appendix B that will safely support the loading shown. The allowable bending stress is $\sigma_{allow} = 150$ MPa and the allowable shear stress is $\tau_{allow} = 84$ MPa.

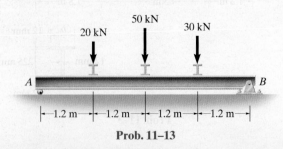

Prob. 11–13

11–14. The beam is used in a railroad yard for loading and unloading cars. If the maximum anticipated hoist load is 60 kN, select the lightest-weight steel wide-flange section from Appendix B that will safely support the loading. The hoist travels along the bottom flange of the beam, $0.3 \text{ m} \leq x \leq 7.5 \text{ m}$ and has negligible size. Assume the beam is pinned to the column at B and roller supported at A. The allowable bending stress is $\sigma_{allow} = 168$ MPa and the allowable shear stress is $\tau_{allow} = 84$ MPa.

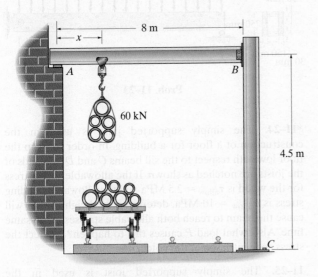

Prob. 11–14

11–15. The simply supported beam is made of timber that has an allowable bending stress of $\sigma_{allow} = 6.72$ MPa and an allowable shear stress of $\tau_{allow} = 0.525$ MPa. Determine its dimensions if it is to be rectangular and have a height-to-width ratio of 1.25.

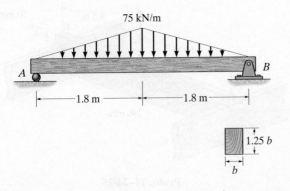

Prob. 11–15

*11–16. The simply supported beam is composed of two W310 × 33 sections built up as shown. Determine the maximum uniform loading w the beam will support if the allowable bending stress is $\sigma_{allow} = 150$ MPa and the allowable shear stress is $\tau_{allow} = 100$ MPa.

•11–17. The simply supported beam is composed of two W310 × 33 sections built up as shown. Determine if the beam will safely support a loading of $w = 30$ kN/m. The allowable bending stress is $\sigma_{allow} = 150$ MPa and the allowable shear stress is $\tau_{allow} = 100$ MPa.

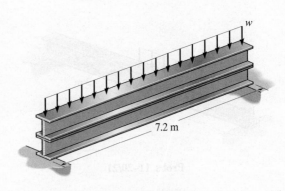

Probs. 11–16/17

11–18. Determine the smallest diameter rod that will safely support the loading shown. The allowable bending stress is $\sigma_{allow} = 167$ MPa and the allowable shear stress is $\tau_{allow} = 97$ MPa.

11–19. The pipe has an outer diameter of 15 mm. Determine the smallest inner diameter so that it will safely support the loading shown. The allowable bending stress is $\sigma_{allow} = 167$ MPa and the allowable shear stress is $\tau_{allow} = 97$ MPa.

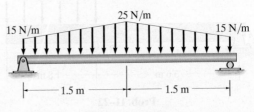

Probs. 11–18/19

11

*11–20. Determine the maximum uniform loading w the W310 × 21 beam will support if the allowable bending stress is $\sigma_{allow} = 150$ MPa and the allowable shear stress is $\tau_{allow} = 84$ MPa.

•11–21. Determine if the W360 × 33 beam will safely support a loading of $w = 25$ kN/m. The allowable bending stress is $\sigma_{allow} = 150$ MPa and the allowable shear stress is $\tau_{allow} = 84$ MPa.

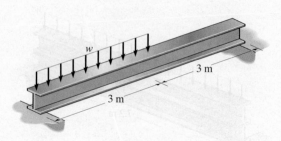

Probs. 11–20/21

11–22. Determine the minimum depth h of the beam to the nearest multiples of 5 mm that will safely support the loading shown. The allowable bending stress is $\sigma_{allow} = 147$ MPa and the allowable shear stress is $\tau_{allow} = 70$ MPa The beam has a uniform thickness of 75 mm.

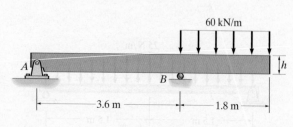

Prob. 11–22

11–23. The box beam has an allowable bending stress of $\sigma_{allow} = 10$ MPa and an allowable shear stress of $\tau_{allow} = 775$ kPa. Determine the maximum intensity w of the distributed loading that it can safely support. Also, determine the maximum safe nail spacing for each third of the length of the beam. Each nail can resist a shear force of 200 N.

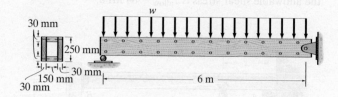

Prob. 11–23

*11–24. The simply supported joist is used in the construction of a floor for a building. In order to keep the floor low with respect to the sill beams C and D, the ends of the joists are notched as shown. If the allowable shear stress for the wood is $\tau_{allow} = 2.5$ MPa and the allowable bending stress is $\sigma_{allow} = 10$ MPa, determine the height h that will cause the beam to reach both allowable stresses at the same time. Also, what load P causes this to happen? Neglect the stress concentration at the notch.

11–25. The simply supported joist is used in the construction of a floor for a building. In order to keep the floor low with respect to the sill beams C and D, the ends of the joists are notched as shown. If the allowable shear stress for the wood is $\tau_{allow} = 2.5$ MPa and the allowable bending stress is $\sigma_{allow} = 13$ MPa, determine the smallest height h so that the beam will support a load of $P = 3$ kN. Also, will the entire joist safely support the load? Neglect the stress concentration at the notch.

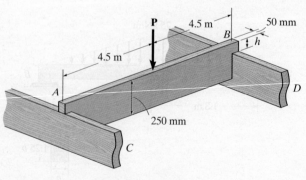

Probs. 11–24/25

11–26. Select the lightest-weight steel wide-flange beam from Appendix B that will safely support the loading shown. The allowable bending stress is $\sigma_{allow} = 150$ MPa and the allowable shear stress is $\tau_{allow} = 84$ MPa.

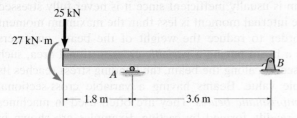

Prob. 11–26

11–27. The T-beam is made from two plates welded together as shown. Determine the maximum uniform distributed load w that can be safely supported on the beam if the allowable bending stress is $\sigma_{allow} = 150$ MPa and the allowable shear stress is $\tau_{allow} = 70$ MPa.

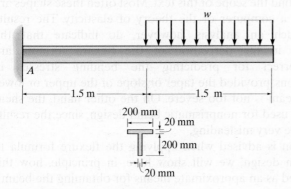

Prob. 11–27

*11–28. The beam is made of a ceramic material having an allowable bending stress of $\sigma_{allow} = 5$ MPa and an allowable shear stress of $\tau_{allow} = 2.8$ MPa. Determine the width b of the beam if the height $h = 2b$.

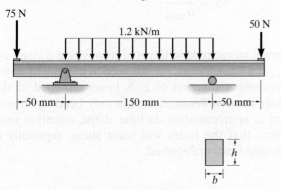

Prob. 11–28

•11–29. The wood beam has a rectangular cross section. Determine its height h so that it simultaneously reaches its allowable bending stress of $\sigma_{allow} = 10$ MPa and an allowable shear stress of $\tau_{allow} = 1$ MPa. Also, what is the maximum load P that the beam can then support?

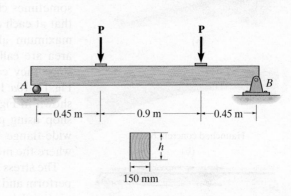

Prob. 11–29

11–30. The beam is constructed from three boards as shown. If each nail can support a shear force of 1.5 kN, determine the maximum allowable spacing of the nails, s, s', s'', for regions AB, BC, and CD respectively. Also, if the allowable bending stress is $\sigma_{allow} = 10$ MPa and the allowable shear stress is $\tau_{allow} = 1$ MPa, determine if it can safely support the load.

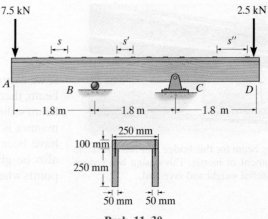

Prob. 11–30

11

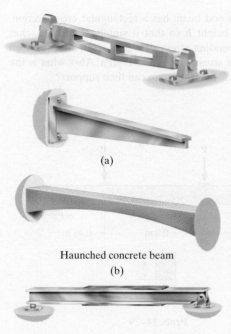

(a)

Haunched concrete beam

(b)

Wide-flange beam with cover plates

(c)

Fig. 11–10

The beam for this bridge pier has a variable moment of inertia. This design will reduce material weight and save cost.

*11.3 Fully Stressed Beams

Since the moment in a beam generally *varies* along its length, the choice of a prismatic beam is usually inefficient since it is never fully stressed at points where the internal moment is less than the maximum moment in the beam. In order to reduce the weight of the beam, engineers sometimes choose a beam having a *variable* cross-sectional area, such that at each cross section along the beam, the bending stress reaches its maximum allowable value. Beams having a variable cross-sectional area are called *nonprismatic beams*. They are often used in machines since they can be readily formed by casting. Examples are shown in Fig. 11–10a. In structures such beams may be "haunched" at their ends as shown in Fig. 11–10b. Also, beams may be "built up" or fabricated in a shop using plates. An example is a girder made from a rolled-shaped wide-flange beam and having cover plates welded to it in the region where the moment is a maximum, Fig. 11–10c.

The stress analysis of a nonprismatic beam is generally very difficult to perform and is beyond the scope of this text. Most often these shapes are analyzed by using a computer or the theory of elasticity. The results obtained from such an analysis, however, do indicate that the assumptions used in the derivation of the flexure formula are approximately correct for predicting the bending stresses in nonprismatic sections, provided the taper or slope of the upper or lower boundary of the beam is not too severe. On the other hand, the shear formula cannot be used for nonprismatic beam design, since the results obtained from it are very misleading.

Although caution is advised when applying the flexure formula to nonprismatic beam design, we will show here, in principle, how this formula can be used as an approximate means for obtaining the beam's general shape. In this regard, the *size* of the cross section of a nonprismatic beam that supports a given loading can be determined using the flexure formula written as

$$S = \frac{M}{\sigma_{\text{allow}}}$$

If we express the internal moment M in terms of its position x along the beam, then since σ_{allow} is a known constant, the section modulus S or the beam's dimensions become a function of x. A beam designed in this manner is called a **fully stressed beam**. Although *only* bending stresses have been considered in approximating its final shape, attention must also be given to ensure that the beam will resist shear, especially at points where concentrated loads are applied.

11

EXAMPLE 11.4

Determine the shape of a fully stressed, simply supported beam that supports a concentrated force at its center, Fig. 11–11a. The beam has a rectangular cross section of constant width b, and the allowable stress is σ_{allow}.

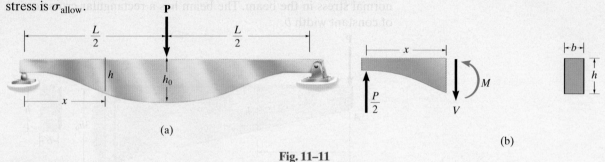

(a)

(b)

Fig. 11–11

SOLUTION

The internal moment in the beam, Fig. 11–11b, expressed as a function of position, $0 \le x < L/2$, is

$$M = \frac{P}{2}x$$

Hence the required section modulus is

$$S = \frac{M}{\sigma_{allow}} = \frac{P}{2\sigma_{allow}}x$$

Since $S = I/c$, then for a cross-sectional area h by b we have

$$\frac{I}{c} = \frac{\frac{1}{12}bh^3}{h/2} = \frac{P}{2\sigma_{allow}}x$$

$$h^2 = \frac{3P}{\sigma_{allow}b}x$$

If $h = h_0$ at $x = L/2$, then

$$h_0^2 = \frac{3PL}{2\sigma_{allow}b}$$

so that

$$h^2 = \left(\frac{2h_0^2}{L}\right)x \qquad Ans.$$

By inspection, the depth h must therefore vary in a *parabolic* manner with the distance x.

NOTE: In practice this *shape* is the basis for the design of leaf springs used to support the rear-end axles of most heavy trucks or train cars as shown in the adjacent photo. Note that although this result indicates that $h = 0$ at $x = 0$, it is necessary that the beam resist shear stress at the supports, and so practically speaking, it must be required that $h > 0$ at the supports, Fig. 11–11a.

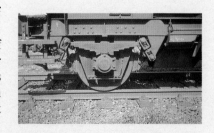

11

EXAMPLE 11.5

The cantilevered beam shown in Fig. 11–12a is formed into a trapezoidal shape having a depth h_0 at A and a depth $3h_0$ at B. If it supports a load **P** at its end, determine the absolute maximum normal stress in the beam. The beam has a rectangular cross section of constant width b.

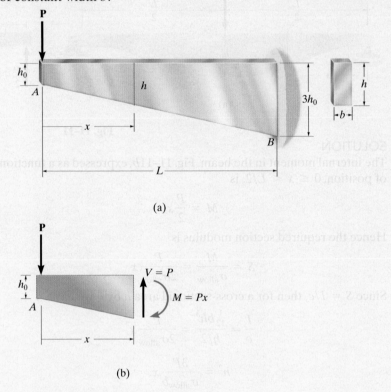

(a)

(b)

Fig. 11–12

SOLUTION

At any cross section, the maximum normal stress occurs at the top and bottom surface of the beam. However, since $\sigma_{max} = M/S$ and the section modulus S increases as x increases, the absolute maximum normal stress does *not* necessarily occur at the wall B, where the moment is maximum. Using the flexure formula, we can express the maximum normal stress at an arbitrary section in terms of its position x, Fig. 11–12b. Here the internal moment has a magnitude of $M = Px$. Since the slope of the bottom of the beam is $2h_0/L$, Fig. 11–12a, the depth of the beam at position x is

$$h = \frac{2h_0}{L}x + h_0 = \frac{h_0}{L}(2x + L)$$

Applying the flexure formula, we have

$$\sigma = \frac{Mc}{I} = \frac{Px(h/2)}{\left(\frac{1}{12}bh^3\right)} = \frac{6PL^2x}{bh_0^2(2x+L)^2} \tag{1}$$

To determine the position x where the absolute maximum normal stress occurs, we must take the derivative of σ with respect to x and set it equal to zero. This gives

$$\frac{d\sigma}{dx} = \left(\frac{6PL^2}{bh_0^2}\right)\frac{1(2x+L)^2 - x(2)(2x+L)(2)}{(2x+L)^4} = 0$$

Thus,

$$4x^2 + 4xL + L^2 - 8x^2 - 4xL = 0$$

$$L^2 - 4x^2 = 0$$

$$x = \frac{1}{2}L$$

Substituting into Eq. 1 and simplifying, the absolute maximum normal stress is therefore

$$\sigma_{\substack{abs \\ max}} = \frac{3}{4}\frac{PL}{bh_0^2} \qquad\qquad Ans.$$

Note that at the wall, B, the maximum normal stress is

$$(\sigma_{max})_B = \frac{Mc}{I} = \frac{PL(1.5h_0)}{\left[\frac{1}{12}b(3h_0)^3\right]} = \frac{2}{3}\frac{PL}{bh_0^2}$$

which is 11.1% smaller than $\sigma_{\substack{abs \\ max}}$.

NOTE: Recall that the flexure formula was derived on the basis of assuming the beam to be *prismatic*. Since this is not the case here, some error is to be expected in this analysis and that of Example 11.4. A more exact mathematical analysis, using the theory of elasticity, reveals that application of the flexure formula as in the above example gives only small errors in the normal stress if the tapered angle of the beam is small. For example, if this angle is 15°, the stress calculated from the formula will be about 5% greater than that calculated by the more exact analysis. It may also be worth noting that the calculation of $(\sigma_{max})_B$ was done only for illustrative purposes, since, by Saint-Venant's principle, the actual stress distribution at the support (wall) is highly irregular.

*11.4 Shaft Design

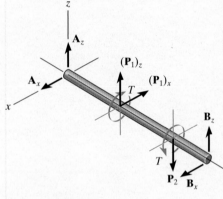

(a)

Shafts that have circular cross sections are often used in the design of mechanical equipment and machinery. As a result, they can be subjected to cyclic or fatigue stress, which is caused by the combined bending and torsional loads they must transmit or resist. In addition to these loadings, stress concentrations may exist on a shaft due to keys, couplings, and sudden transitions in its cross-sectional area (Sec. 5.8). In order to design a shaft properly, it is therefore necessary to take all of these effects into account.

Here we will discuss some of the important aspects of the design of shafts required to transmit power. These shafts are often subjected to loads applied to attached pulleys and gears, such as the one shown in Fig. 11–13a. Since the loads can be applied to the shaft at various angles, the internal bending and torsional moments at any cross section can be determined by first replacing the loads by their statically equivalent counterparts and then resolving these loads into components in two perpendicular planes, Fig. 11–13b. The bending-moment diagrams for the loads *in each plane* can then be drawn, and the resultant internal moment at any section along the shaft is then determined by vector addition, $M = \sqrt{M_x^2 + M_z^2}$, Fig. 11–13c. In addition to the moment, segments of the shaft are also subjected to different internal torques, Fig. 11–13b. To account for this general variation of torque along the shaft, a ***torque diagram*** may also be drawn, Fig. 11–13d.

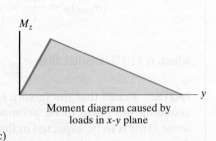

(b)

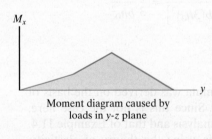

Moment diagram caused by
loads in y-z plane

Moment diagram caused by
loads in x-y plane

(c)

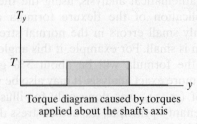

Torque diagram caused by torques
applied about the shaft's axis

(d)

Fig. 11–13

11

Once the moment and torque diagrams have been established, it is then possible to investigate certain critical sections along the shaft where the *combination* of a resultant moment **M** and a torque **T** creates the worst stress situation. Since the moment of inertia of the shaft is the *same* about *any* diametrical axis, we can apply the flexure formula using the *resultant moment* to obtain the maximum bending stress. As shown in Fig. 11–13e, this stress will occur on two elements, *C* and *D*, each located on the outer boundary of the shaft. If a torque **T** is also resisted at this section, then a maximum shear stress is also developed on these elements, Fig. 11–13f. Furthermore, the external forces will also create shear stress in the shaft determined from $\tau = VQ/It$; however, this stress will generally contribute a much smaller stress distribution on the cross section compared with that developed by bending and torsion. In some cases, it must be investigated, but for simplicity, we will neglect its effect in the following analysis. In general, then, the critical element *D* (or *C*) on the shaft is subjected to *plane stress* as shown in Fig. 11–13g, where

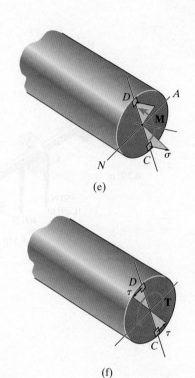

(e)

$$\sigma = \frac{Mc}{I} \quad \text{and} \quad \tau = \frac{Tc}{J}$$

If the allowable normal or shear stress for the material is known, the size of the shaft is then based on the use of these equations and selection of an appropriate theory of failure. For example, if the material is known to be ductile, then the maximum-shear-stress theory may be appropriate. As stated in Sec. 10.7, this theory requires the allowable shear stress, which is determined from the results of a simple tension test, to be equal to the maximum shear stress in the element. Using the stress-transformation equation, Eq. 9–7, for the stress state in Fig. 11–13g, we have

$$\tau_{\text{allow}} = \sqrt{\left(\frac{\sigma}{2}\right)^2 + \tau^2}$$

$$= \sqrt{\left(\frac{Mc}{2I}\right)^2 + \left(\frac{Tc}{J}\right)^2}$$

(f)

Since $I = \pi c^4/4$ and $J = \pi c^4/2$, this equation becomes

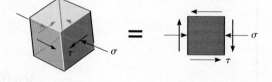

(g)

Fig. 11–13 (cont.)

$$\tau_{\text{allow}} = \frac{2}{\pi c^3}\sqrt{M^2 + T^2}$$

Solving for the radius of the shaft, we get

$$c = \left(\frac{2}{\pi \tau_{\text{allow}}}\sqrt{M^2 + T^2}\right)^{1/3} \qquad (11-2)$$

Application of any other theory of failure will, of course, lead to a different formulation for *c*. However, in all cases it may be necessary to apply this formulation at various "critical sections" along the shaft in order to determine the particular combination of *M* and *T* that gives the largest value for *c*.

The following example illustrates the procedure numerically.

EXAMPLE | 11.6

The shaft in Fig. 11–14a is supported by smooth journal bearings at A and B. Due to the transmission of power to and from the shaft, the belts on the pulleys are subjected to the tensions shown. Determine the smallest diameter of the shaft using the maximum-shear-stress theory, with τ_{allow} = 50 MPa.

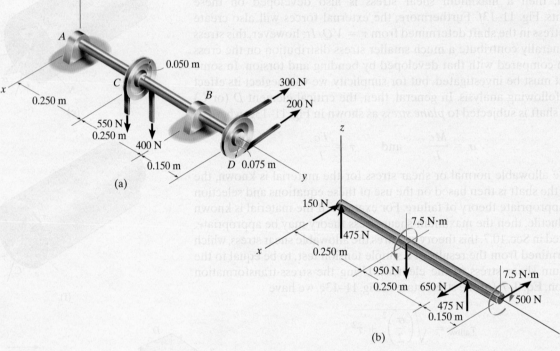

Fig. 11–14

SOLUTION

The support reactions have been calculated and are shown on the free-body diagram of the shaft, Fig. 11–14b. Bending-moment diagrams for M_x and M_z are shown in Figs. 11–14c and 11–14d, respectively. The torque diagram is shown in Fig. 11–14e. By inspection, critical points for bending moment occur either at C or B. Also, just to the right of C and at B the torsional moment is 7.5 N · m. At C, the resultant moment is

$$M_C = \sqrt{(118.75 \text{ N} \cdot \text{m})^2 + (37.5 \text{ N} \cdot \text{m})^2} = 124.5 \text{ N} \cdot \text{m}$$

whereas at B it is smaller, namely

$$M_B = 75 \text{ N} \cdot \text{m}$$

11

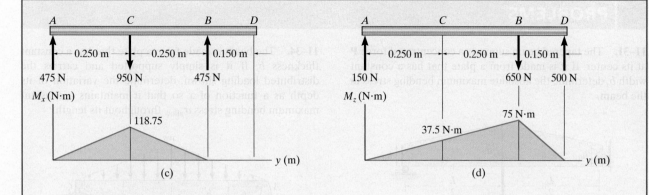

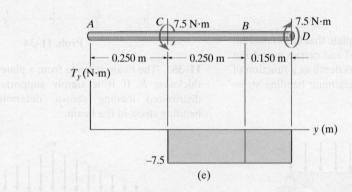

Fig. 11–14 (cont.)

Since the design is based on the maximum-shear-stress theory, Eq. 11–2 applies. The radical $\sqrt{M^2 + T^2}$ will be the largest at a section just to the right of C. We have

$$c = \left(\frac{2}{\pi \tau_{allow}} \sqrt{M^2 + T^2}\right)^{1/3}$$

$$= \left(\frac{2}{\pi (50)(10^6)\ \text{N/m}^2} \sqrt{(124.5\ \text{N} \cdot \text{m})^2 + (7.5\ \text{N} \cdot \text{m})^2}\right)^{1/3}$$

$$= 0.0117\ \text{m}$$

Thus, the smallest allowable diameter is

$$d = 2(0.0117\ \text{m}) = 23.3\ \text{mm} \qquad \textit{Ans.}$$

11

PROBLEMS

11–31.   The tapered beam supports a concentrated force **P** at its center. If it is made from a plate that has a constant width b, determine the absolute maximum bending stress in the beam.

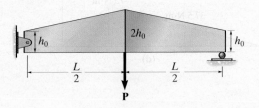

Prob. 11–31

***11–32.**   The beam is made from a plate that has a constant thickness b. If it is simply supported and carries a uniform load w, determine the variation of its depth as a function of x so that it maintains a constant maximum bending stress σ_{allow} throughout its length.

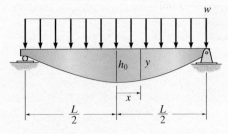

Prob. 11–32

•11–33.   The beam is made from a plate having a constant thickness t and a width that varies as shown. If it supports a concentrated force **P** at its center, determine the absolute maximum bending stress in the beam and specify its location x, $0 < x < L/2$.

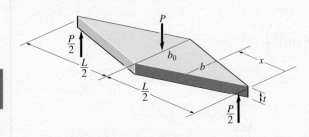

Prob. 11–33

11–34.   The beam is made from a plate that has a constant thickness b. If it is simply supported and carries the distributed loading shown, determine the variation of its depth as a function of x so that it maintains a constant maximum bending stress σ_{allow} throughout its length.

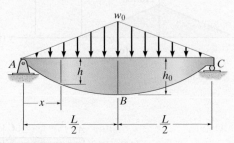

Prob. 11–34

11–35.   The beam is made from a plate that has a constant thickness b. If it is simply supported and carries the distributed loading shown, determine the maximum bending stress in the beam.

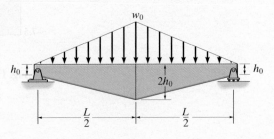

Prob. 11–35

***11–36.**   Determine the variation of the radius r of the cantilevered beam that supports the uniform distributed load so that it has a constant maximum bending stress σ_{max} throughout its length.

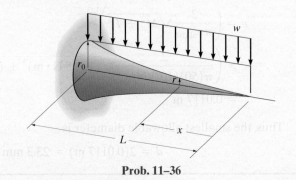

Prob. 11–36

11

•11–37.   Determine the variation in the depth d of a cantilevered beam that supports a concentrated force **P** at its end so that it has a constant maximum bending stress σ_{allow} throughout its length. The beam has a constant width b_0.

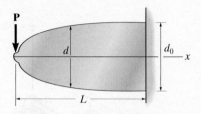

Prob. 11–37

11–38.   Determine the variation in the width b as a function of x for the cantilevered beam that supports a uniform distributed load along its centerline so that it has the same maximum bending stress σ_{allow} throughout its length. The beam has a constant depth t.

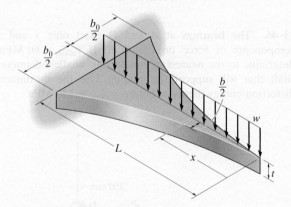

Prob. 11–38

11–39.   The shaft is supported on journal bearings that do not offer resistance to axial load. If the allowable normal stress for the shaft is $\sigma_{allow} = 80$ MPa, determine to the nearest millimeter the smallest diameter of the shaft that will support the loading. Use the maximum-distortion-energy theory of failure.

*11–40.   The shaft is supported on journal bearings that do not offer resistance to axial load. If the allowable shear stress for the shaft is $\tau_{allow} = 35$ MPa, determine to the nearest millimeter the smallest diameter of the shaft that will support the loading. Use the maximum-shear-stress theory of failure.

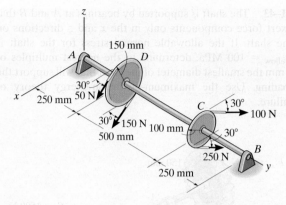

Probs. 11–39/40

•11–41.   The end gear connected to the shaft is subjected to the loading shown. If the bearings at A and B exert only y and z components of force on the shaft, determine the equilibrium torque T at gear C and then determine the smallest diameter of the shaft to the nearest millimeter that will support the loading. Use the maximum-shear-stress theory of failure with $\tau_{allow} = 60$ MPa.

11–42.   The end gear connected to the shaft is subjected to the loading shown. If the bearings at A and B exert only y and z components of force on the shaft, determine the equilibrium torque T at gear C and then determine the smallest diameter of the shaft to the nearest millimeter that will support the loading. Use the maximum-distortion-energy theory of failure with $\sigma_{allow} = 80$ MPa.

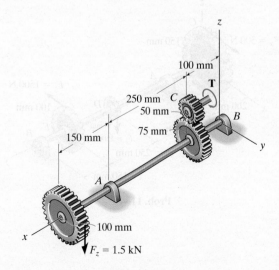

Probs. 11–41/42

11

11–43. The shaft is supported by bearings at A and B that exert force components only in the x and z directions on the shaft. If the allowable normal stress for the shaft is $\sigma_{allow} = 100$ MPa, determine to the nearest multiples of 5 mm the smallest diameter of the shaft that will support the loading. Use the maximum-distortion-energy theory of failure.

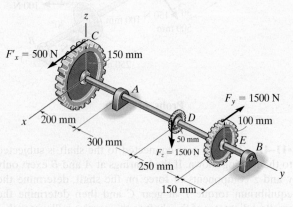

Prob. 11–43

•11–45. The bearings at A and D exert only y and z components of force on the shaft. If $\tau_{allow} = 60$ MPa, determine to the nearest millimeter the smallest-diameter shaft that will support the loading. Use the maximum-shear-stress theory of failure.

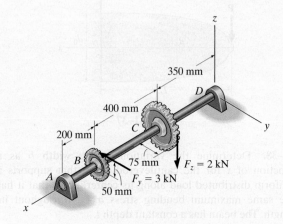

Prob. 11–45

*11–44. The shaft is supported by bearings at A and B that exert force components only in the x and z directions on the shaft. If the allowable normal stress for the shaft is $\sigma_{allow} = 100$ MPa, determine to the nearest multiples of 5 mm the smallest diameter of the shaft that will support the loading. Use the maximum-shear-stress theory of failure. Take $\tau_{allow} = 42$ MPa.

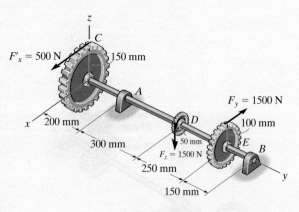

Prob. 11–44

11–46. The bearings at A and D exert only y and z components of force on the shaft. If $\tau_{allow} = 60$ MPa, determine to the nearest millimeter the smallest-diameter shaft that will support the loading. Use the maximum-distortion-energy theory of failure. $\sigma_{allow} = 130$ MPa.

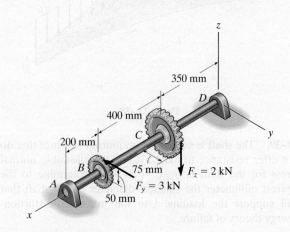

Prob. 11–46

CHAPTER REVIEW

Failure of a beam occurs when the internal shear or moment in the beam is a maximum. To resist these loadings, it is therefore important that the associated maximum shear and bending stress not exceed allowable values as stated in codes. Normally, the cross section of a beam is first designed to resist the allowable bending stress,

$$\sigma_{\text{allow}} = \frac{M_{\max}c}{I}$$

Ref.: Section 11.2

Then the allowable shear stress is checked. For rectangular sections, $\tau_{\text{allow}} \geq 1.5(V_{\max}/A)$ and for wide-flange sections it is appropriate to use $\tau_{\text{allow}} \geq V_{\max}/A_{\text{web}}$. In general, use

$$\tau_{\text{allow}} = \frac{VQ}{It}$$

Ref.: Section 11.2

For built-up beams, the spacing of fasteners or the strength of glue or weld is determined using an allowable shear flow

$$q_{\text{allow}} = \frac{VQ}{I}$$

Ref.: Section 11.2

Fully stressed beams are nonprismatic and designed such that each cross section along the beam will resist the allowable bending stress. This will define the shape of the beam.

Ref.: Section 11.3

A mechanical shaft generally is designed to resist both torsion and bending stresses. Normally, the internal bending moment can be resolved into two planes, and so it is necessary to draw the moment diagrams for each bending-moment component and then select the maximum moment based on vector addition. Once the maximum bending and shear stresses are determined, then depending upon the type of material, an appropriate theory of failure is used to compare the allowable stress to what is required.

Ref.: Section 11.4

REVIEW PROBLEMS

11–47. Draw the shear and moment diagrams for the shaft, and then determine its required diameter to the nearest millimeter if $\sigma_{\text{allow}} = 140$ MPa and $\tau_{\text{allow}} = 80$ MPa. The bearings at A and B exert only vertical reactions on the shaft.

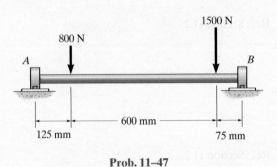

Prob. 11–47

*11–48. The overhang beam is constructed using two 50-mm by 100-mm pieces of wood braced as shown. If the allowable bending stress is $\sigma_{\text{allow}} = 4.2$ MPa, determine the largest load P that can be applied. Also, determine the associated maximum spacing of nails, s, along the beam section AC if each nail can resist a shear force of 4 kN. Assume the beam is pin-connected at A, B, and D. Neglect the axial force developed in the beam along DA.

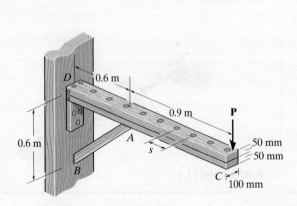

Prob. 11–48

•11–49. The bearings at A and B exert only x and z components of force on the steel shaft. Determine the shaft's diameter to the nearest millimeter so that it can resist the loadings of the gears without exceeding an allowable shear stress of $\tau_{\text{allow}} = 80$ MPa. Use the maximum-shear-stress theory of failure.

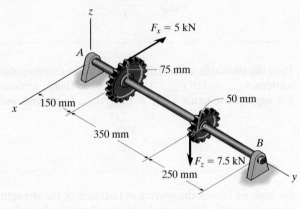

Prob. 11–49

11–50. The bearings at A and B exert only x and z components of force on the steel shaft. Determine the shaft's diameter to the nearest millimeter so that it can resist the loadings of the gears without exceeding an allowable shear stress of $\tau_{\text{allow}} = 80$ MPa. Use the maximum-distortion-energy theory of failure with $\sigma_{\text{allow}} = 200$ MPa.

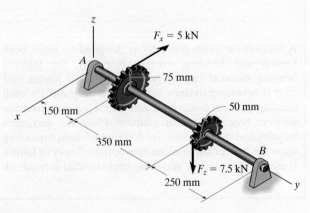

Prob. 11–50

11–51. Draw the shear and moment diagrams for the beam. Then select the lightest-weight steel wide-flange beam from Appendix B that will safely support the loading. Take $\sigma_{allow} = 150$ MPa and $\tau_{allow} = 84$ MPa.

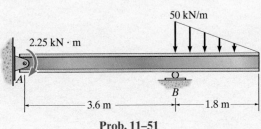

2.25 kN · m

50 kN/m

A

B

3.6 m

1.8 m

Prob. 11–51

*11–52.** The beam is made of cypress having an allowable bending stress of $\sigma_{allow} = 6$ MPa and an allowable shear stress of $\tau_{allow} = 0.56$ MPa. Determine the width b of the beam if the height $h = 1.5b$.

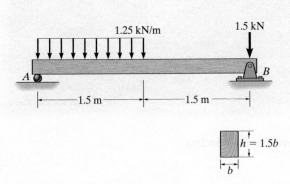

1.25 kN/m

1.5 kN

A

B

1.5 m

1.5 m

$h = 1.5b$

b

Prob. 11–52

•**11–53.** The tapered beam supports a uniform distributed load w. If it is made from a plate and has a constant width b, determine the absolute maximum bending stress in the beam.

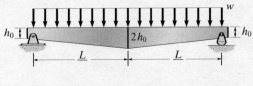

h_0

$2h_0$

h_0

w

L

L

Prob. 11–53

11–54. The tubular shaft has an inner diameter of 15 mm. Determine to the nearest millimeter its outer diameter if it is subjected to the gear loading. The bearings at A and B exert force components only in the y and z directions on the shaft. Use an allowable shear stress of $\tau_{allow} = 70$ MPa, and base the design on the maximum-shear-stress theory of failure.

11–55. Determine to the nearest millimeter the diameter of the solid shaft if it is subjected to the gear loading. The bearings at A and B exert force components only in the y and z directions on the shaft. Base the design on the maximum-distortion-energy theory of failure with $\sigma_{allow} = 150$ MPa.

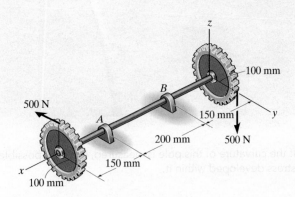

z

100 mm

B

500 N

A

150 mm

y

200 mm

500 N

x

150 mm

100 mm

Probs. 11–54/55

11

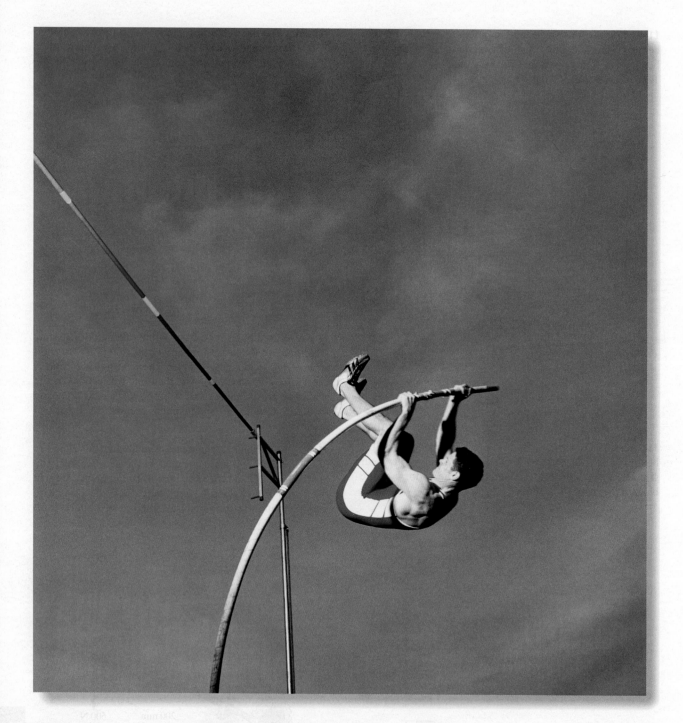

If the curvature of this pole is measured, it is then possible to determine the bending stress developed within it.

Deflection of Beams and Shafts

12

CHAPTER OBJECTIVES

Often limits must be placed on the amount of deflection a beam or shaft may undergo when it is subjected to a load, and so in this chapter we will discuss various methods for determining the deflection and slope at specific points on beams and shafts. The analytical methods include the integration method, the use of discontinuity functions, and the method of superposition. Also, a semigraphical technique, called the moment-area method, will be presented. At the end of the chapter, we will use these methods to solve for the support reactions on a beam or shaft that is statically indeterminate.

Check out the Companion Website for the following video solution(s) that can be found in this chapter.

- Section 5
  Beam Deflection –
  Superpostion Method

12.1 The Elastic Curve

The deflection of a beam or shaft must often be limited in order to provide integrity and stability of a structure or machine, and prevent the cracking of any attached brittle materials such as concrete or glass. Furthermore, code restrictions often require these members not vibrate or deflect severely in order to safely support their intended loading. Most important, though, deflections at specific points on a beam or shaft must be determined if one is to analyze those that are statically indeterminate.

Before the slope or the displacement at a point on a beam (or shaft) is determined, it is often helpful to sketch the deflected shape of the beam when it is loaded, in order to "visualize" any computed results and thereby partially check these results. The deflection curve of the longitudinal axis that passes through the centroid of each cross-sectional area of a beam is called the *elastic curve*. For most beams the elastic curve can be sketched without much difficulty. When doing so, however, it is necessary to know how the slope or displacement is restricted at various types of supports. In general, supports that resist a *force*, such as a pin, restrict *displacement*, and those that resist a *moment*, such as a fixed wall, restrict *rotation* or *slope* as well as displacement. With this in mind, two typical examples of the elastic curves for loaded beams (or shafts), sketched to an exaggerated scale, are shown in Fig. 12–1.

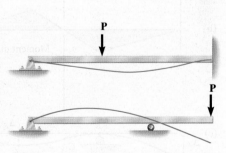

Fig. 12–1

12

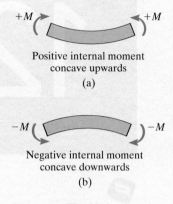

Positive internal moment concave upwards

(a)

Negative internal moment concave downwards

(b)

Fig. 12–2

If the elastic curve for a beam seems difficult to establish, it is suggested that the moment diagram for the beam be drawn first. Using the beam sign convention established in Sec. 6.1, a positive internal moment tends to bend the beam concave upward, Fig. 12–2a. Likewise, a negative moment tends to bend the beam concave downward, Fig. 12–2b. Therefore, if the moment diagram is *known*, it will be easy to construct the elastic curve. For example, consider the beam in Fig. 12–3a with its associated moment diagram shown in Fig. 12–3b. Due to the roller and pin supports, the displacement at B and D must be zero. Within the region of negative moment, AC, Fig. 12–3b, the elastic curve must be concave downward, and within the region of positive moment, CD, the elastic curve must be concave upward. Hence, there must be an *inflection point* at point C, where the curve changes from concave up to concave down, since this is a point of zero moment. Using these facts, the beam's elastic curve is sketched in Fig. 12–3c. It should also be noted that the displacements Δ_A and Δ_E are especially critical. At point E the *slope* of the elastic curve is *zero*, and there the beam's *deflection* may be a *maximum*. Whether Δ_E is actually greater than Δ_A depends on the relative magnitudes of $\mathbf{P}_1$ and $\mathbf{P}_2$ and the location of the roller at B.

Following these same principles, note how the elastic curve in Fig. 12–4 was constructed. Here the beam is cantilevered from a fixed support at A and therefore the elastic curve must have both zero displacement and zero slope at this point. Also, the largest displacement will occur either at D, where the slope is zero, or at C.

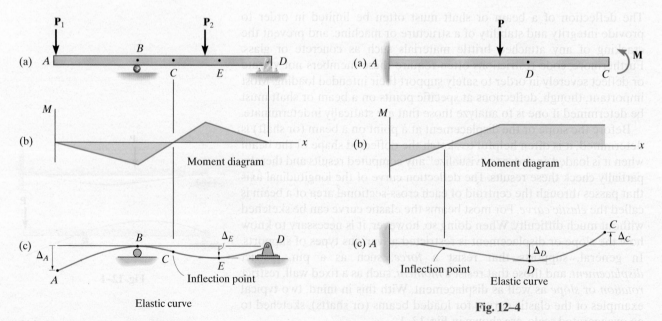

Moment diagram

Elastic curve

Inflection point

Elastic curve

Fig. 12–3

Fig. 12–4

Moment-Curvature Relationship.

We will now develop an important relationship between the internal moment and the radius of curvature ρ (rho) of the elastic curve at a point. The resulting equation will be used for establishing each of the methods presented in the chapter for finding the slope and displacement at points on the elastic curve.

The following analysis, here and in the next section, will require the use of three coordinates. As shown in Fig. 12–5a, the x axis extends positive to the right, along the initially straight longitudinal axis of the beam. It is used to locate the differential element, having an undeformed width dx. The v axis extends *positive upward* from the x axis. It measures the *displacement* of the elastic curve. Lastly, a "localized" y coordinate is used to specify the position of a fiber in the beam element. It is measured *positive upward* from the neutral axis (or elastic curve) as shown in Fig. 12–5b. Recall that this same sign convention for x and y was used in the derivation of the flexure formula.

To derive the relationship between the internal moment and ρ, we will limit the analysis to the most common case of an initially straight beam that is elastically deformed by loads applied perpendicular to the beam's x axis and lying in the x–v plane of symmetry for the beam's cross-sectional area. Due to the loading, the deformation of the beam is caused by both the internal shear force and bending moment. If the beam has a length that is much greater than its depth, the greatest deformation will be caused by bending, and therefore we will direct our attention to its effects. Deflections caused by shear will be discussed in Chapter 14.

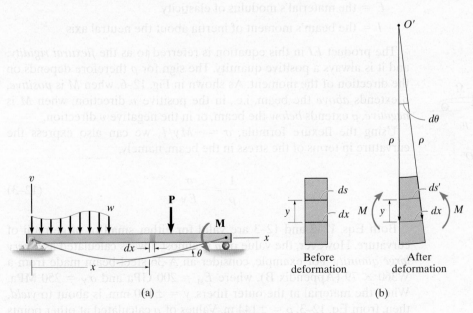

Fig. 12–5

(a)

(b)

Before deformation

After deformation

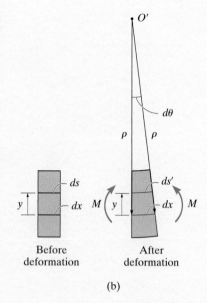

Fig. 12–5 (cont.)

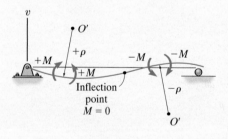

Fig. 12–6

When the internal moment M deforms the element of the beam, the angle between the cross sections becomes $d\theta$, Fig. 12–5b. The arc dx represents a portion of the elastic curve that intersects the neutral axis for each cross section. The *radius of curvature* for this arc is defined as the distance ρ, which is measured from the *center of curvature O'* to dx. Any arc on the element other than dx is subjected to a normal strain. For example, the strain in arc ds, located at a position y from the neutral axis, is $\epsilon = (ds' - ds)/ds$. However, $ds = dx = \rho\, d\theta$ and $ds' = (\rho - y)\, d\theta$, and so $\epsilon = [(\rho - y)\, d\theta - \rho\, d\theta]/\rho\, d\theta$ or

$$\frac{1}{\rho} = -\frac{\epsilon}{y} \tag{12–1}$$

If the material is homogeneous and behaves in a linear-elastic manner, then Hooke's law applies, $\epsilon = \sigma/E$. Also, since the flexure formula applies, $\sigma = -My/I$. Combining these two equations and substituting into the above equation, we have

$$\frac{1}{\rho} = \frac{M}{EI} \tag{12–2}$$

where

ρ = the radius of curvature at the point on the elastic curve
    ($1/\rho$ is referred to as the *curvature*)

M = the internal moment in the beam at the point

E = the material's modulus of elasticity

I = the beam's moment of inertia about the neutral axis

The product EI in this equation is referred to as the *flexural rigidity*, and it is always a positive quantity. The sign for ρ therefore depends on the direction of the moment. As shown in Fig. 12–6, when M is *positive*, ρ extends *above* the beam, i.e., in the positive v direction; when M is *negative*, ρ extends *below* the beam, or in the negative v direction.

Using the flexure formula, $\sigma = -My/I$, we can also express the curvature in terms of the stress in the beam, namely,

$$\frac{1}{\rho} = -\frac{\sigma}{Ey} \tag{12–3}$$

Both Eqs. 12–2 and 12–3 are valid for either small or large radii of curvature. However, the value of ρ is almost always calculated as a *very large quantity*. For example, consider an A-36 steel beam made from a W360 × 79 (Appendix B), where $E_{st} = 200$ GPa and $\sigma_Y = 250$ MPa. When the material at the outer fibers, $y = \pm 180$ mm, is about to *yield*, then, from Eq. 12–3, $\rho = \pm 144$ m. Values of ρ calculated at other points along the beam's elastic curve may be even *larger*, since σ cannot exceed σ_Y at the outer fibers.

12.2   Slope and Displacement by Integration

The equation of the elastic curve for a beam can be expressed mathematically as $v = f(x)$. To obtain this equation, we must first represent the curvature $(1/\rho)$ in terms of v and x. In most calculus books it is shown that this relationship is

$$\frac{1}{\rho} = \frac{d^2v/dx^2}{[1 + (dv/dx)^2]^{3/2}}$$

Substituting into Eq. 12–2, we have

$$\frac{d^2v/dx^2}{[1 + (dv/dx)^2]^{3/2}} = \frac{M}{EI} \qquad (12-4)$$

This equation represents a nonlinear second-order differential equation. Its solution, which is called the *elastica*, gives the exact shape of the elastic curve, assuming, of course, that beam deflections occur only due to bending. Through the use of higher mathematics, elastica solutions have been obtained only for simple cases of beam geometry and loading.

In order to facilitate the solution of a greater number of deflection problems, Eq. 12–4 can be modified. Most engineering design codes specify *limitations* on deflections for tolerance or esthetic purposes, and as a result the elastic deflections for the majority of beams and shafts form a shallow curve. Consequently, the *slope* of the elastic curve, which is determined from dv/dx, will be *very small*, and its square will be negligible compared with unity.* Therefore the curvature, as defined above, can be *approximated* by $1/\rho = d^2v/dx^2$. Using this simplification, Eq. 12–4 can now be written as

$$\frac{d^2v}{dx^2} = \frac{M}{EI} \qquad (12-5)$$

It is also possible to write this equation in two alternative forms. If we differentiate each side with respect to x and substitute $V = dM/dx$ (Eq. 6–2), we get

$$\frac{d}{dx}\left(EI\frac{d^2v}{dx^2}\right) = V(x) \qquad (12-6)$$

Differentiating again, using $w = dV/dx$ (Eq. 6–1), yields

$$\frac{d^2}{dx^2}\left(EI\frac{d^2v}{dx^2}\right) = w(x) \qquad (12-7)$$

*See Example 12.1.

12

For most problems the flexural rigidity (EI) will be constant along the length of the beam. Assuming this to be the case, the above results may be reordered into the following set of three equations:

$$EI\frac{d^4v}{dx^4} = w(x) \qquad (12\text{–}8)$$

$$EI\frac{d^3v}{dx^3} = V(x) \qquad (12\text{–}9)$$

$$EI\frac{d^2v}{dx^2} = M(x) \qquad (12\text{–}10)$$

Solution of any of these equations requires successive integrations to obtain the deflection v of the elastic curve. For each integration it is necessary to introduce a "constant of integration" and then solve for all the constants to obtain a unique solution for a particular problem. For example, if the distributed load w is expressed as a function of x and Eq. 12–8 is used, then four constants of integration must be evaluated; however, if the internal moment M is determined and Eq. 12–10 is used, only two constants of integration must be found. The choice of which equation to start with depends on the problem. Generally, however, it is easier to determine the internal moment M as a function of x, integrate twice, and evaluate only two integration constants.

Recall from Sec. 6.1 that if the loading on a beam is discontinuous, that is, consists of a series of several distributed and concentrated loads, then several functions must be written for the internal moment, each valid within the region between the discontinuities. Also, for convenience in writing each moment expression, *the origin* for each x coordinate can be *selected arbitrarily*. For example, consider the beam shown in Fig. 12–7a. The internal moment in regions AB, BC, and CD can be written in terms of the x_1, x_2, and x_3 coordinates selected, as shown in either Fig. 12–7b or Fig. 12–7c, or in fact in any manner that will yield $M = f(x)$ in as simple a form as possible. Once these functions are integrated twice through the use of Eq. 12–10 and the constants of integration determined, the functions will give the slope and deflection (elastic curve) for each region of the beam for which they are valid.

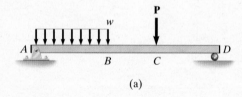

(a)

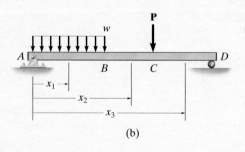

(b)

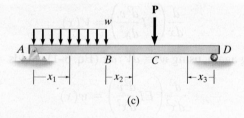

(c)

Fig. 12–7

Sign Convention and Coordinates.

When applying Eqs. 12–8 through 12–10, it is important to use the proper signs for M, V, or w as established by the sign convention that was used in the derivation of these equations. For review, these terms are shown in their *positive directions* in Fig. 12–8a. Furthermore, recall that *positive deflection, v,* is *upward*, and as a result, the *positive slope angle θ* will be measured *counterclockwise* from the x axis when x is *positive to the right*. The reason for this is shown in Fig. 12–8b. Here positive increases dx and dv in x and v create an increased θ that is counterclockwise. If, however, *positive x* is directed to the *left*, then θ will be *positive clockwise*, Fig. 12–8c.

Realize that by assuming dv/dx to be very small, the original horizontal length of the beam's axis and the arc of its elastic curve will be about the same. In other words, ds in Figs. 12–8b and 12–8c is approximately equal to dx, since $ds = \sqrt{(dx)^2 + (dv)^2} = \sqrt{1 + (dv/dx)^2}\, dx \approx dx$. As a result, points on the elastic curve are assumed to be *displaced vertically*, and not horizontally. Also, since the *slope angle θ* will be *very small*, its value in radians can be determined *directly* from $\theta \approx \tan\theta = dv/dx$.

The design of a roof system requires a careful consideration of deflection. For example, rain can accumulate on areas of the roof, which then causes ponding, leading to further deflection, then further ponding, and finally possible failure of the roof.

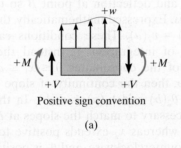

Positive sign convention

(a)

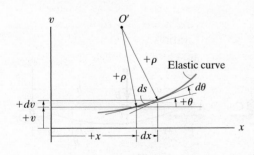

Positive sign convention

(b)

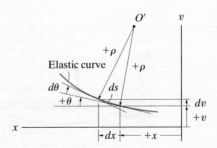

Positive sign convention

(c)

Fig. 12–8

12

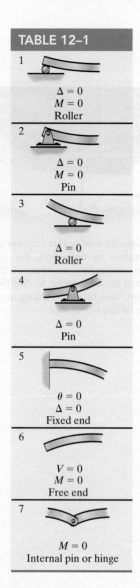

TABLE 12–1

1

$\Delta = 0$
$M = 0$
Roller

2

$\Delta = 0$
$M = 0$
Pin

3

$\Delta = 0$
Roller

4

$\Delta = 0$
Pin

5

$\theta = 0$
$\Delta = 0$
Fixed end

6

$V = 0$
$M = 0$
Free end

7

$M = 0$
Internal pin or hinge

Boundary and Continuity Conditions.

When solving Eqs. 12–8, 12–9, or 12–10, the constants of integration are determined by evaluating the functions for shear, moment, slope, or displacement at a particular point on the beam where the value of the function is known. These values are called **boundary conditions**. Several possible boundary conditions that are often used to solve beam (or shaft) deflection problems are listed in Table 12–1. For example, if the beam is supported by a roller or pin (1, 2, 3, 4), then it is required that the displacement be *zero* at these points. Furthermore, if these supports are located at the *ends of the beam* (1, 2), the internal moment in the beam must also be zero. At the fixed support (5), the slope and displacement are both zero, whereas the free-ended beam (6) has both zero moment and zero shear. Lastly, if two segments of a beam are connected by an "internal" pin or hinge (7), the moment must be zero at this connection.

If the elastic curve cannot be expressed using a single coordinate, then **continuity conditions** must be used to evaluate some of the integration constants. For example, consider the beam in Fig. 12–9a. Here two x coordinates are chosen with origins at A. Each is valid only within the regions $0 \le x_1 \le a$ and $a \le x_2 \le (a + b)$. Once the functions for the slope and deflection are obtained, they must give the *same values* for the slope and deflection at point B so the elastic curve is physically *continuous*. Expressed mathematically, this requires that $\theta_1(a) = \theta_2(a)$ and $v_1(a) = v_2(a)$. These conditions can be used to evaluate two constants of integration. If instead the elastic curve is expressed in terms of the coordinates $0 \le x_1 \le a$ and $0 \le x_2 \le b$, shown in Fig. 12–9b, then the continuity of slope and deflection at B requires $\theta_1(a) = -\theta_2(b)$ and $v_1(a) = v_2(b)$. In this particular case, a *negative* sign is necessary to match the slopes at B since x_1 extends positive to the right, whereas x_2 extends positive to the left. Consequently, θ_1 is positive counterclockwise, and θ_2 is positive clockwise. See Figs. 12–8b and 12–8c.

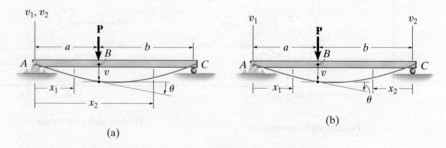

(a)

(b)

Fig. 12–9

Procedure for Analysis

The following procedure provides a method for determining the slope and deflection of a beam (or shaft) using the method of integration.

Elastic Curve.

- Draw an exaggerated view of the beam's elastic curve. Recall that zero slope and zero displacement occur at all fixed supports, and zero displacement occurs at all pin and roller supports.

- Establish the x and v coordinate axes. The x axis must be parallel to the undeflected beam and can have an origin at any point along the beam, with a positive direction either to the right or to the left.

- If several discontinuous loads are present, establish x coordinates that are valid for each region of the beam between the discontinuities. Choose these coordinates so that they will simplify subsequent algebraic work.

- In all cases, the associated positive v axis should be directed upward.

Load or Moment Function.

- For each region in which there is an x coordinate, express the loading w or the internal moment M as a function of x. In particular, *always* assume that M acts in the *positive direction* when applying the equation of moment equilibrium to determine $M = f(x)$.

Slope and Elastic Curve.

- Provided EI is constant, apply either the load equation $EI \, d^4v/dx^4 = w(x)$, which requires four integrations to get $v = v(x)$, or the moment equation $EI \, d^2v/dx^2 = M(x)$, which requires only two integrations. For each integration it is important to include a constant of integration.

- The constants are evaluated using the boundary conditions for the supports (Table 12–1) and the continuity conditions that apply to slope and displacement at points where two functions meet. Once the constants are evaluated and substituted back into the slope and deflection equations, the slope and displacement at *specific points* on the elastic curve can then be determined.

- The numerical values obtained can be checked graphically by comparing them with the sketch of the elastic curve. Realize that *positive* values for *slope* are *counterclockwise* if the x axis extends *positive* to the *right*, and *clockwise* if the x axis extends *positive* to the *left*. In either of these cases, *positive displacement* is *upward*.

12

12

EXAMPLE | **12.1**

The cantilevered beam shown in Fig. 12–10a is subjected to a vertical load **P** at its end. Determine the equation of the elastic curve. EI is constant.

SOLUTION I

Elastic Curve. The load tends to deflect the beam as shown in Fig. 12–10a. By inspection, the internal moment can be represented throughout the beam using a single x coordinate.

Moment Function. From the free-body diagram, with **M** acting in the *positive direction*, Fig. 12–10b, we have

$$M = -Px$$

Slope and Elastic Curve. Applying Eq. 12–10 and integrating twice yields

$$EI\frac{d^2v}{dx^2} = -Px \tag{1}$$

$$EI\frac{dv}{dx} = -\frac{Px^2}{2} + C_1 \tag{2}$$

$$EIv = -\frac{Px^3}{6} + C_1 x + C_2 \tag{3}$$

Using the boundary conditions $dv/dx = 0$ at $x = L$ and $v = 0$ at $x = L$, Eqs. 2 and 3 become

$$0 = -\frac{PL^2}{2} + C_1$$

$$0 = -\frac{PL^3}{6} + C_1 L + C_2$$

Thus, $C_1 = PL^2/2$ and $C_2 = -PL^3/3$. Substituting these results into Eqs. 2 and 3 with $\theta = dv/dx$, we get

$$\theta = \frac{P}{2EI}(L^2 - x^2)$$

$$v = \frac{P}{6EI}(-x^3 + 3L^2x - 2L^3) \qquad \textit{Ans.}$$

Maximum slope and displacement occur at $A(x = 0)$, for which

$$\theta_A = \frac{PL^2}{2EI} \tag{4}$$

$$v_A = -\frac{PL^3}{3EI} \tag{5}$$

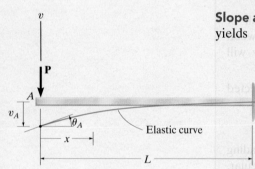

Elastic curve

(a)

(b)

Fig. 12–10

The *positive* result for θ_A indicates *counterclockwise* rotation and the *negative* result for v_A indicates that v_A is *downward*. This agrees with the results sketched in Fig. 12–10a.

In order to obtain some idea as to the actual *magnitude* of the slope and displacement at the end A, consider the beam in Fig. 12–10a to have a length of 5 m, support a load of $P = 30$ kN, and be made of A-36 steel having $E_{st} = 200$ GPa. Using the methods of Sec. 11.2, if this beam was designed without a factor of safety by assuming the allowable normal stress is equal to the yield stress $\sigma_{allow} = 250$ MPa; then a W310 × 39 would be found to be adequate ($I = 84.4(10^6)$ mm^4). From Eqs. 4 and 5 we get

$$\theta_A = \frac{30 \text{ kN}(5 \text{ m})^2(1000 \text{ mm/m})^2}{2[200 \text{ kN/mm}^2](84.4(10^6) \text{ mm}^4)} = 0.0222 \text{ rad}$$

$$v_A = \frac{30 \text{ kN}(5 \text{ m})^3(1000 \text{ mm/m})^3}{3[200 \text{ kN/mm}^2](84.4(10^6) \text{ mm}^4)} = -74.1 \text{ mm}$$

Since $\theta_A^2 = (dv/dx)^2 = 0.000493$ rad$^2 \ll 1$, this justifies the use of Eq. 12–10, rather than applying the more exact Eq. 12–4, for computing the deflection of beams. Also, since this numerical application is for a *cantilevered beam*, we have obtained *larger values* for θ and v than would have been obtained if the beam were supported using pins, rollers, or other fixed supports.

SOLUTION II

This problem can also be solved using Eq. 12–8, $EI\, d^4v/dx^4 = w(x)$. Here $w(x) = 0$ for $0 \le x \le L$, Fig. 12–10a, so that upon integrating once we get the form of Eq. 12–9, i.e.,

$$EI\frac{d^4v}{dx^4} = 0$$

$$EI\frac{d^3v}{dx^3} = C_1' = V$$

The shear constant C_1' can be evaluated at $x = 0$, since $V_A = -P$ (negative according to the beam sign convention, Fig. 12–8a). Thus, $C_1' = -P$. Integrating again yields the form of Eq. 12–10, i.e.,

$$EI\frac{d^3v}{dx^3} = -P$$

$$EI\frac{d^2v}{dx^2} = -Px + C_2' = M$$

Here $M = 0$ at $x = 0$, so $C_2' = 0$, and as a result one obtains Eq. 1 and the solution proceeds as before.

EXAMPLE | 12.2

The simply supported beam shown in Fig. 12–11a supports the triangular distributed loading. Determine its maximum deflection. EI is constant.

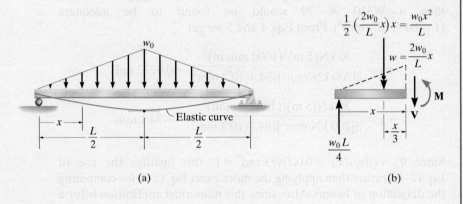

Fig. 12–11

SOLUTION I

Elastic Curve. Due to symmetry, only one x coordinate is needed for the solution, in this case $0 \leq x \leq L/2$. The beam deflects as shown in Fig. 12–11a. The maximum deflection occurs at the center since the slope is zero at this point.

Moment Function. A free-body diagram of the segment on the left is shown in Fig. 12–11b. The equation for the distributed loading is

$$w = \frac{2w_0}{L}x \qquad (1)$$

Hence,

$$\zeta + \Sigma M_{NA} = 0; \qquad M + \frac{w_0 x^2}{L}\left(\frac{x}{3}\right) - \frac{w_0 L}{4}(x) = 0$$

$$M = -\frac{w_0 x^3}{3L} + \frac{w_0 L}{4}x$$

Slope and Elastic Curve.   Using Eq. 12–10 and integrating twice, we have

$$EI\frac{d^2v}{dx^2} = M = -\frac{w_0}{3L}x^3 + \frac{w_0L}{4}x \qquad (2)$$

$$EI\frac{dv}{dx} = -\frac{w_0}{12L}x^4 + \frac{w_0L}{8}x^2 + C_1$$

$$EIv = -\frac{w_0}{60L}x^5 + \frac{w_0L}{24}x^3 + C_1x + C_2$$

The constants of integration are obtained by applying the boundary condition $v = 0$ at $x = 0$ and the symmetry condition that $dv/dx = 0$ at $x = L/2$. This leads to

$$C_1 = -\frac{5w_0L^3}{192} \qquad C_2 = 0$$

Hence,

$$EI\frac{dv}{dx} = -\frac{w_0}{12L}x^4 + \frac{w_0L}{8}x^2 - \frac{5w_0L^3}{192}$$

$$EIv = -\frac{w_0}{60L}x^5 + \frac{w_0L}{24}x^3 - \frac{5w_0L^3}{192}x$$

Determining the maximum deflection at $x = L/2$, we have

$$v_{max} = -\frac{w_0L^4}{120EI} \qquad\qquad Ans.$$

SOLUTION II
Since the distributed loading acts downward, it is negative according to our sign convention. Using Eq. 1 and applying Eq. 12–8, we have

$$EI\frac{d^4v}{dx^4} = -\frac{2w_0}{L}x$$

$$EI\frac{d^3v}{dx^3} = V = -\frac{w_0}{L}x^2 + C_1'$$

Since $V = +w_0L/4$ at $x = 0$, then $C_1' = w_0L/4$. Integrating again yields

$$EI\frac{d^3v}{dx^3} = V = -\frac{w_0}{L}x^2 + \frac{w_0L}{4}$$

$$EI\frac{d^2v}{dx^2} = M = -\frac{w_0}{3L}x^3 + \frac{w_0L}{4}x + C_2'$$

Here $M = 0$ at $x = 0$, so $C_2' = 0$. This yields Eq. 2. The solution now proceeds as before.

12

EXAMPLE | 12.3

The simply supported beam shown in Fig. 12–12*a* is subjected to the concentrated force **P**. Determine the maximum deflection of the beam. *EI* is constant.

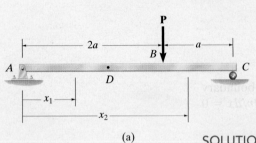

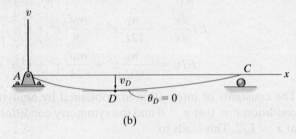

(a)

(b)

SOLUTION

Elastic Curve. The beam deflects as shown in Fig. 12–12*b*. Two coordinates must be used, since the moment function will change at *P*. Here we will take x_1 and x_2, having the *same* origin at *A*.

Moment Function. From the free-body diagrams shown in Fig. 12–12*c*,

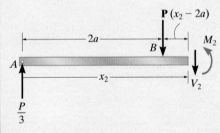

(c)

Fig. 12–12

$$M_1 = \frac{P}{3}x_1$$

$$M_2 = \frac{P}{3}x_2 - P(x_2 - 2a) = \frac{2P}{3}(3a - x_2)$$

Slope and Elastic Curve. Applying Eq. 12–10 for M_1, for $0 \leq x_1 < 2a$, and integrating twice yields

$$EI\frac{d^2v_1}{dx_1{}^2} = \frac{P}{3}x_1$$

$$EI\frac{dv_1}{dx_1} = \frac{P}{6}x_1{}^2 + C_1 \tag{1}$$

$$EIv_1 = \frac{P}{18}x_1{}^3 + C_1 x_1 + C_2 \tag{2}$$

Likewise for M_2, for $2a < x_2 \leq 3a$,

$$EI\frac{d^2v_2}{dx_2{}^2} = \frac{2P}{3}(3a - x_2)$$

$$EI\frac{dv_2}{dx_2} = \frac{2P}{3}\left(3ax_2 - \frac{x_2{}^2}{2}\right) + C_3 \tag{3}$$

$$EIv_2 = \frac{2P}{3}\left(\frac{3}{2}ax_2{}^2 - \frac{x_2{}^3}{6}\right) + C_3 x_2 + C_4 \tag{4}$$

The four constants are evaluated using *two* boundary conditions, namely, $x_1 = 0$, $v_1 = 0$ and $x_2 = 3a$, $v_2 = 0$. Also, *two* continuity conditions must be applied at B, that is, $dv_1/dx_1 = dv_2/dx_2$ at $x_1 = x_2 = 2a$ and $v_1 = v_2$ at $x_1 = x_2 = 2a$. Substitution as specified results in the following four equations:

$v_1 = 0$ at $x_1 = 0$; $0 = 0 + 0 + C_2$

$v_2 = 0$ at $x_2 = 3a$; $0 = \dfrac{2P}{3}\left(\dfrac{3}{2}a(3a)^2 - \dfrac{(3a)^3}{6}\right) + C_3(3a) + C_4$

$\dfrac{dv_1(2a)}{dx_1} = \dfrac{dv_2(2a)}{dx_2}$; $\dfrac{P}{6}(2a)^2 + C_1 = \dfrac{2P}{3}\left(3a(2a) - \dfrac{(2a)^2}{2}\right) + C_3$

$v_1(2a) = v_2(2a)$; $\dfrac{P}{18}(2a)^3 + C_1(2a) + C_2 = \dfrac{2P}{3}\left(\dfrac{3}{2}a(2a)^2 - \dfrac{(2a)^3}{6}\right) + C_3(2a) + C_4$

Solving, we get

$$C_1 = -\frac{4}{9}Pa^2 \qquad C_2 = 0$$

$$C_3 = -\frac{22}{9}Pa^2 \qquad C_4 = \frac{4}{3}Pa^3$$

Thus Eqs. 1–4 become

$$\frac{dv_1}{dx_1} = \frac{P}{6EI}x_1^2 - \frac{4Pa^2}{9EI} \tag{5}$$

$$v_1 = \frac{P}{18EI}x_1^3 - \frac{4Pa^2}{9EI}x_1 \tag{6}$$

$$\frac{dv_2}{dx_2} = \frac{2Pa}{EI}x_2 - \frac{P}{3EI}x_2^2 - \frac{22Pa^2}{9EI} \tag{7}$$

$$v_2 = \frac{Pa}{EI}x_2^2 - \frac{P}{9EI}x_2^3 - \frac{22Pa^2}{9EI}x_2 + \frac{4Pa^3}{3EI} \tag{8}$$

By inspection of the elastic curve, Fig. 12–12b, the maximum deflection occurs at D, somewhere within region AB. Here the slope must be zero. From Eq. 5,

$$\frac{1}{6}x_1^2 - \frac{4}{9}a^2 = 0$$

$$x_1 = 1.633a$$

Substituting into Eq. 6,

$$v_{max} = -0.484\frac{Pa^3}{EI} \qquad\qquad Ans.$$

The negative sign indicates that the deflection is downward.

EXAMPLE | 12.4

The beam in Fig. 12–13a is subjected to a load **P** at its end. Determine the displacement at C. EI is constant.

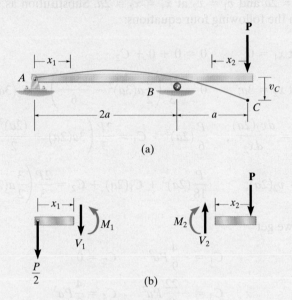

(a)

(b)

Fig. 12–13

SOLUTION

Elastic Curve. The beam deflects into the shape shown in Fig. 12–13a. Due to the loading, two x coordinates will be considered, namely, $0 \leq x_1 < 2a$ and $0 \leq x_2 < a$, where x_2 is directed to the left from C, since the internal moment is easy to formulate.

Moment Functions. Using the free-body diagrams shown in Fig. 12–13b, we have

$$M_1 = -\frac{P}{2}x_1 \qquad M_2 = -Px_2$$

Slope and Elastic Curve. Applying Eq. 12–10,

For $0 \leq x_1 \leq 2a$: $\quad EI\frac{d^2v_1}{dx_1{}^2} = -\frac{P}{2}x_1$

$$EI\frac{dv_1}{dx_1} = -\frac{P}{4}x_1{}^2 + C_1 \qquad (1)$$

$$EIv_1 = -\frac{P}{12}x_1{}^3 + C_1x_1 + C_2 \qquad (2)$$

For $0 \le x_2 \le a$: $EI\dfrac{d^2v_2}{dx_2^2} = -Px_2$

$$EI\dfrac{dv_2}{dx_2} = -\dfrac{P}{2}x_2^2 + C_3 \qquad (3)$$

$$EIv_2 = -\dfrac{P}{6}x_2^3 + C_3x_2 + C_4 \qquad (4)$$

The *four* constants of integration are determined using *three* boundary conditions, namely, $v_1 = 0$ at $x_1 = 0$, $v_1 = 0$ at $x_1 = 2a$, and $v_2 = 0$ at $x_2 = a$, and *one* continuity equation. Here the continuity of slope at the roller requires $dv_1/dx_1 = -dv_2/dx_2$ at $x_1 = 2a$ and $x_2 = a$. Why is there a negative sign in this equation? (Note that continuity of displacement at B has been indirectly considered in the boundary conditions, since $v_1 = v_2 = 0$ at $x_1 = 2a$ and $x_2 = a$.) Applying these four conditions yields

$v_1 = 0$ at $x_1 = 0$; $0 = 0 + 0 + C_2$

$v_1 = 0$ at $x_1 = 2a$; $0 = -\dfrac{P}{12}(2a)^3 + C_1(2a) + C_2$

$v_2 = 0$ at $x_2 = a$; $0 = -\dfrac{P}{6}a^3 + C_3a + C_4$

$\dfrac{dv_1(2a)}{dx_1} = -\dfrac{dv_2(a)}{dx_2}$; $-\dfrac{P}{4}(2a)^2 + C_1 = -\left(-\dfrac{P}{2}(a)^2 + C_3\right)$

Solving, we obtain

$$C_1 = \dfrac{Pa^2}{3} \qquad C_2 = 0 \qquad C_3 = \dfrac{7}{6}Pa^2 \qquad C_4 = -Pa^3$$

Substituting C_3 and C_4 into Eq. 4 gives

$$v_2 = -\dfrac{P}{6EI}x_2^3 + \dfrac{7Pa^2}{6EI}x_2 - \dfrac{Pa^3}{EI}$$

The displacement at C is determined by setting $x_2 = 0$. We get

$$v_C = -\dfrac{Pa^3}{EI} \qquad\qquad Ans.$$

12

FUNDAMENTAL PROBLEMS

F12–1. Determine the slope and deflection of end A of the cantilevered beam. $E = 200$ GPa and $I = 65.0(10^6)$ mm^4.

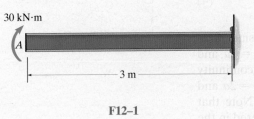

F12–1

F12–2. Determine the slope and deflection of end A of the cantilevered beam. $E = 200$ GPa and $I = 65.0(10^6)$ mm^4.

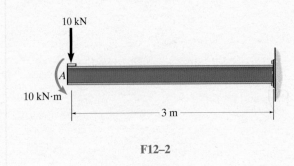

F12–2

F12–3. Determine the slope of end A of the cantilevered beam. $E = 200$ GPa and $I = 65.0(10^6)$ mm^4.

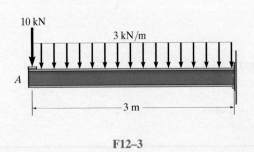

F12–3

F12–4. Determine the maximum deflection of the simply supported beam. The beam is made of wood having a modulus of elasticity of $E_w = 10$ GPa and a rectangular cross section of $b = 60$ mm and $h = 125$ mm.

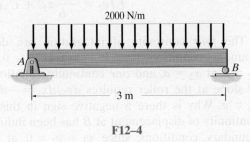

F12–4

F12–5. Determine the maximum deflection of the simply supported beam. $E = 200$ GPa and $I = 39.9(10^{-6})$ m^4.

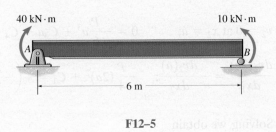

F12–5

F12–6. Determine the slope of the simply supported beam at A. $E = 200$ GPa and $I = 39.9(10^{-6})$ m^4.

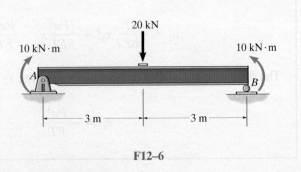

F12–6

PROBLEMS

•12–1. An A-36 steel strap having a thickness of 10 mm and a width of 20 mm is bent into a circular arc of radius $\rho = 10$ m. Determine the maximum bending stress in the strap.

12–2. A picture is taken of a man performing a pole vault, and the minimum radius of curvature of the pole is estimated by measurement to be 4.5 m. If the pole is 40 mm in diameter and it is made of a glass-reinforced plastic for which $E_g = 131$ GPa, determine the maximum bending stress in the pole.

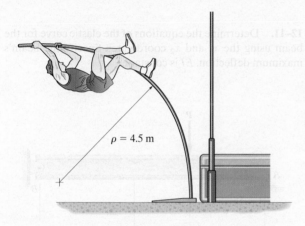

$\rho = 4.5$ m

Prob. 12–2

12–3. When the diver stands at end C of the diving board, it deflects downward 87.5 mm. Determine the weight of the diver. The board is made of material having a modulus of elasticity of $E = 10$ GPa.

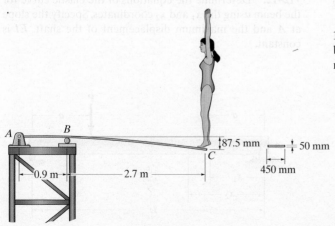

87.5 mm

50 mm

450 mm

0.9 m

2.7 m

Prob. 12–3

*12–4. Determine the equations of the elastic curve using the x_1 and x_2 coordinates. EI is constant.

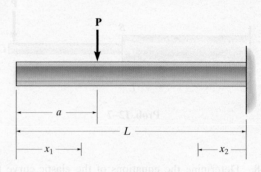

P

a

L

x_1

x_2

Prob. 12–4

•12–5. Determine the equations of the elastic curve for the beam using the x_1 and x_2 coordinates. EI is constant.

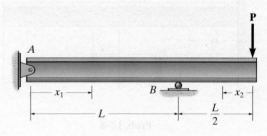

P

A

x_1

B

x_2

L

$\dfrac{L}{2}$

Prob. 12–5

12–6. Determine the equations of the elastic curve for the beam using the x_1 and x_3 coordinates. Specify the beam's maximum deflection. EI is constant.

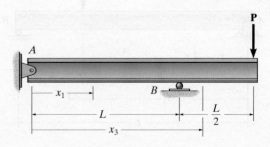

P

A

x_1

B

L

$\dfrac{L}{2}$

x_3

Prob. 12–6

12-7. The beam is made of two rods and is subjected to the concentrated load **P**. Determine the maximum deflection of the beam if the moments of inertia of the rods are I_{AB} and I_{BC}, and the modulus of elasticity is E.

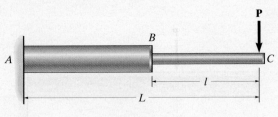

Prob. 12-7

*12-8. Determine the equations of the elastic curve for the beam using the x_1 and x_2 coordinates. EI is constant.

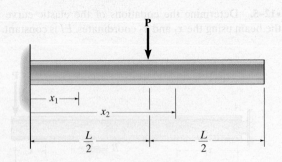

Prob. 12-8

•12-9. Determine the equations of the elastic curve using the x_1 and x_2 coordinates. EI is constant.

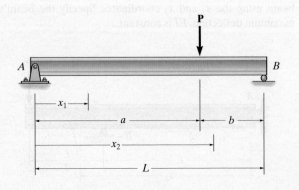

Prob. 12-9

12-10. Determine the maximum slope and maximum deflection of the simply supported beam which is subjected to the couple moment $\mathbf{M}_0$. EI is constant.

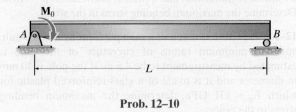

Prob. 12-10

12-11. Determine the equations of the elastic curve for the beam using the x_1 and x_2 coordinates. Specify the beam's maximum deflection. EI is constant.

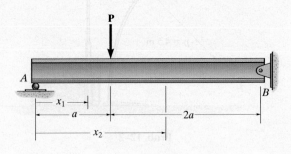

Prob. 12-11

*12-12. Determine the equations of the elastic curve for the beam using the x_1 and x_2 coordinates. Specify the slope at A and the maximum displacement of the shaft. EI is constant.

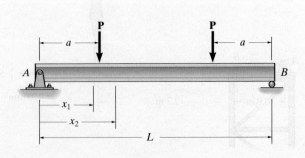

Prob. 12-12

12–13. The bar is supported by a roller constraint at B, which allows vertical displacement but resists axial load and moment. If the bar is subjected to the loading shown, determine the slope at A and the deflection at C. EI is constant

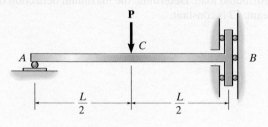

Prob. 12–13

12–14. The simply supported shaft has a moment of inertia of $2I$ for region BC and a moment of inertia I for regions AB and CD. Determine the maximum deflection of the beam due to the load $\mathbf{P}$.

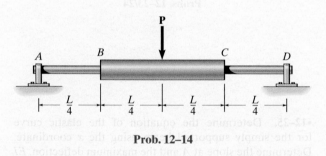

Prob. 12–14

12–15. Determine the equations of the elastic curve for the shaft using the x_1 and x_3 coordinates. Specify the slope at A and the deflection at the center of the shaft. EI is constant.

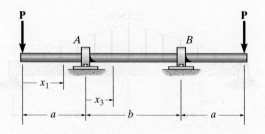

Prob. 12–15

*****12–16.** The fence board weaves between the three smooth fixed posts. If the posts remain along the same line, determine the maximum bending stress in the board. The board has a width of 150 mm and a thickness of 12 mm. $E = 12$ GPa. Assume the displacement of each end of the board relative to its center is 75 mm.

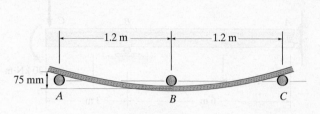

Prob. 12–16

•12–17. Determine the equations of the elastic curve for the shaft using the x_1 and x_2 coordinates. Specify the slope at A and the deflection at C. EI is constant.

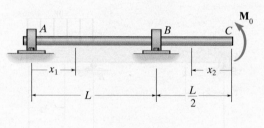

Prob. 12–17

12–18. Determine the equation of the elastic curve for the beam using the x coordinate. Specify the slope at A and the maximum deflection. EI is constant.

12–19. Determine the deflection at the center of the beam and the slope at B. EI is constant.

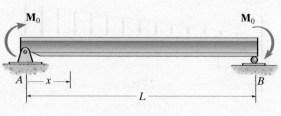

Probs. 12–18/19

*12–20. Determine the equations of the elastic curve using the x_1 and x_2 coordinates, and specify the slope at A and the deflection at C. EI is constant.

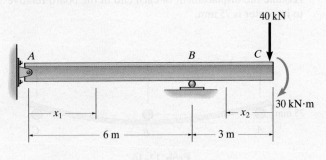

Prob. 12–20

•12–21. Determine the elastic curve in terms of the x_1 and x_2 coordinates and the deflection of end C of the overhang beam. EI is constant.

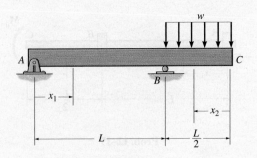

Prob. 12–21

12–22. Determine the elastic curve for the cantilevered W360 × 45 beam using the x coordinate. Specify the maximum slope and maximum deflection. $E = 200$ GPa.

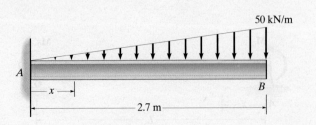

Prob. 12–22

12–23. The beam is subjected to the linearly varying distributed load. Determine the maximum slope of the beam. EI is constant.

*12–24. The beam is subjected to the linearly varying distributed load. Determine the maximum deflection of the beam. EI is constant.

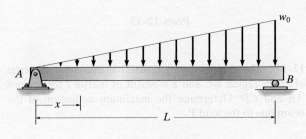

Probs. 12–23/24

•12–25. Determine the equation of the elastic curve for the simply supported beam using the x coordinate. Determine the slope at A and the maximum deflection. EI is constant.

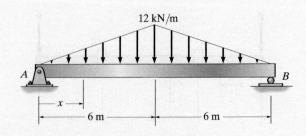

Prob. 12–25

12–26. Determine the equations of the elastic curve using the coordinates x_1 and x_2, and specify the slope and deflection at B. EI is constant.

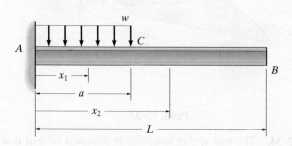

Prob. 12–26

12–27. Wooden posts used for a retaining wall have a diameter of 75 mm. If the soil pressure along a post varies uniformly from zero at the top A to a maximum of 5 kN/m at the bottom B, determine the slope and displacement at the top of the post. $E_w = 12$ GPa.

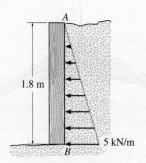

Prob. 12–27

*12–28. Determine the slope at end B and the maximum deflection of the cantilevered triangular plate of constant thickness t. The plate is made of material having a modulus of elasticity E.

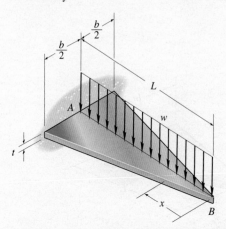

Prob. 12–28

•12–29. The beam is made of a material having a specific weight γ. Determine the displacement and slope at its end A due to its weight. The modulus of elasticity for the material is E.

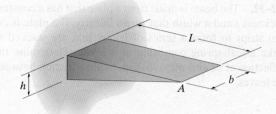

Prob. 12–29

12–30. The beam is made of a material having a specific weight of γ. Determine the displacement and slope at its end A due to its weight. The modulus of elasticity for the material is E.

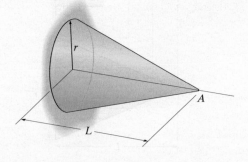

Prob. 12–30

12–31. The tapered beam has a rectangular cross section. Determine the deflection of its free end in terms of the load P, length L, modulus of elasticity E, and the moment of inertia I_0 of its fixed end.

•12–33. The tapered beam has a rectangular cross section. Determine the deflection of its center in terms of the load P, length L, modulus of elasticity E, and the moment of inertia I_c of its center.

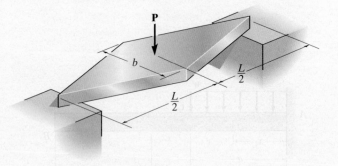

Prob. 12–33

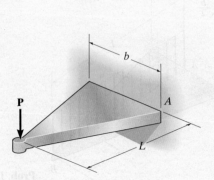

Prob. 12–31

12–34. The leaf spring assembly is designed so that it is subjected to the same maximum stress throughout its length. If the plates of each leaf have a thickness t and can slide freely between each other, show that the spring must be in the form of a circular arc in order that the entire spring becomes flat when a large enough load **P** is applied. What is the maximum normal stress in the spring? Consider the spring to be made by cutting the n strips from the diamond-shaped plate of thickness t and width b. The modulus of elasticity for the material is E. *Hint:* Show that the radius of curvature of the spring is constant.

*12–32. The beam is made from a plate that has a constant thickness t and a width that varies linearly. The plate is cut into strips to form a series of leaves that are stacked to make a leaf spring consisting of n leaves. Determine the deflection at its end when loaded. Neglect friction between the leaves.

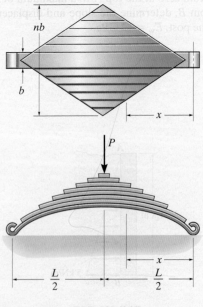

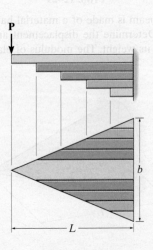

Prob. 12–32

Prob. 12–34

*12.3   Discontinuity Functions

The method of integration, used to find the equation of the elastic curve for a beam or shaft, is convenient if the load or internal moment can be expressed as a continuous function throughout the beam's entire length. If several different loadings act on the beam, however, the method becomes more tedious to apply, because separate loading or moment functions must be written for each region of the beam. Furthermore, integration of these functions requires the evaluation of integration constants using both boundary and continuity conditions. For example, the beam shown in Fig. 12–14 requires four moment functions to be written. They describe the moment in regions AB, BC, CD, and DE. When applying the moment-curvature relationship, $EI\, d^2v/dx^2 = M$, and integrating each moment equation twice, we must evaluate eight constants of integration. These involve two boundary conditions that require zero displacement at points A and E, and six continuity conditions for both slope and displacement at points B, C, and D.

In this section, we will discuss a method for finding the equation of the elastic curve for a *multiply loaded beam* using a *single expression*, either formulated from the loading on the beam, $w = w(x)$, or from the beam's internal moment, $M = M(x)$. If the expression for w is substituted into $EI\, d^4v/dx^4 = w(x)$ and integrated four times, or if the expression for M is substituted into $EI\, d^2v/dx^2 = M(x)$ and integrated twice, the constants of integration will be determined only from the boundary conditions. Since the continuity equations will not be involved, the analysis will be greatly simplified.

For safety purposes these cantilevered beams that support sheets of plywood must be designed for both strength and a restricted amount of deflection.

Discontinuity Functions.   In order to express the load on the beam or the internal moment within it using a single expression, we will use two types of mathematical operators known as *discontinuity functions*.

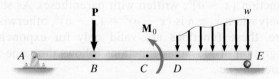

Fig. 12–14

TABLE 12–2

Loading	Loading Function $w = w(x)$	Shear $V = \int w(x)\,dx$	Moment $M = \int V\,dx$
(1) M_0 — x, a	$w = M_0\langle x-a\rangle^{-2}$	$V = M_0\langle x-a\rangle^{-1}$	$M = M_0\langle x-a\rangle^0$
(2) P — x, a	$w = P\langle x-a\rangle^{-1}$	$V = P\langle x-a\rangle^0$	$M = P\langle x-a\rangle^1$
(3) w_0 — x, a	$w = w_0\langle x-a\rangle^0$	$V = w_0\langle x-a\rangle^1$	$M = \dfrac{w_0}{2}\langle x-a\rangle^2$
(4) slope = m — x, a	$w = m\langle x-a\rangle^1$	$V = \dfrac{m}{2}\langle x-a\rangle^2$	$M = \dfrac{m}{6}\langle x-a\rangle^3$

Macaulay Functions. For purposes of beam or shaft deflection, Macaulay functions, named after the mathematician W. H. Macaulay, can be used to describe **distributed loadings**. These functions can be written in general form as

$$\langle x - a\rangle^n = \begin{cases} 0 & \text{for } x < a \\ (x-a)^n & \text{for } x \ge a \end{cases} \qquad n \ge 0 \tag{12–11}$$

Here x represents the coordinate position of a point along the beam, and a is the location on the beam where a "discontinuity" occurs, namely the point where a distributed loading *begins*. Note that the Macaulay function $\langle x - a\rangle^n$ is written with angle brackets to distinguish it from the ordinary function $(x - a)^n$, written with parentheses. As stated by the equation, only when $x \ge a$ is $\langle x - a\rangle^n = (x - a)^n$, otherwise it is zero. Furthermore, these functions are valid only for exponential values $n \ge 0$. Integration of Macaulay functions follows the same rules as for ordinary functions, i.e.,

$$\int \langle x - a\rangle^n \, dx = \frac{\langle x - a\rangle^{n+1}}{n + 1} + C \tag{12–12}$$

Note how the Macaulay functions describe both the *uniform load* $w_0\,(n = 0)$ and *triangular load* $(n = 1)$, shown in Table 12–2, items 3 and 4. This type of description can, of course, be extended to distributed loadings having other forms. Also, it is possible to use superposition with the uniform and triangular loadings to create the Macaulay function for a trapezoidal loading. Using integration, the Macaulay functions for shear, $V = \int w(x)\,dx$, and moment, $M = \int V\,dx$, are also shown in the table.

Singularity Functions. These functions are only used to describe the point location of concentrated forces or couple moments acting on a beam or shaft. Specifically, a concentrated force **P** can be considered as a special case of a distributed loading, where the intensity of the loading is $w = P/\epsilon$ such that its length is ϵ, where $\epsilon \to 0$, Fig. 12–15. The area under this loading diagram is equivalent to P, *positive upward*, and so we will use the singularity function

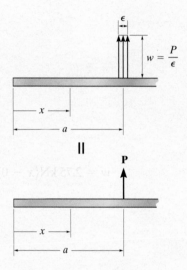

$$w = P\langle x - a \rangle^{-1} = \begin{cases} 0 & \text{for } x \neq a \\ P & \text{for } x = a \end{cases} \qquad (12\text{–}13)$$

to describe the force P. Here $n = -1$ so that the units for w are force per length, as it should be. Furthermore, the function takes on the value of **P** only at the point $x = a$ where the load occurs, otherwise it is zero.

In a similar manner, a couple moment $\mathbf{M}_0$, considered *positive clockwise*, is a limit as $\epsilon \to 0$ of two distributed loadings as shown in Fig. 12–16. Here the following function describes its value.

$$w = M_0 \langle x - a \rangle^{-2} = \begin{cases} 0 & \text{for } x \neq a \\ M_0 & \text{for } x = a \end{cases} \qquad (12\text{–}14)$$

The exponent $n = -2$, in order to ensure that the units of w, force per length, are maintained.

Integration of the above two singularity functions follow the rules of operational calculus and yields results that are *different* from those of Macaulay functions. Specifically,

$$\int \langle x - a \rangle^n dx = \langle x - a \rangle^{n+1},\ n = -1, -2 \qquad (12\text{–}15)$$

Using this formula, notice how M_0 and P, described in Table 12–2, items 1 and 2, are integrated once, then twice, to obtain the internal shear and moment in the beam.

Application of Eqs. 12–11 through 12–15 provides a rather direct means for expressing the loading or the internal moment in a beam as a function of x. When doing so, close attention must be paid to the signs of the external loadings. As stated above, and as shown in Table 12–2, *concentrated forces and distributed loads are positive upward, and couple moments are positive clockwise.* If this sign convention is followed, then the internal shear and moment are in accordance with the beam sign convention established in Sec. 6.1.

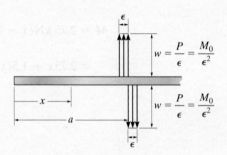

Fig. 12–15

Fig. 12–16

12

As an example of how to apply discontinuity functions to describe the loading or internal moment consider the beam loaded as shown in Fig. 12–17a. Here the reactive 2.75-kN force created by the roller, Fig. 12–17b, is positive since it acts upward, and the 1.5-kN·m couple moment is also positive since it acts clockwise. Finally, the trapezoidal loading is negative and has been separated into triangular and uniform loadings. From Table 12–2, the loading at any point x on the beam is therefore

$$w = 2.75 \text{ kN}\langle x - 0 \rangle^{-1} + 1.5 \text{ kN} \cdot \text{m}\langle x - 3 \text{ m}\rangle^{-2} - 3 \text{ kN/m}\langle x - 3 \text{ m}\rangle^0 - 1 \text{ kN/m}^2\langle x - 3 \text{ m}\rangle^1$$

The reactive force at B is not included here since x is never greater than 6 m, and furthermore, this value is of no consequence in calculating the slope or deflection. We can determine the moment expression directly from Table 12–2, rather than integrating this expression twice. In either case,

$$M = 2.75 \text{ kN}\langle x - 0 \rangle^1 + 1.5 \text{ kN} \cdot \text{m}\langle x - 3 \text{ m}\rangle^0 - \frac{3 \text{ kN/m}}{2}\langle x - 3 \text{ m}\rangle^2 - \frac{1 \text{ kN/m}^2}{6}\langle x - 3 \text{ m}\rangle^3$$

$$= 2.75x + 1.5\langle x - 3 \rangle^0 - 1.5\langle x - 3 \rangle^2 - \frac{1}{6}\langle x - 3 \rangle^3$$

The deflection of the beam can now be determined after this equation is integrated two successive times and the constants of integration are evaluated using the boundary conditions of zero displacement at A and B.

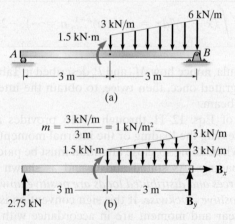

Fig. 12–17

Procedure for Analysis

The following procedure provides a method for using discontinuity functions to determine a beam's elastic curve. This method is particularly advantageous for solving problems involving beams or shafts subjected to *several loadings*, since the constants of integration can be evaluated by using *only* the boundary conditions, while the compatibility conditions are automatically satisfied.

Elastic Curve.

• Sketch the beam's elastic curve and identify the boundary conditions at the supports.

• Zero displacement occurs at all pin and roller supports, and zero slope and zero displacement occur at fixed supports.

• Establish the x axis so that it extends to the right and has its origin at the beam's left end.

Load or Moment Function.

• Calculate the support reactions at $x = 0$ and then use the discontinuity functions in Table 12–2 to express either the loading w or the internal moment M as a function of x. Make sure to follow the sign convention for each loading as it applies for this equation.

• Note that the distributed loadings must extend all the way to the beam's right end to be valid. If this does not occur, use the method of superposition, which is illustrated in Example 12.6.

Slope and Elastic Curve.

• Substitute w into $EI \, d^4v/dx^4 = w(x)$, or M into the moment curvature relation $EI \, d^2v/dx^2 = M$, and integrate to obtain the equations for the beam's slope and deflection.

• Evaluate the constants of integration using the boundary conditions, and substitute these constants into the slope and deflection equations to obtain the final results.

• When the slope and deflection equations are evaluated at any point on the beam, a *positive slope* is *counterclockwise*, and a *positive displacement* is *upward*.

12

EXAMPLE | **12.5**

Determine the maximum deflection of the beam shown in Fig. 12–18a. EI is constant.

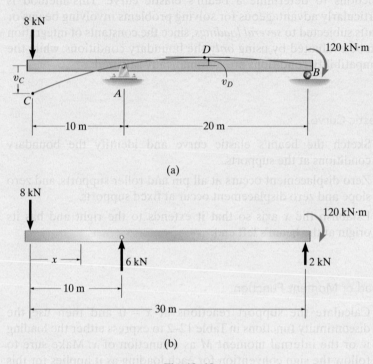

(a)

(b)

Fig. 12–18

SOLUTION

Elastic Curve. The beam deflects as shown in Fig. 12–18a. The boundary conditions require zero displacement at A and B.

Loading Function. The reactions have been calculated and are shown on the free-body diagram in Fig. 12–18b. The loading function for the beam can be written as

$$w = -8 \text{ kN} \langle x - 0 \rangle^{-1} + 6 \text{ kN} \langle x - 10 \text{ m} \rangle^{-1}$$

The couple moment and force at B are not included here, since they are located at the right end of the beam, and x cannot be greater than 30 m. Integrating $dv/dx = w(x)$, we get

$$V = -8 \langle x - 0 \rangle^0 + 6 \langle x - 10 \rangle^0$$

In a similar manner, $dm/dx = V$ yields

$$M = -8 \langle x - 0 \rangle^1 + 6 \langle x - 10 \rangle^1$$
$$= (-8x + 6 \langle x - 10 \rangle^1) \text{ kN·m}$$

Notice how this equation can also be established *directly* using the results of Table 12–2 for moment.

Slope and Elastic Curve.   Integrating twice yields

$$EI\frac{d^2v}{dx^2} = -8x + 6\langle x - 10\rangle^1$$

$$EI\frac{dv}{dx} = -4x^2 + 3\langle x - 10\rangle^2 + C_1$$

$$EIv = -\frac{4}{3}x^3 + \langle x - 10\rangle^3 + C_1x + C_2 \qquad (1)$$

From Eq. 1, the boundary condition $v = 0$ at $x = 10$ m and $v = 0$ at $x = 30$ m gives

$$0 = -1333 + (10 - 10)^3 + C_1(10) + C_2$$
$$0 = -36\,000 + (30 - 10)^3 + C_1(30) + C_2$$

Solving these equations simultaneously for C_1 and C_2, we get $C_1 = 1333$ and $C_2 = -12\,000$. Thus,

$$EI\frac{dv}{dx} = -4x^2 + 3\langle x - 10\rangle^2 + 1333 \qquad (2)$$

$$EIv = -\frac{4}{3}x^3 + \langle x - 10\rangle^3 + 1333x - 12\,000 \qquad (3)$$

From Fig. 12–18a, maximum displacement may occur either at C, or at D, where the slope $dv/dx = 0$. To obtain the displacement of C, set $x = 0$ in Eq. 3. We get

$$v_C = -\frac{12\,000 \text{ kN} \cdot \text{m}^3}{EI} \qquad \qquad Ans.$$

The *negative* sign indicates that the displacement is *downward* as shown in Fig. 12–18a. To locate point D, use Eq. 2 with $x > 10$ m and $dv/dx = 0$. This gives

$$0 = -4x_D^2 + 3(x_D - 10)^2 + 1333$$
$$x_D^2 + 60x_D - 1633 = 0$$

Solving for the positive root,

$$x_D = 20.3 \text{ m}$$

Hence, from Eq. 3,

$$EIv_D = -\frac{4}{3}(20.3)^3 + (20.3 - 10)^3 + 1333(20.3) - 12\,000$$

$$v_D = \frac{5006 \text{ kN} \cdot \text{m}^3}{EI}$$

Comparing this value with v_C, we see that $v_{max} = v_C$.

12

EXAMPLE 12.6

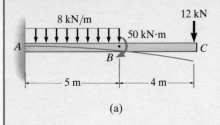

(a)

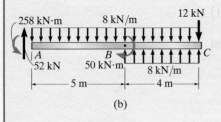

(b)

Fig. 12–19

Determine the equation of the elastic curve for the cantilevered beam shown in Fig. 12–19a. EI is constant.

SOLUTION

Elastic Curve. The loads cause the beam to deflect as shown in Fig. 12–19a. The boundary conditions require zero slope and displacement at A.

Loading Function. The support reactions at A have been calculated and are shown on the free-body diagram in Fig. 12–19b. Since the distributed loading in Fig. 12–19a does not extend to C as required, we can use the superposition of loadings shown in Fig. 12–19b to represent the same effect. By our sign convention, the beam's loading is therefore

$$w = 52 \text{ kN}\langle x - 0 \rangle^{-1} - 258 \text{ kN} \cdot \text{m}\langle x - 0 \rangle^{-2} - 8 \text{ kN/m}\langle x - 0 \rangle^{0}$$

$$+ 50 \text{ kN} \cdot \text{m}\langle x - 5 \text{ m} \rangle^{-2} + 8 \text{ kN/m}\langle x - 5 \text{ m} \rangle^{0}$$

The 12-kN load is *not included* here, since x cannot be greater than 9 m. Because $dV/dx = w(x)$, then by integrating, neglecting the constant of integration since the reactions are included in the load function, we have

$$V = 52\langle x - 0 \rangle^{0} - 258\langle x - 0 \rangle^{-1} - 8\langle x - 0 \rangle^{1} + 50\langle x - 5 \rangle^{-1} + 8\langle x - 5 \rangle^{1}$$

Furthermore, $dM/dx = V$, so that integrating again yields

$$M = -258\langle x - 0 \rangle^{0} + 52\langle x - 0 \rangle^{1} - \frac{1}{2}(8)\langle x - 0 \rangle^{2} + 50\langle x - 5 \rangle^{0} + \frac{1}{2}(8)\langle x - 5 \rangle^{2}$$

$$= (-258 + 52x - 4x^{2} + 50\langle x - 5 \rangle^{0} + 4\langle x - 5 \rangle^{2}) \text{ kN} \cdot \text{m}$$

This same result can be obtained *directly* from Table 12–2.

Slope and Elastic Curve. Applying Eq. 12–10 and integrating twice, we have

$$EI\frac{d^{2}v}{dx^{2}} = -258 + 52x - 4x^{2} + 50\langle x - 5 \rangle^{0} + 4\langle x - 5 \rangle^{2}$$

$$EI\frac{dv}{dx} = -258x + 26x^{2} - \frac{4}{3}x^{3} + 50\langle x - 5 \rangle^{1} + \frac{4}{3}\langle x - 5 \rangle^{3} + C_{1}$$

$$EIv = -129x^{2} + \frac{26}{3}x^{3} - \frac{1}{3}x^{4} + 25\langle x - 5 \rangle^{2} + \frac{1}{3}\langle x - 5 \rangle^{4} + C_{1}x + C_{2}$$

Since $dv/dx = 0$ at $x = 0$, $C_{1} = 0$; and $v = 0$ at $x = 0$, so $C_{2} = 0$. Thus,

$$v = \frac{1}{EI}\left(-129x^{2} + \frac{26}{3}x^{3} - \frac{1}{3}x^{4} + 25\langle x - 5 \rangle^{2} + \frac{1}{3}\langle x - 5 \rangle^{4}\right) \text{m} \quad Ans.$$

PROBLEMS

12–35. The shaft is made of steel and has a diameter of 15 mm. Determine its maximum deflection. The bearings at A and B exert only vertical reactions on the shaft. $E_{st} = 200$ GPa.

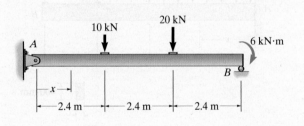

Prob. 12–35

*12–36. The beam is subjected to the loads shown. Determine the equation of the elastic curve. EI is constant.

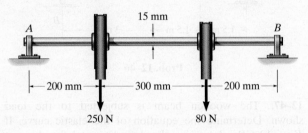

Prob. 12–36

•12–37. Determine the deflection at each of the pulleys C, D, and E. The shaft is made of steel and has a diameter of 30 mm. The bearings at A and B exert only vertical reactions on the shaft. $E_{st} = 200$ GPa.

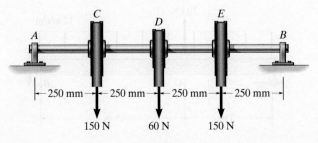

Prob. 12–37

12–38. The shaft supports the two pulley loads shown. Determine the equation of the elastic curve. The bearings at A and B exert only vertical reactions on the shaft. EI is constant.

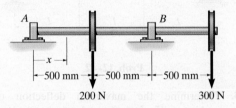

Prob. 12–38

12–39. Determine the maximum deflection of the simply supported beam. $E = 200$ GPa and $I = 65.0(10^6)$ mm^4.

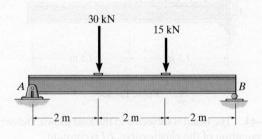

Prob. 12–39

*12–40. Determine the eqution of the elastic curve, the slope at A, and the deflection at B of the simply supported beam. EI is constant.

•12–41. Determine the equation of the elastic curve and the maximum deflection of the simply supported beam. EI is constant.

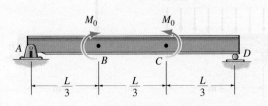

Probs. 12–40/41

12–42. Determine the equation of the elastic curve, the slope at A, and the maximum deflection of the simply supported beam. EI is constant.

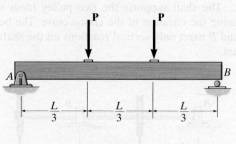

Prob. 12–42

12–43. Determine the maximum deflection of the cantilevered beam. The beam is made of material having an $E = 200$ GPa and $I = 65.0(10^6)$ mm^6.

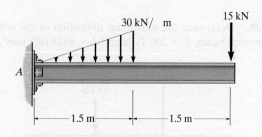

Prob. 12–43

*12–44. The beam is subjected to the load shown. Determine the equation of the elastic curve. EI is constant.

•12–45. The beam is subjected to the load shown. Determine the displacement at $x = 7$ m and the slope at A. EI is constant.

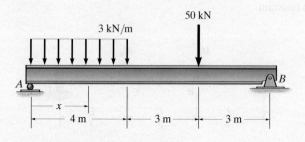

Probs. 12–44/45

12–46. Determine the maximum deflection of the simply supported beam. $E = 200$ GPa and $I = 65.0(10^6)$ mm^4.

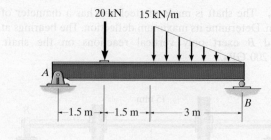

Prob. 12–46

12–47. The wooden beam is subjected to the load shown. Determine the equation of the elastic curve. If $E_w = 12$ GPa, determine the deflection and the slope at end B.

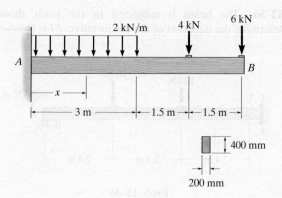

Prob. 12–47

*12–48. The beam is subjected to the load shown. Determine the slopes at A and B and the displacement at C. EI is constant.

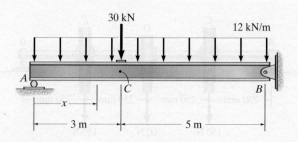

Prob. 12–48

•12–49. Determine the equation of the elastic curve of the simply supported beam and then find the maximum deflection. The beam is made of wood having a modulus of elasticity $E = 10$ GPa.

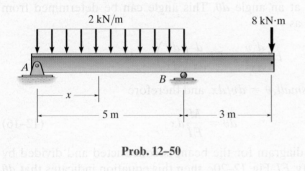

Prob. 12–49

12–50. The beam is subjected to the load shown. Determine the equations of the slope and elastic curve. EI is constant.

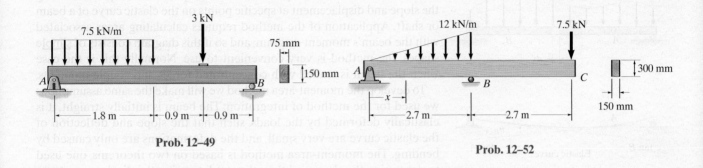

Prob. 12–50

12–51. The beam is subjected to the load shown. Determine the equation of the elastic curve. EI is constant.

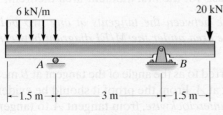

Prob. 12–51

*12–52. The wooden beam is subjected to the load shown. Determine the equation of the elastic curve. Specify the deflection at the end C. $E_w = 12$ GPa.

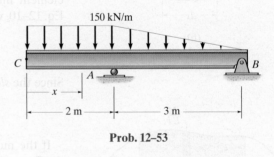

Prob. 12–52

12–53. Determine the displacement at C and the slope at A of the beam.

Prob. 12–53

12–54. The beam is subjected to the load shown. Determine the equation of the elastic curve. EI is constant.

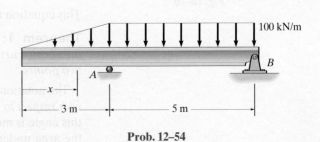

Prob. 12–54

12

*12.4   Slope and Displacement by the Moment-Area Method

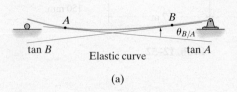

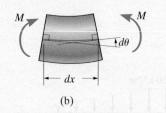

Elastic curve

(a)

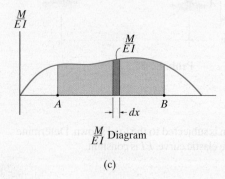

$\dfrac{M}{EI}$ Diagram

(c)

Fig. 12–20

The moment-area method provides a semigraphical technique for finding the slope and displacement at specific points on the elastic curve of a beam or shaft. Application of the method requires calculating areas associated with the beam's moment diagram; and so if this diagram consists of simple shapes, the method is very convenient to use. Normally this is the case when the beam is loaded with concentrated forces and couple moments.

To develop the moment-area method we will make the same assumptions we used for the method of integration: The beam is initially straight, it is elastically deformed by the loads, such that the slope and deflection of the elastic curve are very small, and the deformations are only caused by bending. The moment-area method is based on two theorems, one used to determine the slope and the other to determine the displacement at a point on the elastic curve.

Theorem 1.   Consider the simply supported beam with its associated elastic curve, shown in Fig. 12–20a. A differential segment dx of the beam is isolated in Fig. 12–20b. Here the beam's internal moment M deforms the element such that the *tangents* to the elastic curve at each side of the element intersect at an angle $d\theta$. This angle can be determined from Eq. 12–10, written as

$$EI\frac{d^2v}{dx^2} = EI\frac{d}{dx}\left(\frac{dv}{dx}\right) = M$$

Since the *slope* is *small*, $\theta = dv/dx$, and therefore

$$d\theta = \frac{M}{EI}dx \tag{12–16}$$

If the moment diagram for the beam is constructed and divided by the flexural rigidity, EI, Fig. 12–20c, then this equation indicates that $d\theta$ is equal to the *area* under the "M/EI diagram" for the beam segment dx. Integrating from a selected point A on the elastic curve to another point B, we have

$$\theta_{B/A} = \int_A^B \frac{M}{EI}dx \tag{12–17}$$

This equation forms the basis for the first moment-area theorem.

Theorem 1:   *The angle between the tangents at any two points on the elastic curve equals the area under the M/EI diagram between these two points.*

The notation $\theta_{B/A}$ is referred to as the angle of the tangent at B measured *with respect to* the tangent at A. From the proof it should be evident that this angle is measured *counterclockwise*, from tangent A to tangent B, if the area under the M/EI diagram is *positive*. Conversely, if the area is *negative*, or lies below the *x* axis, the angle $\theta_{B/A}$ is measured clockwise from tangent *A* to tangent *B*. Furthermore, from the dimensions of Eq. 12–17, $\theta_{B/A}$ will be in *radians*.

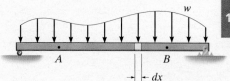

Theorem 2.

The second moment-area theorem is based on the relative deviation of tangents to the elastic curve. Shown in Fig. 12–21*a* is a greatly exaggerated view of the vertical deviation *dt* of the tangents on each side of the differential element *dx*. This deviation is caused by the curvature of the element and has been measured along a vertical line passing through point *A* on the elastic curve. Since the slope of the elastic curve and its deflection are assumed to be very small, it is satisfactory to approximate the length of each tangent line by *x* and the arc *ds'* by *dt*. Using the circular-arc formula $s = \theta r$, where *r* is the length *x* and *s* is *dt*, we can write $dt = x\, d\theta$. Substituting Eq. 12–16 into this equation and integrating from *A* to *B*, the vertical deviation of the tangent at *A* *with respect to* the tangent at *B* can then be determined; that is,

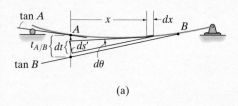

(a)

$$t_{A/B} = \int_A^B x\frac{M}{EI}dx \qquad (12\text{–}18)$$

Since the centroid of an area is found from $\overline{x}\int dA = \int x\, dA$, and $\int (M/EI)\, dx$ represents the area under the *M/EI* diagram, we can also write

$$t_{A/B} = \overline{x}\int_A^B \frac{M}{EI}dx \qquad (12\text{–}19)$$

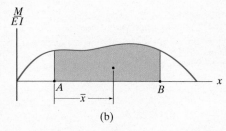

(b)

Here $\overline{x}$ is the distance from *A* to the *centroid* of the area under the *M/EI* diagram between *A* and *B*, Fig. 12–21*b*.

The second moment-area theorem can now be stated in reference to Fig. 12–21*a* as follows:

Theorem 2:   *The vertical distance between the tangent at a point (A) on the elastic curve and the tangent extended from another point (B) equals the moment of the area under the M/EI diagram between these two points (A and B). This moment is calculated about the point (A) where the vertical distance ($t_{A/B}$) is to be determined.*

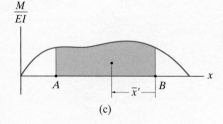

(c)

Fig. 12–21

Note that $t_{A/B}$ is *not* equal to $t_{B/A}$, which is shown in Fig. 12–21*c*. Specifically, the moment of the area under the *M/EI* diagram between *A* and *B* is calculated about point *A* to determine $t_{A/B}$, Fig. 12–21*b*, and it is calculated about point *B* to determine $t_{B/A}$, Fig. 12–21*c*.

If the moment of a *positive M/EI* area between *A* and *B* is found for $t_{A/B}$, it indicates that point *A* is *above* the tangent extended from point *B*, Fig. 12–21*a*. Similarly, *negative M/EI* areas indicate that point *A* is *below* the tangent extended from point *B*. This same rule applies for $t_{B/A}$.

12

Procedure for Analysis

The following procedure provides a method that may be used to apply the two moment-area theorems.

M/EI Diagram.

- Determine the support reactions and draw the beam's M/EI diagram. If the beam is loaded with concentrated forces, the M/EI diagram will consist of a series of straight line segments, and the areas and their moments required for the moment-area theorems will be relatively easy to calculate. If the loading consists of a series of distributed loads, the M/EI diagram will consist of parabolic or perhaps higher-order curves, and it is suggested that the table on the inside front cover be used to locate the area and centroid under each curve.

Elastic Curve.

- Draw an exaggerated view of the beam's elastic curve. Recall that points of zero slope and zero displacement always occur at a fixed support, and zero displacement occurs at all pin and roller supports.

- If it becomes difficult to draw the general shape of the elastic curve, use the moment (or M/EI) diagram. Realize that when the beam is subjected to a *positive moment*, the beam bends *concave up*, whereas *negative moment* bends the beam *concave down*. Furthermore, an inflection point or change in curvature occurs where the moment in the beam (or M/EI) is zero.

- The unknown displacement and slope to be determined should be indicated on the curve.

- Since the moment-area theorems apply *only between two tangents*, attention should be given as to which tangents should be constructed so that the angles or vertical distance between them will lead to the solution of the problem. In this regard, *the tangents at the supports should be considered*, since the beam has zero displacement and/or zero slope at the supports.

Moment-Area Theorems.

- Apply Theorem 1 to determine the *angle* between any two tangents on the elastic curve and Theorem 2 to determine the vertical distance between the tangents.

- The algebraic sign of the answer can be checked from the angle or vertical distance indicated on the elastic curve.

- A *positive* $\theta_{B/A}$ represents a *counterclockwise* rotation of the tangent at B with respect to the tangent at A, and a *positive* $t_{B/A}$ indicates that point B on the elastic curve lies *above* the extended tangent from point A.

EXAMPLE 12.7

Determine the slope of the beam shown in Fig. 12–22a at point B. EI is constant.

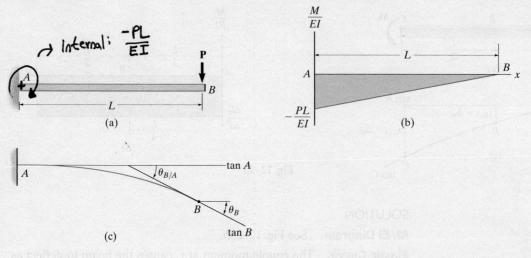

Fig. 12–22

SOLUTION

M/EI Diagram. See Fig. 12–22b.

Elastic Curve. The force P causes the beam to deflect as shown in Fig. 12–22c. (The elastic curve is concave downward, since M/EI is negative.) The tangent at B is indicated since we are required to find θ_B. Also, the tangent at the support (A) is shown. This tangent has a *known* zero slope. By the construction, the angle between tan A and tan B, that is, $\theta_{B/A}$, is equivalent to θ_B, or

$$\theta_B = \theta_{B/A}$$

Moment-Area Theorem. Applying Theorem 1, $\theta_{B/A}$ is equal to the area under the M/EI diagram between points A and B; that is,

$$\theta_B = \theta_{B/A} = \frac{1}{2}\left(-\frac{PL}{EI}\right)L$$

$$= -\frac{PL^2}{2EI} \qquad Ans.$$

The *negative sign* indicates that the angle measured from the tangent at A to the tangent at B is *clockwise*. This checks, since the beam slopes downward at B.

12

EXAMPLE 12.8

Determine the displacement of points B and C of the beam shown in Fig. 12–23a. EI is constant.

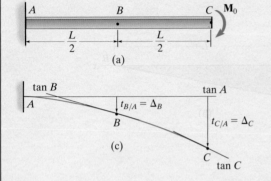

(a)

(c)

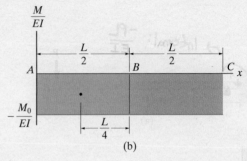

(b)

Fig. 12–23

SOLUTION

M/EI Diagram.   See Fig. 12–23b.

Elastic Curve.   The couple moment at C causes the beam to deflect as shown in Fig. 12–23c. The tangents at B and C are indicated since we are required to find Δ_B and Δ_C. Also, the tangent at the support (A) is shown since it is horizontal. The required displacements can now be related directly to the vertical distance between the tangents at B and A and C and A. Specifically,

$$\Delta_B = t_{B/A}$$

$$\Delta_C = t_{C/A}$$

Moment-Area Theorem.   Applying Theorem 2, $t_{B/A}$ is equal to the moment of the shaded area under the M/EI diagram between A and B calculated about point B (the point on the elastic curve), since this is the point where the vertical distance is to be determined. Hence, from Fig. 12–23b,

$$\Delta_B = t_{B/A} = \left(\frac{L}{4}\right)\left[\left(-\frac{M_0}{EI}\right)\left(\frac{L}{2}\right)\right] = -\frac{M_0 L^2}{8EI} \qquad Ans.$$

Likewise, for $t_{C/A}$ we must determine the moment of the area under the *entire* M/EI diagram from A to C about point C (the point on the elastic curve). We have

$$\Delta_C = t_{C/A} = \left(\frac{L}{2}\right)\left[\left(-\frac{M_0}{EI}\right)(L)\right] = -\frac{M_0 L^2}{2EI} \qquad Ans.$$

NOTE: Since both answers are *negative*, they indicate that points B and C lie *below* the tangent at A. This checks with Fig. 12–23c.

EXAMPLE | 12.9

Determine the slope at point C of the shaft in Fig. 12–24a. EI is constant.

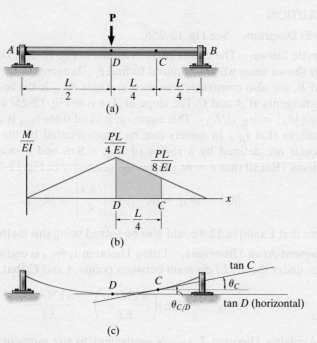

(a)

(b)

(c)

Fig. 12–24

SOLUTION

M/EI Diagram.   See Fig. 12–24b.

Elastic Curve.   Since the loading is applied symmetrically to the beam, the elastic curve is symmetric, and the tangent at D is horizontal, Fig. 12–24c. Also the tangent at C is drawn, since we must find the slope θ_C. By the construction, the angle $\theta_{C/D}$ between the tangents at tan D and C is equal to θ_C; that is,

$$\theta_C = \theta_{C/D}$$

Moment-Area Theorem.   Using Theorem 1, $\theta_{C/D}$ is equal to the shaded area under the M/EI diagram between points D and C. We have

$$\theta_C = \theta_{C/D} = \left(\frac{PL}{8EI}\right)\left(\frac{L}{4}\right) + \frac{1}{2}\left(\frac{PL}{4EI} - \frac{PL}{8EI}\right)\left(\frac{L}{4}\right) = \frac{3PL^2}{64EI} \qquad Ans.$$

What does the positive result indicate?

EXAMPLE | 12.10

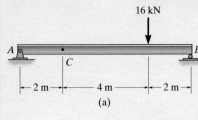

16 kN

A   B

C

|← 2 m →|← 4 m →|← 2 m →|

(a)

Determine the slope at point C for the steel beam in Fig. 12–25a. Take $E_{st} = 200$ GPa, $I = 17(10^6)$ mm^4.

SOLUTION

M/EI Diagram. See Fig. 12–25b.

Elastic Curve. The elastic curve is shown in Fig. 12–25c. The tangent at C is shown since we are required to find θ_C. Tangents at the *supports*, A and B, are also constructed as shown. Angle $\theta_{C/A}$ is the angle between the tangents at A and C. The slope at A, θ_A, in Fig. 12–25c can be found using $|\theta_A| = |t_{B/A}|/L_{AB}$. This equation is valid since $t_{B/A}$ is actually very small, so that $t_{B/A}$ in meters can be approximated by the length of a circular arc defined by a radius of $L_{AB} = 8$ m and a sweep of θ_A in radians. (Recall that $s = \theta r$.) From the geometry of Fig. 12–25c, we have

$$|\theta_C| = |\theta_A| - |\theta_{C/A}| = \left|\frac{t_{B/A}}{8}\right| - |\theta_{C/A}| \qquad (1)$$

Note that Example 12.9 could also be solved using this method.

Moment-Area Theorems. Using Theorem 1, $\theta_{C/A}$ is equivalent to the area under the M/EI diagram between points A and C; that is,

$$\theta_{C/A} = \frac{1}{2}(2\ \text{m})\left(\frac{8\ \text{kN}\cdot\text{m}}{EI}\right) = \frac{8\ \text{kN}\cdot\text{m}^2}{EI}$$

Applying Theorem 2, $t_{B/A}$ is equivalent to the moment of the area under the M/EI diagram between B and A about point B (the point on the elastic curve), since this is the point where the vertical distance is to be determined. We have

$$t_{B/A} = \left(2\ \text{m} + \frac{1}{3}(6\ \text{m})\right)\left[\frac{1}{2}(6\ \text{m})\left(\frac{24\ \text{kN}\cdot\text{m}}{EI}\right)\right]$$

$$+ \left(\frac{2}{3}(2\ \text{m})\right)\left[\frac{1}{2}(2\ \text{m})\left(\frac{24\ \text{kN}\cdot\text{m}}{EI}\right)\right]$$

$$= \frac{320\ \text{kN}\cdot\text{m}^3}{EI}$$

Substituting these results into Eq. 1, we get

$$\theta_C = \frac{320\ \text{kN}\cdot\text{m}^2}{(8\ \text{m})EI} - \frac{8\ \text{kN}\cdot\text{m}^2}{EI} = \frac{32\ \text{kN}\cdot\text{m}^2}{EI}$$

We have calculated this result in units of kN and m, so converting EI into these units, we have

$$\theta_C = \frac{32\ \text{kN}\cdot\text{m}^2}{[200(10^6)\ \text{kN/m}^2][17(10^{-6})\ \text{m}^4]} = 0.00941\ \text{rad} \qquad Ans.$$

$\dfrac{M}{EI}$

$\dfrac{24}{EI}$

$\dfrac{8}{EI}$

A   C   B x

|← 2 m →|← 4 m →|← 2 m →|

(b)

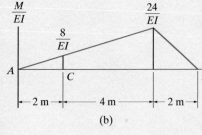

θ_A

A   C

$\theta_{C/A}$ θ_C tan C

B   tan B

$t_{B/A}$

tan A

(c)

Fig. 12–25

EXAMPLE | 12.11

Determine the displacement at C for the beam shown in Fig. 12–26a. EI is constant.

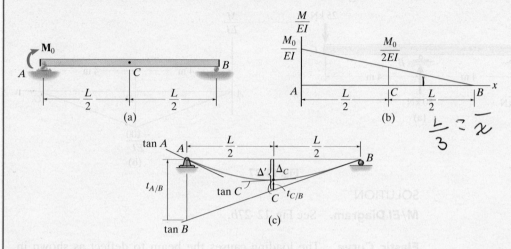

Fig. 12–26

SOLUTION

M/EI Diagram. See Fig. 12–26b.

Elastic Curve. The tangent at C is drawn on the elastic curve since we are required to find Δ_C, Fig. 12–26c. (Note that C is *not* the location of the maximum deflection of the beam, because the loading and hence the elastic curve are *not symmetric*.) Also indicated in Fig. 12–26c are the tangents at the supports A and B. It is seen that $\Delta_C = \Delta' - t_{C/B}$. If $t_{A/B}$ is determined, then Δ' can be found from proportional triangles, that is, $\Delta'/(L/2) = t_{A/B}/L$ or $\Delta' = t_{A/B}/2$. Hence,

$$\Delta_C = \frac{t_{A/B}}{2} - t_{C/B} \qquad (1)$$

Moment-Area Theorem. Applying Theorem 2 to determine $t_{A/B}$ and $t_{C/B}$, we have

$$t_{A/B} = \left(\frac{1}{3}(L)\right)\left[\frac{1}{2}(L)\left(\frac{M_0}{EI}\right)\right] = \frac{M_0 L^2}{6EI}$$

$$t_{C/B} = \left(\frac{1}{3}\left(\frac{L}{2}\right)\right)\left[\frac{1}{2}\left(\frac{L}{2}\right)\left(\frac{M_0}{2EI}\right)\right] = \frac{M_0 L^2}{48EI}$$

Substituting these results into Eq. 1 gives

$$\Delta_C = \frac{1}{2}\left(\frac{M_0 L^2}{6EI}\right) - \left(\frac{M_0 L^2}{48EI}\right)$$

$$= \frac{M_0 L^2}{16EI} \downarrow \qquad\qquad\qquad Ans.$$

12

EXAMPLE | **12.12**

Determine the displacement at point C for the steel overhanging beam shown in Fig. 12–27a. Take $E_{st} = 200$ GPa, $I = 50(10^6)$ mm^4.

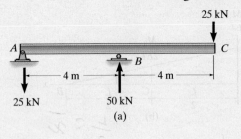

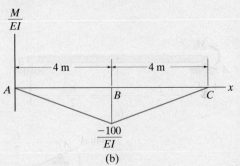

(a)

(b)

Fig. 12–27

SOLUTION

M/EI Diagram. See Fig. 12–27b.

Elastic Curve. The loading causes the beam to deflect as shown in Fig. 12–27c. We are required to find Δ_C. By constructing tangents at C and at the supports A and B, it is seen that $\Delta_C = |t_{C/A}| - \Delta'$. However, Δ' can be related to $t_{B/A}$ by proportional triangles; that is, $\Delta'/8 = |t_{B/A}|/4$ or $\Delta' = 2|t_{B/A}|$. Hence

$$\Delta_C = |t_{C/A}| - 2|t_{B/A}| \qquad (1)$$

Moment-Area Theorem. Applying Theorem 2 to determine $t_{C/A}$ and $t_{B/A}$, we have

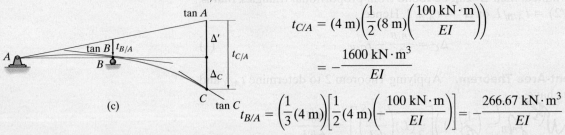

(c)

$$t_{C/A} = (4\text{ m})\left(\frac{1}{2}(8\text{ m})\left(\frac{100\text{ kN}\cdot\text{m}}{EI}\right)\right)$$

$$= -\frac{1600\text{ kN}\cdot\text{m}^3}{EI}$$

$$t_{B/A} = \left(\frac{1}{3}(4\text{ m})\right)\left[\frac{1}{2}(4\text{ m})\left(-\frac{100\text{ kN}\cdot\text{m}}{EI}\right)\right] = -\frac{266.67\text{ kN}\cdot\text{m}^3}{EI}$$

Why are these terms negative? Substituting the results into Eq. 1 yields

$$\Delta_C = \frac{1600\text{ kN}\cdot\text{m}^3}{EI} - 2\left(\frac{266.67\text{ kN}\cdot\text{m}^3}{EI}\right) = \frac{1066.67\text{ kN}\cdot\text{m}^3}{EI} \downarrow$$

Realizing that the calculations were made in units of kip and ft, we have

$$\Delta_C = \frac{1066.67\text{ kN}\cdot\text{m}^3(1000\text{ mm/m})^3}{(200\text{ kN/mm}^2)[50(10^6)\text{ mm}^4]} = 106.7\text{ mm} \downarrow \qquad Ans.$$

FUNDAMENTAL PROBLEMS

F12–7. Determine the slope and deflection of end A of the cantilevered beam. $E = 200$ GPa and $I = 65.0(10^{-6})$ m^4.

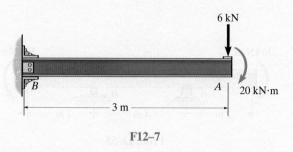

6 kN

B

A 20 kN·m

3 m

F12–7

F12–8. Determine the slope and deflection of end A of the cantilevered beam. $E = 200$ GPa and $I = 126(10^{-6})$ m^4.

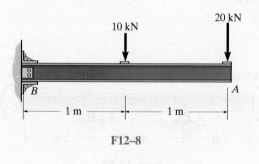

10 kN   20 kN

B

A

1 m   1 m

F12–8

F12–9. Determine the slope and deflection of end A of the cantilevered beam. $E = 200$ GPa and $I = 121(10^{-6})$ m^4.

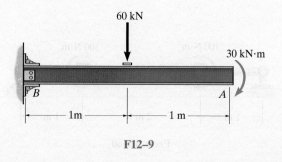

60 kN

30 kN·m

B

A

1m   1 m

F12–9

F12–10. Determine the slope and deflection at A of the cantilevered beam. $E = 200$ GPa, $I = 10(10^6)$ mm^4.

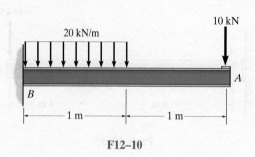

20 kN/m   10 kN

B A

1 m   1 m

F12–10

F12–11. Determine the maximum deflection of the simply supported beam. $E = 200$ GPa and $I = 42.8(10^{-6})$ m^4.

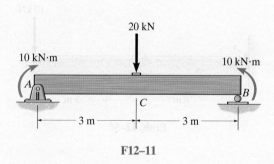

20 kN

10 kN·m   10 kN·m

A C B

3 m   3 m

F12–11

F12–12. Determine the maximum deflection of the simply supported beam. $E = 200$ GPa and $I = 39.9(10^{-6})$ m^4.

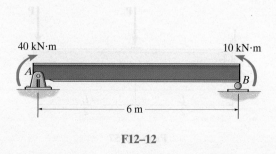

40 kN·m   10 kN·m

A B

6 m

F12–12

12–55. Determine the slope and deflection at C. EI is constant.

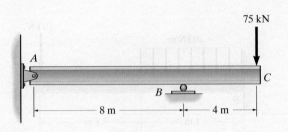

Prob. 12–55

12–58. Determine the slope at A and the maximum deflection. EI is constant.

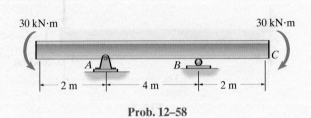

Prob. 12–58

*12–56. Determine the slope and deflection at C. EI is constant.

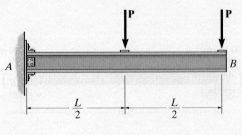

Prob. 12–56

12–59. Determine the slope and deflection at C. EI is constant.

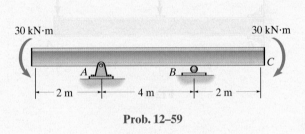

Prob. 12–59

•12–57. Determine the deflection of end B of the cantilever beam. E is constant.

*12–60. If the bearings at A and B exert only vertical reactions on the shaft, determine the slope at A and the maximum deflection of the shaft. EI is constant.

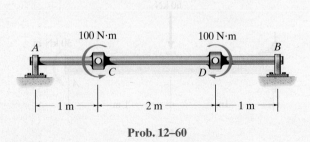

Prob. 12–57

Prob. 12–60

•12–61. Determine the maximum slope and the maximum deflection of the beam. EI is constant.

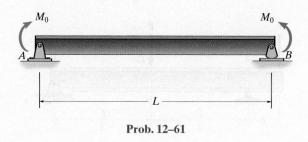

Prob. 12–61

12–62. Determine the deflection and slope at C. EI is constant.

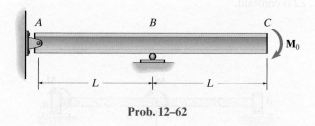

Prob. 12–62

12–63. Determine the slope at A of the overhang beam. $E = 200$ GPa and $I = 45.5(10^6)$ mm^4.

*12–64. Determine the deflection at C of the overhang beam. $E = 200$ GPa and $I = 45.5(10^6)$ mm^4.

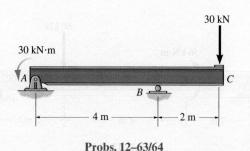

Probs. 12–63/64

•12–65. Determine the position a of roller support B in terms of L so that the deflection at end C is the same as the maximum deflection of region AB of the overhang beam. EI is constant.

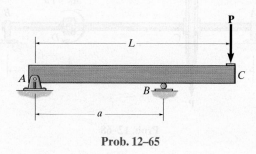

Prob. 12–65

12–66. Determine the slope at A of the simply supported beam. EI is constant.

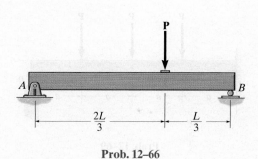

Prob. 12–66

12–67. The beam is subjected to the load P as shown. Determine the magnitude of force F that must be applied at the end of the overhang C so that the deflection at C is zero. EI is constant.

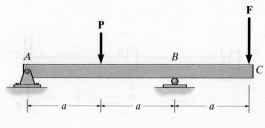

Prob. 12–67

*12–68. If the bearings at A and B exert only vertical reactions on the shaft, determine the slope at A and the maximum deflection.

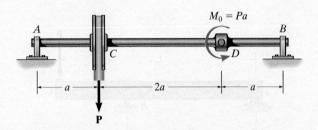

Prob. 12–68

•12–69. The beam is subjected to the loading shown. Determine the slope at A and the displacement at C. Assume the support at A is a pin and B is a roller. EI is constant.

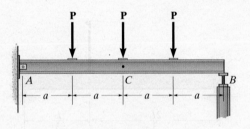

Prob. 12–69

12–70. The shaft supports the gear at its end C. Determine the deflection at C and the slopes at the bearings A and B. EI is constant.

12–71. The shaft supports the gear at its end C. Determine its maximum deflection within region AB. EI is constant. The bearings exert only vertical reactions on the shaft.

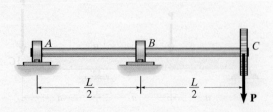

Probs. 12–70/71

*12–72. Determine the value of a so that the displacement at C is equal to zero. EI is constant.

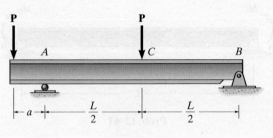

Prob. 12–72

•12–73. The shaft is subjected to the loading shown. If the bearings at A and B only exert vertical reactions on the shaft, determine the slope at A and the displacement at C. EI is constant.

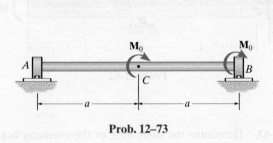

Prob. 12–73

12–74. Determine the slope at A and the maximum deflection in the beam. EI is constant.

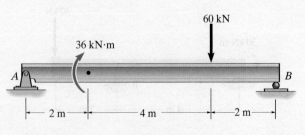

Prob. 12–74

12–75. The beam is made of a ceramic material. In order to obtain its modulus of elasticity, it is subjected to the elastic loading shown. If the moment of inertia is I and the beam has a measured maximum deflection Δ, determine E. The supports at A and D exert only vertical reactions on the beam.

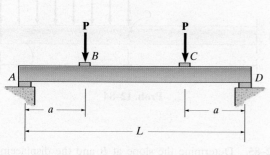

Prob. 12–75

*12–76. The bar is supported by a roller constraint at B, which allows vertical displacement but resists axial load and moment. If the bar is subjected to the loading shown, determine the slope at A and the deflection at C. EI is constant.

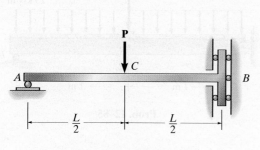

Prob. 12–76

•12–77. The bar is supported by the roller constraint at C, which allows vertical displacement but resists axial load and moment. If the bar is subjected to the loading shown, determine the slope and displacement at A. EI is constant.

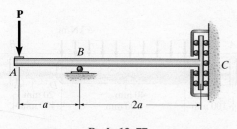

Prob. 12–77

12–78. The rod is constructed from two shafts for which the moment of inertia of AB is I and of BC is $2I$. Determine the maximum slope and deflection of the rod due to the loading. The modulus of elasticity is E.

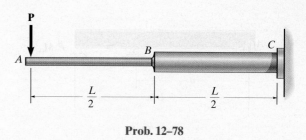

Prob. 12–78

12–79. Determine the slope at point D and the deflection at point C of the simply supported beam. The beam is made of material having a modulus of elasticity E. The moment of inertia of segments AB and CD of the beam is I, while the moment of inertia of segment BC of the beam is $2I$.

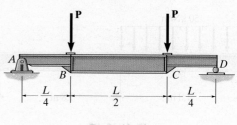

Prob. 12–79

*12–80. Determine the slope at point A and the maximum deflection of the simply supported beam. The beam is made of material having a modulus of elasticity E. The moment of inertia of segments AB and CD of the beam is I, while the moment of inertia of segment BC is $2I$.

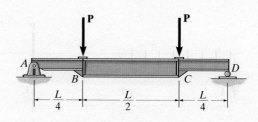

Prob. 12–80

•12–81.   Determine the position a of roller support B in terms of L so that deflection at end C is the same as the maximum deflection of region AB of the simply supported overhang beam. EI is constant.

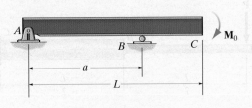

Prob. 12–81

12–82.   The W250 × 22 cantilevered beam is made of A-36 steel and is subjected to the loading shown. Determine the slope and displacement at its end B.

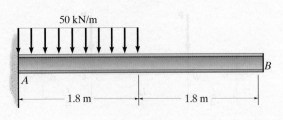

Prob. 12–82

12–83.   The cantilevered beam is subjected to the loading shown. Determine the slope and displacement at C. Assume the support at A is fixed. EI is constant.

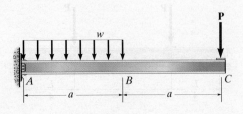

Prob. 12–83

*12–84.   Determine the slope at C and deflection at B. EI is constant.

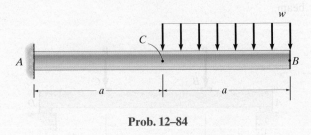

Prob. 12–84

•12–85.   Determine the slope at B and the displacement at C. The member is an A-36 steel structural tee for which $I = 30(10^6)$ mm^4.

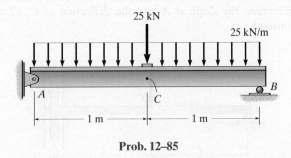

Prob. 12–85

12–86.   The A-36 steel shaft is used to support a rotor that exerts a uniform load of 5 kN/m within the region CD of the shaft. Determine the slope of the shaft at the bearings A and B. The bearings exert only vertical reactions on the shaft.

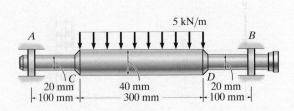

Prob. 12–86

12.5 Method of Superposition

The differential equation $EI\, d^4v/dx^4 = w(x)$ satisfies the two necessary requirements for applying the principle of superposition; i.e., the load $w(x)$ is linearly related to the deflection $v(x)$, and the load is assumed not to change significantly the original geometry of the beam or shaft. As a result, the deflections for a series of separate loadings acting on a beam may be superimposed. For example, if v_1 is the deflection for one load and v_2 is the deflection for another load, the total deflection for both loads acting together is the algebraic sum $v_1 + v_2$. Using tabulated results for various beam loadings, such as the ones listed in Appendix C, or those found in various engineering handbooks, it is therefore possible to find the slope and displacement at a point on a beam subjected to several different loadings by algebraically adding the effects of its various component parts.

The following examples illustrate how to use the method of superposition to solve deflection problems, where the deflection is caused not only by beam deformations, but also by rigid-body displacements, such as those that occur when the beam is supported by springs.

The resultant deflection at any point on this beam can be determined from the superposition of the deflections caused by each of the separate loadings acting on the beam.

12

EXAMPLE   12.13

Determine the displacement at point C and the slope at the support A of the beam shown in Fig. 12–28a. EI is constant.

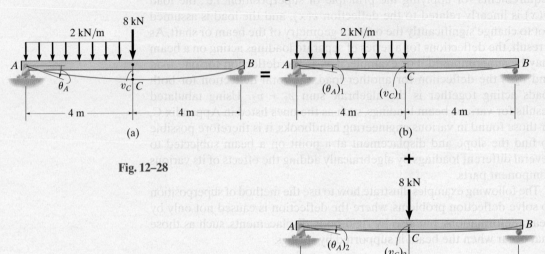

Fig. 12–28

SOLUTION

The loading can be separated into two component parts as shown in Figs. 12–28b and 12–28c. The displacement at C and slope at A are found using the table in Appendix C for each part.

For the distributed loading,

$$(\theta_A)_1 = \frac{3wL^3}{128EI} = \frac{3(2 \text{ kN/m})(8 \text{ m})^3}{128EI} = \frac{24 \text{ kN} \cdot \text{m}^2}{EI} \downarrow$$

$$(v_C)_1 = \frac{5wL^4}{768EI} = \frac{5(2 \text{ kN/m})(8 \text{ m})^4}{768EI} = \frac{53.33 \text{ kN} \cdot \text{m}^3}{EI} \downarrow$$

For the 8-kN concentrated force,

$$(\theta_A)_2 = \frac{PL^2}{16EI} = \frac{8 \text{ kN}(8 \text{ m})^2}{16EI} = \frac{32 \text{ kN} \cdot \text{m}^2}{EI} \downarrow$$

$$(v_C)_2 = \frac{PL^3}{48EI} = \frac{8 \text{ kN}(8 \text{ m})^3}{48EI} = \frac{85.33 \text{ kN} \cdot \text{m}^3}{EI} \downarrow$$

The displacement at C and the slope at A are the algebraic sums of these components. Hence,

$$(+\downarrow) \qquad \theta_A = (\theta_A)_1 + (\theta_A)_2 = \frac{56 \text{ kN} \cdot \text{m}^2}{EI} \downarrow \qquad \textit{Ans.}$$

$$(+\downarrow) \qquad v_C = (v_C)_1 + (v_C)_2 = \frac{139 \text{ kN} \cdot \text{m}^3}{EI} \downarrow \qquad \textit{Ans.}$$

EXAMPLE 12.14

Determine the displacement at the end C of the overhanging beam shown in Fig. 12–29a. EI is constant.

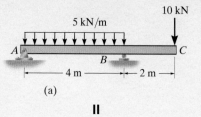

(a)

$\parallel$

SOLUTION

Since the table in Appendix C *does not* include beams with overhangs, the beam will be separated into a simply supported and a cantilevered portion. First we will calculate the slope at B, as caused by the distributed load acting on the simply supported span, Fig. 12–29b.

$$(\theta_B)_1 = \frac{wL^3}{24EI} = \frac{5\text{ kN/m}(4\text{ m})^3}{24EI} = \frac{13.33\text{ kN} \cdot \text{m}^2}{EI} \nwarrow$$

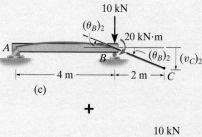

(b)

$+$

Since this angle is *small*, $(\theta_B)_1 \approx \tan(\theta_B)_1$, and the vertical displacement at point C is

$$(v_C)_1 = (2\text{ m})\left(\frac{13.33\text{ kN} \cdot \text{m}^2}{EI}\right) = \frac{26.67\text{ kN} \cdot \text{m}^3}{EI} \uparrow$$

Next, the 10-kN load on the overhang causes a statically equivalent force of 10 kN and couple moment of 20 kN · m at the support B of the simply supported span, Fig. 12–29c. The 10-kN force does not cause a displacement or slope at B; however, the 20-kN · m couple moment does cause a slope. The slope at B due to this moment is

$$(\theta_B)_2 = \frac{M_0 L}{3EI} = \frac{20\text{ kN} \cdot \text{m}(4\text{ m})}{3EI} = \frac{26.67\text{ kN} \cdot \text{m}^2}{EI} \downarrow$$

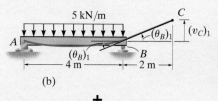

(c)

$+$

so that the extended point C is displaced

$$(v_C)_2 = (2\text{ m})\left(\frac{26.7\text{ kN} \cdot \text{m}^2}{EI}\right) = \frac{53.33\text{ kN} \cdot \text{m}^3}{EI} \downarrow$$

Finally, the cantilevered portion BC is displaced by the 10-kN force, Fig. 12–29d. We have

$$(v_C)_3 = \frac{PL^3}{3EI} = \frac{10\text{ kN}(2\text{ m})^3}{3EI} = \frac{26.67\text{ kN} \cdot \text{m}^3}{EI} \downarrow$$

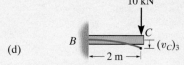

(d)

Fig. 12–29

Summing these results algebraically, we obtain the displacement of point C,

$$(+\downarrow) \qquad v_C = -\frac{26.7}{EI} + \frac{53.3}{EI} + \frac{26.7}{EI} = \frac{53.3\text{ kN} \cdot \text{m}^3}{EI} \downarrow \qquad Ans.$$

12

EXAMPLE | 12.15

Determine the displacement at the end C of the cantilever beam shown in Fig. 12–30. EI is constant.

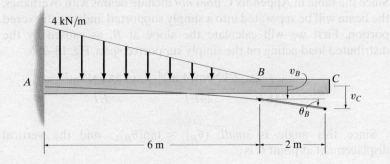

Fig. 12–30

SOLUTION

Using the table in Appendix C for the triangular loading, the slope and displacement at point B are

$$\theta_B = \frac{w_0 L^3}{24EI} = \frac{4 \text{ kN/m}(6 \text{ m})^3}{24EI} = \frac{36 \text{ kN} \cdot \text{m}^2}{EI}$$

$$v_B = \frac{w_0 L^4}{30EI} = \frac{4 \text{ kN/m}(6 \text{ m})^4}{30EI} = \frac{172.8 \text{ kN} \cdot \text{m}^3}{EI}$$

The unloaded region BC of the beam remains straight, as shown in Fig. 12–30. Since θ_B is small, the displacement at C becomes

$$(+\downarrow) \qquad v_C = v_B + \theta_B(L_{BC})$$

$$= \frac{172.8 \text{ kN} \cdot \text{m}^3}{EI} + \frac{36 \text{ kN} \cdot \text{m}^2}{EI}(2 \text{ m})$$

$$= \frac{244.8 \text{ kN} \cdot \text{m}^3}{EI} \downarrow \qquad\qquad\qquad Ans.$$

EXAMPLE 12.16

12

The steel bar shown in Fig. 12–31a is supported by two springs at its ends A and B. Each spring has a stiffness of $k = 45$ kN/m and is originally unstretched. If the bar is loaded with a force of 3 kN at point C, determine the vertical displacement of the force. Neglect the weight of the bar and take $E_{st} = 200$ GPa, $I = 4.6875(10^{-6})$ m^4.

SOLUTION

The end reactions at A and B are calculated and shown in Fig. 12–31b. Each spring deflects by an amount

$$(v_A)_1 = \frac{2 \text{ kN}}{45 \text{ kN/m}} = 0.0444 \text{ m}$$

$$(v_B)_1 = \frac{1 \text{ kN}}{45 \text{ kN/m}} = 0.0222 \text{ m}$$

If the bar is considered to be *rigid*, these displacements cause it to move into the position shown in Fig. 12–31b. For this case, the vertical displacement at C is

$$(v_C)_1 = (v_B)_1 + \frac{2 \text{ m}}{3 \text{ m}}[(v_A)_1 - (v_B)_1]$$

$$= 0.0222 \text{ m} + \frac{2}{3}[0.0444 \text{ m} - 0.0222 \text{ m}] = 0.0370 \text{ m} \downarrow$$

We can find the displacement at C caused by the *deformation* of the bar, Fig. 12–31c, by using the table in Appendix C. We have

$$(v_C)_2 = \frac{Pab}{6EIL}(L^2 - b^2 - a^2)$$

$$= \frac{3 \text{ kN}(1 \text{ m})(2 \text{ m})[(3 \text{ m})^2 - (2 \text{ m})^2 - (1 \text{ m})^2]}{6(200)(10^6) \text{ kN/m}^2[(4.6875(10^{-6}) \text{ m}^4](3 \text{ m})}$$

$$= 0.001422 \text{ m} = 1.422 \text{ mm} \downarrow$$

Adding the two displacement components, we get

$(+\downarrow) \quad v_C = 0.0370 \text{ m} + 0.001422 \text{ m} = 0.0384 \text{ m} = 38.4 \text{ mm} \downarrow$ *Ans.*

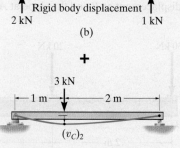

(a)

=

(b)

+

(c)

Deformable body displacement

Fig. 12–31

PROBLEMS

12–87. The W310 × 67 simply supported beam is made of A-36 steel and is subjected to the loading shown. Determine the deflection at its center C.

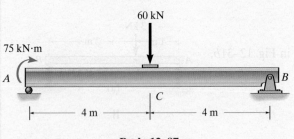

60 kN

75 kN·m

A

B

C

4 m

4 m

Prob. 12–87

*12–88.** The W250 × 22 cantilevered beam is made of A-36 steel and is subjected to the loading shown. Determine the displacement at B and the slope at A.

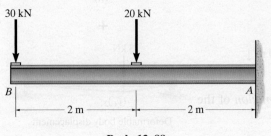

30 kN

20 kN

B

A

2 m

2 m

Prob. 12–88

•**12–89.** Determine the slope and deflection at end C of the overhang beam. EI is constant.

12–90. Determine the slope at A and the deflection at point D of the overhang beam. EI is constant.

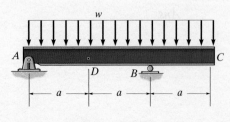

w

A

D

B

C

a

a

a

Probs. 12–89/90

12–91. Determine the slope at B and the deflection at point C of the simply supported beam. $E = 200$ GPa and $I = 45.5(10^6)$ mm^4.

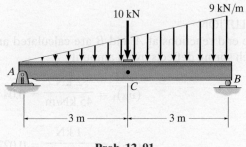

10 kN

9 kN/m

A

C

B

3 m

3 m

Prob. 12–91

*12–92.** Determine the slope at A and the deflection at point C of the simply supported beam. The modulus of elasticity of the wood is $E = 10$ GPa.

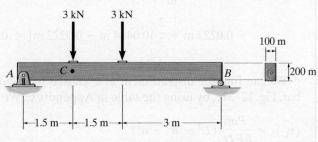

3 kN

3 kN

A

C

B

100 m

200 m

1.5 m

1.5 m

3 m

Prob. 12–92

•**12–93.** The W200 × 36 simply supported beam is made of A-36 steel and is subjected to the loading shown. Determine the deflection at its center C.

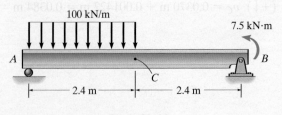

100 kN/m

7.5 kN·m

A

B

C

2.4 m

2.4 m

Prob. 12–93

12–94. Determine the vertical deflection and slope at the end *A* of the bracket. Assume that the bracket is fixed supported at its base, and neglect the axial deformation of segment *AB. EI* is constant.

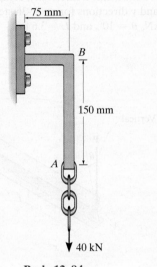

Prob. 12–94

12–95. The simply supported beam is made of A-36 steel and is subjected to the loading shown. Determine the deflection at its center *C. I* = $0.1457(10^{-3})$ m^4.

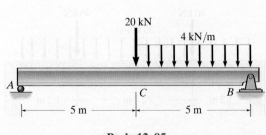

Prob. 12–95

*__12–96.** Determine the deflection at end *E* of beam *CDE.* The beams are made of wood having a modulus of elasticity of *E* = 10 GPa.

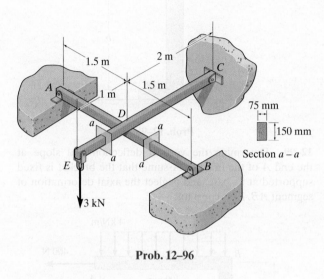

Prob. 12–96

•**12–97.** The pipe assembly consists of three equal-sized pipes with flexibility stiffness *EI* and torsional stiffness *GJ.* Determine the vertical deflection at point *A.*

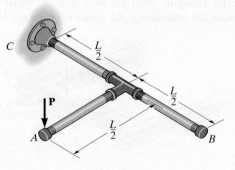

Prob. 12–97

12–98. Determine the vertical deflection at the end A of the bracket. Assume that the bracket is fixed supported at its base B and neglect axial deflection. EI is constant.

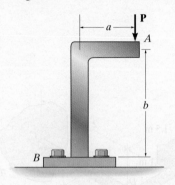

Prob. 12–98

12–99. Determine the vertical deflection and slope at the end A of the bracket. Assume that the bracket is fixed supported at its base, and neglect the axial deformation of segment AB. EI is constant.

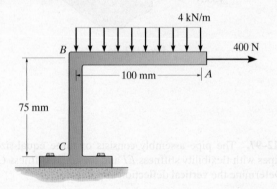

Prob. 12–99

*12–100. The framework consists of two A-36 steel cantilevered beams CD and BA and a simply supported beam CB. If each beam is made of steel and has a moment of inertia about its principal axis of $I_x = 46(10^6)$ mm^4, determine the deflection at the center G of beam CB.

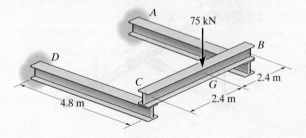

Prob. 12–100

•12–101. The wide-flange beam acts as a cantilever. Due to an error it is installed at an angle θ with the vertical. Determine the ratio of its deflection in the x direction to its deflection in the y direction at A when a load $\mathbf{P}$ is applied at this point. The moments of inertia are I_x and I_y. For the solution, resolve $\mathbf{P}$ into components and use the method of superposition. *Note:* The result indicates that large lateral deflections (x direction) can occur in narrow beams, $I_y \ll I_x$, when they are improperly installed in this manner. To show this numerically, compute the deflections in the x and y directions for an A-36 steel W250 × 22, with $P = 7.5$ kN, $\theta = 10°$, and $L = 3.6$ m.

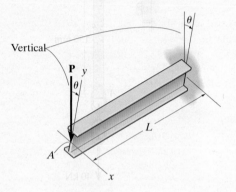

Prob. 12–101

12–102. The simply supported beam carries a uniform load of 30 kN/m. Code restrictions, due to a plaster ceiling, require the maximum deflection not to exceed 1/360 of the span length. Select the lightest-weight A-36 steel wide-flange beam from Appendix B that will satisfy this requirement and safely support the load. The allowable bending stress is $\sigma_{\text{allow}} = 168$ MPa and the allowable shear stress is $\tau_{\text{allow}} = 100$ MPa. Assume A is a pin and B a roller support.

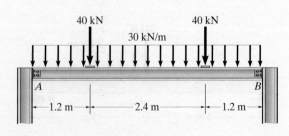

Prob. 12–102

12.6   Statically Indeterminate Beams and Shafts

The analysis of statically indeterminate axially loaded bars and torsionally loaded shafts has been discussed in Secs. 4.4 and 5.5, respectively. In this section we will illustrate a general method for determining the reactions on statically indeterminate beams and shafts. Specifically, a member of any type is classified as *statically indeterminate* if the number of unknown reactions *exceeds* the available number of equilibrium equations.

The additional support reactions on the beam or shaft that are *not needed* to keep it in stable equilibrium are called *redundants*. The number of these redundants is referred to as the *degree of indeterminacy*. For example, consider the beam shown in Fig. 12–32a. If the free-body diagram is drawn, Fig. 12–32b, there will be four unknown support reactions, and since three equilibrium equations are available for solution, the beam is classified as being indeterminate to the first degree. Either A_y, B_y, or M_A can be classified as the redundant, for if any one of these reactions is removed, the beam remains stable and in equilibrium (A_x cannot be classified as the redundant, for if it were removed, $\Sigma F_x = 0$ would not be satisfied.) In a similar manner, the *continuous beam* in Fig. 12–33a is indeterminate to the second degree, since there are five unknown reactions and only three available equilibrium equations, Fig. 12–33b. Here the two redundant support reactions can be chosen among A_y, B_y, C_y, and D_y.

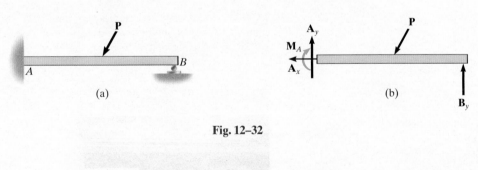

(a)   (b)

Fig. 12–32

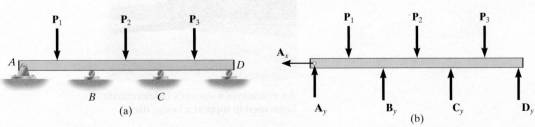

(a)   (b)

Fig. 12–33

12

To determine the reactions on a beam (or shaft) that is statically indeterminate, it is first necessary to specify the redundant reactions. We can determine these redundants from conditions of geometry known as *compatibility conditions*. Once determined, the redundants are then applied to the beam, and the remaining reactions are determined from the equations of equilibrium.

In the following sections we will illustrate this procedure for solution using the method of integration, Sec. 12.7; the moment-area method, Sec. 12.8; and the method of superposition, Sec. 12.9.

12.7   Statically Indeterminate Beams and Shafts—Method of Integration

The method of integration, discussed in Sec. 12.2, requires two integrations of the differential equation $d^2v/dx^2 = M/EI$ once the internal moment M in the beam is expressed as a function of position x. If the beam is statically indeterminate, however, M can also be expressed in terms of the *unknown* redundants. After integrating this equation twice, there will be two constants of integration along with the redundants to be determined. Although this is the case, these unknowns can always be found from the boundary and/or continuity conditions for the problem.

The following example problems illustrate specific applications of this method using the procedure for analysis outlined in Sec. 12.2.

An example of a statically indeterminate beam used to support a bridge deck.

EXAMPLE 12.17

The beam is subjected to the distributed loading shown in Fig. 12–34a. Determine the reaction at A. EI is constant.

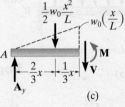

(a)

SOLUTION

Elastic Curve. The beam deflects as shown in Fig. 12–34a. Only one coordinate x is needed. For convenience we will take it directed to the right, since the internal moment is easy to formulate.

Moment Function. The beam is indeterminate to the first degree as indicated from the free-body diagram, Fig. 12–34b. We can express the internal moment M in terms of the redundant force at A using the segment shown in Fig. 12–34c. Here

$$M = A_y x - \frac{1}{6} w_0 \frac{x^3}{L}$$

(b)

Slope and Elastic Curve. Applying Eq. 12–10, we have

$$EI \frac{d^2 v}{dx^2} = A_y x - \frac{1}{6} w_0 \frac{x^3}{L}$$

$$EI \frac{dv}{dx} = \frac{1}{2} A_y x^2 - \frac{1}{24} w_0 \frac{x^4}{L} + C_1$$

$$EI v = \frac{1}{6} A_y x^3 - \frac{1}{120} w_0 \frac{x^5}{L} + C_1 x + C_2$$

(c)

The three unknowns A_y, C_1, and C_2 are determined from the boundary conditions $x = 0$, $v = 0$; $x = L$, $dv/dx = 0$; and $x = L$, $v = 0$. Applying these conditions yields

Fig. 12–34

$x = 0, v = 0;$

$$0 = 0 - 0 + 0 + C_2$$

$x = L, \dfrac{dv}{dx} = 0;$

$$0 = \frac{1}{2} A_y L^2 - \frac{1}{24} w_0 L^3 + C_1$$

$x = L, v = 0;$

$$0 = \frac{1}{6} A_y L^3 - \frac{1}{120} w_0 L^4 + C_1 L + C_2$$

Solving,

$$A_y = \frac{1}{10} w_0 L \qquad \qquad Ans.$$

$$C_1 = -\frac{1}{120} w_0 L^3 \qquad C_2 = 0$$

NOTE: Using the result for A_y, the reactions at B can be determined from the equations of equilibrium, Fig. 12–34b. Show that $B_x = 0$, $B_y = 2w_0 L/5$, and $M_B = w_0 L^2/15$.

12

EXAMPLE | 12.18

The beam in Fig. 12–35a is fixed supported at both ends and is subjected to the uniform loading shown. Determine the reactions at the supports. Neglect the effect of axial load.

SOLUTION

Elastic Curve.   The beam deflects as shown in Fig. 12–35a. As in the previous problem, only one x coordinate is necessary for the solution since the loading is continuous across the span.

Moment Function.   From the free-body diagram, Fig. 12–35b, the respective shear and moment reactions at A and B must be equal, since there is symmetry of both loading and geometry. Because of this, the equation of equilibrium, $\Sigma F_y = 0$, requires

$$V_A = V_B = \frac{wL}{2} \qquad\qquad \textit{Ans.}$$

The beam is indeterminate to the first degree, where M' is redundant. Using the beam segment shown in Fig. 12–35c, the internal moment $\mathbf{M}$ can be expressed in terms of M' as follows:

$$M = \frac{wL}{2}x - \frac{w}{2}x^2 - M'$$

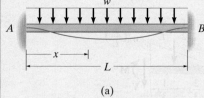

w

A B

x

L

(a)

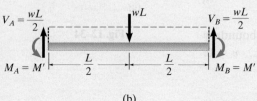

$V_A = \dfrac{wL}{2}$ wL $V_B = \dfrac{wL}{2}$

$M_A = M'$ $\dfrac{L}{2}$ $\dfrac{L}{2}$ $M_B = M'$

(b)

Slope and Elastic Curve.   Applying Eq. 12–10, we have

$$EI\frac{d^2v}{dx} = \frac{wL}{2}x - \frac{w}{2}x^2 - M'$$

$$EI\frac{dv}{dx} = \frac{wL}{4}x^2 - \frac{w}{6}x^3 - M'x + C_1$$

$$EIv = \frac{wL}{12}x^3 - \frac{w}{24}x^4 - \frac{M'}{2}x^2 + C_1x + C_2$$

The three unknowns, M', C_1, and C_2, can be determined from the *three* boundary conditions $v = 0$ at $x = 0$, which yields $C_2 = 0$; $dv/dx = 0$ at $x = 0$, which yields $C_1 = 0$; and $v = 0$ at $x = L$, which yields

$$M' = \frac{wL^2}{12} \qquad\qquad \textit{Ans.}$$

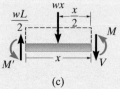

$\dfrac{wL}{2}$ wx $\dfrac{x}{2}$ M

M' x V

(c)

Fig. 12–35

Using these results, notice that because of symmetry the remaining boundary condition $dv/dx = 0$ at $x = L$ is automatically satisfied.

NOTE: It should be realized that this method of solution is generally suitable when only one x coordinate is needed to describe the elastic curve. If several x coordinates are needed, equations of continuity must be written, thus complicating the solution process.

PROBLEMS

12–103. Determine the reactions at the supports A and B, then draw the moment diagram. EI is constant.

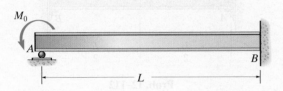

Prob. 12–103

*12–104. Determine the value of a for which the maximum positive moment has the same magnitude as the maximum negative moment. EI is constant.

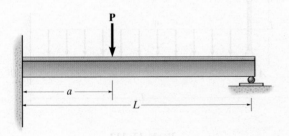

Prob. 12–104

•12–105. Determine the reactions at the supports A, B, and C; then draw the shear and moment diagrams. EI is constant.

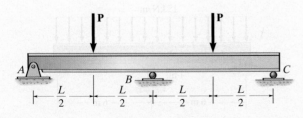

Prob. 12–105

12–106. Determine the reactions at the supports, then draw the shear and moment diagram. EI is constant.

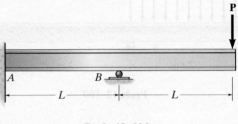

Prob. 12–106

12–107. Determine the moment reactions at the supports A and B. EI is constant.

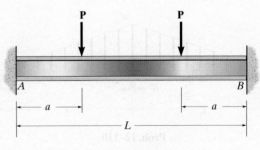

Prob. 12–107

*12–108. Determine the reactions at roller support A and fixed support B.

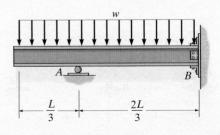

Prob. 12–108

•**12–109.** Use discontinuity functions and determine the reactions at the supports, then draw the shear and moment diagrams. *EI* is constant.

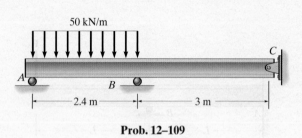

Prob. 12–109

12–110. Determine the reactions at the supports, then draw the shear and moment diagrams. *EI* is constant.

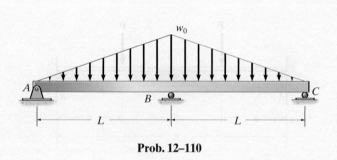

Prob. 12–110

12–111. Determine the reactions at pin support *A* and roller supports *B* and *C*. *EI* is constant.

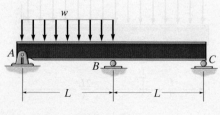

Prob. 12–111

*12–112.** Determine the moment reactions at fixed supports *A* and *B*. *EI* is constant.

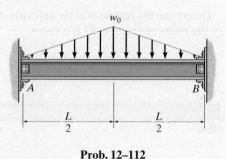

Prob. 12–112

•**12–113.** The beam has a constant E_1I_1 and is supported by the fixed wall at *B* and the rod *AC*. If the rod has a cross-sectional area A_2 and the material has a modulus of elasticity E_2, determine the force in the rod.

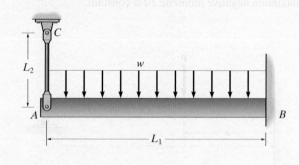

Prob. 12–113

12–114. The beam is supported by a pin at *A*, a roller at *B*, and a post having a diameter of 50 mm at *C*. Determine the support reactions at *A*, *B*, and *C*. The post and the beam are made of the same material having a modulus of elasticity $E = 200$ GPa, and the beam has a constant moment of inertia $I = 255(10^6)$ mm^4.

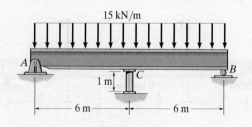

Prob. 12–114

*12.8 Statically Indeterminate Beams and Shafts—Moment-Area Method

If the moment-area method is used to determine the unknown redundants of a statically indeterminate beam or shaft, then the M/EI diagram must be drawn such that the redundants are represented as unknowns on this diagram. Once the M/EI diagram is established, the two moment-area theorems can then be applied to obtain the proper relationships between the tangents on the elastic curve in order to meet the conditions of displacement and/or slope at the supports of the beam. In all cases the number of these compatibility conditions will be equivalent to the number of redundants, and so a solution for the redundants can be obtained.

Moment Diagrams Constructed by the Method of Superposition. Since application of the moment-area theorems requires calculation of both the area under the M/EI diagram and the centroidal location of this area, it is often convenient to use *separate* M/EI diagrams for *each* of the known loads and redundants rather than using the *resultant diagram* to calculate these geometric quantities. This is especially true if the resultant moment diagram has a complicated shape. The method for drawing the moment diagram in parts is based on the principle of superposition.

Most loadings on cantilevered *beams or shafts* will be a combination of the four loadings shown in Fig. 12–36. Construction of the associated moment diagrams, also shown in this figure, has been discussed in the examples of Chapter 6. Based on these results, we will now show how to use the method of superposition to represent the resultant moment diagram by a series of separate moment diagrams for the cantilevered beam shown in Fig. 12–37a. To do this, we will first replace the loads by a system of statically equivalent loads. For example, the three cantilevered beams shown in Fig. 12–37a are statically equivalent to the

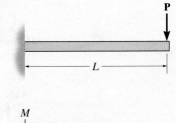

Sloping line

(a)

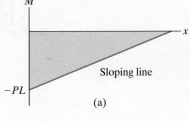

Parabolic curve

(c)

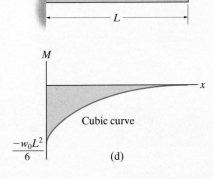

Cubic curve

(d)

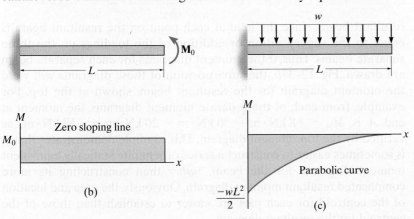

Zero sloping line

(b)

Fig. 12–36

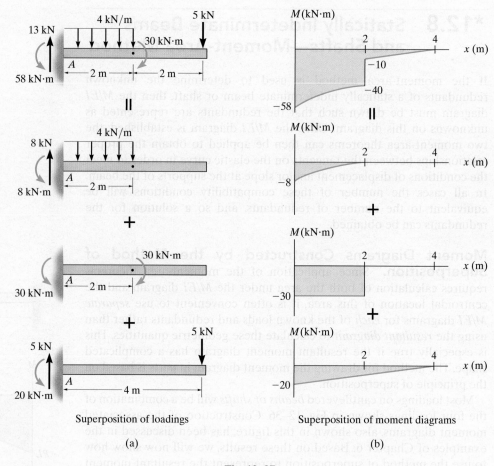

Superposition of loadings
(a)

Superposition of moment diagrams
(b)

Fig. 12–37

resultant beam, since the load at each point on the resultant beam is equal to the superposition or addition of the loadings on the three separate beams. Thus, if the moment diagrams for each separate beam are drawn, Fig. 12–37b, the superposition of these diagrams will yield the moment diagram for the resultant beam, shown at the top. For example, from each of the separate moment diagrams, the moment at end A is $M_A = -8\,\text{kN}\cdot\text{m} - 30\,\text{kN}\cdot\text{m} - 20\,\text{kN}\cdot\text{m} = -58\,\text{kN}\cdot\text{m}$, as verified by the top moment diagram. This example demonstrates that it is sometimes easier to construct a series of separate statically equivalent moment diagrams for the beam, *rather* than constructing its more complicated resultant moment diagram. Obviously, the area and location of the centroid for each part are easier to establish than those of the centroid for the resultant diagram.

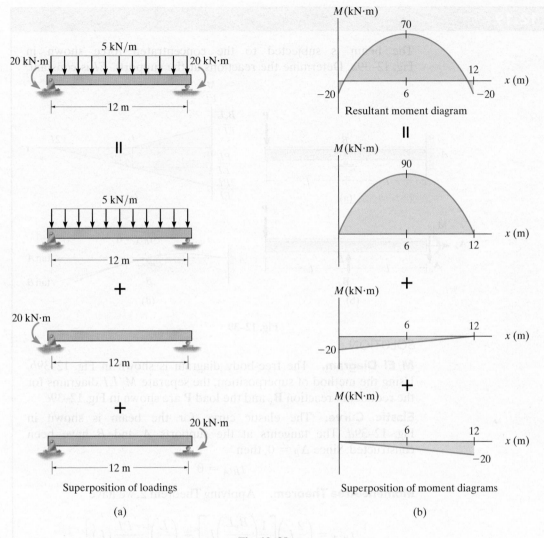

Fig. 12–38

In a similar manner, we can also represent the resultant moment diagram for a *simply supported beam* by using a superposition of moment diagrams for each loading acting on a series of simply supported beams. For example, the beam loading shown at the top of Fig. 12–38a is equivalent to the sum of the beam loadings shown below it. Consequently, the sum of the moment diagrams for each of these three loadings can be used rather than the resultant moment diagram shown at the top of Fig. 12–38b.

The examples that follow should also clarify some of these points and illustrate how to use the moment-area theorems to obtain the redundant reactions on statically indeterminate beams and shafts. The solutions follow the procedure for analysis outlined in Sec. 12.4.

12

EXAMPLE | **12.19**

The beam is subjected to the concentrated force shown in Fig. 12–39a. Determine the reactions at the supports. EI is constant.

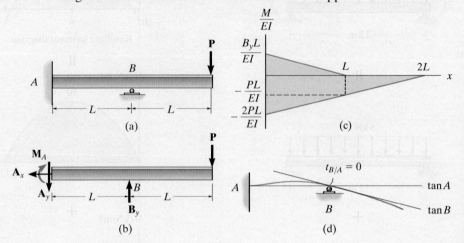

Fig. 12–39

SOLUTION

M/EI Diagram.   The free-body diagram is shown in Fig. 12–39b. Using the method of superposition, the separate M/EI diagrams for the redundant reaction $\mathbf{B}_y$ and the load $\mathbf{P}$ are shown in Fig. 12–39c.

Elastic Curve.   The elastic curve for the beam is shown in Fig. 12–39d. The tangents at the supports A and B have been constructed. Since $\Delta_B = 0$, then

$$t_{B/A} = 0$$

Moment-Area Theorem.   Applying Theorem 2, we have

$$t_{B/A} = \left(\frac{2}{3}L\right)\left[\frac{1}{2}\left(\frac{B_yL}{EI}\right)L\right] + \left(\frac{L}{2}\right)\left[\frac{-PL}{EI}(L)\right]$$

$$+ \left(\frac{2}{3}L\right)\left[\frac{1}{2}\left(\frac{-PL}{EI}\right)(L)\right] = 0$$

$$B_y = 2.5P \qquad\qquad Ans.$$

Equations of Equilibrium.   Using this result, the reactions at A on the free-body diagram, Fig. 12–39b, are

$$\xrightarrow{+} \Sigma F_x = 0; \qquad\qquad A_x = 0 \qquad\qquad Ans.$$

$$+\uparrow \Sigma F_y = 0; \qquad -A_y + 2.5P - P = 0$$

$$A_y = 1.5P \qquad\qquad Ans.$$

$$\zeta + \Sigma M_A = 0; \qquad -M_A + 2.5P(L) - P(2L) = 0$$

$$M_A = 0.5PL \qquad\qquad Ans.$$

EXAMPLE 12.20

The beam is subjected to the couple moment at its end C as shown in Fig. 12–40a. Determine the reaction at B. EI is constant.

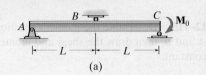
(a)

SOLUTION

M/EI Diagram. The free-body diagram is shown in Fig. 12–40b. By inspection, the beam is indeterminate to the first degree. In order to obtain a direct solution, we will choose $\mathbf{B}_y$ as the redundant. Using superposition, the M/EI diagrams for $\mathbf{B}_y$ and $\mathbf{M}_0$, each applied to a simply supported beam, are shown in Fig. 12–40c. (Note that for such a beam A_x, A_y, and C_y do not contribute to an M/EI diagram.)

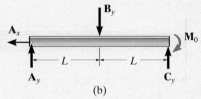
(b)

Elastic Curve. The elastic curve for the beam is shown in Fig. 12–40d. The tangents at A, B, and C have been established. Since $\Delta_A = \Delta_B = \Delta_C = 0$, then the vertical distances shown must be proportional; i.e.,

$$t_{B/C} = \frac{1}{2} t_{A/C} \qquad (1)$$

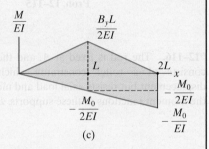
(c)

From Fig. 12–40c, we have

$$t_{B/C} = \left(\frac{1}{3}L\right)\left[\frac{1}{2}\left(\frac{B_yL}{2EI}\right)(L)\right] + \left(\frac{2}{3}L\right)\left[\frac{1}{2}\left(\frac{-M_0}{2EI}\right)(L)\right]$$

$$+ \left(\frac{L}{2}\right)\left[\left(\frac{-M_0}{2EI}\right)(L)\right]$$

$$t_{A/C} = (L)\left[\frac{1}{2}\left(\frac{B_yL}{2EI}\right)(2L)\right] + \left(\frac{2}{3}(2L)\right)\left[\frac{1}{2}\left(\frac{-M_0}{EI}\right)(2L)\right]$$

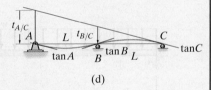
(d)

Substituting into Eq. 1 and simplifying yields

$$B_y = \frac{3M_0}{2L} \qquad \qquad \textit{Ans.}$$

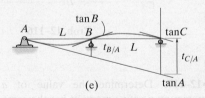
(e)

Fig. 12–40

Equations of Equilibrium. The reactions at A and C can now be determined from the equations of equilibrium, Fig. 12–40b. Show that $A_x = 0$, $C_y = 5M_0/4L$, and $A_y = M_0/4L$.

Note from Fig. 12–40e that this problem can also be worked in terms of the vertical distances,

$$t_{B/A} = \frac{1}{2} t_{C/A}$$

PROBLEMS

12–115. Determine the moment reactions at the supports A and B, then draw the shear and moment diagrams. EI is constant.

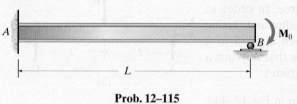

Prob. 12–115

*12–116. The rod is fixed at A, and the connection at B consists of a roller constraint which allows vertical displacement but resists axial load and moment. Determine the moment reactions at these supports. EI is constant.

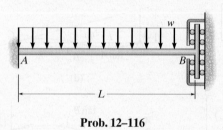

Prob. 12–116

•12–117. Determine the value of a for which the maximum positive moment has the same magnitude as the maximum negative moment. EI is constant.

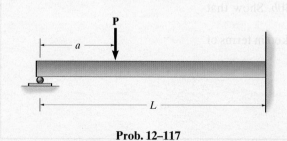

Prob. 12–117

12–118. Determine the reactions at the supports, then draw the shear and moment diagrams. EI is constant.

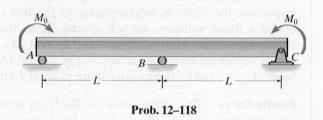

Prob. 12–118

12–119. Determine the reactions at the supports, then draw the shear and moment diagrams. EI is constant. Support B is a thrust bearing.

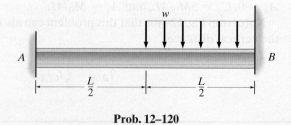

Prob. 12–119

*12–120. Determine the moment reactions at the supports A and B. EI is constant.

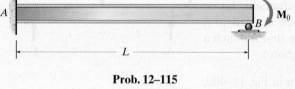

Prob. 12–120

12.9 Statically Indeterminate Beams and Shafts—Method of Superposition

The method of superposition has been used previously to solve for the redundant loading on axially loaded bars and torsionally loaded shafts. In order to apply this method to the solution of statically indeterminate beams (or shafts), it is first necessary to identify the redundant support reactions as explained in Sec. 12.6. By *removing* them from the beam we obtain the so-called *primary beam*, which is statically determinate and stable, and is subjected *only* to the external load. If we add to this beam a succession of similarly supported beams, each loaded with a *separate* redundant, then by the principle of superposition, we obtain the actual loaded beam. Finally, in order to solve for the redundants, we must write the *conditions of compatibility* that exist at the supports where each of the redundants acts. Since the redundant forces are determined directly in this manner, this method of analysis is sometimes called the ***force method***. Once the redundants are obtained, the other reactions on the beam can then be determined from the three equations of equilibrium.

To clarify these concepts, consider the beam shown in Fig. 12–41a. If we choose the reaction $\mathbf{B}_y$ at the roller as the redundant, then the primary beam is shown in Fig. 12–41b, and the beam with the redundant $\mathbf{B}_y$ acting on it is shown in Fig. 12–41c. The displacement at the roller is to be zero, and since the displacement of point B on the primary beam is v_B, and $\mathbf{B}_y$ causes point B to be displaced upward v'_B, we can write the compatibility equation at B as

$$(+\uparrow) \qquad 0 = -v_B + v'_B$$

The displacements v_B and v'_B can be obtained using any one of the methods discussed in Secs. 12.2 through 12.5. Here we will obtain them directly from the table in Appendix C. We have

$$v_B = \frac{5PL^3}{48EI} \quad \text{and} \quad v'_B = \frac{B_yL^3}{3EI}$$

Substituting into the compatibility equation, we get

$$0 = -\frac{5PL^3}{48EI} + \frac{B_yL^3}{3EI}$$

$$B_y = \frac{5}{16}P$$

Now that $\mathbf{B}_y$ is known, the reactions at the wall are determined from the three equations of equilibrium applied to the free-body diagram of the beam, Fig. 12–41d. The results are

$$A_x = 0 \qquad A_y = \frac{11}{16}P$$

$$M_A = \frac{3}{16}PL$$

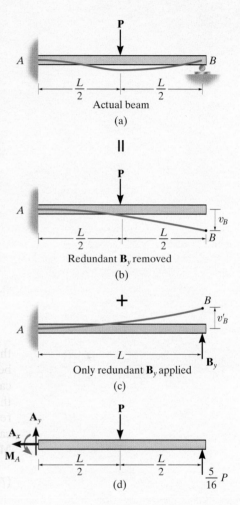

Actual beam

(a)

$\parallel$

Redundant $\mathbf{B}_y$ removed

(b)

$+$

Only redundant $\mathbf{B}_y$ applied

(c)

(d)

Fig. 12–41

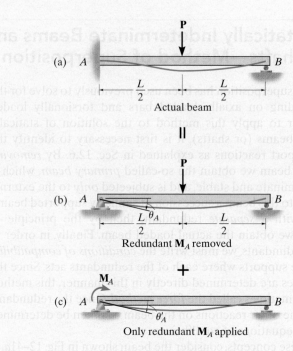

(a)

(b)   Redundant M_A removed

(c)   Only redundant M_A applied

Fig. 12–42

As stated in Sec. 12.6, choice of the redundant is *arbitrary*, provided the primary beam remains stable. For example, the moment at A for the beam in Fig. 12–42a can also be chosen as the redundant. In this case the capacity of the beam to resist M_A is removed, and so the primary beam is then pin supported at A, Fig. 12–42b. To it we add the beam with the redundant at A acting on it, Fig. 12–42c. Referring to the slope at A caused by the load P as θ_A, and the slope at A caused by the redundant M_A as θ'_A, the compatibility equation for the slope at A requires

$$(\uparrow+) \qquad\qquad 0 = \theta_A + \theta'_A$$

Again using the table in Appendix C, we have

$$\theta_A = \frac{PL^2}{16EI} \qquad \text{and} \qquad \theta'_A = \frac{M_A L}{3EI}$$

Thus,

$$0 = \frac{PL^2}{16EI} + \frac{M_A L}{3EI}$$

$$M_A = -\frac{3}{16}PL$$

This is the same result determined previously. Here the negative sign for M_A simply means that M_A acts in the opposite sense of direction of that shown in Fig. 12–42c.

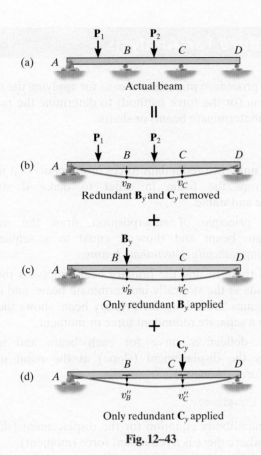

Fig. 12–43

Another example that illustrates this method is given in Fig. 12–43a. In this case the beam is indeterminate to the second degree and therefore *two* compatibility equations will be necessary for the solution. We will choose the forces at the roller supports B and C as redundants. The primary (statically determinate) beam deforms as shown in Fig. 12–43b when the redundants are removed. Each redundant force deforms this beam as shown in Figs. 12–43c and 12–43d, respectively. By superposition, the compatibility equations for the displacements at B and C are

$(+\downarrow)$ $\qquad\qquad\qquad 0 = v_B + v'_B + v''_B$

$(+\downarrow)$ $\qquad\qquad\qquad 0 = v_C + v'_C + v''_C$ $\qquad$ (12–20)

Here the displacement components v'_B and v'_C will be expressed in terms of the unknown $\mathbf{B}_y$, and the components v''_B and v''_C will be expressed in terms of the unknown $\mathbf{C}_y$. When these displacements have been determined and substituted into Eq. 12–20, these equations may then be solved simultaneously for the two unknowns $\mathbf{B}_y$ and $\mathbf{C}_y$.

Procedure for Analysis

The following procedure provides a means for applying the method of superposition (or the force method) to determine the reactions on statically indeterminate beams or shafts.

Elastic Curve.

- Specify the unknown redundant forces or moments that must be removed from the beam in order to make it statically determinate and stable.
- Using the principle of superposition, draw the statically indeterminate beam and show it equal to a sequence of corresponding *statically determinate beams*.
- The first of these beams, the primary beam, supports the same external loads as the statically indeterminate beam, and each of the other beams "added" to the primary beam shows the beam loaded with a separate redundant force or moment.
- Sketch the deflection curve for each beam and indicate symbolically the displacement (slope) at the point of each redundant force (moment).

Compatibility Equations.

- Write a compatibility equation for the displacement (slope) at each point where there is a redundant force (moment).
- Determine all the displacements or slopes using an appropriate method as explained in Secs. 12.2 through 12.5.
- Substitute the results into the compatibility equations and solve for the unknown redundants.
- If a numerical value for a redundant is *positive*, it has the *same sense of direction* as originally assumed. Similarly, a *negative* numerical value indicates the redundant acts *opposite* to its assumed *sense of direction*.

Equilibrium Equations.

- Once the redundant forces and/or moments have been determined, the remaining unknown reactions can be found from the equations of equilibrium applied to the loadings shown on the beam's free-body diagram.

The following examples illustrate application of this procedure. For brevity, all displacements and slopes have been found using the table in Appendix C.

EXAMPLE | 12.21

Determine the reactions at the roller support B of the beam shown in Fig. 12–44a, then draw the shear and moment diagrams. EI is constant.

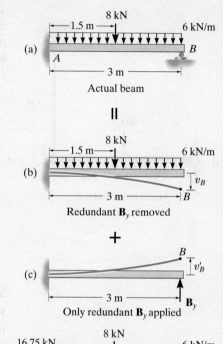

(a)

Actual beam

SOLUTION

Principle of Superposition. By inspection, the beam is statically indeterminate to the first degree. The roller support at B will be chosen as the redundant so that $\mathbf{B}_y$ will be determined *directly*. Figures 12–44b and 12–44c show application of the principle of superposition. Here we have assumed that $\mathbf{B}_y$ acts upward on the beam.

Compatibility Equation. Taking positive displacement as downward, the compatibility equation at B is

$$(+\downarrow) \qquad\qquad 0 = v_B - v_B' \qquad\qquad (1)$$

These displacements can be obtained directly from the table in Appendix C.

$$v_B = \frac{wL^4}{8EI} + \frac{5PL^3}{48EI}$$

$$= \frac{6 \text{ kN/m}(3 \text{ m})^4}{8EI} = \frac{5(8 \text{ kN})(3 \text{ m})^3}{48EI} = \frac{83.25 \text{ kN} \cdot \text{m}^3}{EI}\downarrow$$

$$v_B = \frac{PL^4}{3EI} = \frac{B_y(3 \text{ m})^3}{3EI} = \frac{9 \text{ m}^3 B_y}{EI}\uparrow$$

Substituting into Eq. 1 and solving yields

$$0 = \frac{83.25}{EI} - \frac{9 B_y}{EI}$$

$$By = 9.25 \text{ kN} \qquad\qquad Ans.$$

Equilibrium Equations. Using this result and applying the three equations of equilibrium, we obtain the results shown on the beam's free-body diagram in Fig. 12–44d. The shear and moment diagrams are shown in Fig. 12–44e.

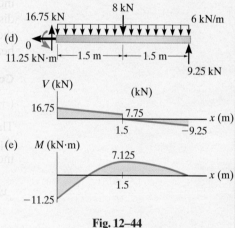

Fig. 12–44

12

EXAMPLE | 12.22

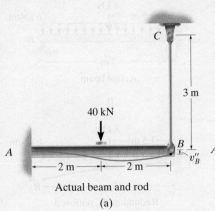

Actual beam and rod

(a)

The beam in Fig. 12–45a is fixed supported to the wall at A and pin connected to a 12-mm-diameter rod BC. If $E = 210$ GPa for both members, determine the force developed in the rod due to the loading. The moment of inertia of the beam about its neutral axis is $I = 186(10^6)$ mm^4.

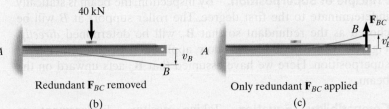

Redundant $\mathbf{F}_{BC}$ removed

(b)

Only redundant $\mathbf{F}_{BC}$ applied

(c)

Fig. 12–45

SOLUTION I

Principle of Superposition. By inspection, this problem is indeterminate to the first degree. Here B will undergo an unknown displacement v''_B, since the rod will stretch. The rod will be treated as the redundant and hence the force of the rod is removed from the beam at B, Fig. 12–45b, and then reapplied, Fig. 12–45c.

Compatibility Equation.   At point B we require

$$(+\downarrow) \qquad\qquad\qquad v''_B = v_B - v'_B \qquad\qquad (1)$$

The displacements v_B and v'_B are determined from the table in Appendix C. v''_B is calculated from Eq. 4–2. Working in kilopounds and inches, we have

$$v''_B = \frac{PL}{AE} = \frac{F_{BC}(3\text{ m})(1000\text{ mm/m})}{(\pi/4)(12\text{ mm})^2[210(10^3)\text{ N/mm}^2]} = 1.26(10^{-4})F_{BC}\downarrow$$

$$v_B = \frac{5PL^3}{48EI} = \frac{5([40(10^3)\text{ N}])(4\text{ m})^3(1000\text{ mm/m})^3}{48[210(10^3)\text{ N/mm}^2][186(10^6)\text{ mm}^4]} = 6.83\text{ mm}\downarrow$$

$$v'_B = \frac{PL^3}{3EI} = \frac{F_{BC}(4\text{ m})^3(1000\text{ mm/m})^3}{3[210(10^3)\text{ N/mm}^2][186(10^6)\text{ mm}^4]} = 1.067(10^{-3})F_{BC}\uparrow$$

Thus, Eq. 1 becomes

$$(+\downarrow) \qquad\qquad 1.26(10^{-4})F_{BC} = 6.83 - 1.067(10^{-3})F_{BC}$$

$$F_{BC} = 5.725(10^3)\text{ N} = 5.725\text{ kN} \qquad\qquad Ans.$$

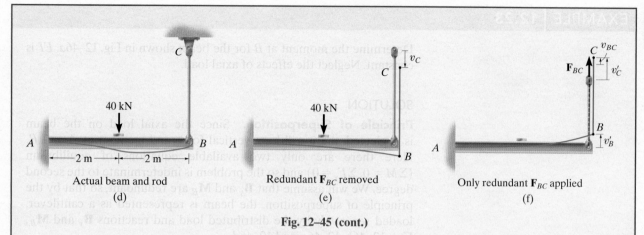

Actual beam and rod
(d)

Redundant $\mathbf{F}_{BC}$ removed
(e)

Only redundant $\mathbf{F}_{BC}$ applied
(f)

Fig. 12–45 (cont.)

SOLUTION II

Principle of Superposition. We can also solve this problem by removing the pin support at C and keeping the rod attached to the beam. In this case the 40-kN load will cause points B and C to be displaced downward the *same amount* v_C, Fig. 12–45e, since no force exists in rod BC. When the redundant force $\mathbf{F}_{BC}$ is applied at point C, it causes the end C of the rod to be displaced upward v'_C and the end B of the beam to be displaced upward v'_B, Fig. 12–45f. The difference in these two displacements, v_{BC}, represents the stretch of the rod due to $\mathbf{F}_{BC}$, so that $v'_C = v_{BC} + v'_B$. Hence, from Figs. 12–45d, 12–45e, and 12–45f, the compatibility of displacement at point C is

$$(+\downarrow) \qquad\qquad 0 = v_C - (v_{BC} + v'_B) \qquad\qquad (2)$$

From Solution I, we have

$$v_C = v_B = 6.83 \text{ mm} \downarrow$$

$$v_{BC} = v''_B = 1.26(10^{-4})F_{BC}\uparrow$$

$$v'_B = 1.067(10^{-3})F_{BC}\uparrow$$

Therefore, Eq. 2 becomes

$$(+\downarrow) \qquad 0 = 6.83 - [1.26(10^{-4})F_{BC} + 1.067(10^{-3})F_{BC}]$$

$$F_{BC} = 5725 \text{ N} = 5.725 \text{ kN} \qquad\qquad\qquad Ans.$$

EXAMPLE | 12.23

Determine the moment at B for the beam shown in Fig. 12–46a. EI is constant. Neglect the effects of axial load.

SOLUTION

Principle of Superposition.   Since the axial load on the beam is neglected, there will be a vertical force and moment at A and B. Here there are only two available equations of equilibrium ($\Sigma M = 0, \Sigma F_y = 0$) and so the problem is indeterminate to the second degree. We will assume that $\mathbf{B}_y$ and $\mathbf{M}_B$ are redundant, so that by the principle of superposition, the beam is represented as a cantilever, loaded *separately* by the distributed load and reactions $\mathbf{B}_y$ and $\mathbf{M}_B$, Figs. 12–46b, 12–46c, and 12–46d.

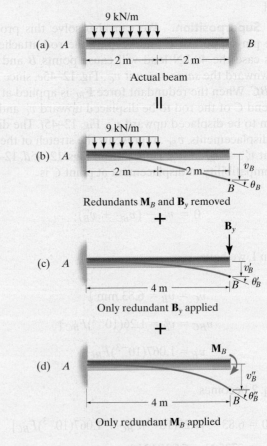

Fig. 12–46

Compatibility Equations.   Referring to the displacement and slope at B, we require

$(\curvearrowleft +)$ $$0 = \theta_B + \theta'_B + \theta''_B \qquad (1)$$

$(+\downarrow)$ $$0 = v_B + v'_B + v''_B \qquad (2)$$

Using the table in Appendix C to calculate the slopes and displacements, we have

$$\theta_B = \frac{wL^3}{48AE} = \frac{9 \text{ kN/m}(4 \text{ m})^3}{48EI} = \frac{12 \text{ kN} \cdot \text{m}^3}{EI} \downarrow$$

$$v_B = \frac{7wL^4}{384EI} = \frac{7(9 \text{ kN/m})(4 \text{ m})^4}{384EI} = \frac{42 \text{ kN} \cdot \text{m}^3}{EI} \downarrow$$

$$\theta'_B = \frac{PL^2}{2EI} = \frac{B_y(4 \text{ m})^2}{2EI} = \frac{8B_y}{EI} \downarrow$$

$$v'_B = \frac{PL^3}{3EI} = \frac{B_y(4 \text{ m})^3}{3EI} = \frac{21.33B_y}{EI} \downarrow$$

$$\theta''_B = \frac{ML}{EI} = \frac{M_B(4 \text{ m})}{EI} = \frac{4M_B}{EI} \downarrow$$

$$v''_B = \frac{ML^2}{2EI} = \frac{M_B(4 \text{ m})^2}{2EI} = \frac{8M_B}{EI} \downarrow$$

Substituting these values into Eqs. 1 and 2 and canceling out the common factor EI, we get

$(\curvearrowleft +)$ $$0 = 12 + 8B_y + 4M_B$$

$(+\downarrow)$ $$0 = 42 + 21.33B_y + 8M_B$$

Solving these equations simultaneously gives

$$B_y = 3.375 \text{ kN}$$

$$M_B = 3.75 \text{ kN} \cdot \text{m} \qquad \qquad Ans.$$

FUNDAMENTAL PROBLEMS

F12–13. Determine the reactions at the fixed support A and the roller B. EI is constant.

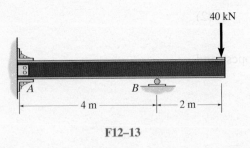

F12–13

F12–14. Determine the reactions at the fixed support A and the roller B. EI is constant.

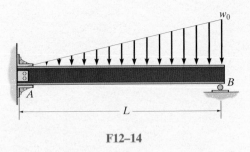

F12–14

F12–15. Determine the reactions at the fixed support A and the roller B. Support B settles 2 mm. $E = 200$ GPa, $I = 65.0(10^{-6})$ m^4.

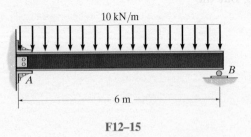

F12–15

F12–16. Determine the reaction at the roller B. EI is constant.

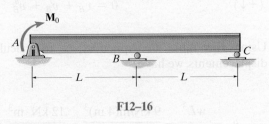

F12–16

F12–17. Determine the reaction at the roller B. EI is constant.

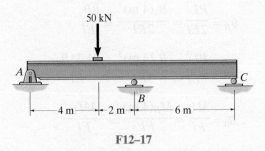

F12–17

F12–18. Determine the reaction at the roller support B if it settles 5 mm. $E = 200$ GPa and $I = 65.0(10^{-6})$ m^4.

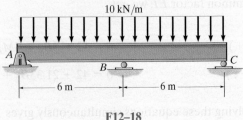

F12–18

PROBLEMS

•12–121. Determine the reactions at the bearing supports A, B, and C of the shaft, then draw the shear and moment diagrams. EI is constant. Each bearing exerts only vertical reactions on the shaft.

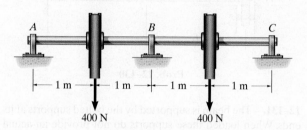

400 N   400 N

Prob. 12–121

12–122. Determine the reactions at the supports A and B. EI is constant.

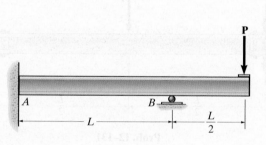

Prob. 12–122

12–123. Determine the reactions at the supports A, B, and C, then draw the shear and moment diagrams. EI is constant.

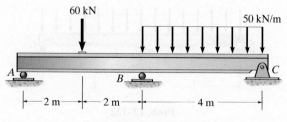

Prob. 12–123

*12–124. The assembly consists of a steel and an aluminum bar, each of which is 25 mm thick, fixed at its ends A and B, and pin connected to the *rigid* short link CD. If a horizontal force of 400 N is applied to the link as shown, determine the moments created at A and B. E_{al} = 200 GPa, E_{al} = 70 GPa.

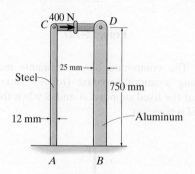

Prob. 12–124

•12–125. Determine the reactions at the supports A, B, and C, then draw the shear and moment diagrams. EI is constant.

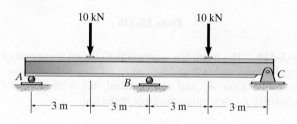

Prob. 12–125

12–126. Determine the reactions at the supports A and B. EI is constant.

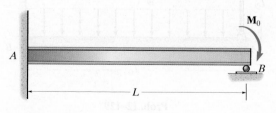

Prob. 12–126

12–127. Determine the reactions at support *C*. *EI* is constant for both beams.

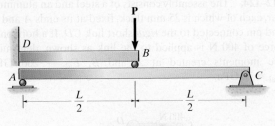

Prob. 12–127

***12–128.** The compound beam segments meet in the center using a smooth contact (roller). Determine the reactions at the fixed supports *A* and *B* when the load **P** is applied. *EI* is constant.

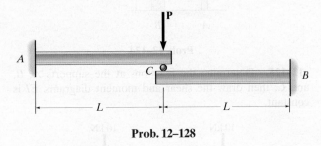

Prob. 12–128

•12–129. The beam has a constant E_1I_1 and is supported by the fixed wall at *B* and the rod *AC*. If the rod has a cross-sectional area A_2 and the material has a modulus of elasticity E_2, determine the force in the rod.

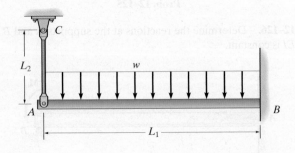

Prob. 12–129

12–130. Determine the reactions at *A* and *B*. Assume the support at *A* only exerts a moment on the beam. *EI* is constant.

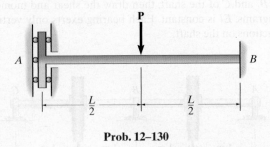

Prob. 12–130

12–131. The beam is supported by the bolted supports at its ends. When loaded these supports do not provide an actual fixed connection, but instead allow a slight rotation α before becoming fixed. Determine the moment at the connections and the maximum deflection of the beam.

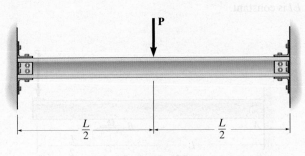

Prob. 12–131

***12–132.** The beam is supported by a pin at *A*, a spring having a stiffness *k* at *B*, and a roller at *C*. Determine the force the spring exerts on the beam. *EI* is constant.

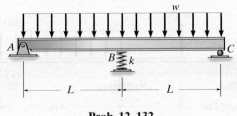

Prob. 12–132

•12–133. The beam is made from a soft linear elastic material having a constant EI. If it is originally a distance Δ from the surface of its end support, determine the distance a at which it rests on this support when it is subjected to the uniform load w_0, which is great enough to cause this to happen.

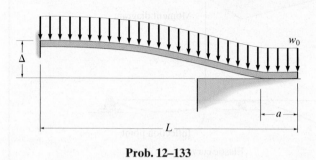

Prob. 12–133

12–134. Before the uniform distributed load is applied on the beam, there is a small gap of 0.2 mm between the beam and the post at B. Determine the support reactions at $A, B,$ and C. The post at B has a diameter of 40 mm, and the moment of inertia of the beam is $I = 875(10^6) \text{ mm}^4$. The post and the beam are made of material having a modulus of elasticity of $E = 200 \text{ GPa}$.

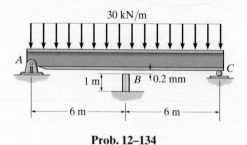

Prob. 12–134

12–135. The 25-mm-diameter A-36 steel shaft is supported by unyielding bearings at A and C. The bearing at B rests on a simply supported steel wide-flange beam having a moment of inertia of $I = 195(10^6) \text{ mm}^4$. If the belt loads on the pulley are 2 kN each, determine the vertical reactions at $A, B,$ and C.

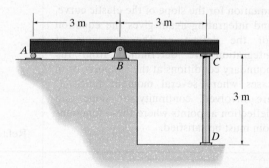

Prob. 12–135

*12–136. If the temperature of the 75-mm-diameter post CD is increased by 60°C, determine the force developed in the post. The post and the beam are made of A-36 steel, and the moment of inertia of the beam is $I = 255(10^6) \text{ mm}^4$.

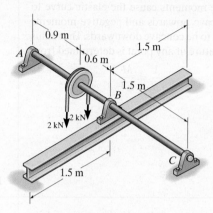

Prob. 12–136

12

CHAPTER REVIEW

The elastic curve represents the centerline deflection of a beam or shaft. Its shape can be determined using the moment diagram. Positive moments cause the elastic curve to be concave upwards and negative moments cause it to be concave downwards. The radius of curvature at any point is determined from

$$\frac{1}{\rho} = \frac{M}{EI}$$

Moment diagram

Inflection point

Elastic curve

Ref.: Section 12.1

The equation of the elastic curve and its slope can be obtained by first finding the internal moment in the member as a function of x. If several loadings act on the member, then separate moment functions must be determined between each of the loadings. Integrating these functions once using $EI(d^2v/dx^2) = M(x)$ gives the equation for the slope of the elastic curve, and integrating again gives the equation for the deflection. The constants of integration are determined from the boundary conditions at the supports, or in cases where several moment functions are involved, continuity of slope and deflection at points where these functions join must be satisfied.

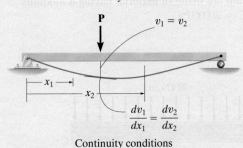

$\theta = 0$
$v = 0$ $v = 0$

Boundary conditions

P $v_1 = v_2$

x_1
x_2
$\dfrac{dv_1}{dx_1} = \dfrac{dv_2}{dx_2}$

Continuity conditions

Ref.: Section 12.2

Discontinuity functions allow one to express the equation of the elastic curve as a continuous function, regardless of the number of loadings on the member. This method eliminates the need to use continuity conditions, since the two constants of integration can be determined solely from the two boundary conditions.

Ref.: Section 12.3

The moment-area method is a semi-graphical technique for finding the slope of tangents or the vertical distance between tangents at specific points on the elastic curve. It requires finding area segments under the M/EI diagram, or the moment of these segments about points on the elastic curve. The method works well for M/EI diagrams composed of simple shapes, such as those produced by concentrated forces and couple moments.

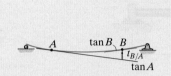

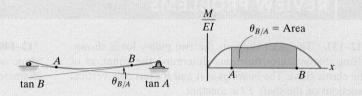

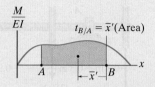

Ref.: Section 12.4

The deflection or slope at a point on a member subjected to combinations of loadings can be determined using the method of superposition. The table in Appendix C is available for this purpose.

Ref.: Section 12.5

Statically indeterminate beams and shafts have more unknown support reactions than available equations of equilibrium. To solve, one first identifies the redundant reactions. The method of integration or the moment-area theorems can then be used to solve for the unknown redundants. It is also possible to determine the redundants by using the method of superposition, where one considers the conditions of continuity at the redundant. Here the displacement due to the external loading is determined with the redundant removed, and again with the redundant applied and the external loading removed. The tables in Appendix C can be used to determine these necessary displacements.

Ref.: Sections 12.6 & 12.7 & 12.8

REVIEW PROBLEMS

•**12–137.** The shaft supports the two pulley loads shown. Using discontinuity functions, determine the equation of the elastic curve. The bearings at A and B exert only vertical reactions on the shaft. EI is constant.

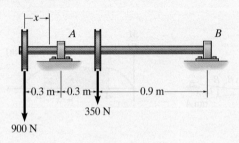

Prob. 12–137

12–138. The shaft is supported by a journal bearing at A, which exerts only vertical reactions on the shaft, and by a thrust bearing at B, which exerts both horizontal and vertical reactions on the shaft. Draw the bending-moment diagram for the shaft and then, from this diagram, sketch the deflection or elastic curve for the shaft's centerline. Determine the equations of the elastic curve using the coordinates x_1 and x_2. EI is constant.

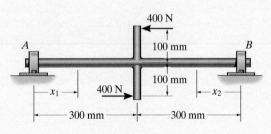

Prob. 12–138

12–139. The $W200 \times 36$ simply supported beam is subjected to the loading shown. Using the method of superposition, determine the deflection at its center C. The beam is made of A-36 steel.

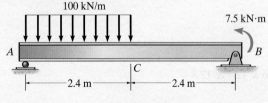

Prob. 12–139

12–140. Using the moment-area method, determine the slope and deflection at end C of the shaft. The 75-mm-diameter shaft is made of material having $E = 200$ GPa.

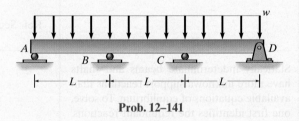

Prob. 12–140

•**12–141.** Determine the reactions at the supports. EI is constant. Use the method of superposition.

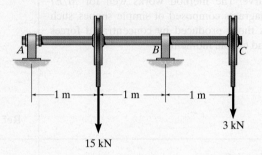

Prob. 12–141

12–142. Determine the moment reactions at the supports A and B. Use the method of integration. EI is constant.

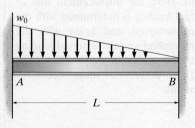

Prob. 12–142

12–143. If the cantilever beam has a constant thickness t, determine the deflection at end A. The beam is made of material having a modulus of elasticity E.

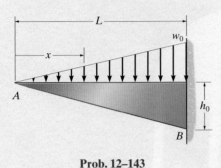

Prob. 12–143

*12–144. Beam ABC is supported by beam DBE and fixed at C. Determine the reactions at B and C. The beams are made of the same material having a modulus of elasticity $E = 200$ GPa, and the moment of inertia of both beams is $I = 25.0(10^6)$ mm^4.

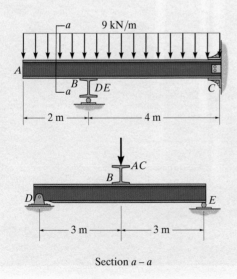

Section a – a

Prob. 12–144

•12–145. Using the method of superposition, determine the deflection at C of beam AB. The beams are made of wood having a modulus of elasticity of $E = 10$ GPa.

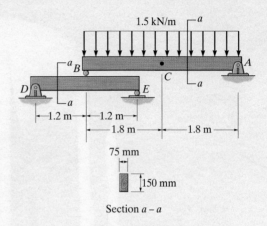

Section a – a

Prob. 12–145

12–146. The rim on the flywheel has a thickness t, width b, and specific weight γ. If the flywheel is rotating at a constant rate of ω, determine the maximum moment developed in the rim. Assume that the spokes do not deform. *Hint:* Due to symmetry of the loading, the slope of the rim at each spoke is zero. Consider the radius to be sufficiently large so that the segment AB can be considered as a straight beam fixed at both ends and loaded with a uniform centrifugal force per unit length. Show that this force is $w = bt\gamma\omega^2 r/g$.

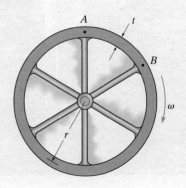

Prob. 12–146

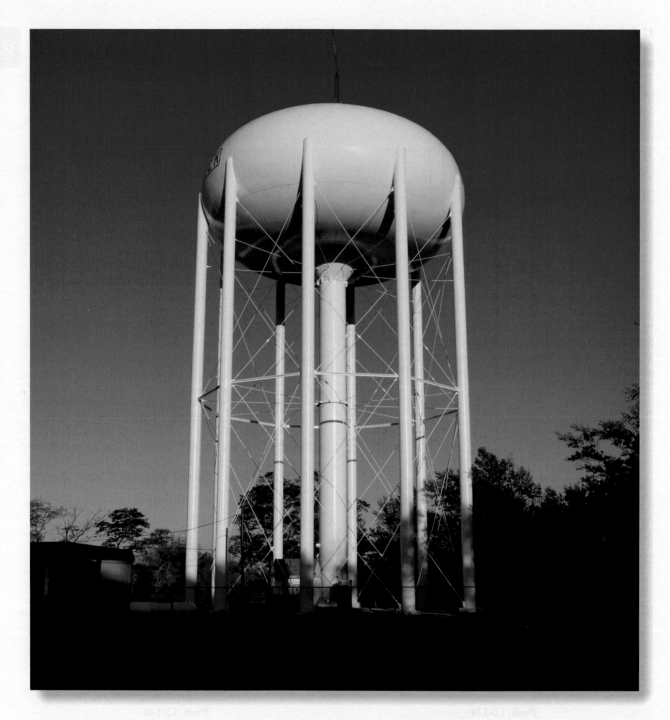

The columns used to support this water tank are braced at their mid-height in order to reduce their chance of buckling.

Buckling of Columns

13

CHAPTER OBJECTIVES

In this chapter, we will discuss the behavior of columns and indicate some of the methods used for their design. The chapter begins with a general discussion of buckling, followed by a determination of the axial load needed to buckle a so-called ideal column. Afterwards, a more realistic analysis is considered, which accounts for any bending of the column. Also, inelastic buckling of a column is presented as a special topic. At the end of the chapter we will discuss some of the methods used to design both concentrically and eccentrically loaded columns made of common engineering materials.

Check out the Companion Website for the following video solution(s) that can be found in this chapter.

- Section 2, 3
  Hollow Pipe Design for Buckling
- Section 4
  Maximum Stress – Secant Formula
- Section 6
  Column Design – Concentric Loading
- Section 7
  Column Design – Eccentric Loading

13.1 Critical Load

Whenever a member is designed, it is necessary that it satisfy specific strength, deflection, and stability requirements. In the preceding chapters we have discussed some of the methods used to determine a member's strength and deflection, while assuming that the member was always in stable equilibrium. Some members, however, may be subjected to compressive loadings, and if these members are long and slender the loading may be large enough to cause the member to deflect laterally or sidesway. To be specific, long slender members subjected to an axial compressive force are called ***columns***, and the lateral deflection that occurs is called ***buckling***. Quite often the buckling of a column can lead to a sudden and dramatic failure of a structure or mechanism, and as a result, special attention must be given to the design of columns so that they can safely support their intended loadings without buckling.

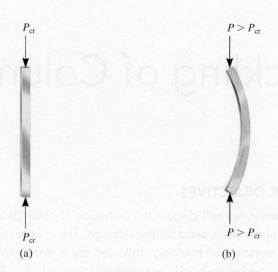

P_{cr}

$P > P_{cr}$

P_{cr} $P > P_{cr}$

(a)   (b)

Fig. 13–1

The maximum axial load that a column can support when it is on the *verge* of buckling is called the ***critical load***, P_{cr}, Fig. 13–1a. Any additional loading will cause the column to buckle and therefore deflect laterally as shown in Fig. 13–1b. In order to better understand the nature of this instability, consider a two-bar mechanism consisting of weightless bars that are rigid and pin connected as shown in Fig. 13–2a. When the bars are in the vertical position, the spring, having a stiffness k, is unstretched, and a *small* vertical force **P** is applied at the top of one of the bars. We can upset this equilibrium position by displacing the pin at A by a small amount Δ, Fig. 13–2b. As shown on the free-body diagram of the pin when the bars are displaced, Fig. 13–2c, the spring will produce a restoring force $F = k\Delta$, while the applied load **P** develops two horizontal components, $P_x = P \tan \theta$, which tend to push the pin (and the bars) further out of equilibrium. Since θ is small, $\Delta \approx \theta(L/2)$ and $\tan \theta \approx \theta$. Thus the *restoring* spring force becomes $F = k\theta L/2$, and the *disturbing* force is $2P_x = 2P\theta$.

If the restoring force is greater than the disturbing force, that is, $k\theta L/2 > 2P\theta$, then, noticing that θ cancels out, we can solve for P, which gives

$$P < \frac{kL}{4} \qquad \text{stable equilibrium}$$

This is a condition for *stable equilibrium* since the force developed by the spring would be adequate to restore the bars back to their vertical position. However, if $kL\theta/2 < 2P\theta$, or

$$P > \frac{kL}{4} \qquad \text{unstable equilibrium}$$

then the mechanism would be in *unstable equilibrium*. In other words, if this load **P** is applied, and a slight displacement occurs at A, the mechanism will tend to move out of equilibrium and not be restored to its original position.

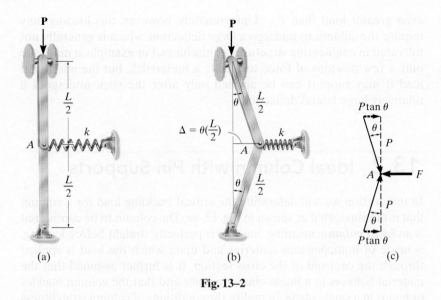

Fig. 13–2

The intermediate value of P, which requires $kL\theta/2 = 2P\theta$, is the *critical load*. Here

$$P_{cr} = \frac{kL}{4} \quad \text{neutral equilibrium}$$

This loading represents a case of the mechanism being in *neutral equilibrium*. Since P_{cr} is *independent* of the (small) displacement θ of the bars, any slight disturbance given to the mechanism will not cause it to move further out of equilibrium, nor will it be restored to its original position. Instead, the bars will *remain* in the deflected position.

These three different states of equilibrium are represented graphically in Fig. 13–3. The transition point where the load is equal to the critical value $P = P_{cr}$ is called the *bifurcation point*. At this point the mechanism will be in equilibrium for any *small value* of θ, measured either to the right or to the left of the vertical. Physically, P_{cr} represents the load for which the mechanism is on the verge of buckling. It is quite reasonable to determine this value by assuming *small displacements* as done here; however, it should be understood that P_{cr} may *not* be the largest value of P that the mechanism can support. Indeed, if a larger load is placed on the bars, then the mechanism may have to undergo a further deflection before the spring is compressed or elongated enough to hold the mechanism in equilibrium.

Like the two-bar mechanism just discussed, the critical buckling loads on columns supported in various ways can be obtained, and the method used to do this will be explained in the next section. Although in engineering design the critical load may be considered to be the largest load the column can support, realize that, like the two-bar mechanism in the deflected or buckled position, a column may actually support an

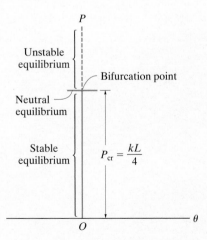

Fig. 13–3

13

even greater load than P_{cr}. Unfortunately, however, this loading may require the column to undergo a *large* deflection, which is generally not tolerated in engineering structures or machines. For example, it may take only a few newtons of force to buckle a meterstick, but the additional load it may support can be applied only after the stick undergoes a relatively large lateral deflection.

13.2 Ideal Column with Pin Supports

Some slender pin-connected members used in moving machinery, such as this short link, are subjected to compressive loads and thus act as columns.

In this section we will determine the critical buckling load for a column that is pin supported as shown in Fig. 13–4a. The column to be considered is an *ideal column*, meaning one that is perfectly straight before loading, is made of homogeneous material, and upon which the load is applied through the centroid of the cross section. It is further assumed that the material behaves in a linear-elastic manner and that the column buckles or bends in a single plane. In reality, the conditions of column straightness and load application are never accomplished; however, the analysis to be performed on an "ideal column" is similar to that used to analyze initially crooked columns or those having an eccentric load application. These more realistic cases will be discussed later in this chapter.

Since an ideal column is straight, theoretically the axial load P could be increased until failure occurs by either fracture or yielding of the material. However, when the critical load P_{cr} is reached, the column will be on the verge of becoming unstable, so that a small lateral force F, Fig. 13–4b, will cause the column to remain in the deflected position when F is removed, Fig. 13–4c. Any slight reduction in the axial load P from P_{cr} will allow the column to straighten out, and any slight increase in P, beyond P_{cr}, will cause further increases in lateral deflection.

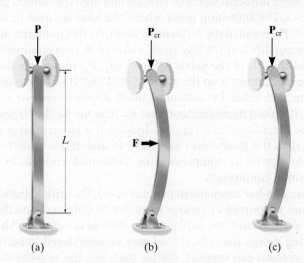

Fig. 13–4

Whether or not a column will remain stable or become unstable when subjected to an axial load will depend on its ability to restore itself, which is based on its resistance to bending. Hence, in order to determine the critical load and the buckled shape of the column, we will apply Eq. 12–10, which relates the internal moment in the column to its deflected shape, i.e.,

$$EI\frac{d^2v}{dx^2} = M \qquad (13\text{–}1)$$

Recall that this equation assumes that the slope of the elastic curve is small and that deflections occur only by bending. When the column is in its deflected position, Fig. 13–5a, the internal bending moment can be determined by using the method of sections. The free-body diagram of a segment in the deflected position is shown in Fig. 13–5b. Here both the deflection v and the internal moment M are shown in the *positive direction* according to the sign convention used to establish Eq. 13–1. Moment equilibrium requires $M = -Pv$. Thus Eq. 13–1 becomes

$$EI\frac{d^2v}{dx^2} = -Pv$$

$$\frac{d^2v}{dx^2} + \left(\frac{P}{EI}\right)v = 0 \qquad (13\text{–}2)$$

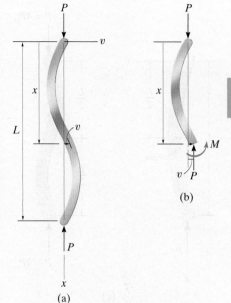

Fig. 13–5

This is a homogeneous, second-order, linear differential equation with constant coefficients. It can be shown by using the methods of differential equations, or by direct substitution into Eq. 13–2, that the general solution is

$$v = C_1\sin\left(\sqrt{\frac{P}{EI}}\,x\right) + C_2\cos\left(\sqrt{\frac{P}{EI}}\,x\right) \qquad (13\text{–}3)$$

The two constants of integration are determined from the boundary conditions at the ends of the column. Since $v = 0$ at $x = 0$, then $C_2 = 0$. And since $v = 0$ at $x = L$, then

$$C_1\sin\left(\sqrt{\frac{P}{EI}}\,L\right) = 0$$

This equation is satisfied if $C_1 = 0$; however, then $v = 0$, which is a *trivial solution* that requires the column to always remain straight, even though the load may cause the column to become unstable. The other possibility is for

$$\sin\left(\sqrt{\frac{P}{EI}}\,L\right) = 0$$

which is satisfied if

$$\sqrt{\frac{P}{EI}}\,L = n\pi$$

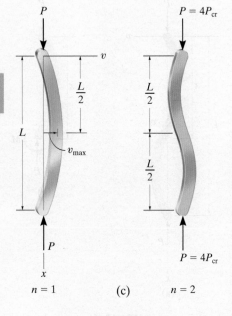

Fig. 13–5 (cont.)

or

$$P = \frac{n^2\pi^2 EI}{L^2} \qquad n = 1, 2, 3, \ldots \qquad (13\text{–}4)$$

The *smallest value* of P is obtained when $n = 1$, so the *critical load* for the column is therefore*

$$P_{cr} = \frac{\pi^2 EI}{L^2}$$

This load is sometimes referred to as the *Euler load*, named after the Swiss mathematician Leonhard Euler, who originally solved this problem in 1757. The corresponding buckled shape is defined by the equation

$$v = C_1 \sin\frac{\pi x}{L}$$

Here the constant C_1 represents the maximum deflection, v_{max}, which occurs at the midpoint of the column, Fig. 13–5c. Specific values for C_1 cannot be obtained, since the exact deflected form for the column is unknown once it has buckled. It has been assumed, however, that this deflection is small.

Note that the critical load is independent of the strength of the material; rather it only depends on the column's dimensions (I and L) and the material's stiffness or modulus of elasticity E. For this reason, as far as elastic buckling is concerned, columns made, for example, of high-strength steel offer no advantage over those made of lower-strength steel, since the modulus of elasticity for both is approximately the same. Also note that the load-carrying capacity of a column will increase as the moment of inertia of the cross section increases. Thus, efficient columns are designed so that most of the column's cross-sectional area is located as far away as possible from the principal centroidal axes for the section. This is why hollow sections such as tubes are more economical than solid sections. Furthermore, wide-flange sections, and columns that are "built up" from channels, angles, plates, etc., are better than sections that are solid and rectangular.

---

*n represents the number of waves in the deflected shape of the column. For example, if $n = 2$, then *two* waves will appear, Fig. 13–5c. Here the critical load is $4 P_{cr}$ just prior to buckling, which practically speaking will not exist.

It is also important to realize that a column will buckle about the principal axis of the cross section having the ***least moment of inertia*** (the weakest axis). For example, a column having a rectangular cross section, like a meter stick, as shown in Fig. 13–6, will buckle about the *a–a* axis, not the *b–b* axis. As a result, engineers usually try to achieve a balance, keeping the moments of inertia the same in all directions. Geometrically, then, circular tubes would make excellent columns. Also, square tubes or those shapes having $I_x \approx I_y$ are often selected for columns.

Summarizing the above discussion, the buckling equation for a pin-supported long slender column can be rewritten, and the terms defined as follows:

$$P_{cr} = \frac{\pi^2 EI}{L^2} \qquad (13\text{--}5)$$

where

P_{cr} = critical or maximum axial load on the column just before it begins to buckle. This load must *not* cause the stress in the column to exceed the proportional limit

E = modulus of elasticity for the material

I = *least* moment of inertia for the column's cross-sectional area

L = unsupported length of the column, whose ends are pinned

Fig. 13–6

For purposes of design, the above equation can also be written in a more useful form by expressing $I = Ar^2$, where A is the cross-sectional area and r is the ***radius of gyration*** of the cross-sectional area. Thus,

$$P_{cr} = \frac{\pi^2 E(Ar^2)}{L^2}$$

$$\left(\frac{P}{A}\right)_{cr} = \frac{\pi^2 E}{(L/r)^2}$$

or

$$\sigma_{cr} = \frac{\pi^2 E}{(L/r)^2} \qquad (13\text{--}6)$$

Here

σ_{cr} = critical stress, which is an average normal stress in the column just before the column buckles. This stress is an *elastic stress* and therefore $\sigma_{cr} \le \sigma_Y$

E = modulus of elasticity for the material

L = unsupported length of the column, whose ends are pinned

r = *smallest* radius of gyration of the column, determined from $r = \sqrt{I/A}$, where I is the *least* moment of inertia of the column's cross-sectional area A

Typical interior steel pipe columns used to support the roof of a single story building.

The geometric ratio L/r in Eq. 13–6 is known as the ***slenderness ratio***. It is a measure of the column's flexibility, and as will be discussed later, it serves to classify columns as long, intermediate, or short.

It is possible to graph Eq. 13–6 using axes that represent the critical stress versus the slenderness ratio. Examples of this graph for columns made of a typical structural steel and aluminum alloy are shown in Fig. 13–7. Note that the curves are hyperbolic and are valid only for critical stresses below the material's yield point (proportional limit), since the material must behave elastically. For the steel the yield stress is $(\sigma_Y)_{st} = 250$ MPa [$E_{st} = 200$ GPa], and for the aluminum it is $(\sigma_Y)_{al} = 190$ MPa [$E_{al} = 70$ GPa]. Substituting $\sigma_{cr} = \sigma_Y$ into Eq. 13–6, the *smallest* allowable slenderness ratios for the steel and aluminum columns are therefore $(L/r)_{st} = 89$ and $(L/r)_{al} = 60.5$, respectively. Thus, for a steel column, if $(L/r)_{st} \geq 89$, Euler's formula can be used to determine the critical load since the stress in the column remains elastic. On the other hand, if $(L/r)_{st} < 89$, the column's stress will exceed the yield point before buckling can occur, and therefore the Euler formula is not valid in this case.

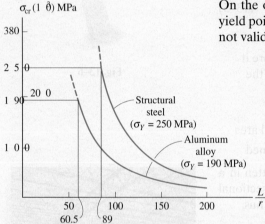

Fig. 13–7

Important Points

- *Columns* are long slender members that are subjected to axial compressive loads.

- The *critical load* is the maximum axial load that a column can support when it is on the verge of buckling. This loading represents a case of *neutral equilibrium*.

- An *ideal column* is initially perfectly straight, made of homogeneous material, and the load is applied through the centroid of the cross section.

- A pin-connected column will buckle about the principal axis of the cross section having the *least* moment of inertia.

- The *slenderness ratio* is L/r, where r is the smallest radius of gyration of the cross section. Buckling will occur about the axis where this ratio gives the greatest value.

EXAMPLE | 13.1

The A-36 steel W200 × 46 member shown in Fig. 13–8 is to be used as a pin-connected column. Determine the largest axial load it can support before it either begins to buckle or the steel yields.

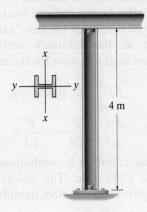

Fig. 13–8

SOLUTION

From the table in Appendix B, the column's cross-sectional area and moments of inertia are $A = 5890$ mm², $I_x = 45.5(10^6)$ mm⁴, and $I_y = 15.3(10^6)$ mm⁴. By inspection, buckling will occur about the y–y axis. Why? Applying Eq. 13–5, we have

$$P_{cr} = \frac{\pi^2 EI}{L^2} = \frac{\pi^2(200 \text{ kN/mm}^2)[15.3(10^6) \text{ mm}^4]}{[4 \text{ m}(1000 \text{ mm/m}]^2} = 1887.6 \text{ kN}$$

When fully loaded, the average compressive stress in the column is

$$\sigma_{cr} = \frac{P_{cr}}{A} = \frac{1887.6(10^3) \text{ N}}{5890 \text{ mm}^2} = 320.5 \text{ N/mm}^2 = 320.5 \text{ MPa}$$

Since this stress exceeds the yield stress (250 MPa), the load P is determined from simple compression:

$$250 \text{ N/mm}^2 = \frac{P}{5890 \text{ mm}^2} \qquad P = 1472.5 \text{ kN} = 1.47 \text{ MN} \qquad Ans.$$

In actual practice, a factor of safety would be placed on this loading.

13

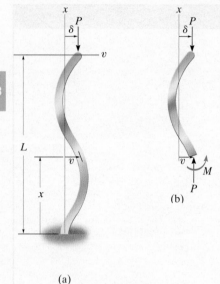

L

x

(a)

(b)

Fig. 13-9

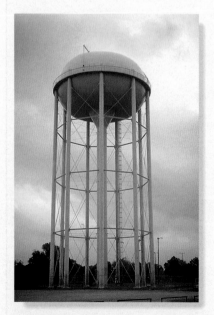

The tubular columns used to support this water tank have been braced at three locations along their length to prevent them from buckling.

13.3 Columns Having Various Types of Supports

The Euler load was derived for a column that is pin connected or free to rotate at its ends. Oftentimes, however, columns may be supported in some other way. For example, consider the case of a column fixed at its base and free at the top, Fig. 13-9a. As the column buckles the load displaces δ and at x the displacement is v. From the free-body diagram in Fig. 13-9b, the internal moment at the arbitrary section is $M = P(\delta - v)$. Consequently, the differential equation for the deflection curve is

$$EI\frac{d^2v}{dx^2} = P(\delta - v)$$

$$\frac{d^2v}{dx^2} + \frac{P}{EI}v = \frac{P}{EI}\delta \qquad (13-7)$$

Unlike Eq. 13-2, this equation is nonhomogeneous because of the nonzero term on the right side. The solution consists of both a complementary and a particular solution, namely,

$$v = C_1 \sin\left(\sqrt{\frac{P}{EI}}x\right) + C_2 \cos\left(\sqrt{\frac{P}{EI}}x\right) + \delta$$

The constants are determined from the boundary conditions. At $x = 0$, $v = 0$, so that $C_2 = -\delta$. Also,

$$\frac{dv}{dx} = C_1\sqrt{\frac{P}{EI}}\cos\left(\sqrt{\frac{P}{EI}}x\right) - C_2\sqrt{\frac{P}{EI}}\sin\left(\sqrt{\frac{P}{EI}}x\right)$$

At $x = 0$, $dv/dx = 0$, so that $C_1 = 0$. The deflection curve is therefore

$$v = \delta\left[1 - \cos\left(\sqrt{\frac{P}{EI}}x\right)\right] \qquad (13-8)$$

Since the deflection at the top of the column is δ, that is, at $x = L$, $v = \delta$, we require

$$\delta \cos\left(\sqrt{\frac{P}{EI}}L\right) = 0$$

The trivial solution $\delta = 0$ indicates that no buckling occurs, regardless of the load P. Instead,

$$\cos\left(\sqrt{\frac{P}{EI}}L\right) = 0 \qquad \text{or} \qquad \sqrt{\frac{P}{EI}}L = \frac{n\pi}{2}, n = 1, 3, 5\ldots$$

The smallest critical load occurs when $n = 1$, so that

$$P_{cr} = \frac{\pi^2 EI}{4L^2} \qquad (13-9)$$

By comparison with Eq. 13-5, it is seen that a column fixed-supported at its base and free at its top will support only one-fourth the critical load that can be applied to a column pin-supported at both ends.

Other types of supported columns are analyzed in much the same way and will not be covered in detail here.* Instead, we will tabulate the results for the most common types of column support and show how to apply these results by writing Euler's formula in a general form.

Effective Length. As stated previously, the Euler formula, Eq. 13–5, was developed for the case of a column having ends that are pinned or free to rotate. In other words, L in the equation represents the unsupported distance between the points of zero moment. This formula can be used to determine the critical load on columns having other types of support provided "L" represents the distance between the zero-moment points. This distance is called the column's **effective length**, L_e. Obviously, for a pin-ended column $L_e = L$, Fig. 13–10a. For the fixed- and free-ended column, the deflection curve, Eq. 13–8, was found to be one-half that of a column that is pin connected and has a length of $2L$, Fig. 13–10b. Thus the effective length between the points of zero moment is $L_e = 2L$. Examples for two other columns with different end supports are also shown in Fig. 13–10. The column fixed at its ends, Fig. 13–10c, has inflection points or points of zero moment $L/4$ from each support. The effective length is therefore represented by the middle half of its length, that is, $L_e = 0.5L$. Lastly, the pin- and fixed-ended column, Fig. 13–10d, has an inflection point at approximately $0.7L$ from its pinned end, so that $L_e = 0.7L$.

Rather than specifying the column's effective length, many design codes provide column formulas that employ a dimensionless coefficient K called the **effective-length factor**. This factor is defined from

$$L_e = KL \qquad (13\text{–}10)$$

Specific values of K are also given in Fig. 13–10. Based on this generality, we can therefore write Euler's formula as

$$P_{cr} = \frac{\pi^2 EI}{(KL)^2} \qquad (13\text{–}11)$$

or

$$\sigma_{cr} = \frac{\pi^2 E}{(KL/r)^2} \qquad (13\text{–}12)$$

Here (KL/r) is the column's **effective-slenderness ratio**. For example, if the column is fixed at its base and free at its end, we have $K = 2$, and therefore Eq. 13–11 gives the same result as Eq. 13–9.

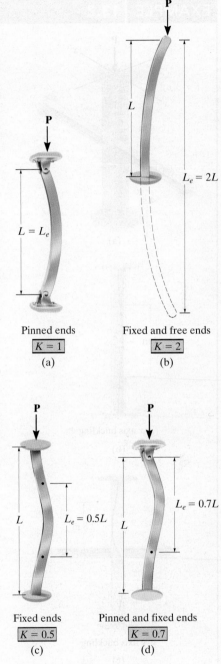

Pinned ends
$K = 1$
(a)

Fixed and free ends
$K = 2$
(b)

Fixed ends
$K = 0.5$
(c)

Pinned and fixed ends
$K = 0.7$
(d)

Fig. 13–10

Refer to the Companion Website for the animation of the concepts that can be found in Figure 13-10.)

13

EXAMPLE | 13.2

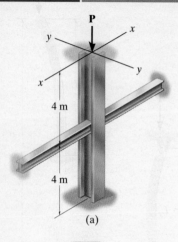

4 m

4 m

(a)

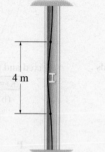

4 m

x–x axis buckling

(b)

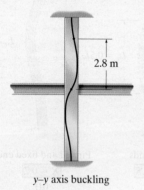

2.8 m

y–y axis buckling

(c)

Fig. 13–11

A W150 × 24 steel column is 8 m long and is fixed at its ends as shown in Fig. 13–11a. Its load-carrying capacity is increased by bracing it about the *y–y* (weak) axis using struts that are assumed to be pin connected to its midheight. Determine the load it can support so that the column does not buckle nor the material exceed the yield stress. Take $E_{st} = 200$ GPa and $\sigma_Y = 410$ MPa.

SOLUTION

The buckling behavior of the column will be *different* about the *x–x* and *y–y* axes due to the bracing. The buckled shape for each of these cases is shown in Figs. 13–11b and 13–11c. From Fig. 13–11b, the effective length for buckling about the *x–x* axis is $(KL)_x = 0.5(8\text{ m}) = 4\text{ m} = 4000$ mm, and from Fig. 13–11c, for buckling about the *y–y* axis, $(KL)_y = 0.7(8\text{ m}/2) = 2.8\text{ m} = 2800$ mm. The moments of inertia for a W150 × 24 are found from the table in Appendix B. We have $I_x = 13.4(10^6)$ mm⁴, $I_y = 1.83(10^6)$ mm⁴.

Applying Eq. 13–11,

$$(P_{cr})_x = \frac{\pi^2 EI}{(KL)_y^2} = \frac{\pi^2 (200 \text{ kN/mm}^2)[13.4(10^6) \text{ mm}^4]}{(4000 \text{ mm})^2} = 1653.2 \text{ kN} \quad (1)$$

$$(P_{cr})_y = \frac{\pi^2 EI}{(KL)_y^2} = \frac{\pi^2 (200 \text{ kN/mm}^2)[1.83(10^6) \text{ mm}^4]}{(2800 \text{ mm})^2} = 460.8 \text{ kN} \quad (2)$$

By comparison, buckling will occur about the *y–y* axis.

The area of the cross section is 3060 mm², so the average compressive stress in the column is

$$\sigma_{cr} = \frac{P_{er}}{A} = \frac{460.8(10^3) \text{ N}}{3060 \text{ mm}^2} = 150.6 \text{ N/mm}^2 = 150.6 \text{ MPa}$$

Since this stress is less than the yield stress, buckling will occur before the material yields. Thus,

$$P_{cr} = 461 \text{ kN} \qquad\qquad Ans.$$

NOTE: From Eq. 13–12 it can be seen that buckling will always occur about the column axis having the *largest* slenderness ratio, since a large slenderness ratio will give a small critical stress. Thus, using the data for the radius of gyration from the table in Appendix B, we have

$$\left(\frac{KL}{r}\right)_x = \frac{4000 \text{ mm}}{66.2 \text{ mm}} = 60.4$$

$$\left(\frac{KL}{r}\right)_y = \frac{2800 \text{ mm}}{24.5 \text{ mm}} = 114.3$$

Hence, *y–y* axis buckling will occur, which is the same conclusion reached by comparing Eqs. 1 and 2.

EXAMPLE | 13.3

The aluminum column is fixed at its bottom and is braced at its top by cables so as to prevent movement at the top along the x axis, Fig. 13–12a. If it is assumed to be fixed at its base, determine the largest allowable load P that can be applied. Use a factor of safety for buckling of F.S. = 3.0. Take E_{al} = 70 GPa, σ_Y = 215 MPa, $A = 7.5(10^{-3})$ m², $I_x = 61.3(10^{-6})$ m⁴, $I_y = 23.2(10^{-6})$ m⁴.

SOLUTION
Buckling about the x and y axes is shown in Figs. 13–12b and 13–12c, respectively. Using Fig. 13–10a, for x–x axis buckling, $K = 2$, so $(KL)_x = 2(5 \text{ m}) = 10$ m. Also, for y–y axis buckling, $K = 0.7$, so $(KL)_y = 0.7(5 \text{ m}) = 3.5$ m.

Applying Eq. 13–11, the critical loads for each case are

$$(P_{cr})_x = \frac{\pi^2 E I_x}{(KL)_x^2} = \frac{\pi^2[70(10^9) \text{ N/m}^2](61.3(10^{-6}) \text{ m}^4)}{(10 \text{ m})^2}$$

$$= 424 \text{ kN}$$

$$(P_{cr})_y = \frac{\pi^2 E I_y}{(KL)_y^2} = \frac{\pi^2[70(10^9) \text{ N/m}^2](23.2(10^{-6}) \text{ m}^4)}{(3.5 \text{ m})^2}$$

$$= 1.31 \text{ MN}$$

By comparison, as P is increased the column will buckle about the x–x axis. The allowable load is therefore

$$P_{allow} = \frac{P_{cr}}{\text{F.S.}} = \frac{424 \text{ kN}}{3.0} = 141 \text{ kN} \qquad Ans.$$

Since

$$\sigma_{cr} = \frac{P_{cr}}{A} = \frac{424 \text{ kN}}{7.5(10^{-3}) \text{ m}^2} = 56.5 \text{ MPa} < 215 \text{ MPa}$$

Euler's equation can be applied.

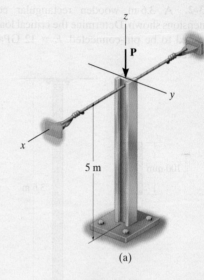

5 m

(a)

Fig. 13–12

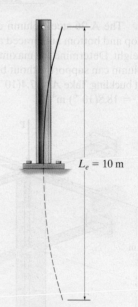

L_e = 10 m

x–x axis buckling

(b)

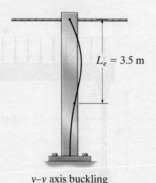

L_e = 3.5 m

y–y axis buckling

(c)

13

FUNDAMENTAL PROBLEMS

F13–1. A 1.25-m-long rod is made from a 25-mm-diameter steel rod. Determine the critical buckling load if the ends are fixed supported. $E = 200$ GPa, $\sigma_Y = 250$ MPa.

F13–2. A 3.6-m wooden rectangular column has the dimensions shown. Determine the critical load if the ends are assumed to be pin-connected. $E = 12$ GPa. Yielding does not occur.

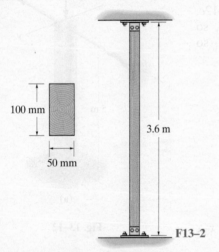

100 mm

3.6 m

50 mm

F13–2

F13–3. The A-36 steel column can be considered pinned at its top and bottom and braced against its weak axis at the mid-height. Determine the maximum allowable force **P** that the column can support without buckling. Apply a F.S. = 2 against buckling. Take $A = 7.4(10^{-3})$ m^2, $I_x = 87.3(10^{-6})$ m^4, and $I_y = 18.8(10^{-6})$ m^4.

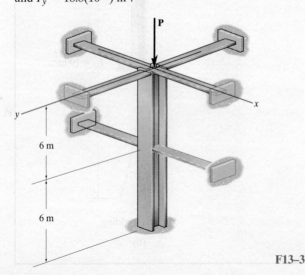

P

y

x

6 m

6 m

F13–3

F13–4. A steel pipe is fixed supported at its ends. If it is 5 m long and has an outer diameter of 50 mm and a thickness of 10 mm, determine the maximum axial load P that it can carry without buckling. $E_{st} = 200$ GPa, $\sigma_Y = 250$ MPa.

F13–5. Determine the maximum force **P** that can be supported by the assembly without causing member AC to buckle. The member is made of A-36 steel and has a diameter of 50 mm. Take F.S. = 2 against buckling.

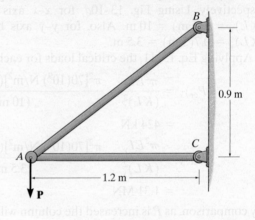

B

0.9 m

C

A

1.2 m

P

F13–5

F13–6. The A-36 steel rod BC has a diameter of 50 mm and is used as a strut to support the beam. Determine the maximum intensity w of the uniform distributed load that can be applied to the beam without causing the strut to buckle. Take F.S. = 2 against buckling.

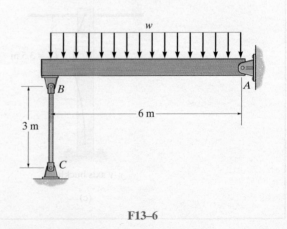

w

B

A

3 m

6 m

C

F13–6

PROBLEMS

•13–1.   Determine the critical buckling load for the column. The material can be assumed rigid.

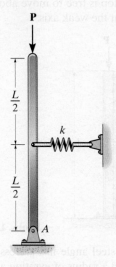

Prob. 13–1

13–2.   Determine the critical load P_{cr} for the rigid bar and spring system. Each spring has a stiffness k.

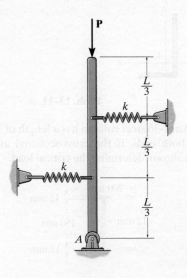

Prob. 13–2

13–3.   The leg in (a) acts as a column and can be modeled (b) by the two pin-connected members that are attached to a torsional spring having a stiffness k (torque/rad). Determine the critical buckling load. Assume the bone material is rigid.

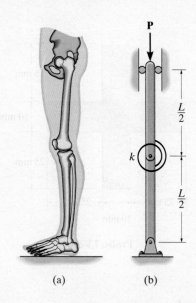

(a)   (b)

Prob. 13–3

*13–4.   Rigid bars AB and BC are pin connected at B. If the spring at D has a stiffness k, determine the critical load P_{cr} for the system.

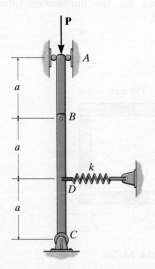

Prob. 13–4

13

•13–5. An A-36 steel column has a length of 4 m and is pinned at both ends. If the cross-sectional area has the dimensions shown, determine the critical load.

13–6. Solve Prob. 13–5 if the column is fixed at its bottom and pinned at its top.

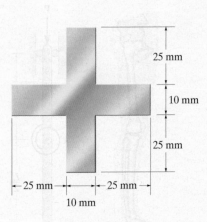

25 mm

10 mm

25 mm

25 mm

25 mm

10 mm

Probs. 13–5/6

13–7. A column is made of A-36 steel, has a length of 6 m, and is pinned at both ends. If the cross-sectional area has the dimensions shown, determine the critical load.

*13–8. A column is made of 2014-T6 aluminum, has a length of 9 m, and is fixed at its bottom and pinned at its top. If the cross-sectional area has the dimensions shown, determine the critical load.

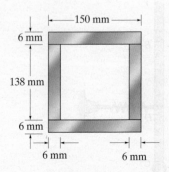

150 mm

6 mm

138 mm

6 mm

6 mm   6 mm

Probs. 13–7/8

•13–9. The W360 × 57 column is made of A-36 steel and is fixed supported at its base. If it is subjected to an axial load of $P = 75$ kN, determine the factor of safety with respect to buckling.

13–10. The W360 × 57 column is made of A-36 steel. Determine the critical load if its bottom end is fixed supported and its top is free to move about the strong axis and is pinned about the weak axis.

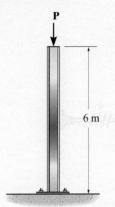

P

6 m

Probs. 13–9/10

13–11. The A-36 steel angle has a cross-sectional area of $A = 1550$ mm^2 and a radius of gyration about the x axis of $r_x = 31.5$ mm and about the y axis of $r_y = 21.975$ mm. The smallest radius of gyration occurs about the z axis and is $r_z = 16.1$ mm. If the angle is to be used as a pin-connected 3-m-long column, determine the largest axial load that can be applied through its centroid C without causing it to buckle.

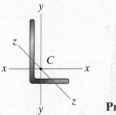

y

z

x ——— C ——— x

z

y   Prob. 13–11

*13–12. An A-36 steel column has a length of 4.5 m and is pinned at both ends. If the cross-sectional area has the dimensions shown, determine the critical load.

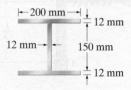

200 mm

12 mm

12 mm   150 mm

12 mm

Prob. 13–12

•13–13. An A-36 steel column has a length of 5 m and is fixed at both ends. If the cross-sectional area has the dimensions shown, determine the critical load.

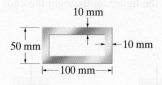

Prob. 13–13

13–14. The two steel channels are to be laced together to form a 9-m-long bridge column assumed to be pin connected at its ends. Each channel has a cross-sectional area of $A = 1950$ mm^2 and moments of inertia $I_x = 21.60(10^6)$ mm^4, $I_y = 0.15(10^6)$ mm^4. The centroid C of its area is located in the figure. Determine the proper distance d between the centroids of the channels so that buckling occurs about the x–x and y'–y' axes due to the same load. What is the value of this critical load? Neglect the effect of the lacing. $E_{st} = 200$ GPa, $\sigma_Y = 350$ MPa.

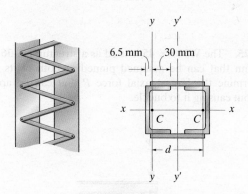

Prob. 13–14

13–15. An A-36-steel W200 × 36 column is fixed at one end and free at its other end. If it is subjected to an axial load of 100 kN, determine the maximum allowable length of the column if F.S. = 2 against buckling is desired.

*13–16. An A-36-steel W200 × 36 column is fixed at one end and pinned at the other end. If it is subjected to an axial load of 300 kN, determine the maximum allowable length of the column if F.S. = 2 against buckling is desired.

•13–17. The 3-m wooden rectangular column has the dimensions shown. Determine the critical load if the ends are assumed to be pin connected. $E_w = 12$ GPa, $\sigma_Y = 35$ MPa.

13–18. The 3-m column has the dimensions shown. Determine the critical load if the bottom is fixed and the top is pinned. $E_w = 12$ GPa, $\sigma_Y = 35$ MPa.

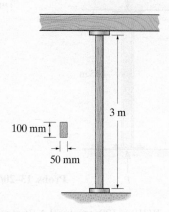

Probs. 13–17/18

13–19. Determine the maximum force P that can be applied to the handle so that the A-36 steel control rod BC does not buckle. The rod has a diameter of 25 mm.

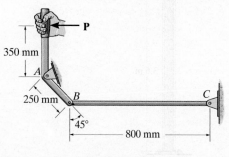

Prob. 13–19

*13–20. The W250 × 67 is made of A-36 steel and is used as a column that has a length of 4.5 m. If its ends are assumed pin supported, and it is subjected to an axial load of 500 kN, determine the factor of safety with respect to buckling.

•13–21. The W250 × 67 is made of A-36 steel and is used as a column that has a length of 4.5 m. If the ends of the column are fixed supported, can the column support the critical load without yielding?

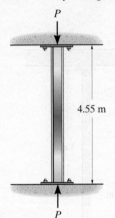

Probs. 13–20/21

13–22. The W310 × 129 structural A-36 steel column has a length of 3.6 m. If its bottom end is fixed supported while its top is free, and it is subjected to an axial load of P = 1900 kN determine the factor of safety with respect to buckling.

13–23. The W310 × 129 structural A-36 steel column has a length of 3.6 m. If its bottom end is fixed supported while its top is free, determine the largest axial load it can support. Use a factor of safety with respect to buckling of 1.75.

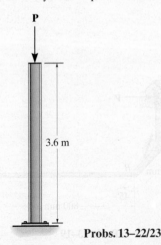

Probs. 13–22/23

*13–24. An L-2 tool steel link in a forging machine is pin connected to the forks at its ends as shown. Determine the maximum load P it can carry without buckling. Use a factor of safety with respect to buckling of F.S. = 1.75. Note from the figure on the left that the ends are pinned for buckling, whereas from the figure on the right the ends are fixed.

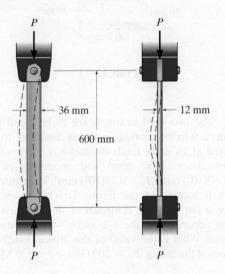

36 mm

12 mm

600 mm

Prob. 13–24

•13–25. The W360 × 45 is used as a structural A-36 steel column that can be assumed pinned at both of its ends. Determine the largest axial force P that can be applied without causing it to buckle.

7.5 m

Prob. 13–25

13–26. The A-36 steel bar AB has a square cross section. If it is pin connected at its ends, determine the maximum allowable load P that can be applied to the frame. Use a factor of safety with respect to buckling of 2.

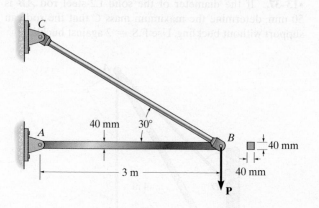

Prob. 13–26

13–27. Determine the maximum allowable intensity w of the distributed load that can be applied to member BC without causing member AB to buckle. Assume that AB is made of steel and is pinned at its ends for x–x axis buckling and fixed at its ends for y–y axis buckling. Use a factor of safety with respect to buckling of 3. $E_{st} = 200$ GPa, $\sigma_Y = 360$ MPa.

***13–28.** Determine if the frame can support a load of $w = 6$ kN/m if the factor of safety with respect to buckling of member AB is 3. Assume that AB is made of steel and is pinned at its ends for x–x axis buckling and fixed at its ends for y–y axis buckling. $E_{st} = 200$ GPa, $\sigma_Y = 360$ MPa.

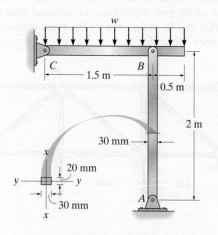

Probs. 13–27/28

•**13–29.** The beam supports the load of $P = 30$ kN. As a result, the A-36 steel member BC is subjected to a compressive load. Due to the forked ends on the member, consider the supports at B and C to act as pins for x–x axis buckling and as fixed supports for y–y axis buckling. Determine the factor of safety with respect to buckling about each of these axes.

13–30. Determine the greatest load P the frame will support without causing the A-36 steel member BC to buckle. Due to the forked ends on the member, consider the supports at B and C to act as pins for x–x axis buckling and as fixed supports for y–y axis buckling.

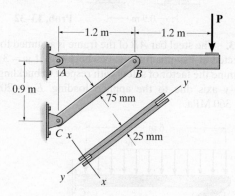

Probs. 13–29/30

13–31. Determine the maximum distributed load that can be applied to the bar so that the A-36 steel strut AB does not buckle. The strut has a diameter of 50 mm. It is pin connected at its ends.

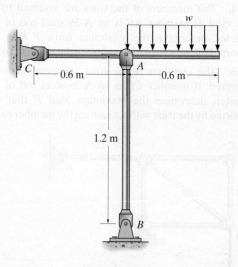

Prob. 13–31

*13–32. The members of the truss are assumed to be pin connected. If member *AC* is an A-36 steel rod of 50 mm diameter, determine the maximum load *P* that can be supported by the truss without causing the member to buckle.

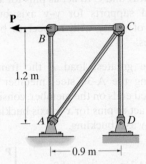

1.2 m

0.9 m

Prob. 13–32

•13–33. The steel bar *AB* of the frame is assumed to be pin connected at its ends for *y–y* axis buckling. If *w* = 3 kN/m, determine the factor of safety with respect to buckling about the *y–y* axis due to the applied loading. E_{st} = 200 GPa, σ_Y = 360 MPa.

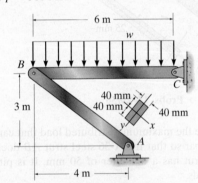

6 m

w

B

3 m

40 mm
40 mm
40 mm

y   *x*

A

4 m

Prob. 13–33

13–34. The members of the truss are assumed to be pin connected. If member *AB* is an A-36 steel rod of 40 mm diameter, determine the maximum force *P* that can be supported by the truss without causing the member to buckle.

13–35. The members of the truss are assumed to be pin connected. If member *CB* is an A-36 steel rod of 40 mm diameter, determine the maximum load *P* that can be supported by the truss without causing the member to buckle.

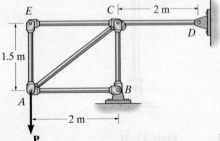

E

C   2 m

D

1.5 m

B

A

2 m

P

Probs. 13–34/35

*13–36. If load *C* has a mass of 500 kg, determine the required minimum diameter of the solid L2-steel rod *AB* to the nearest mm so that it will not buckle. Use F.S. = 2 against buckling.

•13–37. If the diameter of the solid L2-steel rod *AB* is 50 mm, determine the maximum mass *C* that the rod can support without buckling. Use F.S. = 2 against buckling.

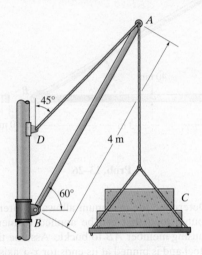

A

45°

D

4 m

60°

B

C

Probs. 13–36/37

13–38. The members of the truss are assumed to be pin connected. If member *GF* is an A-36 steel rod having a diameter of 50 mm, determine the greatest magnitude of load **P** that can be supported by the truss without causing this member to buckle.

13–39. The members of the truss are assumed to be pin connected. If member *AG* is an A-36 steel rod having a diameter of 50 mm, determine the greatest magnitude of load **P** that can be supported by the truss without causing this member to buckle.

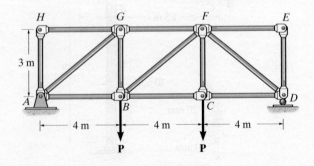

H   *G*   *F*   *E*

3 m

A   *B*   *C*   *D*

4 m   4 m   4 m

P   *P*

Probs. 13–38/39

*13–40. The column is supported at B by a support that does not permit rotation but allows vertical deflection. Determine the critical load P_{cr}. EI is constant.

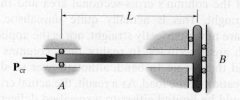

Prob. 13–40

•13–41. The ideal column has a weight w (force/length) and rests in the horizontal position when it is subjected to the axial load P. Determine the maximum moment in the column at midspan. EI is constant. *Hint:* Establish the differential equation for deflection, Eq. 13–1, with the origin at the mid span. The general solution is $v = C_1 \sin kx + C_2 \cos kx + (w/(2P))x^2 - (wL/(2P))x - (wEI/P^2)$ where $k^2 = P/EI$.

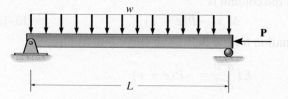

Prob. 13–41

13–42. The ideal column is subjected to the force $\mathbf{F}$ at its midpoint and the axial load $\mathbf{P}$. Determine the maximum moment in the column at midspan. EI is constant. *Hint:* Establish the differential equation for deflection, Eq. 13–1. The general solution is $v = C_1 \sin kx + C_2 \cos kx - c^2 x/k^2$, where $c^2 = F/2EI$, $k^2 = P/EI$.

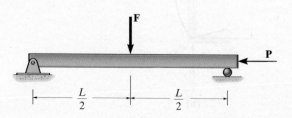

Prob. 13–42

13–43. The column with constant EI has the end constraints shown. Determine the critical load for the column.

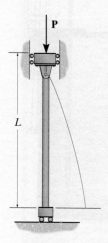

Prob. 13–43

*13–44. Consider an ideal column as in Fig. 13–10c, having both ends fixed. Show that the critical load on the column is given by $P_{cr} = 4\pi^2 EI/L^2$. *Hint:* Due to the vertical deflection of the top of the column, a constant moment M' will be developed at the supports. Show that $d^2v/dx^2 + (P/EI)v = M'/EI$. The solution is of the form $v = C_1 \sin(\sqrt{P/EI}x) + C_2 \cos(\sqrt{P/EI}x) + M'/P$.

•13–45. Consider an ideal column as in Fig. 13–10d, having one end fixed and the other pinned. Show that the critical load on the column is given by $P_{cr} = 20.19EI/L^2$. *Hint:* Due to the vertical deflection at the top of the column, a constant moment M' will be developed at the fixed support and horizontal reactive forces $\mathbf{R}'$ will be developed at both supports. Show that $d^2v/dx^2 + (P/EI)v = (R'/EI)(L - x)$. The solution is of the form $v = C_1 \sin(\sqrt{P/EI}x) + C_2 \cos(\sqrt{P/EI}x) + (R'/P)(L - x)$. After application of the boundary conditions show that $\tan(\sqrt{P/EI}L) = \sqrt{P/EI}\,L$. Solve by trial and error for the smallest nonzero root.

13

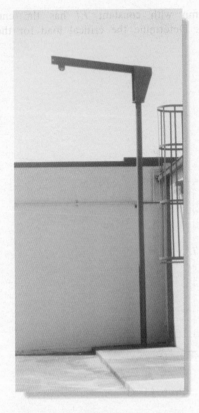

The column supporting this crane is unusually long. It will be subjected not only to uniaxial load, but also a bending moment. To ensure it will not buckle, it should be braced at the roof as a pin connection.

*13.4 The Secant Formula

The Euler formula was derived assuming the load P is always applied through the centroid of the column's cross-sectional area and that the column is perfectly straight. This is actually quite unrealistic, since manufactured columns are never perfectly straight, nor is the application of the load known with great accuracy. In reality, then, columns never suddenly buckle; instead they begin to bend, although ever so slightly, immediately upon application of the load. As a result, the actual criterion for load application should be limited either to a specified deflection of the column or by not allowing the maximum stress in the column to exceed an allowable stress.

To study this effect, we will apply the load P to the column at a short *eccentric distance* e from its centroid, Fig. 13–13a. This loading on the column is statically equivalent to the axial load P and bending moment $M' = Pe$ shown in Fig. 13–13b. As shown, in both cases, the ends A and B are supported so that they are free to rotate (pin supported). As before, we will only consider small slopes and deflections and linear-elastic material behavior. Furthermore, the x–v plane is a plane of symmetry for the cross-sectional area.

From the free-body diagram of the arbitrary section, Fig. 13–13c, the internal moment in the column is

$$M = -P(e + v) \qquad (13\text{–}13)$$

The differential equation for the deflection curve is therefore

$$EI\frac{d^2v}{dx^2} = -P(e + v)$$

Fig. 13–13

or

$$\frac{d^2v}{dx^2} + \frac{P}{EI}v = -\frac{P}{EI}e$$

This equation is similar to Eq. 13–7 and has a general solution consisting of the complementary and particular solutions, namely,

$$v = C_1 \sin\sqrt{\frac{P}{EI}}x + C_2 \cos\sqrt{\frac{P}{EI}}x - e \qquad (13\text{–}14)$$

To evaluate the constants we must apply the boundary conditions. At $x = 0$, $v = 0$, so $C_2 = e$. And at $x = L$, $v = 0$, which gives

$$C_1 = \frac{e[1 - \cos(\sqrt{P/EI}\,L)]}{\sin(\sqrt{P/EI}\,L)}$$

Since $1 - \cos(\sqrt{P/EI}\,L) = 2\sin^2(\sqrt{P/EI}\,L/2)$ and $\sin(\sqrt{P/EI}\,L) = 2\sin(\sqrt{P/EI}\,L/2)\cos(\sqrt{P/EI}\,L/2)$, we have

$$C_1 = e\tan\left(\sqrt{\frac{P}{EI}}\frac{L}{2}\right)$$

Hence, the deflection curve, Eq. 13–14, can be written as

$$v = e\left[\tan\left(\sqrt{\frac{P}{EI}}\frac{L}{2}\right)\sin\left(\sqrt{\frac{P}{EI}}x\right) + \cos\left(\sqrt{\frac{P}{EI}}x\right) - 1\right] \qquad (13\text{–}15)$$

Maximum Deflection. Due to symmetry of loading, both the maximum deflection and maximum stress occur at the column's midpoint. Therefore, when $x = L/2$, $v = v_{max}$, so

$$v_{max} = e\left[\sec\left(\sqrt{\frac{P}{EI}}\frac{L}{2}\right) - 1\right] \qquad (13\text{–}16)$$

Notice that if e approaches zero, then v_{max} approaches zero. However, if the terms in the brackets approach infinity as e approaches zero, then v_{max} will have a nonzero value. Mathematically, this would represent the behavior of an axially loaded column at failure when subjected to the critical load P_{cr}. Therefore, to find P_{cr} we require

$$\sec\left(\sqrt{\frac{P_{cr}}{EI}}\frac{L}{2}\right) = \infty$$

$$\sqrt{\frac{P_{cr}}{EI}}\frac{L}{2} = \frac{\pi}{2}$$

$$P_{cr} = \frac{\pi^2 EI}{L^2} \qquad (13\text{–}17)$$

which is the same result found from the Euler formula, Eq. 13–5.

If Eq. 13–16 is plotted as load P versus deflection v_{max} for various values of eccentricity e, the family of colored curves shown in Fig. 13–14 results. Here the critical load becomes an asymptote to the curves, and of

13

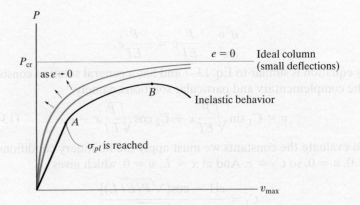

Fig. 13–14

course represents the unrealistic case of an ideal column ($e = 0$). As stated earlier, e is *never zero* due to imperfections in initial column straightness and load application; however, as $e \rightarrow 0$, the curves tend to approach the ideal case. Furthermore, these curves are appropriate only for *small deflections*, since the curvature was approximated by d^2v/dx^2 when Eq. 13–16 was developed. Had a more exact analysis been performed, all these curves would tend to turn upward, intersecting and then rising above the line $P = P_{cr}$. This, of course, indicates that a larger load P is needed to create larger column deflections. We have not considered this analysis here, however, since most often engineering design restricts the deflection of columns to small values.

It should also be noted that the colored curves in Fig. 13–14 apply only for linear-elastic material behavior. Such is the case if the column is long and slender. However, if a short or intermediate-length stocky column is considered, then the applied load, as it is increased, may eventually cause the material to yield, and the column will begin to behave in an *inelastic manner*. This occurs at point A for the black curve in Fig. 13–14. As the load is further increased, the curve never reaches the critical load, and instead the load reaches a maximum value at B. Afterwards, a sudden decrease in load-carrying capacity occurs as the column continues to yield and deflect by larger amounts.

Lastly, the colored curves in Fig. 13–14 also illustrate that a *nonlinear* relationship occurs between the load P and the deflection v. As a result, the principle of superposition *cannot be used* to determine the total deflection of a column caused by applying *successive loads* to the column. Instead, the loads must first be added, and then the corresponding deflection due to their resultant can be determined. Physically, the reason that successive loads and deflections cannot be superimposed is that the column's internal moment *depends* on *both* the *load P* and the *deflection v*, that is, $M = -P(e + v)$, Eq. 13–13.

The Secant Formula.

The maximum stress in the column can be determined by realizing that it is caused by both the axial load and the moment, Fig. 13–15a. Maximum moment occurs at the column's midpoint, and using Eqs. 13–13 and 13–16, it has a magnitude of

$$M = |P(e + v_{max})| \qquad M = Pe \sec\left(\sqrt{\frac{P}{EI}}\frac{L}{2}\right) \qquad (13\text{–}18)$$

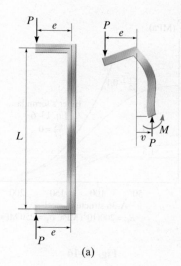

As shown in Fig. 13–15b, the maximum stress in the column is compressive, and it has a value of

$$\sigma_{max} = \frac{P}{A} + \frac{Mc}{I}; \qquad \sigma_{max} = \frac{P}{A} + \frac{Pec}{I} \sec\left(\sqrt{\frac{P}{EI}}\frac{L}{2}\right)$$

Since the radius of gyration is defined as $r^2 = I/A$, the above equation can be written in a form called the *secant formula*:

$$\sigma_{max} = \frac{P}{A}\left[1 + \frac{ec}{r^2} \sec\left(\frac{L}{2r}\sqrt{\frac{P}{EA}}\right)\right] \qquad (13\text{–}19)$$

Here

σ_{max} = maximum *elastic stress* in the column, which occurs at the inner concave side at the column's midpoint. This stress is compressive

P = vertical load applied to the column. $P < P_{cr}$ unless $e = 0$; then $P = P_{cr}$ (Eq. 13–5)

e = eccentricity of the load P, measured from the centroidal axis of the column's cross-sectional area to the line of action of P

c = distance from the centroidal axis to the outer fiber of the column where the maximum compressive stress σ_{max} occurs

A = cross-sectional area of the column

L = unsupported length of the column *in the plane of bending*. For supports other than pins, the effective length $L_e = KL$ should be used. See Fig. 13–10

E = modulus of elasticity for the material

r = radius of gyration, $r = \sqrt{I/A}$, where I is calculated about the centroidal or bending axis

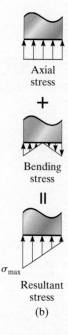

Axial
stress

+

Bending
stress

=

σ_{max}
Resultant
stress
(b)

Fig. 13–15

Like Eq. 13–16, Eq. 13–19 indicates that there is a nonlinear relationship between the load and the stress. Hence, the principle of superposition does not apply, and therefore the loads have to be added *before* the stress is determined. Furthermore, due to this nonlinear relationship, any factor of safety used for design purposes applies to the load and not to the stress.

For a given value of σ_{max}, graphs of Eq. 13–19 can be plotted as the slenderness ratio KL/r versus the average stress P/A for various values of the *eccentricity ratio* ec/r^2. A specific set of graphs for a structural-grade A-36 steel having a yield point of $\sigma_{max} = \sigma_Y = 250$ MPa

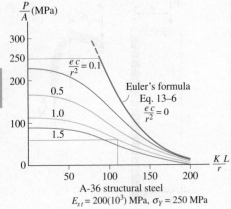

Fig. 13–16

A-36 structural steel
$E_{st} = 200(10^3)$ MPa, $\sigma_Y = 250$ MPa

and a modulus of elasticity of $E_{st} = 200$ GPa is shown in Fig. 13–16. Note that when $e \to 0$, or when $ec/r^2 \to 0$, Eq. 13–19 gives $\sigma_{max} = P/A$, where P is the critical load on the column, defined by Euler's formula. This results in Eq. 13–6, which has been plotted in Fig. 13–7 and repeated in Fig. 13–16. Since both Eqs. 13–6 and 13–19 are valid only for elastic loadings, the stresses shown in Fig. 13–16 cannot exceed $\sigma_Y = 250$ GPa, represented here by the horizontal line.

The curves in Fig. 13–16 indicate that differences in the eccentricity ratio have a marked effect on the load-carrying capacity of columns that have small slenderness ratios. However, columns that have large slenderness ratios tend to fail at or near the Euler critical load regardless of the eccentricity ratio. When using Eq. 13–19 for design purposes, it is therefore important to have a somewhat accurate value for the eccentricity ratio for shorter-length columns.

Design. Once the eccentricity ratio has been determined, the column data can be substituted into Eq. 13–19. If a value of $\sigma_{max} = \sigma_Y$ is chosen, then the corresponding load P_Y can be determined from a trial-and-error procedure, since the equation is transcendental and cannot be solved explicitly for P_Y. As a design aid, computer software, or graphs such as those in Fig. 13–16, can also be used to determine P_Y directly.

Realize that P_Y is the load that will cause the column to develop a maximum compressive stress of σ_Y at its inner concave fibers. Due to the eccentric application of P_Y, this load will always be smaller than the critical load P_{cr}, which is determined from the Euler formula that assumes (unrealistically) that the column is axially loaded. Once P_Y is obtained, an appropriate factor of safety can then be applied in order to specify the column's safe load.

Important Points

- Due to imperfections in manufacturing or specific application of the load, a column will never suddenly buckle; instead, it begins to bend.

- The load applied to a column is related to its deflection in a nonlinear manner, and so the principle of superposition does not apply.

- As the slenderness ratio increases, eccentrically loaded columns tend to fail at or near the Euler buckling load.

EXAMPLE 13.4

The W200 × 59 A-36 steel column shown in Fig. 13–17a is fixed at its base and braced at the top so that it is fixed from displacement, yet free to rotate about the y–y axis. Also, it can sway to the side in the y–z plane. Determine the maximum eccentric load the column can support before it either begins to buckle or the steel yields.

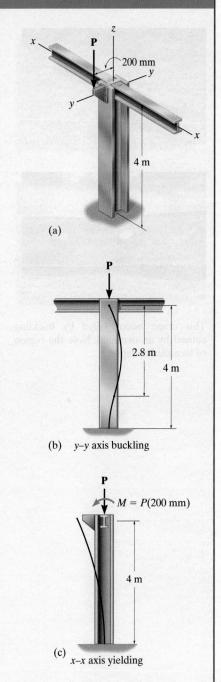

(a)

SOLUTION

From the support conditions it is seen that about the y–y axis the column behaves as if it were pinned at its top and fixed at the bottom and subjected to an axial load P, Fig. 13–17b. About the x–x axis the column is free at the top and fixed at the bottom, and it is subjected to both an axial load P and moment $M = P(200 \text{ mm})$, Fig. 13–17c.

y–y Axis Buckling. From Fig. 13–10d the effective length factor is $K_y = 0.7$, so $(KL)_y = 0.7(4 \text{ m}) = 2.8 \text{ m} = 2800 \text{ mm}$. Using the table in Appendix B to determine I_y for the W200 × 59 section and applying Eq. 13–11, we have

$$(P_{cr})_y = \frac{\pi^2 EI}{(KL)_y^2} = \frac{\pi^2 (200 \text{ kN/mm}^2)[20.4(10^6) \text{ mm}^4]}{(2800 \text{ mm})^2} = 5136 \text{ kN}$$

x–x Axis Yielding. From Fig. 13–10b, $K_x = 2$, so $(KL)_x = 2(4 \text{ m}) = 8 \text{ m} = 8000 \text{ mm}$. Again using the table in Appendix B to determine $A = 7580 \text{ mm}^2$, $c = 210 \text{ mm}/2 = 105 \text{ mm}$, and $r_x = 89.9 \text{ mm}$ and applying the secant formula, we have

$$\sigma_Y = \frac{P_x}{A} \left[1 + \frac{ec}{r_x^2} \sec\left(\frac{(KL)_x}{2r_x} \sqrt{\frac{P_x}{EA}} \right) \right]$$

$$250(10^{-3}) \text{ kN/mm}^2 = \frac{P_x}{7580 \text{ mm}^2} \left[1 + \frac{(200 \text{ mm})(105 \text{ mm})}{(89.9 \text{ mm})^2} \sec \left(\frac{8000 \text{ mm}}{2(89.8 \text{ mm})} \sqrt{\frac{P_x}{200 \text{ kN/mm}^2(7580 \text{ mm}^2)}} \right) \right]$$

(b)   y–y axis buckling

Substituting the data and simplifying yields

$$1895 = P_x[1 + 2.598 \sec(0.03618\sqrt{P_x})]$$

Solving for P_x by trial and error, noting that the argument for the secant is in radians, we get

$$P_x = 419.4 \text{ kN} \qquad\qquad Ans.$$

Since this value is less than $(P_{cr})_y = 5136 \text{ kN}$ failure will occur about the x–x axis.

(c)   x–x axis yielding

Fig. 13–17

*13.5   Inelastic Buckling

In engineering practice, columns are generally classified according to the type of stresses developed within the column at the time of failure. *Long slender columns* will become unstable when the compressive stress remains elastic. The failure that occurs is referred to as *elastic instability*. *Intermediate columns* fail due to *inelastic instability*, meaning that the compressive stress at failure is greater than the material's proportional limit. And *short columns*, sometimes called *posts*, do not become unstable; rather the material simply yields or fractures.

Application of the Euler equation requires that the stress in the column remain *below* the material's yield point (actually the proportional limit) when the column buckles, and so this equation applies only to long columns. In practice, however, most columns are selected to have intermediate lengths. The behavior of these columns can be studied by modifying the Euler equation so that it applies for inelastic buckling. To show how this can be done, consider the material to have a stress–strain diagram as shown in Fig. 13–18a. Here the proportional limit is σ_{pl}, and the modulus of elasticity, or slope of the line AB, is E.

If the column has a slenderness ratio that is *less* than $(KL/r)_{pl}$, then the critical stress in the column must be greater than σ_{pl}. For example, suppose a column has a slenderness ratio of $(KL/r)_1 < (KL/r)_{pl}$, with corresponding critical stress $\sigma_D > \sigma_{pl}$ needed to cause instability. When the column is *about to buckle*, the change in stress and strain that occurs in the column is within a *small range* $\Delta\sigma$ and $\Delta\epsilon$, so that the modulus of elasticity or stiffness for the material can be taken as the ***tangent modulus*** $E_t = \Delta\sigma/\Delta\epsilon$ defined as the slope of the $\sigma-\epsilon$ diagram at point D, Fig. 13–18a. In other words, at the time of failure, the column behaves as if it were made from a material that has a *lower stiffness* than when it behaves elastically, $E_t < E$.

This crane boom failed by buckling caused by an overload. Note the region of localized collapse.

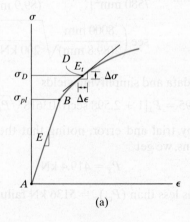

(a)

Fig. 13–18

In general, therefore, as the slenderness ratio (KL/r) decreases, the *critical stress* for a column continues to rise; and from the $\sigma-\epsilon$ diagram, the *tangent modulus* for the material *decreases*. Using this idea, we can modify Euler's equation to include these cases of inelastic buckling by substituting the material's tangent modulus E_t for E, so that

$$\sigma_{cr} = \frac{\pi^2 E_t}{(KL/r)^2} \tag{13–20}$$

This is the so-called *tangent modulus* or *Engesser equation*, proposed by F. Engesser in 1889. A plot of this equation for intermediate and short-length columns of a material defined by the $\sigma-\epsilon$ diagram in Fig. 13–18a is shown in Fig. 13–18b.

No *actual column* can be considered to be either perfectly straight or loaded along its centroidal axis, as assumed here, and therefore it is indeed very difficult to develop an expression that will provide a complete analysis of this phenomenon. As a result, other methods of describing the inelastic buckling of columns have been considered. One of these methods was developed by the aeronautical engineer F. R. Shanley and is called the *Shanley theory* of inelastic buckling. Although it provides a better description of the phenomenon than the tangent modulus theory, as explained here, experimental testing of a large number of columns, each of which approximates the ideal column, has shown that Eq. 13–20 is *reasonably accurate* in predicting the column's critical stress. Furthermore, the tangent modulus approach to modeling inelastic column behavior is relatively easy to apply.

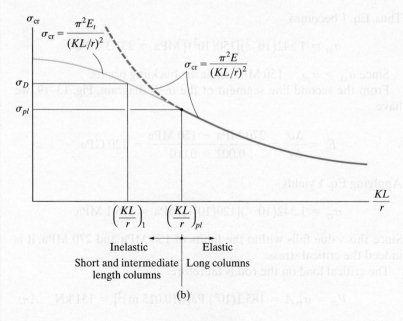

Fig. 13–18 (cont.)

13

EXAMPLE | 13.5

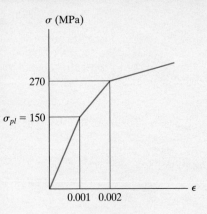

Fig. 13–19

A solid rod has a diameter of 30 mm and is 600 mm long. It is made of a material that can be modeled by the stress–strain diagram shown in Fig. 13–19. If it is used as a pin-supported column, determine the critical load.

SOLUTION
The radius of gyration is

$$r = \sqrt{\frac{I}{A}} = \sqrt{\frac{(\pi/4)(15 \text{ mm})^4}{\pi(15 \text{ mm})^2}} = 7.5 \text{ mm}$$

and therefore the slenderness ratio is

$$\frac{KL}{r} = \frac{1(600 \text{ mm})}{7.5 \text{ mm}} = 80$$

Applying Eq. 13–20 we have,

$$\sigma_{cr} = \frac{\pi^2 E_t}{(KL/r)^2} = \frac{\pi^2 E_t}{(80)^2} = 1.542(10^{-3})E_t \qquad (1)$$

First we will assume that the critical stress is elastic. From Fig. 13–19,

$$E = \frac{150 \text{ MPa}}{0.001} = 150 \text{ GPa}$$

Thus, Eq. 1 becomes

$$\sigma_{cr} = 1.542(10^{-3})[150(10^3)] \text{ MPa} = 231.3 \text{ MPa}$$

Since $\sigma_{cr} > \sigma_{pl} = 150$ MPa, inelastic buckling occurs.
From the second line segment of the $\sigma - \epsilon$ diagram, Fig. 13–19, we have

$$E_t = \frac{\Delta\sigma}{\Delta\epsilon} = \frac{270 \text{ MPa} - 150 \text{ MPa}}{0.002 - 0.001} = 120 \text{ GPa}$$

Applying Eq. 1 yields

$$\sigma_{cr} = 1.542(10^{-3})[120(10^3)] \text{ MPa} = 185.1 \text{ MPa}$$

Since this value falls within the limits of 150 MPa and 270 MPa, it is indeed the critical stress.
The critical load on the rod is therefore

$$P_{cr} = \sigma_{cr}A = 185.1(10^6) \text{ Pa}[\pi(0.015 \text{ m})^2] = 131 \text{ kN} \qquad Ans.$$

PROBLEMS

13–46. Determine the load P required to cause the A-36 steel W200 × 22 column to fail either by buckling or by yielding. The column is fixed at its base and free at its top.

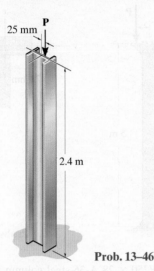

Prob. 13–46

13–47. The hollow red brass C83400 copper alloy shaft is fixed at one end but free at the other end. Determine the maximum eccentric force P the shaft can support without causing it to buckle or yield. Also, find the corresponding maximum deflection of the shaft.

*13–48. The hollow red brass C83400 copper alloy shaft is fixed at one end but free at the other end. If the eccentric force $P = 5$ kN is applied to the shaft as shown, determine the maximum normal stress and the maximum deflection.

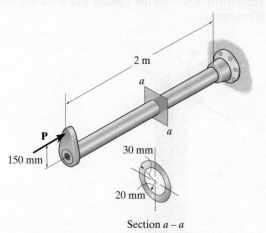

Probs. 13–47/48

•13–49. The tube is made of copper and has an outer diameter of 35 mm and a wall thickness of 7 mm. Using a factor of safety with respect to buckling and yielding of F.S. = 2.5, determine the allowable eccentric load P. The tube is pin supported at its ends. $E_{cu} = 120$ GPa, $\sigma_Y = 750$ MPa.

13–50. The tube is made of copper and has an outer diameter of 35 mm and a wall thickness of 7 mm. Using a factor of safety with respect to buckling and yielding of F.S. = 2.5, determine the allowable eccentric load P that it can support without failure. The tube is fixed supported at its ends. $E_{cu} = 120$ GPa, $\sigma_Y = 750$ MPa.

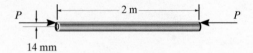

Probs. 13–49/50

13–51. The wood column is fixed at its base and can be assumed pin connected at its top. Determine the maximum eccentric load P that can be applied without causing the column to buckle or yield. $E_w = 12$ GPa, $\sigma_Y = 56$ MPa.

*13–52. The wood column is fixed at its base and can be assumed fixed connected at its top. Determine the maximum eccentric load P that can be applied without causing the column to buckle or yield. $E_w = 12$ GPa, $\sigma_Y = 56$ MPa.

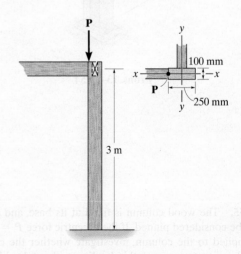

Probs. 13–51/52

13

•**13–53.** The W200 × 22 A-36-steel column is fixed at its base. Its top is constrained to rotate about the y–y axis and free to move along the y–y axis. Also, the column is braced along the x–x axis at its mid-height. Determine the allowable eccentric force P that can be applied without causing the column either to buckle or yield. Use F.S. = 2 against buckling and F.S. = 1.5 against yielding.

13–54. The W200 × 22 A-36-steel column is fixed at its base. Its top is constrained to rotate about the y–y axis and free to move along the y–y axis. Also, the column is braced along the x–x axis at its mid-height. If P = 25 kN, determine the maximum normal stress developed in the column.

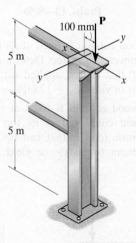

Probs. 13–53/54

*13–56.** The wood column is fixed at its base, and its top can be considered pinned. Determine the maximum eccentric force P the column can support without causing it to either buckle or yield. Take E = 10 GPa and σY = 15 MPa.

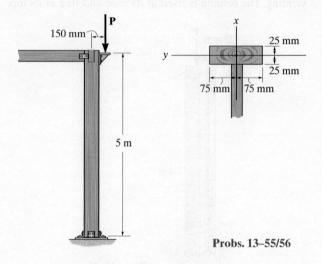

Probs. 13–55/56

•**13–57.** The W250 × 28 A-36-steel column is fixed at its base. Its top is constrained to rotate about the y–y axis and free to move along the y–y axis. If e = 350 mm, determine the allowable eccentric force P that can be applied without causing the column either to buckle or yield. Use F.S. = 2 against buckling and F.S. = 1.5 against yielding.

13–58. The W250 × 28 A-36-steel column is fixed at its base. Its top is constrained to rotate about the y–y axis and free to move along the y–y axis. Determine the force **P** and its eccentricity e so that the column will yield and buckle simultaneously.

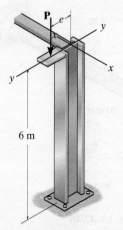

Probs. 13–57/58

13–55. The wood column is fixed at its base, and its top can be considered pinned. If the eccentric force P = 10 kN is applied to the column, investigate whether the column is adequate to support this loading without buckling or yielding. Take E = 10 GPa and σY = 15 MPa.

13–59. The steel column supports the two eccentric loadings. If it is assumed to be pinned at its top, fixed at the bottom, and fully braced against buckling about the y–y axis, determine the maximum deflection of the column and the maximum stress in the column. $E_{st} = 200$ GPa, $\sigma_Y = 360$ MPa.

*13–60. The steel column supports the two eccentric loadings. If it is assumed to be fixed at its top and bottom, and braced against buckling about the y–y axis, determine the maximum deflection of the column and the maximum stress in the column. $E_{st} = 200$ GPa, $\sigma_Y = 360$ MPa.

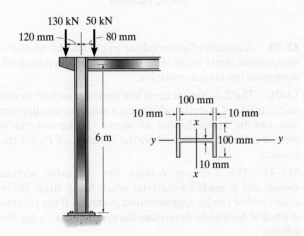

Probs. 13–59/60

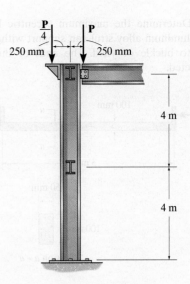

Probs. 13–61/62

13–63. The W360 × 39 structural A-36 steel member is used as a 6-m-long column that is assumed to be fixed at its top and fixed at its bottom. If the 75-kN load is applied at an eccentric distance of 250 mm, determine the maximum stress in the column.

*13–64. The W360 × 39 structural A-36 steel member is used as a column that is assumed to be fixed at its top and pinned at its bottom. If the 75-kN load is applied at an eccentric distance of 250 mm, determine the maximum stress in the column.

13–61. The W250 × 45 A-36-steel column is pinned at its top and fixed at its base. Also, the column is braced along its weak axis at mid-height. If $P = 250$ kN, investigate whether the column is adequate to support this loading. Use F.S. = 2 against buckling and F.S. = 1.5 against yielding.

•13–62. The W250 × 45 A-36-steel column is pinned at its top and fixed at its base. Also, the column is braced along its weak axis at mid-height. Determine the allowable force P that the column can support without causing it either to buckle or yield. Use F.S. = 2 against buckling and F.S. = 1.5 against yielding.

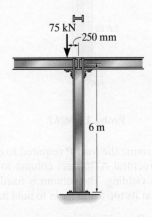

Probs. 13–63/64

•**13–65.** Determine the maximum eccentric load P the 2014-T6-aluminum-alloy strut can support without causing it either to buckle or yield. The ends of the strut are pin-connected.

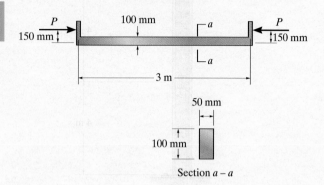

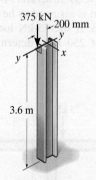

Section $a – a$

Prob. 13–65

13–66. The W200 × 71 structural A-36 steel column is fixed at its bottom and free at its top. If it is subjected to the eccentric load of 375 kN, determine the factor of safety with respect to either the initiation of buckling or yielding.

13–67. The W200 × 71 structural A-36 steel column is fixed at its bottom and pinned at its top. If it is subjected to the eccentric load of 375 kN, determine if the column fails by yielding. The column is braced so that it does not buckle about the $y–y$ axis.

Probs. 13–66/67

****13–68.** Determine the load P required to cause the steel W310 × 74 structural A-36 steel column to fail either by buckling or by yielding. The column is fixed at its bottom and the cables at its top act as a pin to hold it.

•**13–69.** Solve Prob. 13–68 if the column is an A-36 steel W310 × 24 section.

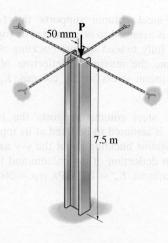

Probs. 13–68/69

13–70. A column of intermediate length buckles when the compressive stress is 280 MPa. If the slenderness ratio is 60, determine the tangent modulus.

13–71. The 2-m-long column has the cross section shown and is made of material which has a stress-strain diagram that can be approximated as shown. If the column is pinned at both ends, determine the critical load P_{cr} for the column.

****13–72.** The 2-m-long column has the cross section shown and is made of material which has a stress-strain diagram that can be approximated as shown. If the column is fixed at both ends, determine the critical load P_{cr} for the column.

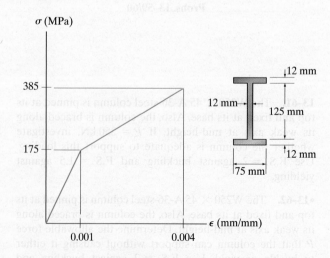

Probs. 13–71/72

•**13–73.** The stress-strain diagram of the material of a column can be approximated as shown. Plot P/A vs. KL/r for the column.

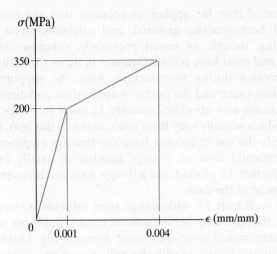

Prob. 13–73

13–74. Construct the buckling curve, P/A versus L/r, for a column that has a bilinear stress–strain curve in compression as shown. The column is pinned at its ends.

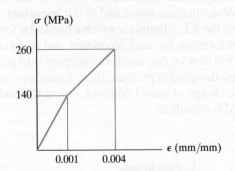

Prob. 13–74

13–75. The stress-strain diagram for a material can be approximated by the two line segments shown. If a bar having a diameter of 80 mm and a length of 1.5 m is made from this material, determine the critical load provided the ends are pinned. Assume that the load acts through the axis of the bar. Use Engesser's equation.

****13–76.** The stress-strain diagram for a material can be approximated by the two line segments shown. If a bar having a diameter of 80 mm and a length of 1.5 m is made from this material, determine the critical load provided the ends are fixed. Assume that the load acts through the axis of the bar. Use Engesser's equation.

•**13–77.** The stress-strain diagram for a material can be approximated by the two line segments shown. If a bar having a diameter of 80 mm and length of 1.5 m is made from this material, determine the critical load provided one end is pinned and the other is fixed. Assume that the load acts through the axis of the bar. Use Engesser's equation.

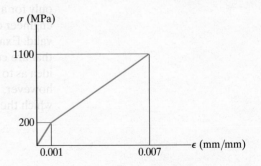

Probs. 13–75/76/77

13

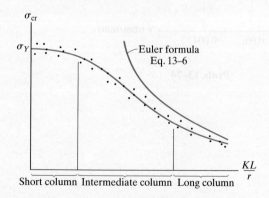

These long unbraced timber columns are used to support the roof of this building.

*13.6   Design of Columns for Concentric Loading

The theory presented thus far applies to columns that are perfectly straight, made of homogeneous material, and originally stress free. Practically speaking, though, as stated previously, columns are not perfectly straight, and most have residual stresses in them, primarily due to nonuniform cooling during manufacture. Also, the supports for columns are less than exact, and the points of application and directions of loads are not known with absolute certainty. In order to compensate for these effects, which actually vary from one column to the next, many design codes specify the use of column formulas that are empirical. By performing experimental tests on a large number of axially loaded columns, the results may be plotted and a design formula developed by curve-fitting the mean of the data.

An example of such tests for wide-flange steel columns is shown in Fig. 13–20. Notice the similarity between these results and those of the family of curves determined from the secant formula, Fig. 13–16. The reason for this similarity has to do with the influence of an "accidental" eccentricity ratio on the column's strength. As stated in Sec. 13.4, this ratio has more of an effect on the strength of short and intermediate-length columns than on those that are long. Tests have indicated that ec/r^2 can range from 0.1 to 0.6 for most axially loaded columns.

In order to account for the behavior of different-length columns, design codes usually specify several formulas that will best fit the data within the short, intermediate, and long column range. Hence, each formula will apply only for a specific *range* of slenderness ratios, and so it is important that the engineer carefully observe the KL/r limits for which a particular formula is valid. Examples of design formulas for steel, aluminum, and wood columns that are currently in use will now be discussed. The purpose is to give some idea as to how columns are designed in practice. These formulas should not, however, be used for the design of actual columns, unless the code from which they are referenced is consulted.

Fig. 13–20

Steel Columns.

Columns made of structural steel can be designed on the basis of formulas proposed by the Structural Stability Research Council (SSRC). Factors of safety have been applied to these formulas and adopted as specifications for building construction by the American Institute of Steel Construction (AISC). Basically these specifications provide two formulas for column design, each of which gives the maximum allowable stress in the column for a specific range of slenderness ratios.*

For long columns the Euler formula is proposed, i.e., $\sigma_{max} = \pi^2 E/(KL/r)^2$.

Application of this formula requires that a factor of safety F.S. $= \frac{23}{12} \approx 1.92$ be applied. Thus, for design,

$$\sigma_{allow} = \frac{12\pi^2 E}{23(KL/r)^2} \qquad \left(\frac{KL}{r}\right)_c \le \frac{KL}{r} \le 200 \qquad (13\text{–}21)$$

As stated, this equation is applicable for a slenderness ratio bounded by 200 and $(KL/r)_c$. A specific value of $(KL/r)_c$ is obtained by requiring the Euler formula to be used only for elastic material behavior. Through experiments it has been determined that compressive residual stresses can exist in rolled-formed steel sections that may be as much as one-half the yield stress. Consequently, if the stress in the Euler formula is greater than $\frac{1}{2}\sigma_Y$, the equation will not apply. Therefore the value of $(KL/r)_c$ is determined as follows:

$$\frac{1}{2}\sigma_Y = \frac{\pi^2 E}{(KL/r)_c^2} \qquad \text{or} \qquad \left(\frac{KL}{r}\right)_c = \sqrt{\frac{2\pi^2 E}{\sigma_Y}} \qquad (13\text{–}22)$$

Columns having slenderness ratios less than $(KL/r)_c$ are designed on the basis of an empirical formula that is parabolic and has the form

$$\sigma_{max} = \left[1 - \frac{(KL/r)^2}{2(KL/r)_c^2}\right]\sigma_Y$$

Since there is more uncertainty in the use of this formula for longer columns, it is divided by a factor of safety defined as follows:

$$\text{F.S.} = \frac{5}{3} + \frac{3}{8}\frac{(KL/r)}{(KL/r)_c} - \frac{(KL/r)^3}{8(KL/r)_c^3}$$

Here it is seen that F.S. $= \frac{5}{3} \approx 1.67$ at $KL/r = 0$ and increases to F.S. $= \frac{23}{12} \approx 1.92$ at $(KL/r)_c$. Hence, for design purposes,

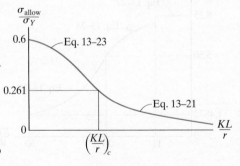

Fig. 13–21

$$\sigma_{allow} = \frac{\left[1 - \dfrac{(KL/r)^2}{2(KL/r)_c^2}\right]\sigma_Y}{(5/3) + [(3/8)(KL/r)/(KL/r)_c] - \left[(KL/r)^3/8(KL/r)_c^3\right]} \qquad (13\text{–}23)$$

Equations 13–21 and 13–23 are plotted in Fig. 13–21. When applying any of these equations, either FPS or SI units can be used for the calculations.

---

*The current AISC code enables engineers to use one of two methods for design, namely, Load and Resistance Factor Design, and Allowable Stress Design. The latter is explained here.

Aluminum Columns.

Column design for structural aluminum is specified by the Aluminum Association using three equations, each applicable for a specific range of slenderness ratios. Since several types of aluminum alloy exist, there is a unique set of formulas for each type. For a common alloy (2014-T6) used in building construction, the formulas are

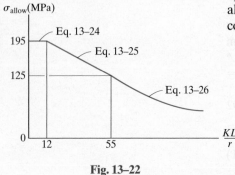

$\sigma_{allow}(MPa)$

Fig. 13–22

$$\sigma_{allow} = 195 \text{ MPa} \quad 0 \le \frac{KL}{r} \le 12 \tag{13–24}$$

$$\sigma_{allow} = \left[214.5 - 1.618\left(\frac{KL}{r}\right)\right] \text{MPa} \quad 12 < \frac{KL}{r} < 55 \tag{13–25}$$

$$\sigma_{allow} = \frac{378125 \text{ MPa}}{(KL/r)^2} \quad 55 \le \frac{KL}{r} \tag{13–26}$$

These equations are plotted in Fig. 13–22. As shown, the first two represent straight lines and are used to model the effects of columns in the short and intermediate range. The third formula has the same form as the Euler formula and is used for long columns.

Timber Columns.

Columns used in timber construction are designed on the basis of formulas published by the National Forest Products Association (NFPA) or the American Institute of Timber Construction (AITC). For example, the NFPA formulas for the allowable stress in short, intermediate, and long columns having a rectangular cross section of dimensions b and d, where d is the *smallest* dimension of the cross section, are

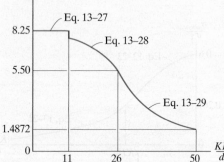

$\sigma_{allow}(MPa)$

Fig. 13–23

$$\sigma_{allow} = 8.25 \text{ MPa} \quad 0 \le \frac{KL}{d} \le 11 \tag{13–27}$$

$$\sigma_{allow} = 8.25\left[1 - \frac{1}{3}\left(\frac{KL/d}{26.0}\right)^2\right] \text{MPa} \quad 11 < \frac{KL}{d} \le 26 \tag{13–28}$$

$$\sigma_{allow} = \frac{3781 \text{ MPa}}{(KL/d)^2} \quad 26 < \frac{KL}{d} \le 50 \tag{13–29}$$

Here wood has a modulus of elasticity of $E_w = 12.4$ GPa and an allowable compressive stress of 8.25 MPa parallel to the grain. In particular, Eq. 13–29 is simply Euler's equation having a factor of safety of 3. These three equations are plotted in Fig. 13–23.

Procedure for Analysis

Column Analysis.

- When using any formula to *analyze* a column, that is, to find its allowable load, it is first necessary to calculate the slenderness ratio in order to determine which column formula applies.

- Once the average allowable stress has been calculated, the allowable load on the column is determined from $P = \sigma_{\text{allow}}A$.

Column Design.

- If a formula is used to *design* a column, that is, to determine the column's cross-sectional area for a given loading and effective length, then a trial-and-check procedure generally must be followed when the column has a composite shape, such as a wide-flange section.

- One possible way to apply a trial-and-check procedure would be to *assume* the column's cross-sectional area, A', and calculate the corresponding stress $\sigma' = P/A'$. Also, use an appropriate design formula to determine the allowable stress σ_{allow}. From this, calculate the *required* column area $A_{\text{req'd}} = P/\sigma_{\text{allow}}$.

- If $A' > A_{\text{req'd}}$, the design is safe. When making the comparison, it is practical to require A' to be close to but greater than $A_{\text{req'd}}$, usually within 2–3%. A redesign is necessary if $A' < A_{\text{req'd}}$.

- Whenever a trial-and-check procedure is repeated, the choice of an area is determined by the previously calculated required area. In engineering practice this method for design is usually shortened through the use of computer software or published tables and graphs.

These timber columns can be considered pinned at their bottom and fixed connected to the beams at their tops.

EXAMPLE | 13.6

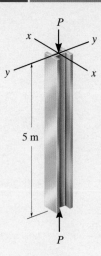

Fig. 13–24

An A-36 steel W250 × 149 member is used as a pin-supported column, Fig. 13–24. Using the AISC column design formulas, determine the largest load that it can safely support.

SOLUTION

The following data for a W250 × 149 is taken from the table in Appendix B.

$$A = 19\,000 \text{ mm}^2 \quad r_x = 117 \text{ mm} \quad r_y = 67.4 \text{ mm}$$

Since $K = 1$ for both x and y axis buckling, the slenderness ratio is largest if r_y is used. Thus,

$$\frac{KL}{r} = \frac{1(5000 \text{ mm})}{67.4 \text{ mm}} = 74.18$$

From Eq. 13–22, we have

$$\left(\frac{KL}{r}\right)_c = \sqrt{\frac{2\pi^2 E}{\sigma_Y}}$$

$$= \sqrt{\frac{2\pi^2 [200(10^3)] \text{ MPa}}{250 \text{ MPa}}}$$

$$= 125.66$$

Here $0 < KL/r < (KL/r)_c$, so Eq. 13–23 applies.

$$\sigma_{\text{allow}} = \frac{\left[1 - \dfrac{(KL/r)^2}{2(KL/r)_c^2}\right]\sigma_Y}{(5/3) + [(3/8)(KL/r)/(KL/r)_c] - \left[(KL/r)^3/8(KL/r_c)^3\right]}$$

$$= \frac{[1 - (74.18)^2/2(125.66)^2]250 \text{ MPa}}{(5/3) + [(3/8)(74.18/125.66)] - [(74.18)^3/8(125.66)^3]}$$

$$= 110.85 \text{ MPa}$$

The allowable load P on the column is therefore

$$\sigma_{\text{allow}} = \frac{P}{A}; \qquad 110.85 \text{ N/mm}^2 = \frac{P}{19\,000 \text{ mm}^2}$$

$$P = 2\,106\,150 \text{ N} = 2106 \text{ kN} \qquad \qquad \textit{Ans.}$$

EXAMPLE | 13.7

The steel rod in Fig. 13–25 is to be used to support an axial load of 80 kN. If $E_{st} = 210(10^3)$ MPa and $\sigma_Y = 300$ MPa, determine the smallest diameter of the rod as allowed by the AISC specification. The rod is fixed at both ends.

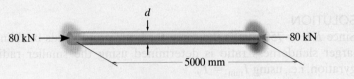

d

80 kN →   ← 80 kN

5000 mm

Fig. 13–25

SOLUTION

For a circular cross section the radius of gyration becomes

$$r = \sqrt{\frac{I}{A}} = \sqrt{\frac{(1/4)\pi(d/2)^4}{(1/4)\pi d^2}} = \frac{d}{4}$$

Applying Eq. 13–22, we have

$$\left(\frac{KL}{r}\right)_c = \sqrt{\frac{2\pi^2 E}{\sigma_Y}} = \sqrt{\frac{2\pi^2[210(10^3)\ \text{MPa}]}{360\ \text{MPa}}} = 107.3$$

Since the rod's radius of gyration is unknown, KL/r is unknown, and therefore a choice must be made as to whether Eq. 13–21 or Eq. 13–23 applies. We will consider Eq. 13–21. For a fixed-end column $K = 0.5$, so

$$\sigma_{\text{allow}} = \frac{12\pi^2 E}{23(KL/r)^2}$$

$$\frac{80\ \text{kN}}{(1/4)\pi d^2} = \frac{12\pi^2[210\ \text{kN/mm}^2]}{23[0.5(5000\ \text{mm})/(d/4)]^2}$$

$$\frac{101.86}{d^2} = 0.0108 d^2$$

$$d = 55.42\ \text{mm}$$

Use

$$d = 56\ \text{mm} \qquad\qquad Ans.$$

For this design, we must check the slenderness-ratio limits; i.e.,

$$\frac{KL}{r} = \frac{0.5(5000\ \text{mm})}{(56\ \text{mm}/4)} = 179$$

Since $107.3 < 179 < 200$, use of Eq. 13–21 is appropriate.

EXAMPLE | 13.8

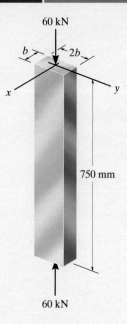

60 kN

b $2b$

x y

750 mm

60 kN

Fig. 13–26

A bar having a length of 750 mm is used to support an axial compressive load of 60 kN, Fig. 13–26. It is pin supported at its ends and made of a 2014-T6 aluminum alloy. Determine the dimensions of its cross-sectional area if its width is to be twice its thickness.

SOLUTION

Since $KL = 750$ mm is the same for both x and y axis buckling, the larger slenderness ratio is determined using the smaller radius of gyration, i.e., using $I_{min} = I_y$:

$$\frac{KL}{r_y} = \frac{KL}{\sqrt{I_y/A}} = \frac{1(750)}{\sqrt{(1/12)2b(b^3)/[2b(b)]}} = \frac{2598.1}{b} \quad (1)$$

Here we must apply Eq. 13–24, 13–25, or 13–26. Since we do not as yet know the slenderness ratio, we will begin by using Eq. 13–24.

$$\frac{P}{A} = 195 \text{ N/mm}^2$$

$$\frac{60(10^3) \text{ N}}{2b(b)} = 195 \text{ N/mm}^2$$

$$b = 12.40 \text{ mm}$$

Checking the slenderness ratio, we have

$$\frac{KL}{r} = \frac{2598.1}{12.40} = 209.5 > 12$$

Try Eq. 13–26, which is valid for $KL/r \geq 55$,

$$\frac{P}{A} = \frac{378\,125 \text{ MPa}}{(KL/r)^2}$$

$$\frac{60(10^3)}{2b(b)} = \frac{378\,125}{(2598.1/b)^2}$$

$$b = 27.05 \text{ mm} \qquad \qquad Ans.$$

From Eq. 1,

$$\frac{KL}{r} = \frac{2598.1}{27.05} = 96.0 > 55 \quad \text{OK}$$

NOTE: It would be satisfactory to choose the cross section with dimensions 27 mm by 54 mm.

EXAMPLE | 13.9

A board having cross-sectional dimensions of 150 mm by 40 mm is used to support an axial load of 20 kN, Fig. 13–27. If the board is assumed to be pin supported at its top and bottom, determine its *greatest* allowable length L as specified by the NFPA.

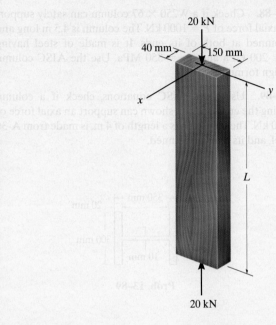

Fig. 13–27

SOLUTION

By inspection, the board will buckle about the y axis. In the NFPA equations, $d = 40$ mm. Assuming that Eq. 13–29 applies, we have

$$\frac{P}{A} = \frac{3781 \text{ MPa}}{(KL/r)^2}$$

$$\frac{20(10^3) \text{ N}}{(150 \text{ mm})(40 \text{ mm})} = \frac{3781 \text{ MPa}}{(1 \ L/40 \text{ mm})^2}$$

$$L = 1336 \text{ mm} \qquad\qquad\qquad Ans.$$

Here

$$\frac{KL}{d} = \frac{1(1336 \text{ mm})}{40 \text{ mm}} = 33.4$$

Since $26 < KL/d \le 50$, the solution is valid.

PROBLEMS

13–78. Determine the largest length of a structural A-36 steel rod if it is fixed supported and subjected to an axial load of 100 kN. The rod has a diameter of 50 mm. Use the AISC equations.

13–79. Determine the largest length of a W250 × 67 structural steel column if it is pin supported and subjected to an axial load of 1450 kN. $E_{st} = 200$ GPa, $\sigma_Y = 350$ MPa. Use the AISC equations.

*13–80. Determine the largest length of a W250 × 18 structural A-36 steel section if it is pin supported and is subjected to an axial load of 140 kN. Use the AISC equations.

•13–81. Using the AISC equations, select from Appendix B the lightest-weight structural A-36 steel column that is 4.2 m long and supports an axial load of 200 kN. The ends are pinned. Take $\sigma_Y = 350$ MPa.

13–82. Using the AISC equations, select from Appendix B the lightest-weight structural A-36 steel column that is 3.6 m long and supports an axial load of 200 kN. The ends are fixed. Take $\sigma_Y = 350$ MPa.

13–83. Using the AISC equations, select from Appendix B the lightest-weight structural A-36 steel column that is 7.2 m long and supports an axial load of 450 kN. The ends are fixed.

*13–84. Using the AISC equations, select from Appendix B the lightest-weight structural A-36 steel column that is 9 m long and supports an axial load of 1000 kN. The ends are fixed.

•13–85. A W200 × 36 A-36-steel column of 9-m length is pinned at both ends and braced against its weak axis at mid-height. Determine the allowable axial force P that can be safely supported by the column. Use the AISC column design formulas.

13–86. Check if a W250 × 58 column can safely support an axial force of $P = 1150$ kN. The column is 6 m long and is pinned at both ends and braced against its weak axis at mid-height. It is made of steel having $E = 200$ GPa and $\sigma_Y = 350$ MPa. Use the AISC column design formulas.

13–87. A 1.5-m-long rod is used in a machine to transmit an axial compressive load of 15 kN. Determine its smallest diameter if it is pin connected at its ends and is made of a 2014-T6 aluminum alloy.

*13–88. Check if a W250 × 67 column can safely support an axial force of $P = 1000$ kN. The column is 4.5 m long and is pinned at both of its ends. It is made of steel having $E = 200$ GPa and $\sigma_Y = 350$ MPa. Use the AISC column design formulas.

•13–89. Using the AISC equations, check if a column having the cross section shown can support an axial force of 1500 kN. The column has a length of 4 m, is made from A-36 steel, and its ends are pinned.

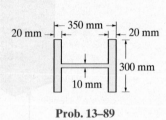

Prob. 13–89

13–90. The A-36-steel tube is pinned at both ends. If it is subjected to an axial force of 150 kN, determine the maximum length that the tube can safely support using the AISC column design formulas.

Prob. 13–90

13–91. The bar is made of a 2014-T6 aluminum alloy. Determine its smallest thickness b if its width is $5b$. Assume that it is pin connected at its ends.

*13–92. The bar is made of a 2014-T6 aluminum alloy. Determine its smallest thickness b if its width is $5b$. Assume that it is fixed connected at its ends.

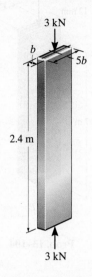

3 kN

b

$5b$

2.4 m

3 kN

Probs. 13–91/92

•13–93. The 2014-T6 aluminum column of 3-m length has the cross section shown. If the column is pinned at both ends and braced against the weak axis at its mid-height, determine the allowable axial force P that can be safely supported by the column.

13–94. The 2014-T6 aluminum column has the cross section shown. If the column is pinned at both ends and subjected to an axial force $P = 100$ kN, determine the maximum length the column can have to safely support the loading.

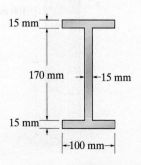

15 mm

170 mm   15 mm

15 mm

100 mm

Probs. 13–93/94

13–95. The 2014-T6 aluminum hollow section has the cross section shown. If the column is 3 m long and is fixed at both ends, determine the allowable axial force P that can be safely supported by the column.

*13–96. The 2014-T6 aluminum hollow section has the cross section shown. If the column is fixed at its base and pinned at its top, and is subjected to the axial force $P = 500$ kN determine the maximum length of the column for it to safely support the load.

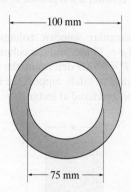

100 mm

75 mm

Probs. 13–95/96

•13–97. The tube is 6 mm thick, is made of a 2014-T6 aluminum alloy, and is fixed at its bottom and pinned at its top. Determine the largest axial load that it can support.

13–98. The tube is 6 mm thick, is made of a 2014-T6 aluminum alloy, and is fixed connected at its ends. Determine the largest axial load that it can support.

13–99. The tube is 6 mm thick, is made of 2014-T6 aluminum alloy and is pin connected at its ends. Determine the largest axial load it can support.

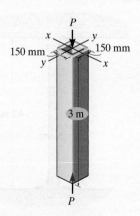

P

x y

150 mm   150 mm

y x

3 m

P

Probs. 13–97/98/99

*13–100. A rectangular wooden column has the cross section shown. If the column is 1.8 m long and subjected to an axial force of $P = 75$ kN, determine the required minimum dimension a of its cross-sectional area to the nearest multiples of 5 mm so that the column can safely support the loading. The column is pinned at both ends.

•13–101. A rectangular wooden column has the cross section shown. If $a = 75$ mm and the column is 3.6 m long, determine the allowable axial force P that can be safely supported by the column if it is pinned at its top and fixed at its base.

13–102. A rectangular wooden column has the cross section shown. If $a = 75$ mm and the column is subjected to an axial force of $P = 75$ kN determine the maximum length the column can have to safely support the load. The column is pinned at its top and fixed at its base.

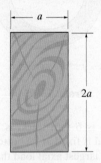

Probs. 13–100/101/102

13–103. The timber column has a square cross section and is assumed to be pin connected at its top and bottom. If it supports an axial load of 250 kN, determine its smallest side dimension a to the nearest multiples of 10 mm. Use the NFPA formulas.

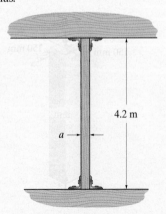

Prob. 13–103

*13–104. The wooden column shown is formed by gluing together the 150 mm × 12 mm boards. If the column is pinned at both ends and is subjected to an axial load $P = 100$ kN determine the required number of boards needed to form the column in order to safely support the loading.

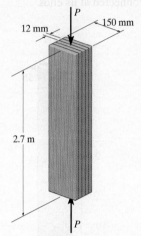

Prob. 13–104

•13–105. The column is made of wood. It is fixed at its bottom and free at its top. Use the NFPA formulas to determine its greatest allowable length if it supports an axial load of $P = 10$ kN.

13–106. The column is made of wood. It is fixed at its bottom and free at its top. Use the NFPA formulas to determine the largest allowable axial load P that it can support if it has a length $L = 1.2$ m.

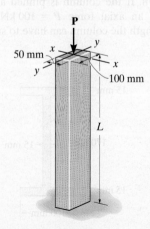

Probs. 13–105/106

*13.7 Design of Columns for Eccentric Loading

Occasionally a column may be required to support a load acting either at its edge or on an angle bracket attached to its side, such as shown in Fig. 13–28a. The bending moment $M = Pe$, which is caused by the eccentric loading, must be accounted for when the column is designed. There are several acceptable ways in which this is done in engineering practice. We will discuss two of the most common methods.

Use of Available Column Formulas. The stress distribution acting over the cross-sectional area of the column shown in Fig. 13–28a is determined from a superposition of both the axial force P and the bending moment $M = Pe$. In particular, the maximum compressive stress is

$$\sigma_{max} = \frac{P}{A} + \frac{Mc}{I} \qquad (13\text{–}30)$$

A typical stress profile is shown in Fig. 13–28b. If we conservatively *assume* that the entire cross section is subjected to the uniform stress σ_{max} as determined from Eq. 13–30, then we can compare σ_{max} with σ_{allow}, which is determined using the formulas given in Sec. 13.6. Calculation of σ_{allow} is usually done using the *largest* slenderness ratio for the column, regardless of the axis about which the column experiences bending. This requirement is normally specified in design codes and will in most cases lead to a conservative design. If

$$\sigma_{max} \leq \sigma_{allow}$$

then the column can carry the specified loading. If this inequality does not hold, then the column's area A must be increased, and a new σ_{max} and σ_{allow} must be calculated. This method of design is rather simple to apply and works well for columns that are short or of intermediate length.

Interaction Formula. When *designing* an eccentrically loaded column it is desirable to see how the bending and axial loads *interact*, so that a balance between these two effects can be achieved. To do this, we will consider the separate contributions made to the total column area by the axial force and moment. If the allowable stress for the axial load is $(\sigma_a)_{allow}$, then the required area for the column needed to support the load P is

$$A_a = \frac{P}{(\sigma_a)_{allow}}$$

Similarly, if the allowable bending stress is $(\sigma_b)_{allow}$, then since $I = Ar^2$, the required area of the column needed to support the eccentric moment is determined from the flexure formula, that is,

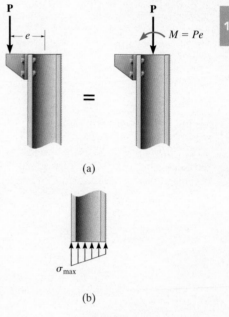

(a)

(b)

Fig. 13–28

13

$$A_b = \frac{Mc}{(\sigma_b)_{\text{allow}} r^2}$$

The total area A for the column needed to resist *both* the axial load and moment requires that

$$A_a + A_b = \frac{P}{(\sigma_a)_{\text{allow}}} + \frac{Mc}{(\sigma_b)_{\text{allow}} r^2} \leq A$$

or

$$\frac{P/A}{(\sigma_a)_{\text{allow}}} + \frac{Mc/Ar^2}{(\sigma_b)_{\text{allow}}} \leq 1$$

$$\frac{\sigma_a}{(\sigma_a)_{\text{allow}}} + \frac{\sigma_b}{(\sigma_b)_{\text{allow}}} \leq 1 \qquad (13\text{–}31)$$

Here

σ_a = axial stress caused by the force P and determined from $\sigma_a = P/A$, where A is the cross-sectional area of the column

σ_b = bending stress caused by an eccentric load or applied moment M; σ_b is found from $\sigma_b = Mc/I$, where I is the moment of inertia of the cross-sectional area calculated about the bending or centroidal axis

$(\sigma_a)_{\text{allow}}$ = allowable axial stress as defined by formulas given in Sec. 13.6 or by other design code specifications. For this purpose, always use the *largest* slenderness ratio for the column, regardless of the axis about which the column experiences bending

$(\sigma_b)_{\text{allow}}$ = allowable bending stress as defined by code specifications

Notice that, if the column is subjected only to an axial load, then the bending-stress ratio in Eq. 13–31 would be equal to zero and the design will be based only on the allowable axial stress. Likewise, when no axial load is present, the axial-stress ratio is zero and the stress requirement will be based on the allowable bending stress. Hence, each stress ratio indicates the contribution of axial load or bending moment. Since Eq. 13–31 shows how these loadings interact, this equation is sometimes referred to as the **interaction formula**. This design approach requires a trial-and-check procedure, where it is required that the designer *pick* an available column and then check to see if the inequality is satisfied. If it is not, a larger section is then picked and the process repeated. An economical choice is made when the left side is close to but less than 1.

The interaction method is often specified in codes for the design of columns made of steel, aluminum, or timber. In particular, for allowable stress design, the American Institute of Steel Construction specifies the use of this equation only when the axial-stress ratio $\sigma_a/(\sigma_a)_{\text{allow}} \leq 0.15$. For other values of this ratio, a modified form of Eq. 13–31 is used.

Typical example of a column used to support an eccentric roof loading.

EXAMPLE | 13.10

The column in Fig. 13–29 is made of aluminum alloy 2014-T6 and is used to support an eccentric load **P**. Determine the maximum magnitude of **P** that can be supported if the column is fixed at its base and free at its top. Use Eq. 13–30.

Refer to the Companion Website for the animation of the concepts that can be found in Figure 13-10.

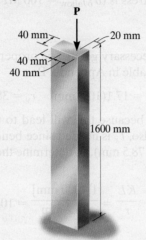

P

40 mm   20 mm
40 mm
40 mm

1600 mm

Fig. 13–29

SOLUTION
From Fig. 13–10b, $K = 2$. The largest slenderness ratio for the column is therefore

$$\frac{KL}{r} = \frac{1600 \text{ mm}}{\sqrt{[(1/12)(80 \text{ mm})(40 \text{ mm})^3]/[(40 \text{ mm}) 80 \text{ mm}]}} = 277.1$$

By inspection, Eq. 13–26 must be used (277.1 > 55). Thus,

$$\sigma_{allow} = \frac{378\,125 \text{ MPa}}{(KL/r)^2} = \frac{378\,125 \text{ MPa}}{(277.1)^2} = 4.92 \text{ MPa}$$

The maximum compressive stress in the column is determined from the combination of axial load and bending. We have

$$\sigma_{max} = \frac{P}{A} + \frac{(Pe)c}{I}$$

$$= \frac{P}{40 \text{ mm}(80 \text{ mm})} + \frac{P(20 \text{ mm})(40 \text{ mm})}{(1/12)(40 \text{ mm})(30 \text{ mm})^3}$$

$$= 0.00078125P$$

Assuming that this stress is *uniform* over the cross section, we require

$$\sigma_{allow} = \sigma_{max}; \qquad 4.92 = 0.00078125\,P$$

$$P = 6297.6 \text{ N} = 6.30 \text{ kN} \qquad\qquad Ans.$$

13

EXAMPLE | 13.11

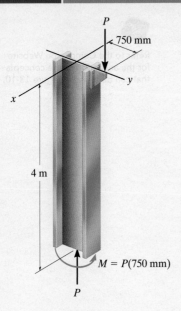

P

750 mm

y

x

4 m

$M = P(750 \text{ mm})$

P

Fig. 13–30

The A-36 steel W150 × 30 column in Fig. 13–30 is pin connected at its ends and is subjected to the eccentric load P. Determine the maximum allowable value of P using the interaction method if the allowable bending stress is $(\sigma_b)_{\text{allow}} = 160$ MPa.

SOLUTION
Here $K = 1$. The necessary geometric properties for the W150 × 30 are taken from the table in Appendix B.

$$A = 3790 \text{ mm}^2 \quad I_x = 17.1(10^6) \text{ mm}^4 \quad r_y = 38.2 \text{ mm} \quad d = 157 \text{ mm}$$

We will consider r_y because this will lead to the *largest* value of the slenderness ratio. Also, I_x is needed since bending occurs about the x axis ($c = 157 \text{ mm}/2 = 78.5 \text{ mm}$). To determine the allowable compressive stress, we have

$$\frac{KL}{r} = \frac{1[4000 \text{ mm}]}{38.2 \text{ mm}} = 104.71$$

Since

$$\left(\frac{KL}{r}\right)_c = \sqrt{\frac{2\pi^2 E}{\sigma_Y}} = \sqrt{\frac{2\pi^2[200(10^3) \text{ MPa}]}{250 \text{ MPa}}} = 125.66$$

then $KL/r < (KL/r)_c$ and so Eq. 13–23 must be used.

$$\sigma_{\text{allow}} = \frac{[1 - (KL/r)^2/2(KL/r)_c{}^2]\sigma_Y}{(5/3) + [(3/8)(KL/r)/(KL/r)_c] - \left[(KL/r)^3/8(KL/r)_c{}^3\right]}$$

$$= \frac{[1 - (104.71)^2/2(125.66)^2]250 \text{ MPa}}{(5/3) + [(3/8)(104.71)/(125.66)] - [(104.71)^3/8(125.66)^3]}$$

$$= 85.59 \text{ MPa}$$

Applying the interaction Eq. 13–31 yields

$$\frac{\sigma_a}{(\sigma_a)_{\text{allow}}} + \frac{\sigma_b}{(\sigma_b)_{\text{allow}}} \leq 1$$

$$\frac{P/3790 \text{ mm}^2}{85.59 \text{ MPa}} + \frac{P(750 \text{ mm})(78.5 \text{ mm})/[17.1(10^6) \text{ mm}^4]}{160 \text{ N/mm}^2} = 1$$

$$P = 40.65 \text{ kN} \qquad \qquad Ans.$$

Checking the application of the interaction method for the steel section, we require

$$\frac{\sigma_a}{(\sigma_a)_{\text{allow}}} = \frac{40.65(10^3) \text{ N}/(3790 \text{ mm}^2)}{85.59 \text{ N/mm}^2} = 0.125 < 0.15 \qquad \text{OK}$$

EXAMPLE | 13.12

The timber column in Fig. 13–31 is made from two boards nailed together so that the cross section has the dimensions shown. If the column is fixed at its base and free at its top, use Eq. 13–30 to determine the eccentric load **P** that can be supported.

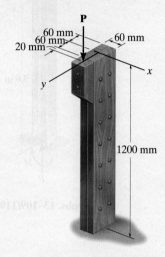

Fig. 13–31

SOLUTION
From Fig. 13–10b, $K = 2$. Here we must calculate KL/d to determine which equation from Eqs. 13–27 through 13–29 should be used. Since σ_{allow} is determined using the largest slenderness ratio, we choose $d = 60$ mm. This is done to make this ratio as large as possible, and thereby yields the lowest possible allowable axial stress. We have

$$\frac{KL}{d} = \frac{2(1200 \text{ mm})}{60 \text{ mm}} = 40$$

Since $26 < KL/d < 50$ the allowable axial stress is determined using Eq. 13–29. Thus,

$$\sigma_{allow} = \frac{3718 \text{ MPa}}{(KL/d)^2} = \frac{3718 \text{ MPa}}{(40)^2} = 2.324 \text{ MPa}$$

Applying Eq. 13–30 with $\sigma_{allow} = \sigma_{max}$, we have

$$\sigma_{allow} = \frac{P}{A} + \frac{Mc}{I}$$

$$2.324 \text{ N/mm}^2 = \frac{P}{60 \text{ mm}(120 \text{ mm})} + \frac{P(80 \text{ mm})(60 \text{ mm})}{(1/12)(60 \text{ mm})(120 \text{ mm})^3}$$

$$P = 3.35 \text{ kN} \qquad Ans.$$

13

PROBLEMS

13–107. The W360 × 79 structural A-36 steel column supports an axial load of 400 kN in addition to an eccentric load *P*. Determine the maximum allowable value of *P* based on the AISC equations of Sec. 13.6 and Eq. 13–30. Assume the column is fixed at its base, and at its top it is free to sway in the *x–z* plane while it is pinned in the *y–z* plane.

***13–108.** The W310 × 67 structural A-36 steel column supports an axial load of 400 kN in addition to an eccentric load of *P* = 300 kN. Determine if the column fails based on the AISC equations of Sec. 13.6 and Eq. 13–30. Assume that the column is fixed at its base, and at its top it is free to sway in the *x–z* plane while it is pinned in the *y–z* plane.

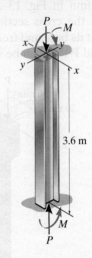

Probs. 13–109/110

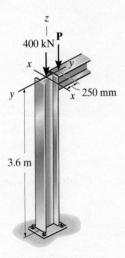

Probs. 13–107/108

•13–109. The W360 × 33 structural A-36 steel column is fixed at its top and bottom. If a horizontal load (not shown) causes it to support end moments of *M* = 15 kN · m, determine the maximum allowable axial force *P* that can be applied. Bending is about the *x–x* axis. Use the AISC equations of Sec. 13.6 and Eq. 13–30.

13–110. The W360 × 33 column is fixed at its top and bottom. If a horizontal load (not shown) causes it to support end moments of *M* = 22.5 kN · m, maximum allowable axial force *P* that can be applied. Bending is about the *x–x* axis. Use the interaction formula with $(\sigma_b)_{allow}$ = 168 MPa.

13–111. The W360 × 57 structural A-36 steel column is fixed at its bottom and free at its top. Determine the greatest eccentric load *P* that can be applied using Eq. 13–30 and the AISC equations of Sec. 13.6.

***13–112.** The W250 × 67 structural A-36 steel column is fixed at its bottom and free at its top. If it is subjected to a load of *P* = 10 kN, determine if it is safe based on the AISC equations of Sec. 13.6 and Eq. 13–30.

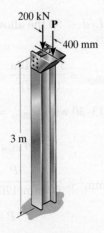

Probs. 13–111/112

•13–113. The A-36-steel W250 × 67 column is fixed at its base. Its top is constrained to move along the x–x axis but free to rotate about and move along the y–y axis. Determine the maximum eccentric force P that can be safely supported by the column using the allowable stress method.

13–114. The A-36-steel W250 × 67 column is fixed at its base. Its top is constrained to move along the x–x axis but free to rotate about and move along the y–y axis. Determine the maximum eccentric force P that can be safely supported by the column using an interaction formula. The allowable bending stress is $(\sigma_b)_{\text{allow}} = 100$ MPa.

13–115. The A-36-steel W310 × 74 column is fixed at its base. Its top is constrained to move along the x–x axis but free to rotate about and move along the y–y axis. If the eccentric force P = 75 kN is applied to the column, investigate if the column is adequate to support the loading. Use the allowable stress method.

*13–116. The A-36-steel W310 × 74 column is fixed at its base. Its top is constrained to move along the x–x axis but free to rotate about and move along the y–y axis. If the eccentric force P = 65 kN is applied to the column, investigate if the column is adequate to support the loading. Use the interaction formula. The allowable bending stress is $(\sigma_b)_{\text{allow}} = 100$ MPa.

•13–117. A 4.8-m-long column is made of aluminum alloy 2014-T6. If it is fixed at its top and bottom, and a compressive load P is applied at point A, determine the maximum allowable magnitude of P using the equations of Sec. 13.6 and Eq. 13–30.

13–118. A 4.8-m-long column is made of aluminum alloy 2014-T6. If it is fixed at its top and bottom, and a compressive load P is applied at point A, determine the maximum allowable magnitude of P using the equations of Sec. 13.6 and the interaction formula with $(\sigma_b)_{\text{allow}} = 140$ MPa.

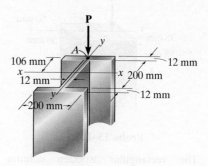

Probs. 13–117/118

13–119. The 2014-T6 hollow column is fixed at its base and free at its top. Determine the maximum eccentric force P that can be safely supported by the column. Use the allowable stress method. The thickness of the wall for the section is t = 12 mm.

*13–120. The 2014-T6 hollow column is fixed at its base and free at its top. Determine the maximum eccentric force P that can be safely supported by the column. Use the interaction formula. The allowable bending stress is $(\sigma_b)_{\text{allow}} = 200$ MPa. The thickness of the wall for the section is t = 12 mm.

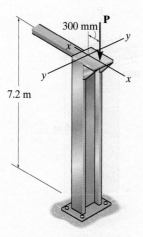

Probs. 13–113/114/115/116

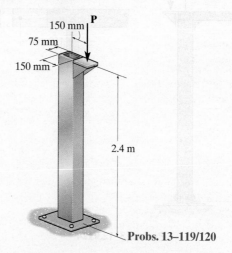

Probs. 13–119/120

13

•**13–121.** The 3-m-long bar is made of aluminum alloy 2014-T6. If it is fixed at its bottom and pinned at the top, determine the maximum allowable eccentric load **P** that can be applied using the formulas in Sec. 13.6 and Eq. 13–30.

13–122. The 3-m-long bar is made of aluminum alloy 2014-T6. If it is fixed at its bottom and pinned at the top, determine the maximum allowable eccentric load **P** that can be applied using the equations of Sec. 13.6 and the interaction formula with $(\sigma_b)_{allow} = 126$ MPa.

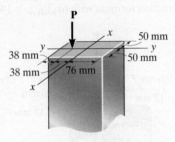

Probs. 13–121/122

13–123. The rectangular wooden column can be considered fixed at its base and pinned at its top. Also, the column is braced at its mid-height against the weak axis. Determine the maximum eccentric force P that can be safely supported by the column using the allowable stress method.

*__13–124.__ The rectangular wooden column can be considered fixed at its base and pinned at its top. Also, the column is braced at its mid-height against the weak axis. Determine the maximum eccentric force P that can be safely supported by the column using the interaction formula. The allowable bending stress is $(\sigma_b)_{allow} = 10$ MPa.

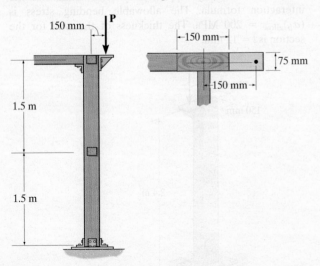

Probs. 13–123/124

•**13–125.** The 250-mm-diameter utility pole supports the transformer that has a weight of 3 kN and center of gravity at G. If the pole is fixed to the ground and free at its top, determine if it is adequate according to the NFPA equations of Sec. 13.6 and Eq. 13–30.

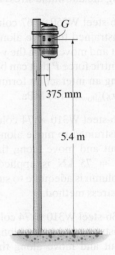

Prob. 13–125

13–126. Using the NFPA equations of Sec. 13.6 and Eq. 13–30, determine the maximum allowable eccentric load P that can be applied to the wood column. Assume that the column is pinned at both its top and bottom.

13–127. Using the NFPA equations of Sec. 13.6 and Eq. 13–30, determine the maximum allowable eccentric load P that can be applied to the wood column. Assume that the column is pinned at the top and fixed at the bottom.

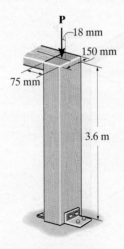

Probs. 13–126/127

CHAPTER REVIEW

Buckling is the sudden instability that occurs in columns or members that support an axial compressive load. The maximum axial load that a member can support just before buckling is called the critical load P_{cr}.	Ref.: Section 13.1	
The critical load for an ideal column is determined from Euler's formula, where $K = 1$ for pin supports, $K = 0.5$ for fixed supports, $K = 0.7$ for a pin and a fixed support, and $K = 2$ for a fixed support and a free end.	$$P_{cr} = \frac{\pi^2 EI}{(KL)^2}$$ Ref.: Sections 13.2 & 13.3	P_{cr}
If the axial loading is applied eccentrically to the column, then the secant formula can be used to determine the maximum stress in the column.	$$\sigma_{max} = \frac{P}{A}\left[1 + \frac{ec}{r^2}\sec\left(\frac{L}{2r}\sqrt{\frac{P}{EA}}\right)\right]$$ Ref.: Section 13.4	
When the axial load causes yielding of the material, then the tangent modulus should be used with Euler's formula to determine the critical load for the column. This is referred to as Engesser's equation.	$$\sigma_{cr} = \frac{\pi^2 E_t}{(KL/r)^2}$$ Ref.: Section 13.5	
Empirical formulas based on experimental data have been developed for use in the design of steel, aluminum, and timber columns.	Ref.: Section 13.6	

REVIEW PROBLEMS

*13–128. The wood column is 4 m long and is required to support the axial load of 25 kN. If the cross section is square, determine the dimension a of each of its sides using a factor of safety against buckling of F.S. = 2.5. The column is assumed to be pinned at its top and bottom. Use the Euler equation. $E_w = 11$ GPa, and $\sigma_Y = 10$ MPa.

13–130. Determine the maximum intensity w of the uniform distributed load that can be applied on the beam without causing the compressive members of the supporting truss to buckle. The members of the truss are made from A-36-steel rods having a 60-mm diameter. Use F.S. = 2 against buckling.

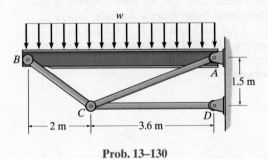

Prob. 13–130

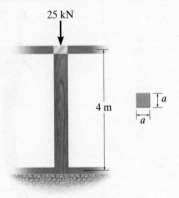

Prob. 13–128

•13–129. If the torsional springs attached to ends A and C of the rigid members AB and BC have a stiffness k, determine the critical load P_{cr}.

13–131. The W250 × 67 steel column supports an axial load of 300 kN in addition to an eccentric load P. Determine the maximum allowable value of P based on the AISC equations of Sec. 13.6 and Eq. 13–30. Assume that in the x–z plane $K_x = 1.0$ and in the y–z plane $K_y = 2.0$. $E_{st} = 200$ GPa, $\sigma_Y = 350$ MPa

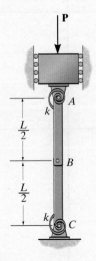

Prob. 13–129

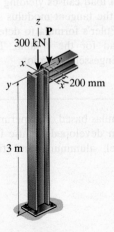

Prob. 13–131

*13–132. The A-36-steel column can be considered pinned at its top and fixed at its base. Also, it is braced at its mid-height along the weak axis. Investigate whether a W250 × 45 section can safely support the loading shown. Use the allowable stress method.

•13–133. The A-36-steel column can be considered pinned at its top and fixed at its base. Also, it is braced at its mid-height along the weak axis. Investigate whether a W250 × 45 section can safely support the loading shown. Use the interaction formula. The allowable bending stress is $(\sigma_b)_{allow} = 100$ MPa.

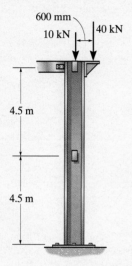

Probs. 13–132/133

13–134. The member has a symmetric cross section. If it is pin connected at its ends, determine the largest force it can support. It is made of 2014-T6 aluminum alloy.

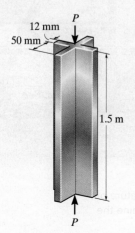

Prob. 13–134

13–135. The W200 × 46 A-36-steel column can be considered pinned at its top and fixed at its base. Also, the column is braced at its mid-height against the weak axis. Determine the maximum axial load the column can support without causing it to buckle.

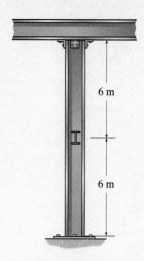

Prob. 13–135

*13–136. The structural A-36 steel column has the cross section shown. If it is fixed at the bottom and free at the top, determine the maximum force P that can be applied at A without causing it to buckle or yield. Use a factor of safety of 3 with respect to buckling and yielding.

•13–137. The structural A-36 steel column has the cross section shown. If it is fixed at the bottom and free at the top, determine if the column will buckle or yield when the load $P = 10$ kN. Use a factor of safety of 3 with respect to buckling and yielding.

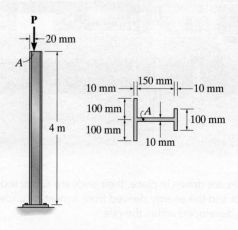

Probs. 13–136/137

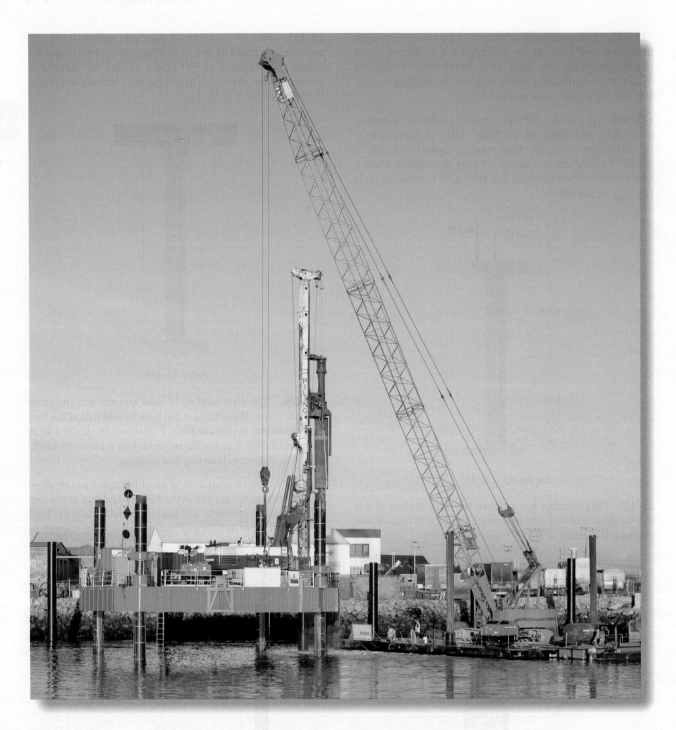

As piles are driven in place, their ends are subjected to impact loading. The nature of impact and the energy derived from it must be understood in order to determine the stress developed within the pile.

Energy Methods

14

CHAPTER OBJECTIVES

In this chapter, we will show how to apply energy methods to solve problems involving deflection. The chapter begins with a discussion of work and strain energy, followed by a development of the principle of conservation of energy. Using this principle, the stress and deflection of a member are determined when the member is subjected to impact. The method of virtual work and Castigliano's theorem are then developed, and these methods are used to determine the displacement and slope at points on structural members and mechanical elements.

14.1 External Work and Strain Energy

The deflection of joints on a truss or points on a beam or shaft can be determined using energy methods. Before developing any of these methods, however, we will first define the work caused by an external force and couple moment and show how to express this work in terms of a body's strain energy. The formulations to be presented here and in the next section will provide the basis for applying the work and energy methods that follow throughout the chapter.

14

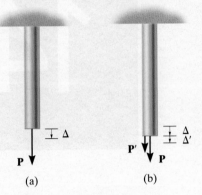

(a)   (b)

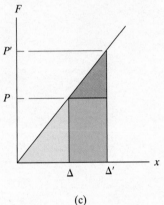

(c)

Fig. 14–1

Work of a Force. In mechanics, a force does *work* when it undergoes a displacement dx that is in the *same direction* as the force. The work done is a scalar, defined as $dU_e = F\,dx$. If the total displacement is Δ, the work becomes

$$U_e = \int_0^{\Delta} F\,dx \qquad (14\text{–}1)$$

To show how to apply this equation, we will calculate the work done by an axial force applied to the end of the bar shown in Fig. 14–1a. As the magnitude of the force is *gradually* increased from zero to some limiting value $F = P$, the final displacement of the end of the bar becomes Δ. If the material behaves in a linear-elastic manner, then the force will be directly proportional to the displacement; that is, $F = (P/\Delta)x$. Substituting into Eq. 14–1 and integrating from 0 to Δ, we get

$$U_e = \frac{1}{2}P\Delta \qquad (14\text{–}2)$$

Therefore, as the force is gradually applied to the bar, its magnitude builds from zero to some value P, and consequently, the work done is equal to the *average force magnitude*, $P/2$, times the total displacement Δ. We can represent this graphically as the light-blue shaded area of the triangle in Fig. 14–1c.

Suppose, however, that $\mathbf{P}$ is already applied to the bar and that *another force* $\mathbf{P}'$ is now applied, so that the end of the bar is displaced *further* by an amount Δ', Fig. 14–1b. The work done by $\mathbf{P}'$ is equal to the gray shaded triangular area, but now the work done by $\mathbf{P}$ when the bar undergoes this further displacement is

$$U_e' = P\Delta' \qquad (14\text{–}3)$$

Here the work represents the dark-blue shaded *rectangular area* in Fig. 14–1c. In this case $\mathbf{P}$ does not change its magnitude, since the bar's displacement Δ' is caused only by $\mathbf{P}'$. Therefore, work here is simply the force magnitude P times the displacement Δ'.

Work of a Couple Moment.

A couple moment **M** does work when it undergoes an angular displacement $d\theta$ along its line of action. The work is defined as $dU_e = M\,d\theta$, Fig. 14–2. If the total angular displacement is θ rad, the work becomes

$$U_e = \int_0^\theta M\,d\theta \qquad (14\text{–}4)$$

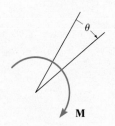

Fig. 14–2

As in the case of force, if the couple moment is applied to a *body* having linear elastic material behavior, such that its magnitude is increased gradually from zero at $\theta = 0$ to M at θ, then the work is

$$U_e = \frac{1}{2}M\theta \qquad (14\text{–}5)$$

However, if the couple moment is already applied to the body and other loadings further rotate the body by an amount θ', then the work is

$$U'_e = M\theta'$$

Strain Energy.

When loads are applied to a body, they will deform the material. Provided no energy is lost in the form of heat, the external work done by the loads will be converted into internal work called **strain energy**. This energy, which is *always positive*, is stored in the body and is caused by the action of either normal or shear stress.

Normal Stress.

If the volume element shown in Fig. 14–3 is subjected to the normal stress σ_z, then the force created on the element's top and bottom faces is $dF_z = \sigma_z\,dA = \sigma_z\,dx\,dy$. If this force is applied gradually to the element, like the force **P** discussed previously, its magnitude is increased from zero to dF_z, while the element undergoes an elongation $d\Delta_z = \epsilon_z\,dz$. The work done by dF_z is therefore $dU_i = \frac{1}{2}dF_z\,d\Delta_z = \frac{1}{2}[\sigma_z\,dx\,dy]\epsilon_z\,dz$. Since the volume of the element is $dV = dx\,dy\,dz$, we have

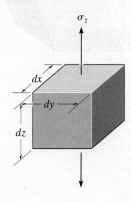

Fig. 14–3

$$dU_i = \frac{1}{2}\sigma_z\epsilon_z\,dV \qquad (14\text{–}6)$$

Notice that dU_i is *always positive*, even if σ_z is compressive, since σ_z and ϵ_z will always be in the same direction.

In general then, if the body is subjected only to a uniaxial *normal stress* σ, the strain energy in the body is then

$$U_i = \int_V \frac{\sigma\epsilon}{2}\,dV \qquad (14\text{–}7)$$

Also, if the material behaves in a linear-elastic manner, then Hooke's law applies, and we can express the strain energy in terms of the normal stress as

$$U_i = \int_V \frac{\sigma^2}{2E} dV \qquad (14\text{–}8)$$

Shear Stress.

A strain-energy expression similar to that for normal stress can also be established for the material when it is subjected to shear stress. Consider the volume element shown in Fig. 14–4. Here the shear stress causes the element to deform such that only the shear force $dF = \tau(dx\,dy)$, acting on the top face of the element, is displaced $\gamma\,dz$ relative to the bottom face. The *vertical faces* only rotate, and therefore the shear forces on these faces do no work. Hence, the strain energy stored in the element is

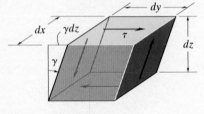

Fig. 14–4

$$dU_i = \frac{1}{2}[\tau(dx\,dy)]\gamma\,dz$$

or since $dV = dx\,dy\,dz$

$$dU_i = \frac{1}{2}\tau\gamma\,dV \qquad (14\text{–}9)$$

The strain energy stored in the body is therefore

$$U_i = \int_V \frac{\tau\gamma}{2} dV \qquad (14\text{–}10)$$

Like the case for normal strain energy, shear strain energy is always positive since τ and γ are always in the same direction. If the material is linear elastic, then, applying Hooke's law, $\gamma = \tau/G$, we can express the strain energy in terms of the shear stress as

$$U_i = \int_V \frac{\tau^2}{2G} dV \qquad (14\text{–}11)$$

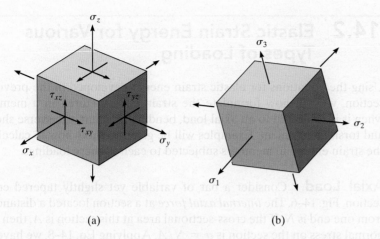

(a)   (b)

Fig. 14–5

In the next section, we will use Eqs. 14–8 and 14–11 to obtain formal expressions for the strain energy stored in members subjected to several types of loads. Once this is done we will then be able to develop the energy methods necessary to determine the displacement and slope at points on a body.

Multiaxial Stress.   The previous development may be expanded to determine the strain energy in a body when it is subjected to a general state of stress, Fig. 14–5a. The strain energies associated with each of the normal and shear stress components can be obtained from Eqs. 14–6 and 14–9. Since energy is a scalar, the total strain energy in the body is therefore

$$U_i = \int_V \left[\frac{1}{2}\sigma_x \epsilon_x + \frac{1}{2}\sigma_y \epsilon_y + \frac{1}{2}\sigma_z \epsilon_z \right.$$

$$\left. + \frac{1}{2}\tau_{xy}\gamma_{xy} + \frac{1}{2}\tau_{yz}\gamma_{yz} + \frac{1}{2}\tau_{xz}\gamma_{xz} \right] dV \qquad (14\text{--}12)$$

The strains can be eliminated by using the generalized form of Hooke's law given by Eqs. 10–18 and 10–19. After substituting and combining terms, we have

$$U_i = \int_V \left[\frac{1}{2E} \left(\sigma_x^2 + \sigma_y^2 + \sigma_z^2 \right) - \frac{\nu}{E} \left(\sigma_x\sigma_y + \sigma_y\sigma_z + \sigma_x\sigma_z \right) \right.$$

$$\left. + \frac{1}{2G} \left(\tau_{xy}^2 + \tau_{yz}^2 + \tau_{xz}^2 \right) \right] dV \qquad (14\text{--}13)$$

If only the principal stresses $\sigma_1, \sigma_2, \sigma_3$ act on the element, Fig. 14–5b, this equation reduces to a simpler form, namely,

$$U_i = \int_V \left[\frac{1}{2E} \left(\sigma_1^2 + \sigma_2^2 + \sigma_3^2 \right) - \frac{\nu}{E} \left(\sigma_1\sigma_2 + \sigma_2\sigma_3 + \sigma_1\sigma_3 \right) \right] dV \quad (14\text{--}14)$$

This equation was used in Sec. 10.7 as a basis for developing the maximum-distortion-energy theory.

14.2 Elastic Strain Energy for Various Types of Loading

Using the equations for elastic strain energy developed in the previous section, we will now formulate the strain energy stored in a member when it is subjected to an axial load, bending moment, transverse shear, and torsional moment. Examples will be given to show how to calculate the strain energy in members subjected to each of these loadings.

Axial Load. Consider a bar of variable yet slightly tapered cross section, Fig. 14–6. The *internal axial force* at a section located a distance x from one end is N. If the cross-sectional area at this section is A, then the normal stress on the section is $\sigma = N/A$. Applying Eq. 14–8, we have

Fig. 14–6

$$U_i = \int_V \frac{\sigma_x^2}{2E}\,dV = \int_V \frac{N^2}{2EA^2}\,dV$$

If we choose an element or differential slice having a volume $dV = A\,dx$, the general formula for the strain energy in the bar is therefore

$$U_i = \int_0^L \frac{N^2}{2AE}\,dx \tag{14–15}$$

For the more common case of a prismatic bar of constant cross-sectional area A, length L, and constant axial load N, Fig. 14–7, Eq. 14–15, when integrated, gives

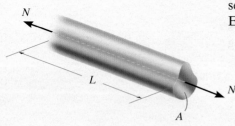

Fig. 14–7

$$U_i = \frac{N^2 L}{2AE} \tag{14–16}$$

Notice that the bar's elastic strain energy will *increase* if the length of the bar is increased, or if the modulus of elasticity or cross-sectional area is decreased. For example, an aluminum rod $[E_{al} = 70 \text{ GPa}]$ will store approximately three times as much energy as a steel rod $[E_{st} = 200 \text{ GPa}]$ having the same size and subjected to the same load. However, doubling the cross-sectional area of a rod will decrease its ability to store energy by one-half. The following example illustrates this point numerically.

EXAMPLE | 14.1

One of the two high-strength steel bolts A and B shown in Fig. 14–8 is to be chosen to support a sudden tensile loading. For the choice it is necessary to determine the greatest amount of elastic strain energy that each bolt can absorb. Bolt A has a diameter of 20 mm for 50 mm of its length and a root (or smallest) diameter of 18 mm within the 6 mm threaded region. Bolt B has "upset" threads, such that the diameter throughout its 56-mm length can be taken as 18 mm. In both cases, neglect the extra material that makes up the threads. Take $E_{st} = 210$ MPa, $\sigma_Y = 310$ MPa.

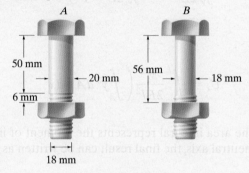

Fig. 14–8

SOLUTION

Bolt A. If the bolt is subjected to its maximum tension, the maximum stress of $\sigma_Y = 310$ MPa will occur within the 6 mm region. This tension force is

$$P_{max} = \sigma_Y A = 310 \text{ N/mm}^2 \left[\pi \left(\frac{18 \text{ mm}}{2} \right)^2 \right] = 78\,886 \text{ N} = 78.89 \text{ kN}$$

Applying Eq. 14–16 to each region of the bolt, we have

$$U_i = \sum \frac{N^2 L}{2AE}$$

$$= \frac{(78\,886 \text{ N})^2(50 \text{ mm})}{2[\pi(20 \text{ mm}/2)^2][210(10^3) \text{ N/mm}^2]} + \frac{(78\,886 \text{ N})^2(6 \text{ mm})}{2[\pi(18 \text{ mm}/2)^2][210(10^3) \text{ N/mm}^2]}$$

$$= 2708 \text{ N} \cdot \text{mm} = 2.708 \text{ N} \cdot \text{m} = 2.708 \text{ J} \qquad \text{Ans.}$$

Bolt B. Here the bolt is assumed to have a uniform diameter of 18 mm throughout its 56-mm length. Also, from the calculation above, it can support a maximum tension force of $P_{max} = 78\,886$ N. Thus,

$$U_i = \frac{N^2 L}{2AE} = \frac{(78\,886 \text{ N})^2(56 \text{ mm})}{2[\pi(18 \text{ mm}/2)^2][210(10^3) \text{ N/mm}^2]} = 3261 \text{ N} \cdot \text{mm} = 3.26 \text{ N} \cdot \text{m} = 3.26 \text{ J} \quad \text{Ans.}$$

NOTE: By comparison, bolt B can absorb 36% more elastic energy than bolt A, because it has a smaller cross section along its shank.

Bending Moment. Since a bending moment applied to a straight prismatic member develops *normal stress* in the member, we can use Eq. 14–8 to determine the strain energy stored in the member due to bending. For example, consider the axisymmetric beam shown in Fig. 14–9. Here the internal moment is M, and the normal stress acting on the arbitrary element a distance y from the neutral axis is $\sigma = My/I$. If the volume of the element is $dV = dA\,dx$, where dA is the area of its exposed face and dx is its length, the elastic strain energy in the beam is

$$U_i = \int_V \frac{\sigma^2}{2E}\,dV = \int_V \frac{1}{2E}\left(\frac{My}{I}\right)^2 dA\,dx$$

or

$$U_i = \int_0^L \frac{M^2}{2EI^2}\left(\int_A y^2\,dA\right)dx$$

Realizing that the area integral represents the moment of inertia of the area about the neutral axis, the final result can be written as

$$U_i = \int_0^L \frac{M^2\,dx}{2EI} \tag{14–17}$$

To evaluate the strain energy, therefore, we must first express the internal moment as a function of its position x along the beam, and then perform the integration over the beam's entire length.* The following examples illustrate this procedure.

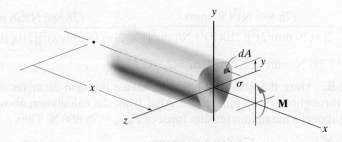

Fig. 14–9

*Recall that the flexure formula, as used here, can also be used with justifiable accuracy to determine the stress in slightly tapered beams. (See Sec. 6.4.) So in the general sense, I in Eq. 14–17 may also have to be expressed as a function of x.

EXAMPLE 14.2

Determine the elastic strain energy due to bending of the cantilevered beam in Fig. 14–10a. *EI* is constant.

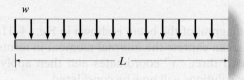

(a)

Fig. 14–10

SOLUTION
The internal moment in the beam is determined by establishing the *x* coordinate with origin at the left side. The left segment of the beam is shown in Fig. 14–10b. We have

$$\zeta + \Sigma M_{NA} = 0; \qquad M + wx\left(\frac{x}{2}\right) = 0$$

$$M = -w\left(\frac{x^2}{2}\right)$$

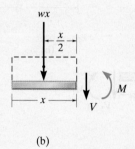

(b)

Applying Eq. 14–17 yields

$$U_i = \int_0^L \frac{M^2 \, dx}{2EI} = \int_0^L \frac{[-w(x^2/2)]^2 \, dx}{2EI} = \frac{w^2}{8EI} \int_0^L x^4 \, dx$$

or

$$U_i = \frac{w^2 L^5}{40EI} \qquad\qquad Ans.$$

   We can also obtain the strain energy using an *x* coordinate having its origin at the right side of the beam and extending positive to the left, Fig. 14–10c. In this case,

$$\zeta + \Sigma M_{NA} = 0; \quad -M - wx\left(\frac{x}{2}\right) + wL(x) - \frac{wL^2}{2} = 0$$

$$M = -\frac{wL^2}{2} + wLx - w\left(\frac{x^2}{2}\right)$$

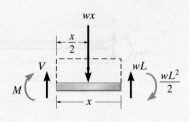

(c)

Applying Eq. 14–17, we obtain the same result as before; however, more calculations are involved in this case.

14

EXAMPLE 14.3

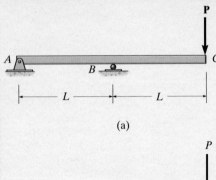

(a)

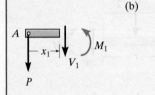

(b)

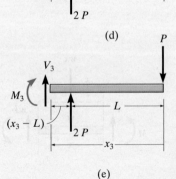

(c)

(d)

(e)

Fig. 14–11

Determine the bending strain energy in region AB of the beam shown in Fig. 14–11a. EI is constant.

SOLUTION

A free-body diagram of the beam is shown in Fig. 14–11b. To obtain the answer we can express the internal moment in terms of any one of the indicated three "x" coordinates and then apply Eq. 14–17. Each of these solutions will now be considered.

$0 \leq x_1 \leq L$. From the free-body diagram of the section in Fig. 14–11c, we have

$$\zeta + \Sigma M_{NA} = 0; \qquad M_1 + Px_1 = 0$$

$$M_1 = -Px_1$$

$$U_i = \int \frac{M^2\,dx}{2EI} = \int_0^L \frac{(-Px_1)^2\,dx_1}{2EI} = \frac{P^2 L^3}{6EI} \qquad Ans.$$

$0 \leq x_2 \leq L$. Using the free-body diagram of the section in Fig. 14–11d gives

$$\zeta + \Sigma M_{NA} = 0; \qquad -M_2 + 2P(x_2) - P(x_2 + L) = 0$$

$$M_2 = P(x_2 - L)$$

$$U_i = \int \frac{M^2\,dx}{2EI} = \int_0^L \frac{[P(x_2 - L)]^2\,dx_2}{2EI} = \frac{P^2 L^3}{6EI} \qquad Ans.$$

$L \leq x_3 \leq 2L$. From the free-body diagram in Fig. 14–11e, we have

$$\zeta + \Sigma M_{NA} = 0; \qquad -M_3 + 2P(x_3 - L) - P(x_3) = 0$$

$$M_3 = P(x_3 - 2L)$$

$$U_i = \int \frac{M^2\,dx}{2EI} = \int_L^{2L} \frac{[P(x_3 - 2L)]^2\,dx_3}{2EI} = \frac{P^2 L^3}{6EI} \qquad Ans.$$

NOTE: This and the previous example indicate that the strain energy for the beam can be found using *any* suitable x coordinate. It is only necessary to integrate over the range of the coordinate where the internal energy is to be determined. Here the choice of x_1 provides the simplest solution.

Transverse Shear. The strain energy due to shear stress in a beam element can be determined by applying Eq. 14–11. Here we will consider the beam to be prismatic and to have an axis of symmetry about the y axis as shown in Fig. 14–12. If the internal shear at the section x is V, then the shear stress acting on the volume element of material, having an area dA and length dx, is $\tau = VQ/It$. Substituting into Eq. 14–11, the strain energy for shear becomes

$$U_i = \int_V \frac{\tau^2}{2G} dV = \int_V \frac{1}{2G}\left(\frac{VQ}{It}\right)^2 dA\, dx$$

$$U_i = \int_0^L \frac{V^2}{2GI^2}\left(\int_A \frac{Q^2}{t^2} dA\right) dx$$

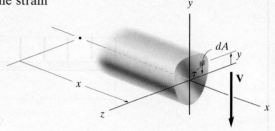

Fig. 14–12

The integral in parentheses can be simplified if we define the *form factor* for shear as

$$f_s = \frac{A}{I^2}\int_A \frac{Q^2}{t^2} dA \qquad (14\text{–}18)$$

Substituting into the above equation, we get

$$\boxed{U_i = \int_0^L \frac{f_s V^2\, dx}{2GA}} \qquad (14\text{–}19)$$

The form factor defined by Eq. 14–18 is a dimensionless number that is unique for each specific cross-sectional area. For example, if the beam has a rectangular cross section of width b and height h, Fig. 14–13, then

$$t = b$$
$$dA = b\,dy$$
$$I = \frac{1}{12}bh^3$$
$$Q = \overline{y}'A' = \left(y + \frac{(h/2) - y}{2}\right)b\left(\frac{h}{2} - y\right) = \frac{b}{2}\left(\frac{h^2}{4} - y^2\right)$$

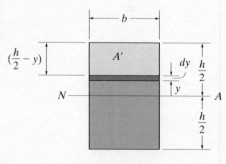

Fig. 14–13

Substituting these terms into Eq. 14–18, we get

$$f_s = \frac{bh}{\left(\frac{1}{12}bh^3\right)^2}\int_{-h/2}^{h/2} \frac{b^2}{4b^2}\left(\frac{h^2}{4} - y^2\right)^2 b\,dy = \frac{6}{5} \qquad (14\text{–}20)$$

The form factor for other sections can be determined in a similar manner. Once obtained, this factor is substituted into Eq. 14–19 and the strain energy for transverse shear can then be evaluated.

EXAMPLE 14.4

Determine the strain energy in the cantilevered beam due to shear if the beam has a square cross section and is subjected to a uniform distributed load w, Fig. 14–14a. EI and G are constant.

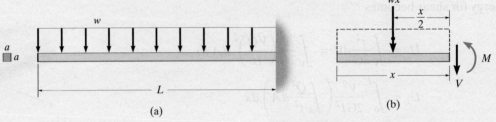

Fig. 14–14

(a)

(b)

SOLUTION

From the free-body diagram of an arbitrary section, Fig. 14–14b, we have

$$+\uparrow \Sigma F_y = 0; \qquad\qquad -V - wx = 0$$

$$V = -wx$$

Since the cross section is square, the form factor $f_s = \frac{6}{5}$ (Eq. 14–20) and therefore Eq. 14–19 becomes

$$(U_i)_s = \int_0^L \frac{\frac{6}{5}(-wx)^2 \, dx}{2GA} = \frac{3w^2}{5GA} \int_0^L x^2 \, dx$$

or

$$(U_i)_s = \frac{w^2 L^3}{5GA} \qquad\qquad Ans.$$

NOTE: Using the results of Example 14.2, with $A = a^2$, $I = \frac{1}{12}a^4$, the ratio of shear to bending strain energy is

$$\frac{(U_i)_s}{(U_i)_b} = \frac{w^2 L^3/5Ga^2}{w^2 L^5/40E\left(\frac{1}{12}a^4\right)} = \frac{2}{3}\left(\frac{a}{L}\right)^2 \frac{E}{G}$$

Since $G = E/2(1 + \nu)$ and $\nu \leq \frac{1}{2}$ (Sec. 10.6), then as an *upper* bound, $E = 3G$, so that

$$\frac{(U_i)_s}{(U_i)_b} = 2\left(\frac{a}{L}\right)^2$$

It can be seen that this ratio will increase as L decreases. However, even for very short beams, where, say, $L = 5a$, the contribution due to shear strain energy is only 8% of the bending strain energy. For this reason, the shear strain energy stored in beams is usually neglected in engineering analysis.

Torsional Moment. To determine the internal strain energy in a circular shaft or tube due to an applied torsional moment, we must apply Eq. 14–11. Consider the slightly tapered shaft in Fig. 14–15. A section of the shaft taken a distance x from one end is subjected to an internal torque T. The shear stress distribution that causes this torque varies linearly from the center of the shaft. On the arbitrary element of area dA and length dx, the stress is $\tau = T\rho/J$. The strain energy stored in the shaft is thus

$$U_i = \int_V \frac{\tau^2}{2G}\,dV = \int_V \frac{1}{2G}\left(\frac{T\rho}{J}\right)^2 dA\,dx = \int_0^L \frac{T^2}{2GJ^2}\left(\int_A \rho^2\,dA\right)dx$$

Since the area integral represents the polar moment of inertia J for the shaft at the section, the final result can be written as

$$U_i = \int_0^L \frac{T^2}{2GJ}\,dx \qquad (14\text{–}21)$$

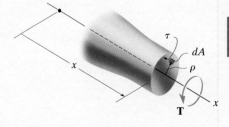

Fig. 14–15

The most common case occurs when the shaft (or tube) has a constant cross-sectional area and the applied torque is constant, Fig. 14–16. Integration of Eq. 14–21 then gives

$$U_i = \frac{T^2 L}{2GJ} \qquad (14\text{–}22)$$

From this equation we may conclude that, like an axially loaded member, the energy-absorbing capacity of a torsionally loaded shaft is *decreased* by increasing the diameter of the shaft, since this increases J.

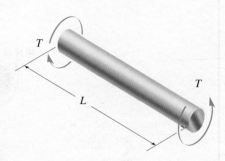

Fig. 14–16

Important Points

- A *force* does work when it moves through a *displacement*. When a force is applied to a body and its magnitude is increased gradually from zero to F, the work is $U = (F/2)\Delta$, whereas if the force is constant when the displacement occurs then $U = F\Delta$.

- A *couple moment* does work when it displaces through a *rotation*.

- *Strain energy* is caused by the internal work of the normal and shear stresses. It is always a *positive* quantity.

- The strain energy can be related to the resultant internal loadings N, V, M, and T.

- As the beam becomes longer, the strain energy due to bending becomes much larger than the strain energy due to shear. For this reason, the *shear strain energy* in beams can generally be *neglected*.

The following example illustrates how to determine the strain energy in a circular shaft due to a torsional loading.

EXAMPLE | 14.5

The tubular shaft in Fig. 14–17a is fixed at the wall and subjected to two torques as shown. Determine the strain energy stored in the shaft due to this loading. $G = 75$ GPa.

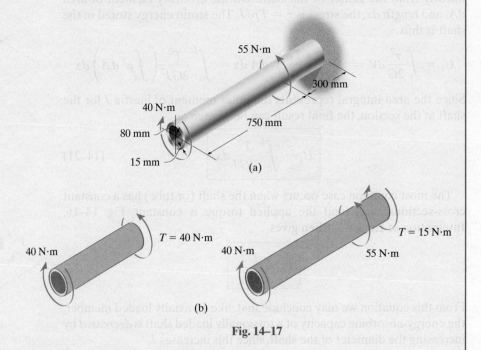

(a)

(b)

Fig. 14–17

SOLUTION

Using the method of sections, the internal torque is first determined within the two regions of the shaft where it is constant, Fig. 14–17b. Although these torques (40 N·m and 15 N·m) are in opposite directions, this will be of no consequence in determining the strain energy, since the torque is squared in Eq. 14–22. In other words, the strain energy is always positive. The polar moment of inertia for the shaft is

$$J = \frac{\pi}{2}[(0.08 \text{ m})^4 - (0.065 \text{ m})^4] = 36.30(10^{-6}) \text{ m}^4$$

Applying Eq. 14–22, we have

$$U_i = \sum \frac{T^2 L}{2GJ}$$

$$= \frac{(40 \text{ N·m})^2(0.750 \text{ m})}{2[75(10^9) \text{ N/m}^2]36.30(10^{-6}) \text{ m}^4} + \frac{(15 \text{ N·m})^2(0.300 \text{ m})}{2[75(10^9) \text{ N/m}^2]36.30(10^{-6}) \text{ m}^4}$$

$$= 233 \ \mu\text{J} \qquad\qquad\qquad Ans.$$

PROBLEMS

14–1. A material is subjected to a general state of plane stress. Express the strain energy density in terms of the elastic constants E, G, and ν and the stress components σ_x, σ_y, and τ_{xy}.

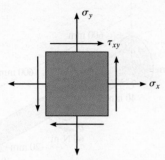

Prob. 14–1

14–2. The strain-energy density must be the same whether the state of stress is represented by σ_x, σ_y, and τ_{xy}, or by the principal stresses σ_1 and σ_2. This being the case, equate the strain–energy expressions for each of these two cases and show that $G = E/[2(1 + \nu)]$.

14–3. Determine the strain energy in the stepped rod assembly. Portion AB is steel and BC is brass. $E_{br} = 101$ GPa, $E_{st} = 200$ GPa, $(\sigma_Y)_{br} = 410$ MPa, $(\sigma_Y)_{st} = 250$ MPa.

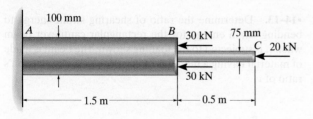

Prob. 14–3

***14–4.** Determine the torsional strain energy in the A-36 steel shaft. The shaft has a diameter of 40 mm.

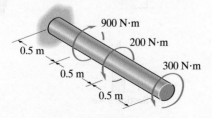

Prob. 14–4

•**14–5.** Determine the strain energy in the rod assembly. Portion AB is steel, BC is brass, and CD is aluminum. $E_{st} = 200$ GPa, $E_{br} = 101$ GPa, and $E_{al} = 73.1$ GPa.

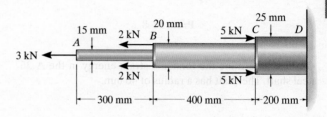

Prob. 14–5

14–6. If $P = 60$ kN, determine the total strain energy stored in the truss. Each member has a cross-sectional area of $2.5(10^3)$ mm^2 and is made of A-36 steel.

14–7. Determine the maximum force **P** and the corresponding maximum total strain energy stored in the truss without causing any of the members to have permanent deformation. Each member has the cross-sectional area of $2.5(10^3)$ mm^2 and is made of A-36 steel.

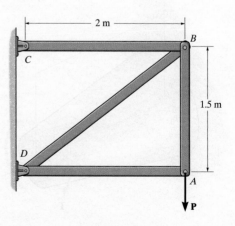

Probs. 14–6/7

*14–8. Determine the torsional strain energy in the A-36 steel shaft. The shaft has a radius of 30 mm.

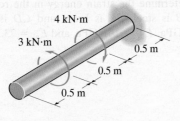

4 kN·m

3 kN·m

0.5 m

0.5 m

0.5 m

Prob. 14–8

•14–9. Determine the torsional strain energy in the A-36 steel shaft. The shaft has a radius of 40 mm.

12 kN·m

6 kN·m

0.5 m

8 kN·m

0.4 m

0.6 m

Prob. 14–9

14–10. Determine the torsional strain energy stored in the tapered rod when it is subjected to the torque **T**. The rod is made of material having a modulus of rigidity of G.

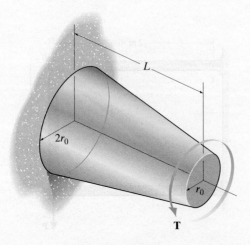

L

$2r_0$

r_0

T

Prob. 14–10

14–11. The shaft assembly is fixed at C. The hollow segment BC has an inner radius of 20 mm and outer radius of 40 mm, while the solid segment AB has a radius of 20 mm. Determine the torsional strain energy stored in the shaft. The shaft is made of 2014-T6 aluminum alloy. The coupling at B is rigid.

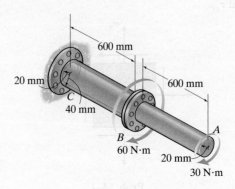

600 mm

20 mm

600 mm

C

40 mm

A

B

60 N·m

20 mm

30 N·m

Prob. 14–11

*14–12. Consider the thin-walled tube of Fig. 5–28. Use the formula for shear stress, $\tau_{avg} = T/2tA_m$, Eq. 5–18, and the general equation of shear strain energy, Eq. 14–11, to show that the twist of the tube is given by Eq. 5–20, *Hint*: Equate the work done by the torque T to the strain energy in the tube, determined from integrating the strain energy for a differential element, Fig. 14–4, over the volume of material.

•14–13. Determine the ratio of shearing strain energy to bending strain energy for the rectangular cantilever beam when it is subjected to the loading shown. The beam is made of material having a modulus of elasticity of E and Poisson's ratio of ν.

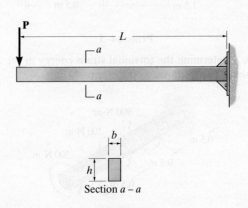

P

L

a

a

b

h

Section $a – a$

Prob. 14–13

14–14. Determine the bending strain-energy in the beam due to the loading shown. EI is constant.

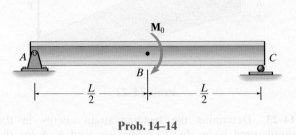

Prob. 14–14

14–15. Determine the bending strain energy in the beam. EI is constant.

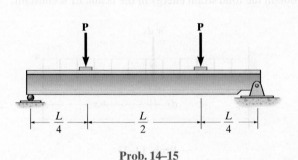

Prob. 14–15

*****14–16.** Determine the bending strain energy in the A-36 structural steel W250 × 18 beam. Obtain the answer using the coordinates (a) x_1 and x_4, and (b) x_2 and x_3.

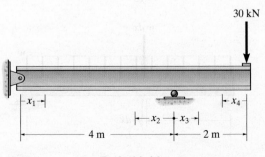

Prob. 14–16

•14–17. Determine the bending strain energy in the A-36 steel beam. $I = 99.2 \, (10^6) \, mm^4$.

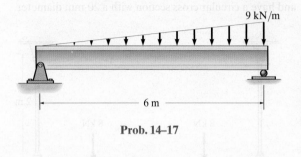

Prob. 14–17

14–18. Determine the bending strain energy in the A-36 steel beam due to the distributed load. $I = 122 \, (10^6) \, mm^4$.

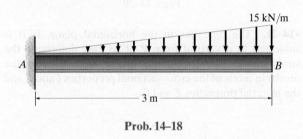

Prob. 14–18

14–19. Determine the strain energy in the *horizontal* curved bar due to torsion. There is a *vertical* force **P** acting at its end. JG is constant.

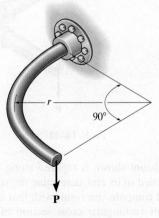

Prob. 14–19

14

*14–20. Determine the bending strain energy in the beam and the axial strain energy in each of the two rods. The beam is made of 2014-T6 aluminum and has a square cross section 50 mm by 50 mm. The rods are made of A-36 steel and have a circular cross section with a 20-mm diameter.

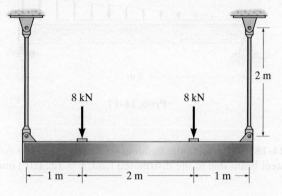

Prob. 14–20

•14–21. The pipe lies in the horizontal plane. If it is subjected to a vertical force **P** at its end, determine the strain energy due to bending and torsion. Express the results in terms of the cross-sectional properties I and J, and the material properties E and G.

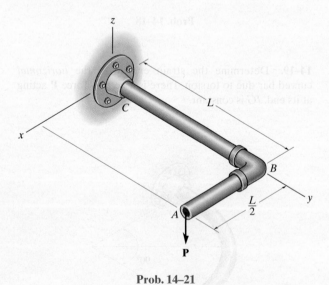

Prob. 14–21

14–22. The beam shown is tapered along its width. If a force **P** is applied to its end, determine the strain energy in the beam and compare this result with that of a beam that has a constant rectangular cross section of width b and height h.

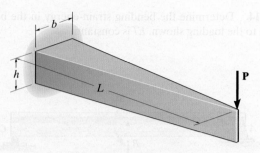

Prob. 14–22

14–23. Determine the bending strain energy in the cantilevered beam due to a uniform load w. Solve the problem two ways. (a) Apply Eq. 14–17. (b) The load $w\,dx$ acting on a segment dx of the beam is displaced a distance y, where $y = w(-x^4 + 4L^3x - 3L^4)/(24EI)$, the equation of the elastic curve. Hence the internal strain energy in the differential segment dx of the beam is equal to the external work, i.e., $dU_i = \frac{1}{2}(w\,dx)(-y)$. Integrate this equation to obtain the total strain energy in the beam. EI is constant.

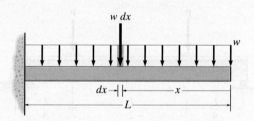

Prob. 14–23

*14–24. Determine the bending strain energy in the simply supported beam due to a uniform load w. Solve the problem two ways. (a) Apply Eq. 14–17. (b) The load $w\,dx$ acting on the segment dx of the beam is displaced a distance y, where $y = w(-x^4 + 2Lx^3 - L^3x)/(24EI)$, the equation of the elastic curve. Hence the internal strain energy in the differential segment dx of the beam is equal to the external work, i.e., $dU_i = \frac{1}{2}(w\,dx)(-y)$. Integrate this equation to obtain the total strain energy in the beam. EI is constant.

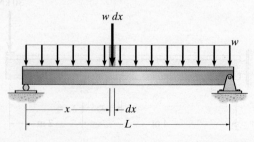

Prob. 14–24

14.3   Conservation of Energy

All energy methods used in mechanics are based on a balance of energy, often referred to as the conservation of energy. In this chapter, only mechanical energy will be considered in the energy balance; that is, the energy developed by heat, chemical reactions, and electromagnetic effects will be neglected. As a result, if a loading is applied *slowly* to a body, then physically the external loads tend to deform the body so that the loads do *external work* U_e as they are displaced. This external work on the body is transformed into *internal work* or strain energy U_i, which is stored in the body. Furthermore, when the loads are removed, the strain energy restores the body back to its original undeformed position, provided the material's elastic limit is not exceeded. The conservation of energy for the body can therefore be stated mathematically as

$$U_e = U_i \qquad (14\text{--}23)$$

We will now show three examples of how this equation can be applied to determine the displacement of a point on a deformable member or structure. As the first example, consider the truss in Fig. 14–18 subjected to the load **P**. Provided **P** is applied gradually, the external work done by **P** is determined from Eq. 14–2, that is, $U_e = \frac{1}{2}P\Delta$, where Δ is the vertical displacement of the truss at the joint where **P** is applied. Assuming that **P** develops an axial force **N** in a particular member, the strain energy stored in this member is determined from Eq. 14–16, that is, $U_i = N^2 L/2AE$. Summing the strain energies for all the members of the truss, we can write Eq. 14–23 as

$$\frac{1}{2}P\Delta = \sum \frac{N^2 L}{2AE} \qquad (14\text{--}24)$$

Once the internal forces (N) in all the members of the truss are determined and the terms on the right calculated, it is then possible to determine the unknown displacement Δ.

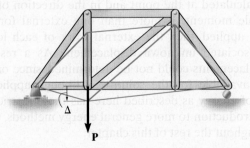

Fig. 14–18

As a second example, consider finding the vertical displacement Δ under the load **P** acting on the beam in Fig. 14–19. Again, the external work is $U_e = \frac{1}{2}P\Delta$. In this case the strain energy is the result of internal shear and moment loadings caused by **P**. In particular, the contribution of strain energy due to shear is generally *neglected* in most beam deflection problems unless the beam is short and supports a very large load. (See Example 14.4.) Consequently, the beam's strain energy will be determined only by the internal bending moment M, and therefore, using Eq. 14–17, Eq. 14–23 can be written symbolically as

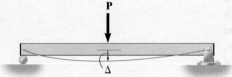

$$\frac{1}{2}P\Delta = \int_0^L \frac{M^2}{2EI}dx \qquad (14\text{–}25)$$

Fig. 14–19

Once M is expressed as a function of position x and the integral is evaluated, Δ can then be determined.

As a last example, we will consider a beam loaded by a couple moment $\mathbf{M}_0$ as shown in Fig. 14–20. This moment causes the rotational displacement θ at the point of application of the couple moment. Since the couple moment only does work when it *rotates*, using Eq. 14–5, the external work is $U_e = \frac{1}{2}M_0\theta$. Therefore Eq. 14–23 becomes

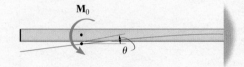

$$\frac{1}{2}M_0\theta = \int_0^L \frac{M^2}{2EI}dx \qquad (14\text{–}26)$$

Fig. 14–20

Here the strain energy is the result of the internal bending moment M caused by application of the couple moment $\mathbf{M}_0$. Once M has been expressed as a function of x and the strain energy evaluated, then θ which measures the slope of the elastic curve can be determined.

In each of the above examples, it should be noted that application of Eq. 14–23 is *quite limited*, because only a *single* external force or couple moment must act on the member or structure. Also, the displacement can *only* be calculated at the point and in the direction of the external force or couple moment. If more than one external force or couple moment were applied, then the external work of each loading would involve its associated unknown displacement. As a result, *all* these unknown displacements could not be determined, since only the single Eq. 14–23 is available for the solution. Although application of the conservation of energy as described here has these restrictions, it does serve as an introduction to more general energy methods, which we will consider throughout the rest of this chapter.

EXAMPLE 14.6

The three-bar truss in Fig. 14–21a is subjected to a horizontal force of 20 kN. If the cross-sectional area of each member is 100 mm^2, determine the horizontal displacement at point B. $E = 200$ GPa.

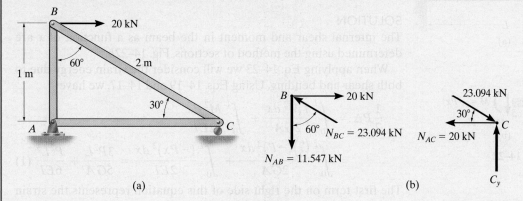

Fig. 14–21

SOLUTION

We can apply the conservation of energy to solve this problem because only a *single* external force acts on the truss and the required displacement happens to be in the *same direction* as the force. Furthermore, the reactive forces on the truss do no work since they are not displaced.

Using the method of joints, the force in each member is determined as shown on the free-body diagrams of the pins at B and C, Fig. 14–21b.

Applying Eq. 14–24, we have

$$\frac{1}{2}P\Delta = \sum \frac{N^2 L}{2AE}$$

$$\frac{1}{2}(20 \text{ kN})(\Delta_B)_h = \frac{(11.547 \text{ kN})^2(1 \text{ m})}{2AE} + \frac{(-23.094 \text{ kN})^2(2 \text{ m})}{2AE}$$

$$+ \frac{(20 \text{ kN})^2(1.732 \text{ m})}{2AE}$$

$$(\Delta_B)_h = \frac{94.64 \text{ kN} \cdot \text{m}}{AE}$$

Notice that since N is squared, it does not matter if a particular member is in tension or compression. Substituting in the numerical data for A and E and solving, we get

$$(\Delta_B)_h = \frac{94.64 \text{ kN} \cdot \text{m}(1000 \text{ mm/m})}{(100 \text{ mm}^2)[200 \text{ kN/mm}^2]}$$

$$= 4.732 \text{ mm} \quad \rightarrow \qquad \qquad Ans.$$

EXAMPLE | 14.7

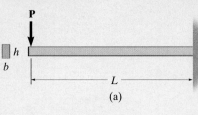

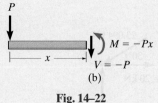

Fig. 14–22

The cantilevered beam in Fig. 14–22a has a rectangular cross section and is subjected to a load **P** at its end. Determine the displacement of the load. EI is constant.

SOLUTION

The internal shear and moment in the beam as a function of x are determined using the method of sections, Fig. 14–22b.

When applying Eq. 14–23 we will consider the strain energy due to both shear and bending. Using Eqs. 14–19 and 14–17, we have

$$\frac{1}{2}P\Delta = \int_0^L \frac{f_s V^2\, dx}{2GA} + \int_0^L \frac{M^2\, dx}{2EI}$$

$$= \int_0^L \frac{\left(\frac{6}{5}\right)(-P)^2\, dx}{2GA} + \int_0^L \frac{(-Px)^2\, dx}{2EI} = \frac{3P^2 L}{5GA} + \frac{P^2 L^3}{6EI} \quad (1)$$

The first term on the right side of this equation represents the strain energy due to shear, while the second is the strain energy due to bending. As stated in Example 14.4, for most beams the shear strain energy is much smaller than the bending strain energy. To show when this is the case for the beam in Fig. 14–22a, we require

$$\frac{3}{5}\frac{P^2 L}{GA} \ll \frac{P^2 L^3}{6EI}$$

$$\frac{3}{5}\frac{P^2 L}{G(bh)} \ll \frac{P^2 L^3}{6E\left[\frac{1}{12}(bh^3)\right]}$$

$$\frac{3}{5G} \ll \frac{2L^2}{Eh^2}$$

Since $E \leq 3G$ (see Example 14.4), then

$$0.9 \ll \left(\frac{L}{h}\right)^2$$

Hence if L is relatively long compared with h, the beam becomes slender and the shear strain energy can be neglected. In other words, the *shear strain energy* becomes important *only* for *short, deep beams*. For example, beams for which $L = 5h$ have approximately 28 times more bending strain energy than shear strain energy, so neglecting the shear strain energy represents an error of about 3.6%. With this in mind, Eq. 1 can be simplified to

$$\frac{1}{2}P\Delta = \frac{P^2 L^3}{6EI}$$

so that

$$\Delta = \frac{PL^3}{3EI} \qquad\qquad\qquad Ans.$$

PROBLEMS

•14–25. Determine the horizontal displacement of joint A. Each bar is made of A-36 steel and has a cross-sectional area of 950 mm².

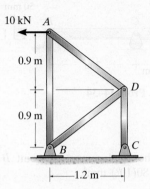

Prob. 14–25

14–26. Determine the horizontal displacement of joint C. AE is constant.

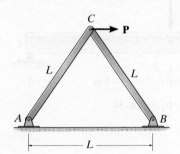

Prob. 14–26

14–27. Determine the vertical displacement of joint C. AE is constant.

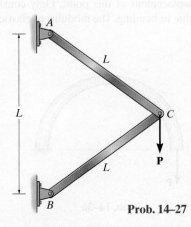

Prob. 14–27

*14–28. Determine the horizontal displacement of joint D. AE is constant.

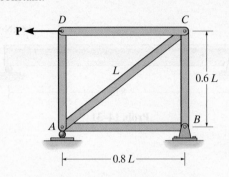

Prob. 14–28

•14–29. The cantilevered beam is subjected to a couple moment M_0 applied at its end. Determine the slope of the beam at B. EI is constant.

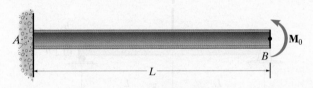

Prob. 14–29

14–30. Determine the vertical displacement of point C of the simply supported 6061-T6 aluminum beam. Consider both shearing and bending strain energy.

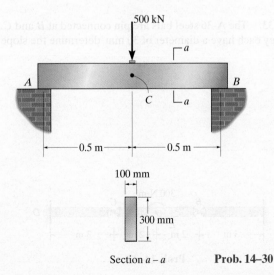

Section a – a   **Prob. 14–30**

14–31. Determine the slope at the end B of the A-36 steel beam. $I = 80(10^6)$ mm^4.

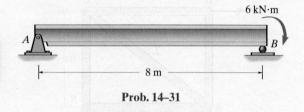

Prob. 14–31

14–34. The A-36 steel bars are pin connected at B. If each has a square cross section, determine the vertical displacement at B.

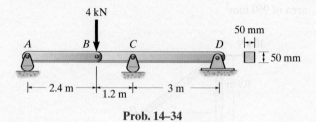

Prob. 14–34

*14–32. Determine the deflection of the beam at its center caused by shear. The shear modulus is G.

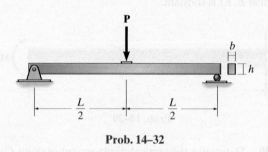

Prob. 14–32

14–35. Determine the displacement of point B on the A-36 steel beam. $I = 80(10^6)$ mm^4.

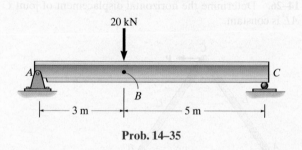

Prob. 14–35

•14–33. The A-36 steel bars are pin connected at B and C. If they each have a diameter of 30 mm, determine the slope at E.

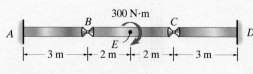

Prob. 14–33

*14–36. The rod has a circular cross section with a moment of inertia I. If a vertical force $\mathbf{P}$ is applied at A, determine the vertical displacement at this point. Only consider the strain energy due to bending. The modulus of elasticity is E.

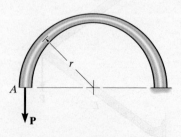

Prob. 14–36

•14–37. The load **P** causes the open coils of the spring to make an angle θ with the horizontal when the spring is stretched. Show that for this position this causes a torque $T = PR \cos \theta$ and a bending moment $M = PR \sin \theta$ at the cross section. Use these results to determine the maximum normal stress in the material.

14–38. The coiled spring has n coils and is made from a material having a shear modulus G. Determine the stretch of the spring when it is subjected to the load **P**. Assume that the coils are close to each other so that $\theta \approx 0°$ and the deflection is caused entirely by the torsional stress in the coil.

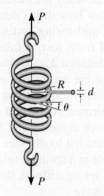

Probs. 14–37/38

*14–40. The rod has a circular cross section with a polar moment of inertia J and moment of inertia I. If a vertical force **P** is applied at A, determine the vertical displacement at this point. Consider the strain energy due to bending and torsion. The material constants are E and G.

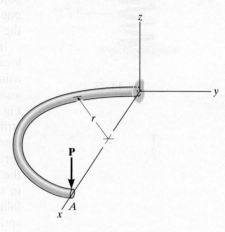

Prob. 14–40

14–39. The pipe assembly is fixed at A. Determine the vertical displacement of end C of the assembly. The pipe has an inner diameter of 40 mm and outer diameter of 60 mm and is made of A-36 steel. Neglect the shearing strain energy.

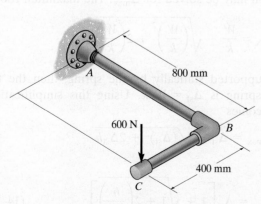

Prob. 14–39

•14–41. Determine the vertical displacement of end B of the frame. Consider only bending strain energy. The frame is made using two A-36 steel W460 × 68 wide-flange sections.

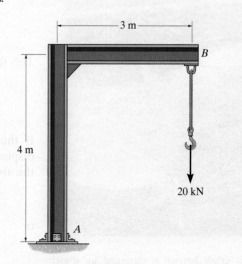

Prob. 14–41

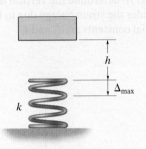

Fig. 14–23

14.4 Impact Loading

Throughout this text we have considered all loadings to be applied to a body in a gradual manner, such that when they reach a maximum value the body remains static. Some loadings, however, are dynamic; that is, they vary with time. A typical example would be caused by the collision of objects. This is called an impact loading. Specifically, *impact* occurs when one object strikes another, such that large forces are developed between the objects during a very short period of time.

If we assume no energy is lost during impact, due to heat, sound or localized plastic deformations, then we can study the mechanics of impact using the conservation of energy. To show how this is done, we will first analyze the motion of a simple block-and-spring system as shown in Fig. 14–23. When the block is released from rest, it falls a distance h, striking the spring and compressing it a distance Δ_{max} before momentarily coming to rest. If we neglect the mass of the spring and assume that the spring responds *elastically*, then the conservation of energy requires that the energy of the falling block be transformed into stored (strain) energy in the spring; or in other words, the work done by the block's weight, falling $h + \Delta_{max}$, is equal to the work needed to displace the end of the spring by an amount Δ_{max}. Since the force in a spring is related to Δ_{max} by the equation $F = k\Delta_{max}$, where k is the spring stiffness, then applying the conservation of energy and Eq. 14–2, we have

$$U_e = U_i$$

$$W(h + \Delta_{max}) = \frac{1}{2}(k\Delta_{max})\,\Delta_{max}$$

$$W(h + \Delta_{max}) = \frac{1}{2}k\Delta_{max}^2 \qquad (14\text{–}27)$$

$$\Delta_{max}^2 - \frac{2W}{k}\Delta_{max} - 2\left(\frac{W}{k}\right)h = 0$$

This quadratic equation may be solved for Δ_{max}. The maximum root is

$$\Delta_{max} = \frac{W}{k} + \sqrt{\left(\frac{W}{k}\right)^2 + 2\left(\frac{W}{k}\right)h}$$

If the weight W is supported statically by the spring, then the top displacement of the spring is $\Delta_{st} = W/k$. Using this simplification, the above equation becomes

$$\Delta_{max} = \Delta_{st} + \sqrt{(\Delta_{st})^2 + 2\Delta_{st}h}$$

or

$$\Delta_{max} = \Delta_{st}\left[1 + \sqrt{1 + 2\left(\frac{h}{\Delta_{st}}\right)}\right] \qquad (14\text{–}28)$$

This crash barrier is designed to absorb the impact energy of moving vehicles.

Once Δ_{max} is calculated, the maximum force applied to the spring can be determined from

$$F_{max} = k\Delta_{max} \qquad (14\text{–}29)$$

It should be realized, however, that this force and associated displacement occur only at an *instant*. Provided the block does not rebound off the spring, it will continue to vibrate until the motion dampens out and the block assumes the static position, Δ_{st}. Note also that if the block is held just above the spring, $h = 0$, and released, then, from Eq. 14–28, the maximum displacement of the block is

$$\Delta_{max} = 2\Delta_{st}$$

In other words, when the block is released from the top of the spring (a dynamic load), the displacement is *twice* what it would be if it were set on the spring (a static load).

Using a similar analysis, it is also possible to determine the maximum displacement of the end of the spring if the block is sliding on a smooth horizontal surface with a known velocity **v** just before it collides with the spring, Fig. 14–24. Here the block's kinetic energy,* $\frac{1}{2}(W/g)v^2$, will be transformed into stored energy in the spring. Hence,

$$U_e = U_i$$

$$\frac{1}{2}\left(\frac{W}{g}\right)v^2 = \frac{1}{2}k\Delta_{max}^2$$

$$\Delta_{max} = \sqrt{\frac{Wv^2}{gk}} \qquad (14\text{–}30)$$

Since the static displacement at the top of the spring caused by the weight W resting on it is $\Delta_{st} = W/k$, then

$$\Delta_{max} = \sqrt{\frac{\Delta_{st}v^2}{g}} \qquad (14\text{–}31)$$

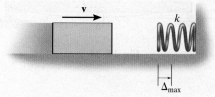

Fig. 14–24

The results of this simplified analysis can be used to determine both the approximate deflection and the stress developed in a deformable member when it is subjected to impact. To do this we must make the necessary assumptions regarding the collision, so that the behavior of the colliding bodies is similar to the response of the block-and-spring models discussed above. Hence we will consider the moving body to be *rigid* like the block and the stationary body to be deformable like the spring. Also, it is assumed that the material behaves in a linear-elastic manner. When collision occurs, the bodies remain in contact until the elastic body reaches its maximum deformation, and during the motion the inertia or mass of the elastic body is neglected. Realize that each of these assumptions will lead to a *conservative* estimate of both the maximum stress and deflection of the elastic body. In other words, their values will be larger than those that actually occur.

---

*Recall from physics that kinetic energy is "energy of motion." For the translation of a body it is determined from $\frac{1}{2}mv^2$, where m is the body's mass, $m = W/g$.

14

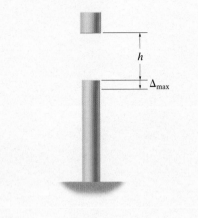

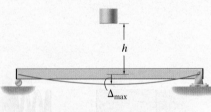

Fig. 14–25

The members of this crash guard must be designed to resist a prescribed impact loading in order to arrest the motion of a rail car.

A few examples of when this theory can be applied are shown in Fig. 14–25. Here a block of known weight is dropped onto a post and a beam, causing them to deform a maximum amount Δ_{max}. The energy of the falling block is transformed momentarily into axial strain energy in the post and bending strain energy in the beam.* In order to determine the deformation Δ_{max}, we could use the same approach as for the block–spring system, and that is to write the conservation-of-energy equation for the block and post or block and beam, and then solve for Δ_{max}. However, we can also solve these problems in a more direct manner by modeling the post and beam by an *equivalent spring*. For example, if a force **P** displaces the top of the post $\Delta = PL/AE$, then a spring having a stiffness $k = AE/L$ would be displaced the same amount by **P**, that is, $\Delta = P/k$. In a similar manner, from Appendix C, a force **P** applied to the center of a simply supported beam displaces the center $\Delta = PL^3/48EI$, and therefore an equivalent spring would have a stiffness of $k = 48EI/L^3$. It is not necessary, however, to actually find the equivalent spring stiffness to apply Eq. 14–28 or 14–30. All that is needed to determine the dynamic displacement, Δ_{max}, is to calculate the *static displacement*, Δ_{st}, due to the weight $P_{st} = W$ of the block resting on the member.

Once Δ_{max} is determined, the maximum dynamic force can then be calculated from $P_{max} = k\Delta_{max}$. If we consider P_{max} to be an *equivalent static load* then the maximum stress in the member can be determined using statics and the theory of mechanics of materials. Recall that this stress acts only for an *instant*. In reality, vibrational waves pass through the material, and the stress in the post or the beam, for example, does not remain constant.

The ratio of the equivalent static load P_{max} to the static load $P_{st} = W$ is called the *impact factor*, n. Since $P_{max} = k\Delta_{max}$ and $P_{st} = k\Delta_{st}$, then from Eq. 14–28, we can express it as

$$n = 1 + \sqrt{1 + 2\left(\frac{h}{\Delta_{st}}\right)} \qquad (14\text{–}32)$$

This factor represents the magnification of a statically applied load so that it can be treated dynamically. Using Eq. 14–32, n can be calculated for any member that has a linear relationship between load and deflection. For a complicated system of connected members, however, impact factors are determined from experience and experimental testing. Once n is determined, the dynamic stress and deflection at the point of impact are easily found from the static stress σ_{st} and static deflection Δ_{st} caused by the load W, that is, $\sigma_{max} = n\sigma_{st}$ and $\Delta_{max} = n\Delta_{st}$.

*Strain energy due to shear is neglected for reasons discussed in Example 14.4.

Important Points

- *Impact* occurs when a large force is developed between two objects which strike one another during a short period of time.

- We can analyze the effects of impact by assuming the moving body is rigid, the material of the stationary body is linear elastic, no energy is lost during collision, the bodies remain in contact during collision, and the inertia of the elastic body is neglected.

- The dynamic load on a body can be determined by multiplying the static load by an *impact factor*.

EXAMPLE | 14.8

The aluminum pipe shown in Fig. 14–26 is used to support a load of 600 kN. Determine the maximum displacement at the top of the pipe if the load is (a) applied gradually, and (b) applied by suddenly releasing it from the top of the pipe when $h = 0$. Take $E_{al} = 70$ GPa and assume that the aluminum behaves elastically.

SOLUTION

Part (a). When the load is applied gradually, the work done by the weight is transformed into elastic strain energy in the pipe. Applying the conservation of energy, we have

$$U_e = U_i$$

$$\frac{1}{2} W \Delta_{st} = \frac{W^2 L}{2AE}$$

$$\Delta_{st} = \frac{WL}{AE} = \frac{600 \text{ kN}(240 \text{ mm})}{\pi[(60 \text{ mm})^2 - (50 \text{ mm})^2]70 \text{ kN/mm}^2}$$

$$= 0.595 \text{ mm} \qquad Ans.$$

Part (b). Here Eq. 14–28 can be applied, with $h = 0$. Hence,

$$\Delta_{max} = \Delta_{st}\left[1 + \sqrt{1 + 2\left(\frac{h}{\Delta_{st}}\right)}\right]$$

$$= 2\Delta_{st} = 2(0.595 \text{ mm})$$

$$= 1.19 \text{ mm} \qquad Ans.$$

Hence, the displacement of the weight when applied dynamically is twice as great as when the load is applied statically. In other words, the impact factor is $n = 2$, Eq. 14–32.

600 kN

$t = 10$ mm

60 mm h

240 mm

Fig. 14–26

EXAMPLE 14.9

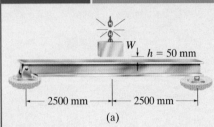

(a)

Fig. 14-27

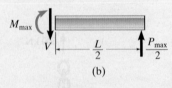

(b)

The A-36 steel beam shown in Fig. 14–27a is a W250 × 58. Determine the maximum bending stress in the beam and the beam's maximum deflection if the weight $W = 6$ kN is dropped from a height $h = 50$ mm onto the beam. $E_{st} = 210$ GPa.

SOLUTION I
We will apply Eq. 14–28. First, however, we must calculate Δ_{st}. Using the table in Appendix C, and the data in Appendix B for the properties of a W250 × 58, we have

$$\Delta_{st} = \frac{WL^3}{48EI} = \frac{(6\text{ kN})(5000\text{ mm})^3}{48(210\text{ kN/mm}^2)[87.3(10^6)\text{ mm}^4]} = 0.8523\text{ mm}$$

$$\Delta_{max} = \Delta_{st}\left[1 + \sqrt{1 + 2\left(\frac{h}{\Delta_{st}}\right)}\right]$$

$$= 0.8523\text{ mm}\left[1 + \sqrt{1 + 2\left(\frac{50\text{ mm}}{0.8523\text{ mm}}\right)}\right] = 10.124\text{ mm}$$
Ans.

The equivalent static load that causes this displacement is therefore

$$P_{max} = \frac{48EI}{L^3} = \frac{48(210\text{ kN/mm}^2)[87.3(10^6)\text{ mm}^4]}{(5000\text{ mm})^3}(10.124\text{ mm}) = 7.040\text{ kN}$$

The internal moment caused by this load is maximum at the center of the beam, such that by the method of sections, Fig. 14–27b, $M_{max} = P_{max}L/4$. Applying the flexure formula to determine the bending stress, we have

$$\sigma_{max} = \frac{M_{max}c}{I} = \frac{P_{max}Lc}{4I} = \frac{12E\,\Delta_{max}c}{L^2}$$

$$= \frac{12[210\text{ kN/mm}^2](10.124\text{ mm})(252\text{ mm}/2)}{(5000\text{ mm})^2} = 0.1286\text{ kN/mm}^2 = 128.6\text{ MPa}$$
Ans.

SOLUTION II
It is also possible to obtain the dynamic or maximum deflection Δ_{max} from first principles. The external work of the falling weight W is $U_e = W(h + \Delta_{max})$. Since the beam deflects Δ_{max}, and $P_{max} = 48EI\Delta_{max}/L^3$, then

$$U_e = U_i$$

$$W(h + \Delta_{max}) = \frac{1}{2}\left(\frac{48EI\Delta_{max}}{L^3}\right)\Delta_{max}$$

$$(6\text{ kN})(50\text{ mm} + \Delta_{max}) = \frac{1}{2}\left[\frac{48(210\text{ kN/mm}^2)[87.3(10^6)\text{ mm}^4]}{(5000\text{ mm})^3}\right]\Delta_{max}^2$$

$$3.520\,\Delta_{max}^2 - 6\Delta_{max}^2 - 300 = 0$$

Solving and choosing the positive root yields

$$\Delta_{max} = 10.12\text{ mm}$$
Ans.

EXAMPLE 14.10

A railroad car that is assumed to be rigid and has a mass of 80 Mg is moving forward at a speed of $v = 0.2$ m/s when it strikes a steel 200-mm by 200-mm post at A, Fig. 14–28a. If the post is fixed to the ground at C, determine the maximum horizontal displacement of its top B due to the impact. Take $E_{st} = 200$ GPa.

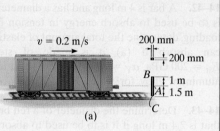

(a)

SOLUTION

Here the kinetic energy of the railroad car is transformed into internal bending strain energy only for region AC of the post. (Region BA is not subjected to an internal loading.) Assuming that point A is displaced $(\Delta_A)_{max}$, then the force P_{max} that causes this displacement can be determined from the table in Appendix C. We have

$$P_{max} = \frac{3EI(\Delta_A)_{max}}{L_{AC}^3} \qquad (1)$$

$$U_e = U_i; \qquad \frac{1}{2}mv^2 = \frac{1}{2}P_{max}(\Delta_A)_{max}$$

$$\frac{1}{2}mv^2 = \frac{1}{2}\frac{3EI}{L_{AC}^3}(\Delta_A)_{max}^2; \qquad (\Delta_A)_{max} = \sqrt{\frac{mv^2 L_{AC}^3}{3EI}}$$

Substituting in the numerical data yields

$$(\Delta_A)_{max} = \sqrt{\frac{80(10^3)\,\text{kg}(0.2\,\text{m/s})^2(1.5\,\text{m})^3}{3[200(10^9)\,\text{N/m}^2][\frac{1}{12}(0.2\,\text{m})^4]}} = 0.01162\,\text{m} = 11.62\,\text{mm}$$

Using Eq. 1, the force P_{max} is therefore

$$P_{max} = \frac{3[200(10^9)\,\text{N/m}^2][\frac{1}{12}(0.2\,\text{m})^4](0.01162\,\text{m})}{(1.5\,\text{m})^3} = 275.4\,\text{kN}$$

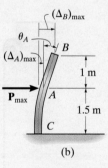

(b)

Fig. 14–28

With reference to Fig. 14–28b, segment AB of the post remains straight. To determine the maximum displacement at B, we must first determine the slope at A. Using the appropriate formula from the table in Appendix C to determine θ_A, we have

$$\theta_A = \frac{P_{max}L_{AC}^2}{2EI} = \frac{275.4(10^3)\,\text{N}\,(1.5\,\text{m})^2}{2[200(10^9)\,\text{N/m}^2][\frac{1}{12}(0.2\,\text{m})^4]} = 0.01162\,\text{rad}$$

The maximum displacement at B is thus

$$(\Delta_B)_{max} = (\Delta_A)_{max} + \theta_A L_{AB}$$

$$= 11.62\,\text{mm} + (0.01162\,\text{rad})\,1(10^3)\,\text{mm} = 23.2\,\text{mm} \qquad Ans.$$

PROBLEMS

14–42. A bar is 4 m long and has a diameter of 30 mm. If it is to be used to absorb energy in tension from an impact loading, determine the total amount of elastic energy that it can absorb if (a) it is made of steel for which E_{st} = 200 GPa, σ_Y = 800 MPa, and (b) it is made from an aluminum alloy for which E_{al} = 70 GPa, σ_Y = 405 MPa.

14–43. Determine the diameter of a red brass C83400 bar that is 2.4 m long if it is to be used to absorb 1200 N · m of energy in tension from an impact loading. No yielding occurs.

*14–44. A steel cable having a diameter of 10 mm wraps over a drum and is used to lower an elevator having a weight of 4 kN (≈ 400 kg). The elevator is 45 m below the drum and is descending at the constant rate of 0.6 m/s when the drum suddenly stops. Determine the maximum stress developed in the cable when this occurs. E_{st} = 200 GPa, σ_Y = 350 MPa.

•14–45. The composite aluminum bar is made from two segments having diameters of 5 mm and 10 mm. Determine the maximum axial stress developed in the bar if the 5-kg collar is dropped from a height of h = 100 mm. E_{al} = 70 GPa, σ_Y = 410 MPa.

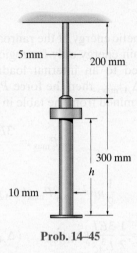

5 mm
200 mm
300 mm
h
10 mm

Prob. 14–45

14–46. The composite aluminum bar is made from two segments having diameters of 5 mm and 10 mm. Determine the maximum height h from which the 5-kg collar should be dropped so that it produces a maximum axial stress in the bar of σ_{max} = 300 MPa, E_{al} = 70 GPa, σ_Y = 410 MPa.

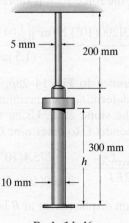

5 mm
200 mm
300 mm
h
10 mm

Prob. 14–46

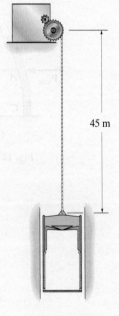

45 m

Prob. 14–44

14–47. The 5-kg block is traveling with the speed of $v = 4$ m/s just before it strikes the 6061-T6 aluminum stepped cylinder. Determine the maximum normal stress developed in the cylinder.

*14–48. Determine the maximum speed v of the 5-kg block without causing the 6061-T6 aluminum stepped cylinder to yield after it is struck by the block.

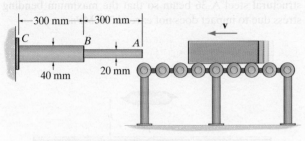

Probs. 14–47/48

•14–49. The steel beam AB acts to stop the oncoming railroad car, which has a mass of 10 Mg and is coasting towards it at $v = 0.5$ m/s. Determine the maximum stress developed in the beam if it is struck at its center by the car. The beam is simply supported and only horizontal forces occur at A and B. Assume that the railroad car and the supporting framework for the beam remains rigid. Also, compute the maximum deflection of the beam. $E_{st} = 200$ GPa, $\sigma_Y = 250$ MPa.

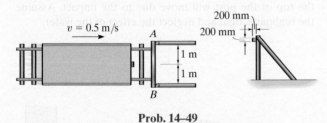

Prob. 14–49

14–50. The aluminum bar assembly is made from two segments having diameters of 40 mm and 20 mm. Determine the maximum axial stress developed in the bar if the 10-kg collar is dropped from a height of $h = 150$ mm. Take $E_{al} = 70$ GPa, $\sigma_Y = 410$ MPa.

14–51. The aluminum bar assembly is made from two segments having diameters of 40 mm and 20 mm. Determine the maximum height h from which the 60-kg collar can be dropped so that it will not cause the bar to yield. Take $E_{al} = 70$ GPa, $\sigma_Y = 410$ MPa.

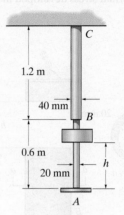

Probs. 14–50/51

*14–52. The 250-N (≈ 25-kg) weight is falling at 0.9 m/s at the instant it is 0.6 m above the spring and post assembly. Determine the maximum stress in the post if the spring has a stiffness of $k = 40$ MN/m. The post has a diameter of 75 mm and a modulus of elasticity of $E = 48$ GPa. Assume the material will not yield.

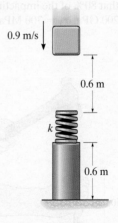

Prob. 14–52

•14–53. The 50-kg block is dropped from $h = 600$ mm onto the bronze C86100 tube. Determine the minimum length L the tube can have without causing the tube to yield.

14–54. The 50-kg block is dropped from $h = 600$ mm onto the bronze C86100 tube. If $L = 900$ mm, determine the maximum normal stress developed in the tube.

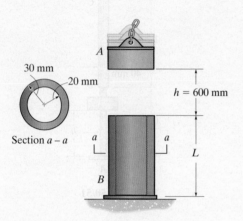

Section a – a

Probs. 14–53/54

14–55. The steel chisel has a diameter of 12 mm and a length of 250 mm. It is struck by a hammer that weighs 15 N(≈ 1.5 kg), and at the instant of impact it is moving at 3.6 m/s. Determine the maximum compressive stress in the chisel, assuming that 80% of the impacting energy goes into the chisel. $E_{st} = 200$ GPa, $\sigma_Y = 700$ MPa.

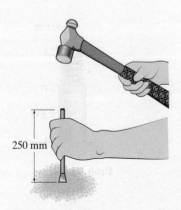

Prob. 14–55

*14–56. The sack of cement has a mass of 45 kg. If it is dropped from rest at a height of $h = 1.2$ m onto the center of the W250 × 58 structural steel A-36 beam, determine the maximum bending stress developed in the beam due to the impact. Also, what is the impact factor?

•14–57. The sack of cement has a mass of 45 kg. Determine the maximum height h from which it can be dropped from rest onto the center of the W250 × 58 structural steel A-36 beam so that the maximum bending stress due to impact does not exceed 210 MPa.

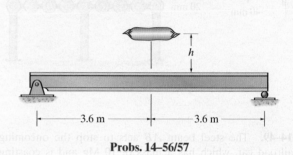

Probs. 14–56/57

14–58. The tugboat has a mass of 60 tonnes and is traveling forward at 0.6 m/s when it strikes the 300-mm-diameter fender post AB used to protect a bridge pier. If the post is made from treated white spruce and is assumed fixed at the river bed, determine the maximum horizontal distance the top of the post will move due to the impact. Assume the tugboat is rigid and neglect the effect of the water.

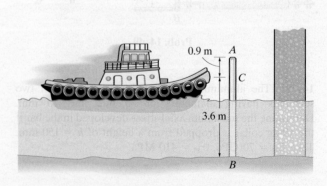

Prob. 14–58

14–59. The wide-flange beam has a length of $2L$, a depth $2c$, and a constant EI. Determine the maximum height h at which a weight W can be dropped on its end without exceeding a maximum elastic stress σ_{max} in the beam.

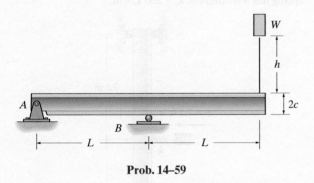

Prob. 14–59

*14–60. The 50-kg block C is dropped from $h = 1.5$ m onto the simply supported beam. If the beam is an A-36 steel W250 × 45 wide-flange section, determine the maximum bending stress developed in the beam.

•14–61. Determine the maximum height h from which the 50-kg block C can be dropped without causing yielding in the A-36 steel W310 × 39 wide flange section when the block strikes the beam.

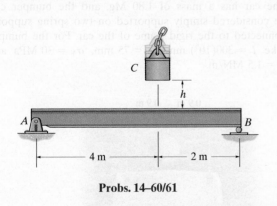

Probs. 14–60/61

14–62. The diver weighs 750 N (≈ 75 kg) and, while holding himself rigid, strikes the end of a wooden diving board ($h = 0$) with a downward velocity of 1.2 m. Determine the maximum bending stress developed in the board. The board has a thickness of 40 mm and width of 450 mm. $E_w = 12.6$ GPa, $\sigma_Y = 56$ MPa.

14–63. The diver weighs 750 N (≈ 75 kg) and, while holding himself rigid, strikes the end of the wooden diving board. Determine the maximum height h from which he can jump onto the board so that the maximum bending stress in the wood does not exceed 42 MPa. The board has a thickness of 40 mm and width of 450 mm. $E_w = 12.6$ GPa.

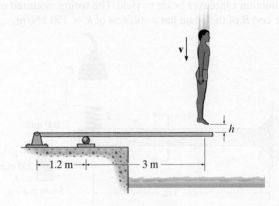

Probs. 14–62/63

*14–64. The weight of 90 kg is dropped from a height of 1.2 m from the top of the A-36 steel beam. Determine the maximum deflection and maximum stress in the beam if the supporting springs at A and B each have a stiffness of $k = 100$ kN/m. The beam is 75 mm thick and 100 mm wide.

•14–65. The weight of 90 kg, is dropped from a height of 1.2 m from the top of the A-36 steel beam. Determine the load factor n if the supporting springs at A and B each have a stiffness of $k = 60$ kN/m. The beam is 75 mm thick and 100 mm wide.

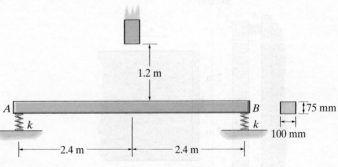

Probs. 14–64/65

14

14–66. Block C of mass 50 kg is dropped from height $h = 0.9$ m onto the spring of stiffness $k = 150$ kN/m mounted on the end B of the 6061-T6 aluminum cantilever beam. Determine the maximum bending stress developed in the beam.

14–67. Determine the maximum height h from which 200-kg block C can be dropped without causing the 6061-T6 aluminum cantilever beam to yield. The spring mounted on the end B of the beam has a stiffness of $k = 150$ kN/m.

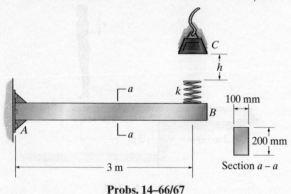

Probs. 14–66/67

*14–68. The 2014-T6 aluminum bar AB can slide freely along the guides mounted on the rigid crash barrier. If the railcar of mass 10 Mg is traveling with a speed of $v = 1.5$ m/s, determine the maximum bending stress developed in the bar. The springs at A and B have a stiffness of $k = 15$ MN/m.

•14–69. The 2014-T6 aluminum bar AB can slide freely along the guides mounted on the rigid crash barrier. Determine the maximum speed v the 10-Mg railcar without causing the bar to yield when it is struck by the railcar. The springs at A and B have a stiffness of $k = 15$ MN/m.

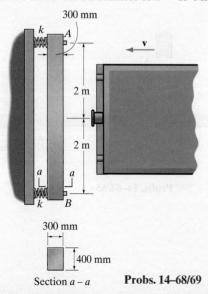

Probs. 14–68/69

14–70. The simply supported W250 × 22 structural A-36 steel beam lies in the horizontal plane and acts as a shock absorber for the 250-kg block which is traveling toward it at 1.5 m/s. Determine the maximum deflection of the beam and the maximum stress in the beam during the impact. The spring has a stiffness of $k = 200$ kN/m.

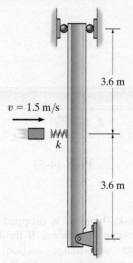

Prob. 14–70

14–71. The car bumper is made of polycarbonate-polybutylene terephthalate. If $E = 2.0$ GPa, determine the maximum deflection and maximum stress in the bumper if it strikes the rigid post when the car is coasting at $v = 0.75$ m/s. The car has a mass of 1.80 Mg, and the bumper can be considered simply supported on two spring supports connected to the rigid frame of the car. For the bumper take $I = 300(10^6)$ mm^4, $c = 75$ mm, $\sigma_Y = 30$ MPa and $k = 1.5$ MN/m.

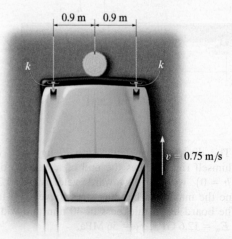

Prob. 14–71

*14.5 Principle of Virtual Work

The principle of virtual work was developed by John Bernoulli in 1717, and like other energy methods of analysis, it is based on the conservation of energy. Although the principle of virtual work has many applications in mechanics, in this text we will use it to obtain the displacement and slope at a point on a deformable body.

To do this, we will consider the body to be of arbitrary shape as shown in Fig. 14–29b, and to be subjected to the "real loads" $\mathbf{P}_1$, $\mathbf{P}_2$, and $\mathbf{P}_3$. It is assumed that these loads cause no movement of the supports; however, in general they can strain the material *beyond* the elastic limit. Suppose that it is necessary to determine the displacement Δ of point A on the body. Since there is no force acting at A, then Δ will *not* be included as an external "work term" in the equation when the conservation of energy principle is applied to the body. In order to get around this limitation, we will place an *imaginary* or "virtual" force $\mathbf{P}'$ on the body at point A, such that $\mathbf{P}'$ acts in the *same direction* as Δ. Furthermore, this load is applied to the body *before* the real loads are applied, Fig. 14–29a. For convenience, which will be made clear later, we will choose $\mathbf{P}'$ to have a "unit" magnitude; that is, $P' = 1$. It is to be emphasized that the term "***virtual***" is used to describe this load because it is *imaginary* and does not actually exist as part of the real loading.

This external virtual load, however, does create an internal virtual load $\mathbf{u}$ in a representative element or fiber of the body, as shown in Fig. 14–29a. As expected, P' and u can be related by the equations of equilibrium. Also, because of P' and u, the body and the element will each undergo a virtual (imaginary) displacement, although we will *not* be

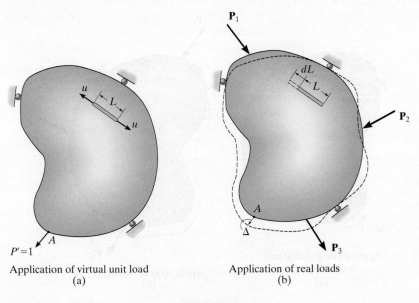

Application of virtual unit load
(a)

Application of real loads
(b)

Fig. 14–29

concerned with their magnitudes. Once the virtual load is applied and *then* the body is subjected to the *real loads* $\mathbf{P}_1$, $\mathbf{P}_2$, and $\mathbf{P}_3$, point A will be displaced a real amount Δ, which causes the element to be displaced dL, Fig. 14–29b. As a result, the external virtual force $\mathbf{P}'$ and internal virtual load $\mathbf{u}$ "ride along" or are displaced by Δ and dL, respectively. Consequently these loads perform *external virtual work* $1 \cdot \Delta$ on the body and *internal virtual work* $u \cdot dL$ on the element. Considering *only* the conservation of *virtual* energy, the external virtual work is then equal to the internal virtual work done on all the elements of the body. Therefore, we can write the virtual-work equation as

$$1 \cdot \Delta = \Sigma u \cdot dL \qquad (14\text{–}34)$$

virtual loadings

real displacements

Here

$P' = 1 =$ external virtual unit load acting in the direction of Δ

$u =$ internal virtual load acting on the element

$\Delta =$ external displacement caused by the real loads

$dL =$ internal displacement of the element in the direction of $\mathbf{u}$, caused by the real loads

By choosing $P' = 1$, it can be seen that the solution for Δ follows directly, since $\Delta = \Sigma u \, dL$.

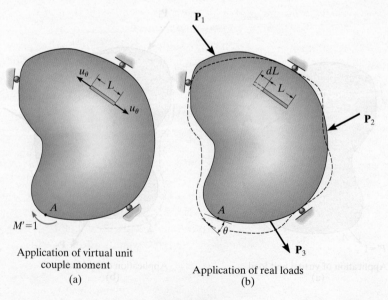

Application of virtual unit
couple moment
(a)

Application of real loads
(b)

Fig. 14–30

In a similar manner, if the angular displacement or slope of the tangent at a point on the body is to be determined at A, Fig. 14–30b, then a virtual *couple moment* $\mathbf{M'}$, having a "unit" magnitude, is applied at the point, Fig. 14–30a. As a result, this couple moment causes a virtual load u_θ in one of the elements of the body. Assuming the real loads $\mathbf{P}_1$, $\mathbf{P}_2$, $\mathbf{P}_3$ deform the element an amount dL, the angular displacement θ can be found from the virtual-work equation

$$\overbrace{1 \cdot \underbrace{\theta = \Sigma u_\theta \, dL}_{\text{real displacements}}}^{\text{virtual loadings}} \qquad (14\text{–}35)$$

Here

$M' = 1 =$ external virtual unit couple moment acting in the direction of θ

$u_\theta =$ internal virtual load acting on an element

$\theta =$ external angular displacement in radians caused by the real loads

$dL =$ internal displacement of the element in the direction of u_θ, caused by the real loads

This method for applying the principle of virtual work is often referred to as the *method of virtual forces*, since a *virtual force* is applied, resulting in a determination of an external *real displacement*. The equation of virtual work in this case represents a statement of *compatibility requirements* for the body. Although it is not important here, realize that we can also apply the principle of virtual work as a *method of virtual displacements*. In this case, *virtual displacements* are imposed on the body when the body is subjected to *real loadings*. This method can be used to determine the external reactive force on the body or an unknown internal loading. When it is used in this manner, the equation of virtual work is a statement of the *equilibrium requirements* for the body.*

Internal Virtual Work.

The terms on the right side of Eqs. 14–34 and 14–35 represent the internal virtual work developed in the body. The real internal displacements dL in these terms can be produced in several different ways. For example, these displacements may result from geometric fabrication errors, from a change in temperature, or more commonly from stress. In particular, no restriction has been placed on the magnitude of the external loading, so the stress may be large enough to cause yielding or even strain hardening of the material.

*See *Engineering Mechanics: Statics*, 12th edition, R.C. Hibbeler, Prentice Hall, Inc., 2009.

TABLE 14–1

Deformation caused by	Strain energy	Internal virtual work
Axial load N	$\int_0^L \dfrac{N^2}{2EA}\,dx$	$\int_0^L \dfrac{nN}{EA}\,dx$
Shear V	$\int_0^L \dfrac{f_s V^2}{2GA}\,dx$	$\int_0^L \dfrac{f_s vV}{GA}\,dx$
Bending moment M	$\int_0^L \dfrac{M^2}{2EI}\,dx$	$\int_0^L \dfrac{mM}{EI}\,dx$
Torsional moment T	$\int_0^L \dfrac{T^2}{2GJ}\,dx$	$\int_0^L \dfrac{tT}{GJ}\,dx$

If we assume that the material behavior is linear elastic and the stress does not exceed the proportional limit, we can then formulate the expressions for internal virtual work caused by stress using the equations of elastic strain energy developed in Sec. 14.2. They are listed in the center column of Table 14–1. Recall that each of these expressions assumes that the internal loading N, V, M, or T was applied gradually from zero to its full value. As a result, the work done by these resultants is shown in these expressions as *one-half* the product of the internal loading and its displacement. In the case of the virtual-force method, however, the "full" virtual internal loading is applied *before* the real loads cause displacements, and therefore the work of the virtual loading is simply the product of the virtual load and its real displacement. Referring to these internal virtual loadings (u) by the corresponding lowercase symbols n, v, m, and t, the virtual work due to axial load, shear, bending moment, and torsional moment is listed in the right-hand column of Table 14–1. Using these results, the virtual-work equation for a body subjected to a general loading can therefore be written as

$$1 \cdot \Delta = \int \frac{nN}{AE}\,dx + \int \frac{mM}{EI}\,dx + \int \frac{f_s vV}{GA}\,dx + \int \frac{tT}{GJ}\,dx \quad (14\text{–}36)$$

In the following sections we will apply the above equation to problems involving the displacement of joints on trusses, and points on beams and mechanical elements. We will also include a discussion of how to handle the effects of fabrication errors and differential temperature. For application it is important that a consistent set of units be used for all the terms. For example, if the real loads are expressed in kilonewtons and the body's dimensions are in meters, a 1-kN virtual force or 1-kN · m virtual couple should be applied to the body. By doing so a calculated displacement Δ will be in meters, and a calculated slope will be in radians.

*14.6 Method of Virtual Forces Applied to Trusses

In this section, we will apply the method of virtual forces to determine the displacement of a truss joint. To illustrate the principles, the vertical displacement of joint A of the truss shown in Fig. 14–31b will be determined. To do this, we must place a virtual unit force at this joint, Fig. 14–31a, so that when the real loads $\mathbf{P}_1$ and $\mathbf{P}_2$ are applied to the truss, they cause the external virtual work $1 \cdot \Delta$. The internal virtual work in each member is $n\Delta L$. Since each member has a constant cross-sectional area A, and n and N are constant throughout the member's length, then from Table 14–1, the internal virtual work for each member is

$$\int_0^L \frac{nN}{AE}dx = \frac{nNL}{AE}$$

Therefore, the virtual-work equation for the entire truss is

$$1 \cdot \Delta = \sum \frac{nNL}{AE} \qquad (14\text{–}37)$$

Here

$1 =$ external virtual unit load acting on the truss joint in the direction of Δ

$\Delta =$ joint displacement caused by the real loads on the truss

$n =$ internal virtual force in a truss member caused by the external virtual unit load

$N =$ internal force in a truss member caused by the real loads

$L =$ length of a member

$A =$ cross-sectional area of a member

$E =$ modulus of elasticity of a member

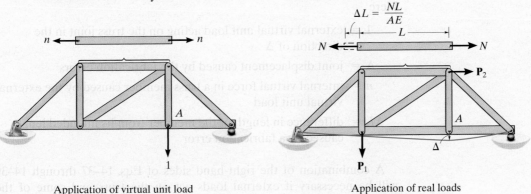

Application of virtual unit load

(a)

Application of real loads

(b)

Fig. 14–31

Temperature Change. Truss members can change their length due to a change in temperature. If α is the coefficient of thermal expansion for a member and ΔT is the change in temperature, the change in length of a member is $\Delta L = \alpha \, \Delta T L$ (Eq. 4–4). Hence, we can determine the displacement of a selected truss joint due to this temperature change from Eq. 14–34, written as

$$1 \cdot \Delta = \Sigma n \alpha \, \Delta T L \qquad (14\text{–}38)$$

Here

    $1 =$ external virtual unit load acting on the truss joint in the direction of Δ

    $\Delta =$ joint displacement caused by the temperature change

    $n =$ internal virtual force in a truss member caused by the external virtual unit load

    $\alpha =$ coefficient of thermal expansion of material

    $\Delta T =$ change in temperature of member

    $L =$ length of member

Fabrication Errors. Occasionally errors in fabricating the lengths of the members of a truss may occur. If this happens, the displacement Δ in a particular direction of a truss joint from its expected position can be determined from direct application of Eq. 14–34 written as

$$1 \cdot \Delta = \Sigma n \, \Delta L \qquad (14\text{–}39)$$

Here

    $1 =$ external virtual unit load acting on the truss joint in the direction of Δ

    $\Delta =$ joint displacement caused by the fabrication errors

    $n =$ internal virtual force in a truss member caused by the external virtual unit load

    $\Delta L =$ difference in length of the member from its intended length caused by a fabrication error

A combination of the right-hand sides of Eqs. 14–37 through 14–39 will be necessary if external loads act on the truss and some of the members undergo a temperature change or have been fabricated with the wrong dimensions.

Procedure for Analysis

The following procedure provides a method that may be used to determine the displacement of any joint on a truss using the method of virtual forces.

Virtual Forces n.

- Place the virtual unit load on the truss at the joint where the displacement is to be determined. The load should be directed along the line of action of the displacement.

- With the unit load so placed and all the real loads *removed* from the truss, calculate the internal n force in each truss member. Assume that tensile forces are positive and compressive forces are negative.

Real Forces N.

- Determine the N forces in each member. These forces are caused only by the real loads acting on the truss. Again, assume that tensile forces are positive and compressive forces are negative.

Virtual-Work Equation.

- Apply the equation of virtual work to determine the desired displacement. It is important to retain the algebraic sign for each of the corresponding n and N forces when substituting these terms into the equation.

- If the resultant sum $\Sigma nNL/AE$ is positive, the displacement Δ is in the same direction as the virtual unit load. If a negative value results, Δ is opposite to the virtual unit load.

- When applying $1 \cdot \Delta = \Sigma n\alpha \Delta T L$, an *increase* in temperature, ΔT, will be *positive*; whereas a *decrease* in temperature will be *negative*.

- For $1 \cdot \Delta = \Sigma n \Delta L$, when a fabrication error causes an increase in the length of a member, ΔL is *positive*, whereas a *decrease* in length is *negative*.

- When applying this method, attention should be paid to the units of each numerical quantity. Notice, however, that the virtual unit load can be assigned any arbitrary unit: pounds, kips, newtons, etc., since the n forces will have these *same* units, and as a result, the units for both the virtual unit load and the n forces will cancel from both sides of the equation.

EXAMPLE 14.11

Determine the vertical displacement of joint C of the steel truss shown in Fig. 14–32a. The cross-sectional area of each member is $A = 400$ mm^2 and $E_{st} = 200$ GPa.

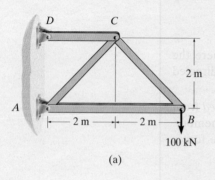

(a)

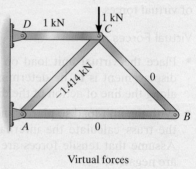

Virtual forces

(b)

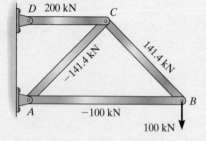

Real forces

(c)

Fig. 14–32

SOLUTION

Virtual Forces n. Since the vertical displacement at joint C is to be determined, *only* a vertical 1-kN virtual load is placed at joint C; and the force in each member is calculated using the method of joints. The results of this analysis are shown in Fig. 14–32b. Using our sign convention, positive numbers indicate tensile forces and negative numbers indicate compressive forces.

Real Forces N. The applied load of 100 kN causes forces in the members that are also calculated using the method of joints. The results of this analysis are shown in Fig. 14–32c.

Virtual-Work Equation. Arranging the data in tabular form, we have

Member	n	N	L	nNL
AB	0	−100	4	0
BC	0	141.4	2.828	0
AC	−1.414	−141.4	2.828	565.7
CD	1	200	2	400
				Σ 965.7 kN^2 · m

Thus,

$$1 \text{ kN} \cdot \Delta_{C_v} = \sum \frac{nNL}{AE} = \frac{965.7 \text{ kN}^2 \cdot \text{m}}{AE}$$

Substituting the numerical values for A and E, we have

$$1 \text{ kN} \cdot \Delta_{C_v} = \frac{965.7 \text{ kN}^2 \cdot \text{m}}{[400(10^{-6}) \text{ m}^2] 200(10^6) \text{ kN/m}^2}$$

$$\Delta_{C_v} = 0.01207 \text{ m} = 12.1 \text{ mm} \qquad Ans.$$

EXAMPLE   14.12

Determine the horizontal displacement of the roller at B of the truss shown in Fig. 14–33a. Due to radiant heating, member AB is subjected to an *increase* in temperature of $\Delta T = +60°C$, and this member has been fabricated 3 mm too short. The members are made of steel, for which $\alpha_{st} = 12(10^{-6})/°C$ and $E_{st} = 200$ GPa. The cross-sectional area of each member is 250 mm^2.

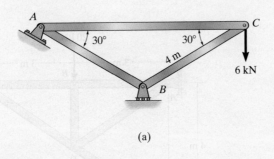

(a)

SOLUTION

Virtual Forces n.   A horizontal 1-kN virtual load is applied to the truss at joint B, and the forces in each member are calculated, Fig. 14–33b.

Real Forces N.   Since the n forces in members AC and BC are *zero*, the N forces in these members do *not* have to be determined. Why? For completeness, though, the entire "real" force analysis is shown in Fig. 14–33c.

Virtual-Work Equation.   The loads, temperature, and the fabrication error all affect the displacement of point B; therefore, Eqs. 14–37, 14–38, and 14–39 must be combined, which gives

$$1 \text{ kN} \cdot \Delta_{B_h} = \sum \frac{nNL}{AE} + \Sigma n\alpha \, \Delta T L + \Sigma n \Delta L$$

$$= 0 + 0 + \frac{(-1.155 \text{ kN})(-12 \text{ kN})(4 \text{ m})}{[250(10^{-6}) \text{ m}^2][200(10^6) \text{ kN/m}^2]}$$

$$+ 0 + 0 + (-1.155 \text{ kN})[12(10^{-6})/°C](60°C)(4 \text{ m})$$

$$+ (-1.155 \text{ kN})(-0.003 \text{ m})$$

$$\Delta_{B_h} = 0.00125 \text{ m}$$

$$= 1.25 \text{ mm} \leftarrow \qquad \qquad \textit{Ans.}$$

Virtual forces

(b)

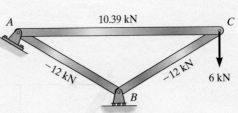

Real forces

(c)

Fig. 14–33

PROBLEMS

*14–72.   Determine the horizontal displacement of joint B on the two-member frame. Each A-36 steel member has a cross-sectional area of 1250 mm^2.

14–75.   Determine the vertical displacement of joint C on the truss. Each A-36 steel member has a cross-sectional area of $A = 300$ mm^2.

*14–76.   Determine the vertical displacement of joint D on the truss. Each A-36 steel member has a cross-sectional area of $A = 300$ mm^2.

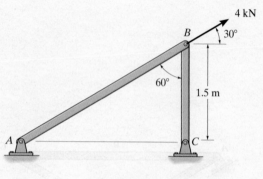

Prob. 14–72

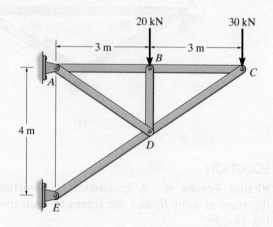

Probs. 14–75/76

•14–73.   Determine the horizontal displacement of point B. Each A-36 steel member has a cross-sectional area of 1250 mm^2.

14–74.   Determine the vertical displacement of point B. Each A-36 steel member has a cross-sectional area of 1250 mm^2.

•14–77.   Determine the vertical displacement of point B. Each A-36 steel member has a cross-sectional area of 2800 mm^2.

14–78.   Determine the vertical displacement of point E. Each A-36 steel member has a cross-sectional area of 2800 mm^2.

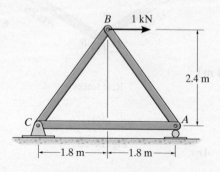

Probs. 14–73/74

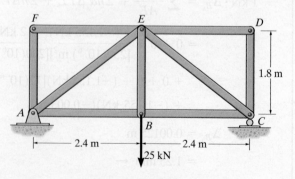

Probs. 14–77/78

14–79. Determine the horizontal displacement of joint B of the truss. Each A-36 steel member has a cross-sectional area of 400 mm^2.

*14–80. Determine the vertical displacement of joint C of the truss. Each A-36 steel member has a cross-sectional area of 400 mm^2.

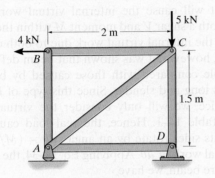

Probs. 14–79/80

14–83. Determine the vertical displacement of joint C. Each A-36 steel member has a cross-sectional area of 2800 mm^2.

*14–84. Determine the vertical displacement of joint H. Each A-36 steel member has a cross-sectional area of 2800 mm^2.

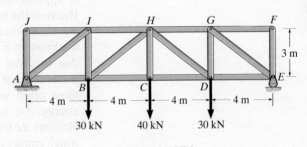

Probs. 14–83/84

•14–81. Determine the vertical displacement of point A. Each A-36 steel member has a cross-sectional area of 400 mm^2.

14–82. Determine the vertical displacement of point B. Each A-36 steel member has a cross-sectional area of 400 mm^2.

•14–85. Determine the vertical displacement of joint C. The truss is made from A-36 steel bars having a cross-sectional area of 150 mm^2.

14–86. Determine the vertical displacement of joint G. The truss is made from A-36 steel bars having a cross-sectional area of 150 mm^2.

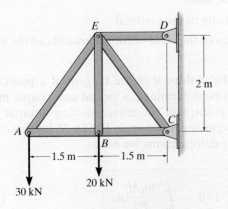

Probs. 14–81/82

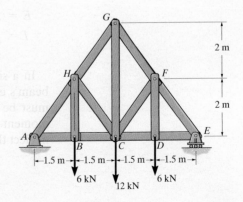

Probs. 14–85/86

*14.7   Method of Virtual Forces Applied to Beams

In this section we will apply the method of virtual forces to determine the displacement and slope at a point on a beam. To illustrate the principles, the vertical displacement Δ of point A on the beam shown in Fig. 14–34b will be determined. To do this we must place a vertical unit load at this point, Fig. 14–34a, so that when the "real" distributed load w is applied to the beam it will cause the internal virtual work $1 \cdot \Delta$. Because the load causes both a shear V and moment M within the beam, we must actually consider the internal virtual work due to both of these loadings. In Example 14.7, however, it was shown that beam deflections due to shear are negligible compared with those caused by bending, particularly if the beam is long and slender. Since this type of beam is most often used in practice, we will only consider the virtual strain energy due to bending, Table 14–1. Hence, the real load causes the element dx to deform so its sides rotate by an angle $d\theta = (M/EI)dx$, which causes internal virtual work $m \, d\theta$. Applying Eq. 14–34, the virtual-work equation for the entire beam, we have

$$1 \cdot \Delta = \int_0^L \frac{mM}{EI} dx \qquad (14\text{–}40)$$

Here

$1 =$ external virtual unit load acting on the beam in the direction of Δ

$\Delta =$ displacement caused by the real loads acting on the beam

$m =$ internal virtual moment in the beam, expressed as a function of x and caused by the external virtual unit load

$M =$ internal moment in the beam, expressed as a function of x and caused by the real loads

$E =$ modulus of elasticity of the material

$I =$ moment of inertia of the cross-sectional area about the neutral axis

In a similar manner, if the slope θ of the tangent at a point on the beam's elastic curve is to be determined, a virtual unit couple moment must be applied at the point, and the corresponding internal virtual moment m_θ has to be determined. If we apply Eq. 14–35 for this case and neglect the effect of shear deformations, we have

$$1 \cdot \theta = \int_0^L \frac{m_\theta M}{EI} dx \qquad (14\text{–}41)$$

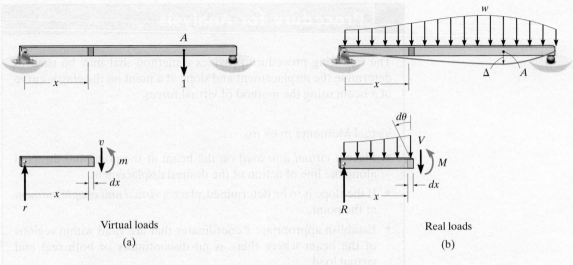

Virtual loads
(a)

Real loads
(b)

Fig. 14–34

14

When applying these equations, keep in mind that the integrals on the right side represent the amount of virtual bending strain energy that is *stored* in the beam. If concentrated forces or couple moments act on the beam or the distributed load is discontinuous, a single integration *cannot* be performed across the beam's entire length. Instead, separate x coordinates must be chosen within regions that have no discontinuity of loading. Also, it is not necessary that each x have the same origin; however, the x selected for determining the real moment M in a particular region must be the *same x* as that selected for determining the virtual moment m or m_θ within the same region. For example, consider the beam shown in Fig. 14–35. In order to determine the displacement at D, we can use x_1 to determine the strain energy in region AB, x_2 for region BC, x_3 for region DE, and x_4 for region DC. In any case, each x coordinate should be selected so that both M and m (or m_θ) can easily be formulated.

Unlike beams, as discussed here, some members may also be subjected to significant virtual strain energy caused by axial load, shear, and torsional moment. When this is the case, we must include in the above equations the energy terms for these loadings as formulated in Eq. 14–36.

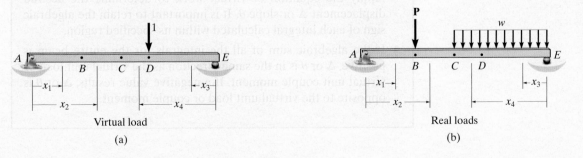

Virtual load
(a)

Real loads
(b)

Fig. 14–35

Procedure for Analysis

The following procedure provides a method that may be used to determine the displacement and slope at a point on the elastic curve of a beam using the method of virtual forces.

Virtual Moments m or m_θ.

- Place a *virtual unit load* on the beam at the point and directed along the line of action of the desired displacement.

- If the slope is to be determined, place a *virtual unit couple moment* at the point.

- Establish appropriate x coordinates that are valid within regions of the beam where there is no discontinuity of both real and virtual load.

- With the virtual load in place, and all the real loads *removed* from the beam, calculate the internal moment m or m_θ as a function of each x coordinate.

- Assume that m or m_θ acts in the positive direction according to the established beam sign convention for positive moment, Fig. 6–3.

Real Moments.

- Using the *same x coordinates* as those established for m or m_θ, determine the internal moments M caused by the real loads.

- Since positive m or m_θ was assumed to act in the conventional "positive direction," it is important that positive M acts in this *same direction*. This is necessary since positive or negative internal virtual work depends on the directional sense of *both* the virtual load, defined by $\pm m$ or $\pm m_\theta$, and displacement, caused by $\pm M$.

Virtual-Work Equation.

- Apply the equation of virtual work to determine the desired displacement Δ or slope θ. It is important to retain the algebraic sign of each integral calculated within its specified region.

- If the algebraic sum of all the integrals for the entire beam is positive, Δ or θ is in the same direction as the virtual unit load or virtual unit couple moment. If a negative value results, Δ or θ is opposite to the virtual unit load or couple moment.

EXAMPLE 14.13

Determine the displacement of point B on the beam shown in Fig. 14–36a. EI is constant.

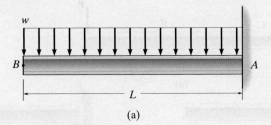

(a)

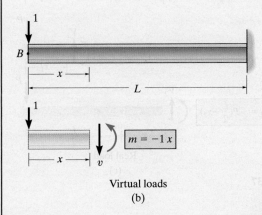

Virtual loads
(b)

$m = -1\,x$

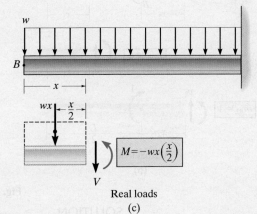

Real loads
(c)

$M = -wx\left(\dfrac{x}{2}\right)$

Fig. 14–36

SOLUTION

Virtual Moment m. The vertical displacement of point B is obtained by placing a virtual unit load at B, Fig. 14–36b. By inspection, there are no discontinuities of loading on the beam for *both* the real and virtual loads. Thus, a *single x* coordinate can be used to determine the virtual strain energy. This coordinate will be selected with its origin at B, so that the reactions at A do not have to be determined in order to find the internal moments m and M. Using the method of sections, the internal moment m is shown in Fig. 14–36b.

Real Moment M. Using the *same x* coordinate, the internal moment M is shown in Fig. 14–36c.

Virtual-Work Equation. The vertical displacement at B is thus

$$1 \cdot \Delta_B = \int \frac{mM}{EI}\,dx = \int_0^L \frac{(-1x)(-wx^2/2)\,dx}{EI}$$

$$\Delta_B = \frac{wL^4}{8EI} \qquad\qquad Ans.$$

EXAMPLE | 14.14

Determine the slope at point B of the beam shown in Fig. 14–37a. EI is constant.

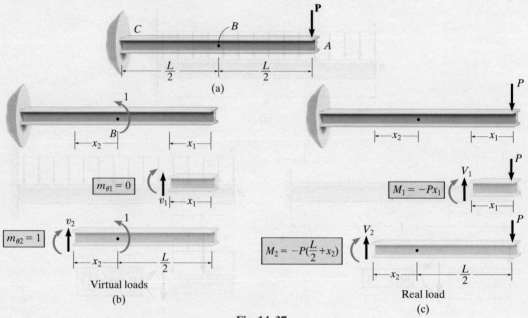

Virtual loads
(b)

Real load
(c)

Fig. 14–37

SOLUTION

Virtual Moments m_θ. The slope at B is determined by placing a virtual unit couple moment at B, Fig. 14–37b. Two x coordinates must be selected in order to determine the total virtual strain energy in the beam. Coordinate x_1 accounts for the strain energy within segment AB, and coordinate x_2 accounts for the strain energy in segment BC. Using the method of sections the internal moments m_θ within each of these segments are shown in Fig. 14–37b.

Real Moments M. Using the *same coordinates* x_1 and x_2 (Why?), the internal moments M are shown in Fig. 14–37c.

Virtual-Work Equation. The slope at B is thus

$$1 \cdot \theta_B = \int \frac{m_\theta M}{EI} dx$$

$$= \int_0^{L/2} \frac{0(-Px_1)\, dx_1}{EI} + \int_0^{L/2} \frac{1\{-P[(L/2) + x_2]\}\, dx_2}{EI}$$

$$\theta_B = -\frac{3PL^2}{8EI} \qquad\qquad Ans.$$

The *negative sign* indicates that θ_B is *opposite* to the direction of the virtual couple moment shown in Fig. 14–37b.

PROBLEMS

14–87. Determine the displacement at point *C*. *EI* is constant.

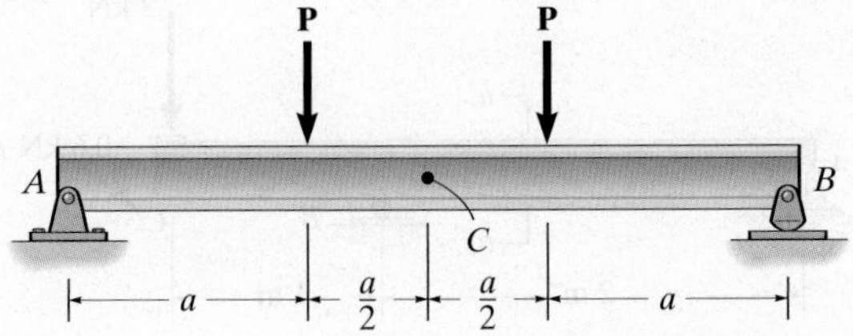

Prob. 14–87

***14–88.** The beam is made of southern pine for which $E_p = 13$ GPa. Determine the displacement at *A*.

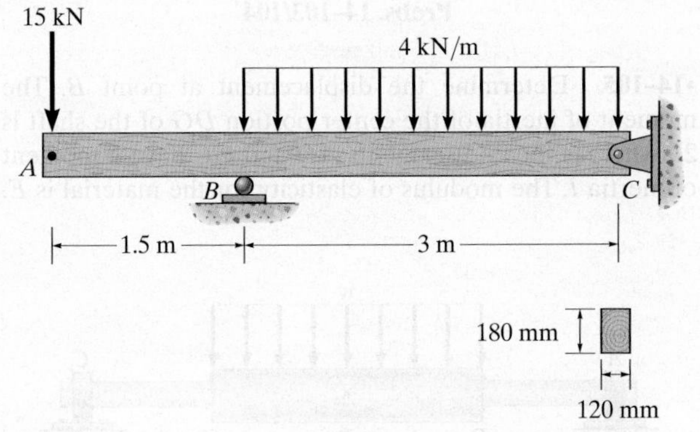

Prob. 14–88

•14–89. Determine the displacement at *C* of the A-36 steel beam. $I = 70(10^6)$ mm^4.

14–90. Determine the slope at *A* of the A-36 steel beam. $I = 70(10^6)$ mm^4.

14–91. Determine the slope at *B* of the A-36 steel beam. $I = 70(10^6)$ mm^4.

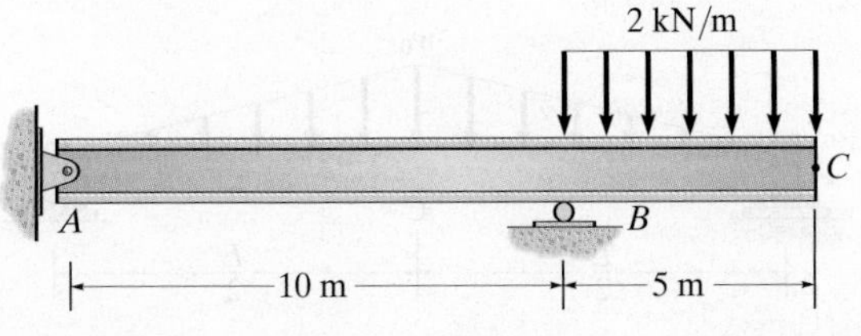

Probs. 14–89/90/91

***14–92.** Determine the displacement at *B* of the 30-mm-diameter A-36 steel shaft.

•14–93. Determine the slope of the 30-mm-diameter A-36 steel shaft at the bearing support *A*.

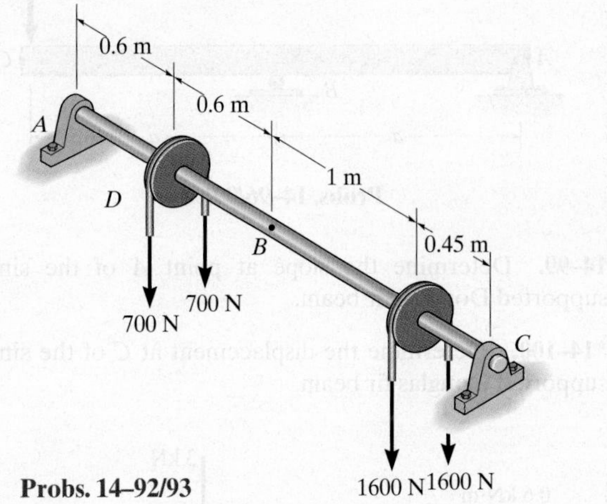

Probs. 14–92/93

14–94. The beam is made of Douglas fir. Determine the slope at *C*.

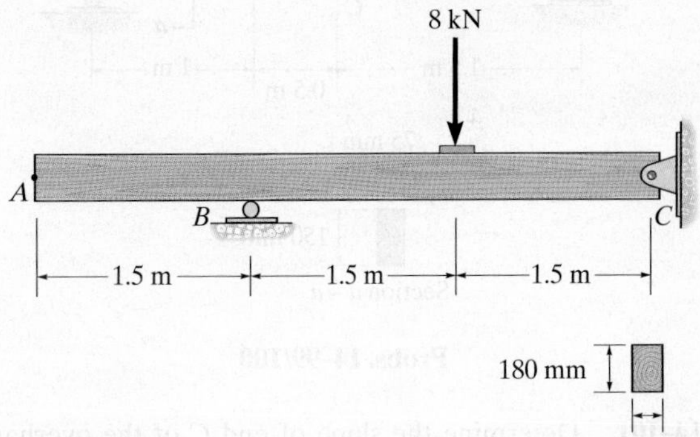

Prob. 14–94

14–95. The beam is made of oak, for which $E_o = 11$ GPa. Determine the slope and displacement at *A*.

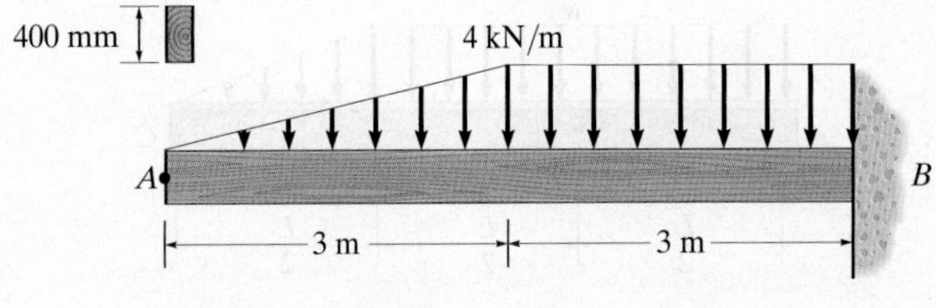

Prob. 14–95

14

*14-96.** Determine the displacement at point *C*. *EI* is constant.

•14-97.** Determine the slope at point *C*. *EI* is constant.

14-98.** Determine the slope at point *A*. *EI* is constant.

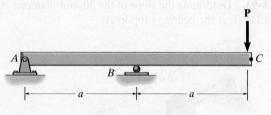

Probs. 14-96/97/98

14-99.** Determine the slope at point *A* of the simply supported Douglas fir beam.

*14-100.** Determine the displacement at *C* of the simply supported Douglas fir beam.

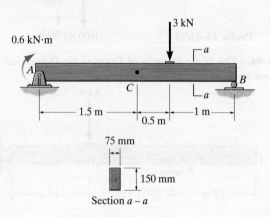

Probs. 14-99/100

•14-101.** Determine the slope of end *C* of the overhang beam. *EI* is constant.

14-102.** Determine the displacement of point *D* of the overhang beam. *EI* is constant.

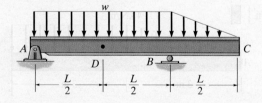

Probs. 14-101/102

14-103.** Determine the displacement of end *C* of the overhang Douglas fir beam.

*14-104.** Determine the slope at *A* of the overhang white spruce beam.

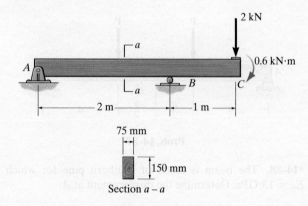

Probs. 14-103/104

•14-105.** Determine the displacement at point *B*. The moment of inertia of the center portion *DG* of the shaft is $2I$, whereas the end segments *AD* and *GC* have a moment of inertia *I*. The modulus of elasticity for the material is *E*.

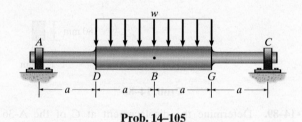

Prob. 14-105

14-106.** Determine the displacement of the shaft at *C*. *EI* is constant.

14-107.** Determine the slope of the shaft at the bearing support *A*. *EI* is constant.

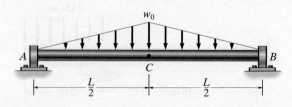

Probs. 14-106/107

*14–108. Determine the slope and displacement of end C of the cantilevered beam. The beam is made of a material having a modulus of elasticity of E. The moments of inertia for segments AB and BC of the beam are $2I$ and I, respectively.

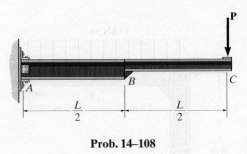

Prob. 14–108

•14–109. Determine the slope at A of the A-36 steel W200 × 46 simply supported beam.

14–110. Determine the displacement at point C of the A-36 steel W200 × 46 simply supported beam.

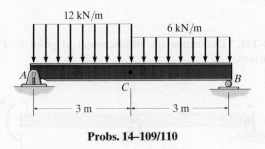

Probs. 14–109/110

14–111. The simply supported beam having a square cross section is subjected to a uniform load w. Determine the maximum deflection of the beam caused only by bending, and caused by bending and shear. Take $E = 3G$.

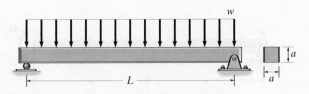

Prob. 14–111

*14–112. The frame is made from two segments, each of length L and flexural stiffness EI. If it is subjected to the uniform distributed load determine the vertical displacement of point C. Consider only the effect of bending.

•14–113. The frame is made from two segments, each of length L and flexural stiffness EI. If it is subjected to the uniform distributed load, determine the horizontal displacement of point B. Consider only the effect of bending.

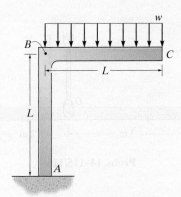

Probs. 14–112/113

14–114. Determine the vertical displacement of point A on the angle bracket due to the concentrated force $\mathbf{P}$. The bracket is fixed connected to its support. EI is constant. Consider only the effect of bending.

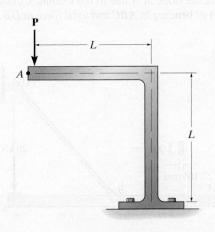

Prob. 14–114

14

14–115. Beam AB has a square cross section of 100 mm by 100 mm. Bar CD has a diameter of 10 mm. If both members are made of A-36 steel, determine the vertical displacement of point B due to the loading of 10 kN.

*14–116. Beam AB has a square cross section of 100 mm by 100 mm. Bar CD has a diameter of 10 mm. If both members are made of A-36 steel, determine the slope at A due to the loading of 10 kN.

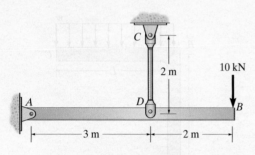

Probs. 14–115/116

14–117. Bar ABC has a rectangular cross section of 300 mm by 100 mm. Attached rod DB has a diameter of 20 mm. If both members are made of A-36 steel, determine the vertical displacement of point C due to the loading. Consider only the effect of bending in ABC and axial force in DB.

14–118. Bar ABC has a rectangular cross section of 300 mm by 100 mm. Attached rod DB has a diameter of 20 mm. If both members are made of A-36 steel, determine the slope at A due to the loading. Consider only the effect of bending in ABC and axial force in DB.

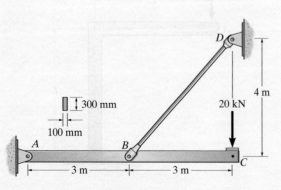

Probs. 14–117/118

14–119. Determine the vertical displacement of point C. The frame is made using A-36 steel W250 × 45 members. Consider only the effects of bending.

*14–120. Determine the horizontal displacement of end B. The frame is made using A-36 steel W250 × 45 members. Consider only the effects of bending.

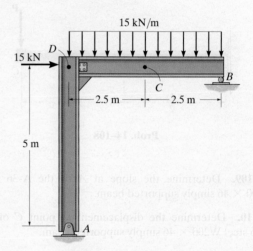

Probs. 14–119/120

•14–121. Determine the displacement at point C. EI is constant.

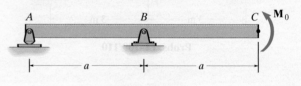

Prob. 14–121

14–122. Determine the slope at B. EI is constant.

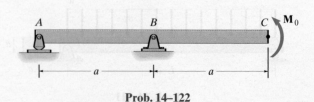

Prob. 14–122

*14.8   Castigliano's Theorem

In 1879, Alberto Castigliano, an Italian railroad engineer, published a book in which he outlined a method for determining the displacement and slope at a point in a body. This method, which is referred to as Castigliano's second theorem, applies only to bodies that have constant temperature and material with linear-elastic behavior. If the displacement at a point is to be determined, the theorem states that the displacement is equal to the first partial derivative of the strain energy in the body with respect to a force acting at the point and in the direction of displacement. In a similar manner, the slope of the tangent at a point in a body is equal to the first partial derivative of the strain energy in the body with respect to a couple moment acting at the point and in the direction of the slope angle.

To derive Castigliano's second theorem, consider a body of any arbitrary shape, which is subjected to a series of n forces $\mathbf{P}_1, \mathbf{P}_2, \ldots, \mathbf{P}_n$, Fig. 14–38. According to the conservation of energy, the external work done by these forces is equal to the internal strain energy stored in the body. However, the external work is a function of the external loads, $U_e = \Sigma \int P \, dx$, Eq. 14–1, so the internal work is also a function of the external loads. Thus,

$$U_i = U_e = f(P_1, P_2, \ldots, P_n) \tag{14–42}$$

Now, if any one of the external forces, say P_j, is increased by a differential amount dP_j, the internal work will also be increased, such that the strain energy becomes

$$U_i + dU_i = U_i + \frac{\partial U_i}{\partial P_j} dP_j \tag{14–43}$$

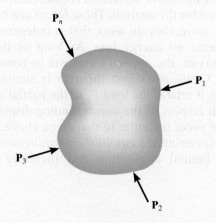

Fig. 14–38

This value, however, will not depend on the sequence in which the n forces are applied to the body. For example, we could apply $d\mathbf{P}_j$ to the body *first*, then apply the loads $\mathbf{P}_1, \mathbf{P}_2, \ldots, \mathbf{P}_n$. In this case, $d\mathbf{P}_j$ would cause the body to displace a differential amount $d\Delta_j$ in the direction of $d\mathbf{P}_j$. By Eq. 14–2 ($U_e = \frac{1}{2} P_j \Delta_j$), the increment of strain energy would be $\frac{1}{2} dP_j d\Delta_j$. This quantity, however, is a second-order differential and may be neglected. Further application of the loads $\mathbf{P}_1, \mathbf{P}_2, \ldots, \mathbf{P}_n$ causes $d\mathbf{P}_j$ to move through the displacement Δ_j so that now the strain energy becomes

$$U_i + dU_i = U_i + dP_j \, \Delta_j \qquad (14\text{–}44)$$

Here, as above, U_i is the internal strain energy in the body, caused by the loads $\mathbf{P}_1, \mathbf{P}_2, \ldots, \mathbf{P}_n$, and $dP_j \Delta_j$ is the *additional* strain energy caused by $d\mathbf{P}_j$.

In summary, Eq. 14–43 represents the strain energy in the body determined by first applying the loads $\mathbf{P}_1, \mathbf{P}_2, \ldots, \mathbf{P}_n$, then $d\mathbf{P}_j$; Eq. 14–44 represents the strain energy determined by first applying $d\mathbf{P}_j$ and then the loads $\mathbf{P}_1, \mathbf{P}_2, \ldots, \mathbf{P}_n$. Since these two equations must be equal, we require

$$\Delta_j = \frac{\partial U_i}{\partial P_j} \qquad (14\text{–}45)$$

which proves the theorem; i.e., the displacement Δ_j in the direction of $\mathbf{P}_j$ is equal to the first partial derivative of the strain energy with respect to $\mathbf{P}_j$.

Castigliano's second theorem, Eq. 14–45, is a statement regarding the body's *compatibility requirements*, since it is a condition related to displacement. Also, the above derivation requires that *only conservative forces* be considered for the analysis. These forces can be applied in any order, and furthermore, they do work that is independent of the path and therefore create no energy loss. As long as the material has linear-elastic behavior, the applied forces will be conservative and the theorem is valid. Castigliano's first theorem is similar to his second theorem; however, it relates the load P_j to the partial derivative of the strain energy with respect to the corresponding displacement, that is, $P_j = \partial U_i / \partial \Delta_j$. The proof is similar to that given above. This theorem is another way of expressing the *equilibrium requirements* for the body; however, it has limited application and therefore it will not be discussed here.

*14.9   Castigliano's Theorem Applied to Trusses

Since a truss member is only subjected to an axial load, the strain energy for the member is given by Eq. 14–16, $U_i = N^2 L / 2AE$. Substituting this equation into Eq. 14–45 and omitting the subscript i, we have

$$\Delta = \frac{\partial}{\partial P} \sum \frac{N^2 L}{2AE}$$

It is generally easier to perform the differentiation prior to summation. Also, L, A, and E are constant for a given member, and therefore we can write

$$\Delta = \sum N \left(\frac{\partial N}{\partial P} \right) \frac{L}{AE} \qquad (14\text{–}46)$$

Here

Δ = displacement of the truss joint

P = an external force of *variable magnitude* applied to the truss joint in the direction of Δ

N = internal axial force in a member caused by *both* force **P** and the actual loads on the truss

L = length of a member

A = cross-sectional area of a member

E = modulus of elasticity of the material

   By comparison, Eq. 14–46 is similar to that used for the method of virtual forces, Eq. 14–37 ($1 \cdot \Delta = \sum nNL/AE$), except that n is replaced by $\partial N/\partial P$. These terms, n and $\partial N/\partial P$, are the *same*, since they represent the change of the member's axial force with respect to the load P or, in other words, the axial force per unit load.

Procedure for Analysis

The following procedure provides a method that may be used to determine the displacement of any joint on a truss using Castigliano's second theorem.

External Force P.

• Place a force $\mathbf{P}$ on the truss at the joint where the displacement is to be determined. This force is assumed to have a *variable magnitude* and should be directed along the line of action of the displacement.

Internal Forces N.

• Determine the force N in each member in terms of both the actual (numerical) loads and the (variable) force P. Assume that tensile forces are positive and compressive forces are negative.

• Find the respective partial derivative $\partial N/\partial P$ for each member.

• *After* N and $\partial N/\partial P$ have been determined, assign P its numerical value if it has actually replaced a real force on the truss. Otherwise, set P equal to zero.

Castigliano's Second Theorem.

• Apply Castigliano's second theorem to determine the desired displacement Δ. It is important to retain the algebraic signs for corresponding values of N and $\partial N/\partial P$ when substituting these terms into the equation.

• If the resultant sum $\Sigma N(\partial N/\partial P)L/AE$ is positive, Δ is in the same direction as $\mathbf{P}$. If a negative value results, Δ is opposite to $\mathbf{P}$.

EXAMPLE | 14.15

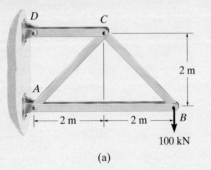

(a)

Fig. 14–39

Determine the vertical displacement of joint C of the steel truss shown in Fig. 14–39a. The cross-sectional area of each member is $A = 400 \text{ mm}^2$, and $E_{st} = 200$ GPa.

SOLUTION

External Force P.   A vertical force $\mathbf{P}$ is applied to the truss at joint C, since this is where the vertical displacement is to be determined, Fig. 14–39b.

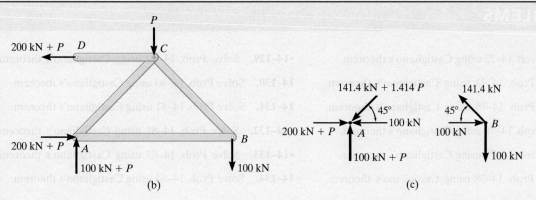

Fig. 14–39 (cont.)

Internal Forces N.

The reactions at the truss supports A and D are calculated and the results are shown in Fig. 14–39b. Using the method of joints, the N forces in each member are determined, Fig. 14–39c.* For convenience, these results along with their partial derivatives $\partial N/\partial P$ are listed in tabular form. Note that since **P** does not actually exist as a real load on the truss, we require $P = 0$.

Member	N	$\dfrac{\partial N}{\partial P}$	$N(P = 0)$	L	$N\left(\dfrac{\partial N}{\partial P}\right)L$
AB	-100	0	-100	4	0
BC	141.4	0	141.4	2.828	0
AC	$-(141.4 + 1.414P)$	-1.414	-141.4	2.828	565.7
CD	$200 + P$	1	200	2	400
					$\Sigma\ 965.7\ \text{kN} \cdot \text{m}$

Castigliano's Second Theorem.

Applying Eq. 14–46, we have

$$\Delta_{C_v} = \Sigma N\left(\frac{\partial N}{\partial P}\right)\frac{L}{AE} = \frac{965.7\ \text{kN} \cdot \text{m}}{AE}$$

Substituting the numerical values for A and E, we get

$$\Delta_{C_v} = \frac{965.7\ \text{kN} \cdot \text{m}}{[400(10^{-6})\ \text{m}^2]\ 200(10^6)\ \text{kN/m}^2}$$

$$= 0.01207\ \text{m} = 12.1\ \text{mm} \qquad \qquad Ans.$$

This solution should be compared with that of Example 14.11, using the virtual-work method.

*It may be more convenient to analyze the truss with just the 100-kN load on it, then analyze the truss with the **P** load on it. The results can then be summed algebraically to give the N forces.

PROBLEMS

14–123. Solve Prob. 14–72 using Castigliano's theorem.

*14–124. Solve Prob. 14–73 using Castigliano's theorem.

•14–125. Solve Prob. 14–75 using Castigliano's theorem.

14–126. Solve Prob. 14–76 using Castigliano's theorem.

14–127. Solve Prob. 14–77 using Castigliano's theorem.

*14–128. Solve Prob. 14–78 using Castigliano's theorem.

•14–129. Solve Prob. 14–79 using Castigliano's theorem.

14–130. Solve Prob. 14–80 using Castigliano's theorem.

14–131. Solve Prob. 14–81 using Castigliano's theorem.

*14–132. Solve Prob. 14–82 using Castigliano's theorem.

•14–133. Solve Prob. 14–83 using Castigliano's theorem.

14–134. Solve Prob. 14–84 using Castigliano's theorem.

*14.10 Castigliano's Theorem Applied to Beams

The internal strain energy for a beam is caused by both bending and shear. However, as pointed out in Example 14.7, if the beam is long and slender, the strain energy due to shear can be neglected compared with that of bending. Assuming this to be the case, the internal strain energy for a beam is given by $U_i = \int M^2\, dx/2EI$, Eq. 14–17. Omitting the subscript i, Castigliano's second theorem, $\Delta_i = \partial U_i/\partial P_i$, becomes

$$\Delta = \frac{\partial}{\partial P} \int_0^L \frac{M^2\, dx}{2EI}$$

Rather than squaring the expression for internal moment, integrating, and then taking the partial derivative, it is generally easier to differentiate prior to integration. Provided E and I are constant, we have

$$\Delta = \int_0^L M\left(\frac{\partial M}{\partial P}\right)\frac{dx}{EI} \qquad (14\text{–}47)$$

Here

Δ = displacement of the point caused by the real loads acting on the beam

P = an external force of *variable magnitude* applied to the beam at the point and in the direction of Δ

M = internal moment in the beam, expressed as a function of x and caused by both the force P and the actual loads on the beam

E = modulus of elasticity of the material

I = moment of inertia of cross-sectional area about the neutral axis

If the slope of the tangent θ at a point on the elastic curve is to be determined, the partial derivative of the internal moment M with respect to an *external couple moment M'* acting at the point must be found. For this case,

$$\theta = \int_0^L M \left(\frac{\partial M}{\partial M'} \right) \frac{dx}{EI} \qquad (14\text{--}48)$$

The above equations are similar to those used for the method of virtual forces, Eqs. 14–40 and 14–41, except m and m_θ replace $\partial M/\partial P$ and $\partial M/\partial M'$, respectively.

In addition, if axial load, shear, and torsion cause significant strain energy within the member, then the effects of all these loadings should be included when applying Castigliano's theorem. To do this we must use the strain-energy functions developed in Sec. 14.2, along with their associated partial derivatives. The result is

$$\Delta = \Sigma N \left(\frac{\partial N}{\partial P} \right) \frac{L}{AE} + \int_0^L f_s V \left(\frac{\partial V}{\partial P} \right) \frac{dx}{GA} + \int_0^L M \left(\frac{\partial M}{\partial P} \right) \frac{dx}{EI} + \int_0^L T \left(\frac{\partial T}{\partial P} \right) \frac{dx}{GJ} \qquad (14\text{--}49)$$

The method of applying this general formulation is similar to that used to apply Eqs. 14–47 and 14–48.

Procedure for Analysis

The following procedure provides a method that may be used to apply Castigliano's second theorem.

External Force P or Couple Moment M'.

- Place a force **P** on the beam at the point and directed along the line of action of the desired displacement.

- If the slope of the tangent is to be determined at the point, place a couple moment **M'** at the point.

- Assume that both **P** and **M'** have a variable magnitude.

Internal Moments M.

- Establish appropriate x coordinates that are valid within regions of the beam where there is no discontinuity of force, distributed load, or couple moment.

- Determine the internal moments M as a function of x, the actual (numerical) loads, and P or M', and then find the partial derivatives $\partial M / \partial P$ or $\partial M / \partial M'$ for each coordinate x.

- *After* M and $\partial M / \partial P$ or $\partial M / \partial M'$ have been determined, assign P or M' its numerical value if it has actually replaced a real force or couple moment. Otherwise, set P or M' equal to zero.

Castigliano's Second Theorem.

- Apply Eq. 14–47 or 14–48 to determine the desired displacement Δ or θ. It is important to retain the algebraic signs for corresponding values of M and $\partial M / \partial P$ or $\partial M / \partial M'$.

- If the resultant sum of all the definite integrals is positive, Δ or θ is in the same direction as **P** or **M'**. If a negative value results, Δ or θ is opposite to **P** or **M'**.

EXAMPLE | 14.16

Determine the displacement of point B on the beam shown in Fig. 14–40a. EI is constant.

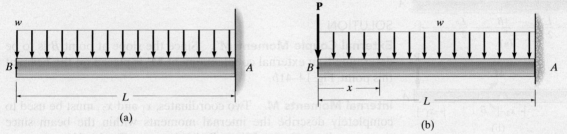

Fig. 14–40

SOLUTION

External Force P.   A vertical force P is placed on the beam at B as shown in Fig. 14–40b.

Internal Moments M.   A single x coordinate is needed for the solution, since there are no discontinuities of loading between A and B. Using the method of sections, Fig. 14–40c, the internal moment and its partial derivative are determined as follows:

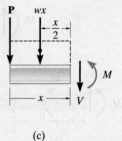

(c)

$$\zeta+\Sigma M_{NA} = 0; \qquad M + wx\left(\frac{x}{2}\right) + P(x) = 0$$

$$M = -\frac{wx^2}{2} - Px$$

$$\frac{\partial M}{\partial P} = -x$$

Setting $P = 0$ gives

$$M = \frac{-wx^2}{2} \qquad \text{and} \qquad \frac{\partial M}{\partial P} = -x$$

Castigliano's Second Theorem.   Applying Eq. 14–47, we have

$$\Delta_B = \int_0^L M\left(\frac{\partial M}{\partial P}\right)\frac{dx}{EI} = \int_0^L \frac{(-wx^2/2)(-x)\, dx}{EI}$$

$$= \frac{wL^4}{8EI} \qquad\qquad\qquad \textit{Ans.}$$

The similarity between this solution and that of the virtual-work method, Example 14.13, should be noted.

EXAMPLE | 14.17

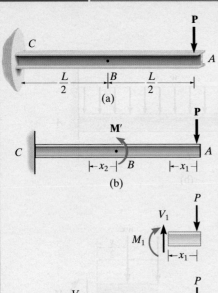

(a)

(b)

(c)

Fig. 14–41

Determine the slope at point B of the beam shown in Fig. 14–41a. EI is constant.

SOLUTION

External Couple Moment M'.   Since the slope at point B is to be determined, an external couple moment M' is placed on the beam at this point, Fig. 14–41b.

Internal Moments M.   Two coordinates, x_1 and x_2, must be used to completely describe the internal moments within the beam since there is a discontinuity, M', at B. As shown in Fig. 14–41b, x_1 ranges from A to B and x_2 ranges from B to C. Using the method of sections, Fig. 14–41c, the internal moments and the partial derivatives for x_1 and x_2 are determined as follows:

$$\zeta + \Sigma M_{NA} = 0; \qquad M_1 = -Px_1, \qquad \frac{\partial M_1}{\partial M'} = 0$$

$$\zeta + \Sigma M_{NA} = 0; \qquad M_2 = M' - P\left(\frac{L}{2} + x_2\right), \qquad \frac{\partial M_2}{\partial M'} = 1$$

Castigliano's Second Theorem.   Setting $M' = 0$ and applying Eq. 14–48, we have

$$\theta_B = \int_0^L M\left(\frac{\partial M}{\partial M'}\right)\frac{dx}{EI} = \int_0^{L/2}\frac{(-Px_1)(0)\,dx_1}{EI} + \int_0^{L/2}\frac{-P[(L/2) + x_2](1)\,dx_2}{EI} = -\frac{3PL^2}{8EI} \quad Ans.$$

Note the similarity between this solution and that of Example 14.14.

PROBLEMS

14–135.   Solve Prob. 14–87 using Castigliano's theorem.

*14–136.   Solve Prob. 14–88 using Castigliano's theorem.

•14–137.   Solve Prob. 14–90 using Castigliano's theorem.

14–138.   Solve Prob. 14–92 using Castigliano's theorem.

14–139.   Solve Prob. 14–93 using Castigliano's theorem.

*14–140.   Solve Prob. 14–96 using Castigliano's theorem.

14–141.   Solve Prob. 14–97 using Castigliano's theorem.

14–142.   Solve Prob. 14–98 using Castigliano's theorem.

14–143.   Solve Prob. 14–112 using Castigliano's theorem.

*14–144.   Solve Prob. 14–114 using Castigliano's theorem.

•14–145.   Solve Prob. 14–121 using Castigliano's theorem.

CHAPTER REVIEW

When a force (couple moment) acts on a deformable body it will do external work when it displaces (rotates). The internal stresses produced in the body also undergo displacement, thereby creating elastic strain energy that is stored in the material. The conservation of energy states that the external work done by the loading is equal to the internal elastic strain energy produced by the stresses in the body.

$$U_e = U_i$$

Ref.: Sections 14.1 & 14.2 & 14.3

The conservation of energy can be used to solve problems involving elastic impact, which assumes the moving body is rigid and all the strain energy is stored in the stationary body. This leads to use of an impact factor n, which is a ratio of the dynamic load to the static load. It is used to determine the maximum stress and displacement of the body at the point of impact.

$$n = 1 + \sqrt{1 + 2\left(\frac{h}{\Delta_{st}}\right)}$$

$$\sigma_{max} = n\sigma_{st}$$

$$\Delta_{max} = n\Delta_{st}$$

Ref.: Section 14.4

The principle of virtual work can be used to determine the displacement of a joint on a truss or the slope and the displacement of points on a beam. It requires placing an external virtual unit force (virtual unit couple moment) at the point where the displacement (rotation) is to be determined. The external virtual work that is produced by the external loading is then equated to the internal virtual strain energy in the structure.

$$1 \cdot \Delta = \sum \frac{nNL}{AE}$$

$$1 \cdot \Delta = \int_0^L \frac{mM}{EI} dx$$

$$1 \cdot \theta = \int_0^L \frac{m_\theta M}{EI} dx$$

Ref.: Sections 14.5 & 14.6 & 14.7

Castigliano's second theorem can also be used to determine the displacement of a joint on a truss or the slope and the displacement at a point on a beam. Here a variable force P (couple moment M') is placed at the point where the displacement (slope) is to be determined. The internal loading is then determined as a function of $P(M')$ and its partial derivative with respect to $P(M')$ is determined. Castigliano's second theorem is then applied to obtain the desired displacement (rotation).

$$\Delta = \sum N\left(\frac{\partial N}{\partial P}\right)\frac{L}{AE}$$

$$\Delta = \int_0^L M\left(\frac{\partial M}{\partial P}\right)\frac{dx}{EI}$$

$$\theta = \int_0^L M\left(\frac{\partial M}{\partial M'}\right)\frac{dx}{EI}$$

Ref.: Sections 14.8 & 14.9 & 14.10

14

REVIEW PROBLEMS

14–146. Determine the bending strain energy in the beam due to the loading shown. *EI* is constant.

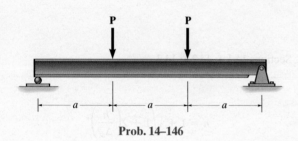

Prob. 14–146

14–147. The 200-kg block *D* is dropped from a height *h* = 1 m onto end *C* of the A-36 steel W200 × 36 overhang beam. If the spring at *B* has a stiffness *k* = 200 kN/m, determine the maximum bending stress developed in the beam.

***14–148.** Determine the maximum height *h* from which the 200-kg block *D* can be dropped without causing the A-36 steel W200 × 36 overhang beam to yield. The spring at *B* has a stiffness *k* = 200 kN/m.

•**14–149.** The L2 steel bolt has a diameter of 5 mm, and the link *AB* has a rectangular cross section that is 10 mm wide by 4 mm thick. Determine the strain energy in the link *AB* due to bending, and in the bolt due to axial force. The bolt is tightened so that it has a tension of 1750 N. Neglect the hole in the link.

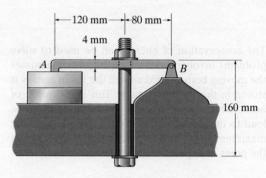

Prob. 14–149

14–150. Determine the vertical displacement of joint *A*. Each bar is made of A-36 steel and has a cross-sectional area of 600 mm². Use the conservation of energy.

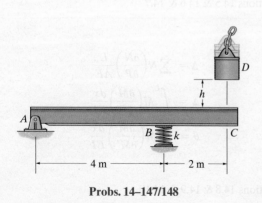

Probs. 14–147/148

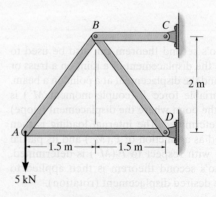

Prob. 14–150

14–151. Determine the total strain energy in the A-36 steel assembly. Consider the axial strain energy in the two 12-mm-diameter rods and the bending strain energy in the beam for which $I = 17.0(10^6)$ mm^4.

14–154. The cantilevered beam is subjected to a couple moment $\mathbf{M}_0$ applied at its end. Determine the slope of the beam at B. EI is constant. Use the method of virtual work.

14–155. Solve Prob. 14–154 using Castigliano's theorem.

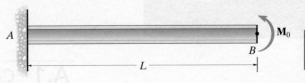

Probs. 14–154/155

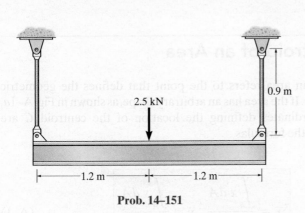

Prob. 14–151

*__14–156.__ Determine the displacement of point B on the aluminum beam. $E_{al} = 75$ GPa. Use the conservation of energy.

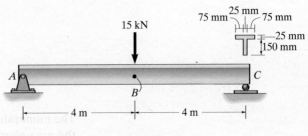

Prob. 14–156

*__14–152.__ Determine the vertical displacement of joint E. For each member $A = 400$ mm^2, $E = 200$ GPa. Use the method of virtual work.

•__14–153.__ Solve Prob. 14–152 using Castigliano's theorem.

14–157. A 10-kg weight is dropped from a height of 1.2 m onto the end of a cantilevered A-36 steel beam. If the beam is a W310 × 74, determine the maximum stress developed in the beam.

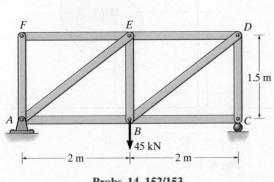

Probs. 14–152/153

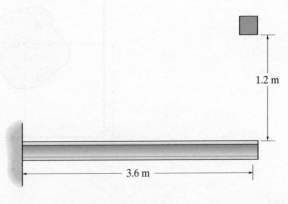

Prob. 14–157

Geometric Properties of an Area

A.1 Centroid of an Area

The *centroid* of an area refers to the point that defines the geometric center for the area. If the area has an arbitrary shape, as shown in Fig. A–1a, the x and y coordinates defining the location of the centroid C are determined using the formulas

$$\bar{x} = \frac{\int_A x\, dA}{\int_A dA} \qquad \bar{y} = \frac{\int_A y\, dA}{\int_A dA} \tag{A–1}$$

The numerators in these equations are formulations of the "moment" of the area element dA about the y and the x axis, respectively, Fig. A–1b; the denominators represent the total area A of the shape.

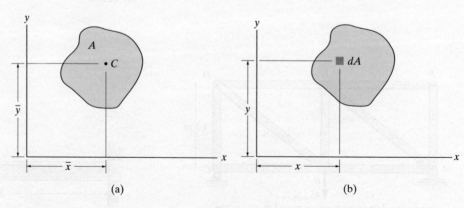

(a)   (b)

Fig. A–1

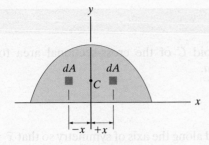

Fig. A–2

The location of the centroid for some areas may be partially or completely specified by using symmetry conditions. In cases where the area has an axis of symmetry, the centroid for the area will lie along this axis. For example, the centroid C for the area shown in Fig. A–2 must lie along the y axis, since for every elemental area dA a distance $+x$ to the right of the y axis, there is an identical element a distance $-x$ to the left. The total moment for all the elements about the axis of symmetry will therefore cancel; that is, $\int x \, dA = 0$ (Eq. A–1), so that $\overline{x} = 0$. In cases where a shape has two axes of symmetry, it follows that the centroid lies at the intersection of these axes, Fig. A–3. Based on the principle of symmetry, or using Eq. A–1, the locations of the centroid for common area shapes are listed on the inside front cover.

Composite Areas.   Often an area can be sectioned or divided into several parts having simpler shapes. Provided the area and location of the centroid of each of these "composite shapes" are known, one can eliminate the need for integration to determine the centroid for the entire area. In this case, equations analogous to Eq. A–1 must be used, except that finite summation signs replace the integrals; i.e.,

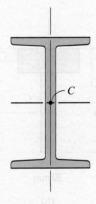

Fig. A–3

$$\overline{x} = \frac{\Sigma \tilde{x} A}{\Sigma A} \qquad \overline{y} = \frac{\Sigma \tilde{y} A}{\Sigma A} \qquad\qquad \text{(A–2)}$$

Here $\tilde{x}$ and $\tilde{y}$ represent the *algebraic distances* or x, y coordinates for the centroid of each composite part, and ΣA represents the sum of the areas of the composite parts or simply the *total area*. In particular, if a hole, or a geometric region having no material, is located within a composite part, the hole is considered as an additional composite part having a *negative* area. Also, as discussed above, if the total area is symmetrical about an axis, the centroid of the area lies on the axis.

The following example illustrates application of Eq. A–2.

A

EXAMPLE A.1

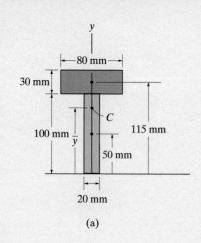

(a)

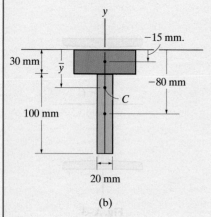

(b)

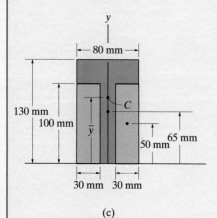

(c)

Fig. A–4

Locate the centroid C of the cross-sectional area for the T-beam shown in Fig. A–4a.

SOLUTION I

The y axis is placed along the axis of symmetry so that $\bar{x} = 0$, Fig. A–4a. To obtain $\bar{y}$ we will establish the x axis (reference axis) through the base of the area. The area is segmented into two rectangles as shown, and the centroidal location $\bar{y}$ for each is established. Applying Eq. A–2, we have

$$\bar{y} = \frac{\Sigma \bar{y}A}{\Sigma A} = \frac{[50 \text{ mm}](100 \text{ mm})(20 \text{ mm}) + [115 \text{ mm}](30 \text{ mm})(80 \text{ mm})}{(100 \text{ mm})(20 \text{ mm}) + (30 \text{ mm})(80 \text{ mm})}$$

$$= 85.5 \text{ mm} \qquad\qquad Ans.$$

SOLUTION II

Using the same two segments, the x axis can be located at the top of the area, Fig. A–4b. Here

$$\bar{y} = \frac{\Sigma \tilde{y}A}{\Sigma A} = \frac{[-15 \text{ mm}](30 \text{ mm})(80 \text{ mm}) + [-80 \text{ mm}](100 \text{ mm})(20 \text{ mm})}{(30 \text{ mm})(80 \text{ mm}) + (100 \text{ mm})(20 \text{ mm})}$$

$$= -44.5 \text{ mm} \qquad\qquad Ans.$$

The negative sign indicates that C is located *below* the x axis, which is to be expected. Also note that from the two answers 85.5 mm + 44.5 mm + 130 mm, which is the depth of the beam.

SOLUTION III

It is also possible to consider the cross-sectional area to be one large rectangle *less* two small rectangles shown shaded in Fig A–4c. Here we have

$$\bar{y} = \frac{\Sigma \tilde{y}A}{\Sigma A} = \frac{[65 \text{ mm}](130 \text{ mm})(80 \text{ mm}) - 2[50 \text{ mm}](100 \text{ mm})(30 \text{ mm})}{(130 \text{ mm})(80 \text{ mm}) - 2(100 \text{ mm})(30 \text{ mm})}$$

$$= 85.5 \text{ mm} \qquad\qquad Ans.$$

A

A.2   Moment of Inertia for an Area

The moment of inertia of an area often appears in formulas used in mechanics of materials. It is a geometric property that is calculated about an axis, and for the x and y axes shown in Fig. A–5, it is defined as

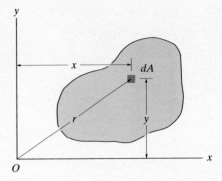

$$I_x = \int_A y^2 \, dA$$
$$I_y = \int_A x^2 \, dA$$

(A–3)

Fig. A–5

These integrals have no physical meaning, but they are so named because they are similar to the formulation of the moment of inertia of a mass, which is a dynamical property of matter.

We can also calculate the moment of inertia of an area about the pole O or z axis, Fig. A–5. This is referred to as the *polar moment of inertia*,

$$J_O = \int_A r^2 \, dA = I_x + I_y$$

(A–4)

Here r is the perpendicular distance from the pole (z axis) to the element dA. The relationship between J_O and I_x, I_y is possible since $r^2 = x^2 + y^2$, Fig. A–5.

From the above formulations it is seen that I_x, I_y, and J_O will *always* be *positive*, since they involve the product of distance squared and area. Furthermore, the units for moment of inertia involve length raised to the fourth power, e.g., m^4, mm^4.

Using the above equations, the moments of inertia for some common area shapes have been calculated about their *centroidal axes* and are listed on the inside front cover.

Parallel-Axis Theorem for an Area.

If the moment of inertia for an area is known about a centroidal axis, we can determine the moment of inertia of the area about a corresponding parallel axis using the *parallel-axis theorem*. To derive this theorem, consider finding the moment of inertia of the shaded area shown in Fig. A–6 about the x axis. In this case, a differential element dA is located at the arbitrary distance y' from the centroidal x' axis, whereas the *fixed distance* between the

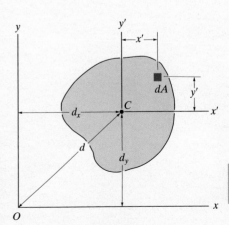

Fig. A–6

parallel x and x' axes is d_y. Since the moment of inertia of dA about the x axis is $dI_x = (y' + d_y)^2 \, dA$, then for the entire area,

$$I_x = \int_A (y' + d_y)^2 \, dA = \int_A y'^2 \, dA + 2d_y \int_A y' \, dA + d_y^2 \int_A dA$$

The first term on the right represents the moment of inertia of the area about the x' axis, $\bar{I}_{x'}$. The second term is zero since the x' axis passes through the area's centroid C, that is, $\int y' \, dA = \bar{y}'A = 0$ since $\bar{y}' = 0$. The final result is therefore

$$\boxed{I_x = \bar{I}_{x'} + A d_y^2} \tag{A–5}$$

A similar expression can be written for I_y, that is,

$$\boxed{I_y = \bar{I}_{y'} + A d_x^2} \tag{A–6}$$

And finally, for the polar moment of inertia about an axis perpendicular to the x–y plane and passing through the pole O (z axis), Fig. A–6, we have

$$\boxed{J_O = \bar{J}_C + A d^2} \tag{A–7}$$

The form of each of the above equations states that *the moment of inertia of an area about an axis is equal to the area's moment of inertia about a parallel axis passing through the "centroid" plus the product of the area and the square of the perpendicular distance between the axes.*

Composite Areas.

Many cross-sectional areas consist of a series of connected simpler shapes, such as rectangles, triangles, and semicircles. In order to properly determine the moment of inertia of such an area about a specified axis, it is first necessary to divide the area into its composite parts and indicate the perpendicular distance from the axis to the parallel centroidal axis for each part. Using the table on the inside front cover of the book, the moment of inertia of each part is determined about the centroidal axis. If this axis does not coincide with the specified axis, the parallel-axis theorem, $I = \bar{I} + A d^2$, should be used to determine the moment of inertia of the part about the specified axis. The moment of inertia of the entire area about this axis is then determined by summing the results of its composite parts. In particular, if a composite part has a "hole," the moment of inertia for the composite is found by "subtracting" the moment of inertia for the hole from the moment of inertia of the entire area including the hole.

A

EXAMPLE | A.2

Determine the moment of inertia of the cross-sectional area of the T-beam shown in Fig. A–7a about the centroidal x' axis.

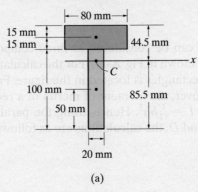

(a)

Fig. A–7

SOLUTION I

The area is segmented into two rectangles as shown in Fig. A–7a, and the distance from the x' axis and each centroidal axis is determined. Using the table on the inside front cover, the moment of inertia of a rectangle about its centroidal axis is $I = \frac{1}{12}bh^3$. Applying the parallel-axis theorem, Eq. A–5, to each rectangle and adding the results, we have

$$I = \Sigma(\bar{I}_{x'} + Ad_y^2)$$

$$= \left[\frac{1}{12}(20 \text{ mm})(100 \text{ mm})^3 + (20 \text{ mm})(100 \text{ mm})(85.5 \text{ mm} - 50 \text{ mm})^2\right]$$

$$+ \left[\frac{1}{12}(80 \text{ mm})(30 \text{ mm})^3 + (80 \text{ mm})(30 \text{ mm})(44.5 \text{ mm} - 15 \text{ mm})^2\right]$$

$$I = 6.46(10^6) \text{ mm}^4 \qquad\qquad\qquad Ans.$$

SOLUTION II

The area can be considered as one large rectangle less two small rectangles, shown shaded in Fig. A–7b. We have

$$I = \Sigma(\bar{I}_{x'} + Ad_y^2)$$

$$= \left[\frac{1}{12}(80 \text{ mm})(130 \text{ mm})^3 + (80 \text{ mm})(130 \text{ mm})(85.5 \text{ mm} - 65 \text{ mm})^2\right]$$

$$- 2\left[\frac{1}{12}(30 \text{ mm})(100 \text{ mm})^3 + (30 \text{ mm})(100 \text{ mm})(85.5 \text{ mm} - 50 \text{ mm})^2\right]$$

$$I = 6.46(10^6) \text{ mm}^4 \qquad\qquad\qquad Ans.$$

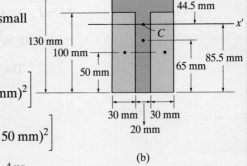

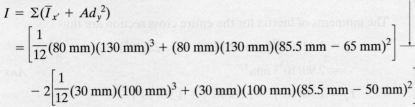

(b)

A

EXAMPLE A.3

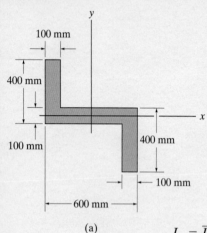

(a)

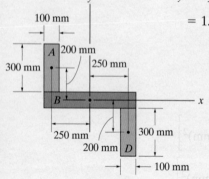

(b)

Fig. A–8

Determine the moments of inertia of the beam's cross-sectional area shown in Fig. A–8a about the x and y centroidal axes.

SOLUTION

The cross section can be considered as three composite rectangular areas A, B, and D shown in Fig. A–8b. For the calculation, the centroid of each of these rectangles is located in the figure. From the table on the inside front cover, the moment of inertia of a rectangle about its centroidal axis is $I = \frac{1}{12}bh^3$. Hence, using the parallel-axis theorem for rectangles A and D, the calculations are as follows:

Rectangle A:

$$I_x = \bar{I}_{x'} + Ad_y^2 = \frac{1}{12}(100 \text{ mm})(300 \text{ mm})^3 + (100 \text{ mm})(300 \text{ mm})(200 \text{ mm})^2$$

$$= 1.425(10^9) \text{ mm}^4$$

$$I_y = \bar{I}_{y'} + Ad_x^2 = \frac{1}{12}(300 \text{ mm})(100 \text{ mm})^3 + (100 \text{ mm})(300 \text{ mm})(250 \text{ mm})^2$$

$$= 1.90(10^9) \text{ mm}^4$$

Rectangle B:

$$I_x = \frac{1}{12}(600 \text{ mm})(100 \text{ mm})^3 = 0.05(10^9) \text{ mm}^4$$

$$I_y = \frac{1}{12}(100 \text{ mm})(600 \text{ mm})^3 = 1.80(10^9) \text{ mm}^4$$

Rectangle D:

$$I_x = \bar{I}_{x'} + Ad_y^2 = \frac{1}{12}(100 \text{ mm})(300 \text{ mm})^3 + (100 \text{ mm})(300 \text{ mm})(200 \text{ mm})^2$$

$$= 1.425(10^9) \text{ mm}^4$$

$$I_y = \bar{I}_{y'} + Ad_x^2 = \frac{1}{12}(300 \text{ mm})(100 \text{ mm})^3 + (100 \text{ mm})(300 \text{ mm})(250 \text{ mm})^2$$

$$= 1.90(10^9) \text{ mm}^4$$

The moments of inertia for the entire cross section are thus

$$I_x = 1.425(10^9) + 0.05(10^9) + 1.425(10^9)$$

$$= 2.90(10^9) \text{ mm}^4 \qquad \qquad Ans.$$

$$I_y = 1.90(10^9) + 1.80(10^9) + 1.90(10^9)$$

$$= 5.60(10^9) \text{ mm}^4 \qquad \qquad Ans.$$

A.3   Product of Inertia for an Area

In general, the moment of inertia for an area is different for every axis about which it is computed. In some applications of mechanical or structural design it is necessary to know the orientation of those axes that give, respectively, the maximum and minimum moments of inertia for the area. The method for determining this is discussed in Sec. A.4. To use this method, however, one must first determine the product of inertia for the area as well as its moments of inertia for given x, y axes.

The **product of inertia** for the area A shown in Fig. A–9 is defined as

$$I_{xy} = \int_A xy \, dA \qquad\qquad (\text{A–8})$$

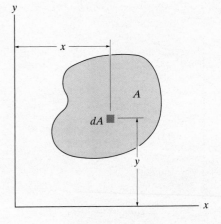

Fig. A–9

Like the moment of inertia, the product of inertia has units of length raised to the fourth power, e.g., m^4, mm^4. However, since x or y may be a negative quantity, while dA is always positive, the product of inertia may be positive, negative, or zero, depending on the location and orientation of the coordinate axes. For example, the product of inertia I_{xy} for an area will be *zero* if either the x or y axis is an axis of *symmetry* for the area. To show this, consider the shaded area in Fig. A–10, where for every element dA located at point (x, y) there is a corresponding element dA located at $(x, -y)$. Since the products of inertia for these elements are, respectively, $xy \, dA$ and $-xy \, dA$, their algebraic sum or the integration of all the elements of area chosen in this way will cancel each other. Consequently, the product of inertia for the total area becomes zero.

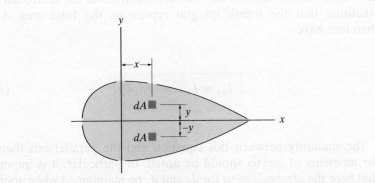

Fig. A–10

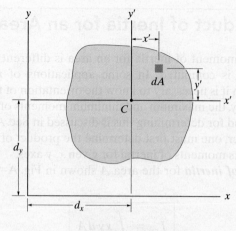

Fig. A–11

Parallel-Axis Theorem. Consider the shaded area shown in Fig. A–11, where x' and y' represent a set of centroidal axes, and x and y represent a corresponding set of parallel axes. Since the product of inertia of dA with respect to the x and y axes is $dI_{xy} = (x' + d_x)(y' + d_y)\,dA$, then for the entire area,

$$I_{xy} = \int_A (x' + d_x)(y' + d_y)\,dA$$

$$= \int_A x'y'\,dA + d_x \int_A y'\,dA + d_y \int_A x'\,dA + d_x d_y \int_A dA$$

The first term on the right represents the product of inertia of the area with respect to the centroidal axis, $\overline{I}_{x'y'}$. The second and third terms are zero since the moments of the area are taken about the centroidal axis. Realizing that the fourth integral represents the total area A, we therefore have

$$\boxed{I_{xy} = \overline{I}_{x'y'} + A d_x d_y} \tag{A–9}$$

The similarity between this equation and the parallel-axis theorem for moments of inertia should be noted. In particular, it is important that here the *algebraic signs* for d_x and d_y be maintained when applying Eq. A–9.

EXAMPLE A.4

Determine the product of inertia of the beam's cross-sectional area, shown in Fig. A–12a, about the x and y centroidal axes.

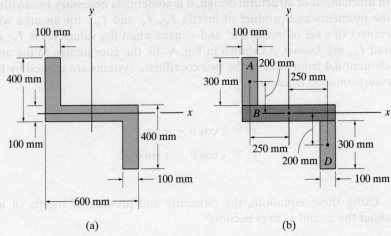

(a)   (b)

Fig. A–12

SOLUTION

As in Example A.3, the cross section can be considered as three composite rectangular areas A, B, and D, Fig. A–12b. The coordinates for the centroid of each of these rectangles are shown in the figure. Due to symmetry, the product of inertia of *each rectangle* is *zero* about a set of x', y' axes that pass through the rectangle's centroid. Hence, application of the parallel-axis theorem to each of the rectangles yields

Rectangle A:

$$I_{xy} = \bar{I}_{x'y'} + Ad_xd_y$$
$$= 0 + (300 \text{ mm})(100 \text{ mm})(-250 \text{ mm})(200 \text{ mm})$$
$$= -1.50(10^9) \text{ mm}^4$$

Rectangle B:

$$I_{xy} = \bar{I}_{x'y'} + Ad_xd_y$$
$$= 0 + 0$$
$$= 0$$

Rectangle D:

$$I_{xy} = \bar{I}_{x'y'} + Ad_xd_y$$
$$= 0 + (300 \text{ mm})(100 \text{ mm})(250 \text{ mm})(-200 \text{ mm})$$
$$= -1.50(10^9) \text{ mm}^4$$

The product of inertia for the entire cross section is thus

$$I_{xy} = [-1.50(10^9) \text{ mm}^4] + 0 + [-1.50(10^9) \text{ mm}^4]$$
$$= -3.00(10^9) \text{ mm}^4 \qquad\qquad Ans.$$

A

A.4   Moments of Inertia for an Area about Inclined Axes

In mechanical or structural design, it is sometimes necessary to calculate the moments and product of inertia $I_{x'}$, $I_{y'}$ and $I_{x'y'}$ for an area with respect to a set of inclined x' and y' axes when the values for θ, I_x, I_y, and I_{xy} are *known*. As shown in Fig. A–13, the coordinates to the area element dA from each of the two coordinate systems are related by the *transformation equations*

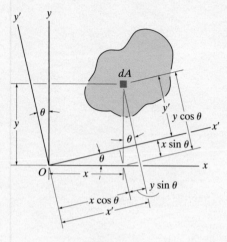

Fig. A–13

$$x' = x \cos \theta + y \sin \theta$$
$$y' = y \cos \theta - x \sin \theta$$

Using these equations, the moments and product of inertia of dA about the x' and y' axes become

$$dI_{x'} = y'^2 \, dA = (y \cos \theta - x \sin \theta)^2 \, dA$$
$$dI_{y'} = x'^2 \, dA = (x \cos \theta + y \sin \theta)^2 \, dA$$
$$dI_{x'y'} = x'y' \, dA = (x \cos \theta + y \sin \theta)(y \cos \theta - x \sin \theta) \, dA$$

Expanding each expression and integrating, realizing that $I_x = \int y^2 \, dA$, $I_y = \int x^2 \, dA$, and $I_{xy} = \int xy \, dA$, we obtain

$$I_{x'} = I_x \cos^2 \theta + I_y \sin^2 \theta - 2I_{xy} \sin \theta \cos \theta$$
$$I_{y'} = I_x \sin^2 \theta + I_y \cos^2 \theta + 2I_{xy} \sin \theta \cos \theta$$
$$I_{x'y'} = I_x \sin \theta \cos \theta - I_y \sin \theta \cos \theta + I_{xy}(\cos^2 \theta - \sin^2 \theta)$$

These equations may be simplified by using the trigonometric identities $\sin 2\theta = 2 \sin \theta \cos \theta$ and $\cos 2\theta = \cos^2 \theta - \sin^2 \theta$, in which case

$$
\boxed{
\begin{aligned}
I_{x'} &= \frac{I_x + I_y}{2} + \frac{I_x - I_y}{2} \cos 2\theta - I_{xy} \sin 2\theta \\[1em]
I_{y'} &= \frac{I_x + I_y}{2} - \frac{I_x - I_y}{2} \cos 2\theta + I_{xy} \sin 2\theta \\[1em]
I_{x'y'} &= \frac{I_x - I_y}{2} \sin 2\theta + I_{xy} \cos 2\theta
\end{aligned}
}
\qquad \text{(A–10)}
$$

Principal Moments of Inertia. Note that $I_{x'}$, $I_{y'}$, and $I_{x'y'}$ depend on the angle of inclination, θ, of the x', y' axes. We will now determine the orientation of these axes about which the moments of inertia for the area, $I_{x'}$ and $I_{y'}$, are maximum and minimum. This particular set of axes is called the *principal axes* of inertia for the area, and the corresponding moments of inertia with respect to these axes are called the *principal moments of inertia*. In general, there is a set of principal axes for every chosen origin O; however, in mechanics of materials the area's centroid is the most important location for O.

The angle $\theta = \theta_p$, which defines the orientation of the principal axes for the area, can be found by differentiating the first of Eq. A–10 with respect to θ and setting the result equal to zero. Thus,

$$\frac{dI_{x'}}{d\theta} = -2\left(\frac{I_x - I_y}{2}\right)\sin 2\theta - 2I_{xy}\cos 2\theta = 0$$

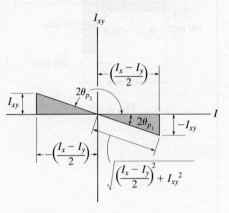

Therefore, at $\theta = \theta_p$,

$$\boxed{\tan 2\theta_p = \frac{-I_{xy}}{(I_x - I_y)/2}} \qquad \text{(A–11)}$$

Fig. A–14

This equation has two roots, θ_{p_1} and θ_{p_2}, which are 90° apart and so specify the inclination of each principal axis.

The sine and cosine of $2\theta_{p_1}$ and $2\theta_{p_2}$ can be obtained from the triangles shown in Fig. A–14, which are based on Eq. A–11. If these trigonometric relations are substituted into the first or second of Eq. A–10 and simplified, the result is

$$\boxed{I_{\substack{max \\ min}} = \frac{I_x + I_y}{2} \pm \sqrt{\left(\frac{I_x - I_y}{2}\right)^2 + I_{xy}^2}} \qquad \text{(A–12)}$$

Depending on the sign chosen, this result gives the maximum or minimum moment of inertia for the area. Furthermore, if the above trigonometric relations for θ_{p_1} and θ_{p_2} are substituted into the third of Eq. A–10, it will be seen that $I_{x'y'} = 0$; that is, the *product of inertia with respect to the principal axes is zero*. Since it was indicated in Sec. A.3 that the product of inertia is zero with respect to any symmetrical axis, it therefore follows that *any symmetrical axis and the one perpendicular to it represent principal axes of inertia for the area*. Also, notice that the equations derived in this section are similar to those for stress and strain transformation developed in Chapters 9 and 10, respectively.

EXAMPLE | A.5

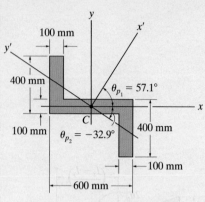

Fig. A–15

Determine the principal moments of inertia for the beam's cross-sectional area shown in Fig. A–15 with respect to an axis passing through the centroid C.

SOLUTION
The moments and product of inertia of the cross section with respect to the x, y axes have been determined in Examples A.3 and A.4. The results are

$$I_x = 2.90(10^9) \text{ mm}^4 \qquad I_y = 5.60(10^9) \text{ mm}^4 \qquad I_{xy} = -3.00(10^9) \text{ mm}^4$$

Using Eq. A–11, the angles of inclination of the principal axes x' and y' are

$$\tan 2\theta_p = \frac{-I_{xy}}{(I_x - I_y)/2} = \frac{3.00(10^9)}{[2.90(10^9) - 5.60(10^9)]/2} = -2.22$$

$$2\theta_{p_1} = 114.2° \qquad \text{and} \qquad 2\theta_{p_2} = -65.8°$$

Thus, as shown in Fig. A–15,

$$\theta_{p1} = 57.1° \qquad \text{and} \qquad \theta_{p2} = -32.9°$$

The principal moments of inertia with respect to the x' and y' axes are determined by using Eq. A–12.

$$I_{\substack{\max \\ \min}} = \frac{I_x + I_y}{2} \pm \sqrt{\left(\frac{I_x - I_y}{2}\right)^2 + I_{xy}^2}$$

$$= \frac{2.90(10^9) + 5.60(10^9)}{2} \pm \sqrt{\left[\frac{2.90(10^9) - 5.60(10^9)}{2}\right]^2 + [-3.00(10^9)]^2}$$

$$= 4.25(10^9) \pm 3.29(10^9)$$

or

$$I_{\max} = 7.54(10^9) \text{ mm}^4 \qquad I_{\min} = 0.960(10^9) \text{ mm}^4 \qquad \textit{Ans.}$$

Specifically, the maximum moment of inertia, $I_{\max} = 7.54(10^9) \text{ mm}^4$, occurs with respect to the x' axis (major axis), since *by inspection* most of the cross-sectional area is farthest away from this axis. To show this, substitute the data with $\theta = 57.1°$ into the first of Eq. A–10.

A

A.5   Mohr's Circle for Moments of Inertia

Equations A–10 through A–12 have a semi-graphical solution that is convenient to use and generally easy to remember. Squaring the first and third of Eq. A–10 and adding, it is found that

$$\left(I_{x'} - \frac{I_x + I_y}{2}\right)^2 + I_{x'y'}^2 = \left(\frac{I_x - I_y}{2}\right)^2 + I_{xy}^2 \qquad \text{(A–13)}$$

In any given problem, $I_{x'}$ and $I_{x'y'}$ are *variables*, and I_x, I_y, and I_{xy} are *known constants*. Thus, the above equation may be written in compact form as

$$(I_{x'} - a)^2 + I_{x'y'}^2 = R^2$$

When this equation is plotted, the resulting graph represents a *circle* of radius

$$R = \sqrt{\left(\frac{I_x - I_y}{2}\right)^2 + I_{xy}^2}$$

and having its center located at point $(a, 0)$, where $a = (I_x + I_y)/2$. The circle so constructed is called *Mohr's circle*. Its application is similar to that used for stress and strain transformation developed in Chapters 9 and 10, respectively.

Procedure for Analysis

The main purpose for using Mohr's circle here is to have a convenient means of transforming I_x, I_y, and I_{xy} into the principal moments of inertia for the area. The following procedure provides a method for doing this.

Calculate I_x, I_y, I_{xy}.

Establish the x, y axes for the area, with the origin located at the point P of interest, usually the centroid, and determine I_x, I_y, and I_{xy}, Fig. A–16a.

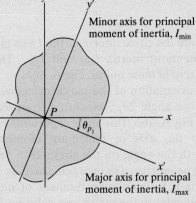

(a)   Fig. A–16

Procedure for Analysis (continued)

Construct the Circle.

Establish a rectangular coordinate system such that the horizontal axis represents the moment of inertia I, and the vertical axis represents the product of inertia I_{xy}, Fig. A–16b. Determine the center of the circle, C, which is located a distance $(I_x + I_y)/2$ from the origin, and plot the "reference point" A having coordinates (I_x, I_{xy}). By definition, I_x is always positive, whereas I_{xy} can either be positive or negative. Connect the reference point A with the center of the circle, and determine the distance CA by trigonometry. This distance represents the radius of the circle, Fig. A–16b. Finally, draw the circle.

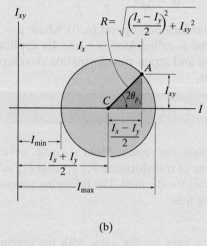

$$R = \sqrt{\left(\frac{I_x - I_y}{2}\right)^2 + I_{xy}^2}$$

(b)

Fig. A–16 (cont.)

Principal Moments of Inertia.

The points where the circle intersects the I axis give the values of the principal moments of inertia I_{min} and I_{max}. Here the *product of inertia will be zero at these points*, Fig. A–16b.

To find the orientation of the major principal axis, determine by trigonometry the angle $2\theta_{p_1}$, *measured from the radius CA to the positive I axis*, Fig. A–16b. This angle represents twice the angle from the x axis to the axis of maximum moment of inertia I_{max}, Fig. A–16a. Both the angle on the circle, $2\theta_{p_1}$, and the angle on the area, θ_{p_1}, *must be measured in the same sense*, as shown in Fig. A–16. The minor axis is for minimum moment of inertia I_{min}, which is always perpendicular to the major axis defining I_{max}.

EXAMPLE A.6

Use Mohr's circle to determine the principal moments of inertia for the beam's cross-sectional area, shown in Fig. A–17a, with respect to principal axes passing through the centroid C.

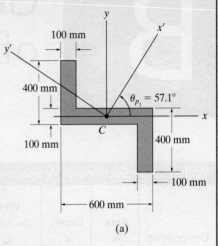

(a)

SOLUTION

Compute I_x, I_y, I_{xy}. The moments of inertia and the product of inertia have been determined in Examples A.3 and A.4 with respect to the x, y axes shown in Fig. A–17a. The results are $I_x = 2.90(10^9)$ mm^4, $I_y = 5.60(10^9)$ mm^4, and $I_{xy} = -3.00(10^9)$ mm^4.

Construct the Circle. The I and I_{xy} axes are shown in Fig. A–17b. The center of the circle, C, lies at a distance $(I_x + I_y)/2 = (2.90 + 5.60)/2 = 4.25$ from the origin. When the reference point $A(2.90, -3.00)$ is connected to point C, the radius CA is determined from the shaded triangle CBA using the Pythagorean theorem:

$$CA = \sqrt{(1.35)^2 + (-3.00)^2} = 3.29$$

The circle is constructed in Fig. A–17c.

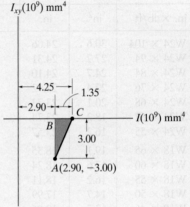

(b)

Principal Moments of Inertia. The circle intersects the I axis at points $(7.54, 0)$ and $(0.960, 0)$. Hence

$$I_{max} = 7.54(10^9) \text{ mm}^4 \qquad Ans.$$

$$I_{min} = 0.960(10^9) \text{ mm}^4 \qquad Ans.$$

As shown in Fig. A–17c, the angle $2\theta_{p_1}$ is determined from the circle by measuring *counterclockwise* from CA to the *positive I* axis. Hence,

$$2\theta_{p_1} = 180° - \tan^{-1}\left(\frac{|BA|}{|BC|}\right) = 180° - \tan^{-1}\left(\frac{3.00}{1.35}\right) = 114.2°$$

$$\theta_{p_1} = 57.1°$$

The major principal axis (for $I_{max} = 7.54(10^9)$ mm^4) is therefore oriented at an angle $\theta_{p_1} = 57.1°$, measured *counterclockwise*, from the *positive x* axis. The minor axis is perpendicular to this axis. The results are shown in Fig. A–17a.

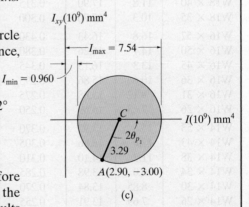

(c)

Fig. A–17

A

APPENDIX

B

Geometric Properties of Structural Shapes

Wide-Flange Sections or W Shapes   FPS Units

Designation	Area A	Depth d	Web thickness t_w	Flange width b_f	Flange thickness t_f	x–x axis I	x–x axis S	x–x axis r	y–y axis I	y–y axis S	y–y axis r
in. $\times$ lb/ft	in^2	in.	in.	in.	in.	in^4	in^3	in.	in^4	in^3	in.
W24 $\times$ 104	30.6	24.06	0.500	12.750	0.750	3100	258	10.1	259	40.7	2.91
W24 $\times$ 94	27.7	24.31	0.515	9.065	0.875	2700	222	9.87	109	24.0	1.98
W24 $\times$ 84	24.7	24.10	0.470	9.020	0.770	2370	196	9.79	94.4	20.9	1.95
W24 $\times$ 76	22.4	23.92	0.440	8.990	0.680	2100	176	9.69	82.5	18.4	1.92
W24 $\times$ 68	20.1	23.73	0.415	8.965	0.585	1830	154	9.55	70.4	15.7	1.87
W24 $\times$ 62	18.2	23.74	0.430	7.040	0.590	1550	131	9.23	34.5	9.80	1.38
W24 $\times$ 55	16.2	23.57	0.395	7.005	0.505	1350	114	9.11	29.1	8.30	1.34
W18 $\times$ 65	19.1	18.35	0.450	7.590	0.750	1070	117	7.49	54.8	14.4	1.69
W18 $\times$ 60	17.6	18.24	0.415	7.555	0.695	984	108	7.47	50.1	13.3	1.69
W18 $\times$ 55	16.2	18.11	0.390	7.530	0.630	890	98.3	7.41	44.9	11.9	1.67
W18 $\times$ 50	14.7	17.99	0.355	7.495	0.570	800	88.9	7.38	40.1	10.7	1.65
W18 $\times$ 46	13.5	18.06	0.360	6.060	0.605	712	78.8	7.25	22.5	7.43	1.29
W18 $\times$ 40	11.8	17.90	0.315	6.015	0.525	612	68.4	7.21	19.1	6.35	1.27
W18 $\times$ 35	10.3	17.70	0.300	6.000	0.425	510	57.6	7.04	15.3	5.12	1.22
W16 $\times$ 57	16.8	16.43	0.430	7.120	0.715	758	92.2	6.72	43.1	12.1	1.60
W16 $\times$ 50	14.7	16.26	0.380	7.070	0.630	659	81.0	6.68	37.2	10.5	1.59
W16 $\times$ 45	13.3	16.13	0.345	7.035	0.565	586	72.7	6.65	32.8	9.34	1.57
W16 $\times$ 36	10.6	15.86	0.295	6.985	0.430	448	56.5	6.51	24.5	7.00	1.52
W16 $\times$ 31	9.12	15.88	0.275	5.525	0.440	375	47.2	6.41	12.4	4.49	1.17
W16 $\times$ 26	7.68	15.69	0.250	5.500	0.345	301	38.4	6.26	9.59	3.49	1.12
W14 $\times$ 53	15.6	13.92	0.370	8.060	0.660	541	77.8	5.89	57.7	14.3	1.92
W14 $\times$ 43	12.6	13.66	0.305	7.995	0.530	428	62.7	5.82	45.2	11.3	1.89
W14 $\times$ 38	11.2	14.10	0.310	6.770	0.515	385	54.6	5.87	26.7	7.88	1.55
W14 $\times$ 34	10.0	13.98	0.285	6.745	0.455	340	48.6	5.83	23.3	6.91	1.53
W14 $\times$ 30	8.85	13.84	0.270	6.730	0.385	291	42.0	5.73	19.6	5.82	1.49
W14 $\times$ 26	7.69	13.91	0.255	5.025	0.420	245	35.3	5.65	8.91	3.54	1.08
W14 $\times$ 22	6.49	13.74	0.230	5.000	0.335	199	29.0	5.54	7.00	2.80	1.04

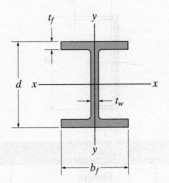

Wide-Flange Sections or W Shapes   FPS Units

Designation	Area A	Depth d	Web thickness t_w	Flange width b_f	Flange thickness t_f	x–x axis I	x–x axis S	x–x axis r	y–y axis I	y–y axis S	y–y axis r
in. × lb/ft	in²	in.	in.	in.	in.	in⁴	in³	in.	in⁴	in³	in.
W12 × 87	25.6	12.53	0.515	12.125	0.810	740	118	5.38	241	39.7	3.07
W12 × 50	14.7	12.19	0.370	8.080	0.640	394	64.7	5.18	56.3	13.9	1.96
W12 × 45	13.2	12.06	0.335	8.045	0.575	350	58.1	5.15	50.0	12.4	1.94
W12 × 26	7.65	12.22	0.230	6.490	0.380	204	33.4	5.17	17.3	5.34	1.51
W12 × 22	6.48	12.31	0.260	4.030	0.425	156	25.4	4.91	4.66	2.31	0.847
W12 × 16	4.71	11.99	0.220	3.990	0.265	103	17.1	4.67	2.82	1.41	0.773
W12 × 14	4.16	11.91	0.200	3.970	0.225	88.6	14.9	4.62	2.36	1.19	0.753
W10 × 100	29.4	11.10	0.680	10.340	1.120	623	112	4.60	207	40.0	2.65
W10 × 54	15.8	10.09	0.370	10.030	0.615	303	60.0	4.37	103	20.6	2.56
W10 × 45	13.3	10.10	0.350	8.020	0.620	248	49.1	4.32	53.4	13.3	2.01
W10 × 39	11.5	9.92	0.315	7.985	0.530	209	42.1	4.27	45.0	11.3	1.98
W10 × 30	8.84	10.47	0.300	5.810	0.510	170	32.4	4.38	16.7	5.75	1.37
W10 × 19	5.62	10.24	0.250	4.020	0.395	96.3	18.8	4.14	4.29	2.14	0.874
W10 × 15	4.41	9.99	0.230	4.000	0.270	68.9	13.8	3.95	2.89	1.45	0.810
W10 × 12	3.54	9.87	0.190	3.960	0.210	53.8	10.9	3.90	2.18	1.10	0.785
W8 × 67	19.7	9.00	0.570	8.280	0.935	272	60.4	3.72	88.6	21.4	2.12
W8 × 58	17.1	8.75	0.510	8.220	0.810	228	52.0	3.65	75.1	18.3	2.10
W8 × 48	14.1	8.50	0.400	8.110	0.685	184	43.3	3.61	60.9	15.0	2.08
W8 × 40	11.7	8.25	0.360	8.070	0.560	146	35.5	3.53	49.1	12.2	2.04
W8 × 31	9.13	8.00	0.285	7.995	0.435	110	27.5	3.47	37.1	9.27	2.02
W8 × 24	7.08	7.93	0.245	6.495	0.400	82.8	20.9	3.42	18.3	5.63	1.61
W8 × 15	4.44	8.11	0.245	4.015	0.315	48.0	11.8	3.29	3.41	1.70	0.876
W6 × 25	7.34	6.38	0.320	6.080	0.455	53.4	16.7	2.70	17.1	5.61	1.52
W6 × 20	5.87	6.20	0.260	6.020	0.365	41.4	13.4	2.66	13.3	4.41	1.50
W6 × 16	4.74	6.28	0.260	4.030	0.405	32.1	10.2	2.60	4.43	2.20	0.966
W6 × 15	4.43	5.99	0.230	5.990	0.260	29.1	9.72	2.56	9.32	3.11	1.46
W6 × 12	3.55	6.03	0.230	4.000	0.280	22.1	7.31	2.49	2.99	1.50	0.918
W6 × 9	2.68	5.90	0.170	3.940	0.215	16.4	5.56	2.47	2.19	1.11	0.905

B

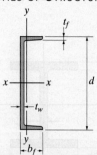

American Standard Channels or C Shapes   FPS Units

Designation	Area A	Depth d	Web thickness t_w		Flange width b_f		Flange thickness t_f		x–x axis I	x–x axis S	x–x axis r	y–y axis I	y–y axis S	y–y axis r
in. × lb/ft	in^2	in.	in.		in.		in.		in^4	in^3	in.	in^4	in^3	in.
C15 × 50	14.7	15.00	0.716	11/16	3.716	$3\frac{3}{4}$	0.650	5/8	404	53.8	5.24	11.0	3.78	0.867
C15 × 40	11.8	15.00	0.520	1/2	3.520	$3\frac{1}{2}$	0.650	5/8	349	46.5	5.44	9.23	3.37	0.886
C15 × 33.9	9.96	15.00	0.400	3/8	3.400	$3\frac{3}{8}$	0.650	5/8	315	42.0	5.62	8.13	3.11	0.904
C12 × 30	8.82	12.00	0.510	1/2	3.170	$3\frac{1}{8}$	0.501	1/2	162	27.0	4.29	5.14	2.06	0.763
C12 × 25	7.35	12.00	0.387	3/8	3.047	3	0.501	1/2	144	24.1	4.43	4.47	1.88	0.780
C12 × 20.7	6.09	12.00	0.282	5/16	2.942	3	0.501	1/2	129	21.5	4.61	3.88	1.73	0.799
C10 × 30	8.82	10.00	0.673	11/16	3.033	3	0.436	7/16	103	20.7	3.42	3.94	1.65	0.669
C10 × 25	7.35	10.00	0.526	1/2	2.886	$2\frac{7}{8}$	0.436	7/16	91.2	18.2	3.52	3.36	1.48	0.676
C10 × 20	5.88	10.00	0.379	3/8	2.739	$2\frac{3}{4}$	0.436	7/16	78.9	15.8	3.66	2.81	1.32	0.692
C10 × 15.3	4.49	10.00	0.240	1/4	2.600	$2\frac{5}{8}$	0.436	7/16	67.4	13.5	3.87	2.28	1.16	0.713
C9 × 20	5.88	9.00	0.448	7/16	2.648	$2\frac{5}{8}$	0.413	7/16	60.9	13.5	3.22	2.42	1.17	0.642
C9 × 15	4.41	9.00	0.285	5/16	2.485	$2\frac{1}{2}$	0.413	7/16	51.0	11.3	3.40	1.93	1.01	0.661
C9 × 13.4	3.94	9.00	0.233	1/4	2.433	$2\frac{3}{8}$	0.413	7/16	47.9	10.6	3.48	1.76	0.962	0.669
C8 × 18.75	5.51	8.00	0.487	1/2	2.527	$2\frac{1}{2}$	0.390	3/8	44.0	11.0	2.82	1.98	1.01	0.599
C8 × 13.75	4.04	8.00	0.303	5/16	2.343	$2\frac{3}{8}$	0.390	3/8	36.1	9.03	2.99	1.53	0.854	0.615
C8 × 11.5	3.38	8.00	0.220	1/4	2.260	$2\frac{1}{4}$	0.390	3/8	32.6	8.14	3.11	1.32	0.781	0.625
C7 × 14.75	4.33	7.00	0.419	7/16	2.299	$2\frac{1}{4}$	0.366	3/8	27.2	7.78	2.51	1.38	0.779	0.564
C7 × 12.25	3.60	7.00	0.314	5/16	2.194	$2\frac{1}{4}$	0.366	3/8	24.2	6.93	2.60	1.17	0.703	0.571
C7 × 9.8	2.87	7.00	0.210	3/16	2.090	$2\frac{1}{8}$	0.366	3/8	21.3	6.08	2.72	0.968	0.625	0.581
C6 × 13	3.83	6.00	0.437	7/16	2.157	$2\frac{1}{8}$	0.343	5/16	17.4	5.80	2.13	1.05	0.642	0.525
C6 × 10.5	3.09	6.00	0.314	5/16	2.034	2	0.343	5/16	15.2	5.06	2.22	0.866	0.564	0.529
C6 × 8.2	2.40	6.00	0.200	3/16	1.920	$1\frac{7}{8}$	0.343	5/16	13.1	4.38	2.34	0.693	0.492	0.537
C5 × 9	2.64	5.00	0.325	5/16	1.885	$1\frac{7}{8}$	0.320	5/16	8.90	3.56	1.83	0.632	0.450	0.489
C5 × 6.7	1.97	5.00	0.190	3/16	1.750	$1\frac{3}{4}$	0.320	5/16	7.49	3.00	1.95	0.479	0.378	0.493
C4 × 7.25	2.13	4.00	0.321	5/16	1.721	$1\frac{3}{4}$	0.296	5/16	4.59	2.29	1.47	0.433	0.343	0.450
C4 × 5.4	1.59	4.00	0.184	3/16	1.584	$1\frac{5}{8}$	0.296	5/16	3.85	1.93	1.56	0.319	0.283	0.449
C3 × 6	1.76	3.00	0.356	3/8	1.596	$1\frac{5}{8}$	0.273	1/4	2.07	1.38	1.08	0.305	0.268	0.416
C3 × 5	1.47	3.00	0.258	1/4	1.498	$1\frac{1}{2}$	0.273	1/4	1.85	1.24	1.12	0.247	0.233	0.410
C3 × 4.1	1.21	3.00	0.170	3/16	1.410	$1\frac{3}{8}$	0.273	1/4	1.66	1.10	1.17	0.197	0.202	0.404

B

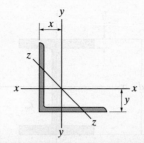

Angles Having Equal Legs   FPS Units

Size and thickness	Weight per foot	Area A	x–x axis				y–y axis				z–z axis
			I	S	r	y	I	S	r	x	r
in.	lb	in²	in⁴	in³	in.	in.	in⁴	in³	in.	in.	in.
L8 × 8 × 1	51.0	15.0	89.0	15.8	2.44	2.37	89.0	15.8	2.44	2.37	1.56
L8 × 8 × $\frac{3}{4}$	38.9	11.4	69.7	12.2	2.47	2.28	69.7	12.2	2.47	2.28	1.58
L8 × 8 × $\frac{1}{2}$	26.4	7.75	48.6	8.36	2.50	2.19	48.6	8.36	2.50	2.19	1.59
L6 × 6 × 1	37.4	11.0	35.5	8.57	1.80	1.86	35.5	8.57	1.80	1.86	1.17
L6 × 6 × $\frac{3}{4}$	28.7	8.44	28.2	6.66	1.83	1.78	28.2	6.66	1.83	1.78	1.17
L6 × 6 × $\frac{1}{2}$	19.6	5.75	19.9	4.61	1.86	1.68	19.9	4.61	1.86	1.68	1.18
L6 × 6 × $\frac{3}{8}$	14.9	4.36	15.4	3.53	1.88	1.64	15.4	3.53	1.88	1.64	1.19
L5 × 5 × $\frac{3}{4}$	23.6	6.94	15.7	4.53	1.51	1.52	15.7	4.53	1.51	1.52	0.975
L5 × 5 × $\frac{1}{2}$	16.2	4.75	11.3	3.16	1.54	1.43	11.3	3.16	1.54	1.43	0.983
L5 × 5 × $\frac{3}{8}$	12.3	3.61	8.74	2.42	1.56	1.39	8.74	2.42	1.56	1.39	0.990
L4 × 4 × $\frac{3}{4}$	18.5	5.44	7.67	2.81	1.19	1.27	7.67	2.81	1.19	1.27	0.778
L4 × 4 × $\frac{1}{2}$	12.8	3.75	5.56	1.97	1.22	1.18	5.56	1.97	1.22	1.18	0.782
L4 × 4 × $\frac{3}{8}$	9.8	2.86	4.36	1.52	1.23	1.14	4.36	1.52	1.23	1.14	0.788
L4 × 4 × $\frac{1}{4}$	6.6	1.94	3.04	1.05	1.25	1.09	3.04	1.05	1.25	1.09	0.795
L$3\frac{1}{2}$ × $3\frac{1}{2}$ × $\frac{1}{2}$	11.1	3.25	3.64	1.49	1.06	1.06	3.64	1.49	1.06	1.06	0.683
L$3\frac{1}{2}$ × $3\frac{1}{2}$ × $\frac{1}{2}$	8.5	2.48	2.87	1.15	1.07	1.01	2.87	1.15	1.07	1.01	0.687
L$3\frac{1}{2}$ × $3\frac{1}{2}$ × $\frac{1}{4}$	5.8	1.69	2.01	0.794	1.09	0.968	2.01	0.794	1.09	0.968	0.694
L3 × 3 × $\frac{1}{2}$	9.4	2.75	2.22	1.07	0.898	0.932	2.22	1.07	0.898	0.932	0.584
L3 × 3 × $\frac{3}{8}$	7.2	2.11	1.76	0.833	0.913	0.888	1.76	0.833	0.913	0.888	0.587
L3 × 3 × $\frac{1}{4}$	4.9	1.44	1.24	0.577	0.930	0.842	1.24	0.577	0.930	0.842	0.592
L$2\frac{1}{2}$ × $2\frac{1}{2}$ × $\frac{1}{2}$	7.7	2.25	1.23	0.724	0.739	0.806	1.23	0.724	0.739	0.806	0.487
L$2\frac{1}{2}$ × $2\frac{1}{2}$ × $\frac{3}{8}$	5.9	1.73	0.984	0.566	0.753	0.762	0.984	0.566	0.753	0.762	0.487
L$2\frac{1}{2}$ × $2\frac{1}{2}$ × $\frac{1}{4}$	4.1	1.19	0.703	0.394	0.769	0.717	0.703	0.394	0.769	0.717	0.491
L2 × 2 × $\frac{3}{8}$	4.7	1.36	0.479	0.351	0.594	0.636	0.479	0.351	0.594	0.636	0.389
L2 × 2 × $\frac{1}{4}$	3.19	0.938	0.348	0.247	0.609	0.592	0.348	0.247	0.609	0.592	0.391
L2 × 2 × $\frac{1}{8}$	1.65	0.484	0.190	0.131	0.626	0.546	0.190	0.131	0.626	0.546	0.398

B

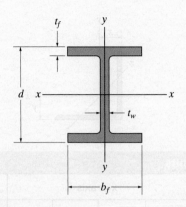

Wide-Flange Sections or W Shapes   SI Units

Designation	Area A	Depth d	Web thickness t_w	Flange width b_f	Flange thickness t_f	x–x axis I	x–x axis S	x–x axis r	y–y axis I	y–y axis S	y–y axis r
mm × kg/m	mm²	mm	mm	mm	mm	10^6 mm⁴	10^3 mm³	mm	10^6 mm⁴	10^3 mm³	mm
W610 × 155	19 800	611	12.70	324.0	19.0	1 290	4 220	255	108	667	73.9
W610 × 140	17 900	617	13.10	230.0	22.2	1 120	3 630	250	45.1	392	50.2
W610 × 125	15 900	612	11.90	229.0	19.6	985	3 220	249	39.3	343	49.7
W610 × 113	14 400	608	11.20	228.0	17.3	875	2 880	247	34.3	301	48.8
W610 × 101	12 900	603	10.50	228.0	14.9	764	2 530	243	29.5	259	47.8
W610 × 92	11 800	603	10.90	179.0	15.0	646	2 140	234	14.4	161	34.9
W610 × 82	10 500	599	10.00	178.0	12.8	560	1 870	231	12.1	136	33.9
W460 × 97	12 300	466	11.40	193.0	19.0	445	1 910	190	22.8	236	43.1
W460 × 89	11 400	463	10.50	192.0	17.7	410	1 770	190	20.9	218	42.8
W460 × 82	10 400	460	9.91	191.0	16.0	370	1 610	189	18.6	195	42.3
W460 × 74	9 460	457	9.02	190.0	14.5	333	1 460	188	16.6	175	41.9
W460 × 68	8 730	459	9.14	154.0	15.4	297	1 290	184	9.41	122	32.8
W460 × 60	7 590	455	8.00	153.0	13.3	255	1 120	183	7.96	104	32.4
W460 × 52	6 640	450	7.62	152.0	10.8	212	942	179	6.34	83.4	30.9
W410 × 85	10 800	417	10.90	181.0	18.2	315	1 510	171	18.0	199	40.8
W410 × 74	9 510	413	9.65	180.0	16.0	275	1 330	170	15.6	173	40.5
W410 × 67	8 560	410	8.76	179.0	14.4	245	1 200	169	13.8	154	40.2
W410 × 53	6 820	403	7.49	177.0	10.9	186	923	165	10.1	114	38.5
W410 × 46	5 890	403	6.99	140.0	11.2	156	774	163	5.14	73.4	29.5
W410 × 39	4 960	399	6.35	140.0	8.8	126	632	159	4.02	57.4	28.5
W360 × 79	10 100	354	9.40	205.0	16.8	227	1 280	150	24.2	236	48.9
W360 × 64	8 150	347	7.75	203.0	13.5	179	1 030	148	18.8	185	48.0
W360 × 57	7 200	358	7.87	172.0	13.1	160	894	149	11.1	129	39.3
W360 × 51	6 450	355	7.24	171.0	11.6	141	794	148	9.68	113	38.7
W360 × 45	5 710	352	6.86	171.0	9.8	121	688	146	8.16	95.4	37.8
W360 × 39	4 960	353	6.48	128.0	10.7	102	578	143	3.75	58.6	27.5
W360 × 33	4 190	349	5.84	127.0	8.5	82.9	475	141	2.91	45.8	26.4

B

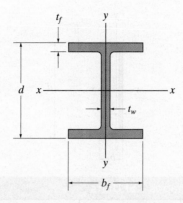

Wide-Flange Sections or W Shapes   SI Units

Designation	Area A	Depth d	Web thickness t_w	Flange width b_f	Flange thickness t_f	x–x axis I	x–x axis S	x–x axis r	y–y axis I	y–y axis S	y–y axis r
mm × kg/m	mm^2	mm	mm	mm	mm	10^6 mm^4	10^3 mm^3	mm	10^6 mm^4	10^3 mm^3	mm
W310 × 129	16 500	318	13.10	308.0	20.6	308	1940	137	100	649	77.8
W310 × 74	9 480	310	9.40	205.0	16.3	165	1060	132	23.4	228	49.7
W310 × 67	8 530	306	8.51	204.0	14.6	145	948	130	20.7	203	49.3
W310 × 39	4 930	310	5.84	165.0	9.7	84.8	547	131	7.23	87.6	38.3
W310 × 33	4 180	313	6.60	102.0	10.8	65.0	415	125	1.92	37.6	21.4
W310 × 24	3 040	305	5.59	101.0	6.7	42.8	281	119	1.16	23.0	19.5
W310 × 21	2 680	303	5.08	101.0	5.7	37.0	244	117	0.986	19.5	19.2
W250 × 149	19 000	282	17.30	263.0	28.4	259	1840	117	86.2	656	67.4
W250 × 80	10 200	256	9.40	255.0	15.6	126	984	111	43.1	338	65.0
W250 × 67	8 560	257	8.89	204.0	15.7	104	809	110	22.2	218	50.9
W250 × 58	7 400	252	8.00	203.0	13.5	87.3	693	109	18.8	185	50.4
W250 × 45	5 700	266	7.62	148.0	13.0	71.1	535	112	7.03	95	35.1
W250 × 28	3 620	260	6.35	102.0	10.0	39.9	307	105	1.78	34.9	22.2
W250 × 22	2 850	254	5.84	102.0	6.9	28.8	227	101	1.22	23.9	20.7
W250 × 18	2 280	251	4.83	101.0	5.3	22.5	179	99.3	0.919	18.2	20.1
W200 × 100	12 700	229	14.50	210.0	23.7	113	987	94.3	36.6	349	53.7
W200 × 86	11 000	222	13.00	209.0	20.6	94.7	853	92.8	31.4	300	53.4
W200 × 71	9 100	216	10.20	206.0	17.4	76.6	709	91.7	25.4	247	52.8
W200 × 59	7 580	210	9.14	205.0	14.2	61.2	583	89.9	20.4	199	51.9
W200 × 46	5 890	203	7.24	203.0	11.0	45.5	448	87.9	15.3	151	51.0
W200 × 36	4 570	201	6.22	165.0	10.2	34.4	342	86.8	7.64	92.6	40.9
W200 × 22	2 860	206	6.22	102.0	8.0	20.0	194	83.6	1.42	27.8	22.3
W150 × 37	4 730	162	8.13	154.0	11.6	22.2	274	68.5	7.07	91.8	38.7
W150 × 30	3 790	157	6.60	153.0	9.3	17.1	218	67.2	5.54	72.4	38.2
W150 × 22	2 860	152	5.84	152.0	6.6	12.1	159	65.0	3.87	50.9	36.8
W150 × 24	3 060	160	6.60	102.0	10.3	13.4	168	66.2	1.83	35.9	24.5
W150 × 18	2 290	153	5.84	102.0	7.1	9.19	120	63.3	1.26	24.7	23.5
W150 × 14	1 730	150	4.32	100.0	5.5	6.84	91.2	62.9	0.912	18.2	23.0

B

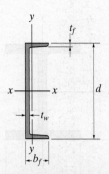

American Standard Channels or C Shapes SI Units											
Designation	Area A	Depth d	Web thickness t_w	Flange width b_f	Flange thickness t_f	x–x axis I	x–x axis S	x–x axis r	y–y axis I	y–y axis S	y–y axis r
mm × kg/m	mm²	mm	mm	mm	mm	10^6 mm⁴	10^3 mm³	mm	10^6 mm⁴	10^3 mm³	mm
C380 × 74	9 480	381.0	18.20	94.4	16.50	168	882	133	4.58	61.8	22.0
C380 × 60	7 610	381.0	13.20	89.4	16.50	145	761	138	3.84	55.1	22.5
C380 × 50	6 430	381.0	10.20	86.4	16.50	131	688	143	3.38	50.9	22.9
C310 × 45	5 690	305.0	13.00	80.5	12.70	67.4	442	109	2.14	33.8	19.4
C310 × 37	4 740	305.0	9.83	77.4	12.70	59.9	393	112	1.86	30.9	19.8
C310 × 31	3 930	305.0	7.16	74.7	12.70	53.7	352	117	1.61	28.3	20.2
C250 × 45	5 690	254.0	17.10	77.0	11.10	42.9	338	86.8	1.61	27.1	17.0
C250 × 37	4 740	254.0	13.40	73.3	11.10	38.0	299	89.5	1.40	24.3	17.2
C250 × 30	3 790	254.0	9.63	69.6	11.10	32.8	258	93.0	1.17	21.6	17.6
C250 × 23	2 900	254.0	6.10	66.0	11.10	28.1	221	98.4	0.949	19.0	18.1
C230 × 30	3 790	229.0	11.40	67.3	10.50	25.3	221	81.7	1.01	19.2	16.3
C230 × 22	2 850	229.0	7.24	63.1	10.50	21.2	185	86.2	0.803	16.7	16.8
C230 × 20	2 540	229.0	5.92	61.8	10.50	19.9	174	88.5	0.733	15.8	17.0
C200 × 28	3 550	203.0	12.40	64.2	9.90	18.3	180	71.8	0.824	16.5	15.2
C200 × 20	2 610	203.0	7.70	59.5	9.90	15.0	148	75.8	0.637	14.0	15.6
C200 × 17	2 180	203.0	5.59	57.4	9.90	13.6	134	79.0	0.549	12.8	15.9
C180 × 22	2 790	178.0	10.60	58.4	9.30	11.3	127	63.6	0.574	12.8	14.3
C180 × 18	2 320	178.0	7.98	55.7	9.30	10.1	113	66.0	0.487	11.5	14.5
C180 × 15	1 850	178.0	5.33	53.1	9.30	8.87	99.7	69.2	0.403	10.2	14.8
C150 × 19	2 470	152.0	11.10	54.8	8.70	7.24	95.3	54.1	0.437	10.5	13.3
C150 × 16	1 990	152.0	7.98	51.7	8.70	6.33	83.3	56.4	0.360	9.22	13.5
C150 × 12	1 550	152.0	5.08	48.8	8.70	5.45	71.7	59.3	0.288	8.04	13.6
C130 × 13	1 700	127.0	8.25	47.9	8.10	3.70	58.3	46.7	0.263	7.35	12.4
C130 × 10	1 270	127.0	4.83	44.5	8.10	3.12	49.1	49.6	0.199	6.18	12.5
C100 × 11	1 370	102.0	8.15	43.7	7.50	1.91	37.5	37.3	0.180	5.62	11.5
C100 × 8	1 030	102.0	4.67	40.2	7.50	1.60	31.4	39.4	0.133	4.65	11.4
C75 × 9	1 140	76.2	9.04	40.5	6.90	0.862	22.6	27.5	0.127	4.39	10.6
C75 × 7	948	76.2	6.55	38.0	6.90	0.770	20.2	28.5	0.103	3.83	10.4
C75 × 6	781	76.2	4.32	35.8	6.90	0.691	18.1	29.8	0.082	3.32	10.2

B

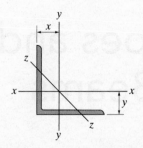

Angles Having Equal Legs   SI Units

Size and thickness	Mass per Meter	Area	x–x axis				y–y axis				z–z axis
			I	S	r	y	I	S	r	x	r
mm	kg	mm^2	$10^6 mm^4$	$10^6 mm^3$	mm	mm	$10^6 mm^4$	$10^6 mm^3$	mm	mm	mm
L203 × 203 × 25.4	75.9	9 680	36.9	258	61.7	60.1	36.9	258	61.7	60.1	39.6
L203 × 203 × 19.0	57.9	7 380	28.9	199	62.6	57.8	28.9	199	62.6	57.8	40.1
L203 × 203 × 12.7	39.3	5 000	20.2	137	63.6	55.5	20.2	137	63.6	55.5	40.4
L152 × 152 × 25.4	55.7	7 100	14.6	139	45.3	47.2	14.6	139	45.3	47.2	29.7
L152 × 152 × 19.0	42.7	5 440	11.6	108	46.2	45.0	11.6	108	46.2	45.0	29.7
L152 × 152 × 12.7	29.2	3 710	8.22	75.1	47.1	42.7	8.22	75.1	47.1	42.7	30.0
L152 × 152 × 9.5	22.2	2 810	6.35	57.4	47.5	41.5	6.35	57.4	47.5	41.5	30.2
L127 × 127 × 19.0	35.1	4 480	6.54	73.9	38.2	38.7	6.54	73.9	38.2	38.7	24.8
L127 × 127 × 12.7	24.1	3 060	4.68	51.7	39.1	36.4	4.68	51.7	39.1	36.4	25.0
L127 × 127 × 9.5	18.3	2 330	3.64	39.7	39.5	35.3	3.64	39.7	39.5	35.3	25.1
L102 × 102 × 19.0	27.5	3 510	3.23	46.4	30.3	32.4	3.23	46.4	30.3	32.4	19.8
L102 × 102 × 12.7	19.0	2 420	2.34	32.6	31.1	30.2	2.34	32.6	31.1	30.2	19.9
L102 × 102 × 9.5	14.6	1 840	1.84	25.3	31.6	29.0	1.84	25.3	31.6	29.0	20.0
L102 × 102 × 6.4	9.8	1 250	1.28	17.3	32.0	27.9	1.28	17.3	32.0	27.9	20.2
L89 × 89 × 12.7	16.5	2 100	1.52	24.5	26.9	26.9	1.52	24.5	26.9	26.9	17.3
L89 × 89 × 9.5	12.6	1 600	1.20	19.0	27.4	25.8	1.20	19.0	27.4	25.8	17.4
L89 × 89 × 6.4	8.6	1 090	0.840	13.0	27.8	24.6	0.840	13.0	27.8	24.6	17.6
L76 × 76 × 12.7	14.0	1 770	0.915	17.5	22.7	23.6	0.915	17.5	22.7	23.6	14.8
L76 × 76 × 9.5	10.7	1 360	0.726	13.6	23.1	22.5	0.726	13.6	23.1	22.5	14.9
L76 × 76 × 6.4	7.3	927	0.514	9.39	23.5	21.3	0.514	9.39	23.5	21.3	15.0
L64 × 64 × 12.7	11.5	1 450	0.524	12.1	19.0	20.6	0.524	12.1	19.0	20.6	12.4
L64 × 64 × 9.5	8.8	1 120	0.420	9.46	19.4	19.5	0.420	9.46	19.4	19.5	12.4
L64 × 64 × 6.4	6.1	766	0.300	6.59	19.8	18.2	0.300	6.59	19.8	18.2	12.5
L51 × 51 × 9.5	7.0	877	0.202	5.82	15.2	16.2	0.202	5.82	15.2	16.2	9.88
L51 × 51 × 6.4	4.7	605	0.146	4.09	15.6	15.1	0.146	4.09	15.6	15.1	9.93
L51 × 51 × 3.2	2.5	312	0.080	2.16	16.0	13.9	0.080	2.16	16.0	13.9	10.1

Slopes and Deflections of Beams

Simply Supported Beam Slopes and Deflections

Beam	Slope	Deflection	Elastic Curve
	$\theta_{max} = \dfrac{-PL^2}{16EI}$	$v_{max} = \dfrac{-PL^3}{48EI}$	$v = \dfrac{-Px}{48EI}(3L^2 - 4x^2)$ $0 \le x \le L/2$
	$\theta_1 = \dfrac{-Pab(L + b)}{6EIL}$ $\theta_2 = \dfrac{Pab(L + a)}{6EIL}$	$v\Big|_{x=a} = \dfrac{-Pba}{6EIL}(L^2 - b^2 - a^2)$	$v = \dfrac{-Pbx}{6EIL}(L^2 - b^2 - x^2)$ $0 \le x \le a$
	$\theta_1 = \dfrac{-M_0 L}{6EI}$ $\theta_2 = \dfrac{M_0 L}{3EI}$	$v_{max} = \dfrac{-M_0 L^2}{\sqrt{243}EI}$ at $x = 0.5774L$	$v = \dfrac{-M_0 x}{6EIL}(L^2 - x^2)$
	$\theta_{max} = \dfrac{-wL^3}{24EI}$	$v_{max} = \dfrac{-5wL^4}{384EI}$	$v = \dfrac{-wx}{24EI}(x^3 - 2Lx^2 + L^3)$
	$\theta_1 = \dfrac{-3wL^3}{128EI}$ $\theta_2 = \dfrac{7wL^3}{384EI}$	$v\Big|_{x=L/2} = \dfrac{-5wL^4}{768EI}$ $v_{max} = -0.006563\dfrac{wL^4}{EI}$ at $x = 0.4598L$	$v = \dfrac{-wx}{384EI}(16x^3 - 24Lx^2 + 9L^3)$ $0 \le x \le L/2$ $v = \dfrac{-wL}{384EI}(8x^3 - 24Lx^2$ $\qquad + 17L^2 x - L^3)$ $L/2 \le x < L$
	$\theta_1 = \dfrac{-7w_0 L^3}{360EI}$ $\theta_2 = \dfrac{w_0 L^3}{45EI}$	$v_{max} = -0.00652\dfrac{w_0 L^4}{EI}$ at $x = 0.5193L$	$v = \dfrac{-w_0 x}{360EIL}(3x^4 - 10L^2 x^2 + 7L^4)$

Cantilevered Beam Slopes and Deflections

Beam	Slope	Deflection	Elastic Curve
	$\theta_{max} = \dfrac{-PL^2}{2EI}$	$v_{max} = \dfrac{-PL^3}{3EI}$	$v = \dfrac{-Px^2}{6EI}(3L - x)$
	$\theta_{max} = \dfrac{-PL^2}{8EI}$	$v_{max} = \dfrac{-5PL^3}{48EI}$	$v = \dfrac{-Px^2}{6EI}\left(\dfrac{3}{2}L - x\right) \quad 0 \le x \le L/2$ $v = \dfrac{-PL^2}{24EI}\left(3x - \dfrac{1}{2}L\right) \quad L/2 \le x \le L$
	$\theta_{max} = \dfrac{-wL^3}{6EI}$	$v_{max} = \dfrac{-wL^4}{8EI}$	$v = \dfrac{-wx^2}{24EI}(x^2 - 4Lx + 6L^2)$
	$\theta_{max} = \dfrac{M_0 L}{EI}$	$v_{max} = \dfrac{M_0 L^2}{2EI}$	$v = \dfrac{M_0 x^2}{2EI}$
	$\theta_{max} = \dfrac{-wL^3}{48EI}$	$v_{max} = \dfrac{-7wL^4}{384EI}$	$v = \dfrac{-wx^2}{24EI}\left(x^2 - 2Lx + \dfrac{3}{2}L^2\right)$ $0 \le x \le L/2$ $v = \dfrac{-wL^3}{192EI}(4x - L/2)$ $L/2 \le x \le L$
	$\theta_{max} = \dfrac{-w_0 L^3}{24EI}$	$v_{max} = \dfrac{-w_0 L^4}{30EI}$	$v = \dfrac{-w_0 x^2}{120EIL}(10L^3 - 10L^2 x + 5Lx^2 - x^3)$

C

Fundamental Problems Partial Solutions and Answers

Chapter 1

F1–1 Entire beam:

$\zeta + \Sigma M_B = 0;$ $60 - 10(2) - A_y(2) = 0$ $A_y = 20$ kN

Left segment:

$\xrightarrow{+} \Sigma F_x = 0;$ $N_C = 0$ *Ans.*

$+\uparrow \Sigma F_y = 0;$ $20 - V_C = 0$ $V_C = 20$ kN   *Ans.*

$\zeta + \Sigma M_C = 0;$ $M_C + 60 - 20(1) = 0$ $M_C = -40$ kN $\cdot$ m   *Ans.*

F1–2 Entire beam:

$\zeta + \Sigma M_A = 0;$ $B_y(3) - 100(1.5)(0.75) - 200(1.5)(2.25) = 0$

                 $B_y = 262.5$ N

Right segment:

$\xrightarrow{+} \Sigma F_x = 0;$ $N_C = 0$ *Ans.*

$+\uparrow \Sigma F_y = 0;$ $V_C + 262.5 - 200(1.5) = 0$ $V_C = 37.5$ N   *Ans.*

$\zeta + \Sigma M_C = 0;$ $262.5(1.5) - 200(1.5)(0.75) - M_C = 0$

                 $M_C = 169$ N $\cdot$ m   *Ans.*

F1–3 Entire beam:

$\xrightarrow{+} \Sigma F_x = 0;$ $B_x = 0$

$\zeta + \Sigma M_A = 0;$ $20(2)(1) - B_y(4) = 0$ $B_y = 10$ kN

Right segment:

$\xrightarrow{+} \Sigma F_x = 0;$ $N_C = 0$ *Ans.*

$+\uparrow \Sigma F_y = 0;$ $V_C - 10 = 0$ $V_C = 10$ kN   *Ans.*

$\zeta + \Sigma M_C = 0;$ $-M_C - 10(2) = 0$ $M_C = -20$ kN $\cdot$ m   *Ans.*

F1–4 Entire beam:

$\zeta + \Sigma M_B = 0;$ $\dfrac{1}{2}(10)(3)(2) + 10(3)(4.5) - A_y(6) = 0$ $A_y = 27.5$ kN

Left segment:

$\xrightarrow{+} \Sigma F_x = 0;$ $N_C = 0$ *Ans.*

$+\uparrow \Sigma F_y = 0;$ $27.5 - 10(3) - V_C = 0$ $V_C = -2.5$ kN   *Ans.*

$\zeta + \Sigma M_C = 0;$ $M_C + 10(3)(1.5) - 27.5(3) = 0$ $M_C = 37.5$ kN $\cdot$ m   *Ans.*

F1–5 Entire beam:

$\xrightarrow{+} \Sigma F_x = 0;$ $A_x = 0$

$\zeta + \Sigma M_B = 0;$ $5(2)(1) - \dfrac{1}{2}(5)(1)\left(\dfrac{1}{3}\right) - A_y(2) = 0$ $A_y = 4.583$ kN

Left segment:

$\xrightarrow{+} \Sigma F_x = 0;$ $N_C = 0$ *Ans.*

$+\uparrow \Sigma F_y = 0;$ $4.583 - 5(1) - V_C = 0$ $V_C = -0.417$ kN   *Ans.*

$\zeta + \Sigma M_C = 0;$ $M_C + 5(1)(0.5) - 4.583(1) = 0$ $M_C = 2.083$ kN $\cdot$ m   *Ans.*

F1–6   Entire beam:

$\zeta + \Sigma M_A = 0;$ $F_{BD}\left(\dfrac{3}{5}\right)(4) - 5(6)(3) = 0$ $F_{BD} = 37.5$ kN

$\overset{+}{\rightarrow} \Sigma F_x = 0;$ $37.5\left(\dfrac{4}{5}\right) - A_x = 0$ $A_x = 30$ kN

$+\uparrow \Sigma F_y = 0;$ $A_y + 37.5\left(\dfrac{3}{5}\right) - 5(6) = 0$ $A_y = 7.5$ kN

Left segment:

$\overset{+}{\rightarrow} \Sigma F_x = 0;$ $N_C - 30 = 0$ $N_C = 30$ kN   *Ans.*

$+\uparrow \Sigma F_y = 0;$ $7.5 - 5(2) - V_C = 0$ $V_C = -2.5$ kN   *Ans.*

$\zeta + \Sigma M_C = 0;$ $M_C + 5(2)(1) - 7.5(2) = 0$ $M_C = 5$ kN·m   *Ans.*

F1–7   Beam:

$\Sigma M_A = 0; T_{CD} = 2w$

$\Sigma F_y = 0; T_{AB} = w$

Rod AB:

$\sigma = \dfrac{P}{A}; 300(10^3) = \dfrac{w}{10};$

$w = 3$ N/m

Rod CD:

$\sigma = \dfrac{P}{A}; 300(10^3) = \dfrac{2w}{15};$

$w = 2.25$ N/m   *Ans.*

F1–8 $A = \pi(0.1^2 - 0.08^2) = 3.6(10^{-3})\pi$ m^2

$\sigma_{\text{avg}} = \dfrac{P}{A} = \dfrac{300(10^3)}{3.6(10^{-3})\pi} = 26.5$ MPa   *Ans.*

F1–9 $A = 3[100(25)] = 7500$ mm^2

$\sigma_{\text{avg}} = \dfrac{P}{A} = \dfrac{75(10^3)}{7500} = 10$ N/mm^2 = 10 MPa   *Ans.*

F1–10   Consider the cross section to be a rectangle and two triangles.

$\bar{y} = \dfrac{\Sigma \tilde{y} A}{\Sigma A} = \dfrac{0.15[(0.3)(0.12)] + (0.1)\left[\dfrac{1}{2}(0.16)(0.3)\right]}{0.3(0.12) + \dfrac{1}{2}(0.16)(0.3)}$

$= 0.13$ m $= 130$ mm   *Ans.*

$\sigma_{\text{avg}} = \dfrac{P}{A} = \dfrac{600(10^3)}{0.06} = 10$ MPa   *Ans.*

F1–11

$A_A = A_C = \dfrac{\pi}{4}(5^2) = 6.25\pi$ mm^2, $A_B = \dfrac{\pi}{4}(10^2) = 25\pi$ mm^2

$\sigma_A = \dfrac{N_A}{A_A} = \dfrac{300}{6.25\pi} = 15.3$ N/mm^2 (T)   *Ans.*

$\sigma_B = \dfrac{N_B}{A_B} = \dfrac{-600}{25\pi} = -7.64$ N/mm^2 = -7.64 N/mm^2 (C) *Ans.*

$\sigma_C = \dfrac{N_C}{A_C} = \dfrac{200}{6.25\pi} = 10.2$ N/mm^2 (T)   *Ans.*

F1–12

$F_{AD} = 50(9.81)$ N $= 490.5$ N

$+\uparrow \Sigma F_y = 0;$ $F_{AC}\left(\dfrac{3}{5}\right) - 490.5 = 0$ $F_{AC} = 817.5$ N

$\overset{+}{\rightarrow} \Sigma F_x = 0;$ $817.5\left(\dfrac{4}{5}\right) - F_{AB} = 0$ $F_{AB} = 654$ N

$A_{AB} = \dfrac{\pi}{4}(0.008^2) = 16(10^{-6})\pi$ m^2

$(\sigma_{AB})_{\text{avg}} = \dfrac{F_{AB}}{A_{AB}} = \dfrac{654}{16(10^{-6})\pi} = 13.0$ MPa   *Ans.*

F1–13   Ring C:

$+\uparrow \Sigma F_y = 0;$ $2F\cos 60° - 200(9.81) = 0$ $F = 1962$ N

$(\sigma_{\text{allow}})_{\text{avg}} = \dfrac{F}{A};$ $150(10^6) = \dfrac{1962}{\dfrac{\pi}{4}d^2}$

$d = 0.00408$ m $= 4.08$ mm

Use $d = 5$ mm   *Ans.*

F1–14   Entire frame:

$\Sigma F_y = 0;\ A_y = 600$ kN

$\Sigma M_B = 0;\ A_y = 800$ kN

$F_A = \sqrt{(600)^2 + (800)^2} = 1000$ kN

$(\tau_A)_{avg} = \dfrac{F_A/2}{A} = \dfrac{1000/2}{\frac{\pi}{4}(50)^2} \equiv 0.255$ kN/mm^2 *Ans.*

$= 255$ N/mm^2

F1–15   Double shear:

$\Sigma F_x = 0;\ \ 4V - 10 = 0 \ \ \ \ V = 2.5$ kN

$A = \dfrac{\pi}{4}(12)^2 = 113.1$ mm^2

$\tau_{avg} = \dfrac{V}{A} = \dfrac{2.5(10^3)}{113.1} = 22.1$ N/mm^2 *Ans.*

F1–16   Single shear:

$\Sigma F_x = 0;\ \ \ \ P - 3V = 0 \ \ \ \ V = \dfrac{P}{3}$

$A = \dfrac{\pi}{4}(0.004^2) = 4(10^{-6})\pi$ m^2

$(\tau_{avg})_{allow} = \dfrac{V}{A};\ \ \ \ 60(10^6) = \dfrac{\frac{P}{3}}{4(10^{-6})\pi}$

$P = 2.262(10^3)$ N $= 2.26$ kN   *Ans.*

F1–17 $\xrightarrow{+} \Sigma F_x = 0;\ \ \ \ V - P\cos 60° = 0 \ \ \ \ V = 0.5P$

$A = \left(\dfrac{0.05}{\sin 60°}\right)(0.025) = 1.4434(10^{-3})$ m^2

$(\tau_{avg})_{allow} = \dfrac{V}{A};\ \ \ \ 600(10^3) = \dfrac{0.5P}{1.4434(10^{-3})}$

$P = 1.732(10^3)$ N $= 1.73$ kN   *Ans.*

F1–18   The resultant force on the pin is
$F = \sqrt{30^2 + 40^2} = 50$ kN.

Here we have double shear:

$V = \dfrac{F}{2} = \dfrac{50}{2} = 25$ kN

$A = \dfrac{\pi}{4}(0.03^2) = 0.225(10^{-3})\pi$ m^2

$\tau_{avg} = \dfrac{V}{A} = \dfrac{25(10^3)}{0.225(10^{-3})\pi} = 35.4$ MPa   *Ans.*

F1–19 $\xrightarrow{+} \Sigma F_x = 0;\ \ \ \ 30 - N = 0 \ \ \ \ N = 30$ kN

$\sigma_{allow} = \dfrac{\sigma_Y}{\text{F.S.}} = \dfrac{250}{1.5} = 166.67$ MPa

$\sigma_{allow} = \dfrac{N}{A};\ \ \ \ 166.67(10^6) = \dfrac{30(10^3)}{\frac{\pi}{4}d^2}$

$d = 15.14$ mm

Use $d = 16$ mm   *Ans.*

F1–20

$\xrightarrow{+} \Sigma F_x = 0;\ \ \ N_{AB} - 150 = 0 \ \ \ \ \ \ \ \ N_{AB} = 150$ kN

$\xrightarrow{+} \Sigma F_x = 0;\ \ \ N_{BC} - 75 - 75 - 150 = 0 \ \ \ N_{BC} = 300$ kN

$\sigma_{allow} = \dfrac{\sigma_Y}{\text{F.S.}} = \dfrac{350}{1.5} = 233.3$ MPa

Segment AB:

$\sigma_{allow} = \dfrac{N_{AB}}{A_{AB}};\ \ \ \ 233.3 = \dfrac{150(10^3)}{h_1(12)}$

$h_1 = 54.0$ mm

Segment BC:

$\sigma_{allow} = \dfrac{N_{BC}}{A_{BC}};\ \ \ \ 233.3 = \dfrac{300(10^3)}{h_2(12)}$

$h_2 = 108$ mm

Use $h_1 = 54$ mm and $h_2 = 108$ mm.   *Ans.*

F1–21 $N = P$

$\sigma_{allow} = \dfrac{\sigma_Y}{\text{F.S.}} = \dfrac{250}{2} = 125$ MPa

$A_r = \dfrac{\pi}{4}(0.04^2) = 1.2566(10^{-3})$ m^2

$A_{a-a} = 2(0.06 - 0.03)(0.05) = 3(10^{-3})$ m^2

The rod will fail first.

$\sigma_{allow} = \dfrac{N}{A_r};\ \ \ \ 125(10^6) = \dfrac{P}{1.2566(10^{-3})}$

$P = 157.08(10^3)$ N $= 157$ kN   *Ans.*

F1–22 $\xrightarrow{+} \Sigma F_x = 0;\ \ \ \ \ 80 - 2V = 0 \ \ \ \ V = 40$ kN

$\tau_{allow} = \dfrac{\tau_{fail}}{\text{F.S.}} = \dfrac{100}{2.5} = 40$ MPa

$\tau_{allow} = \dfrac{V}{A};\ \ \ \ \ 40(10^6) = \dfrac{40(10^3)}{\frac{\pi}{4}d^2}$

$d = 0.03568$ m $= 35.68$ mm

Use $d = 36$ mm   *Ans.*

F1–23 $V = P$

$\tau_{allow} = \dfrac{\tau_{fail}}{\text{F.S.}} = \dfrac{120}{2.5} = 48$ MPa

Area of shear plane for bolt head and plate:

$A_b = \pi dt = \pi(0.04)(0.075) = 0.003\pi$ m^2

$A_p = \pi dt = \pi(0.08)(0.03) = 0.0024\pi$ m^2

Since the area of shear plane for the plate is smaller,

$$\tau_{allow} = \frac{V}{A_p}; \qquad 48(10^6) = \frac{P}{0.0024\pi}$$

$$P = 361.91(10^3) \text{ N} = 362 \text{ kN} \qquad Ans.$$

F1–24

$$\downarrow + \Sigma M_B = 0; \qquad \frac{1}{2}(5)(3)(2) - 6V(3) = 0 \qquad V = 0.833 \text{ kN}$$

$$\tau_{allow} = \frac{\tau_{fail}}{\text{F.S.}} = \frac{112}{2} = 56 \text{ MPa} = 56 \text{ N/mm}^2$$

$$\tau_{allow} = \frac{V}{A}; \qquad 56 = \frac{0.833(10^3)}{\frac{\pi}{4}d^2}$$

$$d = 4.35 \text{ mm}$$

Use $d = 5$ mm $\qquad Ans.$

Chapter 2

F2–1 $\quad \dfrac{\delta_C}{600} = \dfrac{0.2}{400}; \qquad \delta_C = 0.3$ mm

$$\epsilon_{CD} = \frac{\delta_C}{L_{CD}} = \frac{0.3}{300} = 0.001 \text{ mm/mm} \qquad Ans.$$

F2–2

$$\theta = \left(\frac{0.02°}{180°}\right)\pi \text{ rad} = 0.3491(10^{-3}) \text{ rad}$$

$$\delta_B = \theta L_{AB} = 0.3491(10^{-3})(600) = 0.2094 \text{ mm}$$

$$\delta_C = \theta L_{AC} = 0.3491(10^{-3})(1200) = 0.4189 \text{ mm}$$

$$\epsilon_{BD} = \frac{\delta_B}{L_{BD}} = \frac{0.2094}{400} = 0.524(10^{-3}) \text{ mm/mm} \qquad Ans.$$

$$\epsilon_{CE} = \frac{\delta_C}{L_{CE}} = \frac{0.4189}{600} = 0.698(10^{-3}) \text{ mm/mm} \qquad Ans.$$

F2–3

$$\alpha = \frac{2}{400} = 0.005 \text{ rad} \qquad \beta = \frac{4}{300} = 0.01333 \text{ rad}$$

$$(\gamma_A)_{xy} = \frac{\pi}{2} - \theta$$

$$= \frac{\pi}{2} - \left(\frac{\pi}{2} - \alpha + \beta\right)$$

$$= \alpha - \beta$$

$$= 0.005 - 0.01333$$

$$= -0.00833 \text{ rad} \qquad Ans.$$

F2–4

$$L_{BC} = \sqrt{300^2 + 400^2} = 500 \text{ mm}$$

$$L_{B'C} = \sqrt{(300 - 3)^2 + (400 + 5)^2} = 502.2290 \text{ mm}$$

$$\alpha = \frac{3}{405} = 0.007407 \text{ rad}$$

$$(\epsilon_{BC})_{avg} = \frac{L_{B'C} - L_{BC}}{L_{BC}} = \frac{502.2290 - 500}{500}$$

$$= 0.00446 \text{ mm/mm} \qquad Ans.$$

$$(\gamma_A)_{xy} = \frac{\pi}{2} - \theta = \frac{\pi}{2} - \left(\frac{\pi}{2} + \alpha\right) = -\alpha = -0.00741 \text{ rad} \qquad Ans.$$

F2–5

$$L_{AC} = \sqrt{L_{CD}^2 + L_{AD}^2} = \sqrt{300^2 + 300^2} = 424.2641 \text{ mm}$$

$$L_{A'C'} = \sqrt{L_{C'D'}^2 + L_{A'D'}^2} = \sqrt{306^2 + 296^2} = 425.7370 \text{ mm}$$

$$\frac{\theta}{2} = \tan^{-1}\left(\frac{L_{C'D'}}{L_{A'D'}}\right); \theta = 2\tan^{-1}\left(\frac{306}{296}\right) = 1.6040 \text{ rad}$$

$$(\epsilon_{AC})_{avg} = \frac{L_{A'C'} - L_{AC}}{L_{AC}} = \frac{425.7370 - 424.2641}{424.2641}$$

$$= 0.00347 \text{ mm/mm} \qquad Ans.$$

$$(\gamma_E)_{xy} = \frac{\pi}{2} - \theta = \frac{\pi}{2} - 1.6040 = -0.0332 \text{ rad} \qquad Ans.$$

Chapter 3

F3–1   Material has uniform properties throughout. $\quad Ans.$

F3–2   Proportional limit is A. $\qquad Ans.$

Ultimate stress is D. $\qquad Ans.$

F3–3   The initial slope of the $\sigma - \epsilon$ diagram. $\quad Ans.$

F3–4   True. $\qquad Ans.$

F3–5   False. Use the *original* cross-sectional area and length. $\qquad Ans.$

F3–6   False. It will normally decrease. $\qquad Ans.$

F3–7 $\quad \epsilon = \dfrac{\sigma}{E} = \dfrac{P}{AE}$

$$\delta = \epsilon L = \frac{PL}{AE} = \frac{100(10^3)(0.100)}{\frac{\pi}{4}(0.015)^2 200(10^9)}$$

$$= 0.283 \text{ mm} \qquad Ans.$$

F3–8 $\quad \epsilon = \dfrac{\sigma}{E} = \dfrac{P}{AE}$

$$\delta = \epsilon L = \frac{PL}{AE};$$

$$0.075 = \frac{(50)(200)}{7500E}$$

$$E = 17.78 \text{ kN/mm}^2 = 17.78 \text{ GPa} \qquad Ans.$$

F3–9 $\epsilon = \dfrac{\sigma}{E} = \dfrac{P}{AE}$

$\delta = \epsilon L = \dfrac{PL}{AE} = \dfrac{6(10^3)4}{\frac{\pi}{4}(0.01)^2\,100(10^9)}$

$= 3.06 \text{ mm}$ *Ans.*

F3–10 $\sigma = \dfrac{P}{A} = \dfrac{100(10^3)}{\frac{\pi}{4}(0.02^2)} = 318.31 \text{ MPa}$

Since $\sigma < \sigma_Y = 450$ MPa, Hooke's Law is applicable.

$E = \dfrac{\sigma_Y}{\epsilon_Y} = \dfrac{450(10^6)}{0.00225} = 200 \text{ GPa}$

$\epsilon = \dfrac{\sigma}{E} = \dfrac{318.31(10^6)}{200(10^9)} = 0.001592 \text{ mm/mm}$

$\delta = \epsilon L = 0.001592(50) = 0.0796 \text{ mm}$ *Ans.*

F3–11 $\sigma = \dfrac{P}{A} = \dfrac{150(10^3)}{\frac{\pi}{4}(0.02^2)} = 477.46 \text{ MPa}$

Since $\sigma > \sigma_Y = 450$ MPa, Hooke's Law is not applicable. From the geometry of the stress-strain diagram,

$\dfrac{\epsilon - 0.00225}{0.03 - 0.00225} = \dfrac{477.46 - 450}{500 - 450}$

$\epsilon = 0.017493$

When the load is removed, the strain recovers along a line parallel to the original elastic line. Here $E = \dfrac{\sigma_Y}{\epsilon_Y} = \dfrac{450(10^6)}{0.00225} = 200 \text{ GPa}$.

The elastic recovery is

$\epsilon_r = \dfrac{\sigma}{E} = \dfrac{477.46(10^6)}{200(10^9)} = 0.002387 \text{ mm/mm}$

$\epsilon_p = \epsilon - \epsilon_r = 0.017493 - 0.002387$

$= 0.01511 \text{ mm/mm}$

$\delta_p = \epsilon_p L = 0.01511(50) = 0.755 \text{ mm}$ *Ans.*

F3–12 $\epsilon_{BC} = \dfrac{\delta_{BC}}{L_{BC}} = \dfrac{0.2}{300} = 0.6667(10^{-3}) \text{ mm/mm}$

$\sigma_{BC} = E\epsilon_{BC} = 200(10^9)[0.6667(10^{-3})]$

$= 133.33 \text{ MPa}$

Since $\sigma_{BC} < \sigma_Y = 250$ MPa, Hooke's Law is valid.

$\sigma_{BC} = \dfrac{F_{BC}}{A_{BC}}; \qquad 133.33(10^6) = \dfrac{F_{BC}}{\frac{\pi}{4}(0.003^2)}$

$F_{BC} = 942.48 \text{ N}$

$\zeta + \Sigma M_A = 0; \qquad 942.48(0.4) - P(0.6) = 0$

$P = 628.31 \text{ N} = 628 \text{ N}$ *Ans.*

F3–13 $\sigma = \dfrac{P}{A} = \dfrac{10(10^3)}{\frac{\pi}{4}(0.015)^2} = 56.59 \text{ MPa}$

$\epsilon_{\text{long}} = \dfrac{\sigma}{E} = \dfrac{56.59(10^6)}{70(10^9)} = 0.808(10^{-3})$

$\epsilon_{\text{lat}} = -\nu\epsilon_{\text{long}} = -0.35(0.808(10^{-3}))$

$= -0.283(10^{-3})$

$\delta d = (-0.283(10^{-3}))(15 \text{ mm}) = -4.24(10^{-3}) \text{ mm}$

Ans.

F3–14 $\sigma = \dfrac{P}{A} = \dfrac{50(10^3)}{\frac{\pi}{4}(0.02^2)} = 159.15 \text{ MPa}$

$\epsilon_a = \dfrac{\delta}{L} = \dfrac{1.40}{600} = 0.002333 \text{ mm/mm}$

$E = \dfrac{\sigma}{\epsilon_a} = \dfrac{159.15(10^6)}{0.002333} = 68.2 \text{ GPa}$ *Ans.*

$\epsilon_e = \dfrac{d' - d}{d} = \dfrac{19.9837 - 20}{20} = -0.815(10^{-3}) \text{ mm/mm}$

$\nu = -\dfrac{\epsilon_e}{\epsilon_a} = -\dfrac{-0.815(10^{-3})}{0.002333} = 0.3493 = 0.349$

$G = \dfrac{E}{2(1 + \nu)} = \dfrac{68.21}{2(1 + 0.3493)} = 25.3 \text{ GPa}$ *Ans.*

F3–15 $\alpha = \dfrac{0.5}{150} = 0.003333 \text{ rad}$

$\gamma = \dfrac{\pi}{2} - \theta = \dfrac{\pi}{2} - \left(\dfrac{\pi}{2} - \alpha\right)$

$= \alpha = 0.003333 \text{ rad}$

$\tau = G\gamma = [26(10^9)](0.003333) = 86.67 \text{ MPa}$

$\tau = \dfrac{V}{A}; \qquad 86.67(10^6) = \dfrac{P}{0.15(0.02)}$

$P = 260 \text{ kN}$ *Ans.*

F3–16 $\alpha = \dfrac{3}{150} = 0.02 \text{ rad}$

$\gamma = \dfrac{\pi}{2} - \theta = \dfrac{\pi}{2} - \left(\dfrac{\pi}{2} - \alpha\right) = \alpha = 0.02 \text{ rad}$

When P is removed, the shear strain recovers along a line parallel to the original elastic line.

$\gamma_r = \gamma_Y = 0.005 \text{ rad}$

$\gamma_p = \gamma - \gamma_r = 0.02 - 0.005 = 0.015 \text{ rad}$ *Ans.*

Chapter 4

F4–1 $A = \dfrac{\pi}{4}(0.02^2) = 0.1(10^{-3})\pi \text{ m}^2$

$\delta_C = \dfrac{1}{AE}\{40(10^3)(400) + [-60(10^3)(600)]\}$

$= \dfrac{-20(10^6) \text{ N} \cdot \text{mm}}{AE}$

$= -0.318 \text{ mm}$ *Ans.*

F4–2 $A_{AB} = A_{CD} = \dfrac{\pi}{4}(0.02^2) = 0.1(10^{-3})\pi$ m^2

$A_{BC} = \dfrac{\pi}{4}(0.04^2 - 0.03^2) = 0.175(10^{-3})\pi$ m^2

$\delta_{D/A} = \dfrac{[-10(10^3)](400)}{[0.1(10^{-3})\pi][68.9(10^9)]}$

$+ \dfrac{[10(10^3)](400)}{[0.175(10^{-3})\pi][68.9(10^9)]}$

$+ \dfrac{[-20(10^3)](400)}{[0.1(10^{-3})\pi][68.9(10^9)]}$

$= -0.449$ mm   *Ans.*

F4–3 $A = \dfrac{\pi}{4}(0.03^2) = 0.225(10^{-3})\pi$ m^2

$\delta_C = \dfrac{1}{0.225(10^{-3})\pi[200(10^9)]} \left\{ \left[-90(10^3) \right.\right.$

$\left.\left. - 2\left(\dfrac{4}{5}\right)30(10^3) \right](0.4) + [-90(10^3)(0.6)] \right\}$

$= -0.772(10^{-3})$ m $= -0.772$ mm   *Ans.*

F4–4 $\delta_{A/B} = \dfrac{PL}{AE} = \dfrac{[60(10^3)](0.8)}{[0.1(10^{-3})\pi][200(10^9)]}$

$= 0.7639(10^{-3})$ m $\downarrow$

$\delta_B = \dfrac{F_{sp}}{k} = \dfrac{60(10^3)}{50(10^6)} = 1.2(10^{-3})$ m $\downarrow$

$+\downarrow \quad \delta_A = \delta_B + \delta_{A/B}$

$\delta_A = 1.2(10^{-3}) + 0.7639(10^{-3})$

$= 1.9639(10^{-3})$ m $= 1.96$ mm $\downarrow$ *Ans.*

F4–5 $A = \dfrac{\pi}{4}(0.02^2) = 0.1(10^{-3})\pi$ m^2

Internal load $P(x) = 30(10^3)x$

$\delta_A = \displaystyle\int \dfrac{P(x)dx}{AE}$

$= \dfrac{1}{[0.1(10^{-3})\pi][73.1(10^9)]} \displaystyle\int_0^{0.9\,\text{m}} 30(10^3)x\,dx$

$= 0.529(10^{-3})$ m $= 0.529$ mm   *Ans.*

F4–6   Distributed load $P(x) = \dfrac{45(10^3)}{0.9}x = 50(10^3)x$ N/m

Internal load $P(x) = \frac{1}{2}(50(10^3))x(x) = 25(10^3)x^2$

$\delta_A = \displaystyle\int_0^L \dfrac{P(x)dx}{AE}$

$= \dfrac{1}{[0.1(10^{-3})\pi][73.1(10^9)]} \displaystyle\int_0^{0.9\,\text{m}} [25(10^3)x^2]dx$

$= 0.265$ mm   *Ans.*

Chapter 5

F5–1 $J = \dfrac{\pi}{2}(0.04^4) = 1.28(10^{-6})\pi$ m^4

$\tau_A = \tau_{\max} = \dfrac{T_C}{J} = \dfrac{5(10^3)(0.04)}{1.28(10^{-6})\pi} = 49.7$ MPa   *Ans.*

$\tau_B = \dfrac{T\rho_B}{J} = \dfrac{5(10^3)(0.03)}{1.28(10^{-6})\pi} = 37.3$ MPa   *Ans.*

F5–2 $J = \dfrac{\pi}{2}(0.06^4 - 0.04^4) = 5.2(10^{-6})\pi$ m^4

$\tau_B = \tau_{\max} = \dfrac{Tc}{J} = \dfrac{10(10^3)(0.06)}{5.2(10^{-6})\pi} = 36.7$ MPa   *Ans.*

$\tau_A = \dfrac{T\rho_A}{J} = \dfrac{10(10^3)(0.04)}{5.2(10^{-6})\pi} = 24.5$ MPa   *Ans.*

F5–3 $J_{AB} = \dfrac{\pi}{2}(0.04^4 - 0.03^4) = 0.875(10^{-6})\pi$ m^4

$J_{BC} = \dfrac{\pi}{2}(0.04^4) = 1.28(10^{-6})\pi$ m^4

$(\tau_{AB})_{\max} = \dfrac{T_{AB}c_{AB}}{J_{AB}} = \dfrac{[2(10^3)](0.04)}{0.875(10^{-6})\pi} = 29.1$ MPa

$(\tau_{BC})_{\max} = \dfrac{T_{BC}c_{BC}}{J_{BC}} = \dfrac{[6(10^3)](0.04)}{1.28(10^{-6})\pi}$

$= 59.7$ MPa   *Ans.*

F5–4 $T_{AB} = 0,\ T_{BC} = 600$ N·m, $T_{CD} = 0$

$J = \dfrac{\pi}{2}(0.02^4) = 80(10^{-9})\pi$ m^4

$\tau_{\max} = \dfrac{Tc}{J} = \dfrac{600(0.02)}{80(10^{-9})\pi} = 47.7$ MPa   *Ans.*

F5–5 $J_{BC} = \dfrac{\pi}{2}(0.04^4 - 0.03^4) = 0.875(10^{-6})\pi$ m^4

$(\tau_{BC})_{\max} = \dfrac{T_{BC}c_{BC}}{J_{BC}} = \dfrac{2100(0.04)}{0.875(10^{-6})\pi}$

$= 30.6$ MPa   *Ans.*

F5–6 $t = 5(10^3)$ N·m/m

Internal torque is $T = 5(10^3)(0.8) = 4000$ N·m

$J = \dfrac{\pi}{2}(0.04^4) = 1.28(10^{-6})\pi$ m^4

$\tau_A = \dfrac{T_Ac}{J} = \dfrac{4000(0.04)}{1.28(10^{-6})\pi} = 39.8$ MPa   *Ans.*

F5–7 $J = \dfrac{\pi}{2}(0.03^4) = 0.405(10^{-6})\pi \ m^4$

$\phi_{A/C} = \dfrac{1}{[0.405(10^{-6})\pi][75(10^9)]}\{[-2(10^3)](0.6)$

$\qquad + 1(10^3)(0.4)\}$

$\qquad = -0.00838 \ rad = -0.480°$ *Ans.*

F5–8 $J = \dfrac{\pi}{2}(0.02^4) = 80(10^{-9})\pi \ m^4$

$\phi_{B/A} = \dfrac{600(0.45)}{[80(10^{-9})\pi][75(10^9)]}$

$\qquad = 0.01432 \ rad = 0.821°$ *Ans.*

F5–9 $J = \dfrac{\pi}{2}(0.04^4 - 0.03^4) = 0.875(10^{-6})\pi \ m^4$

$\phi_{A/B} = \dfrac{T_{AB}L_{AB}}{JG} = \dfrac{3(10^3)(0.9)}{[0.875(10^{-6})\pi][26(10^9)]}$

$\qquad = 0.03778 \ rad$

$\phi_B = \dfrac{T_B}{k} = \dfrac{3(10^3)}{90(10^3)} = 0.03333 \ rad$

$\phi_A = \phi_B + \phi_{A/B}$

$\qquad = 0.03333 + 0.03778$

$\qquad = 0.07111 \ rad = 4.07°$ *Ans.*

F5–10 $J = \dfrac{\pi}{2}(0.02^4) = 80(10^{-9})\pi \ m^4$

$\phi_{B/A} = \dfrac{0.2}{[80(10^{-9})\pi][75(10^9)]}[600 + (-300)$

$\qquad + 200 + 500]$

$\qquad = 0.01061 \ rad = 0.608°$ *Ans.*

F5–11 $J = \dfrac{\pi}{2}(0.04^4) = 1.28(10^{-6})\pi \ m^4$

$t = 5(10^3) \ N·m/m$

Internal torque is $5(10^3)x \ N·m$

$\phi_{A/B} = \displaystyle\int_0^L \dfrac{T(x)dx}{JG}$

$\qquad = \dfrac{1}{[1.28(10^{-6})\pi][75(10^9)]}\displaystyle\int_0^{0.8\,m} 5(10^3)x\,dx$

$\qquad = 0.00531 \ rad = 0.304°$ *Ans.*

F5–12 $J = \dfrac{\pi}{2}(0.04^4) = 1.28(10^{-6})\pi \ m^4$

Distributed torque is $t = \dfrac{15(10^3)}{0.6}(x)$

$\qquad = 25(10^3)x \ N·m/m$

Internal torque is $T(x) = \dfrac{1}{2}(25x)(x)$

$\qquad = 12.5(10^3)x^2 \ N·m$

$\phi_{A/C} = \displaystyle\int_0^L \dfrac{T(x)dx}{JG} + \dfrac{T_{BC}L_{BC}}{JG}$

$= \dfrac{1}{[1.28(10^{-6})\pi][75(10^9)]}\left[\displaystyle\int_0^{0.6\,m} 12.5(10^3)x^2 dx + 4500(0.4)\right]$

$= 0.008952 \ rad = 0.513°$ *Ans.*

Chapter 6

F6–1 $+\uparrow \Sigma F_y = 0; \quad -V - 9 = 0 \quad V = -9 \ kN \quad$ *Ans.*

$\quad\;\; \zeta + \Sigma M_O = 0; \quad M + 9x = 0 \quad M = \{-9x\} \ kN·m$

$\hspace{10cm}$ *Ans.*

F6–2 $+\uparrow \Sigma F_y = 0; \quad -V - 50x = 0 \quad V = \{-50x\} \ kN \quad$ *Ans.*

$\quad\;\; \zeta + \Sigma M_O = 0; \quad M + 50x\left(\dfrac{x}{2}\right) - 20 = 0$

$\qquad\qquad M = \{20 - 25x^2\} \ kN·m \quad$ *Ans.*

F6–3 $+\uparrow \Sigma F_y = 0; \quad -V - \dfrac{1}{2}(4x)(x) = 0$

$\qquad\qquad V = \{-2x^2\} \ kN \quad$ *Ans.*

$\Sigma M_O = 0; \ M + \left[\dfrac{1}{2}(4x)(x)\right]\left(\dfrac{x}{3}\right) = 0$

$\qquad\qquad M = \left\{-\dfrac{2}{3}x^3\right\} \ kN·m \quad$ *Ans.*

F6–4 $0 \le x < 1.5 \ m$

$+\uparrow \Sigma F_y = 0; \quad V = 0 \qquad\qquad$ *Ans.*

$\zeta + \Sigma M_O = 0; \quad M - 4 = 0 \quad M = 4 \ kN·m \;$ *Ans.*

$1.5 \ m < x \le 3 \ m$

$+\uparrow \Sigma F_y = 0; \quad -V - 9 = 0 \quad V = -9 \ kN \quad$ *Ans.*

$\zeta + \Sigma M_O = 0; \quad M + 9(x - 1.5) - 4 = 0$

$\qquad\qquad M = \{17.5 - 9x\} \ kN·m \quad$ *Ans.*

F6–5 $\zeta + \Sigma M_B = 0; \quad A_y(6) - 30 = 0 \quad A_y = 5 \ kN$

$+\uparrow \Sigma F_y = 0; \quad -V - 5 = 0 \quad V = -5 \ kN$

$\hspace{9cm}$ *Ans.*

$\zeta + \Sigma M_O = 0; \quad M + 5x = 0 \quad M = \{-5x\} \ kN·m$

$\hspace{9cm}$ *Ans.*

F6–6 $\zeta + \Sigma M_B = 0; \quad A_y(6) + 20 - 50 = 0$

$\qquad\qquad\qquad A_y = 5 \ kN$

$+\uparrow \Sigma F_y = 0; \quad -V - 5 = 0 \quad V = -5 \ kN$

$\hspace{9cm}$ *Ans.*

$\zeta + \Sigma M_O = 0; \quad M + 5x - 50 = 0$

$\qquad\qquad M = \{50 - 5x\} \ kN·m \quad$ *Ans.*

F6–7 *Shear diagram.* $V = -4$, $x = 0$. Zero slope to $x = 6$.

Moment diagram. $M = 0$, $x = 0$. Constant negative slope to $M = -16$, $x = 4^-$, $M = 8$, $x = 4^+$. Constant negative slope to $M = 0$, $x = 6$.

F6–8 *Shear diagram.* $V = -6$, $x = 0$. Zero slope to $x = 3$.

Moment diagram. $M = 0$, $x = 0$. Constant negative slope to $M = -9$, $x = 1.5^-$, $M = -21$, $x = 1.5^+$. Constant negative slope to $M = -30$, $x = 3$.

F6–9 *Shear diagram.* $V = 0$, $x = 0$. Zero slope to $x = 1.5^-$. $V = 4$, $x = 1.5^+$. Zero slope to $x = 4.5^-$, $V = 0$, $x = 4.5^+$. Zero slope to $V = 0$, $x = 6$.

Moment diagram. $M = 6$, $x = 0$. Zero slope to $x = 1.5$. $M = 6$, $x = 1.5$. Constant positive slope to $x = 4.5$. $M = 18$, $x = 4.5$. Zero slope to $M = 18$, $x = 6$.

F6–10 *Shear diagram.* $V = 16.5$, $x = 0$. Constant negative slope to $x = 3$. $V = 0$, $x = 2.75$, $V = -1.5$, $x = 3$. Negative decreasing slope, $V = -10.5$, $x = 6$.

Moment diagram. $M = 0$, $x = 0$. Positive decreasing slope. $M = 22.7$, $x = 2.75$. Negative decreasing slope to $M = 0$, $x = 6$.

F6–11 *Shear diagram.* $V = 0$, $x = 0$. Negative constant slope, $V = -6$, $x = 1.5^-$, $V = 0$, $x = 1.5^+$. Zero slope to $x = 4.5$. $V = 0$, $x = 4.5^-$. $V = 6$, $x = 4.5^+$. Constant negative slope, $V = 0$, $x = 6$.

Moment diagram. $M = 0$, $x = 0$. Negative increasing slope, $M = -4.5$, $x = 1.5$. Zero slope, $M = -4.5$, $x = 4.5$. Positive decreasing slope, $M = 0$, $x = 6$.

F6–12 *Shear diagram.* $V = 15$, $x = 0$. Negative decreasing slope to zero slope at $x = 3$. $V = 0$, $x = 0$. Negative increasing slope to $V = -15$, $x = 6$.

Moment diagram. $M = 0$, $x = 0$. Positive decreasing slope to zero slope at $x = 3$. $M = 15$, $x = 3$. Negative increasing slope. $M = 0$, $x = 6$.

F6–13 *Shear diagram.* $V = 1050$, $x = 0$. Constant negative slope. $V = 0$, $x = 5.25$, $V = -150$, $x = 6$. Zero slope. $V = -150$, $x = 9^-$, $V = -750$, $x = 9^+$. Zero slope. $V = -750$, $x = 12$.

Moment diagram. $M = 0$, $x = 0$. Positive decreasing slope to $x = 5.25$. $M = 2756$, $x = 5.25$. Negative increasing slope to $M = 2700$, $x = 6$.

Constant negative slope. $M = 2250$, $x = 9$. Constant negative slope. $M = 0$, $x = 12$.

F6–14 *Shear diagram.* $V = 30$, $x = 0$. Constant negative slope, $V = 0$, $x = 1.5$, $V = -50$, $x = 4^-$. $V = 20$, $x = 4^+$. Zero slope, $V = 20$, $x = 6$.

Moment diagram. $M = 0$, $x = 0$. Positive decreasing slope, to zero slope at $x = 1.5$. $M = 22.5$, $x = 1.5$. Negative increasing slope, $M = -40$, $x = 4$. Positive constant slope, $M = 0$, $x = 6$.

F6–15 $I = 2\left[\dfrac{1}{12}(0.02)(0.2^3)\right] + \dfrac{1}{12}(0.26)(0.02^3)$

$= 26.84(10^{-6})$ m^4

$\sigma_{max} = \dfrac{Mc}{I} = \dfrac{20(10^3)(0.1)}{26.84(10^{-6})} = 74.5$ MPa   *Ans.*

F6–16 $\bar{y} = \dfrac{0.3}{3} = 0.1$ m

$I = \dfrac{1}{36}(0.3)(0.3^3) = 0.225(10^{-3})$ m^4

$(\sigma_{max})_c = \dfrac{Mc}{I} = \dfrac{50(10^3)(0.3 - 0.1)}{0.225(10^{-3})}$

$= 44.4$ MPa (C)   *Ans.*

$(\sigma_{max})_t = \dfrac{My}{I} = \dfrac{50(10^3)(0.1)}{0.225(10^{-3})} = 22.2$ MPa (T)   *Ans.*

F6–17 $I = \dfrac{1}{12}(0.2)(0.3^3) - \dfrac{1}{12}(0.18)(0.26^3)$

$= 0.18636(10^{-3})$ m^4

$\sigma_{max} = \dfrac{Mc}{I} = \dfrac{50(10^3)(0.15)}{0.18636(10^{-3})} = 40.2$ MPa   *Ans.*

F6–18

$I = 2\left[\dfrac{1}{12}(0.03)(0.4^3)\right] + 2\left[\dfrac{1}{12}(0.14)(0.03^3) + 0.14(0.03)(0.15^2)\right]$

$= 0.50963(10^{-3})$ m^4

$\sigma_{max} = \dfrac{Mc}{I} = \dfrac{10(10^3)(0.2)}{0.50963(10^{-3})} = 3.92$ MPa   *Ans.*

F6–19

$I = \dfrac{1}{12}(0.05)(0.4)^3 + 2\left[\dfrac{1}{12}(0.025)(0.3)^3\right]$

$= 0.37917(10^{-3})$ m^4

$\sigma_A = \dfrac{My_A}{I} = -\dfrac{5(10^3)(-0.15)}{0.37917(10^{-3})} = 1.98$ MPa (T)   *Ans.*

F6–20 $M_y = 50\left(\dfrac{4}{5}\right) = 40 \text{ kN} \cdot \text{m}$

$M_z = 50\left(\dfrac{3}{5}\right) = 30 \text{ kN} \cdot \text{m}$

$I_y = \dfrac{1}{12}(0.3)(0.2^3) = 0.2(10^{-3}) \text{ m}^4$

$I_z = \dfrac{1}{12}(0.2)(0.3^3) = 0.45(10^{-3}) \text{ m}^4$

$\sigma = -\dfrac{M_z y}{I_z} + \dfrac{M_y z}{I_y}$

$\sigma_A = -\dfrac{[30(10^3)](-0.15)}{0.45(10^{-3})} + \dfrac{[40(10^3)](0.1)}{0.2(10^{-3})}$

$= 30 \text{ MPa (T)}$ *Ans.*

$\sigma_B = -\dfrac{[30(10^3)](0.15)}{0.45(10^{-3})} + \dfrac{[40(10^3)](0.1)}{0.2(10^{-3})}$

$= 10 \text{ MPa (T)}$ *Ans.*

$\tan \alpha = \dfrac{I_z}{I_y} \tan \theta$

$\tan \alpha = \left[\dfrac{0.45(10^{-3})}{0.2(10^{-3})}\right]\left(\dfrac{4}{3}\right)$

$\alpha = 71.6°$ *Ans.*

F6–21   Maximum stress occurs at D or A.

$(\sigma_{\max})_D = \dfrac{(75 \cos 30°)(10^3)(75)}{\frac{1}{12}(100)(150)^3} + \dfrac{(75 \cos 30°)(10^3)(50)}{\frac{1}{12}(150)(100)^3}$

$= 0.323 \text{ N/mm}^2$ *Ans.*

Chapter 7

F7–1 $I = 2\left[\dfrac{1}{12}(0.02)(0.2^3)\right] + \dfrac{1}{12}(0.26)(0.02^3)$

$= 26.84(10^{-6})$

$Q_A = 0.055(0.09)(0.02) = 99(10^{-6}) \text{ m}^3$

$\tau_A = \dfrac{VQ_A}{It} = \dfrac{100(10^3)[99(10^{-6})]}{[26.84(10^{-6})](0.02)}$

$= 18.4 \text{ MPa}$ *Ans.*

F7–2 $I = \dfrac{1}{12}(0.1)(0.3^3) + \dfrac{1}{12}(0.2)(0.1^3) = 0.24167(10^{-3}) \text{ m}^4$

$Q_A = y'_1 A'_1 + y'_2 A'_2$

$= \left[\dfrac{1}{2}(0.05)\right](0.05)(0.3) + 0.1(0.1)(0.1)$

$= 1.375(10^{-3}) \text{ m}^3$

$Q_B = y'_3 A'_3 = 0.1(0.1)(0.1) = 1(10^{-3}) \text{ m}^3$

$\tau_A = \dfrac{VQ}{It} = \dfrac{600(10^3)[1.375(10^{-3})]}{[0.24167(10^{-3})](0.3)} = 11.4 \text{ MPa}$ *Ans.*

$\tau_B = \dfrac{VQ}{It} = \dfrac{600(10^3)[1(10^{-3})]}{[0.24167(10^{-3})](0.1)} = 24.8 \text{ MPa}$ *Ans.*

F7–3 $V_{\max} = 22.5 \text{ kN}$

$I = \dfrac{1}{12}(75)(150^3) = 21\,093\,750 \text{ mm}^4$

$Q_{\max} = y'A' = 37.5(75)(75) = 210\,937.5 \text{ mm}^3$

$(\tau_{\max})_{\text{abs}} = \dfrac{V_{\max} Q_{\max}}{It} = \dfrac{22.5(10^3)(210\,937.5)}{21\,093\,750(75)} = 3.0 \text{ N/mm}^2$

Ans.

F7–4 $I = 2\left[\dfrac{1}{12}(0.03)(0.4)^3\right] + 2\left[\dfrac{1}{12}(0.14)(0.03)^3\right]$

$+ 0.14(0.03)(0.15^2)\right] = 0.50963(10^{-3}) \text{ m}^4$

$Q_{\max} = 2y'_1 A'_1 + y'_2 A'_2 = 2(0.1)(0.2)(0.03)$

$+ (0.15)(0.14)(0.03) = 1.83(10^{-3}) \text{ m}^3$

$\tau_{\max} = \dfrac{VQ_{\max}}{It} = \dfrac{20(10^3)[1.83(10^{-3})]}{0.50963(10^{-3})[2(0.3)]} = 1.20 \text{ MPa}$

Ans.

F7–5 $I = \dfrac{1}{12}(0.05)(0.4)^3 + 2\left[\dfrac{1}{12}(0.025)(0.3)^3\right]$

$= 0.37917(10^3) \text{ m}^4$

$Q_{\max} = 2y'_1 A'_1 + y'_2 A'_2 = 2(0.075)(0.025)(0.15)$

$+ (0.1)(0.05)(0.2) = 1.5625(10^{-3}) \text{ m}^3$

$\tau_{\max} = \dfrac{VQ_{\max}}{It} = \dfrac{20(10^3)[1.5625(10^{-3})]}{[0.37917(10^{-3})][2(0.025)]}$

$= 1.65 \text{ MPa}$ *Ans.*

F7–6 $I = \dfrac{1}{12}(0.3)(0.2^3) = 0.2(10^{-3}) \text{ m}^4$

$Q = y'A' = 0.05(0.1)(0.3) = 1.5(10^{-3}) \text{ m}^3$

$q_{\text{allow}} = 2\left(\dfrac{F}{s}\right) = \dfrac{2[15(10^3)]}{s} = \dfrac{30(10^3)}{s}$

$q_{\text{allow}} = \dfrac{VQ}{I}; \quad \dfrac{30(10^3)}{s} = \dfrac{50(10^3)[1.5(10^{-3})]}{0.2(10^{-3})}$

$s = 0.08 \text{ m} = 80 \text{ mm}$ *Ans.*

F7–7 $I = \dfrac{1}{12}(0.3)(0.2^3) = 0.2(10^{-3})\text{ m}^4$

$Q = y'A' = 0.05(0.1)(0.3) = 1.5(10^{-3})\text{ m}^3$

$q_{\text{allow}} = 2\left(\dfrac{F}{s}\right) = \dfrac{2[15(10^3)]}{0.1} = 300(10^3)\text{ N/m}$

$q_{\text{allow}} = \dfrac{VQ}{I}; \qquad 300(10^3) = \dfrac{V[1.5(10^{-3})]}{0.2(10^{-3})}$

$\qquad\qquad\qquad V = 40(10^3)\text{ N} = 40\text{ kN} \qquad Ans.$

F7–8 $I = \dfrac{1}{12}(0.2)(0.34^3) - \dfrac{1}{12}(0.19)(0.28^3)$

$\qquad = 0.3075(10^{-3})\text{ m}^4$

$Q = y'A' = 0.16(0.02)(0.2) = 0.64(10^{-3})\text{ m}^3$

$q_{\text{allow}} = 2\left(\dfrac{F}{s}\right) = \dfrac{2[30(10^3)]}{s} = \dfrac{60(10^3)}{s}$

$q_{\text{allow}} = \dfrac{VQ}{I}; \qquad \dfrac{60(10^3)}{s} = \dfrac{300(10^3)[0.64(10^{-3})]}{0.3075(10^{-3})}$

$s = 0.09609\text{ m} = 96.1\text{ mm}$

Use $s = 96$ mm $\qquad Ans.$

F7–9

$I = 2\left[\dfrac{1}{12}(0.025)(0.3^3)\right] + 2\left[\dfrac{1}{12}(0.05)(0.2^3) + 0.05(0.2)(0.15^2)\right]$

$\qquad = 0.62917(10^{-3})\text{ m}^4$

$Q = y'A' = 0.15(0.2)(0.05) = 1.5(10^{-3})\text{ m}^3$

$q_{\text{allow}} = 2\left(\dfrac{F}{s}\right) = \dfrac{2[8(10^3)]}{s} = \dfrac{16(10^3)}{s}$

$q_{\text{allow}} = \dfrac{VQ}{I}; \qquad \dfrac{16(10^3)}{s} = \dfrac{20(10^3)[1.5(10^{-3})]}{0.62917(10^{-3})}$

$\qquad\qquad\qquad s = 0.3356\text{ m} = 335.56\text{ mm}$

Use $s = 335$ mm $\qquad Ans.$

F7–10 $I = \dfrac{1}{12}(25)(150^3) + 4\left[\dfrac{1}{12}(12)(100^3) + 12(100)(75^2)\right]$

$\qquad = 38\ 031\ 250\text{ mm}^4$

$Q = y'A' = 75(100)(12) = 90\ 000\text{ mm}^3$

$q_{\text{allow}} = \dfrac{F}{s} = \dfrac{30(10^3)}{s}$

$q_{\text{allow}} = \dfrac{VQ}{I}; \qquad \dfrac{30(10^3)}{s} = \dfrac{75(10^3)(90\ 000)}{38\ 031\ 250}$

$\qquad\qquad\qquad s = 169.03\text{ mm}$

Use $s = 170$ mm $\qquad Ans.$

Chapter 8

F8–1 $+\uparrow\Sigma F_z = (F_R)_z; \quad -500 - 300 = P$

$P = -800$ kN

$\Sigma M_x = 0; \qquad 300(0.05) - 500(0.1) = M_x$

$M_x = -35\text{ kN}\cdot\text{m}$

$\Sigma M_y = 0; \qquad 300(0.1) - 500(0.1) = M_y$

$M_y = -20\text{ kN}\cdot\text{m}$

$A = 0.3(0.3) = 0.09\text{ m}^2$

$I_x = I_y = \dfrac{1}{12}(0.3)(0.3^3) = 0.675(10^{-3})\text{ m}^4$

$\sigma_A = \dfrac{-800(10^3)}{0.09} + \dfrac{[20(10^3)](0.15)}{0.675(10^{-3})} + \dfrac{[35(10^3)](0.15)}{0.675(10^{-3})}$

$\qquad = 3.3333\text{ MPa} = 3.33\text{ MPa (T)} \qquad Ans.$

$\sigma_B = \dfrac{-800(10^3)}{0.09} + \dfrac{[20(10^3)](0.15)}{0.675(10^{-3})} - \dfrac{[35(10^3)](0.15)}{0.675(10^{-3})}$

$\qquad = -12.22\text{ MPa} = 12.2\text{ MPa (C)} \qquad Ans.$

F8–2 $+\uparrow\Sigma F_y = 0; \qquad V - 400 = 0 \qquad V = 400\text{ kN}$

$\downarrow+\Sigma M_A = 0; \ -M - 400(0.5) = 0 \ \ M = -200\text{ kN}\cdot\text{m}$

$I = \dfrac{1}{12}(0.1)(0.3^3) = 0.225(10^{-3})\text{ m}^4$

$Q_A = y'A' = 0.1(0.1)(0.1) = 1(10^{-3})\text{ m}^3$

$\sigma_A = \dfrac{My}{I} = \dfrac{[200(10^3)](-0.05)}{0.225(10^{-3})}$

$\qquad = -44.44\text{ MPa} = 44.4\text{ MPa (C)} \qquad Ans.$

$\tau_A = \dfrac{VQ}{It} = \dfrac{400(10^3)[1(10^{-3})]}{0.225(10^{-3})(0.1)} = 17.8\text{ MPa} \qquad Ans.$

F8–3   Left reaction is 20 kN.

Left segment:

$+\uparrow\Sigma F_y = 0; \qquad 20 - V = 0 \qquad V = 20\text{ kN}$

$\downarrow+\Sigma M_s = 0; \qquad M - 20(0.5) = 0 \ \ M = 10\text{ kN}\cdot\text{m}$

$I = \dfrac{1}{12}(0.1)(0.2^3) - \dfrac{1}{12}(0.09)(0.18^3)$

$\qquad = 22.9267(10^{-6})\text{ m}^4$

$Q_A = y_1'A_1' + y_2'A_2' = 0.07(0.04)(0.01)$

$\qquad + 0.095(0.1)(0.01) = 0.123(10^{-3})\text{ m}^3$

$\sigma_A = -\dfrac{My_A}{I} = -\dfrac{[10(10^3)](0.05)}{22.9267(10^{-6})}$

$\qquad = -21.81\text{ MPa} = 21.8\text{ MPa (C)} \qquad Ans.$

$\tau_A = \dfrac{VQ_A}{It} = \dfrac{20(10^3)[0.123(10^{-3})]}{[22.9267(10^{-6})](0.01)}$

$\qquad = 10.7\text{ MPa} \qquad Ans.$

F8–4   At the section through centroidal axis:

$N = P$

$V = 0$

$M = (50 + 25)P = 75P$

$\sigma = \dfrac{P}{A} + \dfrac{Mc}{I}$

$210 = \dfrac{P}{50(12)} = \dfrac{(75P)(25)}{\frac{1}{12}(12)(50)^3}$

$P = 12.6$ kN   *Ans.*

F8–5   At section through B:

$N = 1000$ N, $V = 800$ N

$M = 800(100) = 80\,000$ N · m

Axial load:

$\sigma_x = \dfrac{P}{A} = \dfrac{1000}{40(30)} = 0.833$ N/mm^2 (T)

Shear load:

$\tau_{xy} = \dfrac{VQ}{It} = \dfrac{800[(15)(30)(10)]}{[\frac{1}{12}(30)(40)^3]30} = 0.75$ N/mm^2

Bending moment:

$\sigma_x = \dfrac{My}{I} = \dfrac{(80\,000)(10)}{\frac{1}{12}(30)(40)^3} = 5.0$ N/mm^2 (C)

Thus

$\sigma_x = 0.833 - 5.0 = 4.167$ N/mm^2 (C)   *Ans.*

$\sigma_y = 0$ *Ans.*

$\tau_{xy} = 0.75$ N/mm^2 *Ans.*

F8–6   Top segment:

$\Sigma F_y = 0;$ $V_y + 1000 = 0$ $V_y = -1000$ N

$\Sigma F_x = 0;$ $V_x - 1500 = 0$ $V_x = 1500$ N

$\Sigma M_z = 0;$ $T_z - 1500(0.4) = 0$ $T_z = 600$ N · m

$\Sigma M_y = 0;$ $M_y - 1500(0.2) = 0$ $M_y = 300$ N · m

$\Sigma M_x = 0;$ $M_x - 1000(0.2) = 0$ $M_x = 200$ N · m

$I_y = I_x = \dfrac{\pi}{4}(0.02^4) = 40(10^{-9})\pi$ m^4

$J = \dfrac{\pi}{2}(0.02^4) = 80(10^{-9})\pi$ m^4

$(Q_y)_A = \dfrac{4(0.02)}{3\pi}\left[\dfrac{\pi}{2}(0.02^2)\right] = 5.3333(10^{-6})$ m^3

$\sigma_A = \dfrac{M_x y}{I_x} + \dfrac{M_y z}{I_y} = \dfrac{200(0)}{40(10^{-9})\pi} + \dfrac{300(0.02)}{40(10^{-9})\pi}$

$= 47.7$ MPa (T)   *Ans.*

$[(\tau_{zy})_T]_A = \dfrac{T_z c}{J} = \dfrac{600(0.02)}{80(10^{-9})\pi} = 47.746$ MPa

$[(\tau_{zy})_V]_A = \dfrac{V_y(Q_y)_A}{I_x t} = \dfrac{1000[5.3333(10^{-6})]}{[40(10^{-9})\pi](0.04)}$

$= 1.061$ MPa

Combining these two shear stress components,

$(\tau_{zy})_A = 47.746 + 1.061 = 48.8$ MPa   *Ans.*

F8–7   Right Segment:

$\Sigma F_z = 0;$ $V_z - 6 = 0$ $V_z = 6$ kN

$\Sigma M_y = 0;$ $T_y - 6(0.3) = 0$ $T_y = 1.8$ kN · m

$\Sigma M_x = 0;$ $M_x - 6(0.3) = 0$ $M_x = 1.8$ kN · m

$I_x = \dfrac{\pi}{4}(0.05^4 - 0.04^4) = 0.9225(10^{-6})\pi$ m^4

$J = \dfrac{\pi}{2}(0.05^4 - 0.04^4) = 1.845(10^{-6})\pi$ m^4

$(Q_z)_A = y_2' A_2' - y_1' A_1'$

$= \dfrac{4(0.05)}{3\pi}\left[\dfrac{\pi}{2}(0.05^2)\right] - \dfrac{4(0.04)}{3\pi}\left[\dfrac{\pi}{2}(0.04^2)\right]$

$= 40.6667(10^{-6})$ m^3

$\sigma_A = \dfrac{M_x z}{I_x} = \dfrac{1.8(10^3)}{0.9225(10^{-6})\pi} = 0$ *Ans.*

$[(\tau_{yz})_T]_A = \dfrac{T_y c}{J} = \dfrac{[1.8(10^3)](0.05)}{1.845(10^{-6})\pi} = 15.53$ MPa

$[(\tau_{yz})_V]_A = \dfrac{V_z(Q_z)_A}{I_x t} = \dfrac{6(10^3)[40.6667(10^{-6})]}{[0.9225(10^{-6})\pi](0.02)}$

$= 4.210$ MPa

Combining these two shear stress components,

$(\tau_{yz})_A = 15.53 - 4.210 = 11.3$ MPa   *Ans.*

F8–8   Left Segment:

$\Sigma F_z = 0;$ $V_z - 900 - 300 = 0$ $V_z = 1200$ N

$\Sigma M_y = 0;$ $T_y + 300(0.1) - 900(0.1) = 0$ $T_y = 60$ N · m

$\Sigma M_x = 0;$ $M_x + (900 + 300)0.3 = 0$ $M_x = -360$ N · m

$I_x = \dfrac{\pi}{4}(0.025^4 - 0.02^4) = 57.65625(10^{-9})\pi$ m^4

$J = \dfrac{\pi}{2}(0.025^4 - 0.02^4) = 0.1153125(10^{-6})\pi$ m^4

$(Q_y)_A = 0$

$\sigma_A = \dfrac{M_x y}{I_x} = \dfrac{(360)(0.025)}{57.65625(10^{-9})\pi} = 49.7$ MPa   *Ans.*

$[(\tau_{xy})_T]_A = \dfrac{T_y \rho_A}{J} = \dfrac{60(0.025)}{0.1153125(10^{-6})\pi} = 4.14$ MPa *Ans.*

$[(\tau_{yz})_V]_A = \dfrac{V_z(Q_z)_A}{I_x t} = 0$ *Ans.*

Chapter 9

F9–1 $\theta = 120°$ $\sigma_x = 500$ kPa $\sigma_y = 0$ $\tau_{xy} = 0$
Apply Eqs. 9–1, 9–2.
$\sigma_{x'} = 125$ kPa   *Ans.*
$\tau_{x'y'} = 217$ kPa   *Ans.*

F9–2 $\theta = -45°$ $\sigma_x = 0$ $\sigma_y = -400$ kPa
$\tau_{xy} = -300$ kPa
Apply Eqs. 9–1, 9–3, 9–2.
$\sigma_{x'} = 100$ kPa   *Ans.*
$\sigma_{y'} = -500$ kPa   *Ans.*
$\tau_{x'y'} = 200$ kPa   *Ans.*

F9–3 $\sigma_x = 80$ MPa $\sigma_y = 0$ $\tau_{xy} = 30$ MPa
Apply Eqs. 9–5, 9–4.
$\sigma_1 = 90$ MPa $\sigma_2 = -10$ MPa   *Ans.*
$\theta_p = 18.43°$ and $108.43°$
From Eq. 9–1,
$$\sigma_{x'} = \frac{80 + 0}{2} + \frac{80 - 0}{2}\cos 2(18.43°)$$
$$+ 30\sin 2(18.43°)$$
$$= 90 \text{ MPa} = \sigma_1$$
Thus,
$(\theta_p)_1 = 18.4°$ and $(\theta_p)_2 = 108°$ *Ans.*

F9–4 $\sigma_x = 100$ kPa $\sigma_y = 700$ kPa
$\tau_{xy} = -400$ kPa
Apply Eqs. 9–7, 9–8.
$\tau_{\max}$ in-plane $= 500$ kPa   *Ans.*
$\sigma_{\text{avg}} = 400$ kPa   *Ans.*

F9–5 At the cross section through B:
$N = 4$ kN $V = 2$ kN
$M = 2(2) = 4$ kN·m
$$\sigma_B = \frac{P}{A} + \frac{Mc}{I} = \frac{4(10^3)}{0.03(0.06)} + \frac{4(10^3)(0.03)}{\frac{1}{12}(0.03)(0.06)^3}$$
$$= 224 \text{ MPa (T)}$$
Note $\tau_B = 0$ since $Q = 0$.
Thus
$\sigma_1 = 224$ MPa   *Ans.*
$\sigma_2 = 0$

F9–6 $A_y = B_y = 12$ kN
Segment AC:
$V_C = 0$ $M_C = 24$ kN·m
$\tau_C = 0$ (since $V_C = 0$)
$\sigma_C = 0$ (since C is on neutral axis)
$\sigma_1 = \sigma_2 = 0$ *Ans.*

F9–7 $$\sigma_{\text{avg}} = \frac{\sigma_x + \sigma_y}{2} = \frac{500 + 0}{2} = 250 \text{ kPa}$$
The coordinates of the center C of the circle and the reference point A are
$$A(500, 0) \qquad C(250, 0)$$
$R = CA = 500 - 250 = 250$ kPa
$\theta = 120°$ (counterclockwise). Rotate the radial line CA counterclockwise $2\theta = 240°$ to the coordinates of point $P(\sigma_{x'}, \tau_{x'y'})$.
$\alpha = 240° - 180° = 60°$
$\sigma_{x'} = 250 - 250\cos 60° = 125$ kPa   *Ans.*
$\tau_{x'y'} = 250\sin 60° = 217$ kPa   *Ans.*

F9–8 $$\sigma_{\text{avg}} = \frac{\sigma_x + \sigma_y}{2} = \frac{80 + 0}{2} = 40 \text{ kPa}$$
The coordinates of the center C of the circle and the reference point A are
$$A(80, 30) \qquad C(40, 0)$$
$R = CA = \sqrt{(80 - 40)^2 + 30^2} = 50$ MPa
$\sigma_1 = 40 + 50 = 90$ MPa   *Ans.*
$\sigma_2 = 40 - 50 = -10$ MPa   *Ans.*
$$\tan 2(\theta_p)_1 = \frac{30}{80 - 40} = 0.75$$
$(\theta_p)_1 = 18.4°$ (counterclockwise)   *Ans.*

F9–9 $J = \dfrac{\pi}{2}(0.04^4 - 0.03^4) = 0.875(10^{-6})\pi$ m^4
$$\tau = \frac{Tc}{J} = \frac{4(10^3)(0.04)}{0.875(10^{-6})\pi} = 58.21 \text{ MPa}$$
$\sigma_x = \sigma_y = 0$ and $\tau_{xy} = -58.21$ MPa
$$\sigma_{\text{avg}} = \frac{\sigma_x + \sigma_y}{2} = 0$$
The coordinates of the reference point A and the center C of the circle are
$$A(0, -58.21) \qquad C(0, 0)$$
$R = CA = 58.21$ MPa
$\sigma_1 = 0 + 58.21 = 58.2$ MPa   *Ans.*
$\sigma_2 = 0 - 58.21 = -58.2$ MPa   *Ans.*

F9–10
$+\uparrow \Sigma F_y = 0;$ $V - 30 = 0$ $V = 30$ kN
$\downarrow+\Sigma M_O = 0;$ $-M - 30(0.3) = 0$ $M = -9$ kN·m

$$I = \frac{1}{12}(0.05)(0.15^3) = 14.0625(10^{-6}) \text{ m}^4$$

$$Q_A = y'A' = 0.05(0.05)(0.05) = 0.125(10^{-3}) \text{ m}^3$$

$$\sigma_A = -\frac{My_A}{I} = \frac{[-9(10^3)](0.025)}{14.0625(10^{-6})} = 16 \text{ MPa (T)}$$

$$\tau_A = \frac{VQ_A}{It} = \frac{30(10^3)[0.125(10^{-3})]}{14.0625(10^{-6})(0.05)} = 5.333 \text{ MPa}$$

$\sigma_x = 16 \text{ MPa}, \sigma_y = 0, \text{ and } \tau_{xy} = -5.333 \text{ MPa}$

$$\sigma_{avg} = \frac{\sigma_x + \sigma_y}{2} = \frac{16 + 0}{2} = 8 \text{ MPa}$$

The coordinates of the reference point A and the center C of the circle are

$$A\ (16, -5.333) \qquad C(8, 0)$$

$$R = CA = \sqrt{(16 - 8)^2 + (-5.333)^2} = 9.615 \text{ MPa}$$

$$\sigma_1 = 8 + 9.615 = 17.6 \text{ MPa} \qquad\qquad Ans.$$

$$\sigma_2 = 8 - 9.615 = -1.61 \text{ MPa} \qquad\qquad Ans.$$

F9–11

$$\downarrow + \Sigma M_B = 0; \qquad 60(1) - A_y(1.5) = 0 \qquad A_y = 40 \text{ kN}$$

$$+\uparrow \Sigma F_y = 0; \qquad 40 - V = 0 \qquad\qquad V = 40 \text{ kN}$$

$$\downarrow + \Sigma M_O = 0; \qquad M - 40(0.5) = 0 \qquad M = 20 \text{ kN} \cdot \text{m}$$

$$I = \frac{1}{12}(0.1)(0.2^3) - \frac{1}{12}(0.09)(0.18^3) = 22.9267(10^{-6}) \text{ m}^4$$

$$Q_A = y'A' = 0.095(0.01)(0.1) = 95(10^{-6}) \text{ m}^3$$

$$\sigma_A = -\frac{My_A}{I} = -\frac{[20(10^3)](0.09)}{22.9267(10^{-6})} = -78.51 \text{ MPa}$$

$$= 78.51 \text{ MPa (C)}$$

$$\tau_A = \frac{VQ_A}{It} = \frac{40(10^3)[95(10^{-6})]}{[22.9267(10^{-6})](0.01)} = 16.57 \text{ MPa}$$

$\sigma_x = -78.51 \text{ MPa}, \sigma_y = 0, \text{ and } \tau_{xy} = -16.57 \text{ MPa}$

$$\sigma_{avg} = \frac{\sigma_x + \sigma_y}{2} = \frac{-78.51 + 0}{2} = -39.26 \text{ MPa}$$

The coordinates of the reference point A and the center C of the circle are

$$A(-78.51, -16.57) \quad C(-39.26, 0)$$

$$R = CA = \sqrt{[-78.51 - (-39.26)]^2 + (-16.57)^2}$$

$$= 42.61 \text{ MPa}$$

$$\tau_{max}^{\text{in-plane}} = |R| = 42.6 \text{ MPa} \qquad\qquad Ans.$$

Chapter 11

F11–1

$$V_{max} = 12 \text{ kN} \qquad M_{max} = 18 \text{ kN} \cdot \text{m}$$

$$\sigma_{allow} = \frac{M_{max}c}{I}; \qquad 10(10^6) = \frac{18(10^3)(a)}{\frac{2}{3}a^4}$$

$$a = 0.1392 \text{ m} = 139.2 \text{ mm}$$

Use $a = 140 \text{ mm}$ \qquad\qquad Ans.

$$I = \frac{2}{3}(0.14^4) = 0.2561(10^{-3}) \text{ m}^4$$

$$Q_{max} = \frac{0.14}{2}(0.14)(0.14) = 1.372(10^{-3}) \text{ m}^3$$

$$\tau_{max} = \frac{V_{max}Q_{max}}{It} = \frac{12(10^3)[1.372(10^{-3})]}{[0.2561(10^{-3})](0.14)}$$

$$= 0.459 \text{ MPa} < \tau_{allow} = 1 \text{ MPa (OK)}$$

F11–2

$$V_{max} = 15 \text{ kN} \qquad M_{max} = 20 \text{ kN} \cdot \text{m}$$

$$I = \frac{\pi}{4}\left(\frac{d}{2}\right)^4 = \frac{\pi d^4}{64}$$

$$\sigma_{allow} = \frac{M_{max}c}{I}; \qquad 100 = \frac{20(10^6)\left(\frac{d}{2}\right)}{\frac{\pi d^4}{64}}$$

$$d = 126.8 \text{ mm}$$

Use $d = 127 \text{ mm}$ \qquad\qquad Ans.

$$I = \frac{\pi}{64}(127^4) = 12.77(10^6) \text{ mm}^4$$

$$Q_{max} = \frac{4(127/2)}{3\pi}\left[\frac{1}{2}\left(\frac{\pi}{4}\right)(127^2)\right] = 170\ 699 \text{ mm}^3$$

$$\tau_{max} = \frac{V_{max}Q_{max}}{It} = \frac{15(10^3)(170\ 699)}{12.77(10^6)(127)}$$

$$= 1.58 \text{ N/mm}^2 < \tau_{allow} = 50 \text{ N/mm}^2 \text{ (OK)}$$

F11–3

$$V_{max} = 10 \text{ kN} \qquad\qquad M_{max} = 5 \text{ kN} \cdot \text{m}$$

$$I = \frac{1}{12}(a)(2a)^3 = \frac{2}{3}a^4$$

$$\sigma_{allow} = \frac{M_{max}c}{I}; \qquad 12(10^6) = \frac{5(10^3)(a)}{\frac{2}{3}a^4}$$

$$a = 0.0855 \text{ m} = 85.5 \text{ mm}$$

Use $a = 86 \text{ mm}$ \qquad\qquad Ans.

$$I = \frac{2}{3}(0.086^4) = 36.4672(10^{-6}) \text{ m}^4$$

$$Q_{max} = \frac{0.086}{2}(0.086)(0.086)$$

$$= 0.318028(10^{-3}) \text{ m}^3$$

$$\tau_{max} = \frac{V_{max}Q_{max}}{It} = \frac{10(10^3)[0.318028(10^{-3})]}{[36.4672(10^{-6})](0.086)}$$

$$= 1.01 \text{ MPa} < \tau_{allow} = 1.5 \text{ MPa (OK)}$$

F11–4

$$V_{max} = 25 \text{ kN} \qquad M_{max} = 12.5 \text{ kN} \cdot \text{m}$$

$$I = \frac{1}{12}(100)(h^3) = \frac{h^3}{0.12}$$

$$\sigma_{allow} = \frac{M_{max}c}{I}; \qquad 15 = \frac{12.5(10^6)\left(\frac{h}{3}\right)}{\frac{h^3}{0.12}}$$

$$h = 223.6 \text{ mm}$$

$$Q_{max} = y'A' = \frac{h}{4}\left(\frac{h}{2}\right)(100) = \frac{h^2}{0.08}$$

$$\tau_{max} = \frac{V_{max}Q_{max}}{It}; \quad 1.5 = \frac{25(10^3)\left(\frac{h^2}{0.08}\right)}{\frac{h^3}{0.12}(100)}$$

$$h = 250 \text{ mm (controls)}$$

Use $h = 250$ mm   *Ans.*

F11–5

$$V_{max} = 25 \text{ kN} \qquad M_{max} = 20 \text{ kN} \cdot \text{m}$$

$$I = \frac{1}{12}(b)(3b)^3 = 2.25b^4$$

$$\sigma_{allow} = \frac{M_{max}c}{I}; \qquad 12(10^6) = \frac{20(10^3)(1.5b)}{2.25b^4}$$

$$b = 0.1036 \text{ m} = 103.6 \text{ mm}$$

Use $b = 104$ mm   *Ans.*

$$I = 2.25(0.104^4) = 0.2632(10^{-3}) \text{ m}^4$$

$$Q_{max} = 0.75(0.104)[1.5(0.104)(0.104)] = 1.2655(10^{-3}) \text{ m}^3$$

$$\tau_{max} = \frac{V_{max}Q_{max}}{It} = \frac{25(10^3)[1.2655(10^{-3})]}{[0.2632(10^{-3})](0.104)}$$

$$= 1.156 \text{ MPa} < \tau_{allow} = 1.5 \text{ MPa (OK)}.$$

F11–6

$$V_{max} = 150 \text{ kN} \qquad M_{max} = 150 \text{ kN} \cdot \text{m}$$

$$S_{req'd} = \frac{M_{max}}{\sigma_{allow}} = \frac{150(10^3)}{150(10^6)} = 0.001 \text{ m}^3 = 1000(10^3) \text{ mm}^3$$

Select W410 × 67 [$S_x = 1200(10^3)$ mm^3, $d = 410$ mm, and $t_w = 8.76$ mm].   *Ans.*

$$\tau_{max} = \frac{V}{t_w d} = \frac{150(10^3)}{0.00876(0.41)}$$

$$= 41.76 \text{ MPa} < \tau_{allow} = 75 \text{ MPa (OK)}$$

Chapter 12

F12–1

$$\zeta + \Sigma M_O = 0; \qquad M(x) = 30 \text{ kN} \cdot \text{m}$$

$$EI\frac{d^2v}{dx^2} = 30$$

$$EI\frac{dv}{dx} = 30x + C_1$$

$$EIv = 15x^2 + C_1x + C_2$$

At $x = 3$ m, $\frac{dv}{dx} = 0$.

$$C_1 = -90 \text{ kN} \cdot \text{m}^2$$

At $x = 3$ m, $v = 0$.

$$C_2 = 135 \text{ kN} \cdot \text{m}^3$$

$$\frac{dv}{dx} = \frac{1}{EI}(30x - 90)$$

$$v = \frac{1}{EI}(15x^2 - 90x + 135)$$

For end $A, x = 0$

$$\theta_A = \frac{dv}{dx}\bigg|_{x=0} = -\frac{90(10^3)}{200(10^9)[65.0(10^{-6})]} = -0.00692 \text{ rad}$$

Ans.

$$v_A = v|_{x=0} = \frac{135(10^3)}{200(10^9)[65.0(10^{-6})]} = 0.01038 \text{ m} = 10.4 \text{ mm}$$

Ans.

F12–2

$$\zeta + \Sigma M_O = 0; \qquad M(x) = (-10x - 10) \text{ kN} \cdot \text{m}$$

$$EI\frac{d^2x}{dx^2} = -10x - 10$$

$$EI\frac{dv}{dx} = -5x^2 - 10x + C_1$$

$$EIv = -\frac{5}{3}x^3 - 5x^2 + C_1x + C_2$$

At $x = 3$ m, $\frac{dv}{dx} = 0$.

$$EI(0) = -5(3^2) - 10(3) + C_1 \qquad C_1 = 75 \text{ kN} \cdot \text{m}^2$$

At $x = 3$ m, $v = 0$.

$$EI(0) = -\frac{5}{3}(3^3) - 5(3^2) + 75(3) + C_2 \quad C_2 = -135 \text{ kN} \cdot \text{m}^3$$

$$\frac{dv}{dx} = \frac{1}{EI}(-5x^2 - 10x + 75)$$

$$v = \frac{1}{EI}\left(-\frac{5}{3}x^3 - 5x^2 + 75x - 135\right)$$

For end $A, x = 0$

$$\theta_A = \frac{dv}{dx}\bigg|_{x=0} = \frac{1}{EI}[-5(0) - 10(0) + 75]$$

$$= \frac{75(10^3)}{200(10^9)[65.0(10^{-6})]} = 0.00577 \text{ rad} \qquad Ans.$$

$$v_A = v|_{x=0} = \frac{1}{EI}\left[-\frac{5}{3}(0^3) - 5(0^2) + 75(0) - 135\right]$$

$$= -\frac{135(10^3)}{200(10^9)[65.0(10^{-6})]} = -0.01038 \text{ m} = -10.4 \text{ mm}$$

$$Ans.$$

F12–3

$$\downarrow + \Sigma M_O = 0; \qquad M(x) = \left(-\frac{3}{2}x^2 - 10x\right) \text{kN} \cdot \text{m}$$

$$EI\frac{d^2x}{dx^2} = -\frac{3}{2}x^2 - 10x$$

$$EI\frac{dv}{dx} = -\frac{1}{2}x^3 - 5x^2 + C_1$$

At $x = 3$ m, $\dfrac{dv}{dx} = 0$.

$$EI(0) = -\frac{1}{2}(3^3) - 5(3^2) + C_1 \qquad C_1 = 58.5 \text{ kN} \cdot \text{m}^2$$

$$\frac{dv}{dx} = \frac{1}{EI}\left(-\frac{1}{2}x^3 - 5x^2 + 58.5\right)$$

For end $A, x = 0$

$$\theta_A = \frac{dv}{dx}\bigg|_{x=0} = \frac{58.5(10^3)}{200(10^9)[65.0(10^{-6})]} = 0.0045 \text{ rad} \qquad Ans.$$

F12–4

$$A_y = 3000 \text{ N}$$

$$\downarrow + \Sigma M_O = 0; \qquad M(x) = (3000x - 1000x^2) \text{ N} \cdot \text{m}$$

$$EI\frac{d^2x}{dx^2} = 3000x - 1000x^2$$

$$EI\frac{dv}{dx} = 1500x^2 - 333.33x^3 + C_1$$

$$EIv = 500x^3 - 83.333x^4 + C_1x + C_2$$

At $x = 0, v = 0$.

$$EI(0) = 500(0^3) - 83.333(0^4) + C_1(0) + C_2 \qquad C_2 = 0$$

At $x = 3$ m, $v = 0$.

$$EI(0) = 500(3^3) - 83.333(3^4) + C_1(3)$$

$$C_1 = -2250 \text{ N} \cdot \text{m}^2$$

$$\frac{dv}{dx} = \frac{1}{EI}(1500x^2 - 333.33x^3 - 2250)$$

$$v = \frac{1}{EI}(500x^3 - 83.333x^4 - 2250)$$

v_{max} occurs where $\dfrac{dv}{dx} = 0$.

$$1500x^2 - 333.33x^3 - 2250 = 0$$

$$x = 1.5 \text{ m} \qquad Ans.$$

$$v = \frac{1}{EI}[500(1.5^3) - 83.333(1.5^4) - 2250(1.5)]$$

$$= \frac{-2109.373}{10(10^9)\left[\frac{1}{12}(0.060)(0.125^3)\right]} = -0.0216 \text{ m} = -21.6 \text{ mm}$$

$$Ans.$$

F12–5

$$\downarrow + \Sigma M_O = 0; \qquad M(x) = (40 - 5x) \text{ kN} \cdot \text{m}$$

$$EI\frac{d^2x}{dx^2} = 40 - 5x$$

$$EI\frac{dv}{dx} = 40x - 2.5x^2 + C_1$$

$$EIv = 20x^2 - 0.8333x^3 + C_1x + C_2$$

At $x = 0, v = 0$.

$$EI(0) = 20(0^2) - 0.8333(0^3) + C_1(0) + C_2 \qquad C_2 = 0$$

At $x = 6$ m, $v = 0$.

$$EI(0) = 20(6^2) - 0.8333(6^3) + C_1(6) + 0$$

$$C_1 = -90 \text{ kN} \cdot \text{m}^2$$

$$\frac{dv}{dx} = \frac{1}{EI}(40x - 2.5x^2 - 90)$$

$$v = \frac{1}{EI}(20x^2 - 0.8333x^3 - 90x)$$

v_{max} occurs where $\dfrac{dv}{dx} = 0$.

$$40x - 2.5x^2 - 90 = 0$$

$$x = 2.7085 \text{ m}$$

$$v = \frac{1}{EI}[20(2.7085^2) - 0.83333(2.7085^3) - 90(2.7085)]$$

$$= -\frac{113.60(10^3)}{200(10^9)[39.9(10^{-6})]} = -0.01424 \text{ m} = -14.2 \text{ mm}$$

$$Ans.$$

F12–6

$$\downarrow + \Sigma M_O = 0; \qquad M(x) = (10x + 10) \text{ kN} \cdot \text{m}$$

$$EI\frac{d^2x}{dx^2} = 10x + 10$$

$$EI\frac{dv}{dx} = 5x^2 + 10x + C_1$$

Due to symmetry, $\dfrac{dv}{dx} = 0$ at $x = 3$ m.

$EI(0) = 5(3^2) + 10(3) + C_1$ $C_1 = -75$ kN·m^2

$\dfrac{dv}{dx} = \dfrac{1}{EI}[5x^2 + 10x - 75]$

At $x = 0$,

$\dfrac{dv}{dx} = \dfrac{-75(10^3)}{200(10^9)(39.9(10^{-6}))} = -9.40(10^{-3})$ rad   *Ans.*

F12–7

Since B is a fixed support, $\theta_B = 0$.

$\theta_A = |\theta_{A/B}| = \dfrac{1}{2}\left(\dfrac{38}{EI} + \dfrac{20}{EI}\right)(3) = \dfrac{87 \text{ kN·m}^2}{EI}$

$= \dfrac{87(10^3)}{200(10^9)[65(10^{-6})]} = 0.00669$ rad   *Ans.*

$\Delta_A = |t_{A/B}| = (1.5)\left[\dfrac{20}{EI}(3)\right] + 2\left[\dfrac{1}{2}\left(\dfrac{18}{EI}\right)(3)\right]$

$= \dfrac{144(10^3)}{200(10^9)[65(10^{-6})]} = 0.01108$ m $= 11.1$ mm   *Ans.*

F12–8

Since B is a fixed support, $\theta_B = 0$.

$\theta_A = |\theta_{A/B}| = \dfrac{1}{2}\left(\dfrac{50}{EI} + \dfrac{20}{EI}\right)(1) + \dfrac{1}{2}\left(\dfrac{20}{EI}\right)(1) = \dfrac{45 \text{ kN·m}^2}{EI}$

$= \dfrac{45(10^3)}{200(10^9)[126(10^{-6})]} = 0.00179$ rad   *Ans.*

$\Delta_A = |t_{A/B}| =$

$(1.6667)\left[\dfrac{1}{2}\left(\dfrac{30}{EI}\right)(1)\right] + 1.5\left[\dfrac{20}{EI}(1)\right] + 0.6667\left[\dfrac{1}{2}\left(\dfrac{20}{EI}\right)(1)\right]$

$= \dfrac{61.667 \text{ kN·m}^3}{EI} = \dfrac{61.667(10^3)}{200(10^9)[126(10^{-6})]}$

$= 0.002447$ m $= 2.48$ mm   *Ans.*

F12–9

Since B is a fixed support, $\theta_B = 0$.

$\theta_A = |\theta_{A/B}| = \dfrac{1}{2}\left[\dfrac{60}{EI}(1)\right] + \dfrac{30}{EI}(2) = \dfrac{90 \text{ kN·m}^2}{EI}$

$= \dfrac{90(10^3)}{200(10^9)[121(10^{-6})]} = 0.00372$ rad   *Ans.*

$\Delta_A = |t_{A/B}| = 1.6667\left[\dfrac{1}{2}\left(\dfrac{60}{EI}\right)(1)\right] + (1)\left[\dfrac{30}{EI}(2)\right]$

$= \dfrac{110 \text{ kN·m}^3}{EI}$

$= \dfrac{110(10^3)}{200(10^9)[121(10^{-6})]} = 0.004545$ m $= 4.55$ mm   *Ans.*

F12–10

Since B is a fixed support, $\theta_B = 0$.

$\theta_A = |\theta_{A/B}| = \dfrac{1}{2}\left(\dfrac{20}{EI}\right)(2) + \dfrac{1}{3}\left(\dfrac{10}{EI}\right)(1) = \dfrac{23.333 \text{ kN·m}^2}{EI}$

$= \dfrac{23.333}{200(10^6)[10(10^{-6})]} = 0.01167$ rad   *Ans.*

$\Delta_A = |t_{A/B}| = \dfrac{4}{3}\left[\dfrac{1}{2}\left(\dfrac{20}{EI}\right)(2)\right] + (1 + 0.75)\left[\dfrac{1}{3}\left(\dfrac{10}{EI}\right)(1)\right]$

$= \dfrac{32.5 \text{ kN·m}^3}{EI} = \dfrac{32.5}{200(10^6)[10(10^{-6})]} = 0.01625$ m

$= 16.25$ mm   *Ans.*

F12–11

Due to symmetry, the slope at the midspan of the beam (point C) is zero, i.e., $\theta_C = 0$.

$\Delta_{\max} = \Delta_C = |t_{A/C}| = (2)\left[\dfrac{1}{2}\left(\dfrac{30}{EI}\right)(3)\right] + 1.5\left[\dfrac{10}{EI}(3)\right]$

$= \dfrac{135 \text{ kN·m}^3}{EI}$

$= \dfrac{135(10^3)}{200(10^9)[42.8(10^{-6})]} = 0.0158$ m $= 15.8$ mm $\downarrow$ *Ans.*

F12–12

$t_{A/B} = 2\left[\dfrac{1}{2}\left(\dfrac{30}{EI}\right)(6)\right] + 3\left[\dfrac{10}{EI}(6)\right] = \dfrac{360}{EI}$

$\theta_B = \dfrac{|t_{A/B}|}{L} = \dfrac{\frac{360}{EI}}{6} = \dfrac{60}{EI}$

The maximum deflection occurs at point C where the slope of the elastic curve is zero.

$\theta_B = \theta_{B/C}$

$\dfrac{60}{EI} = \left(\dfrac{10}{EI}\right)x + \dfrac{1}{2}\left(\dfrac{5x}{EI}\right)x$

$2.5x^2 + 10x - 60 = 0$

$x = 3.2915$ m

$\Delta_{\max} = |t_{B/C}| =$

$\dfrac{2}{3}(3.2915)\left\{\dfrac{1}{2}\left[\dfrac{5(3.2915)}{EI}\right](3.2915)\right\} + \dfrac{1}{2}(3.2915)\left[\dfrac{10}{EI}(3.2915)\right]$

$= \dfrac{113.60 \text{ kN·m}^3}{EI}$

$= \dfrac{113.60(10^3)}{200(10^9)[39.9(10^{-6})]} = 0.01424$ m $= 14.2$ mm $\downarrow$ *Ans.*

F12–13

$$(v_B)_1 = \frac{Px^2}{6EI}(3L - x) = \frac{40(4^2)}{6EI}[3(6) - 4] = \frac{1493.33}{EI}\downarrow$$

$$(v_B)_2 = \frac{PL^3}{3EI} = \frac{B_y(4^3)}{3EI} = \frac{21.33B_y}{EI}\uparrow$$

$$(+\uparrow)v_B = 0 = (v_B)_1 + (v_B)_2$$

$$0 = -\frac{1493.33}{EI} + \frac{21.33B_y}{EI}$$

$$B_y = 70 \text{ kN} \qquad\qquad Ans.$$

$$\overset{+}{\to} \Sigma F_x = 0; \qquad A_x = 0 \qquad\qquad Ans.$$

$$+\uparrow\Sigma F_y = 0; \qquad 70 - 40 - A_y = 0 \quad A_y = 30 \text{ kN} \qquad Ans.$$

$$\downarrow+\Sigma M_A = 0; \qquad 70(4) - 40(6) - M_A = 0$$

$$M_A = 40 \text{ kN} \cdot \text{m} \qquad\qquad Ans.$$

F12–14

To use the deflection tables, consider loading as a superposition of uniform distributed load minus a triangular load.

$$(v_B)_1 = \frac{w_0L^4}{8EI}\downarrow \quad (v_B)_2 = \frac{w_0L^4}{30EI}\uparrow \quad (v_B)_3 = \frac{B_yL^3}{3EI}\uparrow$$

$$(+\uparrow) \quad v_B = 0 = (v_B)_1 + (v_B)_2 + (v_B)_3$$

$$0 = -\frac{w_0L^4}{8EI} + \frac{w_0L^4}{30EI} + \frac{B_yL^3}{3EI}$$

$$B_y = \frac{11w_0L}{40} \qquad\qquad Ans.$$

$$\overset{+}{\to} \Sigma F_x = 0; \qquad A_x = 0 \qquad\qquad Ans.$$

$$+\uparrow\Sigma F_y = 0; \qquad A_y + \frac{11w_0L}{40} - \frac{1}{2}w_0L = 0$$

$$A_y = \frac{9w_0L}{40} \qquad\qquad Ans.$$

$$\downarrow+\Sigma M_A = 0; \qquad M_A + \frac{11w_0L}{40}(L) - \frac{1}{2}w_0L\left(\frac{2}{3}L\right) = 0$$

$$M_A = \frac{7w_0L^2}{120} \qquad\qquad Ans.$$

F12–15

$$(v_B)_1 = \frac{wL^4}{8EI} = \frac{[10(10^3)](6^4)}{8[200(10^9)][65.0(10^{-6})]} = 0.12461 \text{ m}\downarrow$$

$$(v_B)_2 = \frac{B_yL^3}{3EI} = \frac{B_y(6^3)}{3[200(10^9)][65.0(10^{-6})]} = 5.5385(10^{-6})B_y\uparrow$$

$$(+\downarrow) \quad v_B = (v_B)_1 + (v_B)_2$$

$$0.002 = 0.12461 - 5.5385(10^{-6})B_y$$

$$B_y = 22.14(10^3) \text{ N} = 22.1 \text{ kN} \qquad Ans.$$

$$\overset{+}{\to} \Sigma F_x = 0; \qquad A_x = 0 \qquad\qquad Ans.$$

$$+\uparrow\Sigma F_y = 0; \qquad A_y + 22.14 - 10(6) = 0 \quad A_y = 37.9 \text{ kN}$$
$$\qquad\qquad Ans.$$

$$\downarrow+\Sigma M_A = 0; \qquad M_A + 22.14(6) - 10(6)(3) = 0$$

$$M_A = 47.2 \text{ kN} \cdot \text{m} \qquad\qquad Ans.$$

F12–16

$$(v_B)_1 = \frac{M_OL}{6EI(2L)}[(2L)^2 - L^2] = \frac{M_OL^2}{4EI}\downarrow$$

$$(v_B)_2 = \frac{B_y(2L)^3}{48EI} = \frac{B_yL^3}{6EI}\uparrow$$

$$(+\uparrow) \quad v_B = 0 = (v_B)_1 + (v_B)_2$$

$$0 = -\frac{M_OL^2}{4EI} + \frac{B_yL^3}{6EI}$$

$$B_y = \frac{3M_O}{2L} \qquad\qquad Ans.$$

F12–17

$$(v_B)_1 = \frac{Pbx}{6EIL}(L^2 - b^2 - x^2) = \frac{50(4)(6)}{6EI(12)}(12^2 - 4^2 - 6^2)$$

$$= \frac{1533.3 \text{ kN} \cdot \text{m}^3}{EI}\downarrow$$

$$(v_B)_2 = \frac{B_yL^3}{48EI} = \frac{B_y(12^3)}{48EI} = \frac{36B_y}{EI}\uparrow$$

$$(+\uparrow) \quad v_B = 0 = (v_B)_1 + (v_B)_2$$

$$0 = -\frac{1533.3 \text{ kN} \cdot \text{m}^3}{EI} + \frac{36B_y}{EI}$$

$$B_y = 42.6 \text{ kN} \qquad\qquad Ans.$$

F12–18

$$(v_B)_1 = \frac{5wL^4}{384EI} = \frac{5[10(10^3)](12^4)}{384[200(10^9)][65.0(10^{-6})]} = 0.20769\downarrow$$

$$(v_B)_2 = \frac{B_yL^3}{48EI} = \frac{B_y(12^3)}{48[200(10^9)][65.0(10^{-6})]} = 2.7692(10^{-6})B_y\uparrow$$

$$(+\uparrow) \quad v_B = (v_B)_1 + (v_B)_2$$

$$0.005 = 0.20769 - 2.7692(10^{-6})B_y$$

$$B_y = 73.19(10^3) \text{ N} = 73.2 \text{ kN} \qquad Ans.$$

Chapter 13

F13–1

$$P = \frac{\pi^2EI}{(KL)^2} = \frac{\pi^2(200)[\frac{\pi}{4}(12.5)^4]}{[0.5(1250)]^2} = 96.89 \text{ kN} \qquad Ans.$$

$$\sigma = \frac{P}{A} = \frac{96.89(10^3)}{\pi(12.5)^2} = 197.4 \text{ MPa} < \sigma_Y \qquad OK$$

F13–2

$$P = \frac{\pi^2 EI}{(KL)^2} = \frac{\pi^2 (12)[\frac{1}{12}(100)(50)^3]}{[1(3600)]^2}$$

$$= 9.52 \text{ kN} \qquad\qquad Ans.$$

F13–3

For buckling about the x axis, $K_x = 1$ and $L_x = 12$ m.

$$P_{cr} = \frac{\pi^2 EI_x}{(K_x L_x)^2} = \frac{\pi^2 [200(10^9)][87.3(10^{-6})]}{[1(12)]^2} = 1.197(10^6) \text{ N}$$

For buckling about the y axis, $L = 6$ m and $K_y = 1$

$$P_{cr} = \frac{\pi^2 EI_y}{(K_y L_y)^2} = \frac{\pi^2 [200(10^9)][18.8(10^{-6})]}{[1(6)]^2}$$

$$= 1.031(10^6) \text{ N (controls)}$$

$$P_{allow} = \frac{P_{cr}}{\text{F.S.}} = \frac{1.031(10^6)}{2} = 515 \text{ kN} \qquad Ans.$$

$$\sigma_{cr} = \frac{P_{cr}}{A} = \frac{1.031(10^6)}{7.4(10^{-3})} = 139.30 \text{ MPa} < \sigma_Y = 250 \text{ MPa}$$
$$\text{(OK)}$$

F13–4

$$A = \pi((0.025)^2 - (0.015)^2) = 1.257 (10^{-3}) \text{ m}^2$$

$$I = \frac{1}{4}\pi((0.025)^4 - (0.015)^4) = 267.04(10^{-9}) \text{ m}^4$$

$$P = \frac{\pi^2 EI}{(KL)^2} = \frac{\pi^2 (200(10^9))(267.04)(10^{-9})}{[0.5(5)]^2} = 84.3 \text{ kN} \quad Ans.$$

$$\sigma = \frac{P}{A} = \frac{84.3(10^3)}{1.257(10^{-3})} = 67.1 \text{ MPa} < 250 \text{ MPa} \qquad \text{(OK)}$$

F13–5

$$+\uparrow \Sigma F_y = 0; \qquad F_{AB}\left(\frac{3}{5}\right) - P = 0 \qquad F_{AB} = 1.6667P \text{ (T)}$$

$$\stackrel{+}{\rightarrow} \Sigma F_x = 0; \qquad 1.6667P\left(\frac{4}{5}\right) - F_{AC} = 0$$

$$F_{AC} = 1.3333P \text{ (C)}$$

$$A = \frac{\pi}{4}(50^2) = 625\pi \text{ mm}^2 \quad I = \frac{\pi}{4}(25^4) = \frac{390\,625\pi}{4} \text{ mm}^2$$

$$P_{cr} = F_{AC}(\text{F.S.}) = 1.3333P(2) = 2.6667P$$

$$P_{cr} = \frac{\pi^2 EI}{(KL)^2}$$

$$2.6667P = \frac{\pi^2 [(200)]\left[\dfrac{390\,625\pi}{4}\right]}{[1(1200)]^2}$$

$$P = 157.7 \text{ kN} \qquad\qquad Ans.$$

$$\sigma_{cr} = \frac{P_{cr}}{A} = \frac{2.6667[157.7(10^3)]}{625\pi} = 214.2 \text{ N/mm}^2$$

$$= 214.2 \text{ MPa} < \sigma_Y = 250 \text{ MPa} \qquad \text{(OK)}$$

F13–6

$$\downarrow + \Sigma M_A = 0; \qquad w(6)(3) - F_{BC}(6) = 0 \qquad F_{BC} = 3w$$

$$A = \frac{\pi}{4}(0.05^2) = 0.625(10^{-3})\pi \text{ m}^2 \quad I = \frac{\pi}{4}(0.025^4)$$

$$= 97.65625(10^{-9})\pi \text{ m}^4$$

$$P_{cr} = F_{BC}(\text{F.S.}) = 3w(2) = 6w$$

$$P_{cr} = \frac{\pi^2 EI}{(KL)^2}$$

$$6w = \frac{\pi^2 [200(10^9)][97.65625(10^{-9})\pi]}{[1(3)]^2}$$

$$w = 11.215(10^3) \text{ N/m} = 11.2 \text{ kN/m} \qquad Ans.$$

$$\sigma_{cr} = \frac{P_{cr}}{A} = \frac{6[11.215(10^3)]}{0.625(10^{-3})\pi} = 34.27 \text{ MPa} < \sigma_Y = 250 \text{ MPa}$$
$$\text{(OK)}$$

Answers to Selected Problems

Chapter 1

1–1. (a) $F_A = 24.5$ kN, (b) $F_A = 34.9$ kN

1–2. $T_C = 250$ N·m, $T_D = 0$

1–3. $T_B = 150$ N·m, $T_C = 500$ N·m

1–5. $50(4/3) - A_y(4) = 0, A_y = 16.67$ kN,
$B_y = 33.33$ kN, $N_D = 0, V_D = 4.17$ kN,
$M_D = 25.17$ kN·m, $N_E = 0, V_E = -48.33$ kN,
$M_E = -50$ kN·m

1–6. $N_C = -30.0$ kN, $V_C = -8.00$ kN,
$M_C = 6.00$ kN·m

1–7. $P = 0.533$ kN, $N_C = -2.00$ kN,
$V_C = -0.533$ kN, $M_C = 0.400$ kN·m

1–9. $B_y = 3.00$ kN, $N_D = 0, V_D = -1.875$ kN,
$M_D = 3.94$ kN·m

1–10. $N_A = 0, V_A = 2.17$ kN, $M_A = -1.654$ kN·m,
$N_B = 0, V_B = 3.98$ kN, $M_B = -9.034$ kN·m,
$V_C = 0, N_C = -5.55$ kN, $M_C = -11.554$ kN·m

1–11. $V_A = 386.37$ N, $N_A = 103.53$ N, $M_A = 1.808$ N·m

1–13. $N_B = -2$ kN, $V_B = 4.72$ kN,
$-M_B - 0.72(0.6) - 4(1.275) + 2(0.45) = 0$,
$M_B = -4.632$ kN·m

1–14. $N_C = -2$ kN, $V_C = 5.26$ kN
$M_C = -9.123$ kN·m, $N_D = 0, V_D = 6.97$ kN
$M_D = -22.933$ kN·m

1–15. $V_C = 60$ N, $N_C = 0, M_C = 0.9$ N·m

1–17. $N_B = 5.303$ kN, $N_{a-a} = -3.75$ kN,
$V_{a-a} = 1.25$ kN, $M_{a-a} = 3.75$ kN·m,
$N_{b-b} = -1.77$ kN, $V_{b-b} = 3.54$ kN,
$M_{b-b} = 3.75$ kN·m

1–18. $N_C = -400$ N, $V_C = 0, M_C = -60$ N·m

1–19. $N_C = 0$. $V_C = 25$ kN, $M_C = 58.33$ kN·m

1–21. $N_{a-a} = 779$ N, $V_{a-a} = 450$ N,
$900(0.2) - M_{a-a} = 0, M_{a-a} = 180$ N·m

1–22. $N_G = 9.81$ kN, $V_G = 0, M_G = 0$

1–23. $N_H = -2.71$ kN, $V_H = -20.6$ kN,
$M_H = -4.12$ kN·m

1–25. $(V_B)_x = 7.5$ kN, $(V_B)_y = 0, (N_B)_z = 0$,
$(M_B)_x = 0, (M_B)_y - 7.5(7.5) = 0$;
$(M_B)_y = 56.25$ kN·m, $(T_B)_z - 7.5(0.5) = 0$;
$(T_B)_z = 3.75$ kN·m

1–26. $(V_C)_x = -250$ N, $(N_C)_y = 0$,
$(V_C)_z = -240$ N,
$(M_C)_x = -108$ N·m,
$(T_C)_y = 0, (M_C)_z = -138$ N·m

1–27. $(N_B)_x = 0, (V_B)_y = 0, (V_B)_z = 70.6$ N,
$(T_B)_x = 9.42$ N·m, $(M_B)_y = 6.23$ N·m,
$(M_B)_z = 0$

1–29. $P\cos\theta - N_A = 0, N_A = P\cos\theta$,
$V_A - P\sin\theta = 0, V_A = P\sin\theta$,
$M_A - P[r(1 - \cos\theta)] = 0$,
$M_A = Pr(1 - \cos\theta)$

1–31. $\sigma = 1.82$ MPa

1–33. $V = P\cos\theta, N = P\sin\theta$,
$$\sigma = \frac{P}{A}\sin^2\theta, \tau_{\text{avg}} = \frac{P}{2A}\sin 2\theta$$

1–34. $\sigma_D = 13.3$ MPa (C), $\sigma_E = 70.7$ MPa (T)

1–35. Joint A: $\sigma_{AB} = 85.47$ MPa (T),
$\sigma_{AE} = 68.376$ MPa (C)
Joint E: $\sigma_{ED} = 68.376$ MPa (C),
$\sigma_{EB} = 38.462$ MPa (T)
Joint B: $\sigma_{BC} = 188.034$ MPa (T),
$\sigma_{BD} = 149.573$ MPa (C)

1–37. $dF = 7.5(10^6)\, x^{1/2}\, dx$,
$P = 40$ MN, $d = 2.40$ m

1–38. $\sigma = 533.33$ kPa, $\tau = 923.76$ kPa

1–39. $\sigma_{\text{avg}} = 5$ MPa

1–41. $D_y = 3.25$ kN, $E_x = 2.5$ kN, $E_y = 1.75$ kN,
$C_y = 0.75$ kN, $B_y = 0.75$ kN,
$F_B = F_C = 2.971$ kN
$(\tau_B)_{\text{avg}} = 105.1$ N/mm^2

1–42. $(\tau_D)_{\text{avg}} = 57.47$ MPa, $(\tau_E)_{\text{avg}} = 53.9$ MPa

1–43. $(\tau_D)_{\text{avg}} = 114.9$ MPa, $(\tau_E)_{\text{avg}} = 107.9$ MPa

1–45. Joint B: $\sigma_{AB} = \dfrac{F_{AB}}{A_{AB}} = \dfrac{3125}{1000} = 3.125$ MPa (C),
$\sigma_{BC} = 3.516$ MPa (T)
Joint A: $\sigma_{AC} = 6.25$ MPa

1–46. $\sigma = 339$ MPa

1–47. $\tau_A = 138$ MPa

1–49. $A_x = 9.5263P$, $A_y = 5.5P$, $F_A = 11P$,
$P = 3.70$ kN

1–50. $(\sigma_{a-a})_{\text{avg}} = 66.7$ kPa, $(\tau_{a-a})_{\text{avg}} = 115$ kPa

1–51. $P = 18.75$ kN, $(\tau_{a-a})_{\text{avg}} = 1.875$ MPa

1–53. $V_b = P/4$, $V_p = P/4$, $P = 9.05$ kN

1–54. $(\sigma_{\text{avg}})_s = 56.6$ MPa, $(\sigma_{\text{avg}})_b = 31.8$ MPa

1–55. $(\sigma_{\text{avg}})_{AB} = 118$ MPa, $(\sigma_{\text{avg}})_{BC} = 58.8$ MPa

1–57. $V = 61.57$ kN, $N = 78.80$ kN
Inclined plane: $\sigma' = 549.05$ MPa,
$\tau'_{\text{avg}} = 428.96$ MPa,
Cross section: $\sigma = 884.2$ MPa, $\tau_{\text{avg}} = 0$
$\tau_{\text{avg}} = 0$

1–58. $P = 68.3$ kN

1–59. $\sigma = 25$ MPa, $\tau_{\text{avg}} = 14.434$ MPa

1–61. $A_x = 1.732P$, $A_y = P$, $F_A = 2P$,
$P = 15.3$ kN

1–62. $(\tau_A)_{\text{avg}} = 29.709$ MPa

1–63. $(\tau_B)_{\text{avg}} = 12.732$ MPa

1–65. $V = 636.40$ N, $\tau_{\text{avg}} = 509$ kPa

1–66. $P = 62.5$ kN

1–67. $\sigma = \dfrac{w_0}{2aA}(2a^2 - x^2)$

1–69. $A = 447.6$ mm^2, $N = 3600$ N, $\sigma = 8.04$ MPa

1–70. $r = r_1 e^{\left(\frac{\rho g \pi r_1^2}{2P}\right)z}$

1–71. $\sigma_{a-a} = 12.0$ MPa, $\tau_{b-b} = 5.21$ MPa

1–73. $h = 73.2$ mm, use $h = 75$ mm

1–74. $d = 5.71$ mm

1–75. $d = 13.5$ mm

1–77. Shear limitation, $t = 167$ mm,
Tension limitation, $b = 33.3$ mm

1–78. $a = 171.4$ mm. Use $a = 175$ mm

1–79. $d_A = 27.6$ mm

1–81. $P = 14.28$ kN (Controls)

1–82. $d_{BD} = 7.00$ mm, $d_{AB} = 6.50$ mm,
$d_{BC} = 6.00$ mm

1–83. $P = 4.43$ kN

1–85. $W = 1.739$ kN

1–86. Use $d = 30$ mm

1–87. $P = 90$ kN, $A = 6.19(10^{-3})$ m^2,
$P_{\text{max}} = 155$ kN

1–89. $150 = \dfrac{25(10^3)}{\frac{\pi}{4}d^2}$, use $d = 15$ mm,

$35 = \dfrac{25(10^3)}{\pi(25)(h)}$, use $h = 10$ mm

1–90. $d_B = 7.08$ mm, $d_C = 6.29$ mm

1–91. $(\text{F.S.})_B = 2.24$, $(\text{F.S.})_C = 2.13$

1–93. $\sigma_{rod} = 75.34$ MPa, $(\text{F.S.})_{rod} = 3.32$,
$\tau_{pins} = 63.66$ MPa, $(\text{F.S.})_{pins} = 1.96$

1–94. $w = 14.14$ kN/m

1–95. $a_{A'} = 130$ mm, $a_{B'} = 300$ mm

1–97. $F_{CD} = 6.70$ kN, $F_{AB} = 8.30$ kN,
$d_{AB} = 6.02$ mm, $d_{CD} = 5.41$ mm

1–98. $h = 52.1$ mm

1–99. $P = 55.0$ kN

1–101. $N_{a-a} = 259.81$ kN, $V_{a-a} = 150$ kN,
$(\sigma_{a-a})_{\text{avg}} = 7.16$ MPa, $(\tau_{a-a})_{\text{avg}} = 4.13$ MPa

1–102. $\sigma_s = 208$ MPa, $(\tau_{\text{avg}})_a = 4.72$ MPa,
$(\tau_{\text{avg}})_b = 45.5$ MPa

1–103. $t = 6$ mm, $d_A = 28$ mm, $d_B = 20$ mm

1–105. $F = 3678.75$ N, $\tau_{\text{avg}} = 61.3$ MPa

1–106. $\sigma_{a-a} = 200$ kPa, $\tau_{a-a} = 115$ kPa

1–107. $\sigma_{40} = 3.98$ MPa, $\sigma_{30} = 7.07$ MPa,
$\tau_{\text{avg}} = 5.09$ MPa

Chapter 2

2–1. $\epsilon = 0.167$ mm/mm

2–2. $\epsilon = 0.0472$ mm/mm

2–3. $\epsilon_{CE} = 0.00250$ mm/mm, $\epsilon_{BD} = 0.00107$ mm/mm

2–5. $\delta_B = 4$ mm, $(\epsilon_{\text{avg}})_{BD} = 0.00267$ mm/mm,
$(\epsilon_{\text{avg}})_{CE} = 0.005$ mm/mm

2–6. $\gamma = 0.197$ rad

2–7. $\epsilon_{\text{avg}} = 0.0689$ mm/mm

2–9. $AB = 500$ mm, $AB' = 501.75$ mm,
$\Delta_D = 4.38$ mm

2–10. $(\gamma_{xy})_A = 0.206$ rad, $(\gamma_{xy})_B = -0.206$ rad

2–11. $(\epsilon_{\text{avg}})_{AB} = -0.0889$ mm/mm,
$(\epsilon_{\text{avg}})_{BD} = -0.1875$ mm/mm

2–13. $AD' = 400.01125$ mm, $AB' = 300.00667$,
$D'B' = 496.6014$ mm, $DB = 500$ mm,
$\epsilon_{DB} = -0.00680$ mm/mm,
$\epsilon_{AD} = 0.0281(10^{-3})$ mm/mm

2–14. $x = -4.81$ mm, $y = -5.45$ mm

2–15. $\epsilon_{AB} = 0.152$ mm/mm, $\epsilon_{AC} = 0.0274$ mm/mm

2–17. $L_{DB'} = L\sqrt{1 + (2\epsilon_{AB}\cos^2\theta + 2\epsilon_{CB}\sin^2\theta)}$,
$\epsilon_{DB} = \epsilon_{AB}\cos^2\theta + \epsilon_{CB}\sin^2\theta$

2–18. $(\gamma_B)_{xy} = 11.6(10^{-3})$ rad,
$(\gamma_A)_{xy} = -11.6(10^{-3})$ rad

2–19. $(\gamma_C)_{xy} = -11.6(10^{-3})$ rad,
$(\gamma_D)_{xy} = 11.6(10^{-3})$ rad

2–21. $L_{B'D} = 0.6155$ m, $\epsilon_{avg} = 0.0258$ mm/mm

2–22. $(\gamma_A)_{xy} = 5.24(10^{-3})$ rad

2–23. $\epsilon_{AC} = 16.7(10^{-3})$ mm/mm,
$\epsilon_{BD} = 11.3(10^{-3})$ mm/mm

2–25. $A'B' = 5.08416$ m, $AB = 5.00$ m,
$\epsilon_{AB} = 16.8(10^{-3})$ mm/mm

2–26. $(\epsilon_{avg})_{AC} = 0.0112$ mm/mm,
$(\epsilon_{avg})_{CD} = 0.125$ mm/mm,
$(\gamma_{xy})_F = 0.245$ rad,
$(\epsilon_{avg})_{BE} = 0.0635$ mm/mm

2–27. $(\epsilon_{avg})_{AD} = 0.132$ mm/mm,
$(\epsilon_{avg})_{CF} = -0.0687$ mm/mm

2–29. $\epsilon = 0.05\cos\theta$, $\Delta L = \int_0^{90°}(0.05\cos\theta)(0.6\,d\theta)$,
$\Delta L = 0.03$ m

2–30. $\Delta L = 0.048$ m

2–31. $\epsilon_{avg} = 0.479$ m/m

2–33. $\epsilon_{AB} = \left[1 + \dfrac{2(v_B\sin\theta - u_A\cos\theta)}{L}\right]^{\frac{1}{2}} - 1$,
$\epsilon_{AB} = \dfrac{v_B\sin\theta}{L} - \dfrac{u_A\cos\theta}{L}$

Chapter 3

3–1. $E_{approx} = \dfrac{8.0\text{ MPa} - 0}{0.0003 - 0} = 26.67$ GPa

3–2. $E = 387.3$ GPa, $u_r = 0.0697$ MJ/m^3

3–3. $(u_r)_{approx} = 0.595$ MJ/m^3

3–5. $(u_t)_{approx} = 117$ MJ/m^3

3–6. $E = 76.64$ GPa

3–7. $A = 150$ mm^2, $P = 7.5$ kN

3–9. $\sigma = 11.85$ MPa, $\epsilon = 0.035$ mm/mm, $\delta = 5.775$ mm

3–10. $E = 290$ GPa, $P_Y = 32.80$ kN, $P_{ult} = 62.20$ kN

3–11. Elastic Recovery = 0.08621 mm
$\Delta L = 3.91379$ mm

3–13. $\sigma = 95.81$ MPa, $\epsilon = 0.000400$ mm/mm
$E = 239.5$ GPa

3–14. $\delta_{BD} = 1.98975$ mm

3–15. $P = 2.26$ kN

3–17. $\sigma_{pl} = 308$ MPa, $\sigma_Y = 420$ MPa, $E = 77.0$ GPa

3–18. $(U_i) = 0.616$ MJ/m^3, $[(U_i)_t]_{approx} = 45.5$ MJ/m^3

3–19. $\sigma = 2.22$ MPa

3–21. $\sigma_{AB} = 31.83$ MPa, $\epsilon_{AB} = 0.009885$ mm/mm,
$\sigma_{CD} = 7.958$ MPa, $\epsilon_{CD} = 0.002471$ mm/mm,
$\alpha = 0.708°$

3–22. $P = 11.3$ kN

3–23. From the stress–strain diagram, the *copolymer* will satisfy both stress and strain requirements.

3–25. $\sigma = 1.697$ MPa,
$\delta = 0.126$ mm, $\Delta d = -0.00377$ mm

3–26. (a) $\delta = -0.0181$ mm
(b) $d' = 12.002032$ mm

3–27. $\nu = 0.300$

3–29. $\epsilon_{long} = -0.00030476$, $\epsilon_{lat} = 0.0000880$, $\nu = 0.289$,
$h' = 50.0044$ mm

3–30. $\epsilon_y = -0.0150$ mm/mm, $\epsilon_x = 0.00540$ mm/mm,
$\gamma_{xy} = -0.00524$ rad

3–31. $P = 263.89$ kN, $E = 200.4$ GPa

3–33. $\tau_{avg} = 4166.67$ Pa, $\gamma = 0.02083$ rad,
$\delta = 0.833$ mm

3–34. $\delta = \dfrac{Pa}{2bhG}$

3–35. $G_{al} = 31.80$ GPa

3–37. $E = 38.5$ MPa, $u_t = 134.75$ MJ/m^3, $u_r = 77$ MJ/m^3

3–38. $\delta = -0.0173$ mm, $d' = 20.0016$ mm

3–39. $x = 1.53$ m, $d_A' = 30.008$ mm

3–41. $\tau = 148.89$ kPa, $G = 1.481$ MPa, $\delta_h = 3.02$ mm

3–42. $\epsilon_{DE} = 0.0011574$ mm/mm, $W = 482.25$ N,
$\epsilon_{BC} = 0.00193$ mm/mm

3–43. $\epsilon_b = 0.00227$ mm/mm,
$\epsilon_r = 0.000884$ mm/mm

Chapter 4

4–1. $\delta_A = \dfrac{-5.00(10^3)(8)}{\frac{\pi}{4}(0.4^2 - 0.3^2)200(10^9)} = -3.64(10^{-3})$ mm

4–2. $\delta_{AD} = 0.02189$ mm

4–3. $\delta_D = 0.850$ mm

4–5. $\delta_A = 6.14$ mm

4–6. $\delta_A = 0.00092$ mm

4–7. $\delta_P = 1.0572$ mm $\downarrow$

4–9. $\delta_C = 0.166667$ mm, $\delta_A = 0.333333$ mm,

$\delta_{F/E} = 0.0625$ mm, $\delta'_E = 0.111111$ mm

$\delta_F = 0.340$ mm

4–10. $\theta = 0.01061°$

4–11. $\delta_t = 0.736$ mm

4–13. $\delta = \dfrac{1}{AE}\displaystyle\int_0^L (\gamma Ax + P)\, dx = \dfrac{\gamma L^2}{2E} + \dfrac{PL}{AE}$

4–14. $\delta_{A/B} = -0.864$ mm

4–15. $\delta_{A/B} = -1.03$ mm

4–17. $\delta = -2.34375(10^{-6})P, P = 213.3$ kN

4–18. $\delta_F = 0.5856$ mm $\downarrow$

4–19. $\theta = 0.4476(10^{-3})$ rad

4–21. $\delta_D = 0.1374$ mm, $\delta_{A/B} = 0.3958$ mm,

$\delta_C = 0.5332$ mm, $\delta_{tot} = 33.9$ mm

4–22. $W = 9.69$ kN

4–25. $\delta = 0.360$ mm

4–26. $\delta = \dfrac{\gamma L^2}{6E}$

4–27. $\delta = -\dfrac{P}{2a\pi r_0^2 E}\left(1 - e^{-2aL}\right)$

4–29. $A = \pi r^2 = \pi(r_0 \cos \theta)^2 = \pi r_0^2 \cos^2 \theta,$

$y = r_0 \sin \theta; \quad dy = r_0 \cos \theta\, d\theta,$

When $y = \dfrac{r_0}{4}$: $\theta = 14.48°,$

$\delta = \dfrac{0.511 P}{\pi r_0 E}$

4–30. $p_0 = 250$ kN/m, $\delta = 2.93$ mm

4–31. $\sigma_{st} = 24.323$ MPa, $\sigma_{con} = 3.527$ MPa

4–33. $P_{st} = 57.47$ kN, $P_{con} = 22.53$ kN,

$\sigma_{st} = 48.8$ MPa, $\sigma_{con} = 5.85$ MPa

4–34. $\sigma_{br} = 2.792$ MPa, $\sigma_{st} = 5.335$ MPa

4–35. $d = 58.88$ mm

4–37. $F_D = 107.89$ kN, $\delta_{A/B} = 0.335$ mm

4–38. $\sigma_{st} = 13.23$ MPa,

$\sigma_{con} = 1.92$ MPa

$\delta = 0.159$ mm

4–39. $A_{st} = 11397.4$ mm^2, $\delta = 0.1579$ mm

4–41. $P_{con} = 36.552\, P_{st}$,

$\sigma_{con} = 8.42$ MPa, $\sigma_{st} = 67.3$ MPa

4–42. $d = 24.6$ mm

4–43. $\sigma_{AB} = 26.5$ MPa, $\sigma_{EF} = 33.8$ MPa

4–45. $F_b = 10.17\,(10^3)$ N, $F_t = 29.83\,(10^3)$ N,

$\sigma_b = 32.4$ MPa, $\sigma_t = 34.5$ MPa

4–46. $F_D = 20.4$ kN, $F_A = 180$ kN

4–47. $T_{AB} = 1.469$ kN, $T_{A'B'} = 1.781$ kN

4–49. $y = 75 - 0.025x, F_A = 20.47$ kN, $F_B = 14.53$ kN

4–50. $x = 721.7$ mm, $P = 264.5$ kN

4–51. $T_{CD} = 135.84$ kN, $T_{CB} = 45.28$ kN

4–53. $F_{st} = 7.37$ kN, $F_{al} = 14.74$ kN

$\sigma_{rod} = 65.2$ MPa, $\sigma_{cyl} = 7.5$ MPa

4–54. $\theta = 690°$

4–55. $\sigma_{BE} = 96.3$ MPa, $\sigma_{AD} = 79.6$ MPa,

$\sigma_{CF} = 113$ MPa

4–57. $F_{CD} = 3.0737$ kN, $F_{BC} = 2.02734$ kN, $\theta = 0.0734°$

4–58. $F_A = 5.79$ kN, $F_B = 9.64$ kN, $F_C = 11.6$ kN

4–59. $\theta = 1.14(10^{-3})°$

4–61. Assume failure of AB and EF: $F = 42\,300$ N,

Assume failure of CD: $F_{CD} = 81\,000$ N,

$w = 45.9$ kN/m

4–62. $\sigma_D = 13.4$ MPa, $\sigma_{BC} = 9.55$ MPa

4–63. $\theta = 0.00365°$

4–65. $0.02 = \delta_t + \delta_b, P = 1.16$ kN

4–66. $a = 0.120$ mm

4–67. $\sigma_{AB} = \dfrac{7P}{12A}, \sigma_{CD} = \dfrac{P}{3A}, \sigma_{EF} = \dfrac{P}{12A}$

4–69. $0 = \Delta_T - \delta, F = 4.20$ kN

4–70. $F = 2.442$ kN

4–71. $F = 489.03$ kN

4–73. $0 = \delta_T - \delta_F, F = 134.40A, \sigma = 134.40$ MPa

4–74. $F = 18.566$ kN

4–75. $\delta = 8.640$ mm, $F = 76.80$ kN

4–77. $0 = \Delta_T - \delta_F, F = \dfrac{\alpha AE}{2}(T_B - T_A)$

4–78. $F_B = 183$ kN, $F_A = 383$ kN

4–79. $P = 188$ kN

4–81. $66825 = 1.875F_{AB} + 1.2F_{AD}$,

$F_{AD} = 28.17$ kN, $F_{AC} = F_{AB} = 17.61$ kN

4–82. $\delta_A = 1.027$ mm $\uparrow$

4–83. $F_{AC} = 243.6$ N, $F_{AD} = 405.5$ N

4–85. $F = 107442.47$ N, $T = 172°$ C

4–86. $\sigma_s = 40.1$ MPa, $\sigma_b = 29.5$ MPa

4–87. $\sigma_{max} = 190$ MPa

4–89. $K = 2.50, w = 71.53$ mm

4–90. $P = 77.1$ kN, $\delta = 0.429$ mm

4–91. $P = 5.4$ kN

4–93. Maximum normal stress at fillet: $K = 1.4$,

Maximum normal stress at the hole: $K = 2.65$,

$\sigma_{max} = 88.3$ MPa

4–94. $P = 36.0$ kN, $K = 1.60$

4–95. $P = 73.5$ kN, $K = 1.29$

4–97. $\sigma_{st} = 250$ MPa, $\sigma_{al} = 171.31$ MPa

4–98. $\delta_C = 18.286$ mm

4–99. $w = 21.9$ kN/m, $\delta_G = 4.24$ mm

4–101. $F_{CD} = 1800$ N, $F_{AB} = 3600$ N,
$F_{AB} = 3.14$ kN, $F_{CD} = 2.72$ kN,
$\delta_{CD} = 0.324$ mm, $\delta_{AB} = 0.649$ mm

4–102. (a) $P = 2.62$ kN, (b) $P = 3.14$ kN

4–103. $P = \sigma_Y A (2 \cos \theta + 1)$, $\delta_A = \dfrac{\sigma_Y L}{E \cos \theta}$

4–105. $\sigma_{CF} = 250$ MPa (T), $F_{CF} = 122\,718.46$ N,
$F_{BE} = 91\,844.61$ N, $F_{AD} = 15\,436.93$ N,
$(\sigma_{CF})_r = 17.7$ MPa (C),
$(\sigma_{BE})_r = 53.2$ MPa (T),
$(\sigma_{AD})_r = 35.5$ MPa (C)

4–106. $\sigma_A = 410.3$ MPa, $\delta = 282.8$ mm

4–107. $w = 159.2$ kN/m

4–109. $(F_{al})_Y = 56.55$ kN, $F_{st} = 146.9$ kN,
$d_B = 17.8$ mm

4–110. (a) $\delta_D = 9.062$ mm
(b) $\delta_D = 111.2$ mm

4–111. $P = 549.78$ kN, $\Delta \delta = 0.180$ mm ←

4–113. $\delta = \dfrac{1}{A^2 c^2} \displaystyle\int_0^L (\gamma A x)^2 \, dx$, $\delta = \dfrac{\gamma^3 L^3}{3c^2}$

4–114. $F_B = 8.526$ kN, $F_A = 8.606$ kN

4–115. $P = 19.776$ kN

4–117. $P = 200$ kN, $\sigma_{AB} = 1000$ MPa

4–118. $P = 56.5$ kN, $\delta_{B/A} = 0.0918$ mm

4–119. $\theta = \dfrac{3E_2 L (T_2 - T_1)(\alpha_2 - \alpha_1)}{d(5E_2 + E_1)}$

Chapter 5

5–1. (a) $T = 0.87$ kN $\cdot$ m, (b) $T' = 0.698$ kN $\cdot$ m
$\tau_\rho = \dfrac{69781(12.5)}{\frac{\pi}{2}(18.75^4 - 12.5^4)} = 56.0$ MPa

5–2. (a) $r' = 0.841r$, (b) $r' = 0.841r$

5–3. $\tau_B = 6.04$ MPa, $\tau_A = 6.04$ MPa

5–5. $\tau_{max} = 26.7$ MPa

5–6. $(\tau_{BC})_{max} = 43.5$ MPa, $(\tau_{DE})_{max} = 31.1$ MPa

5–7. $(\tau_{EF})_{max} = 0$, $(\tau_{CD})_{max} = 18.65$ MPa

5–9. $J = 2.545(10^{-6})$ m^4,
$\tau_{max} = 11.9$ MPa

5–10. $n = \dfrac{2r^3}{Rd^2}$

5–11. $\tau_{AB} = 62.5$ MPa, $\tau_{BC} = 18.9$ MPa

5–13. $F = 600$ N, $T_A = 30.0$ N $\cdot$ m,
$(\tau_{EA})_{max} = 5.66$ MPa, $(\tau_{CD})_{max} = 8.91$ MPa

5–14. $(\tau_{max})_{abs} = 10.2$ MPa

5–15. $d = 33$ mm

5–17. $T_A = 108$ N $\cdot$ m, $J = 12207\pi$ mm^4,
$\tau_{max} = 35.2$ MPa

5–18. $\tau_{max} = 52.8$ MPa

5–19. $(\tau_{max})_{AB} = 41.4$ MPa, $(\tau_{max})_{BC} = 82.8$ MPa

5–21. $T_{AB} = (2000x - 1200)$ N $\cdot$ m,
$d = 0.9$ m, $\tau_{min} = 0$,
$d = 0$, $\tau_{max} = 42.4$ MPa

5–22. $d = 57$ mm

5–25. $T_{max} = 390.63$ N $\cdot$ m, $\tau_{abs\,max} = 28.73$ MPa

5–26. $\tau_{max} = \dfrac{T}{2\pi r_i^2 h}$

5–27. $(\tau_{AB})_{max} = 23.9$ MPa, $(\tau_{BC})_{max} = 15.9$ MPa

5–29. $T_A + \dfrac{1}{2} t_A L - T_B = 0$, $T_B = \dfrac{2T_A + t_A L}{2}$,
$\tau_{max} = \dfrac{(2T_A + t_A L) r_o}{\pi (r_o^4 - r_i^4)}$

5–30. $c = (2.98\,x)$ mm

5–31. $(\tau_{AB})_{max} = 1.04$ MPa, $(\tau_{BC})_{max} = 3.11$ MPa

5–33. $P = 1600$ N $\cdot$ m/s,
$T = 339.5$ N $\cdot$ m,
$\tau_{max} = 110.7$ MPa

5–34. Use $d = 25$ mm

5–35. $\omega = 21.7$ rad/s

5–37. $P = 1500(10^3)$ N $\cdot$ m/s,
$T = 9.55$ kN $\cdot$ m,
$\tau_{max} = 48.6$ MPa

5–38. $d_A = 12.4$ mm, $d_B = 16.8$ mm

5–39. $(\tau_{max})_{CF} = 12.5$ MPa,
$(\tau_{max})_{BC} = 7.26$ MPa

5–41. $T = 625$ N $\cdot$ m,
$t = 2.5$ mm

5–42. $\omega = 17.7$ rad/s

5–43. Use $d = 65$ mm

5–45. $T = 795.77$ J,
$r_i = 28.25$ mm,
$t = 2.998$ mm

5–46. Use $d = 20$ mm

5–47. $\tau_{max} = 44.3$ MPa, $\phi = 11.9°$

5–49. $T_{AB} = -85$ N·m,
$T_{BC} = -85$ N·m,
$\phi_{A/D} = 0.879°$

5–50. $\tau_{max} = 20.797$ MPa
$\phi = 4.766°$

5–51. Use $d = 70$ mm

5–53. $T_{BC} = -80$ N·m,
$T_{CD} = -60$ N·m,
$T_{DA} = -90$ N·m,
$\phi_B = |5.74°|$

5–54. $\phi_D = 1.01°$

5–55. $\phi_C = 0.227°$

5–57. Use $d = 30$ mm

5–58. $\tau_{\substack{abs \\ max}} = 24.59$ MPa
$\phi_{C/D} = 0.075152°$

5–59. $\phi_{B/D} = 1.34°$

5–61. $\phi_E = 0.020861$ rad
$\phi_F = 0.031291$ rad
$\phi_B = 1.793°$

5–62. $\phi_A = 2.092°$

5–63. $(\tau_{BC})_{max} = 81.49$ MPa,
$(\tau_{BA})_{max} = 14.9$ MPa,
$\phi_C = 3.125°$

5–65. $\phi_B = 0.002173$ rad,
$\phi_{C/B} = -0.0001357$ rad,
$\phi_C = 0.132°$

5–66. $\phi_A = 2.66°$,
$\phi_C = 2.30°$

5–67. $\phi_A = \phi_B + \phi_{A/B}$,
$\phi_C = \phi_B + \phi_{C/B}$,
$T_1 = 2.19$ kN·m,
$T_2 = 3.28$ kN·m

5–69. $\phi_B = 0.01194$ rad,
$\phi_C = 0.008952$ rad,
$\phi_E = 1.20°$

5–70. $\phi_D = 1.42°$

5–73. $J(x) = \dfrac{\pi r^4}{2L^4}(L + x)^4$,
$\phi = \dfrac{7TL}{12\pi r^4 G}$

5–74. $t_o = \dfrac{4\,pd}{L}$,
$\phi = \dfrac{4PLd}{3\pi r^4 G}$

5–75. $\phi = \dfrac{2L(t_0 L + 3T_A)}{3\pi(r_o^4 - r_i^4)G}$

5–77. $T_A = 200$ N·m,
$T_B = 100$ N·m,
$(\tau_{AC})_{max} = 8.15$ MPa,
$(\tau_{CB})_{max} = 4.07$ MPa

5–78. $\tau_{AC} = 9.77$ MPa

5–79. $\tau_{AC} = 232.3$ MPa

5–81. $T_A = 1.498$ kN·m,
$T_B = 0.502$ kN·m,
$\tau_{CD} = 24.9$ MPa

5–82. $\phi_C = 0.142°$
$(\tau_{st})_{max} = 3.153$ MPa,
$(\gamma_{st})_{max} = 42(10^{-6})$ rad,
$(\tau_{br})_{max} = 0.799$ MPa,
$(\gamma_{bt})_{max} = 21.0(10^{-6})$ rad

5–83. $(\tau_{BC})_{max} = 11.8$ MPa,
$(\tau_{BD})_{max} = 15.7$ MPa,
$\phi = 0.386°$

5–85. $T_R = (1.5x - 0.625x^2)$ kN·m, $\phi_B = 3.208°$,
$\tau_{\substack{abs \\ max}} = 105.66$ MPa

5–86. $T_B = 222$ N·m,
$T_A = 55.6$ N·m

5–87. $\phi_E = 1.66°$

5–89. $F = 24.51$ kN,
$T_E = 7.353$ kN·m,
$T_A = 21.32$ kN·m,
$\phi_B = 1.24°$

5–90. $(\tau_{BD})_{max} = 67.91$ MPa,
$(\tau_{AC})_{max} = 33.95$ MPa

5–91. $\tau_{\substack{abs \\ max}} = 68.75$ MPa

5–93. $J(x) = \dfrac{\pi c^4}{2L^4}(L + x)^4$,
$T_B = \dfrac{37}{189}T$,
$T_A = \dfrac{152}{189}T$

5–94. $T_B = \dfrac{7t_0 L}{12}$,

$T_A = \dfrac{3t_0 L}{4}$

5–95. $(\tau_c)_{max} = 4.23$ MPa,

$(\tau_r)_{max} = 5.74$ MPa,

$\phi_c = 0.0689°$,

$\phi_r = 0.0778°$

5–97. $(\tau_{max})_c = \dfrac{16T}{\pi d^3}$,

$(\tau_{max})_c = \dfrac{16T}{\pi k^2 d^3}$,

Factor of increase in shear stress $= \dfrac{1}{k^2}$

5–98. $(\tau_{BC})_{max} = 0.955$ MPa,

$(\tau_{AC})_{max} = 1.59$ MPa,

$\phi_{B/A} = 0.207°$

5–99. $(\tau_{BC})_{max} = 0.955$ MPa,

$(\tau_{AC})_{max} = 1.59$ MPa,

$\phi_{B/C} = |0.0643°|$

5–101. For segment AB, $T = 3180.86$ N·m,

For segment BC, $T = 11\ 366.94$ N·m,

$T = 2.80$ kN·m

5–102. $T_B = 48$ N·m,

$T_A = 72$ N·m,

$\phi_C = 0.104°$

5–103. $(\tau_{max})_A = 308$ MPa

5–105. $T = 181.91$ N·m, $F = 454.78$ N

5–106. $\tau_{max} = 18.47$ MPa,

$\delta_F = 0.872$ mm

5–107. $t = 2.98$ mm

5–109. $A_m = 906.15$ mm^2,

$A_m' = 1500.10$ mm^2,

Factor = 1.66

5–110. The factor of increase = 2.85

5–111. $q_{st} = \dfrac{\pi}{4} q_{ct}$

5–113. $A_m = 4.97469$ m^2,

use $t = 8$ mm,

$\phi = 0.1333°$

5–114. $T = 7.462$ MN·m

5–115. $(\tau_{avg})_A = 15.6$ MPa,

$(\tau_{avg})_B = 10.4$ MPa

5–117. $A_m = 1.8927$ m^2,

$T = 4.73$ MN·m,

$\phi = 0.428°/$m

5–118. $\tau_{avg} = 119$ MPa,

$\phi = 0.407°/$m

5–119. $(\tau_{avg})_A = (\tau_{avg})_B = 357$ kPa

5–121. $K = 1.28$,

$r = 7.98$ mm,

No, it is not possible.

5–122. $P = 101$ kW

5–123. $(\tau_{max})_f = 50.6$ MPa

5–125. $K = 1.466, r = 2$ mm

5–127. $T = 3.551$ kN·m,

$T_P = 3.665$ kN·m,

5–129. $T = 20.8$ kN·m,

$\phi = 34.4°$,

$G = 40$ GPa,

$\phi' = 0.3875$ rad,

$\phi_r = 12.2°$

5–130. $T = 18.84$ kN·m

5–131. $T_Y = 12.6$ kN·m,

$T_P = 16.8$ kN·m

5–133. $T = 175.9$ kN·m

5–134. $T_P = \dfrac{2}{3}\pi\tau_Y(c_o^3 - c_i^3)$,

$\phi = \dfrac{\tau_Y L}{c_i G}$,

$\gamma_{max} = \dfrac{c_o \tau_Y}{c_i G}$

5–135. $T_C = 9.3$ kN·m, $T_A = 5.70$ kN·m

5–137. $\rho_\gamma = 0.00625$ m,

$\tau_1 = 8(10^9)\rho$,

$\tau_2 = 4(10^9)\rho + 25(10^6)$,

$T = 3.27$ kN·m,

$\phi = 34.4°$

5–138. $T = 51.46$ kN·m, $\phi_P = 0.413°$

5–139. $T = 54.12$ kN·m at 75 mm, $\tau = 87.1$ MPa,

at 37.5 mm, $\tau = 43.6$ MPa

5–141. Elastic, $T_t = 9256.95$ N·m, $T_c = 5743.05$ N·m,

$T_t > (T_Y)_t$, Plastic, $T_t = 7.39$ kN·m,

$T_c = 7.61$ kN·m

5–142. $\tau_{max} = \dfrac{19T}{12\pi r^3}$

5–145. $r_o = 0.0625$ m, $r_i = 0.0575$ m,

Eq. 5–7: $\tau_{\rho=0.06\ m} = 88.27$ MPa,

Eq. 5–18: $\tau_{avg} = 88.42$ MPa,

Eq. 5–15: $\phi = 4.495°$,

Eq. 5–20: $\phi = 4.503°$

5–146. $T = 331 \text{ N} \cdot \text{m}$

5–147. $t = 8 \text{ mm}$

5–149. $T = 71.5 \text{ N} \cdot \text{m}, \tau_{max} = 23.3 \text{ MPa}$

5–150. $\tau_{max} = 82.0 \text{ MPa}$

5–151. $F = 26.2 \text{ N}, \phi = 1.86°$

Chapter 6

6–1. $x = 0.25^-, V = -24, M = -6$

6–2. $x = 2^-, V = 1, M = 2, x = 4^-, V = 1, M = 6$

6–3. $x = 0.9^-, V = -10, M = -9$

6–5. $x = 2^+, V = 8, M = -39$

6–6. $x = 1.5, V = 0, M = 9, x = 4^-,$
$V = -20, M = -16$

6–7. $x = 1^-, V = -30, M = -30$

6–9. $x = 1^+, V = -16.67, M = 58.33$

6–10. $x = 0.45^-, V = 750, M = 337.5$

6–11. $x = 3^-, V = -4, M = -12$

6–13. $x = 3a^-, V = -P, M = -Pa$

6–14. $x = 350 \text{ mm}^+, V = 575, M = -484.4$

6–17. $x = 2, V = -4.5, M = -5,$
$V = \{-1.5 - 0.75x^2\} \text{ kN},$
$M = \{-1.5x - 0.25x^3\} \text{ kN} \cdot \text{m}$

6–18. $V = \{144 - 30x\} \text{ kN},$
$M = \{-1.5x^2 + 144x - 458.6\} \text{ kN} \cdot \text{m},$
$V = 40 \text{ kN},$
$M = \{40x - 320\} \text{ kN} \cdot \text{m}$

6–19. $x = 1.5^-, V = -45, M = -33.75$

6–21. $x = 0.75, V = 0, M = 0.5625,$
$F_{BC} = 7.5 \text{ kN},$
$A_y = 1.5 \text{ kN}$

6–22. $x = 3^-, V = -10, M = -18$

6–23. $x = L, V = -wL, M = 0$

6–25. $x = L, V = 0, M = 0$

6–27. $x = (L/3)^-, V = -w_0 L/18,$
$M = -0.00617 w_0 L^2$

6–29. $x = 4.11, V = 0, M = 25.7,$
At $x = 4.108 \text{ m}, M = 25.67 \text{ kN} \cdot \text{m},$
At $x = 4.5 \text{ m}, M = 25.31 \text{ kN} \cdot \text{m}$

6–30. $x = 0.845, V = 0, M = 0.616$

6–31. $V = \dfrac{w_0}{4}(3L - 4x),$

$M = \dfrac{w_0}{24}(-12x^2 + 18Lx - 7L^2),$

$V = \dfrac{w_0}{L}(L - x)^2, M = -\dfrac{w_0}{3L}(L - x)^3$

6–33. $w = 600 \text{ N/m}$
$V = 150 \text{ N}$
$M = 25 \text{ N} \cdot \text{m}$

6–34. $x = 3^-, V = -11.5, M = -21$

6–35. $V = 200 \text{ N}, M = (200 x) \text{ N} \cdot \text{m},$

$V = \left\{-\dfrac{100}{3}x^2 + 500\right\} \text{ N},$

$M = \left\{-\dfrac{100}{9}x^3 + 500x - 600\right\} \text{ N} \cdot \text{m}$

6–37. $x = 4.5, V = 0, M = 169$

6–38. $A_y = 48.1875 \text{ kN}, A_x = 0$

6–39. $x = L, V = -\dfrac{23}{54}wL, M = -\dfrac{5}{54}wL^2$

6–41. $x = 4^-, V = -2.8, M = -2.4$

6–42. $x = 1, V = 0, M = 2.50$

6–43. $x = 3.6, V = 0, M = 36$

6–45. $x = 0.630L, V = 0, M = 0.0394 w_0 L^2,$

$M = \dfrac{w_0 L x}{12} - \dfrac{w_0 x^4}{12L^2}$

6–46. $x = 0, V = 2w_0 L/\pi, M = -w_0 L^2/\pi$

6–47. $\sigma_{max} = 120 \text{ MPa},$
$\sigma_{max} = 90 \text{ MPa}$

6–49. $\bar{y} = 84.7 \text{ mm},$
$I_{NA} = 34.277(10^6) \text{ mm}^4,$
$(\sigma_t)_{max} = 31.0 \text{ MPa}, (\sigma_c)_{max} = 14.8 \text{ MPa}$

6–50. $\sigma_A = 6.81 \text{ MPa},$
$\sigma_B = 1.01 \text{ MPa},$
$\sigma_C = 4.14 \text{ MPa}$

6–51. $M = 771 \text{ N} \cdot \text{m}$

6–53. $I = 91.14583(10^{-6}) \text{ m}^4,$
$M = 36.5 \text{ kN} \cdot \text{m}, \sigma_{max} = 40.0 \text{ MPa}$

6–54. $\sigma_{max} = 2.06 \text{ MPa}$

6–55. $F = 4.56 \text{ kN}$

6–57. $I = 17.8133(10^{-6}) \text{ m}^4,$
$\sigma_{max} = 49.4 \text{ MPa}$

6–58. $(\sigma_{max})_T = 190.6 \text{ MPa (T)},$
$(\sigma_{max})_C = 160.3 \text{ MPa (C)}$

6–59. $M = 132.2 \text{ kN} \cdot \text{m}$

6–61. $\bar{y} = 232.61 \text{ mm}, I = 426.981(10^6) \text{ mm}^4,$
$\sigma_A = 16.43 \text{ MPa}, \sigma_D = 2.38 \text{ MPa},$
$(F_R)_C = 58.78 \text{ kN}$

6–62. $\sigma_A = 6.21 \text{ MPa (C)},$
$\sigma_B = 5.17 \text{ MPa (T)}$

6–63. $a = 1.68r$

6–65. $I = 786.8(10^6)$ mm^4, $\sigma_{max} = 1.45$ MPa,
$\sigma = 1.14$ MPa

6–66. $F_R = 15.54$ kN

6–67. $\sigma_{max} = 158$ MPa

6–69. $I_a = 0.21645(10^{-3})$ m^4,
$I_b = 0.36135(10^{-3})$ m^4,
$\sigma_{max} = 74.7$ MPa

6–70. $\sigma_{max} = 175.0$ MPa

6–71. $\sigma_{max} = 98.0$ MPa

6–73. $I = 63.685(10^6)$ mm^4, $\sigma_{max} = 111.4$ MPa

6–74. $\sigma_{max} = 166.2$ MPa

6–75. $\sigma_{allow} = 52.8$ MPa

6–77. $I = 63.685(10^6)$ mm^4, $w = 24.94$ kN/m

6–78. $\sigma_{max} = 451.1$ MPa

6–79. $\sigma_{max} = 103.7$ MPa

6–81. $w = 62.5$ kN/m, $\sigma_{max} = 10$ MPa

6–82. Use $t = 150$ mm

6–83. $\sigma_{max} = 129$ MPa

6–85. $b = 53.1$ mm

6–86. $\sigma_{max} = 152.8$ MPa

6–87. $d = 50.3$ mm

6–89. $M_{max} = 7.50$ kN $\cdot$ m, $a = 66.9$ mm

6–90. $\sigma_{max} = \dfrac{23w_0L^2}{36bh^2}$

6–91. $\sigma_{max} = 119$ MPa

6–93. $\bar{y} = 0.012848$ m, $I = 0.79925(10^{-6})$ m^4,
$\epsilon_{max} = 0.711(10^{-3})$ mm/mm

6–94. $d = 116$ mm

6–95. $d = 199$ mm

6–97. $M_{max} = 0.05P$, $P = 558.6$ N

6–98. $\sigma_{max} = 86.0$ MPa

6–99. $\sigma_{max} = 40.3$ MPa

6–101. $I = 69.6888(10^6)$ mm^4, $\sigma_{max} = 201.5$ MPa

6–102. $\sigma_B = 101.5$ MPa, $\sigma_A = 89.6$ MPa,

6–103. $\omega = 11.25$ kN/m

6–105. $I = \dfrac{2}{3}b^4$, $b = 195$ mm

6–106. $\sigma_{max} = 0.873$ MPa

6–107. $c = \dfrac{h\sqrt{E_c}}{\sqrt{E_t} + \sqrt{E_c}}$,
$(\sigma_{max})_t = \dfrac{3M}{b\,h^2}\left(\dfrac{\sqrt{E_t} + \sqrt{E_c}}{\sqrt{E_c}}\right)$

6–109. $M_y = -21.21$ kN $\cdot$ m, $M_z = -21.21$ kN $\cdot$ m,
$I_y = 287.5(10^6)$ mm^4, $I_z = 618.75(10^6)$ mm^4,
$\sigma_{max} = 16.08$ MPa (T), $\sigma_{max} = 16.08$ MPa (C),
$\alpha = 65.1°$

6–110. $M = 186.6$ kN $\cdot$ m

6–111. $\bar{y} = 57.4$ mm, $\sigma_A = 1.30$ MPa (C),
$\sigma_B = 0.587$ MPa (T),
$\alpha = -3.74°$

6–114. $\sigma_A = 89.1$ MPa

6–115. $\sigma_B = 17.1$ MPa

6–117. $M_z = -1039.23$ N $\cdot$ m, $M_y = -600.0$ N $\cdot$ m,
$I_z = 28.44583(10^{-6})$ m^4,
$I_y = 13.34583(10^{-6})$ m^4, $\sigma_A = 7.60$ MPa (T)

6–118. $\sigma_B = 131$ MPa (C), $\alpha = -66.5°$

6–119. $M = 1186$ kN $\cdot$ m

6–121. $M_{max} = 427.2$ N $\cdot$ m, $\sigma_{max} = 161$ MPa

6–122. $\sigma_A = 293$ kPa (C)

6–123. $\sigma_A = 293$ kPa (C)

6–125. $z'_A = 29.133$ mm, $y'_A = -70.711$ mm,
$\sigma_A = 165.4$ MPa

6–126. $\sigma_A = 165.4$ MPa

6–127. $h = 41.3$ mm,
$M = 6.60$ kN $\cdot$ m

6–129. $M_{max} = 37.97$ kN $\cdot$ m, $\bar{y} = 57.575$ mm,
$I = 12.0696(10^6)$ mm^4, $(\sigma_{max})_{st} = 181.1$ MPa,
$(\sigma_{max})_{al} = 106.3$ MPa

6–130. $w = 12.42$ kN/m

6–131. $(\sigma_{max})_{st} = 62.7$ MPa, $(\sigma_{max})_w = 4.1$ MPa

6–133. $\bar{y} = 63.077$ mm, $I_{NA} = 31.821(10^6)$ mm^4,
$M = 20.5$ kN $\cdot$ m

6–134. $(\sigma_{st})_{max} = 20.1$ MPa

6–135. $(\sigma_{st})_{max} = 10.37$ MPa, $(\sigma_w)_{max} = 0.62$ MPa

6–137. $\bar{y} = 0.1882$ m, $I = 18.08(10^{-6})$ m^4,
$(\sigma_{max})_{st} = 154$ MPa, $(\sigma_{max})_{al} = 171$ MPa

6–138. $d = 531$ mm, $M = 98.6$ kN $\cdot$ m

6–139. $(\sigma_{max})_{pvc} = 12.26$ MPa

6–141. $M_{max} = 60$ kN $\cdot$ m, $A_{st} = 1472.6$ mm^2,
$h' = 137.86$ mm, $I = 530.35(10^6)$ mm^4,
$(\sigma_{con})_{max} = 15.60$ MPa, $(\sigma_{st})_{max} = 146.0$ MPa

6–142. $M = 127.98$ kN·m

6–145. $A = 0.0028125\pi$ m², $\int_A \dfrac{dA}{r} = 0.053049301$ m, $M = 14.0$ kN·m

6–146. $P = 55.2$ kN

6–147. $(\sigma_{max})_t = 4.51$ MPa, $(\sigma_{max})_c = -5.44$ MPa

6–149. $\Sigma A = 0.00425$ m², $\bar{r} = 0.5150$ m, $\Sigma \int_A \dfrac{dA}{r} = 8.348614(10^{-3})$ m, $\sigma_C = 2.66$ MPa (T)

6–150. $(\sigma_{max})_t = 1.37$ MPa (T), $(\sigma_{max})_c = 0.79$ MPa (C)

6–151. $\sigma_A = 86.0$ MPa (T), $\sigma_B = 104.1$ MPa (C),

6–153. $\bar{r} = 1.235$ m, $\Sigma \int_A \dfrac{dA}{r} = 6.479051(10^{-3})$ m, $A = 0.008$ m², $\sigma_B = 26.2$ MPa (C)

6–154. $\sigma_t = 2.01$ MPa (T)

6–155. $P = 3.09$ N

6–157. $K = 2.60$, $M = 19.56$ kN·m

6–158. $\sigma_{max} = 95.8$ MPa

6–159. $r = 5.00$ mm

6–161. $K = 1.92$, $P = 0.469$ kN

6–162. $\sigma_{max} = 266.7$ MPa

6–163. $L = 950$ mm

6–165. $I_z = 82.78333(10^{-6})$ m⁴, $M_p = 211.25$ kN·m, $\sigma_{top} = \sigma_{bottom} = 43.5$ MPa

6–166. $k = \dfrac{3h}{2}\left[\dfrac{4bt(h-t)+t(h-2t)^2}{bh^3-(b-t)(h-2t)^3}\right]$

6–167. $k = 1.71$

6–169. $M_p = 289\,062.5$ N·m, $I = 91.14583(10^{-6})$ m⁴, $\sigma'_{top} = \sigma'_{bottom} = 67.1$ MPa

6–170. $k = 1.17$

6–171. $k = 1.70$

6–173. $I_x = 26.8(10^{-6})$ m⁴, $M_p = 0.00042\sigma_Y$, $M_Y = 0.000268\sigma_Y$, $k = 1.57$

6–174. $\sigma_T = \sigma_B = 142$ MPa

6–175. $k = 1.707$

6–177. $M_Y = 1.3724(10^6)\sigma_Y$, $M_p = 1.89583(10^6)\sigma_Y$, $k = 1.38$

6–178. $k = \dfrac{16r_o(r_o^3 - r_i^3)}{3\pi(r_o^4 - r_i^4)}$

6–179. $k = 2$

6–181. $d = \dfrac{2}{3}h$, $M = \dfrac{11ah^2}{54}\sigma_Y$

6–182. (a) $w_0 = 212.67$ kN/m (b) $w_0 = 269.79$ kN/m

6–183. (a) $P = 161.8$ kN (b) $P = 197.5$ kN

6–185. $\sigma - 50\sigma d - 3500(10^6)d = 0$, $M = 94.7$ N·m

6–186. (a) $M = 45.94$ kN·m, (b) $M = 78.54$ kN·m

6–187. $M = 251$ N·m

6–189. $\sigma = 574$ MPa, $M = 96.45$ kN·m

6–190. $F_R = 5.88$ kN

6–191. $(\sigma_{max})_t = 3.43$ MPa (T), $(\sigma_{max})_c = 1.62$ MPa (C)

6–193. $n = 18.182$, $I = 0.130578(10^{-3})$ m⁴, $M = 14.9$ kN·m

6–194. $M = 26.4$ kN·m

6–195. (a) $\sigma_{max} = 0.410$ MPa, (b) $\sigma_{max} = 0.410$ MPa

6–197. $\int_A \dfrac{dA}{r} = 0.012908358$ m, $A = 6.25(10^{-3})$ m², $\sigma_A = 225$ kPa (C), $\sigma_B = 265$ kPa (T)

6–198. $V = 94 - 30x$, $M = -15x^2 + 94x - 243.6$

6–199. $x = 0.6^-$, $V = -233$, $M = -50$

6–201. $\sigma = \dfrac{6M}{a^3}(\cos\theta + \sin\theta)$, $\dfrac{d\sigma}{d\theta} = 0$, $\theta = 45°$, $\alpha = 45°$

Chapter 7

7–1. $I = 0.2501(10^{-3})$ m⁴, $Q_A = 0.64(10^{-3})$ m³, $\tau_A = 2.56$ MPa

7–2. $\tau_{max} = 3.46$ MPa

7–3. $V_w = 19.0$ kN

7–5. $\bar{y} = 82.50$ mm, $I_{NA} = 152.578(10^6)$ mm⁴, $V_f = 19.08$ kN

7–6. $\tau_A = 1.99$ MPa, $\tau_B = 1.65$ MPa

7–7. $\tau_{max} = 4.62$ MPa

7–9. $\bar{y} = 29.167$ mm, $I = 2.6367(10^6)$ mm⁴, $V = 140.58$ kN

7–10. $\tau_{max} = 35.85$ MPa

7–11. $V = 100$ kN

7–13. $\bar{y} = 0.080196$ m, $I = 4.8646(10^{-6})$ m⁴, $\tau_{max} = 4.22$ MPa

7–14. $V = 190$ kN

7–15. The factor $= \dfrac{4}{3}$

7–17. $I = 0.175275(10^{-3})$ m⁴, $Q_{max} = 1.09125(10^{-3})$ m³, $\tau_{max} = 37.4$ MPa

7–18. $V = 723$ kN

7–19. $(\tau_f)_{max} = 9.24$ MPa, $(\tau_w)_{max} = 37.4$ MPa

7–21. $I = 1.5625\pi$ mm^4, $Q = \dfrac{2}{3}(2500 - y^2)^{3/2}$,

$\tau_A = 19.1$ MPa

7–22. $\tau_B = 4.41$ MPa

7–23. $\tau_{max} = 4.85$ MPa

7–25. $V_C = -13.75$ kN, $I = 27.0(10^{-6})$ m^4,

$Q_{max} = 0.216(10^{-3})$ m^3, $\tau_{max} = 3.67$ MPa

7–26. $\tau_{max} = 2.553$ MPa

7–27. $\tau_C = 11.9$ MPa, $\tau_D = 9.83$ MPa

7–33. $I = 12.5(10^6)$ mm^4, $Q = 187500$ mm^3, $F = 3.37$ kN

7–34. $V = 7.20$ kN

7–35. $V = 8.0$ kN, $s = 65$ mm

7–37. $I_{NA} = 34.871(10^6)$ mm^4, $Q = 151650$ mm^3,

$V = 172.5$ kN

7–38. $\tau_n = 35.2$ MPa

7–39. $F = 12.5$ kN

7–41. $I_{NA} = 1126(10^6)$ mm^4, $Q = 2.625(10^6)$ mm^3,

$Q_{max} = 3.2325(10^6)$ mm^4, $P = 467$ kN

7–42. $V = 36.7$ kN, use $s = 35$ mm

7–43. $(\tau_{nail})_{avg} = 119$ MPa

7–45. $I_{NA} = 72.0(10^{-6})$ m^4, $Q = 0.450(10^{-3})$ m^3,

$P = 6.60$ kN

7–47. $s = 216.6$ mm, $s' = 30.7$ mm

7–50. $q_A = 228$ kN/m, $q_B = 462$ kN/m

7–51. $q_C = 0$, $q_D = 601$ kN/m

7–53. $I_{NA} = 125.17(10^{-6})$ m^4, $Q_C = 0.5375(10^{-3})$ m^3,

$q_C = 38.6$ kN/m

7–54. $q_A = 1.39$ kN/m, $q_B = 1.25$ kN/m

7–55. $q_{max} = 1.63$ kN/m

7–57. $\bar{y} = 92.47$ mm, $I = 54.599(10^6)$ mm^4,

$q_{max} = 82.88$ kN/m

7–58. $q_A = 215$ kN/m

7–59. $q_{max} = 232$ kN/m

7–61. $\bar{y} = 70.641$ mm, $I = 34.628(10^6)$ mm^4,

$q_A = 39.2$ kN/m, $q_B = 90.1$ kN/m,

$q_{max} = 122.7$ kN/m

7–62. $\tau = \dfrac{V}{\pi R^2 t}\sqrt{R^2 - y^2}$

7–63. $e = \dfrac{3(b_2^2 - b_1^2)}{h + 6(b_1 + b_2)}$

7–65. $I = \dfrac{10}{3}a^3 t$, $Q_1 = \dfrac{t}{2}y^2$,

$Q_2 = \dfrac{at}{2}(a + 2x)$, $e = \dfrac{7}{10}a$

7–66. $e = \dfrac{2\sqrt{3}}{3}a$

7–67. $e = 0$

7–69. $Pe = F(h) + 2V(b)$,

$I = \dfrac{t}{12}(2h^3 + 6bh^2 - (h - 2h_1)^3)$,

$e = \dfrac{b(6h_1 h^2 + 3h^2 b - 8h_1^3)}{2h^3 + 6bh^2 - (h - 2h_1)^3}$

7–70. $e = \dfrac{4r(\sin \alpha - \alpha \cos \alpha)}{2\alpha - \sin 2\alpha}$

7–71. $V_{AB} = 49.78$ kN

7–73. $\bar{y} = 0.08798$ m,

$I_{NA} = 86.93913(10^{-6})$ m^4,

$Q_A = 0$, $Q_C = 0.16424(10^{-3})$ m^3,

$q_A = 0$, $q_B = 1.21$ kN/m,

$q_C = 3.78$ kN/m

7–74. $V = 20.37$ kN

7–75. $V = 3.731$ kN

Chapter 8

8–1. $t = 18.8$ mm

8–2. $r_o = 1.812$ m

8–3. Case (a): $\sigma_1 = 8.33$ MPa; $\sigma_2 = 0$

Case (b): $\sigma_1 = 8.33$ MPa; $\sigma_2 = 4.17$ MPa

8–5. $\sigma = 133$ MPa, $P_b = 35.56(10^3)\pi$ N,

$\sigma_b = 228$ MPa

8–6. $t = 26.7$ mm, $n = 820$ bolts

8–7. (a) $\sigma_1 = 127$ MPa, (b) $\sigma_1' = 79.1$ MPa,

(c) $(\tau_{avg})_r = 322$ MPa

8–9. $t_c = 40$ mm, $t_s = 20$ mm,

$(P_b)_{allow} = 122.72(10^3)$ N, $n_s = 308$ bolts

8–10. $s = 833.33$ m

8–11. $\sigma_h = 3.15$ MPa, $\sigma_b = 66.8$ MPa

8–13. $\delta_F - \delta_T = 0$, $\sigma_c = 19.69$ MPa

8–14. $\delta_{r_i} = \dfrac{pr_i^2}{E(r_o - r_i)}$

8–15. $p = \dfrac{E(r_2 - r_3)}{\dfrac{r_2^2}{r_2 - r_1} + \dfrac{r_3^2}{r_4 - r_3}}$

8–17. $\sigma_{fil} = \dfrac{pr}{t + t'} + \dfrac{T}{wt'}$,

$\sigma_w = \dfrac{pr}{t + t'} - \dfrac{T}{wt}$

8–18. $d = 66.7$ mm

8–19. $\sigma_L = 66.7$ MPa (C), $\sigma_R = 33.3$ MPa (T)

8–21. $\sigma_A = -\dfrac{P}{A} + \dfrac{Mc}{I} = 123$ MPa, $\sigma_B = 62.5$ MPa

8–22. $\sigma_{max} = 1.07$ MPa

8–23. $\sigma_{max} = 1.07$ MPa

8–25. $N = 3.031$ kN, $V = 1.75$ kN,
$M = 24.5$ N $\cdot$ m,
$\sigma_B = 51.84$ MPa, $\tau_B = 0$

8–26. $w = 79.7$ mm

8–27. $P = 109$ kN

8–29. $A = 216$ mm^2, $Q_A = 324$ mm^3,
$I = 2592$ mm^4
$\sigma_A = 4.63$ MPa, $\sigma_B = 8.10$ MPa,
$\tau_A = 5.21$ MPa, $\tau_B = 0$

8–30. $\sigma_A = 504$ kPa (C), $\tau_A = 14.9$ kPa

8–31. $d = 66.7$ mm

8–33. $I = 416.67$ mm^4, $Q_B = 0$
$Q_C = 62.5$ mm^3, $\sigma_B = 45.5$ MPa (T),
$\tau_B = 0$, $\sigma_C = 0.5$ MPa (C), $\tau_C = 1.30$ MPa

8–34. $\sigma_D = 0$, $\tau_D = 5.33$ MPa, $\sigma_E = 187.7$ MPa, $\tau_E = 0$

8–35. $\sigma_A = -65.31$ MPa, $\tau_A = 0$, $\sigma_B = 18.77$ MPa,
$\tau_B = 7.265$ MPa

8–37. $A = 25\pi(10^{-6})$ m^2, $I_z = 0.15625\pi(10^{-9})$ m^4,
$J = 0.3125\pi(10^{-9})$ m^4, $\sigma_B = 1.53$ MPa (C),
$\tau_B = 100$ MPa

8–38. $T = 9.331$ kN

8–39. $T = 9.343$ kN

8–41. $A = 9.00\,(10^{-3})$ m^2, $I = 82.8\,(10^{-6})$ m,
$Q_B = 0$, $\sigma_B = 0.522$ MPa (C), $\tau_B = 0$

8–42. $\sigma = 17.9$ MPa (C), $\tau = 1.06$ MPa

8–43. $\sigma = 23.9$ MPa (C), $\tau = 0.796$ MPa

8–45. $A = 11250$ mm^2, $I_y = 5273437.5$ mm^4,
$I_z = 21093750$ mm^4, $\sigma_C = 8$ MPa (C),
$\sigma_D = 8$ MPa (C)

8–46. $\sigma_{max} = \dfrac{1.33P}{a^2}$ (C), $\sigma_{min} = \dfrac{P}{3a^2}$ (T)

8–47. $\sigma_{max} = \dfrac{0.368P}{r^2}$ (C), $\sigma_{min} = \dfrac{0.0796P}{r^2}$ (T)

8–49. $R = 0.080889$ m, $N = -24.525$ N,
$M = 14.2463$ N $\cdot$ m,
$\sigma_A = 89.1$ MPa (C), $\sigma_B = 79.3$ kPa (T)

8–50. $(\sigma_t)_{max} = 70.1$ MPa, $(\sigma_c)_{max} = -58.2$ MPa

8–51. $6e_y + 18e_z < 5a$

8–53. $A = 13.5$ m^2, $I_x = 22.78125$ m^4,
$I_y = 10.125$ m^4, $y = 0.75 - 1.5x$

8–54. $\sigma_A = 9.88$ kPa (T), $\sigma_B = 49.4$ kPa (C),
$\sigma_C = 128$ kPa (C), $\sigma_D = 69.1$ kPa (C)

8–55. $\sigma_A = 11.9$ MPa (T), $\tau_A = -0.318$ MPa

8–57. $N_y = -4$ kN, $V_z = -3$ kN
$V_x = 2.5$ kN, $T_y = 0.9$ kN $\cdot$ m,
$M_x = -0.6$ kN $\cdot$ m, $\sigma = 140.6$ MPa (T),
$\tau = 38.7$ MPa

8–58. $\sigma = 46.86$ MPa (C), $\tau = 38.4$ MPa

8–59. $\sigma_{max} = 71.0$ MPa (C)

8–61. $A = 490.87$ mm^2, $J = 38349.5$ mm^4
$I = 19164.8$ mm^4, $(Q_C)_x = 0$,
$(Q_A)_z = 1302.1$ mm^3, $\sigma_A = 129.6$ MPa (T),
$\tau_A = -22.75$ MPa

8–62. $\sigma_B = 103.1$ MPa (T), $\tau_B = 27.16$ MPa

8–63. $\sigma_C = 111.5$ MPa (T), $\sigma_D = 866.2$ MPa (T),
$\tau_D = 434.3$ MPa, $\tau_C = -360.8$ MPa

8–65. $T = -64.95$ N $\cdot$ m, $M_y = 31.25$ N $\cdot$ m,
$M_z = -54.13$ N $\cdot$ m, $I_y = I_z = 204442$ mm^4,
$J = 408884$ mm^4, $(Q_y)_A = 0$,
$(Q_z)_A = 5844$ mm^3,
$\sigma_A = 5.03$ MPa (T), $\tau_A = 2.72$ MPa

8–66. $\sigma_B = 3.82$ MPa (C), $\tau_B = 3.45$ MPa

8–67. $\sigma = -\dfrac{2P}{bh^3}(h^2 + 18e_y y)$, $-\dfrac{h}{6} \le e_y \le \dfrac{h}{12}$

8–69. $I = 0.1256637\,(10^{-6})$ m^4,
$A = 1.256637\,(10^{-3})$ m^2, $Q_B = 0$,
$\sigma_B = -21.7$ MPa, $\tau_B = 0$

8–70. $\tau_A = 0$, $\sigma_A = 262.0$ MPa (C)

8–71. $\sigma_B = 0$, $\tau_B = 3.14$ MPa

8–73. $R = 45.8032$ mm, $e = 0.1968$ mm,
$I = 324\pi$ mm^4, $Q_B = 144$ mm^3,
$\sigma = 0.011$ MPa (T), $\tau = 3.33$ MPa

8–74. $\sigma_A = -0.1703$ MPa, $\sigma_B = -0.0977$ MPa

8–75. $\sigma_E = 802$ kPa, $\tau_E = 69.8$ kPa

8–77. $\displaystyle\int_A \dfrac{dA}{r} = 0.8586$ mm, $A = 28.27$ mm^2,
$(\sigma_t)_{max} = 419.3$ MPa (T), $(\sigma_c)_{max} = 349.4$ MPa (C)

8–78. $(\sigma_t)_{max} = 250.1$ MPa (T), $(\sigma_c)_{max} = 208.4$ MPa (C)

8–79. $\sigma_{max} = 2.12$ MPa (C)

8–81. $p = 12(10^6)$ MPa, $F = 30(10^3)\pi$,
$P = 94.2$ kN

8–82. $\sigma_{max} = 397.4$ MPa (T)

8–83. $\sigma_1 = 7.07$ MPa, $\sigma_2 = 0$

8–85. $p = 3.60$ MPa, $F_b = 6.3617(10^6)$ N,
$n = 113$ bolts

8–86. $\sigma_1 = 50.0$ MPa, $\sigma_2 = 25.0$ MPa, $F_b = 133$ kN

Chapter 9

9–2. $\sigma_{x'} = -3.48$ MPa, $\tau_{x'y'} = 4.63$ MPa

9–3. $\sigma_{x'} = -678$ MPa, $\tau_{x'y'} = 41.5$ MPa

9–5. $\sigma_x = -650$ MPa, $\sigma_y = 400$ MPa, $\tau_{xy} = 0, \theta = 30°$,
$\sigma_{x'} = -388$ MPa, $\tau_{x'y'} = 455$ MPa

9–6. $\sigma_{x'} = 49.7$ MPa, $\tau_{x'y'} = -34.8$ MPa

9–7. $\sigma_{x'} = 49.7$ MPa, $\tau_{x'y'} = -34.8$ MPa

9–9. $\theta = +135°, \sigma_x = 80$ MPa,
$\sigma_y = 0, \tau_{xy} = 45$ MPa,
$\sigma_{x'} = -5$ MPa, $\tau_{x'y'} = 40$ MPa

9–10. $\sigma_{x'} = -2.71$ MPa, $\tau_{x'y'} = 4.17$ MPa

9–11. $\sigma_{x'} = -2.71$ MPa, $\tau_{x'y'} = 4.17$ MPa

9–13. $\theta = 60°, \sigma_x = 200$ MPa,
$\sigma_y = -350$ MPa, $\tau_{xy} = 75$ MPa,
$\sigma_{x'} = -277$ MPa, $\sigma_{y'} = 127$ MPa, $\tau_{x'y'} = 201$ MPa

9–14. (a) $\sigma_1 = 4.21$ MPa, $\sigma_2 = -34.2$ MPa,
$\theta_{p_2} = 19.3°, \theta_{p_1} = -70.7°$

(b) $\tau_{\max}_{\text{in-plane}} = 19.2$ MPa, $\sigma_{\text{avg}} = -15$ MPa,
$\theta_s = -25.7°, 64.3°$

9–15. $\sigma_1 = -19.0$ MPa, $\sigma_2 = -121$ MPa,
$(\theta_p)_1 = 39.3° \ (\theta_p)_2 = -50.7°$,
$\tau_{\max}_{\text{in-plane}} = 51.0$ MPa, $\theta_s = -5.65°, 84.3°$

9–17. $\sigma_x = 125$ MPa, $\sigma_y = -75$ MPa,
$\tau_{xy} = -50$ MPa,
$\sigma_1 = 137$ MPa, $\sigma_2 = -86.8$ MPa,
$(\theta_p)_1 = -13.3°, (\theta_p)_2 = 76.7°$,
$\tau_{\max}_{\text{in-plane}} = 112$ MPa, $\sigma_{\text{avg}} = 25$ MPa

9–18. $\sigma_x = -193$ MPa, $\sigma_y = -357$ MPa,
$\tau_{xy} = 102$ MPa

9–19. $\sigma_1 = 224$ MPa, $\sigma_2 = -64.2$ MPa,
$(\theta_p)_1 = -61.8°, (\theta_p)_2 = 28.2°$,
$\tau_{\max}_{\text{in-plane}} = 144$ MPa, $\theta_s = -16.8°, 73.2°$,
$\sigma_{\text{avg}} = 80$ MPa

9–21. $\sigma_x = 61.962$ MPa, $\tau_{xy} = 30$ MPa
$\sigma_1 = 80.1$ MPa, $\sigma_2 = 19.9$ MPa

9–22. $\sigma_1 = 4.93$ MPa, $\sigma_2 = -111$ MPa,
$(\theta_p)_1 = 78.1°, (\theta_p)_2 = -11.9°$

9–23. $I = 0.45(10^{-3})$ m^4, $Q_A = 1.6875(10^{-3})$ m^3,
$\sigma_{x'} = 0.507$ MPa, $\tau_{x'y'} = 0.958$ MPa

9–25. $N = 400$ N, $M = 100$ N·m,
$A = 0.1(10^{-3})\pi$ m^2, $I = 2.5(10^{-9})\pi$ m^4,
$\sigma_1 = 0, \sigma_2 = -126$ MPa,
$\tau_{\max}_{\text{in-plane}} = 63.0$ MPa, $\sigma_{\text{avg}} = -63.0$ MPa

9–26. $\sigma_1 = 244.5$ MPa, $\sigma_2 = 0$,
$\tau_{\max}_{\text{in-plane}} = 122.25$ MPa, $\theta_s = -45°, 45°$,
$\sigma_{\text{avg}} = 122.25$ MPa

9–27. $\sigma_1 = 0, \sigma_2 = -187.68$ MPa,
$\tau_{\max}_{\text{in-plane}} = 93.84$ MPa, $\theta_s = 45°, 135°$,
$\sigma_{\text{avg}} = -93.84$ MPa

9–29. $V = 70.5$ kN, $M = 39.15$ kN·m,
$I = 49.175(10^{-6})$ m^4, $Q_A = 0.255(10^{-3})$ m^3,
$\sigma_1 = 64.9$ MPa, $\sigma_2 = -5.15$ MPa,
$(\theta_p)_1 = 15.7°, (\theta_p)_2 = -74.3°$

9–30. Point A: $\sigma_1 = 15.0$ MPa, $\sigma_2 = -0.235$ MPa,
Point B: $\sigma_1 = 0.723$ MPa, $\sigma_2 = -6.83$ MPa

9–31. $\sigma_1 = 6.38$ MPa, $\sigma_2 = -0.360$ MPa,
$(\theta_p)_1 = 13.4°, (\theta_p)_2 = 103°$

9–33. $I = 0.3125(10^{-6})$ m^4, $Q_A = 0$,
$Q_B = 9.375(10^{-6})$ m^3,
$\sigma_1 = 0, \sigma_2 = -192$ MPa,
$\sigma_1 = 24.0, \sigma_2 = -24.0$ MPa,
$\theta_{p_1} = -15.0°, \theta_{p_2} = 45.0°$

9–34. $\sigma_1 = 0, \sigma_2 = -10.70$ MPa,
$\tau_{\max}_{\text{in-plane}} = 5.35$ MPa, $\theta_s = 45°, -45°$

9–35. $\tau_{\max}_{\text{in-plane}} = 5$ kPa, $\sigma_{\text{avg}} = 0$

9–37. $A = \dfrac{\pi}{4}d^2, I = \dfrac{\pi}{64}d^4, Q_A = 0$,
$\sigma_1 = \dfrac{4}{\pi d^2}\left(\dfrac{2PL}{d} - F\right), \sigma_2 = 0$,
$\tau_{\max}_{\text{in-plane}} = \dfrac{2}{\pi d^2}\left(\dfrac{2PL}{d} - F\right)$

9–38. $\tau_{x'y'} = -47.5$ kPa

9–39. $\sigma_x = 82.3$ kPa

9–41. $\bar{y} = 0.0991$ m, $I = 7.4862(10^{-6})$ m^4,
$A = 3.75(10^{-3})$ m^2,
$\sigma_1 = 21.2$ MPa, $\sigma_2 = -0.380$ MPa, $\theta_p = -7.63°$

9–42. $\sigma_1 = 24.50$ MPa, $\sigma_2 = -33.96$ MPa,
$\tau_{\max}_{\text{in-plane}} = 29.23$ MPa

9–43. $\sigma_1 = 1.27$ MPa, $\sigma_2 = -62.4$ MPa,
$(\theta_p)_1 = 81.9°, (\theta_p)_2 = -8.11°$

9–45. $V = 10$ kN, $M = 19.5$ kN $\cdot$ m,

$I = 33.854(10^6)$ mm^4, $\tau_{\max_{\text{in-plane}}} = 21.60$ MPa,

$\sigma_{\text{avg}} = -21.60$ MPa, $\theta_s = 45°, -45°$

9–46. $\sigma_1 = 1.754$ MPa, $\sigma_2 = -1.754$ MPa,

$(\theta_p)_1 = 45°, (\theta_p)_2 = -45°$

9–47. $\sigma_1 = 5.50$ MPa, $\sigma_2 = -0.611$ MPa

9–49. $I_z = 0.350(10^{-3})$ m^4, $I_y = 68.75(10^{-6})$ m^4,

$(Q_A)_y = 0$, $\sigma_1 = 0$, $\sigma_2 = -77.4$ MPa,

$\tau_{\max_{\text{in-plane}}} = 38.7$ MPa

9–50. $\sigma_1 = 0$, $\sigma_2 = -20$ kPa, $\tau_{\max_{\text{in-plane}}} = 10$ kPa

9–51. $\sigma_{x'} = -388$ psi, $\tau_{x'y'} = 455$ psi

9–53. $R = 19.21$ MPa,

$\sigma_1 = 4.21$ MPa, $\sigma_2 = -34.2$ MPa, $\theta_{p2} = 19.3°$,

$\tau_{\max_{\text{in-plane}}} = 19.2$ MPa, $\sigma_{\text{avg}} = -15$ MPa, $\theta_{s_2} = 45°$

9–54. $\sigma_1 = 53.0$ MPa, $\sigma_2 = -68.0$ MPa,

$\theta_{p1} = 14.9°$ counterclockwise,

$\tau_{\max_{\text{in-plane}}} = 60.5$ MPa, $\sigma_{\text{avg}} = -7.50$ MPa,

$\theta_{s1} = 30.1°$ clockwise

9–55. $\sigma_{x'} = -19.9$ MPa, $\tau_{x'y'} = 7.70$ MPa,

$\sigma_{y'} = 9.89$ MPa

9–58. $\sigma_{x'} = -421$ MPa, $\tau_{x'y'} = -354$ MPa,

$\sigma_{y'} = 421$ MPa

9–59. $\sigma_{x'} = 4.99$ MPa, $\tau_{x'y'} = -1.46$ MPa,

$\sigma_{y'} = -3.99$ MPa

9–61. $\sigma_{\text{avg}} = -155$ MPa, $R = 569.23$ MPa,

$\sigma_{x'} = -299$ MPa, $\tau_{x'y'} = 551$ MPa,

$\sigma_{y'} = -11.1$ MPa

9–62. $\sigma_{x'} = 0.250$ MPa, $\tau_{x'y'} = 3.03$ MPa,

$\sigma_{y'} = -3.25$ MPa

9–63. $\sigma_{\text{avg}} = 52.5$ MPa

(a) $\sigma_1 = 115.60$ MPa, $\sigma_2 = -10.60$ MPa,

$\theta_{p_1} = 16.8°$ (*Clockwise*)

(b) $\tau_{\max_{\text{in-plane}}} = -63.1$ MPa,

$\theta_s = 28.2°$ (*Counterclockwise*)

9–65. $R = 192.094$, $\sigma_1 = 342$ MPa,

$\sigma_2 = -42.1$ MPa, $\theta_P = 19.3°$,

$\sigma_{\text{avg}} = 150$ MPa, $\tau_{\max_{\text{in-plane}}} = 19.2$ MPa,

$\theta_s = -25.7°$

9–66. (a) $\sigma_1 = 88.1$ MPa,

$\sigma_2 = -13.1$ MPa

(b) $\tau_{\max_{\text{in-plane}}} = 50.6$ MPa,

$\sigma_{\text{avg}} = 37.5$ MPa,

$\theta_2 = 4.27°$

9–67. (a) $\sigma_1 = 646$ MPa,

$\sigma_2 = -496$ MPa,

$\theta_{p_1} = 30.6°$ (*Counterclockwise*)

(b) $\tau_{\max_{\text{in-plane}}} = 571$ MPa,

$\theta_s = 14.4°$ (*Clockwise*)

9–69. $\sigma_{x'} = 11.0$ kPa,

$\tau_{x'y'} = -22.6$ kPa

9–70. $\sigma_{x'} = -12.5$ kPa,

$\tau_{x'y'} = 21.7$ kPa

9–71. $\sigma_1 = 1.144$ MPa, $\sigma_2 = -3.432$ MPa

9–73. $A = 7200$ mm^2, $I = 8640000$ mm^4, $Q_A = 8100$ mm^3,

$\sigma_1 = 15.0$ MPa, $\sigma_2 = -0.235$ MPa

9–74. $\sigma_1 = 0.723$ MPa, $\sigma_2 = -6.83$ MPa

9–75. $\sigma_1 = 35$ MPa, $\sigma_2 = -9.6$ MPa

$\tau_{\max_{\text{in-plane}}} = 22.3$ MPa

9–77. $\sigma_1 = \sigma_2 = 35$ MPa

9–78. $\sigma_{x'} = 500$ MPa, $\tau_{x'y'} = 167$ MPa

9–79. $\tau = 0.2222$ MPa, $\sigma_{\text{avg}} = 0.5556$ MPa,

$R = 0.5984$ MPa,

$\sigma_{x'} = 470$ kPa,

$\tau_{x'y'} = 592$ kPa

9–81. $N = 900$ N, $V = 900$ N, $M = 675$ N $\cdot$ m,

$A = 1.4(10^{-3})$ m^2, $I = 1.7367(10^{-6})$ m^4,

$\sigma_1 = 9.18$ MPa, $\sigma_2 = -0.104$ MPa,

$(\theta_p)_1 = 6.08°$ (*Counterclockwise*)

9–82. $\sigma_1 = 32.5$ MPa, $\sigma_2 = -0.118$ MPa,

$(\theta_p)_1 = 3.44°$ (*Counterclockwise*)

9–83. $\sigma_1 = 7.52$ MPa, $\sigma_2 = 0$,

$\tau_{\max_{\text{in-plane}}} = 3.76$ MPa,

$\theta_s = 45°$ (*Counterclockwise*)

9–85. $\sigma_{\min} = -300$ MPa, $\sigma_{\text{int}} = 0$, $\sigma_{\max} = 400$ MPa

9–86. $\sigma_1 = 0$, $\sigma_2 = 137$ MPa,

$\sigma_3 = -46.8$ MPa,

$\tau_{\text{abs}_{\max}} = 91.8$ MPa

9–87. $\sigma_{\max} = 158$ MPa, $\sigma_{\min} = -8.22$ MPa,

$\sigma_{\text{int}} = 0$ MPa, $\tau_{\text{abs}_{\max}} = 83.2$ MPa

9–89. $\sigma_1 = 222$ MPa, $\sigma_2 = 0$ MPa,
$\sigma_3 = -102$ MPa,
$\tau_{\text{abs}\atop\text{max}} = 162$ MPa

9–90. $\sigma_1 = 6.73$ kPa, $\sigma_2 = 0$, $\sigma_3 = -4.23$ kPa,
$\tau_{\text{abs}\atop\text{max}} = 5.48$ kPa

9–93. $\sigma_{\text{max}} = 100$ MPa, $\sigma_{\text{int}} = 50$ MPa, $\sigma_{\text{min}} = 0$,
$\tau_{\text{abs}\atop\text{m ax}} = 50$ MPa

9–94. $\sigma_{\text{max}} = 5.06$ MPa, $\sigma_{\text{int}} = 0$, $\sigma_{\text{min}} = -8.04$ MPa,
$\tau_{\text{abs}\atop\text{m ax}} = 6.55$ MPa

9–95. $\sigma_{\text{max}} = 98.8$ MPa, $\sigma_{\text{int}} = \sigma_{\text{min}} = 0$,
$\tau_{\text{abs}\atop\text{max}} = 49.4$ MPa

9–97. $P = 0.900(10^6)$ N·m/s, $T_0 = 60.0(10^3)$ N·m,
$A = 0.015625\pi$ m^2, $J = 0.3835(10^{-3})$ m^4,
$\tau_{\text{max}\atop\text{in-plane}} = 23.2$ MPa

9–98. $\sigma_1 = 0.983$ MPa, $\sigma_2 = -0.983$ MPa

9–99. $\sigma_1 = 2.726$ MPa, $\sigma_2 - 0.598$ MPa

9–101. $\sigma_1 = \dfrac{2}{\pi d^2}\left(-F + \sqrt{F^2 + \dfrac{64T_0^2}{d^2}}\right)$,

$\sigma_2 = -\dfrac{2}{\pi d^2}\left(F + \sqrt{F^2 + \dfrac{64T_0^2}{d^2}}\right)$,

$\tau_{\text{max}\atop\text{in-plane}} = \dfrac{2}{\pi d^2}\sqrt{F^2 + \dfrac{64T_0^2}{d^2}}$

9–102. $\sigma_{x'} = -63.3$ MPa, $\tau_{x'y'} = 35.7$ MPa

9–103. $\sigma_1 = 3.29$ MPa,
$\sigma_2 = -4.30$ MPa,
$(\theta_p)_2 = 41.1°$ (*Clockwise*)

9–105. $Q_C = 31.25(10^{-6})$ m^3, $I = 2.0833(10^{-6})$ m^4,
$\tau = 26.4$ kPa, $\sigma_1 = 26.4$ kPa, $\sigma_2 = -26.4$ kPa,
$\theta_{p_1} = -45°$; $\theta_{p_2} = 45°$

9–106. $\sigma_{x'} = -22.9$ kPa, $\tau_{x'y'} = -13.2$ kPa

Chapter 10

10–2. $\epsilon_{x'} = 248(10^{-6})$, $\gamma_{x'y'} = -233(10^{-6})$,
$\epsilon_{y'} = -348(10^{-6})$

10–3. $P = 78.4$ kW

10–5. (a) $\epsilon_1 = 441(10^{-6})$, $\epsilon_2 = -641(10^{-6})$,
$\theta_{p1} = -24.8°$, $\theta_{p2} = 65.2°$

(b) $\gamma_{\text{max}\atop\text{in-plane}} = 1.08(10^{-3})$,

$\epsilon_{\text{avg}} = -100(10^{-6})$

10–6. $\epsilon_{x'} = 145(10^{-6})$, $\gamma_{x'y'} = 583(10^{-6})$,
$\epsilon_{y'} = 155(10^{-6})$

10–7. $\epsilon_{x'} = 215(10^{-6})$, $\gamma_{x'y'} = -248(10^{-6})$,
$\epsilon_{y'} = 185(10^{-6})$

10–9. $\epsilon_1 = 188(10^{-6})$, $\epsilon_2 = -128(10^{-6})$,
$(\theta_p)_1 = -9.22°$, $(\theta_p)_2 = 80.8°$,

$\gamma_{\text{max}\atop\text{in-plane}} = 316(10^{-6})$,

$\theta_s = 35.8°, -54.2°$, $\epsilon_{\text{avg}} = 30(10^{-6})$

10–10. $\epsilon_{x'} = 103(10^{-6})$, $\epsilon_{y'} = 46.7(10^{-6})$,
$\gamma_{x'y'} = 718(10^{-6})$

10–11. $\epsilon_1 = -79.3(10^{-6})$, $\epsilon_2 = -221(10^{-6})$,
$(\theta_p)_1 = 22.5°$, $(\theta_p)_2 = -67.5°$,

$\gamma_{\text{max}\atop\text{in-plane}} = 141(10^{-6})$, $\theta_s = -22.5°, 67.5°$,

$\epsilon_{\text{avg}} = -150(10^{-6})$

10–13. $\epsilon_1 = 17.7(10^{-6})$, $\epsilon_2 = -318(10^{-6})$,
$(\theta_p)_1 = 76.7°$, $(\theta_p)_2 = -13.3°$,

$\gamma_{\text{max}\atop\text{in-plane}} = 335(10^{-6})$, $\theta_s = 31.7°, 122°$,

$\epsilon_{\text{avg}} = -150(10^{-6})$

10–14. (a) $\epsilon_1 = 368(10^{-6})$, $\epsilon_2 = 182(10^{-6})$,
$\theta_{p1} = -52.8°$, $\theta_{p2} = 37.2°$

(b) $\gamma_{\text{max}\atop\text{in-plane}} = 187(10^{-6})$,

$\theta_s = -7.76°, \ 82.2°$, $\epsilon_{\text{avg}} = 275(10^{-6})$

10–17. $R = 167.71(10^{-6})$, $\epsilon_1 = 138(10^{-6})$,
$\epsilon_2 = -198(10^{-6})$, $\theta_p = 13.3°$

10–18. $(\gamma_{xy})_{\text{max}\atop\text{in-plane}} = 335(10^{-6})$, $\epsilon_{\text{avg}} = -30(10^{-6})$,

$\theta_s = -31.7°$

10–19. $\epsilon_{x'} = -309(10^{-6})$, $\epsilon_{y'} = -541(10^{-6})$,
$\gamma_{x'y'} = -423(10^{-6})$

10–21. (a) $R = 93.408$, $\epsilon_1 = 368(10^{-6})$,
$\epsilon_2 = 182(10^{-6})$, $\theta_{p1} = -52.8°$, $\theta_{p2} = 37.2°$

(b) $\gamma_{\text{max}\atop\text{in-plane}} = 187(10^{-6})$,

$\theta_s = -7.76°, \ 82.2°$, $\epsilon_{\text{avg}} = 275(10^{-6})$

10–22. (a) $\epsilon_1 = 773(10^{-6})$, $\epsilon_2 = 76.8(10^{-6})$

(b) $\gamma_{\text{max}\atop\text{in-plane}} = 696(10^{-6})$

(c) $\gamma_{\text{abs}\atop\text{max}} = 773(10^{-6})$

10–23. (a) $\epsilon_1 = 192(10^{-6})$, $\epsilon_2 = -152(10^{-6})$

(b, c) $\gamma_{\substack{abs \\ max}} = \gamma_{\substack{max \\ in\text{-}plane}} = 344(10^{-6})$

10–25. $\epsilon_1 = 308(10^{-6})$, $\epsilon_2 = -108(10^{-6})$,

$(\theta_p)_2 = 8.05°$ (Clockwise), $R = 208.17(10^{-6})$,

$\gamma_{\substack{max \\ in\text{-}plane}} = 416(10^{-6})$, $\epsilon_{avg} = 100(10^{-6})$,

$\theta_s = 36.9°$ (Counterclockwise)

10–26. $\epsilon_1 = 451(10^{-6})$, $\epsilon_2 = -451(10^{-6})$,

$(\theta_p)_1 = 28.2°$ (Counterclockwise),

$\gamma_{\substack{max \\ in\text{-}plane}} = 902(10^{-6})$, $\epsilon_{avg} = 0$,

$\theta_s = 16.8°$ (Clockwise)

10–27. $\epsilon_1 = 339(10^{-6})$, $\epsilon_2 = -489(10^{-6})$,

$(\theta_p)_1 = 32.5°$ (Counterclockwise),

$\gamma_{\substack{max \\ in\text{-}plane}} = 828(10^{-6})$,

$\epsilon_{avg} = -75(10^{-6})$,

$\theta_s = 12.5°$ (Clockwise)

10–33. $\sigma_3 = 0$, $\sigma_1 = 71.64$ MPa, $\sigma_2 = 51.64$ MPa

10–34. $\gamma_{\substack{abs \\ max}} = 41.1(10^{-6})$

10–35. $\epsilon_{max} = 30.5(10^{-6})$, $\epsilon_{int} = \epsilon_{min} = -10.7(10^{-6})$

10–37. (a) $K_r = 23.33$ MPa

(b) $K_g = 35.9$ GPa

10–38. $\epsilon_{max} = 554(10^{-6})$, $\epsilon_{int} = 370(10^{-6})$,

$\epsilon_{min} = -924(10^{-6})$

10–39. $\rho = 3.43$ MPa,

$\tau_{\substack{max \\ in\text{-}plane}} = 0$,

$\tau_{\substack{abs \\ max}} = 85.7$ MPa

10–41. $\sigma_z = -\dfrac{12\,My}{bh^3}$, $\epsilon_y = \dfrac{12\,\nu My}{Ebh^3}$,

$\Delta L_{AB} = \dfrac{3\,\nu M}{2Ebh}$,

$\Delta L_{CD} = \dfrac{6\,\nu M}{Eh^2}$

10–42. $\epsilon_x = 2.35(10^{-3})$, $\epsilon_y = -0.972(10^{-3})$,

$\epsilon_z = -2.44(10^{-3})$

10–43. $P = 1.662$ kN

10–45. $t_h = 4.94$ mm

10–46. $\sigma_1 = 58.56$ MPa, $\sigma_2 = 43.83$ MPa

10–47. $\epsilon_1 = 833(10^{-6})$, $\epsilon_2 = 168(10^{-6})$,

$\epsilon_3 = -763(10^{-6})$

10–49. $\epsilon_x = 0.3125(10^{-3})$, and $\epsilon_y = 0.25(10^{-3})$, $\sigma_z = 0$,

$\sigma_x = 106.7$ MPa (C),

$\sigma_y = 116.1$ MPa (C)

10–50. $w_y = -184$ kN/m, $w_x = 723$ kN/m

10–51. $E = 67.7$ GPa, $\nu = 0.335$,

$G = 25.0$ GPa

10–53. $0 = \sigma_x - 0.35\sigma_y - 0.35\sigma_z + 181.896$,

$0 = \sigma_y - 0.35\sigma_z - 0.35\sigma_x + 181.896$,

$0 = \sigma_z - 0.35\sigma_x - 0.35\sigma_y + 44.096$,

$\sigma_x = \sigma_y = -487.2$ MPa, $\sigma_z = -385.2$ MPa

10–54. $\epsilon_x = \epsilon_y = 0$, $\epsilon_z = 5.48(10^{-3})$

10–57. $E_{eff} = \dfrac{E}{1 - \nu^2}$

10–58. $k = 1.35$

10–59. $\sigma_x^2 + \sigma_y^2 - \sigma_x\sigma_y + 3\tau_{xy}^2 = \sigma_Y^2$

10–61. $\omega = 80\,\pi$ rad/s, $T = \dfrac{3300}{8\pi}$ N·m,

$d = 21.89$ mm

10–62. $d = 20.87$ mm

10–63. $d = 48.25$ mm

10–65. $\sigma = \dfrac{60(10^6)}{a^3}$, $a = 47.50$ mm

10–66. $T_e = \sqrt{\dfrac{4}{3}M^2 + T^2}$

10–67. $M_e = \sqrt{M^2 + \dfrac{3}{4}T^2}$

10–69. $\sigma_1 = 50 + 197.23 = 247$ MPa,

$\sigma_2 = 50 - 197.23 = -147$ MPa,

Yes.

10–70. $M_e = \sqrt{M^2 + T^2}$

10–71. No.

10–73. $\sigma = -76.4$ MPa, $\tau = 244.5$ MPa,

Yes.

10–74. No.

10–75. F.S. = 1.43

10–77. $\sigma_1 = 51.20$ MPa, $\sigma_2 = -107.20$ MPa,

F.S. = 1.58

10–78. F.S. = 1.79

10–79. $\sigma_2 = 255.1$ MPa

10–81. $\sigma_2 = \dfrac{\sigma_x}{2}$, $\sigma_x = 848.7$ MPa

10–82. $\sigma_Y = 660.4$ MPa

10–83. $\sigma_Y = 636.8$ MPa

10–85. $\sigma_1 = 61.94$ MPa, $\sigma_2 = -75.94$ MPa,
$\sigma_Y = 137.9$ MPa

10–86. (a) $p = \dfrac{1}{r}\sigma_Y$

(b) $p = \dfrac{2t}{\sqrt{3}\,r}\sigma_Y$

10–89. $(\tau_{max})_h = 8626.28T$, $(\tau_{max})_s = 9947.18T$,
$\sigma_1 = 9947.18T$, $\sigma_2 = -9947.18T$,
$T = 838$ kN $\cdot$ m

10–90. $T = 9.67$ kN $\cdot$ m

10–91. $d = 39.35$ mm

10–93. $\sigma_{allow} = 166.67(10^6)$ Pa,

(a) $t = 22.5$ mm

(b) $t = 19.5$ mm

10–94. $\epsilon = \dfrac{pr}{2Et}(1 - \nu)$

10–95. (a) $\epsilon_1 = 482(10^{-6})$, $\epsilon_2 = 168(10^{-6})$

(b) $\gamma_{\substack{max \\ in\text{-}plane}} = 313(10^{-6})$

(c) $\gamma_{\substack{abs \\ max}} = 482(10^{-6})$

10–97. $\sigma_1 = 350.42$ MPa, $\sigma_2 = -65.42$ MPa, No.

10–98. $\epsilon_{avg} = 83.3(10^{-6})$,

(a) $\epsilon_1 = 880(10^{-6})$, $\epsilon_2 = -713(10^{-6})$,
$\theta_p = 54.8°$ (*Clockwise*)

(b) $\gamma_{\substack{max \\ in\text{-}plane}} = -1593(10^{-6})$,
$\theta_s = 9.78°$ (*Clockwise*)

10–99. $T = 736$ N $\cdot$ m

10–101. $\epsilon_1 = 996(10^{-6})$, $\epsilon_2 = 374(10^{-6})$,
$(\theta_p)_1 = 15.8°$ (*Clockwise*),
$\gamma_{\substack{max \\ in\text{-}plane}} = -622(10^{-6})$, $\epsilon_{avg} = 685(10^{-6})$,
$\theta_s = 29.2°$ (*Counterclockwise*)

10–102. $\epsilon_{x'} = 480(10^{-6})$, $\gamma_{x'y'} = 23.2(10^{-6})$,
$\epsilon_{y'} = 120(10^{-6})$

10–103. $\epsilon_{avg} = 300(10^{-6})$, $\epsilon_1 = 480(10^{-6})$,
$\epsilon_2 = 120(10^{-6})$, $(\theta_p)_1 = 28.2°$ (*Clockwise*),
$\gamma_{\substack{max \\ in\text{-}plane}} = -361(10^{-6})$,
$\theta_s = 16.8°$ (*Counterclockwise*)

Chapter 11

11–1. $I_x = 0.16276b^4$, $Q_{max} = 0.1953125b^3$,
$b = 211$ mm, $h = 264$ mm

11–2. Use W360 $\times$ 33

11–3. Use $b = 130$ mm

11–5. $S_{req'd} = 267.9(10^3)$ mm^3,
Use W310 $\times$ 24

11–6. Member AB: Use W250 $\times$ 18,
Member BC: Use W150 $\times$ 14

11–7. $w = 6.12$ kN/m

11–9. $S_{req'd} = 540(10^3)$ mm^3,
$\tau_{max} = 20.7$ MPa,
Use W310 $\times$ 39

11–10. Use W360 $\times$ 79

11–11. $P = 2.49$ kN

11–13. $S_{req'd} = 600\,000$ mm^3,
Use W250 $\times$ 58

11–14. Use W410 $\times$ 46

11–15. $b = 393$ mm

11–17. $S = 1069.5(10^3)$ mm^3, $\sigma_{max} = 181.8$ MPa, No, the
beam fails due to bending stress criteria.

11–18. $d = 11.4$ mm

11–19. $d_i = 13.0$ mm

11–21. $\sigma_{max} = 133.2$ MPa, Yes.

11–22. Use $h = 230$ mm

11–23. $w = 3.02$ kN/m,
$s_{ends} = 16.7$ mm,
$s_{mid} = 50.2$ mm

11–25. $h = 18$ mm,
Yes, the joist will safely support the load.

11–26. Use W310 $\times$ 39

11–27. $w = 10.8$ kN/m

11–29. $P = \dfrac{h^2}{1800}$, $h = 180$ mm,
$P = 18$ kN

11–30. Use $s = 100$ mm, $s' = 150$ mm,
$s'' = 300$ mm, Yes, it can support the load.

11–31. $\sigma_{max} = \dfrac{3PL}{8bh_0^2}$

11–33. $S = \dfrac{b_0 r^2}{3L}x$, $\sigma = \dfrac{3PL}{2b_0 t^2}$. The bending stress
is constant throughout the span.

11–34. $h = \dfrac{h_0}{L^{3/2}}(3L^2x - 4x^3)^{1/2}$

11–35. $\sigma_{\substack{abs \\ max}} = \dfrac{0.155w_0L^2}{bh_0^2}$

11–37. $\sigma_{allow} = \dfrac{Px}{b_0d^2/6}, \; d = d_0\sqrt{\dfrac{x}{L}}$

11–38. $b = \dfrac{b_0}{L^2}x^2$

11–39. Use $d = 20$ mm

11–41. $T = 100 \, \text{N} \cdot \text{m}, \; c = 0.01421,$
Use $d = 29$ mm

11–42. $T = 100 \, \text{N} \cdot \text{m},$ Use $d = 33$ mm

11–43. Use $d = 35$ mm

11–45. $M = 496.1 \, \text{N} \cdot \text{m}, c = 0.0176 \, \text{m},$ Use $d = 36$ mm

11–46. $d = 34.3$ mm

11–47. Use $d = 21$ mm

11–49. $M = 1274.75 \, \text{N} \cdot \text{m},$ Use $d = 44$ mm

11–50. Use $d = 41$ mm

11–51. Use W310 $\times$ 21

11–53. $S = \dfrac{bh_0^2}{6L^2}(2x + L)^2, \; \sigma_{max} = \dfrac{wL^2}{4bh_0^2}$

11–54. Use $d = 21$ mm

11–55. Use $d = 19$ mm

Chapter 12

12–1. $\sigma_{max} = \dfrac{c}{\rho}E = 100$ MPa

12–2. $\sigma = 582$ MPa

12–3. $W = 468.9$ N

12–5. $M(x_1) = -\dfrac{P}{2}x_1, \; M(x_2) = -Px_2,$

$v_1 = \dfrac{P}{12EI}(-x_1^3 + L^2x_1),$

$v_2 = \dfrac{P}{24EI}(-4x_2^3 + 7L^2x_2 - 3L^3)$

12–6. $v_1 = \dfrac{Px_1}{12EI}(-x_1^3 + L^2),$

$v_3 = \dfrac{P}{12EI}(2x_3^3 - 9Lx_3^3 + 10L^2x_3 - 3L^3),$

$v_{max} = \dfrac{PL^3}{8EI}$

12–7. $v_{max} = \dfrac{P}{3EI_{AB}}\left\{\left(1 - \dfrac{I_{AB}}{I_{AC}}\right)l^3 - L^3\right\}$

12–9. $M_1 = \dfrac{Pb}{L}x_1, \; M_2 = Pa\left(1 - \dfrac{x_2}{L}\right),$

$v_1 = \dfrac{Pb}{6EIL}\left(x_1^3 - (L^2 - b^2)x_1\right),$

$v_2 = \dfrac{Pa}{6EIL}(3x_2^2L - x_2^3 - (2L^2 + a^2)x_2 + a^2L)$

12–10. $\theta_{max} = \dfrac{M_0L}{3EI},$

$v_{max} = -\dfrac{\sqrt{3}M_0L^2}{27EI}$

12–11. $v_1 = \dfrac{P}{9EI}(x_1^3 - 5a^2x_1),$

$v_2 = \dfrac{P}{18EI}(-x_2^3 + 9ax_2^2 - 19a^2x_2 + 3a^3),$

$v_{max} = \dfrac{0.484 \, Pa^3}{EI}\downarrow$

12–13. $\theta_A = -\dfrac{3}{8}\dfrac{PL^2}{EI}, \; v_C = \dfrac{-PL^3}{6EI}$

12–14. $v_{max} = \dfrac{3PL^3}{256EI}\downarrow$

12–15. $\theta_A = \dfrac{Pab}{2EI}, \; v_1 = \dfrac{P}{6EI}[-x_1^3 + 3a(a + b)x_1$
$- a^2(2a + 3b)],$

$v_3 = \dfrac{Pax_3}{2EI}(-x_3 + b), \; v_C = \dfrac{Pab^2}{8\,EI}$

12–17. $M(x_1) = \dfrac{M_0}{L}x_1, \; M(x_2) = M_0,$

$\theta_A = \dfrac{M_0L}{6EI}, \; \triangledown, \; v_1 = \dfrac{M_0}{6EIL}(x_1^3 - L^2x_1),$

$v_2 = \dfrac{M_0}{24EI}(12x_2^2 - 20Lx_2 + 7L^2),$

$v_C = \dfrac{7M_0L^2}{24EI}\uparrow$

12–18. $\theta_A = -\dfrac{M_0L}{6EI}, \; v = \dfrac{M_0}{6EIL}(3Lx^2 - 2x^3 - L^2x),$

$v_{max} = \dfrac{0.0160 \, M_0L^2}{EI}\downarrow, \; v_{max} = \dfrac{0.0160 \, M_0L^2}{EI}\uparrow$

12–19. $\theta_B = \dfrac{M_0L}{6EI} \; \triangledown, \; v|_{x=\frac{L}{2}} = 0$

12–21. $M(x_1) = -\dfrac{wL}{8}x_1, \; M(x_2) = -\dfrac{w}{2}x_2^2,$

$v_C = \dfrac{11wL^4}{384EI}\downarrow$

12–22. $\theta_{max} = 0.00508$ rad, $v_{max} = 10.1$ mm $\downarrow$

12–23. $\theta_{max} = \dfrac{w_0 L^3}{45EI}$

12–25. $M(x) = \left(36x - \dfrac{1}{3}x^3\right) kN \cdot m$,

$\theta_A = \dfrac{540\ kN \cdot m^2}{EI}$ ⟍ ,

$v = \dfrac{1}{EI}\left(6x^3 - \dfrac{1}{60}x^5 - 540x\right) kN \cdot m^3$,

$v_{max} = \dfrac{2074\ kN \cdot m^3}{EI}$ ↓

12–26. $\theta_B = -\dfrac{wa^3}{6EI}$,

$v_1 = \dfrac{w}{24EI}\left(-x_1^4 + 4ax_1^3 - 6a^2x_1^2\right)$,

$v_2 = \dfrac{wa^3}{24EI}\left(-4x_2 + a\right)$,

$v_B = \dfrac{wa^3}{24EI}\left(-4L + a\right)$

12–27. $\theta_A = 0.0652\ rad$, $v_A = -93.9\ mm$

12–29. $I(x) = \dfrac{bh^3}{12L^3}x^3$, $\theta_A = \dfrac{2\gamma L^3}{h^2 E}$,

$v_A|_{x=0} = \dfrac{\gamma L^4}{h^2 E}$

12–30. $I = \left(\dfrac{2Lc}{L}\right)x$,

$\theta_A = \dfrac{\gamma L^3}{3r^2 E}$,

$v_A|_{x=0} = -\dfrac{\gamma L^4}{6r^2 E}$

12–31. $v_{max} = -\dfrac{PL^3}{2EI_0}$

12–33. $I = \left(\dfrac{2Ic}{L}\right)x$, $v_C = -\dfrac{PL^3}{32EI_C}$

12–34. $\sigma_{max} = \dfrac{3PL}{2nbt^2}$

12–35. $v_{max} = -3.64\ mm$

12–37. $M = 180x - 1.50\langle x - 0.25\rangle - 60\langle x - 0.5\rangle$

$- 150\langle x - 0.75\rangle$, $v_C = v_E = -0.501\ mm$,

$v_D = -0.698\ mm$, $v_E = -0.501\ mm$

12–38. $v = \dfrac{1}{EI}\left[-0.833x^3 - 33.33\langle x - 0.5\rangle^3 + \right.$

$\left. 91.67\langle x - 1.0\rangle^3 + 12.50x\right] N \cdot m^3$

12–39. $v = \dfrac{1}{EI}[4.1667x^3 - 5\langle x - 2\rangle^3$

$- 2.5\langle x - 4\rangle^3 - 93.333x]$,

$v_{max} = 13.3\ mm$ ↓

12–41. $M = M_0\left\langle x - \dfrac{L}{3}\right\rangle^0 - M_0\left\langle x - \dfrac{2}{3}L\right\rangle$,

$v = \dfrac{M_0}{6EI}\left[3\left\langle x - \dfrac{L}{3}\right\rangle^2 \right.$

$\left. - 3\left\langle x - \dfrac{2}{3}L\right\rangle^2 - Lx\right]$,

$v_{max} = \dfrac{5M_0 L^2}{72EI}$ ↓

12–42. $\theta_A = \dfrac{PL^2}{9EI}$,

$v = \dfrac{P}{18EI}\left[3x^3 - 3\left\langle x - \dfrac{L}{3}\right\rangle^3 \right.$

$\left. - 3\left\langle x - \dfrac{2}{3}L\right\rangle^3 - 2L^2 x\right]$,

$v_{max} = \dfrac{23PL^3}{648\ EI}$ ↓

12–43. $v = \dfrac{1}{EI}\left[6.25x^3 - 33.75x^2 - \dfrac{1}{6}x^5 \right.$

$\left. + \dfrac{1}{6}\langle x - 1.5\rangle^5 + \dfrac{5}{4}\langle x - 1.5\rangle^4\right]$,

$v_{max} = 12.9\ mm$ ↓

12–45. $M = 24.6x - 1.5x^2 + 1.5\langle x-4\rangle^2 - 50\langle x-7\rangle$,

$\theta_A = -\dfrac{279\ kN \cdot m^2}{EI}$,

$v|_{x=7\ m} = -\dfrac{835\ kN \cdot m^3}{EI}$

12–46. $v = \dfrac{1}{EI}\left[3.75x^3 - \dfrac{10}{3}\langle x - 1.5\rangle^3 \right.$

$\left. - 0.625\langle x - 3\rangle^4 + \dfrac{1}{24}\langle x - 3\rangle^5 - 77.625x\right]$,

$v_{max} = 11.0\ mm$ ↓

12–47. $\theta_B = -0.705°$, $v_B = -51.7\ mm$

12–49. $M = 10.875x - 3\langle x - 2.7\rangle$

$- 3.75x^2 + 3.75\langle x - 1.8\rangle^2$,

$v = \dfrac{1}{EI}\left[1.8125x^3 - 0.5\langle x - 2.7\rangle^3 \right.$

$\left. -0.3125\langle x - 1.8\rangle^4 - 9.72x\right] N \cdot m^3$

$v_{max} = 48.5\ mm$ ↓

12–50. $\dfrac{dv}{dx} = \dfrac{1}{EI}\{0.100x^2 - 0.333x^3$

$\qquad + 0.333\langle x - 5\rangle^3 + 8.90\langle x - 5\rangle^2$

$\qquad + 9.58\}\ \text{kN} \cdot \text{m}^2,$

$\qquad v = \dfrac{1}{EI}\{0.0333x^3 - 0.0833x^4$

$\qquad + 0.833\langle x - 5\rangle^4 + 2.97\langle x - 5\rangle^2$

$\qquad + 9.58x\}\ \text{kN} \cdot \text{m}^3$

12–51. $v = \dfrac{1}{EI}[-0.25x^4 + 0.208\langle x - 1.5\rangle^3$

$\qquad + 0.25\langle x - 1.5\rangle^4 + 4.625\langle x - 4.5\rangle^3$

$\qquad + 25.1x - 36.4]\ \text{kN} \cdot \text{m}^3$

12–53. $\theta_A = \dfrac{210\ \text{kN} \cdot \text{m}^2}{EI},$

$\qquad \theta_A = -\dfrac{720\ \text{kN} \cdot \text{m}^3}{EI}$

12–54. $v = \dfrac{1}{EI}[-0.2778x^5 + 71.6667\langle x - 3\rangle^3$

$\qquad + 0.2778\langle x - 3\rangle^5$

$\qquad - 158.33 + 542.5]\ \text{kN} \cdot \text{m}^3$

12–55. $\theta_C = \dfrac{1400}{EI}, \Delta_C = \dfrac{4800}{EI}$

12–57. $\theta_B = |\theta_{B/A}| = \dfrac{5PL^2}{8EI},$

$\qquad \Delta_B = |t_{B/A}| = \dfrac{7PL^3}{16EI}\downarrow,$

12–58. $\theta_A = \dfrac{60\ \text{kN} \cdot \text{m}^2}{EI}\ \measuredangle,$

$\qquad v_{\max} = v_C = \dfrac{180\ \text{kN} \cdot \text{m}^3}{EI}\downarrow$

12–59. $\theta_C = \dfrac{120\ \text{kN} \cdot \text{m}^2}{EI}\ \measuredangle,$

$\qquad v_C = \dfrac{180\ \text{kN} \cdot \text{m}^3}{EI}\downarrow$

12–61. $\theta_{C/A} = \dfrac{M_0 L}{2EI}, \theta_{\max} = \theta_A = \dfrac{M_0 L}{2EI},$

$\qquad \Delta_{\max} = |t_{B/C}| = \dfrac{M_0 L^2}{8EI}$

12–62. $\Delta_C = \dfrac{5M_0 L^2}{6EI}, \theta_C = \dfrac{4M_0 L}{3EI}$

12–63. $\theta_A = 0.00879\ \text{rad}$

12–65. $\Delta_C = |t_{C/A}| - \dfrac{|t_{B/A}|}{a}L, \theta_A = \dfrac{|t_{B/A}|}{a},$

$\qquad x = \dfrac{\sqrt{3}}{3}a, a = 0.858L$

12–66. $\theta_A = \dfrac{4PL^2}{81EI}$

12–67. $F = \dfrac{P}{4}$

12–69. $\theta_A = \theta_{A/C} = \dfrac{5Pa^2}{2EI}, \Delta_C = t_{A/C} = \dfrac{19Pa^3}{6EI}\downarrow$

12–70. $\Delta_C = \dfrac{PL^3}{12EI}, \theta_A = \dfrac{PL^2}{24EI}, \theta_B = \dfrac{PL^2}{12EI}$

12–71. $\Delta_{\max} = \dfrac{0.00802PL^3}{EI}$

12–73. $t_{B/A} = -\dfrac{5M_0 a^2}{6EI}, t_{C/A} = -\dfrac{M_0 a^2}{6EI},$

$\qquad \theta_A = \dfrac{|t_{B/A}|}{L} = \dfrac{5M_0 a}{12EI},$

$\qquad \Delta_C = \left|\dfrac{1}{2}t_{B/A}\right| - |t_{C/A}| = \dfrac{M_0 a^2}{4EI}\uparrow$

12–74. $v_{\max} = \dfrac{625.8\ \text{kN} \cdot \text{m}^3}{EI}\downarrow$

12–75. $E = \dfrac{Pa}{24\Delta I}(3L^2 - 4a^2)$

12–77. $\theta_{A/C} = -\dfrac{5Pa^2}{2EI}, t_{B/C} = -\dfrac{2Pa^3}{EI},$

$\qquad t_{A/C} = -\dfrac{13Pa^3}{3EI}, \theta_A = |\theta_{A/C}| = \dfrac{5Pa^2}{2EI},$

$\qquad \Delta_A = |t_{A/C}| - |t_{B/C}| = \dfrac{7Pa^3}{3EI}\downarrow$

12–78. $\theta_{\max} = \dfrac{5PL^2}{16EI}, \Delta_{\max} = \dfrac{3PL^3}{16EI}$

12–79. $\theta_D = \dfrac{PL^2}{16EI}, \Delta_C = \dfrac{5PL^3}{384EI}$

12–81. $\Delta_C = |t_{C/A}| - \dfrac{|t_{B/A}|}{a}L, \theta_A = \dfrac{|t_{B/A}|}{a} = \dfrac{M_0 a}{6EI},$

$\qquad x = \dfrac{\sqrt{3}}{3}a, \Delta_D = |t_{A/D}| = \dfrac{\sqrt{3}M_0 a^2}{27EI},$

$\qquad a = 0.865L$

12–82. $\theta_B = 0.0070833$ rad $\triangledown$, $v_B = 26.6$ mm $\downarrow$

12–83. $\theta_C = \dfrac{a^2}{6EI}(12P + wa)$,

$\Delta_C = \dfrac{a^3}{24EI}(64P + 7wa)\downarrow$

12–85. $\theta_B = |\theta_{B/C}| = 0.00243$ rad,

$\Delta_C = |t_{A/C}| = 1.56$ mm $\downarrow$

12–86. $\theta_A = 0.175°$

12–87. $\Delta_C = 13.1$ mm $\downarrow$

12–89. $(\theta_C)_1 = (\theta_B)_1 = \dfrac{wa^3}{3EI}$, $(\Delta_C)_1 = \dfrac{wa^4}{3EI}\uparrow$,

$(\theta_C)_2 = \dfrac{wa^3}{6EI}$, $(\Delta_C)_2 = \dfrac{wa^4}{8EI}\downarrow$,

$(\Delta_C)_3 = (\theta_B)_3(a) = \dfrac{wa^4}{3EI}\downarrow$,

$\theta_C = \dfrac{wa^3}{6EI}$, $\Delta_C = \dfrac{wa^4}{8EI}\downarrow$

12–90. $\theta_A = \dfrac{wa^3}{6EI}$, $\Delta_D = \dfrac{wa^4}{12EI}\downarrow$

12–91. $\theta_B = 0.00722$ rad, $\Delta_C = 13.3$ mm $\downarrow$

12–93. $(\Delta_C)_1 = \dfrac{345.6}{EI}\downarrow$,

$\Delta_2(x) = \dfrac{Mx}{6LEI}(L^2 - x^2)$,

$(\Delta_C)_2 = \dfrac{10.8}{EI}\downarrow$, $\Delta_C = 51.8$ mm

12–94. $\Delta_A = \dfrac{5.625 \text{ N} \cdot \text{m}^3}{EI}$, $\theta_A = \dfrac{112.5 \text{ N} \cdot \text{m}^2}{EI}$

12–95. $\Delta_C = 23.2$ m $\downarrow$

12–97. $\Delta_B = \dfrac{PL^3}{24EI}$, $(\Delta_A)_1 = \dfrac{PL^3}{24EI}$, $\theta = \dfrac{PL^2}{4JG}$,

$(\Delta_A)_2 = \dfrac{PL^3}{8JG}$, $\Delta_A = PL^3\left(\dfrac{1}{12EI} + \dfrac{1}{8JG}\right)$

12–98. $\Delta_A = \dfrac{Pa^2(3b + a)}{3EI}$

12–99. $\theta_A = \dfrac{3.29 \text{ N} \cdot \text{m}^2}{EI}$,

$(\Delta_A)_v = \dfrac{0.313 \text{ N} \cdot \text{m}^3}{EI}\downarrow$

12–101. $y_{max} = \dfrac{P\cos\theta \, L^3}{3EI_x}$; $x_{max} = \dfrac{P\sin\theta \, L^3}{3EI_y}$,

$\dfrac{x_{max}}{y_{max}} = \dfrac{I_x}{I_y}\tan\theta$, $y_{max} = 19.94$ mm,

$x_{max} = 83.01$ mm

12–102. Use W410 × 53

12–103. $A_y = \dfrac{3M_0}{2L}$, $B_y = \dfrac{3M_0}{2L}$, $M_B = \dfrac{M_0}{2}$

12–105. $M(x_1) = C_y x_1$, $M(x_2) = C_y x_2 - Px_2 + \dfrac{PL}{2}$,

$C_y = \dfrac{5}{16}P$, $B_y = \dfrac{11}{8}P$, $A_y = \dfrac{5}{16}P$

12–106. $M_A = \dfrac{PL}{2}$, $A_y = \dfrac{3P}{2}$, $B_y = \dfrac{5P}{2}$

12–107. $M_A = \dfrac{Pa}{L}(L - a)$, $M_B = \dfrac{Pa}{L}(L - a)$

12–109. $M(x) = C_y x + B_y\langle x - 3\rangle - 25\langle x - 3\rangle^3$,

$B_y = 72$ kN, $C_y = 5.333$ kN $\downarrow$, $A_y = 53.333$ kN

12–110. $A_x = 0$, $C_y = \dfrac{w_0 L}{10}$, $B_y = \dfrac{4w_0 L}{5}$, $A_y = \dfrac{w_0 L}{10}$

12–111. $A_x = 0$, $A_y = \dfrac{7wL}{16}$, $C_y = -\dfrac{wL}{16}$, $B_y = \dfrac{5wL}{8}$

12–113. $M_a = \dfrac{wL_1^2}{2} - T_{AC}L_1$,

$T_{AC} = \dfrac{3A_2 E_2 w L_1^4}{8(A_2 E_2 L_1^3 + 3E_1 I_1 L_2)}$

12–114. $A_x = 0$, $F_C = 112$ kN,

$A_y = 34.0$ kN, $B_y = 34.0$ kN

12–115. $A_x = 0$, $B_y = \dfrac{3M_0}{2L}$, $A_y = \dfrac{3M_0}{2L}$, $M_A = \dfrac{M_0}{2}$

12–117. $(t_{A/B})_1 = \dfrac{-P(L - a)^2(2L + a)}{6EI}$,

$(t_{A/B})_2 = \dfrac{A_y L^3}{3EI}$, $A_y = \dfrac{P(L - a)^2(2L + a)}{2L^3}$,

$a = 0.414L$

12–118. $A_y = \dfrac{3M_0}{2L}$, $C_y = \dfrac{3M_0}{2L}$, $B_y = \dfrac{3M_0}{L}$, $C_x = 0$

12–119. $B_x = 0$, $B_y = \dfrac{11P}{16}$, $C_y = \dfrac{13P}{32}$, $A_y = \dfrac{3P}{32}$

12–121. $v_B' = \dfrac{366.67 \text{ N} \cdot \text{m}^3}{EI}\downarrow$,

$v_B'' = \dfrac{1.3333B_y \text{ m}^3}{EI}\uparrow$,

$B_y = 550$ N, $A_y = 125$ N, $C_y = 125$ N

12–122. $B_y = \dfrac{7P}{4}$, $A_y = \dfrac{3P}{4}$, $M_A = \dfrac{PL}{4}$

12–123. $C_x = 0$, $B_y = 166.24$ kN,

$A_y = 11.88$ kN, $C_y = 81.88$ kN

12–125. $C_x = 0, (v_p)_1 = (v_p)_2 = \dfrac{247.5 \text{ kN} \cdot \text{m}^3}{EI} \downarrow,$

$(v_B)_y = \dfrac{36B_y}{EI} \uparrow, B_y = 13.75 \text{ kN},$

$A_y = C_y = 3.125 \text{ kN}$

12–126. $A_x = 0, B_y = \dfrac{3M_0}{2L}, A_y = \dfrac{3M_0}{2L}, M_A = \dfrac{M_0}{2}$

12–127. $C_x = 0, \ C_y = \dfrac{P}{3}$

12–129. $(\Delta_A)' = \dfrac{wL_1^4}{8E_1I_1}; \Delta_A = \dfrac{T_{AC}L_2}{A_2E_2}, \delta_A = \dfrac{T_{AC}L_1^3}{3E_1I_1},$

$T_{AC} = \dfrac{3wA_2E_2L_1^4}{8[3E_1I_1L_2 + A_2E_2L_1^3]}$

12–130. $M_A = \dfrac{PL}{8}, M_B = \dfrac{3PL}{8}, B_x = 0, B_y = P$

12–131. $M = \left(\dfrac{PL}{8} - \dfrac{2EI}{L}\alpha\right), \Delta_{\max} = \dfrac{PL^3}{192EI} + \dfrac{\alpha L}{4}$

12–133. $\Delta = \dfrac{w_0(L - a)^4}{8EI} - \dfrac{R(L - a)^3}{3EI},$

$R = \left(\dfrac{8\Delta EIw_0^3}{9}\right)^{\frac{1}{4}}, a = L - \left(\dfrac{72\Delta EI}{w_0}\right)^{\frac{1}{4}}$

12–134. $A_x = 0, F_B = 220 \text{ kN},$

$A_y = C_y = 70.1 \text{ kN}$

12–135. $B_y = 3.17 \text{ kN}, A_y = 1.22 \text{ kN}, C_y = 0.39 \text{ kN}$

12–137. $M = -900 + 1387.5\langle x - 0.3\rangle - 350\langle x - 0.6\rangle,$

$v = \dfrac{1}{EI}[-150x^3 + 231.25\langle x - 0.3\rangle^3$

$- 58.333\langle x - 0.6\rangle^3 + 120.94x - 32.23]$

12–138. $v_1 = \dfrac{1}{EI}(22.22x_1^3 - 2x_1) \text{ N} \cdot \text{m}^3,$

$v_2 = \dfrac{1}{EI}(-22.22x_2^3 + 2x_2) \text{ N} \cdot \text{m}^3$

12–139. $\Delta_C = 51.80 \text{ mm} \downarrow$

12–141. $\Delta_B = \Delta_C = \dfrac{11wL^4}{12EI}, \Delta_{BB} = \Delta_{CC} = \dfrac{4B_yL^3}{9EI},$

$\Delta_{BC} = \Delta_{CB} = \dfrac{7B_yL^3}{18EI}, B_y = C_y = \dfrac{11wL}{10},$

$A_y = \dfrac{2wL}{5}, D_y = \dfrac{2wL}{5}, D_x = 0$

12–142. $M_B = \dfrac{w_0L^2}{30}, M_A = \dfrac{w_0L^2}{20}$

12–143. $v_A = \dfrac{w_0L^4}{Eth_0^3} \downarrow$

12–145. $\Delta_B = \dfrac{777.6 \text{ N} \cdot \text{m}^3}{EI} \downarrow, (\Delta_C)_1 = \dfrac{388.8 \text{ N} \cdot \text{m}^3}{EI} \downarrow,$

$(\Delta_C)_2 = \dfrac{3280.5 \text{ N} \cdot \text{m}^3}{EI} \downarrow, \Delta_C = 17.4 \text{ mm} \downarrow$

12–146. $M_{\max} = \dfrac{\pi^2bt\gamma\omega^2r^3}{108g}$

Chapter 13

13–1. $F = 2P\theta, F_s = \dfrac{kL\theta}{2}, P_{cr} = \dfrac{kL}{4}$

13–2. $P_{cr} = \dfrac{5}{9}kL$

13–3. $P_{cr} = \dfrac{4k}{L}$

13–5. $A = 1.10(10^{-3}) \text{ m}^2,$

$I_x = I_y = 0.184167(10^{-6}) \text{ m}^4, P_{cr} = 22.7 \text{ kN}$

13–6. $P_{cr} = 46.4 \text{ kN}$

13–7. $P_{cr} = 656 \text{ kN}$

13–9. $P_{cr} = 152 \text{ kN}, \text{F.S.} = 2.03$

13–10. $P_{cr} = 1242 \text{ kN}$

13–11. $P_{cr} = 88.1 \text{ kN}$

13–13. $I = 0.86167(10^{-6}) \text{ m}^4, P_{cr} = 272 \text{ kN}$

13–14. $d = 209.76 \text{ mm}, P_{cr} = 1052.8 \text{ kN}$

13–15. $L = 4.34 \text{ m}$

13–17. $A = 5000 \text{ mm}^2, I_x = 4.1667(10^6) \text{ mm}^4,$

$I_y = 1.041667(10^6) \text{ mm}^4, P_{cr} = 13.7 \text{ kN}$

13–18. $P_{cr} = 28 \text{ kN}$

13–19. $P = 29.9 \text{ kN}$

13–21. $P_{cr} = 8467 \text{ kN}, \text{No.}$

13–22. $\text{F.S.} = 2.00$

13–23. $P = 2176 \text{ kN}$

13–25. $A = 5710 \text{ mm}^2, I_y = 8.16(10^6) \text{ mm}^4, P = 286 \text{ kN}$

13–26. $P = 13.51 \text{ kN}$

13–27. $w = 5.55 \text{ kN/m}$

13–29. $F_{BC} = 100 \text{ kN}, P_{cr} = 771 \text{ kN},$

$x–x \text{ axis: F.S.} = 7.71, y–y \text{ axis: F.S.} = 3.43$

13–30. $P = 102.8 \text{ kN}$

13–31. $w = 467.3 \text{ kN/m}$

13–33. $F_{AB} = 15 \text{ kN}, A = 3.2(10^{-3}) \text{ m}^2,$

$I_y = 0.4267(10^{-6}) \text{ m}^4, P_{cr} = 33.69 \text{ kN},$

$\text{F.S.} = 2.25$

13–34. $P = 46.5$ kN

13–35. $P = 110$ kN

13–37. $F_{AB} = 26.8014m$, $A = 0.625(10^{-3})\pi$ m^2,
$I = 97.65625(10^{-9})\pi$m^4, $m = 7.06$ kg

13–38. $P = 28.39$ kN

13–39. $P = 14.53$ kN

13–41. $M(x) = \dfrac{w}{2}(x^2 - Lx) - Pv$,

$$v_{max} = \dfrac{wEI}{P^2}\left[\sec\left(\sqrt{\dfrac{P}{EI}}\dfrac{L}{2}\right) - \dfrac{PL^2}{8EI} - 1\right],$$

$$M_{max} = -\dfrac{wEI}{P}\left[\sec\left(\sqrt{\dfrac{P}{EI}}\dfrac{L}{2}\right) - 1\right]$$

13–42. $v_{max} = \dfrac{F}{2P}\left[\sqrt{\dfrac{EI}{P}}\tan\left(\sqrt{\dfrac{P}{EI}}\dfrac{L}{2}\right) - \dfrac{L}{2}\right]$,

$$M_{max} = -\dfrac{F}{2}\sqrt{\dfrac{EI}{P}}\tan\left(\sqrt{\dfrac{P}{EI}}\dfrac{L}{2}\right)$$

13–43. $P_{cr} = \dfrac{\pi^2 EI}{4L^2}$

13–45. $M(x) = R'(L - x) - Pv$

13–46. $P = 121.66$ kN

13–47. $P = 5.87$ kN, $v_{max} = 42.1$ mm

13–49. $A = 0.61575(10^{-3})$ m^2,
$I = 64.1152(10^{-9})$ m^4, $P_{max} = P_{cr} = 18.98$ kN,
$P = 6.75$ kN

13–50. $P = 20.1$ kN

13–51. $P = 320.08$ kN

13–53. $P_{cr} = 98.70$ kN, $P_{allow} = 49.35$ kN,
$P_{allow} = 26.3$ kN

13–54. $\sigma_{max} = 130$ MPa

13–55. The column is adequate.

13–57. Strong axis yielding controls.
$P_{allow} = 89.0$ kN

13–58. $P_{cr} = 199$ kN, $e = 175$ mm

13–59. $\sigma_{max} = 199$ MPa, $v_{max} = 24.3$ mm

13–61. The column is adequate.

13–62. $e = 0.15$ m, $P_{allow} = 346.92$ kN,
$P_{allow} = 268$ kN

13–63. $\sigma_{max} = 47.88$ MPa

13–65. $(KL)_x = (KL)_y = 3$ m, $P_{cr} = 83.5$ kN

13–66. F.S. = 1.42

13–67. The column does not fail by yielding.

13–69. $(KL)_y = (KL)_y = 5.25$ m, $P = 83.07$ kN

13–70. $E_t = 102.13$ GPa

13–71. $P_{cr} = 372.2$ kN

13–73. $\dfrac{KL}{r} < 49.7$: $\dfrac{P}{A} = \dfrac{0.4935(10^6)}{\left(\dfrac{KL}{r}\right)^2}$ MPa,

$49.7 < \dfrac{KL}{r} < 99.3$: $\dfrac{P}{A} = 200$ MPa,

$\dfrac{KL}{r} > 99.3$: $\dfrac{P}{A} = \dfrac{1.974(10^6)}{\left(\dfrac{KL}{r}\right)^2}$ MPa

13–75. $P_{cr} = 1323$ kN

13–77. $E_1 = 200$ GPa, $E_2 = 150$ GPa,
$r = 0.02$ m, $P_{cr} = 2700$ kN

13–78. $L = 3.56$ m

13–79. $L = 2.575$ m

13–81. $\sigma_{allow} = 79.1$ MPa, Use W150 × 22

13–82. Use W150 × 14

13–83. Use W200 × 36

13–85. $\left(\dfrac{KL}{r}\right)_x = 103.69$, $\left(\dfrac{KL}{r}\right)_y = 111.02$,

$P_{allow} = 368.7$ kN

13–86. Yes.

13–87. $d = 36.72$ mm

13–89. $A = 0.0151$ m^2, $I_y = 90.025833(10^{-6})$ m^4.
The column is adequate.

13–90. $L = 4.46$ m

13–91. $b = 18.20$ mm

13–93. $A = 5.55(10^{-3})$ m^2, $I_x = 31.86625(10^{-6})$ m^4,
$I_y = 2.5478(10^{-6})$ m^4, $P_{allow} = 422$ kN

13–94. $L = 3.08$ m

13–95. $P_{allow} = 468.7$ kN

13–97. $A = 3456$ mm^2, $I = 11964672$ mm^2,
$P_{allow} = 540.5$ kN

13–98. $P_{allow} = 597.9$ kN

13–99. $P_{allow} = 454.4$ kN

13–101. $\sigma_{allow} = 3.35$ MPa, $P_{allow} = 37.7$ kN

13–102. $L = 2.11$ kN

13–103. Use $a = 200$ kN

13–105. $L = 1.078$ m

13–106. $P_{allow} = 8.05$ kN

13–107. $P = 26.90$ kN

13–109. $\dfrac{KL}{r_y} = 69.182$, $(\sigma_a)_{\text{allow}} = 115.2 \text{ MPa}$

$P = 350.4 \text{ kN}$

13–110. $P = 346.6 \text{ kN}$

13–111. $P = 5.07 \text{ kN}$

13–113. $\left(\dfrac{KL}{r}\right)_x = 130.91$ (controls),

$\left(\dfrac{KL}{r}\right)_y = 99.02$, $P = 40.2 \text{ kN}$

13–114. $P = 63.53 \text{ kN}$

13–115. The column is not adequate.

13–117. $A = 7200 \text{ mm}^2$, $I_x = 61.99(10^6) \text{ mm}^4$,

$I_y = 16.0288(10^6) \text{ mm}^4$, $P = 398.6 \text{ kN}$

13–118. $P = 412.5 \text{ kN}$

13–119. $P = 12.0 \text{ kN}$

13–121. $A = 15200 \text{ mm}^2$, $I_x = 29.27(10^6) \text{ mm}^4$,

$I_y = 12.67(10^6) \text{ mm}^4$, $P = 434.4 \text{ kN}$

13–122. $P = 586.9 \text{ kN}$

13–123. $P = 10.6 \text{ kN}$

13–125. $\dfrac{KL}{d} = 43.2$, Yes.

13–126. $P = 7.42 \text{ kN}$

13–127. $P = 15.17 \text{ kN}$

13–129. $B_x = 0$, $B_y = \dfrac{2M}{L\sin\theta}$, $M = \dfrac{PL}{2}\sin\theta$,

$M = \dfrac{PL}{2}\theta$, $P_{cr} = \dfrac{2k}{L}$

13–130. $w = 4.63 \text{ kN/m}$

13–131. $P = 107.1 \text{ kN}$

13–133. $\left(\dfrac{KL}{r}\right)_x = 56.25$, $\left(\dfrac{KL}{r}\right)_y = 128.21$ (controls),

Yes.

13–134. $P_{\text{allow}} = 238.6 \text{ kN}$

13–135. $P_{cr} = 839 \text{ kN}$

13–137. $\bar{x} = 0.06722 \text{ m}$, $I_y = 20.615278(10^{-6}) \text{ m}^4$,

$I_z = 7.5125(10^{-4}) \text{ m}^4$, No.

Chapter 14

14–1. $\dfrac{U_i}{V} = \dfrac{1}{2E}(\sigma_x^2 + \sigma_y^2 - 2\nu\sigma_x\sigma_y) + \dfrac{\tau_{xy}^2}{2G}$

14–3. $(U_i)_a = 3.28 \text{ J}$

14–5. $N_{AB} = 3 \text{ kN}$, $N_{BC} = 7 \text{ kN}$, $N_{CD} = -3 \text{ kN}$,

$U_i = 0.372 \text{ J}$

14–6. $(U_i)_a = 43.2 \text{ J}$

14–7. $P = 375 \text{ kN}$, $(U_i)_a = 1.69 \text{ kJ}$

14–9. $T = 8.0 \text{ kN} \cdot \text{m}$, $T = 2.0 \text{ kN} \cdot \text{m}$,

$T = -10.0 \text{ kN} \cdot \text{m}$, $U_i = 149 \text{ J}$

14–10. $(U_i)_t = \dfrac{7T^2L}{24\pi r_0^4 G}$

14–11. $(U_i)_t = 0.0637 \text{ J}$

14–13. $V = -P$, $M = -Px$, $\dfrac{(U_i)_y}{(U_i)_b} = \dfrac{3(1+\nu)}{5}\left(\dfrac{h}{L}\right)^2$

14–14. $U_i = \dfrac{M_0^2 L}{24EI}$

14–15. $(U_i)_b = \dfrac{P^2 L^3}{48EI}$

14–17. $M(x) = \left(9x - \dfrac{1}{4}x^3\right) \text{kN}\cdot\text{m}$, $(U_i)_b = 33.6 \text{ J}$

14–18. $(U_i)_b = 29.3 \text{ J}$

14–19. $U_i = \dfrac{P^2 r^3}{JG}\left(\dfrac{3\pi}{8} - 1\right)$

14–21. $T_y = \dfrac{PL}{2}$, $M_x = Py$, $U_i = P^2 L^3\left[\dfrac{3}{16EI} + \dfrac{1}{8JG}\right]$

14–22. $U_i = \dfrac{3P^2 L^3}{bh^3 E}$, 1.5 times as great as for a uniform

cross section

14–23. (a) $U_i = \dfrac{w^2 L^5}{40EI}$, (b) $U_i = \dfrac{w^2 L^5}{40EI}$

14–25. $F_{AD} = 12.5 \text{ kN (T)}$, $F_{AB} = 7.50 \text{ kN (C)}$,

$F_{DB} = 12.5 \text{ kN (C)}$, $F_{DC} = 15.0 \text{ kN (C)}$,

$(\Delta_A)_h = 0.407 \text{ mm}$

14–26. $(\Delta_C)_h = \dfrac{2PL}{AE}$

14–27. $\Delta_C = \dfrac{2PL}{AE}$

14–29. $U_e = \dfrac{1}{2}(M_0\theta_B)$, $\theta_B = \dfrac{M_0 L}{EI}$

14–30. $U_i = 216.1 \text{ J}$, $\Delta_C = 0.8644 \text{ mm}$

14–31. $\theta_B = 0.00100 \text{ rad}$

14–33. $U_i = \dfrac{65\,625}{EI}$, $U_e = 150\,\theta_E$,

$\theta_E = 3.15°$

14–34. $\Delta_B = 77.41 \text{ mm}$

14–35. $\Delta_B = 11.7 \text{ mm}$

14–37. $T = PR \cos \theta, \ M = PR \sin \theta,$

$$\sigma_{max} = \frac{16PR}{\pi d^3} [\sin \theta + 1]$$

14–38. $\Delta = \dfrac{64nPR^3}{d^4 G}$

14–39. $\Delta_C = 2.13$ mm

14–41. $M_1 = -20(10^3)x_1, \ M_2 = 60(10^3) \ \text{N} \cdot \text{m},$

$\Delta_B = 15.2$ mm

14–42. (a) $U_i = 4.52$ kJ, (b) $U_i = 3.31$ kJ

14–43. $d = 5.35$ in.

14–45. $\Delta_{st} = 9.8139(10^{-6})$ m, $\sigma_{max} = 359$ MPa

14–46. $h = 69.6$ mm

14–47. $\sigma_{max} = 216$ MPa

14–49. $\Delta_{st} = 0.613125(10^{-3})$ m, $k = 160(10^6)$ N/m,

$\sigma_{max} = 237$ MPa, $\Delta_{max} = 3.95$ mm

14–50. $\sigma_{max} = 85.7$ MPa

14–51. $h = 0.571$ m

14–53. $\sigma_{st} = 0.3123$ MPa, $L = 325$ mm

14–54. $\sigma_{max} = 207$ MPa

14–55. $\sigma_{max} = 335$ MPa

14–57. $\sigma_{st} = 1.147$ MPa, $n = 183.086, h = 3.259$ m

14–58. $(\Delta_A)_{max} = 406.9$ mm

14–59. $h = \dfrac{\sigma_{max}L^2}{3Ec}\left[\dfrac{\sigma_{max}I}{WLc} - 2\right]$

14–61. $n = 209.13, h = 2.23$ m

14–62. $\sigma_{max} = 41.5$ MPa

14–63. $h = 84$ m

14–65. $k_{beam} = 305.176$ N/mm, $n = 16.11$

14–66. $\sigma_{max} = 47.8$ MPa

14–67. $h = 6.57$ m

14–69. $k_b = 49.3425(10^6)$ N/m, $\Delta_b = 0.050342$ m,

$v = 5.75$ m/s

14–70. $\Delta_{beam} = 12.71$ mm, $\sigma_{max} = 74.7$ MPa

14–71. $\Delta_{max} = 23.3$ mm, $\sigma_{max} = 4.89$ MPa

14–73. $1 \ \text{kN} \cdot (\Delta_B)_h = \dfrac{5.0667(10^6) \ \text{kN}^2 \cdot \text{m}}{AE},$

$(\Delta_B)_h = 20.27(10^{-3})$ mm $\rightarrow$

14–74. $(\Delta_B)_v = 2.70(10^{-3})$ mm $\downarrow$

14–75. $\Delta_{C_v} = 20.4$ mm

14–77. $1 \ \text{kN} \cdot (\Delta_B)_v = \dfrac{202.5(10^6) \ \text{kN}^2 \cdot \text{m}}{AE},$

$(\Delta_B)_v = 0.3616$ mm $\downarrow$

14–78. $(\Delta_E)_v = 0.2813$ mm $\downarrow$

14–79. $\Delta_{B_h} = 0.367$ mm

14–81. $1 \ \text{N} \cdot (\Delta_A)_v = \dfrac{498.125(10^3) \ \text{N}^2 \cdot \text{m}}{AE},$

$(\Delta_A)_v = 6.23$ mm $\downarrow$

14–82. $(\Delta_B)_v = 3.79$ mm $\downarrow$

14–83. $\Delta_{C_v} = 5.266$ mm

14–85. $1 \ \text{N} \cdot (\Delta_C)_v = \dfrac{174.28125(10^3)}{0.15(10^{-3})[200(10^9)]},$

$(\Delta_C)_v = 5.81$ mm $\downarrow$

14–86. $(\Delta_G)_v = 3.41$ mm $\downarrow$

14–87. $\Delta_C = \dfrac{23Pa^3}{24EI}$

14–89. $\Delta_C = \dfrac{572.92 \ \text{kN} \cdot \text{m}^3}{EI},$

$M(x_1) = 2.50x_1, M(x_2) = x_2^2,$

$m(x_1) = 0.50x_1, m(x_2) = x_2,$

$\Delta_C = 40.9$ mm $\downarrow$

14–90. $\theta_A = 0.00298$ rad

14–91. $\theta_B = 0.00595$ rad

14–93. $M = 1.626x_1, M = 0.9756 + 0.226x_1,$

$m_\theta = 1 - 0.377x_1, m_\theta = 1 = 0.377x_3,$

$\theta_A = 8.118°$

14–94. $\theta_C = 5.89(10^{-3})$ rad

14–95. $\Delta_A = 27.4$ mm $\downarrow, \theta_A = 5.75(10^{-3})$ rad

14–97. $M_1 = Px, M_2 = Px_2,$

$m_\theta = \dfrac{x_1}{a}, m_\theta = 1,$

$\theta_C = \dfrac{5Pa^2}{6EI}$

14–98. $\theta_A = \dfrac{Pa^2}{6EI}$

14–99. $\theta_A = 0.00700$ rad

14–101. $M(x_1) = \dfrac{w}{24}(11Lx_1 - 12x_1^2), M(x_2) = \dfrac{w}{3L}x_2^3,$

$m_\theta(x_1) = -\dfrac{x_1}{L}, m_\theta(x_2) = 1, \theta_C = \dfrac{13wL^3}{576EI}$

14–102. $\Delta_D = \dfrac{wL^4}{96EI} \downarrow$

14–103. $\Delta_C = 9.77$ mm $\downarrow$

14-105. $M(x_2) = wa(a + x_2) - \frac{w}{2}x_2^2$, $M(x_1) = wax_1$,

$$m(x_2) = \frac{1}{2}(x_2 + a), m(x_1) = \frac{x_1}{2}, \Delta_B = \frac{65wa^4}{48EI} \downarrow$$

14-106. $\Delta_C = \dfrac{w_0 L^4}{120EI}$

14-107. $\theta_A = \dfrac{5w_0 L^3}{192EI}$

14-109. $M(x_1) = (31.5x_1 - 6x_1^2) \text{ kN} \cdot \text{m}$,

$M(x_2) = (22.5x_2 - 3x_2^2) \text{ kN} \cdot \text{m}$,

$m_\theta(x_1) = (1 - 0.1667x_1) \text{ kN} \cdot \text{m}$,

$m_\theta(x_2) = (0.1667x_2) \text{ kN} \cdot \text{m}$,

$\theta_A = 0.00927 \text{ rad}$

14-110. $\Delta_C = 16.7 \text{ mm} \downarrow$

14-111. Bending and shear:

$$\Delta = \left(\frac{w}{G}\right)\left(\frac{L}{a}\right)^2\left[\left(\frac{5}{96}\right)\left(\frac{L}{a}\right)^2 + \frac{3}{20}\right],$$

Bending only: $\Delta = \dfrac{5w}{96G}\left(\dfrac{L}{a}\right)^4$

14-113. $M(x_1) = \dfrac{w}{2}x_1^2$, $M(x_2) = \dfrac{wL^2}{2}$,

$m(x_1) = 0, m(x_2) = 1.0L - 1.0x_4$,

$$(\Delta_B)_h = \frac{wL^4}{4EI} \rightarrow$$

14-114. $\Delta_{A_v} = \dfrac{4PL^3}{3EI}$

14-115. $\Delta_B = 43.5 \text{ mm} \downarrow$

14-117. $\Delta_C = 17.9 \text{ mm} \downarrow$

14-118. $\theta_A = 0.991(10^{-3}) \text{ rad}$

14-119. $(\Delta_C)_v = 16.8 \text{ mm} \downarrow$

14-121. $M = \dfrac{M_0}{a}, M = M_0$,

$$m = 1x, m = 1x, \Delta_C = \frac{5M_0 a^2}{6EI}$$

14-122. $\theta_B = \dfrac{M_0 a}{3EI}$

14-123. $\Delta_{B_h} = 0.0554 \text{ mm}$

14-125. $\Delta_{C_v} = \dfrac{1225.26(10^3)}{300(10^{-6})(200)(10^9)} = 20.4 \text{ mm}$

14-126. $\Delta_{D_v} = 4.88 \text{ mm}$

14-127. $(\Delta_B)_v = 0.3616 \text{ mm} \downarrow$

14-129. $\Delta_{B_h} = \dfrac{29.375(10^3)}{400(10^{-6})(200)(10^9)} = 0.367 \text{ mm}$

14-130. $\Delta_{C_v} = 0.0375 \text{ mm}$

14-131. $(\Delta_A)_v = 6.23 \text{ mm} \downarrow$

14-133. $\Delta_{C_v} = \dfrac{2.948889(10^9)}{(2800)(200)(10^3)} = 5.266 \text{ mm}$

14-134. $\Delta_{H_v} = 5.052 \text{ mm}$

14-135. $\Delta_C = \dfrac{23Pa^3}{24EI}$

14-137. $\theta_A = \dfrac{41.667(10^3)}{200(10^9)\left[70(10^{-6})\right]} = 0.00298 \text{ rad}$

14-138. $\Delta_B = 117.45 \text{ mm}$

14-139. $\theta_A = 8.118° \text{ (clockwise)}$

14-141. $\theta_C = \dfrac{5Pa^2}{6EI}$

14-142. $\theta_A = \dfrac{Pa^2}{6EI}$

14-143. $(\Delta_C)_v = \dfrac{5wL^4}{8EI} \downarrow$

14-145. $M = \dfrac{M_0}{a}x, M = M_0, M = Px, M = Px$,

$$\Delta_C = \frac{5M_0 a^2}{6EI}$$

14-146. $U_i = \dfrac{5P^2 a^3}{6EI}$

14-147. $\sigma_{max} = 116 \text{ MPa}$

14-149. Bending strain energy:

$(U_b)_i = 22.05 \text{ J}$,

Axial force strain energy:

$(U_a)_i = 0.0624 \text{ J}$

14-150. $\Delta_A = 0.536 \text{ mm}$

14-151. $(U_i)_t = 0.274 \text{ J}$

14-153. $\Delta_{E_v} = \dfrac{236.25(10^3)}{400(10^{-6})(200)(10^9)} = 2.95 \text{ mm}$

14-154. $\theta_B = \dfrac{M_0 L}{EI}$

14-155. $\theta_B = \dfrac{M_0 L}{EI}$

14-157. $\sigma_{max} = 75.9 \text{ MPa}$

Index

Average Mechanical Properties of Typical Engineering Materials[a]
(U.S. Customary Units)

Materials	Specific Weight γ (lb/in³)	Modulus of Elasticity E (10³) ksi	Modulus of Rigidity G (10³) ksi	Yield Strength (ksi) σ_Y Tens.	Comp.[b]	Shear	Ultimate Strength (ksi) σ_u Tens.	Comp.[b]	Shear	% Elongation in 2 in. specimen	Poisson's Ratio ν	Coef. of Therm. Expansion α (10⁻⁶)/°F
Metallic												
Aluminum Wrought Alloys ⎡ 2014-T6	0.101	10.6	3.9	60	60	25	68	68	42	10	0.35	12.8
⎣ 6061-T6	0.098	10.0	3.7	37	37	19	42	42	27	12	0.35	13.1
Cast Iron Alloys ⎡ Gray ASTM 20	0.260	10.0	3.9	–	–	–	26	97	–	0.6	0.28	6.70
⎣ Malleable ASTM A-197	0.263	25.0	9.8	–	–	–	40	83	–	5	0.28	6.60
Copper Alloys ⎡ Red Brass C83400	0.316	14.6	5.4	11.4	11.4	–	35	35	–	35	0.35	9.80
⎣ Bronze C86100	0.319	15.0	5.6	50	50	–	95	95	–	20	0.34	9.60
Magnesium Alloy [Am 1004-T61]	0.066	6.48	2.5	22	22	–	40	40	22	1	0.30	14.3
Steel Alloys ⎡ Structural A36	0.284	29.0	11.0	36	36	–	58	58	–	30	0.32	6.60
⎢ Stainless 304	0.284	28.0	11.0	30	30	–	75	75	–	40	0.27	9.60
⎣ Tool L2	0.295	29.0	11.0	102	102	–	116	116	–	22	0.32	6.50
Titanium Alloy [Ti-6Al-4V]	0.160	17.4	6.4	134	134	–	145	145	–	16	0.36	5.20
Nonmetallic												
Concrete ⎡ Low Strength	0.086	3.20	–	–	–	1.8	–	–	–	–	0.15	6.0
⎣ High Strength	0.086	4.20	–	–	–	5.5	–	–	–	–	0.15	6.0
Plastic Reinforced ⎡ Kevlar 49	0.0524	19.0	–	–	–	–	104	70	10.2	2.8	0.34	–
⎣ 30% Glass	0.0524	10.5	–	–	–	–	13	19	–	–	0.34	–
Wood Select Structural Grade ⎡ Douglas Fir	0.017	1.90	–	–	–	–	0.30[c]	3.78[d]	0.90[d]	–	0.29[e]	–
⎣ White Spruce	0.130	1.40	–	–	–	–	0.36[c]	5.18[d]	0.97[d]	–	0.31[e]	–

[a] Specific values may vary for a particular material due to alloy or mineral composition, mechanical working of the specimen, or heat treatment. For a more exact value reference books for the material should be consulted.

[b] The yield and ultimate strengths for ductile materials can be assumed equal for both tension and compression.

[c] Measured perpendicular to the grain.

[d] Measured parallel to the grain.

[e] Deformation measured perpendicular to the grain when the load is applied along the grain.

Average Mechanical Properties of Typical Engineering Materials[a] (SI Units)

Materials	Density ρ (Mg/m³)	Modulus of Elasticity E (GPa)	Modulus of Rigidity G (GPa)	Yield Strength (MPa) σ_Y			Ultimate Strength (MPa) σ_u			% Elongation in 50 mm specimen	Poisson's Ratio ν	Coef. of Therm. Expansion α $(10^{-6})/°C$
				Tens.	Comp.[b]	Shear	Tens.	Comp.[b]	Shear			
Metallic												
Aluminum Wrought Alloys — 2014-T6	2.79	73.1	27	414	414	172	469	469	290	10	0.35	23
Aluminum Wrought Alloys — 6061-T6	2.71	68.9	26	255	255	131	290	290	186	12	0.35	24
Cast Iron Alloys — Gray ASTM 20	7.19	67.0	27	–	–	–	179	669	–	0.6	0.28	12
Cast Iron Alloys — Malleable ASTM A-197	7.28	172	68	–	–	–	276	572	–	5	0.28	12
Copper Alloys — Red Brass C83400	8.74	101	37	70.0	70.0	–	241	241	–	35	0.35	18
Copper Alloys — Bronze C86100	8.83	103	38	345	345	–	655	655	–	20	0.34	17
Magnesium Alloy [Am 1004-T61]	1.83	44.7	18	152	152	–	276	276	152	1	0.30	26
Steel Alloys — Structural A36	7.85	200	75	250	250	–	400	400	–	30	0.32	12
Steel Alloys — Stainless 304	7.86	193	75	207	207	–	517	517	–	40	0.27	17
Steel Alloys — Tool L2	8.16	200	75	703	703	–	800	800	–	22	0.32	12
Titanium Alloy [Ti-6Al-4V]	4.43	120	44	924	924	–	1,000	1,000	–	16	0.36	9.4
Nonmetallic												
Concrete — Low Strength	2.38	22.1	–	–	–	12	–	–	–	–	0.15	11
Concrete — High Strength	2.38	29.0	–	–	–	38	–	–	–	–	0.15	11
Plastic Reinforced — Kevlar 49	1.45	131	–	–	–	–	717	483	20.3	2.8	0.34	–
Plastic Reinforced — 30% Glass	1.45	72.4	–	–	–	–	90	131	–	–	0.34	–
Wood Select Structural Grade — Douglas Fir	0.47	13.1	–	–	–	–	2.1[c]	26[d]	6.2[d]	–	0.29[e]	–
Wood Select Structural Grade — White Spruce	3.60	9.65	–	–	–	–	2.5[c]	36[d]	6.7[d]	–	0.31[e]	–

[a] Specific values may vary for a particular material due to alloy or mineral composition, mechanical working of the specimen, or heat treatment. For a more exact value reference books for the material should be consulted.

[b] The yield and ultimate strengths for ductile materials can be assumed equal for both tension and compression.

[c] Measured perpendicular to the grain.

[d] Measured parallel to the grain.

[e] Deformation measured perpendicular to the grain when the load is applied along the grain.

Fundamental Equations of Mechanics of Materials

Axial Load

Normal Stress

$$\sigma = \frac{P}{A}$$

Displacement

$$\delta = \int_0^L \frac{P(x)dx}{A(x)E}$$

$$\delta = \Sigma \frac{PL}{AE}$$

$$\delta_T = \alpha \, \Delta T L$$

Torsion

Shear stress in circular shaft

$$\tau = \frac{T\rho}{J}$$

where

$$J = \frac{\pi}{2}c^4 \text{ solid cross section}$$

$$J = \frac{\pi}{2}\left(c_o^{\,4} - c_i^{\,4}\right) \text{ tubular cross section}$$

Power

$$P = T\omega = 2\pi f T$$

Angle of twist

$$\phi = \int_0^L \frac{T(x)dx}{J(x)G}$$

$$\phi = \Sigma \frac{TL}{JG}$$

Average shear stress in a thin-walled tube

$$\tau_{\text{avg}} = \frac{T}{2tA_m}$$

Shear Flow

$$q = \tau_{\text{avg}}t = \frac{T}{2A_m}$$

Bending

Normal stress

$$\sigma = \frac{My}{I}$$

Unsymmetric bending

$$\sigma = -\frac{M_z y}{I_z} + \frac{M_y z}{I_y}, \qquad \tan \alpha = \frac{I_z}{I_y} \tan \theta$$

Shear

Average direct shear stress

$$\tau_{\text{avg}} = \frac{V}{A}$$

Transverse shear stress

$$\tau = \frac{VQ}{It}$$

Shear flow

$$q = \tau t = \frac{VQ}{I}$$

Stress in Thin-Walled Pressure Vessel

Cylinder

$$\sigma_1 = \frac{pr}{t} \qquad \sigma_2 = \frac{pr}{2t}$$

Sphere

$$\sigma_1 = \sigma_2 = \frac{pr}{2t}$$

Stress Transformation Equations

$$\sigma_{x'} = \frac{\sigma_x + \sigma_y}{2} + \frac{\sigma_x - \sigma_y}{2}\cos 2\theta + \tau_{xy}\sin 2\theta$$

$$\tau_{x'y'} = -\frac{\sigma_x - \sigma_y}{2}\sin 2\theta + \tau_{xy}\cos 2\theta$$

Principal Stress

$$\tan 2\theta_p = \frac{\tau_{xy}}{(\sigma_x - \sigma_y)/2}$$

$$\sigma_{1,2} = \frac{\sigma_x + \sigma_y}{2} \pm \sqrt{\left(\frac{\sigma_x - \sigma_y}{2}\right)^2 + \tau_{xy}^2}$$

Maximum in-plane shear stress

$$\tan 2\theta_s = -\frac{(\sigma_x - \sigma_y)/2}{\tau_{xy}}$$

$$\tau_{\max} = \sqrt{\left(\frac{\sigma_x - \sigma_y}{2}\right)^2 + \tau_{xy}^2}$$

$$\sigma_{\text{avg}} = \frac{\sigma_x + \sigma_y}{2}$$

Absolute maximum shear stress

$$\tau_{\substack{\text{abs} \\ \max}} = \frac{\sigma_{\max} - \sigma_{\min}}{2}$$

$$\sigma_{\text{avg}} = \frac{\sigma_{\max} + \sigma_{\min}}{2}$$

Geometric Properties of Area Elements

...operty Relations

...s ratio

$$\nu = -\frac{\epsilon_{lat}}{\epsilon_{long}}$$

...eneralized Hooke's Law

$$\epsilon_x = \frac{1}{E}\left[\sigma_x - \nu(\sigma_y + \sigma_z)\right]$$

$$\epsilon_y = \frac{1}{E}\left[\sigma_y - \nu(\sigma_x + \sigma_z)\right]$$

$$\epsilon_z = \frac{1}{E}\left[\sigma_z - \nu(\sigma_x + \sigma_y)\right]$$

$$\gamma_{xy} = \frac{1}{G}\tau_{xy}, \quad \gamma_{yz} = \frac{1}{G}\tau_{yz}, \quad \gamma_{zx} = \frac{1}{G}\tau_{zx}$$

where

$$G = \frac{E}{2(1+\nu)}$$

Relations Between w, V, M

$$\frac{dV}{dx} = -w(x), \qquad \frac{dM}{dx} = V$$

Elastic Curve

$$\frac{1}{\rho} = \frac{M}{EI}$$

$$EI\frac{d^4v}{dx^4} = -w(x)$$

$$EI\frac{d^3v}{dx^3} = V(x)$$

$$EI\frac{d^2v}{dx^2} = M(x)$$

Buckling
Critical axial load

$$P_{cr} = \frac{\pi^2 EI}{(KL)^2}$$

Critical stress

$$\sigma_{cr} = \frac{\pi^2 E}{(KL/r)^2}, \quad r = \sqrt{I/A}$$

Secant formula

$$\sigma_{max} = \frac{P}{A}\left[1 + \frac{ec}{r^2}\sec\left(\frac{L}{2r}\sqrt{\frac{P}{EA}}\right)\right]$$

Energy Methods
Conservation of energy

$$U_e = U_i$$

Strain energy

$$U_i = \frac{N^2 L}{2AE} \quad \text{constant axial load}$$

$$U_i = \int_0^L \frac{M^2 dx}{EI} \quad \text{bending moment}$$

$$U_i = \int_0^L \frac{f_s V^2 dx}{2GA} \quad \text{transverse shear}$$

$$U_i = \int_0^L \frac{T^2 dx}{2GJ} \quad \text{torsional moment}$$

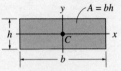

Rectangular area

$$A = bh$$
$$I_x = \frac{1}{12}bh^3$$
$$I_y = \frac{1}{12}hb^3$$

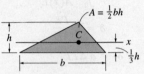

Triangular area

$$A = \frac{1}{2}bh$$
$$I_x = \frac{1}{36}bh^3$$

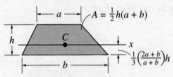

Trapezoidal area

$$A = \frac{1}{2}h(a+b)$$
$$\frac{1}{3}\left(\frac{2a+b}{a+b}\right)h$$

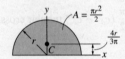

Semicircular area

$$A = \frac{\pi r^2}{2}$$
$$I_x = \frac{1}{8}\pi r^4$$
$$I_y = \frac{1}{8}\pi r^4$$

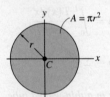

Circular area

$$A = \pi r^2$$
$$I_x = \frac{1}{4}\pi r^4$$
$$I_y = \frac{1}{4}\pi r^4$$

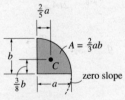

Semiparabolic area

$$A = \frac{2}{3}ab$$

zero slope

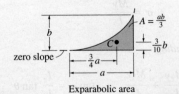

Exparabolic area

$$A = \frac{ab}{3}$$